THE RUMEN MICROBIAL ECOSYSTEM

THE RUMEN MICROBIAL ECOSYSTEM

Edited by

P. N. HOBSON

*Honorary Research Fellow, Biochemistry Department,
Marischal College, Aberdeen University,
Aberdeen, UK*

*Formerly Head, Microbial Chemistry Department,
Rowett Research Institute, Aberdeen, UK*

ELSEVIER APPLIED SCIENCE
LONDON and NEW YORK

ELSEVIER SCIENCE PUBLISHERS LTD
Crown House, Linton Road, Barking, Essex IG11 8JU, England

Sole Distributor in the USA and Canada
ELSEVIER SCIENCE PUBLISHING CO., INC.
52 Vanderbilt Avenue, New York, NY 10017, USA

WITH 68 TABLES AND 48 ILLUSTRATIONS

© 1988 ELSEVIER SCIENCE PUBLISHERS
(except Chapter 10: © Government of Canada)

British Library Cataloguing in Publication Data

The Rumen microbial ecosystem.
 1. Ruminants. Rumen. Microflora
 I. Hobson, P.N.
599.73'504132

Library of Congress Cataloging in Publication Data

The Rumen microbial ecosystem.
 1. Rumen—Microbiology. I. Hobson, P. N.
QR171.R85R86 1988   599.73'504132   88-16424

ISBN 1-85166-188-3

No responsibility is assumed by the Publisher for any injury and/or damage to persons or property as a matter of products liability, negligence or otherwise, or from any use or operation of any methods, products, instructions or ideas contained in the material herein.

Special regulations for readers in the USA

This publication has been registered with the Copyright Clearance Center Inc. (CCC), Salem, Massachusetts. Information can be obtained from the CCC about conditions under which photocopies of parts of this publication may be made in the USA. All other copyright questions, including photocopying outside the USA, should be referred to the publisher.

All rights reserved. No part of this publication may be reproduced, stored in a retrieval system, or transmitted in any form or by any means, electronic, mechanical, photocopying, recording, or otherwise, without the prior written permission of the publisher.

Printed in Great Britain by Galliard (Printers) Ltd, Great Yarmouth

Preface

Ruminants have played a major role in farming production for thousands of years, and have provided mankind with meat, milk and clothing. They can adapt to all kinds of climates and, above all, they can feed on all kinds of temperate and tropical vegetation from lichens to trees.

The ability to digest vegetation is conferred by a part of the digestive tract, the rumen, and this book is about the processes that go on in that organ. Although it was known 100 or more years ago that microorganisms inhabited the rumen, it was not until 40 or 50 years ago that a picture of the true role of the microbes in the digestion of vegetation began to emerge.

In the late 1940s not only was the function of the main microbial products in the nutrition of the animal more clearly defined, but the first bacteria which could be said to have a definite part in feed digestion were cultured.

Progress was fast in the next decade or so; biochemical pathways in the conversion of vegetation into products used by the animal were defined, and bacteria and protozoa which could cause these biochemical reactions were isolated or investigated in other ways. By the mid-1960s the overall scheme of the reactions in the rumen and some, at least, of the microbes which could be causing these reactions were known. To Robert E. Hungate, who might be known as the 'father' of rumen microbiology, the time seemed ripe to put all this knowledge together and *The Rumen and Its Microbes* was published.

Since then, publications on 'the rumen and its microbes' have escalated, along with other scientific literature. The basic pattern presented in 1965 has not altered, but new techniques have added detail to this and discovered new pathways and microorganisms, and the presence of rumen-like organisms in the digestive tracts of non-ruminant herbivores and omnivores, including humans, has been demonstrated.

The rumen is a complex *anaerobic* microbial ecosystem, with many microbes and many reactions. But the system is not unique. Similar anaerobic ecosystems exist in habitats other than the rumen and the intestinal contents of non-ruminant animals, in soils, muds, river silts and elsewhere where they are part of the natural recycling of biomass but can also cause pollution, and in anaerobic digesters and sanitary landfills where they help to abate pollution and form a useful fuel, a fuel that can also be derived from energy crops by the anaerobes. The experience gained in the investigation of the rumen has been used in the investigation of these other ecosystems, to increase knowledge of the causes and control of pollution and the dangers to human life that the metabolism of the anaerobic bacteria can sometimes bring. And the rumen investigations led to the discovery of the roles of anaerobes in human ailments, a branch of medical bacteriology that has grown over the last decade.

Since *The Rumen and Its Microbes*, there have been reviews of rumen, and intestinal, microbiology and biochemistry, and the anaerobes have found their way into bacteriological textbooks, but there has not been a comprehensive book on the rumen. It seemed to us that there was a place for such a book but, as in other subjects, it is the age of the specialist rumen investigator. So it seemed best to draw together a team of authors, and more than one author to each chapter, to contribute their own expertise to the topics under review. The growth of knowledge and papers in the last 10–15 years has meant that not all early works can be referred to; these can be found in the previous rumen book and journal reviews etc. Some selection has also been made in more recent work. However, we hope that the book will give a comprehensive explanation of the rumen ecosystem.

The reader is assumed to have some basic knowledge of microbiology and biochemistry, but we hope that the book will help to extend the information available to the students and teachers of these subjects, and particularly to the agricultural microbiologist. The high productions sought from farm ruminants and the use of feeds and feed additives never encountered by the wild animal make a knowledge of rumen function an essential for many agricultural research workers and advisers.

The information in the book is not only on a particular subject, the rumen, but is also on how a microbial ecosystem has been investigated and the expertise in numerous branches of the biological sciences has been integrated—an example of use in the investigation of other ecosystems. The connection of the anaerobes with recycling in the soils and waters, with pollution and pollution control, and with human ills and human nutrition should enable the worker in these areas to find useful information in the book.

P. N. HOBSON

Contents

Preface v

List of Contributors ix

1. Introduction: The Ruminant and the Rumen 1
   R. E. HUNGATE

2. The Rumen Bacteria 21
   C. S. STEWART and M. P. BRYANT

3. The Rumen Protozoa 77
   A. G. WILLIAMS and G. S. COLEMAN

4. The Rumen Anaerobic Fungi 129
   C. G. ORPIN and K. N. JOBLIN

5. Development of, and Natural Fluctuations in, Rumen Microbial Populations 151
   B. A. DEHORITY and C. G. ORPIN

6. Energy Yielding and Consuming Reactions 185
   J. B. RUSSELL and R. J. WALLACE

7. Metabolism of Nitrogen-Containing Compounds 217
   R. J. WALLACE and M. A. COTTA

8. Polysaccharide Degradation by Rumen Microorganisms . . 251
   A. CHESSON and C. W. FORSBERG

Contents

9. Lipid Metabolism in the Rumen 285
   C. G. HARFOOT and G. P. HAZLEWOOD

10. The Genetics of Rumen Bacteria 323
    G. P. HAZLEWOOD and R. M. TEATHER

11. Microbe–Microbe Interactions 343
    M. J. WOLIN and T. L. MILLER

12. Compartmentation in the Rumen 361
    J. W. CZERKAWSKI and K.-J. CHENG

13. Manipulation of Rumen Fermentation 387
    C. J. VAN NEVEL and D. I. DEMEYER

14. Digestive Disorders and Nutritional Toxicity 445
    K. A. DAWSON and M. J. ALLISON

15. Models, Mathematical and Biological, of the Rumen Function   461
    P. N. HOBSON and J.-P. JOUANY

16. The Future 513
    P. N. HOBSON

Index 519

List of Contributors

M. J. ALLISON

   USDA Agricultural Research Service, National Animal Disease Centre, PO Box 70, Ames, Iowa, USA

M. P. BRYANT

   Departments of Animal Science and Microbiology, University of Illinois, Urbana, Illinois 61801, USA

K.-J. CHENG

   Animal Science Section, Research Station, PO Box 3000 Main, Lethbridge, Alberta, Canada T1J 4B1

A. CHESSON

   Rowett Research Centre, Greenburn Road, Bucksburn, Aberdeen AB2 9SB, Scotland, UK

G. S. COLEMAN

   Biochemistry Department, AFRC Institute of Animal Physiology and Genetics Research, Babraham, Cambridge CB2 4AT, UK

M. A. COTTA

   US Department of Agriculture, 1815 North University Street, Peoria, Illinois 61604, USA

J. W. CZERKAWSKI

   6 Elms Way, Maybole, Ayrshire KA19 SBB, Scotland, UK

K. A. DAWSON

Department of Animal Sciences, University of Kentucky, 907 Agricultural Science Building South, Lexington, Kentucky 40546-0215, USA

B. A. DEHORITY

Department of Animal Sciences, Ohio State University, Wooster, Ohio 44691-6900, USA

D. I. DEMEYER

Research Centre for Nutrition, Animal Production and Meat Science, Faculty of Agricultural Sciences, State University of Ghent, Proefhoevestraat 10, B-9230 Melle, Belgium

C. W. FORSBERG

University of Guelph, Guelph, Ontario, Canada N1G 2W1

C. G. HARFOOT

Department of Biological Sciences, University of Waikato, Private Bag, Hamilton, New Zealand

G. P. HAZLEWOOD

Biochemistry Department, AFRC Institute of Animal Physiology and Genetics Research, Babraham, Cambridge CB2 4AT, UK

P. N. HOBSON

4 North Deeside Road, Aberdeen, Scotland AB1 7PL

R. E. HUNGATE

Department of Bacteriology, University of California, Davis, California 95616, USA

K. N. JOBLIN

Applied Biochemistry Division, Department of Scientific and Industrial Research, Private Bag, Palmerston North, New Zealand

J.-P. JOUANY

Laboratoire de la Digestion des Ruminants, INRA Digestion, Theix, 63122 Ceyrat, France

T. L. MILLER

Office of Public Health, Corning Tower, Empire State Plaza, Albany, New York 12201, USA

C. G. ORPIN

> AFRC Institute of Animal Physiology and Genetics Research, Babraham, Cambridge CB2 4AT, UK and Department of Arctic Biology, Institute of Medical Biology, University of Tromsø, 9001 Tromsø, Norway

J. B. RUSSELL

> ISDA Agricultural Research Center, Cornell University, 3221 Morrison Hall, Ithaca, New York 14853, USA

C. S. STEWART

> Nutrition Division, Rowett Research Institute, Greenburn Road, Bucksburn, Aberdeen AB2 9SB, Scotland, UK

R. M. TEATHER

> Animal Research Centre, Research Branch, Agriculture Canada, Ottawa, Ontario, Canada K1A 0C6

C. J. VAN NEVEL

> Research Centre for Nutrition, Animal Production and Meat Science, Faculty of Agricultural Sciences, State University of Ghent, Proefhoevestraat 10, B-9230 Melle, Belgium

R. J. WALLACE

> Rowett Research Institute, Greenburn Road, Bucksburn, Aberdeen AB2 9SB, Scotland, UK

A. G. WILLIAMS

> Hannah Research Institute, Ayr KA6 5HL, Scotland, UK

M. J. WOLIN

> Office of Public Health, Corning Tower, Empire State Plaza, Albany, New York 12201, USA

THE MAIN MICROBIAL REACTIONS IN THE RUMEN ECOSYSTEM

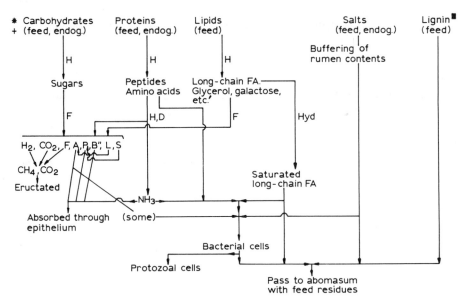

* Primary substrates for microorganisms.
+ Feed, substrate from feed. Endog., substrate from endogenous sources: salivary–mucous carbohydrates; epithelial-cell proteins; urea in saliva and secreted through epithelium; salts, including bicarbonate and phosphate, in saliva.
■ A plant structural material linked with cellulose and hemicellulose and limiting carbohydrate hydrolysis.
H, hydrolysis of polymers.
F, fermentation of mono- and di-saccharides from hydrolysis of polysaccharides or lipids.
D, deamination of amino acids.
FA, fatty acids. F, formic acid; A, acetic acid; P, propionic acid; B″, butyric acid plus C_5 and C_6 straight- and branched-chain acids; L, lactic acid; S, succinic acid.
Hyd, hydrogenation of unsaturated long-chain fatty acids.
′, Residues from phospholipids.
Part of the carbohydrate carbon is also used in microbial-cell synthesis.

1
Introduction: The Ruminant and the Rumen

R. E. Hungate
Department of Bacteriology, University of California, Davis, California, USA

The continuing increase in the number of humans causes them to consume an ever greater fraction of Earth's foods, among which proteins are extremely important. Some protein is formed in plants, especially as reserve food in the seeds, but the bodies and milks of grazing mammals have long been an important source of protein, and among these animals the ruminants predominate. Most of their protein depends on microbial activity.

An essential role of microorganisms in the alimentary tracts of grazing mammals has been increasingly appreciated ever since Sprengel (1832) reported that acetic and butyric acids were products of the breakdown of plant materials in the rumen. Protozoa and bacteria were soon identified as abundant inhabitants of the alimentary tracts of cattle and horses, and, after Pasteur recognized the alcoholic fermentation by yeasts as an anaerobic equivalent of respiration, the rumen acids were similarly interpreted as metabolites of an anaerobic rumen microbiota. Zuntz (1879) cited evidence that the acids were absorbed and oxidized by the animal to meet its energy requirements, and formulated the fermentation hypothesis to explain the host–microbe symbiosis.

This was confirmed experimentally when an animal physiology unit was set up in the 1940s at Cambridge, UK under the leadership of Sir Joseph Barcroft who, with McAnally and Phillipson, demonstrated (Barcroft *et al.*, 1944) that the quantities of acetic, propionic and butyric acids (Elsden, 1946) were sufficient to meet the energy requirements of the ruminant. Later the work was located at the Rowett Research Institute where numerous workers have continued to report important new knowledge of rumen microbiology and physiology and its significance in ruminant production. Innumerable other laboratories and investigators over the world, including the authors of this volume, have similarly contributed new facts and insights to the biology of the rumen. Their review is timely and appropriate.

The bodies of the microorganisms leaving the rumen constitute the proteinaceous food of the host; the fermentation acids are the source of its energy. We shall return to a consideration of the way in which these two important products are related, but first let us trace the evolution of microbial communities as inhabitants of higher animals.

EVOLUTION OF MICROBIAL COMMUNITIES IN ANIMAL GUTS

Many aquatic animals have long used microorganisms as food, filtering them from large volumes of water containing low concentrations of microbial cells. A unique example is *Urechis caupo* (Fisher & MacGinitie, 1928), an archaic, marine, worm-like invertebrate which secretes a membrane between the wall of its burrow and itself and pumps seawater through it, filtering out the small planktonic organisms. At intervals it detaches the membrane and swallows it together with the collected microbes, and then secretes a new membrane.

Land animals cannot easily concentrate microbes from the soil or other terrestrial habitats containing them, and most microbes do not accumulate sufficiently in soil to warrant direct ingestion (Parle, 1968). However, some of the higher fungi seasonally form large fruiting-bodies which are often consumed by higher animals, and certain higher termites cultivate fungi as food (Rohmann, 1978). Many small herbivorous mammals subsist on easily digested fruiting-structures of plants, but most large herbivores consume the stems and leaves of forages. The starch, lipids and proteins in these vegetative structures can be digested by the animal's enzymes, but they are a relatively small fraction of the total nutrients in the plant body as compared to the materials composing the plant skeleton, the celluloses, hemicelluloses and lignins, all indigestible by the enzymes of higher animals.

Cellulose occurs in aquatic plants as a component of the cell walls, strengthening them to withstand internal osmotic pressures. Even greater quantities are needed in land plants to support the stems and leaves in air and to prevent collapse of the water-conducting cells due to the high suction-tensions developed in the water in transport from soil to leaves during periods of water stress. There are thus huge supplies of this carbohydrate on Earth, together with hemicelluloses which occur along with cellulose in structural tissues of plants. Lignins similarly are vital for the support of plants in conjunction with the cellulose, particularly in woody plants, including forages.

Celluloses and hemicelluloses are, like starches, composed of sugars in a polymerized state containing relatively little water. All can be hydrolysed into soluble sugars but their structures are different. Starch serves to store carbohydrate for use as food, and its formation and digestion occur readily and rapidly in accordance with nutritional needs. Cellulose, as a supporting polymer, requires a less easily digested structure with more strength. It may be digested at local points by plant enzymes effecting formation of new buds, but its support function requires relative resistance to enzymatic attack. The hemicelluloses, polymers of a number of 5- and 6-carbon sugars, together with some uronic acids, are generally weaker than cellulose and less resistant to hydrolysis. However, most studies of the hemicelluloses have been on those occurring in a relatively pure state or extracted chemically from complexes with other polymers in plant tissues.

The lignins, complex polymers of variously substituted phenylpropane units, are even more resistant to enzymatic digestion than are the structural carbohydrates. Lignin reduces the rate and completeness of digestion of the carbohydrate components of supporting tissues, particularly under anaerobic conditions, though aerobically a number of terrestrial fungi can almost completely digest and oxidize even highly lignified wood.

Accumulation of plant bodies on Earth is prevented by the ability of fungi and many other microbes in soil and other free-living natural habitats to digest the plant cell walls. The herbage consumed by land animals does not contain a sufficient concentration of these microbes to digest its fibre at a rate sufficient to meet the carbohydrate needs of the animal. Further, in aerobic habitats such as plant bodies, the chief waste products of carbohydrate utilization by microbes are carbon dioxide and water, which would have little nutrient value for animals.

Prior to the evolution of living systems for the photosynthetic reduction of carbon dioxide with hydrogen from water, the Earth's atmosphere lacked dioxygen. Most primitive life was anaerobic. After O_2 appeared, its relative insolubility in water, together with its great chemical reactivity, prevented its penetration into many habitats. In these, anaerobic microbes survived and, as plant structural carbohydrate polymers evolved, concomitant microbial evolution produced species capable of hydrolysing them to their constituent sugars. As anaerobes they could not oxidize foods with O_2 but did possess chemical reactions for fermenting them, synthesizing microbial protoplasm under anoxic conditions.

CHEMICAL WORK IN BIOLOGICAL SYSTEMS

Such chemical reactions constitute metabolism, sometimes distinguished as catabolic reactions yielding energy which is used in anabolic reactions synthesizing protoplasm. Actually, energy is degraded in all chemical reactions, all are exergonic, including those directly concerned in the formation of protoplasm. The relationships between energy and matter during evolution of the universe are pertinent to the concept of energy degradation as an inherent feature of biological growth.

A continuous supply of energy arrives on Earth in the form of sunlight, but it cannot accumulate as such. A fraction is reflected into space, some is absorbed as heat, but much of it is converted into other forms of energy that can be stored more efficiently. In Fig. 1 a diagram shows the various forms in which energy interconversions can occur. Although arrows do not show it in the figure, heat energy can to some extent be converted into other forms, but the conversion cannot be complete, whereas the other forms can be completely converted into heat. On this basis they may be classed as 'high' forms of energy, in contrast to heat, the 'low' form. Heat is an expression of random directions of motion, the others involve more order. In most conversions of high forms of energy, some at least is converted into heat, though in a very few chemical conversions there may be a slight transformation of heat into chemical energy, but in these and all other

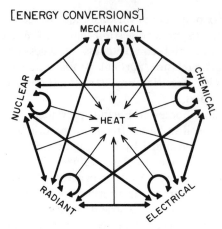

Fig. 1. Energy conversions including work (wide lines) and degradation (narrow lines). All interconversions involve degradation. From Hungate (1985) by permission of Plenum Press.

transformations of energy there is a net degradation, i.e. a decrease in the capacity to do work; the system cannot be returned to its state prior to the transformation without application of additional energy.

Energy has been defined as the capacity to do work, and one definition of work is application of force to cause motion. Work can also be defined as the extent to which any form of energy is converted into a high form. Oparin (1953) and Winfield (1978) have postulated that energy transformations have occurred in the universe following a primitive condition of infinitely vast energy but little or no material, and that during transformations of energy, always involving degradation, there was a concomitant increase in the amount and complexity of matter. Initially, electrons and protons are postulated to have formed in vast quantities, with much degradation of total energy, followed by further degradation during which the atoms spontaneously formed. These first steps in synthesis of matter were followed by others. The atoms combined spontaneously to form a universe of molecules of all sorts, both inorganic and organic. Within this plethora of kinds of molecules, energy degradation continued, synthesizing ever more complex material.

Molecules containing carbon, nitrogen and phosphorus, all atoms with multiple bonds available for three-dimensional structural arrangements, possessed special potential for synthesis of an almost infinite number of very large and complex molecules, including nucleotides. Conceivably, there appeared spontaneously information systems for self-replication from decomposition of simpler molecules, in which chemical patterns of synthesis through degradation were a basic feature (Broda, 1975). This favoured their continuance.

From this perspective, metabolism includes the successive chemical reactions, each involving a degradation of energy, which redistribute the atoms and energy in relatively simply organized materials in the 'food' into even simpler products, 'wastes', unable to react further, with the concomitant synthesis of the living material for which the whole process of replication, including metabolism, preserves the informational pattern. This chemical work forming living material is the surviving expression of the inherent tendency in the universe toward synthesis of more complex forms of matter in reactions involving decrements in free energy which occur spontaneously in the presence of appropriate catalysts. The ultimate problem in biology is to identify the actual reactants and products in the chains of synthesizing reactions, all of them exergonic, in which protoplasm is formed.

EXAMPLES OF GUT MUTUALISMS IN ANIMALS

Coincident with the appearance of higher plants on Earth, animals using them as food evolved, and anaerobic microbes capable of using the plant structural carbohydrates as food must have been taken into the animals along with the consumed plants. Moisture, temperature and food conditions favoured their development in the animal. Initially it must have been the more easily digested portions of the plants that were metabolized, but as the advantages conferred by the nutritional value of the microbial fermentation products selected animals with ever larger holding-chambers, the retention time of the digesta became such that the products of the fermentation of plant fibre supplied most of the nutrients essential to the animal. Various anatomical and physiological adaptations selected mutualistic microbial populations well fitted to the animal gut and unable to survive elsewhere. Adaptations for dissemination of the microbes among individual animals within a species included also social behaviour, as in grooming and licking (Troyer, 1984).

The particular conditions developed in the herbivores using microbes to ferment plant fibres varied from one animal species to another (Moir, 1968), leading to selection of microbial populations unique for each animal type. Bacteria occur in all the gut mutualistic populations, protozoa are prominent in many, and even aquatic fungi may be important (Orpin, 1977). In some animals the microbial digestion occurs in a compartment preceding the stomach in which pepsin and acid are secreted (pre-peptic), in others the microbial attack follows the action of the host enzymes (post-peptic). In both instances the acidity of the animal's peptic secretions is sufficient to kill most of the microbes subjected to it. Some of the higher animals in which a mutualistic relationship with a gut microbial population has developed are listed in Table 1.

The pre-peptic location of microbes makes the host completely dependent on its symbionts because they ferment all ingested foods, whereas with the post-peptic location any products from the animal's digestive enzymes can be absorbed prior to the microbial action. The rumen is the capacious portion of the stomach in which the ingesta are held for pre-peptic microbial action; an enlarged caecum, or a capacious large-intestine, retains the digesta during post-peptic fermentations.

The interior of the mass of digesta in the fermentation chamber provided by the animal host is completely devoid of dioxygen; any slight amounts diffusing in from the gas above the solids digesta are rapidly absorbed. Even in small termites with only a cubic millimetre of digesta, a low oxidation–reduction potential ($-160\,\text{mV}$) can be continuously maintained (Bignell, 1984). A relatively small fraction of gut

Table 1
Animals with Nutritionally Mutualistic Microbes (including Fungi)

Animals	Microbes	Location (see text)
Wood roach, *Cryptocercus punctulatus*	Flagellate protozoa	Post-peptic
Termites	Flagellates in some, bacteria in some, fungi in some	Post-peptic, the fungi cultivated
Birds, grouse and ptarmigan	Bacteria	Post-peptic
Marsupial mammals (quokka, kangaroo and wallaby)	Ciliate protozoa and bacteria	Pre-peptic
Leaf-eating colobid and langur monkeys	Bacteria	Pre-peptic
Sloths	Bacteria	Pre-peptic
Rodents	Bacteria	Post-peptic, with coprophagy
Rabbit and hare	Bacteria	Post-peptic with coprophagy
Elephant and hyrax	Bacteria and ciliates	Post-peptic
Hippopotamus and peccary	Bacteria and ciliates	Pre-peptic
Dugong and manatee	Bacteria	Post-peptic
Camel, llama and alpaca	Bacteria and ciliates	Pre-peptic
Horse, zebra, tapir and rhinoceros	Bacteria and ciliates	Post-peptic
Cattle, sheep and other ruminants	Bacteria, ciliates and microscopic fungi	Pre-peptic

bacteria can absorb dioxygen without harm, and their activities, aided perhaps by fermentation products such as sulphide, completely and rapidly remove any traces of dissolved O_2. Almost all mutualistic gut microbes are obligately anaerobic, and most of them are oxy-labile.

UTILITY OF CARBOHYDRATES FOR ANAEROBIC METABOLISM

Anaerobes, unable to oxidize food through hydration of carbon atoms and removal of the hydrogen to combine it with dioxygen, must obtain energy by transferring hydrogens between carbon atoms. For this type of anaerobic oxido-reduction, carbohydrate is the ideal substrate, and probably evolved as an important food because of this characteristic (Hungate, 1955). Its carbons, at an average redox state of zero (CH_2O), have more potential for anaerobic chemical work, i.e. for more hydrogen transfers between carbons, than have most other organic molecules. The maximum number of transfers occurs in a hexose when its carbon atoms are converted to their most oxidized or most reduced state, $C_6H_{12}O_6 \rightarrow 3CO_2 + 3CH_4$, with a standard free-energy decrement of 102 cal.

No single species can accomplish this complete oxido-reduction. Also, the equation disregards the work done by the conversion. As an inherent part of the breakdown, part of the substrate plus additional chemicals in other foods are transformed into the bodies of the participating microbes. Bacteria are cited as having the composition $C_{4.2}NH_{10}O_{1.25}$ (Luria, 1960) or, as estimated from measurements on microbes from a sheep rumen, $C_{5.66}NH_{11.3}O_{3.43}$ (Hungate et al., 1971).

The innate metabolic advantages of carbohydrate ensured its place in anaerobic life, and its capacities for polymerization made it an important component of the plant skeleton when photosynthetic reduction of CO_2 with water made cellulose abundantly available and the greatest mass of carbohydrate on Earth. As starch, carbohydrate was also a polymerized storage product, readily digested to its component sugar.

THE RUMINANT

Further consideration of the alimentary microbial mutualisms concerned in the utilization by animals of plant bodies as food will be mainly limited to ruminants. The great economic importance of these animals, coupled with the inherent interest in their symbionts, has caused more study to be devoted to them than to other similar mutualisms in higher animals. They are relatively easy to modify operatively, so that the interior of the rumen is easily accessible to sampling and to artificial introduction of experimental treatments with little disturbance of normal functions. Many of the phenomena of the rumen microbial ecosystem are sufficiently similar to those in other herbivores to make much of the knowledge gained from the ruminant pertinent to other anaerobic habitats.

The anatomical and physiological adaptations of ruminants to accommodate the large quantity of plants consumed and to maintain the concentrated population of microbes needed for processing these plants are highly specialized. The consumption of forage stimulates submaxillary and sublingual glands to secrete into the mouth large volumes of viscous saliva containing a relatively high concentration of sodium bicarbonate. The latter glands secrete continuously at a low basal rate, reducing rumen acidity as fermentation acids are produced, and the rate increases during rumination.

The masticated mixture of forage and saliva is swallowed and passes down a fairly long oesophagous, past the cardiac sphincter into the reticulum, which is an anterior sac communicating directly with the rumen (Fig. 2) but separated from it on the ventral side by the rumino-reticular fold. The rumen is much larger than the reticulum, and is actually a blind sac extending from it posteriorly. The exit from the reticulum into the omasum is a few inches ventral to the entrance into it from the oesophagous. The two openings are connected by an oesophageal groove in the inner wall of the reticulum, between two muscular ridges. These are under reflex control and in the young animal can contract in a way that converts the groove into a closed channel leading directly from the oesophagous to the omasum and through it to the acid stomach, the abomasum (Fig. 3). Suckling induces this closure, allowing

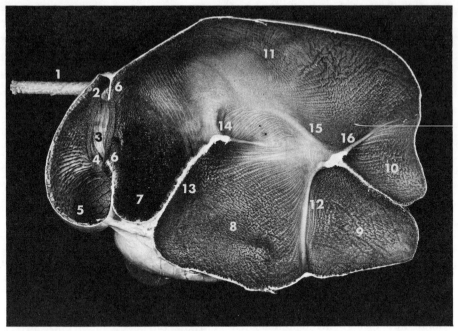

Fig. 2. The right half of a dried rumen prepared by N. J. Benevenga, as seen from the inside: (1) oesophagus; (2) cardia; (3) oesophageal groove; (4) reticulo-omasal opening; (5) reticulum; (6) rumino-reticular fold; (7) anterior sac of the rumen; (8) ventral sac of the rumen; (9) ventral blind sac; (10) dorsal blind sac; (11) dorsal sac; (12) ventral coronary pillar; (13) anterior transverse fold; (14) anterior transverse pillar; (15) longitudinal pillar; (16) dorsal coronary pillar. From Hungate (1966) by permission of Academic Press.

the swallowed milk to bypass the fermentation in the rumen-reticulum, pass by the edges of the omasal leaves, and enter the acid stomach unaffected by microbial activity; this is an exception to the usual pre-peptic microbial attack on ingested food. Peptic digestion in the abomasum is followed by tryptic digestion of protein when the digesta enter the duodenum.

The complex structural adaptations of this mutualistic digestive system are matched by the intricacies of their functions. The fermentation occurs in the rumen-reticulum. The muscles in the walls of this chamber undergo a series of coordinated contractions which mix the contents and cause the rumen liquid to well up and spill over the solids digesta which are slowly rotated and mixed by the contractions. An observer can note through an open fistula the movement of the solids mass caused by the contractions. Some of the solids particles spill over the rumino-reticular fold into the reticulum. If they are sufficiently abundant and stimulatory, the reticulum may contract as the cardiac sphincter into the oesophagous opens, and a reverse peristalsis propels liquid and particles into the mouth, initiating rumination. Much of the liquid is immediately swallowed, and the particles are masticated and mixed with saliva before deglutition. This act of rumination occurs more frequently in

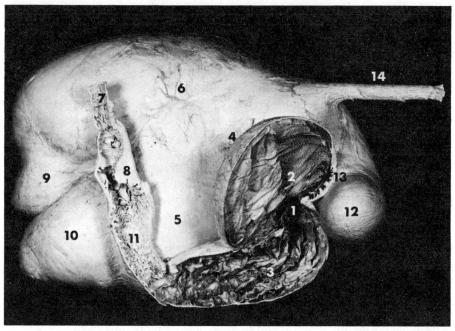

Fig. 3. An exterior view of the right half of the dried rumen of Fig. 2: (1) omasal-abomasal orifice; (2) leaves of the omasum; (3) longitudinal spiral folds of the abomasum; (4) greater curvature of the omasum; (5) region of the ventral sac of the rumen; (6) region of the dorsal sac of the rumen; (7) duodenum; (8) pylorus; (9) dorsal blind sac; (10) ventral blind sac; (11) pyloric portion of the abomasum; (12) reticulum; (13) vestibule of the omasum; (14) oesophagus. From Hungate (1966) by permission of Academic Press.

animals on a diet of coarse hay than on fresh, leafy forages. It hastens digestion and reduces particle size to the point where all solids ultimately suspend in the rumen liquid and pass with it between the omasal leaves (Fig. 3). The large surface of the omasal leaves favours absorption of water and fermentation products, being especially permeable to the undissociated lower volatile fatty acids. The microbial bodies produced in the rumen fermentation are digested in the abomasum and small intestine.

Features of the rumen microcosm

The microbial population required to metabolize the resistant components of forage at a rate sufficient to supply the animal's needs is considerably more concentrated than that usually achieved in batch cultures *in vitro* of similar substrates inoculated with rumen contents. Here, accumulation of acid products of fermentation usually inhibits growth if more than a few tenths of a per cent dry weight of fermentable substrate are provided (Giallo *et al.*, 1983). Such a substrate concentration produces a population much too small to solubilize structural plant carbohydrates at the rate required. The animal's problem is to obtain maximal microbial growth, i.e. to maintain conditions in which the product of the size of the microbial population times its growth rate is greatest (Adams & Hungate, 1950). This is accomplished in the ruminant by important additional physiological adaptations.

The alkaline saliva, copiously secreted into the mouth during feeding and at a somewhat lesser rate during other periods, keeps the rumen contents at a pH level of 6·0–6·7, partially neutralizing the acid wastes from the formation of the microbial cells. The redox reactions producing the acids involve greater decrements in free energy than do reactions capable of further conversion of the acids into methane and carbon dioxide. The microbial growth rate on fibre carbohydrates in the rumen, determined by the turnover rate of the rumen liquid and solids, is faster than the growth rate of microbes able to convert the acids into methane. The rumen methane comes from H_2 reduction of CO_2, not from the volatile fatty acids.

Free-energy decrements and rates of microbial growth

Events associated with sudden changes in the rumen fermentation when excess soluble carbohydrates are available provide insight into the factors concerned in the continuous production of the volatile fatty acids in the rumen fermentation. The work done during the initial stages of anaerobic carbohydrate metabolism is the synthesis of adenosine triphosphate (ATP) or similar molecules high in energy content, and capable of participating in a wide variety of further exergonic reactions of cell synthesis. The energy decrements in the fermentation reactions leading to the formation of the lower volatile fatty acids as waste products permit synthesis of more than two ATP molecules per hexose molecule fermented, as compared to only two ATP when lactic acid is the chief fermentation waste product. This accounts for the preponderance of the volatile fatty acids over lactic acid as the usual product of the rumen fermentation. However, if a large amount of starchy grain is suddenly administered to a ruminant normally fed on forage, there is a sudden change from the usual fermentation waste products to solely lactic acid, produced so rapidly that the ruminant cannot adjust and suffers an acute, and even fatal, acid indigestion (Hungate et al., 1951).

Forage usually contains a small concentration of starch, and the rumen a relatively small number of lactic acid bacteria capable of rapidly converting starch into soluble sugar which they ferment to form lactic acid. These bacteria are greatly outnumbered by the rumen bacteria digesting the abundant structural carbohydrates, but this digestion occurs more slowly than does digestion of starch, and even though the fibrolytic microbes can synthesize more ATP per hexose molecule the lactic acid bacteria can solubilize sugar from starch sufficiently rapidly to offset this advantage. With excess starch available they can synthesize more ATP per unit time, and almost explosively outgrow the fibre digesters. This can be interpreted as indicating that ATP per unit time, i.e. power, is a basic factor determining the rate of microbial growth. Under conditions of limited availability of soluble carbohydrate, power is a function not only of carbohydrate concentration but also of the amount of chemical work accomplished by the fermentation reaction, in this case the number of ATP that can be synthesized per hexose, the ATP in turn synthesizing cell material.

The free-energy decrement in the rumen conversions of solubilized structural carbohydrates into lower volatile fatty acids, carbon dioxide and dihydrogen is approximately 60 cal per mole of hexose; at a conversion efficiency of 30% (derived from studies on oxidative phosphorylation) this is sufficient for the synthesis of 2·3 moles of ATP, each storing about 7·7 cal. Under anaerobic conditions the efficiency may be somewhat greater, but at least this calculation indicates that the ATP yield per hexose fermented in the rumen is greater than two.

Successive cell divisions of the rumen microbes must occur within 0·69 ($=\ln 2$) of the turnover time in a continuously mixed rumen pool in order to avoid washout. Turnover time in hours is calculated as the total material of any sort in a pool divided by the rate per hour at which it enters or leaves. A rumen pool of liquid and small particles (LSP) with a turnover time of 10·6 hours would require a minimum cell division interval of 7·3 hours for microbes suspended in it. Actually, many fibrolytic microbes may be attached; their growth rate could considerably exceed the turnover rate of the fibre.

The 32 cal energy decrement in the reaction, $4H_2 + CO_2 \rightarrow CH_4 + 2H_2O$, enables hydrogenotrophic methanogens to grow at rumen liquid turnover rates and maintain a population sufficiently dense to keep the concentration of H_2 in rumen contents at its measured very low level.

The reactions by which the rumen volatile acids can be further converted into methane do not provide sufficient power to enable an aciditrophic methanogenic consortium to maintain itself against washout at rumen LSP turnover rates. The turnover rate in anaerobic sludge-digesters is much slower, and the free energy decrements in the anaerobic oxidation of propionate and butyrate to form acetate plus CO_2 and CH_4 are sufficient to grow the microbes concerned at rates avoiding washout in sewage and other sludge fermentations. The free-energy decrement in the split of acetate into methane and CO_2 is barely sufficient to synthesize one ATP; it is not enough to support growth at rumen turnover rates. Acetate removal by methanogens in sewage sludge increases the free-energy decrement in breakdown of propionate and butyrate.

These considerations lead to the realization that the adaptations in the ruminant include a microbial ecosystem that obtains maximum power from fermentation of fibre, which in turn is tied to maximum production not only of the required amounts of microbial protein but also of non-nitrogenous products in quantities sufficient, through oxidative phosphorylation in the ruminant, to meet the animal's non-protein energy needs.

Adaptations in the rumen microcosm have led not to a single species accomplishing all the chemical work but to a complex community of specialists. Apparently, in the intense competition for survival in the rumen, a species cannot afford to synthesize more metabolic machinery than is required for catalysing and guiding the work of cell synthesis from the particular foods and environment for which the microbe is specifically adapted.

Rumen fermentation products: Acids

The proportions in which the principal volatile fatty acids (VFA) are formed in the rumen are about 63 acetic, 21 propionic, 16 butyric and higher acids. For example, in a lactating cow on a hay-concentrate ration the fermentation can produce as much as 2·4 kg acetic, 0·95 kg propionic and 0·92 kg butyric acid in a day (Hungate, 1966). On a hay ration the amounts are less: 0·86, 0·37 and 0·39 kg, respectively. The butyrate is specifically important for the development of the papillation and musculature of the wall of the rumen, and appears to serve a similar function in the wall of the caecum of rodents; in germ-free mice the wall of the caecum is very thin and greatly distended. In the rumen the large papillae lining its wall greatly increase the surface for absorption of the fermentation acids.

Reliable samples of rumen contents can be obtained via a fistula, an opening operatively introduced into the rumen through the wall of the abdomen. The edges of the abdominal opening are first sutured to the surface of the exposed wall of the rumen. The apposed tissues fuse and, after healing, the exposed rumen wall is removed and a plug placed in the opening to prevent loss of contents. The appetite remains normal and the rumen fermentation resembles that in unoperated animals even though the sequences of the muscular contractions mixing the contents may be somewhat modified. It is possible to make a rather large fistula in cattle, and samples representative of the rumen contents can be easily and repeatedly withdrawn.

Reliable comparative estimates of the rates of fermentation within the rumen can be obtained in the field by rapidly removing a sample of the contents through a 2 cm diameter tube by applying suction to the receiving vessel. Part of the liquid may need to be removed to make the solids/liquid ratio of the sample resemble that in the rumen contents *in vivo*. During this process, and later, the sample is held at 39°C and the air above it is displaced with O_2-free N_2 containing 5% CO_2 to balance the rumen bicarbonate at a pH of 6·7. The fermentation acids release CO_2 from the bicarbonate and this, together with the CO_2 and CH_4 produced in the fermentation, can be measured with a water-lubricated hypodermic syringe, the needle being inserted and withdrawn directly through the stopper closing the incubating container.

The rate of gas production diminishes with time. By plotting gas production at frequent intervals following removal of the sample, and extrapolating the resulting curve back to zero time, a good estimate of the rate in the rumen contents *in situ* can be obtained. More accurate and complete estimates can be made with more elaborate equipment, but with any *in vitro* incubation method it is important to make measurements as soon as possible after sampling. The products of the high rate of fermentation quickly accumulate and diminish the rate. Although the total fermentation rate is high, so also are the numbers of microbes; the rate per unit of microbes is much less than maximal in the rumen, and rapidly diminishes when the sample is separated from the supply and removal factors operating *in vivo*.

In the usual laboratory-broth pure culture the rate of fermentation per microbial cell is maximal for a short period after inoculation, with a constant generation time until inhibitory products and diminished food concentrations limit the specific growth-rate as the population density increases. In the rumen, following utilization of relatively soluble and easily digested fractions of the ingested forage, the plant cell-wall components become the source of fermentable carbohydrate, and the rate of their solubilization by microbial enzymes limits carbohydrate availability. A very large population is needed in order to produce enough enzymes to digest fibre sufficiently rapidly to meet the animal's needs for fermentation products. The most efficient way to maintain such a concentrated microbial population is by a steady-state continuous fermentation system. The rumen approximates to this.

Cattle spend each day about 8 hours feeding, 8 hours ruminating and 8 hours resting. Rumination increases the rate at which ingested fibre is comminuted, digested and leaves the rumen. There are fluctuations in the rate of intake and outflow, volume, solids/liquid ratio and other parameters, but interpretation on the basis of a steady-state continuous fermentation can provide a meaningful quantification.

Rumen fermentation products: Microbes

The rumen microbes are not only the agents producing the forage-digesting enzymes and the fermentation acids oxidized by the host, they themselves constitute the host's supply of proteins and other essential nutrients. Estimation of the quantity and quality of the microbial bodies supplied to the ruminant is one of the most difficult problems in rumen microbiology, yet thorough understanding of ruminant nutrition requires it. Identification of the numbers and activities of each kind of rumen microbe may ultimately be obtainable through culture counts and pure cultures, but this path to assessment of the quantitative significance to the host of the entire rumen microcosm would be difficult, and at best very time-consuming. An attempt has been made (Hungate *et al.*, 1971) to estimate indirectly the microbial yield from the rumen fermentation.

To simulate a steady-state continuous fermentation in a sheep, a 55 kg Corriedale wether with a rumen fistula was fed 90 g of alfalfa pellets at 2 hour intervals (976 g dry wt/day) by an automatic machine, with water available *ad lib.*, over a period of 7 months. After 1 month, faeces were collected each day for ten consecutive days, dried, and a pooled representative sample was subjected to elemental and proximate analyses. An average daily feed–faeces difference in dry weight (607 g) was used as the reactant in a chemical equation, the products being the amounts of volatile fatty acids, methane, carbon dioxide, ammonia and microbial cells formed per day. Their quantities were obtained from elemental analysis or from the chemical formulae and the measured VFA, methane, carbon dioxide and ammonia production rates. The carbon missing from the measured products as compared to the feed–faeces difference was assumed to represent that in the microbial cells digested and absorbed. The cell content in oxygen, hydrogen and nitrogen was assigned from published analyses for bacteria (Mayberry *et al.*, 1968) and protozoa (Weller, 1957). By these procedures the values in the following equation were obtained:

$$C_{20}H_{37}O_{17.4}N_{1.345} \rightarrow C_{12}H_{24}O_{10} + 0.83CH_4 + 2.76CO_2 + NH_3$$
$$\text{feed–faeces diff.} \qquad\qquad + C_{4.44}H_{8.88}O_{2.35}N_{0.785} \quad (1)$$
$$\text{microbial cells}$$

The agreement between the quantity of each chemical element in the reactants as compared to the products is surprisingly good considering the estimates involved. The percentages by which the total gram-atoms in the products differ from those in the reactants are about 2% for the H, 6% for the O and 5% for the N. The weight of the microbial cells is 110·75 g dry weight; 116·5 g including expected inorganic elements.

An independent estimate of the quantity of microbes was obtained from 6 samplings of rumen contents at spaced intervals between March and May. Samples of the liquid rumen contents (LSP) containing suspended microbes and other small particles were fixed and stained. The contained bacteria were counted in a Petroff–Hauser counting chamber, while at the same time the dimensions and volume of each cell were determined. From the average count in each size (volume) range the total rumen bacterial volume was calculated to be 93 μl/ml, with a wet weight of 103 mg/ml (assuming a specific gravity of 1·1 for the wet cells). A known volume of the LSP was mounted under a coverslip, the protozoa were counted and sized and their volume was estimated at 105 μl/ml (115 mg/ml). The total microbial dry weight was estimated to be 10% of the wet weight, giving 21·8 mg/ml.

Samples of the total rumen contents (TR) were removed at the same samplings as the LSP material. They contained an average total microbial volume of 218 μl/ml as compared to the 198 for the LSP pool. The latter had been sampled by inserting into the rumen contents a half-inch copper tube with numerous $\frac{1}{8}$ inch holes bored through 6 inches of the wall of the tubing just above the closed lower end. The liquid and suspended small particles draining into the tube were withdrawn through a $\frac{1}{4}$ inch plastic tube with attached rubber bulb to apply suction. This provided a sample of the LSP pool.

The microbial volume of the LSP pool in the rumen was used instead of that obtained from samples of the total rumen contents, in part because it was more conservative, but chiefly because a turnover rate constant could be estimated for the LSP pool. A known quantity of polyethylene glycol (PEG) was added through the fistula as a material miscible with the liquid, not metabolizable by the microbes, and easily determined. On the premise that the rate at which added PEG leaves the rumen is a first-order function of its concentration, the log of the concentration against time yields a straight line with a slope showing the rate constant at which PEG and the LSP pool leave the rumen. This was determined six different times between February and June and the data from all these experiments were plotted together, giving a linear regression with a slope of 0·0946 per hour, i.e. a turnover time of 10·6 hours or 2·27 times per day.

It is taken that all of the microbes in the rumen are viable. Viability is easily confirmed in the case of the protozoa by their active motility and intact intracellular structures, but the bacteria are only assumed to be viable. PEG does not enter living cells, and its initial concentration after mixing in the rumen indicates the liquid volume with which it mixes. This averaged 3·77 litres, or 4·7 litres including the 197 μl/ml of suspended microbial bodies, with a microbial dry weight of 93 g in the entire LSP volume. If the LSP pool turned over at the PEG-measured rate the ruminant would have been supplied with 211 g of microbes per day. This considerably exceeds the estimate of 116·5 g from eqn (1). Part of the discrepancy can be due to incomplete digestion of the microbial cells in the gut posterior to the rumen. However, even if all of the 4·7 g of nitrogen in the faeces represented undigested microbial nitrogen, it would still indicate production of 167 g of microbes per day in the rumen of this animal. Other possible errors in the 211 g estimate are that the microbes in the LSP pool did not turn over at the PEG rate, and that the microscopic measurements overestimated the actual microbial volumes.

In summary the indirect estimate of 116·5 g dry weight from eqn (1) appears to give the best estimate of the weight of the microbes produced in the rumen fermentation of this sheep. It underestimates to the extent that the microbes were not completely digested.

The proximate analyses of feed and faeces showed a disappearance of 306 g of cellulose, hemicellulose and water-soluble carbohydrates. It also showed 126 g of 'other carbohydrates' not recovered in the faeces. If the total 432 g be regarded as the substrate supporting the rumen fermentation, and is assigned an average molecular weight of 162 g for its component sugars, it contained 2·66 moles of sugars. To account for the 116·5 g of microbial cells produced in the rumen fermentation, each mole must have produced an average of 43·7 g of cells. On the basis of 11 g of cells from each mole of ATP, an average of 4 ATP were formed from each sugar molecule fermented. This estimate, probably minimal, represents the efficiency of microbial-cell production by the rumen fermentation in this sheep.

Continuous feeding of cellobiose on a 10-hour turnover time to a pure culture of a rumen cellulolytic anaerobe, *Ruminococcus albus*, gave a dry cell yield of 45·5 mg per mole hexose, after correction to a nitrogen content of 10·5% to avoid including storage polysaccharide which can be abundant in cultures fed on soluble sugar (Hungate, 1963). This better controlled, laboratory, continuous pure culture of a fibrolytic rumen bacterium gave a result closely approximating to that found with the total microbiota in a live sheep.

Reconciliation with the biochemistry of ATP synthesis in fermentations

In most microbial fermentations of carbohydrates two biochemical steps generating ATP are well known. These are its formation from both 1,3-diphosphoglyceric acid and acetyl phosphate. Also, in propionic acid formation an ATP is generated when fumarate is reduced to succinate (Macy et al., 1975). Thus biochemical pathways are known for synthesis of 4 ATP per sugar when acetic or propionic acid is formed. The known steps in formation of butyric and higher acids account for 3 ATP per sugar. Species interactions in the rumen may increase this. It might seem that microbial species in a habitat would each need the same efficiency in ATP production in order to compete successfully. However, species adapted to attachment to the rumination pool with its slower turnover would not need to grow as rapidly as those turning over more rapidly. Or, is it possible that the attachment and elaboration of fibrolytic enzymes increases the work to be done in the synthesis of these organisms? These questions cannot be answered with our present knowledge but they suggest variations in the amount of work to be done in the synthesis of the cells of the various microbial species, and also variation in the efficiency with which ATP can be generated.

Enzymatic digestion of fibre

The success exhibited by ruminants and other animals using mutualistic microbes to digest fibre has excited much interest in the mechanisms involved. All celluloses are not necessarily alike (Van Soest, 1973). The defatted fibres of uniform diameter growing on the seeds of cotton have been most commonly used in enzyme studies because their major component is cellulose, the carbohydrate polymer most resistant to digestion. Many cellulose-digesting bacteria isolated from the rumen show complete clearing of finely divided cellulose dispersed in the agar medium adjacent to a colony, an indication that an effective extracellular cellulase is released. In other instances in which the microbes can move freely, either in liquid or through the agar, they appear to attach directly to the surface of the cellulose; no cleared area is evident.

A soluble derivative of cotton cellulose, carboxymethylcellulose (CMC), has been used as a substrate for the investigation of cellulase activity, as it gives definitive results in a shorter time than most substrates. This rapidity may be due in part to the dissolved condition, allowing sufficient movement at the point of enzyme attachment to change the configuration, releasing the enzyme and the separated chains. The digestion of insoluble structural carbohydrates probably involves considerably more than the breaking of a single bond. It would seem difficult for an enzyme placing breaking-stress on a single bond in a solid matrix to disengage without the bond re-forming.

Some of the enzymes most effective in fibre digestion are very large, of the order of 2×10^6 daltons per soluble unit (Ljungdahl & Eriksson, 1985; Yu & Hungate, 1979; Coughlan et al., 1985). Such large units must contain many active sites capable of breaking numerous structural bonds on the surfaces of fibres composing the cell wall, their combined action releasing soluble products. The structure of an adsorbed giant enzyme could conceivably be sufficiently porous and flexible to allow diffusion of solubilized products through the enzyme, and into the cell if the enzyme was part of the cell surface (Stack & Hungate, 1984). The demonstration (Eriksson & Wood, 1984) of a marked synergism in cellulose digestion by mixed solutions of several kinds of cellulases suggests that inclusion of multiple types of active sites in a giant enzyme or on a microbial cell surface might be even more effective. In addition to glucose, other 6- or 5-carbon sugars, as well as mannuronic and galacturonic acids, are components of the cell walls of alfalfa (Greve et al., 1984; McNeil et al., 1980). Linkages involving them could require many additional types of specificities.

CHALLENGES IN RUMEN MICROBIOLOGY

Microbial ecosystems in the guts of animals are unique in several respects. Each is basically an enrichment culture continuously sustained over millions of years, at constant or variable temperature, degrading photosynthetically renewable plant bodies at a rapid rate. Historically, the microbes themselves and the degradation of the structural polymers have excited the most interest, and will continue to be fruitful subjects for research. However, each of the less abundant components of plants, including ring compounds (Tsai & Jones, 1975), may also have selected for a microbe, or mixture of microbes, capable of supporting growth, and analyses of their metabolic pathways may disclose new mechanisms of chemical work.

Bryant and his students (Bryant, 1979; McInerney & Bryant, 1981) have shown that mixtures of microbial species can ferment substances, including phenolic compounds, undegradable by a single species alone. This syntrophy often involves methanogenic or sulphate-reducing strains oxidizing hydrogen with carbon dioxide or incompletely reduced sulphate compounds. Presumably this removal of hydrogen increases the available free-energy to the point that ATP can be generated sufficiently rapidly to maintain the species in the system.

Some of the problems in sorting out the enzymes concerned in fibre digestion may be solved by immunogenetic techniques for producing each specific enzyme in quantities sufficient to permit identification of its role alone and in conjunction with others. Other rumen phenomena will be similarly aided by application of these new procedures.

CONCLUSION

The aim of this introduction is to stimulate interest in reading the results, approaches and interpretations of the remaining contributors to this volume portraying the current state of knowledge of rumen microbiology. The analyses of any ecosystem are of interest. That in the rumen is complex but it is easily accessible, inexpensive to maintain, and suitable for statistically significant measurements. Its further description will be aided by application of quantitative methods describing and interlinking the rates of the many interacting processes concerned in the life together of so many microbial species using as food so many complex materials, processes made possible by the physiological adaptations of the ruminant harbouring them, and they in turn responsible for the health and growth of the host.

REFERENCES

Adams, S. L. & Hungate, R. E. (1950). Continuous fermentation cycle times. Prediction from growth curve analysis. *Ind. Eng. Chem.*, **42**, 1815–18.

Barcroft, J., McAnally, R. A. & Phillipson, A. T. (1944). Absorption of volatile acids from the alimentary tract of the sheep and other animals. *J. Exp. Biol.*, **20**, 120–9.

Bignell, D. E. (1984). Direct potentiometric determination of redox potentials of the gut contents in the termites, *Zootermopsis nevadensis* and *Cubitermes severans*, and three other arthropods. *J. Insect Physiol.*, **30**, 169–74.

Broda, E. (1975). *The Evolution of the Bioenergetic Processes*. Pergamon Press, Oxford.

Bryant, M. P. (1979). Microbial methane production: theoretical aspects. *J. Anim. Sci.*, **48**, 193–201.

Coughlan, M. P., Hon-nami, K., Hon-nami, H., Ljungdahl, L. G., Paulin, J. J. & Rigsby, W. E. (1985). The cellulolytic enzyme complex of *Clostridium thermocellum* is very large. *Biochem. Biophys. Res. Comm.*, **130**, 904–9.

Elsden, S. R. (1946). The application of silica gel partition chromatography to the estimation of volatile fatty acids. *Biochem. J.*, **40**, 252–6.

Eriksson, K.-E. & Wood, T. M. (1984). Biodegradation of cellulose. In *Biosynthesis and Biodegradation of Wood Components*, ed. T. Higuchi. Academic Press, New York, pp. 469–503.

Fisher, W. K. & MacGinitie, G. E. (1928). The natural history of an echiuroid worm. *Annals and Magazine of Natural History*, S.10, **1**, 199–204.

Giallo, J. C., Gaudin, C., Belaich, J. P., Petildemange, E. & Caillet-Mangin, F. (1983). Metabolism of glucose and cellobiose by cellulolytic mesophilic *Clostridium* sp. Strain H10. *Appl. Environ. Microbiol.*, **45**, 843–9.

Greve, C., Labavitch, J. M. & Hungate, R. E. (1984). Xylanase action on alfalfa cell walls. *Proc. 7th Ann. Symp. Botany* (Univ. Calif., Riverside), ed. W. M. Duggar & S. Bartnicki-Garcia, pp. 150–66.

Hungate, R. E. (1955). Why carbohydrates? In *Biochemistry and Physiology of Protozoa*, Vol. II, ed. S. H. Hutner & A. Lwoff. Academic Press, New York, Appendix II, pp. 195–7.

Hungate, R. E. (1963). Polysaccharide storage and growth efficiency in *Ruminococcus albus*. *J. Bacteriol.*, **86**, 848–64.

Hungate, R. E. (1966). *The Rumen and Its Microbes*. Academic Press, New York.

Hungate, R. E. (1985). Anaerobic biotransformations of organic matter. In *Bacteria in Nature*, Vol. 1, ed. E. R. Leadbetter & J. S. Poindexter. Plenum Press, New York, pp. 39–95.

Hungate, R. E., Dougherty, R. W., Bryant, M. P. & Cello, R. M. (1951). Microbiological and physiological changes associated with acute indigestion in sheep. *Cornell Veterinarian*, **42**, 423–49.

Hungate, R. E., Reichl, J. & Prins, R. A. (1971). Parameters of rumen fermentation in a continuously fed sheep: evidence of a microbial rumination pool. *Appl. Microbiol.*, **22**, 1104–13.

Ljungdahl, L. G. & Eriksson, K.-E. (1985). Ecology of microbial cellulose degradation. *Adv. Microbial Ecol.*, **8**, 237–99.

Luria, S. E. (1960). The bacterial protoplasm: composition and organization. In *The Bacteria*, Vol. I, ed. I. C. Gunsalus & R. Y. Stanier. Academic Press, New York, Ch. 1, pp. 1–34.

Macy, J. M., Probst, I. & Gottschalk, G. (1975). Evidence for cytochrome involvement in fumarate reduction and adenosine-5'-triphosphate synthesis by *Bacteroides fragilis* growth in the presence of hemin. *J. Bacteriol.*, **123**, 436–42.

Mayberry, W. R., Prochazka, G. J. & Payne, W. J. (1968). Factors derived from studies of aerobic grown in minimal media. *J. Bacteriol.*, **96**, 1424–6.

McInerney, M. J. & Bryant, M. P. (1981). Anaerobic degradation of lactate by syntrophic association of *Methanosarcina barkeri* and *Desulfovibrio* species and effect of H_2 on acetate degradation. *Appl. Environ. Microbiol.*, **41**, 346–54.

McNeil, M., Darvill, A. G. & Albersheim, P. (1980). The structural polymers of the primary cell walls of dicots. *Progr. Chem. Nat. Prod.*, **37**, 191–250.

Moir, R. J. (1968). Ruminal digestion and evolution. In *Handbook of Physiology*, V, ed. C. F. Code. Ch. 126, pp. 2673–94.

Oparin, A. I. (1953). *Origin of Life*, 2nd edn. Dover, New York (translation into English of the 1st (1938) edn, in Russian).

Orpin, C. G. (1977). On the induction of zoosporogenesis in the rumen phycomycete *Neocallimastix frontalis, Piromonas communis* and *Sphaeromonas communis. J. Gen. Microbiol.*, **101**, 181–90.
Parle, J. (1968). Micro-organisms in the intestine of earthworms. *J. Gen. Microbiol.*, **31**, 1–11.
Rohmann, G. F. (1978). The origin, structure and nutritional importance of the comb in two species of Macrotermitinae (Insecta, Isoptera). *Pedobiologia*, **18**, 89–98.
Sprengel, C. (1832). *Chemie für Landwirte, Forstmänner und Cameralisten*, Part I. Vandenhouk and Ruprecht, Göttingen.
Stack, R. J. & Hungate, R. E. (1984). Effect of 3-phenylpropanoic acid on capsule and cellulases of *Ruminococcus albus* 8. *Appl. Environ. Microbiol.*, **48**, 218–23.
Troyer, K. (1984). Microbes, herbivory and the evolution of social behavior. *J. Theor. Biol.*, **106**, 157–70.
Tsai, C. G. & Jones, G. A. (1975). Isolation and identification of rumen bacteria capable of anaerobic phloroglucinol degradation. *Can. J. Microbiol.*, **21**, 749–801.
Van Soest, P. J. (1973). The uniformity and nutritive availability of cellulose. *Fed. Proc.*, **32**, 1804–8.
Weller, R. A. (1957). The amino acid composition of hydrolyzates of microbial preparations from the rumen of sheep. *Austral. J. Biol. Sci.*, **10**, 384–9.
Winfield, M. E. (1978). *From Massless to Material Universe*. Craftsman Press, Spit Junction, Australia.
Yu, I. & Hungate, R. E. (1979). The extracellular cellulases of *Ruminococcus albus*. *Ann. Recherche Veterin.*, **10**, 251–4.
Zuntz, N. (1879). Gesichtspunkte zum kritischen Studien der neueren Arbeiten auf dem Gebiete der Ernährung. *Landwirtschaft. Jahrbuch.*, **8**, 65–117.

2
The Rumen Bacteria

C. S. Stewart

Rowett Research Institute, Aberdeen, UK

&

M. P. Bryant

Departments of Animal Science and Microbiology, University of Illinois, Urbana, Illinois, USA

The free-swimming ciliates and flagellates are the most immediately obvious inhabitants of rumen contents examined by light microscopy. Among the bacteria, the large ovals, rods and sheets of cocci up to around 50 μm in dimension, although present in small numbers, are often at first sight more distinctive than the teeming populations of small bacteria, the study of which has provided most of our understanding of rumen microbiology. This understanding has been gained slowly. Most rumen bacteria require strictly oxygen-free conditions for growth, and at first the spread of investigations on these bacteria was delayed by the limited availability of rubber stoppers that were sufficiently impermeable to oxygen to allow the maintenance of anaerobic conditions in test-tubes (Hungate, 1966). Anaerobic culture conditions are now provided by the roll-tube technique (Hungate, 1969) or by the use of anaerobic chambers (Leedle & Hespell, 1980).

The very large numbers of bacteria present in the rumen (up to 10^{11} viable cells/ml) were recognised in the early years of rumen studies (Hungate, 1966). More recently, scanning- and transmission-electron microscopy have enabled us to see that the rumen microflora is compartmentalised, different populations being associated with feed particles, the rumen wall and the liquid phase of the rumen contents (see Chapter 12). The distribution of feed particles in the rumen is dictated by the contraction and relaxation of the rumen wall (Wyburn, 1980), by the selection of regurgitated material for chewing during rumination (the 'Up' bolus; Ulyatt *et al.*, 1986) and by the fact that, as microbial digestion proceeds, many feed particles accumulate to form a buoyant mat on the liquid surface (Chapter 12). In a study by Bryant & Robinson (1968) it was found that the numbers of bacteria were highest in the dorsal area of the rumen while the relative numbers of bacteria present in the reticulum and the ventral rumen varied according to the diet. Bryant & Robinson took particular care to ensure that their sub-samples taken for counting contained the same proportion of liquid and solids as the whole sample. The difficulty of truly representative sampling probably distorts our perception of the relative importance of the different species isolated. More significantly, we do not know if we have now isolated all of the species of small bacteria present in the rumen. The very large bacteria are at least highly distinctive; we are all too aware of our inability to grow most of them in pure culture. However, the rumen contains representatives of all of the major morphological forms of small bacteria, with Gram-positive and -negative rods, cocci, crescents, vibrios and helices occurring singly, in chains and in clumps, as illustrated by Ogimoto & Imai (1981). Although representatives of all of these morphological types have been isolated in pure culture, each type could embrace several different species, some of which may not be cultivable. In addition, bacteria that require for their growth the maintenance of syntrophic associations, such as the presence of a suitable hydrogen-acceptor, may never be isolated in pure culture and will be missed in conventional isolation strategies involving picking single colonies from roll-tubes or plates.

The isolation of rumen bacteria can be carried out using habitat-simulating media containing clarified rumen fluid and a range of fermentable substrates, or media tailored to the nutrient requirements of particular functional groups or species. (A brief summary of media and methods for the isolation and cultivation of small rumen bacteria is presented at the end of this chapter.) Having obtained a culture, identification usually begins with assessment of (1) Gram reaction and cell morphology, (2) motility, (3) ability to grow aerobically, (4) major fermentation products, and (5) the range of substrates fermented, and reactions (such as nitrate reduction, production of H_2S or indole) catalysed. Methods for the performance of these and other identification tests are described by Holdeman et al. (1977) and by Ogimoto & Imai (1981). Not all of these tests give unequivocal results. The colour reaction during Gram staining does not always correlate with the cell-wall ultrastructure as seen by electron microscopy (see for example *Butyrivibrio* and *Ruminococcus*, below). The presence of spores can be difficult to determine and has caused some confusion in the past (Van Gylswyk et al., 1980). Bacterial motility can be influenced by the culture conditions (Sharpe et al., 1973; Hudman, 1984) but, provided that relatively young cultures grown on low ($<0.1\%$) concentrations of substrates are used, motility is easy to detect in 'hanging drops' (Ogimoto & Imai, 1981). Bryant (1972) found that motility was easiest to detect in wet mounts of the water of syneresis of slants containing low concentrations of substrates and sampled at early stages of growth. The position and arrangement of flagella, particularly diagnostic in the identification of Gram-negative rods, can be determined most accurately by electron microscopy (Henderson & Hodgkiss, 1973), although the flagella-staining methods for light microscopy (Holdeman et al., 1977) are very successful in the hands of some investigators. The oxygen sensitivity of some strains of rumen bacteria has been measured (Loesche, 1969), and it was concluded that, even within the group of organisms that are normally considered as anaerobic, bacteria such as *Megasphaera* and *Veillonella* were more tolerant of oxygen than the strains of *Butyrivibrio*, *Selenomonas*, *Succinimonas*, *Succinivibrio* and *Lachnospira* that were studied. In the absence of oxygen, a number of rumen bacteria have been shown to be capable of growth at a wide range of Eh values, from -290 mV or below, up to $+100$ mV (*Selenomonas ruminantium*) to $+414$ mV (*Streptococcus bovis*) (Marounek & Wallace, 1984).

Of the chemotaxonomic methods for bacterial identification, those which have been most successfully applied to rumen bacteria include: determination of the guanine plus cytosine content of the DNA; determination of the fatty acid composition of the cells; detection of cell-wall components such as diaminopimelic acid (Sharpe et al., 1973) and 2-keto-3-deoxyoctonate (KDO; Ogimoto & Imai, 1981); characterisation of the teichoic acids (Sharpe et al., 1975); and determination of the presence or absence of cytochromes (Reddy & Bryant, 1977) and other electron carriers (Shah & Collins, 1983). Techniques so far less widely used for rumen bacteria, but of obvious potential value, include determination of nucleic acid homologies (Johnson, 1978; Johnson & Harich, 1986; Rogosa, 1984), including short-sequence homologies using labelled probes; comparison of polypeptide patterns upon electrophoresis of cell proteins (Begbie & Stewart, 1984); comparison of the electrophoretic mobility of enzymes (Shah & Collins, 1983). In addition to such generally recognised chemotaxonomic criteria, other tests could be used as an aid to identification. Essers (1982) has suggested that antibiotic sensitivity patterns could be used to identify anaerobes, at least to genus level. The responses of rumen bacteria to the animal-feed antibiotics avoparcin, monensin and lasalocid could also be an aid to identification, especially for those bacteria like *Butyrivibrio* that stain Gram-negative but are physiologically and cytologically Gram-positive. However, since antibiotic resistance genes are sometimes mobilisable (Foster, 1983), such tests must be used with caution.

THE DIVERSITY OF THE RUMEN FLORA

The activity of the bacterial flora of the rumen is not constant, but varies according to changing conditions within the gut. In the wild, the biggest cause of changes in activity is the seasonal changes that occur in the profusion and composition of the vegetation. These changes occur globally, and to different patterns according to geographic conditions, but their general nature is illustrated by the studies of Hobson and his colleagues in the Scottish Highlands. They reported peaks of fermentative activity in the rumens of deer feeding on fresh spring and summer vegetation, and troughs of activity in autumn and winter as the diet changed to the woody stalks of windblown heaths and heathers (Hobson et al., 1975/76a,b). The study of domesticated animals has illustrated diurnal changes in the activity and composition of the rumen flora (Leedle et al., 1982), particularly in relation to the time after feeding. Some of the most dramatic changes in the bacterial flora occur in response to changes in the composition of the diet (Eadie & Mann, 1970; Allison et al., 1975; Mackie et al., 1978) and its physical form (Thorley et al., 1968); these and other sources of change are reviewed in Chapter 5.

The rumen is an open system; although certain organisms are repeatedly found to be present and clearly occupy a secure ecological niche, many of the bacteria capable of growth in the rumen are not generally regarded as true rumen bacteria, or they have been isolated only on relatively few occasions under specialised conditions. Bryant et al. (1958b), Jayne-Williams (1979), Fonty et al. (1984) and Stewart et al. (1988) have reviewed the very wide range of aerobic, facultative and anaerobic organisms which is found, especially in young ruminants, and this topic is also

Table 1
Some Occasional Rumen Isolates

Property, niche or group	Bacteria and reference
Methanogens	*Methanomicrobium mobile* (Paynter & Hungate, 1968); CoM-synthesising methanogens (Lovley et al., 1984); *Methanobacterium formicicum* (Oppermann et al., 1957)
Eubacterium	*Eubacterium limosum* (Genthner et al., 1981)
Plasmid-borne urease	*Streptococcus faecium* (Cook, 1976)
Plant cell wall digestion	*Bacillus licheniformis, B. circulans, B. coagulans, B. laterosporus* (Williams & Withers, 1983); *Cellulomonas fimi, Clostridium longisporum, C. lochheadii* (Hungate, 1966); *Micromonospora ruminantium* (Maluzynska & Janota-Bassalik, 1974); unidentified irregular coccus (Macy et al., 1984)
Young ruminants[a]	*Bifidobacterium longum, B. globosum, B. adolescentis, B. thermophyllum, B. boum.* (Scardovi, 1981); *B. ruminale* (Trovatelli & Matteuzzi, 1976); *Propionibacterium acnes* (Gutierrez, 1953); *Clostridium hastiforme, C. butyricum, C. perfringens, C. sartagoformum, C. lentoputrescens, C. malenominatum, C. cochlearum, C. oroticum, C. clostridioforme, Bacteroides corrodens, Bact. pneumosintes, Bact. capillosus, Peptostreptococcus intermedius, P. anaerobius, Escherichia coli, Fusobacterium necrophorum* (Jayne-Williams, 1979); *Alysiella filiformis* (Mueller et al., 1984); *Peptostreptococcus productus* (subsequent identification of cultures from Bryant et al., 1958b)
Nitrogen fixation	*Bacillus macerans* (Jones & Thomas, 1974)
Ethanol utilisation	*Clostridium kluyverii* (Hobson & Howard, 1969)
Sulphate reduction	*Clostridium nigrificans* (Hobson & Howard, 1969); *Desulphatomaculum ruminis* (Postgate, 1970); *Desulphovibrio* (McInerney et al. 1979)
Proteolysis	*Clostridium bifermentans* (Hobson & Howard, 1969)
High starch and acidic conditions	*Clostridium perfringens* (Allison et al., 1975); *C. butyricum* (Hobson & Howard, 1969)
Lactobacilli	*Lactobacillus lactis, L. bifidus, L. brevis, L. buchneri, L. cellobiosus, L. plantarum, L. fermentum, L. acidophilus* (Hobson & Mann, 1957; Hungate, 1966)
Yeasts	Various species (Krogh, 1960)

Table 1 (*continued*)

Property, niche or group	Bacteria and reference
Oxalate degradation	*Oxalobacter formigenes* (Dawson *et al.*, 1980; Allison *et al.*, 1985); *Pseudomonas oxalaticus* (Khambata & Bhat, 1953)
Hydrogenation	'*Fusocillus*' sp. (Kemp *et al.*, 1975, 1984*a*,*b*)
trans-Aconitic acid reduction	Gram-negative rods (Russell, 1985*a*)
3-Hydroxy-4-1(H)-pyridone degradation	Unidentified Gram-negative rods (Allison *et al.*, 1983)
Isolated non-selectively	*Bacteroides* (*Mitsuokella*) *multiacidus* (Flint & Stewart, 1987); *B. levii* (Holdeman *et al.*, 1977); *Coprococcus* sp. (Latham *et al.*, 1979)

[a] For further examples see Bryant *et al.* (1958*b*) and Jayne-Williams (1979).

covered in Chapter 5. Perhaps one of the most remarkable occurrences is the presence of the strict aerobe *Alysiella filiformis* on the rumen epithelium of young lambs (Mueller *et al.*, 1984). This bacterium is more commonly found in the buccal cavity, but can occasionally establish on the rumen epithelium as a transient coloniser. Among other aerobic organisms isolated from ruminants are *Acinetobacter* species, *Pseudomonas aeruginosa*, *Alkaligenes faecalis*, *Micrococcus varians* and *Flavobacterium* spp. (Jayne-Williams, 1979). Of the facultative anaerobes, various staphylococci and streptococci are the genera most frequently found in the rumen. Coliforms are present in large numbers in the gut flora of newborn lambs (Jayne-Williams, 1979) and calves (Bryant *et al.*, 1958*b*); although their numbers decline as animals mature, they are invariably present, albeit in low numbers (Jayne-Williams, 1979; Flint *et al.*, 1987). Some of the bacteria isolated from, or observed in, the rumen but not considered in detail in this chapter are listed in Table 1. Many other types have been isolated, but this list illustrates the diversity of physiological types and taxonomic groups that inhabit the rumen, at least on occasions. The groupings in Table 1 are not mutually exclusive and are primarily for convenience; many of the organisms could occupy several different niches. In particular, many of the types listed as occurring in young animals will also occur in older ruminants. Some of the less frequently isolated species, especially those that degrade toxic substances like 3 hydroxy-4-1(H)-pyridone, may be of considerable importance under some conditions (Chapter 14), and it has been suggested that other toxic amino compounds such as indospicine, a toxic compound found in the leguminous shrub *Indigofera*, may also be degraded by rumen microbes that remain to be isolated (Jones, 1985).

PHYLOGENETIC RELATIONSHIPS OF RUMEN BACTERIA

Although the anaerobic conditions in the rumen are highly selective, the microbes that inhabit the system are taxonomically and phylogenetically very diverse. The phylogenetic relationships of some of these bacteria have been elucidated principally from oligonucleotide sequencing of 16S rRNA. Since 1977, the methanogens have been recognised as belonging to a kingdom, the Archaebacteria, which 'appear to be no more related to typical bacteria than they are to eukaryotic cytoplasms' (Woese & Fox, 1977) (see also Chapter 11). The essential uniqueness of the methanogens has been confirmed in more recent studies (Fox et al., 1980), and the rumen species have been shown to belong to very diverse groupings within this kingdom (Balch et al., 1979). The true *Bacteroides* belong to a prokaryotic phylum that also contains the flavobacteria and cytophagas (Paster et al., 1985). Of the rumen species, only *B. ruminicola* belongs in this grouping, the taxonomic and phylogenetic relationships of *B. amylophilus* and of *B. succinogenes* being uncertain (Paster et al., 1985; Stackebrandt & Hippe, 1986). *Selenomonas ruminantium* and *Megasphaera elsdenii* have 16S rRNA oligonucleotide sequences that suggest closer phylogenetic relationships to Gram-positive bacteria, and particularly to the *Bacillus* cluster, than to any of the Gram-negative bacteria tested (Stackebrandt et al., 1985). The phylogenetic relationship of *Ruminococcus* to other species of the Gram-positive bacteria is not clear. The non-cellulolytic, starch-fermenting species of the rumen and human gut, *R. bromii*, and the cellulolytic species *R. flavefaciens* and *R. albus* are closely related (David Stahl & Bryan White, personal communication). In turn, these bacteria are thought to be most closely related to *Peptococcus glycinophilus*, *Clostridium sphenoides* and *C. amino-valericum* (Stackebrandt & Woese, 1981).

The rumen spirochaete *Treponema bryantii* shows closest phylogenetic relationship, on the basis of rRNA oligonucleotide cataloguing, to *T. succinifaciens*, an unidentified oral strain, and three rumen strains, one of which (PB) has been designated as the type strain of the new species *T. saccharophilum* (Paster et al., 1984; Paster & Canale-Parola, 1985).

CHARACTERISTICS OF SOME TYPICAL RUMEN BACTERIA

The properties of some of the rumen bacterial species which are either isolated frequently, or for which the rumen appears to provide a particularly important habitat, are described below. A list of type species is presented in Table 2, together with values (where known) for the mol % G + C in the DNA. Some of the main features of the fatty acid composition of selected species are shown in Table 3, and the pattern of response to monensin, lasalocid and avoparcin is summarised in Table 4.

The rumen *Bacteroides*

A chemotaxonomic review of the *Bacteroides* collated by Shah & Collins (1983) suggested that there are six important chemotaxonomic features of true *Bacteroides*. In summary these are: the production of large amounts of succinic or butyric acid, with acetate or other short-chain acids; the presence of MDH and

Table 2
Type Strains and % Guanine + Cytosine in DNA of Some Rumen Bacteria

Bacterium	Type strain	Mol % G+C in DNA
Bacteroides ruminicola		
ss. brevis	ATCC 19188	50
ss. ruminicola	ATCC 19189	49
Bacteroides (Ruminobacter) amylophilus	ATCC 29744	40–42
Bacteroides succinogenes	ATCC 19169	47–49
Selenomonas ruminantium		
ss. ruminantium	ATCC 12561	54
ss. lactilytica	ATCC 19205	NR
Butyrivibrio fibrisolvens	ATCC 19171	36–41
Anaerovibrio lipolytica	VPI 7553	NR
Treponema bryantii	ATCC 33254	35–37
Treponema saccharophilum	DSM 2985	54
Wolinella succinogenes	ATCC 29543	47
Succinimonas amylolytica	ATCC 19206	NR
Succinivibrio dextrinosolvens	ATCC 19716	NR
Megasphaera elsdenii	ATCC 25940	53–54
Veillonella parvula	ATCC 10790[a]	38–41
Syntrophococcus sucromutans	S195 (Krumholz & Bryant, 1986a)	52
Ruminococcus bromii	ATCC 27255[a]	39–40
Ruminococcus flavefaciens	ATCC 19208	39–44
Ruminococcus albus	ATCC 27210	42–46
Clostridium polysaccharolyticum	ATCC 33142	42
Lactobacillus ruminis	ATCC 27780	44–47
Lactobacillus vitulinus	ATCC 27783	34–37
Lachnospira multiparus	ATCC 19207	NR
Eubacterium ruminantium	ATCC 17233	NR
Eubacterium oxidoreducens	G41 (Krumholz & Bryant, 1986b)	36
Streptococcus bovis	NCDO 597	37–39
Anaeroplasma abactoclasticum	ATCC 27879	29
Anaeroplasma bactoclasticum	ATCC 27112	32–34
Methanobrevibacter ruminantium	M1 (Smith & Hungate, 1958)	30·6
Methanosarcina barkeri	DSM 800[a]	38·8
Methanomicrobium mobile	BP (Paynter & Hungate, 1968)	48·8
Methanobacterium formicicum	DSM 1535[a]	40·7
Oxalobacter formigenes	ATCC 35274	48–51

[a] Not a rumen strain. NR, not recorded.

GDH; G+C content of DNA from 40% to 52% (Table 2); the presence of sphingolipids; the presence of straight-chain saturated and iso- and anteiso-methyl branched long-chain fatty acids (Table 3); and the presence of menaquinones. From both the phylogenetic and chemotaxonomic standpoints, it seems that only one rumen species, *Bacteroides ruminicola*, is a true member of the genus *Bacteroides*. Although these species are very diverse in their characteristics, they are in general relatively resistant to the antibiotics avoparcin, monensin and lasalocid (Table 4).

Table 3
Major Long-chain Fatty Acids and Fatty Aldehydes of Some Rumen Bacteria

Succinivibrio	18:1 (ω-7), 16:0, 14:0, 16:1, 3-OH 14:0	Miyagawa (1982)
Lachnospira multiparus	18:0, 16:0, 18:1 (ω-7), anteiso 15:0, anteiso 17:0	Miyagawa (1982)
Butyrivibrio group 1	anteiso 15:0, anteiso 15:0 aldehyde	Miyagawa (1982)
Butyrivibrio strain 3U8	16:0, 18:1 aldehyde, 16:0 aldehyde	Miyagawa (1982)
Butyrivibrio strain 2U26	15:0, 16:0, iso 15:0, iso 15:0 aldehyde	Miyagawa (1982)
Selenomonas ruminantium	16:1, 14:0, 3-OH 14:0, 16:1 aldehyde	Miyagawa (1982)
Lactobacillus	16:0, 16:1, 16:0 aldehyde, 16:1 aldehyde	Miyagawa (1982)
Bacteroides ruminicola subsp. brevis (GA 33)	anteiso 15:0, 15:0, iso 15:0, iso 16:0, iso 17:0, anteiso 17:0	Shah & Collins (1983)
B. ruminicola subsp. ruminicola (19189)	anteiso 15:0, 16:0, 18:0, 18:1, 3-OH 16:0	Miyagawa et al. (1979)
B. succinogenes	16:0, 15:0, 14:0, 18:0	Miyagawa et al. (1979)
B. amylophilus	16:0, 14:0, 16:1, 18:1	Miyagawa et al. (1979)
Selenomonas ruminantium[a]	17:1, 15:1, 15:0	Watanabe et al. (1982)
Veillonella parvula[a]	17:1, 15:0, 17:0, 18:1	Verkley et al. (1975)
Anaerovibrio lipolytica[a]	15:0, 17:1, 17:0, 15:1	Verkley et al. (1975)
Megasphaera elsdenii[a]	iso 20:0, 16:0, 17:1	Verkley et al. (1975)

[a] Major fatty acids of phosphatidylserine and phosphatidylethanolamine.

Table 4
Minimum Inhibitory Concentrations of Monensin, Lasalocid and Avoparcin towards Rumen Bacteria in Pure Culture

	Monensin[a,b]	Lasalocid[b]	Avoparcin[c]
Anaerovibrio lipolytica	>48	>48	>50
Bacteroides ruminicola	>20[d]	>10[d]	>50
B. amylophilus	>48	>48	
B. succinogenes	>20[d]	>10[d]	32
Selenomonas ruminantium	>48	>48	>50
Megasphaera elsdenii	>48	>48	>50
Veillonella parvula	>48	24	>50
Lactobacillus acidophilus			50–>100[e]
L. vitulinus	0·38–1·5[e]	0·38–1·5[e]	
L. ruminis	1·5–3·0	1·5	
Lachnospira multiparus			4
Streptococcus bovis	0·75–>48[e]	0·38–0·75[e]	8
Ruminococcus albus	0·38	0·38	16
R. flavefaciens	0·38	0·38	0·5
Butyrivibrio fibrisolvens	0·38	0·38	4·0
Succinimonas amylolytica	>48	>48	
Succinivibrio dextrinosolvens	>48	>48	

[a] Minimum inhibitory concentration (lowest tested concentration which prevented growth) in μg/ml.
[b] Dennis et al. (1981).   [c] Stewart et al. (1983).   [d] Chen & Wolin (1979).
[e] Variations between strains tested: data of Chen & Wolin (1979) and Stewart et al. (1983) obtained after adaptation to the drug; Dennis et al. (1981), acute response.

Bacteroides ruminicola

First described by Bryant et al. (1958a), B. ruminicola is one of the most numerous rumen bacteria and is found in ruminants fed on many different diets. Many strains utilise starch and/or the products of plant cell-wall breakdown (Holdeman et al., 1984; Williams & Withers, 1982a,b) and the organism is ecologically ubiquitous. The general characteristics of B. ruminicola were reviewed by Holdeman et al. (1984). The cells are Gram-negative, commonly 1–6 μm long. Substrates fermented and fermentation products are summarised in Table 5, but depend to some extent on the subspecies (below). Propionate, when formed, is produced via the acrylate pathway (Holdeman et al., 1984; Wallnöfer & Baldwin, 1967). Acetate, 2-methylbutyrate and isobutyrate are essential for growth of some strains; NH_3 and peptides serve as nitrogen source (Hungate, 1966). Sphingolipids are present (Kunsman & Caldwell, 1974; reviewed by Shah & Collins, 1983). Respiratory electron carriers include menaquinones with side chains composed of 11, 12 and 13 prenyl (C_3) units (Shah & Collins, 1983). The fatty acids of B. ruminicola are predominantly anteiso-methyl branched acids with iso-methyl branched- and straight-chain saturated acids (Table 3; Shah & Collins, 1983). Biosynthetic pathways of amino acid synthesis, especially by routes involving carboxylation and amination of fatty acids, have been described (Allison et al., 1984; see also Chapter 7).

Table 5
Culture Characteristics of Rumen Gram-negative Rods

	Bacteroides ruminicola	Bacteroides amylophilus	Bacteroides succinogenes	Selenomonas ruminantium	Butyrivibrio fibrisolvens	Anaerovibrio lipolytica	Succinivibrio dextrinosolvens	Succinimonas amylolytica	Wolinella succinogenes	Treponema bryantii
Acid from:[a]										
starch	d	+	d	d	d	−	−	+	−	−
cellulose	−	−	+	−	d	−	−	−	−	−
xylan	d	−	−	−	+	−	−	−	−	+
pectin	d	+	d	−	+	−	+	−	−	−
maltose	d	+	d	+	d	−	d	+	−	+
cellobiose	+	−	+	+	d	−	d	−	−	+
sucrose	d	−	−	d	d	−	d	−	−	+
D-xylose	d	−	−	+	d	−	d	+	−	+
L-arabinose	d	−	+	+	+	−	d	−	−	+
glucose	+	−	−	+	+	+	+	+	−	−
fructose	+	−	−	+	+	−	d	−	−	+
galactose	+	−	−	+	+	−	d	−	−	+
mannose	+	−	d	+	+	−	d	−	−	+
lactose	+	−	−	+	−	−	−	−	−	−
mannitol	−	−	−	d	−	+	d	−	−	−
glycerol	−	−	−	d	−	+	−	−	−	−
lactate	−	−	−	d	−	−	d	−	−	−
Aesculin hydrolysis	d	−	−		d					
Growth on formate/fumarate	−	−	−	d	−	−	−	−	+	−
Nitrate reduction	−	−	−	d	−	−	−	−	+	−
H_2S production	d	−	−	+	d	+	−	−	+	−
Fermentation products:										
Major[b]	AS	FAS	AS	LPA	FBA	PSA	AS	S	S	FAS
Minor/some strains	FPiB, BiV L	L	FPiV	S	LS	L	FL	AP		
Gas				H_2 CO_2	H_2 CO_2	H_2 CO_2			H_2 CO_2	

Abbreviations: A = acetate; B = n-butyrate; F = formate; iB = isobutyrate; iV = isovalerate; L = lactate; P = propionate; S = succinate; V = n-valerate. + = positive reaction; − = negative reaction; ; d = reaction varies between strains; blank = not tested.

[a] From Hungate (1966), Holdeman et al. (1977) and the relevant entry in *Bergey's Manual*, referred to in the text. *T. bryantii*, Stanton & Canale-Parola (1980).
[b] Major products may vary according to substrate, especially for *Selenomonas* and *Anaerovibrio* (see text).

Bacteroides ruminicola is of particular interest because of its presumed role in proteolysis (Hazlewood & Edwards, 1981; Wallace & Brammall, 1985) and the uptake and fermentation of peptides (Munn *et al.*, 1983; Russell, 1983). The hemicellulases of *B. ruminicola* (Williams & Withers, 1985) are presumably mainly used for the further breakdown of oligosaccharides released by the action of cellulolytic bacteria on plant cell walls.

The B. ruminicola *subspecies*
Two subspecies of *B. ruminicola* are distinguished on the basis of morphology, substrates fermented and requirement for haeme. Cultures of *B. ruminicola* subsp. *brevis* normally contain a high proportion of short coccobacilli, whereas most of the cells of subsp. *ruminicola* are long rods. Each subspecies is represented by a number of biovars (Bryant *et al.*, 1958a). Isolates of subsp. *brevis* do not require haeme, but their growth may be stimulated by it, whereas most strains of *B. ruminicola* subsp. *ruminicola* require haeme. A study by Gardner *et al.* (1983) demonstrated that *B. ruminicola* subsp. *brevis* is able to utilise pre-formed tetrapyrroles for cytochrome-b synthesis, unlike *B. succinogenes* or *Selenomonas ruminantium*. When grown in the presence of magnesium and manganese chelates of deuteroporphyrin, a deuteroporphyrin cytochrome-b was formed, indicating the utilisation of the specific metal chelates. The type strain of *B. ruminicola* subsp. *ruminicola* lacks the enzyme superoxide dismutase (SOD) which is present in the type strain of *B. ruminicola* subsp. *brevis* and many other rumen bacteria tested (Fulghum & Worthington, 1984). Xylan and pectin are more frequently and vigorously fermented by representatives of subsp. *ruminicola* than by subsp. *brevis*. Unpublished data on DNA homologies by Johnson's group (reported by Holdeman *et al.*, 1984) suggest no homology between the type strains of these two subspecies. The polypeptide band patterns of the subspecies type-strains, revealed by polyacrylamide gel electrophoresis of sodium dodecyl sulphate extracts of whole cells (SDS-PAGE), are also markedly different (Begbie & Stewart, 1984). It seems probable that these two subspecies and many other strains will eventually be elevated to species rank (Holdeman *et al.*, 1984).

Bacteroides (Ruminobacter) amylophilus
Hamlin & Hungate (1956) described this Gram-negative amylolytic bacterium from the rumen of cattle. Although not always detectable in rumen contents, on occasions *B. amylophilus* is thought to be the predominant starch digester (Holdeman *et al.*, 1984). Phylogenetically and chemotaxonomically, *B. amylophilus* is not closely related to the true *Bacteroides*; Stackebrandt & Hippe (1986) proposed, on the basis of 16S rRNA oligonucleotide sequencing studies, that it should be transferred to a new species, *Ruminobacter amylophilus*. The type strain consists of pleomorphic oval-to-long rods approximately 1 by 1–3 μm. Acid is produced from few substrates; in addition to those listed in Table 5, glycogen and dextrin are fermented (Holdeman *et al.*, 1984). Major fermentation products are acetate and succinate (Table 5). Carbon dioxide and ammonia are essential for growth (Hungate, 1966).

Bacteroides amylophilus differs from the true *Bacteroides* in that it lacks sphingolipids and menaquinones (Shah & Collins, 1983). Cytochromes are not present (Reddy & Bryant, 1977) and haemin is not stimulatory to growth (Macy & Probst, 1979). The fatty acids of *B. amylophilus* are mainly straight-chain saturated and mono-unsaturated acids (Table 3; Shah & Collins, 1983). Sodium is required in high concentration for optimal growth (Caldwell *et al.*, 1973). In the presence of 1 mM Na^+ the cells demonstrated a 15 hour lag period followed by growth with a doubling time of 135 minutes. With 90 mM Na^+ there was no lag and the doubling time was reduced to 54–60 minutes (Wetzstein & Gottschalk, 1985). An apparently unique membrane-bound NADH-fumarate–reductase system has been described; remarkably, compounds normally regarded as key electron transport components (cytochromes, menaquinones) are not present, and the fumarate–reductase system appears to generate a proton-motive force across the cell membrane (Wetzstein & Gottschalk, 1985). This is discussed in Chapter 6.

Bacteroides succinogenes

Bacteroides succinogenes, first described by Hungate (1950), is now considered to be one of the most widespread cellulolytic bacteria of the rumen. The numbers of *B. succinogenes* present are underestimated when cellulolytic populations are enumerated using cellulose roll-tubes containing 1·2% or more agar, as under these conditions this organism is normally unable to migrate through the agar to digest the cellulose, and clear zones are difficult to detect. For *B. succinogenes*, the concentration of agar has to be low enough (around 0·5% v/v) to allow cell migration (Montgomery & Macy, 1982; Varel *et al.*, 1984). On first isolation, the cells are predominantly rod-shaped, but on maintenance in the laboratory they tend to become coccoid to lemon-shaped or oval, from around 0·8 to 1·6 μm in diameter. Most cells occur singly, but short chains and even rosette formations may be observed. Old cultures, particularly of recently isolated strains, lyse rapidly, with loss of viability, and strains may readily be lost if not subcultured frequently. The major substrates fermented by, and principal products of, *B. succinogenes* are shown in Table 5. Valerate, isobutyrate, CO_2, NH_4^+, Na^+, K^+, Ca^{2+}, Mg^{2+} and PO_4^{3-} are essential for growth; biotin is usually essential and PABA may be required (Bryant *et al.*, 1959). The fatty acids are used mainly in the synthesis of phospholipids, which consist predominantly of ethanolamine plasmalogens (Wegner & Foster, 1963). Isobutyric acid is incorporated mainly into fatty aldehydes and branched-chain C_{14} and C_{16} acids, and valerate into fatty aldehydes and straight-chain C_{13} and C_{15} fatty acids (Wegner & Foster, 1963).

Bacteroides succinogenes differs from the Shah & Collins definition of *Bacteroides* in that it does not possess sphingolipids; in addition, the fatty acids of *B. succinogenes* are very predominantly straight-chain saturated acids (Table 3; Miyagawa *et al.*, 1979; Shah & Collins, 1983). Evidence from 16S rRNA oligonucleotide sequencing suggests that *B. succinogenes* does not belong to the *Bacteroides–Flavobacterium–Cytophaga* subgroup (Paster *et al.*, 1985). Indeed, the fatty-acid composition of *B. succinogenes* is more similar to that of *Bacteroides (Ruminobacter) amylophilus* than to many representatives of the true *Bacteroides*, including *B. ruminicola* (Miyagawa *et al.*, 1979). Montgomery & Stahl (pers.

comm.) have suggested a new name, *Fibrobacter*, for this species. They have also suggested that the strains that remain rod-like on cultivation may belong to a different sub-species to those strains with highly pleomorphic cells.

The cellulases of *B. succinogenes* and its mode of attack on plant cell walls have been widely investigated (see Chapter 8). Following the observation that *B. succinogenes* is particularly active in the hydrolysis of highly ordered celluloses such as cotton fibres (Halliwell & Bryant, 1963), numerous studies have established the ability of *B. succinogenes* to solubilise and digest plant material (Dehority & Scott, 1967; Stewart *et al.*, 1979; Morris & Van Gylswyk, 1980; Morris, 1984; Kolankaya *et al.*, 1985; Graham *et al.*, 1985; Chesson *et al.*, 1986). In these tests, *B. succinogenes* has normally achieved the most extensive solubilisation attained by axenic cultures of rumen bacteria, although the performance of ruminococci in such tests may be hindered by their accumulation of hydrogen, which would normally be removed by methanogens in the rumen. Despite extensive solubilisation of plant material, *B. succinogenes* does not utilise pentoses for growth (Morris & Van Gylswyk, 1980). In common with *B. ruminicola* strain $B_1 4$ and *Butyrivibrio fibrisolvens* 49, and in contrast to *Selenomonas ruminantium* HD4, *Megasphaera elsdenii* B159 and *Streptococcus bovis* JB1, *B. succinogenes* strain S85 does not appear to depend upon a phosphoenolpyruvate–phosphotransferase system for the phosphorylation and uptake of glucose, under the experimental conditions studied by Martin & Russell (1986) and Franklund & Glass (1987).

Bacteroides succinogenes produces succinate from fumarate using an FMN-linked fumarate reductase (Miller, 1978). Experiments with uncouplers of electron transport (Dawson *et al.*, 1979) suggest that ATP is partly derived from electron transport-linked phosphorylation. Cytochrome-b was detected in this species by Reddy & Bryant (1977).

Among the cellulolytic bacteria, *B. succinogenes* is relatively resistant to feed antibiotics (Table 4), and has been found as the predominant cellulolytic bacterium in sheep receiving avoparcin (Stewart & Duncan, 1985).

Selenomonas ruminantium

Like several other species, *S. ruminantium* has been detected in highest numbers in the rumens of animals fed on cereal grains. In steers fed on cracked corn and urea, Caldwell & Bryant (1966) also reported that *S. ruminantium* constituted 22–51% of the total viable count of rumen bacteria. The cells are Gram-negative, curved, crescentic or 'banana-shaped' rods, $0.9–1.1 \times 3.0–6.0\,\mu$m and motile by means of a linear array of up to 16 flagella attached to the middle of the concave side of the cells (reviewed by Bryant, 1984*d*). Cultivation of *S. ruminantium* in a medium containing excess glucose under phosphate-limited conditions resulted in the loss of flagella, and the assumption of a spiral growth form (Hudman, 1984).

The major substrates for growth and the fermentation products of *S. ruminantium* are shown in Table 5. In axenic culture very small amounts of hydrogen accumulate; however, in co-culture with methanogens, substantial methanogenesis is supported (Scheifinger *et al.*, 1975; see Chapter 11). In addition to the substrates listed in

Table 5, others, including melezitose, raffinose, salicin and sucrose, may be fermented. Strains that ferment glycerol and lactate are placed in a different subspecies (*S. ruminantium* subsp. *lactilytica*) from those that do not (*S. ruminantium* subsp. *ruminantium*). The products of fermentation of lactate and glycerol differ; propionate, acetate and CO_2 are formed from lactate but propionate, lactate, succinate and acetate are formed from glycerol (reviewed by Bryant, 1984*d*). A third subspecies, *S. ruminantium* subsp. *bryanti* has been designated to accommodate large isolates, 2–3 × 5–10 µm, that typically do not produce H_2S from cysteine, or ferment arabinose, xylose, galactose, lactose or dulcitol (Prins, 1971). *S. ruminantium* subsp. *lactilytica* requires *n*-valerate when grown on glucose but not on lactate; when grown on lactate, PABA and aspartate may be essential. NH_3 is utilised as N source, but most strains can use certain amino acids as nitrogen and energy sources (Bryant, 1984*d*). The rumen selenomonads are closely related taxonomically to the bacterium *S. ruminantium* subsp. *pseudocataligenes*, isolated from ditchwater (Harborth & Hanert, 1982).

The predominant cellular fatty acids of *S. ruminantium* are shown in Table 3. Phosphatidylserine (PS) and phosphatidylethanolamine (PE) and their plasmalogen analogues account for a major part of the phospholipids.

The incorporation of volatile fatty acids into long-chain fatty acids of the phospholipids of *S. ruminantium* subsp. *lactilytica* has been described by Watanabe *et al.* (1984). PE and PS were found to be synthesised via pathways similar to those of *Escherichia coli*, PS being synthesised from CDP-diglyceride and serine, and PE from PS. It was hypothesised that the diglyceride moieties of a precursor pool of diacyl phospholipids are used for the synthesis of plasmalogens. Diaminopimelic acid is the diagnostic amino acid of the peptidoglycan. Cytochromes-b are present, but menaquinones and ubiquinones absent (summarised by Stackebrandt *et al.*, 1985). The diamine cadaverine is covalently bound to the D-glutamic acid residue of the peptidoglycan of *S. ruminantium*. Kamio *et al.* (1986) have described the synthesis of cadaverine from L-lysine by lysine decarboxylase, and transfer by a particulate enzyme to the α-carboxyl of D-glutamic acid. Cadaverine appears to maintain the structural integrity of the cell envelope, and inhibition of its synthesis prevents normal growth.

The urease of *S. ruminantium* has been purified by Hausinger (1986) and shown to have a molecular mass of 360 000 daltons and subunit molecular mass of 70 000. The enzyme contains 2 nickel ions per subunit, which is similar to the Ni content of urease from jack beans. The kinetic properties of *S. ruminantium* urease differ from those of isolated rumen urease (Hausinger, 1986).

Although generally a strict anaerobe (Loesche, 1969), some strains of *S. ruminantium* are known to tolerate exposure to small amounts of oxygen. Samah & Wimpenny (1982) demonstrated the presence in one (possibly atypical) strain of a soluble NADH-oxidase presumed to reduce oxygen to water or H_2O_2. Superoxide produced in this reaction would be further metabolised by superoxide dismutase (Fulghum & Worthington, 1984). The *Selenomonas* strains so far studied are resistant to monensin, lasalocid and avoparcin (Table 4).

Butyrivibrio fibrisolvens

A large number of isolates of Gram-negative, butyric acid-producing rods were obtained by Bryant & Burkey (1953a) from cows fed on different rations, including alfalfa hay with or without a grain supplement, green alfalfa, soybean hay with grain, blue grass pasture with grain, and grain mixture alone. Further observations on these isolates led to their description as *Butyrivibrio fibrisolvens* (Bryant & Small, 1956a). These bacteria were among the predominant species present and were isolated from the 10^8 dilution of rumen contents. The characteristics of *B. fibrisolvens* vary considerably between isolates, and there is a clear need for a reassessment of the taxonomy of the genus (Bryant, 1984a).

The type strain, D1 of Bryant & Small (1956a), stains Gram-negative, though electron microscopic studies have revealed that the cells are ultrastructurally Gram-positive (Cheng & Costerton, 1977). Motility is by means of a single polar or subpolar flagellum. The cells are curved rods 0·4–0·6 μm wide and 2–5 μm in length, with bluntly tapered ends, found singly, in pairs, and in chains. *B. fibrisolvens* isolates ferment a wide range of sugars (Table 5) with considerable variation from strain to strain (Bryant & Small, 1956a; Hungate, 1966). Shane *et al.* (1969) described 19 cellulolytic isolates from the rumens of sheep fed on low quality teff hay. Within this group of isolates, two *B. fibrisolvens* biotypes were designated. In a later study, Van der Toorn & Van Gylswyk (1985) found representatives of these two biotypes (acetate-utilisers and -producers) and a third biotype represented by propionate-producing isolates. A study of the effects of volatile fatty acids on *Butyrivibrio* also suggested the existence of at least two major biotypes (Roché *et al.*, 1973). Apart from variations in fermentation patterns, strains of *B. fibrisolvens* vary substantially in antigenic properties and in their content of lipoteichoic acids (reviewed by Bryant, 1984a). A study of the fatty acid and fatty aldehyde compositions of *Butyrivibrio* strains showed that the isolates could be divided into three groups; the fatty acid components of some representative types are shown in Table 3. All of the isolates lacked hydroxy fatty acids, supporting the view that these are physiologically Gram-positive organisms (Miyagawa, 1982). The pattern of antibiotic sensitivity of *Butyrivibrio* (Table 4) also suggests that this bacterium is physiologically Gram-positive.

The major fermentation products are shown in Table 5. The very wide range of substrates fermented by *B. fibrisolvens* make it very common in the rumen, where various isolates are thought to play a part in plant cell wall hydrolysis, the hydrogenation of lipids, and proteolysis. Like *Selenomonas ruminantium*, *Butyrivibrio fibrisolvens* may also play a role in selenium metabolism, as ^{75}Se can be shown to be incorporated into selenium analogues of sulphur amino acids by these species (Hudman & Glenn, 1984, 1985). *B. fibrisolvens* is not restricted to the rumen, however, isolates having also been obtained from the hindgut of sheep (Ørskov *et al.*, 1970; Mann & Ørskov, 1973; Lewis & Dehority, 1985) and from the pig caecum (Robinson *et al.*, 1981).

In axenic culture, *Butyrivibrio* is normally less actively cellulolytic than the cellulolytic *Bacteroides* and ruminococci, though this may partly be due to the loss of activity on continued maintenance in the laboratory. Hemicellulose in plant cell walls is more extensively solubilised by *B. fibrisolvens* than is cellulose (Dehority & Scott, 1967; Morris & Van Gylswyk, 1980). The attack on bermuda grass and orchard grass fibres *in vitro* occurs without obvious attachment of the bacterial cells to the substrate (Akin & Rigsby, 1985). *B. fibrisolvens* has been shown to ferment cellodextrins (Russell, 1985*b*), and a range of polysaccharide depolymerases and glycoside hydrolases has been detected in cell extracts (Howard *et al.*, 1960; Williams & Withers, 1982*a,b*; Wojciechowicz *et al.*, 1982; Williams *et al.*, 1984; Heinrichova *et al.*, 1985; Hespell *et al.*, 1987).

Butyrivibrio fibrisolvens hydrogenates certain unsaturated fatty acids (Kepler *et al.*, 1966). Linoleic acid is hydrogenated to *trans*-11-octadecenoic acid. The reductase which catalyses the hydrogenation step has been purified (Hughes *et al.*, 1982; see also Chapter 9). Bacteria of this genus possess proteases which accumulate in culture supernatants (Cotta & Hespell, 1986). The contribution of *B. fibrisolvens* to proteolysis in the rumen has been studied by Wallace & Brammall (1985) and is discussed further in Chapter 7.

Large bacteria with lophotrichous flagella (the 'B-835 like group') often replace *B. fibrisolvens* under acid conditions in the rumen (Bryant, 1959, 1984*a*; Bryant *et al.*, 1960, 1961). These bacteria are physiologically similar to *B. fibrisolvens*, and their physiological properties and systematic relationships are currently under study (R. B. Hespell, personal communication).

Anaerovibrio lipolytica

The cultural properties of *Anaerovibrio lipolytica* suggest two major ecological roles for this bacterium, the hydrolysis of lipids and utilisation of lactate. Hobson & Mann (1961) isolated *A. lipolytica* in the 10^8 dilution of rumen contents of a sheep fed on a ration containing linseed cake meal. Subsequently, Slyter *et al.* (1976) isolated *A. lipolytica* during the transition from forage to concentrate feeding when lactic acid utilisation may have supported enhanced growth of this organism; however, Prins *et al.* (1975) found *A. lipolytica* in the rumens of sheep and a cow fed on hay, and Henderson (1975) isolated 21 strains of lipolytic, curved rods, apparently *A. lipolytica* or a closely related organism, from the rumens of sheep fed on hay and dried grass. These bacteria were present at around 2.7×10^7/ml, approximately 5% of the total viable count.

The cultural and physiological properties of *A. lipolytica* have been reviewed by Prins (1984). Cells are curved Gram-negative rods, about 0.5×1.2–$3.6\,\mu$m, and normally motile by a single polar flagellum. Henderson (1975) illustrated cells of lipolytic bacteria very similar to *A. lipolytica*, but with the flagellum apparently located on the concave side of the cells. Some isolates had two or more flagella. The main substrates for growth are shown in Table 5; in addition, ribose, triacylglycerides and phospholipids may be used and some amino acids are fermented (Prins *et al.*, 1975). The major fermentation products depend upon the substrate. Glycerol is fermented mainly to propionate and succinate, with small amounts of H_2 and L-lactate; ribose and fructose are fermented to acetate, propionate and CO_2, with small amounts of succinate, H_2 and lactate; DL-lactate is fermented mainly to acetate, propionate and CO_2, with small amounts of succinate and H_2 (Henderson, 1975; Prins, 1984). A source of amino acids, and the vitamins folic acid, pantothenate and pyridoxal HCl are essential for growth (Hungate, 1966).

A constitutive extracellular lipase is released during growth of *A. lipolytica*; this enzyme is associated with membranous blebs apparently released from the surface of the cells. The proportions of protein, nucleic acid and lipid in these blebs are similar to those in intact cells (Henderson & Hodgkiss, 1973). When strain 5S was grown in batch culture on glycerol, lipolytic activity in the culture supernatant declined sharply after about 7 hours; this decrease was correlated with the fall in the pH of the medium during growth, and could be prevented by maintaining the pH above 6·3 (Henderson *et al.*, 1969). When grown in continuous culture under glycerol-limited conditions, extracellular lipase production showed two maxima, at dilution rates of around $0.08\,h^{-1}$ and $0.14\,h^{-1}$ (Henderson *et al.*, 1969).

Anaerovibrio lipolytica is able to utilise extracellular hydrogen to reduce fumarate to succinate, a reaction of potential importance in the rumen under conditions in which methanogenesis is reduced (Henderson, 1980). Fumarate can oxidise a membrane-bound cytochrome-b which can be reduced by hydrogen from NADH (Henderson, 1980).

The presence of phosphatidylserine in the membranes of *A. lipolytica* and the composition of the membrane lipids has been described by Van Golde *et al.* (1975) and Verkley *et al.* (1975); the principal fatty acids of phosphatidyl-serine (PS) and -ethanolamine (PE) are shown in Table 3. The composition of plasmalogen analogues of the PE and PS is described by Van Golde *et al.* (1975).

Spirochaetes

Some of the earliest studies on rumen microbes showed that spirochaetes are present in the rumen, and such organisms were often seen in the course of isolation of cellulolytic bacteria (Bryant, 1952; Hungate, 1947). Electron microscopic examinations of plant cell walls during their degradation by rumen microbes often revealed the presence of helical cells growing in close proximity to the cell wall-degrading population (Cheng *et al.*, 1984).

Stanton & Canale-Parola (1979) enumerated spirochaetes in bovine rumen liquor and found that they accounted for between 1% and 6% of the total viable count.

Treponema bryantii
These are Gram-negative helical rods, typically 3–8 μm in length, motile, with one periplasmic flagellum inserted at each end of the cell. Substrates fermented and major products are summarised in Table 5. Carbon dioxide, isobutyrate, 2-methylbutyrate and B vitamins are required for growth, while riboflavin is stimulatory (Smibert, 1984).

Treponema saccharophilum
Treponema saccharophilum was isolated from the rumen of a cow, and described by Paster & Canale-Parola (1985). The cells are helical and large, up to 0·7 μm wide and

20 μm long, with a bundle of (usually 16) periplasmic flagella. A wide range of substrates is utilised, those supporting greatest growth including polygalacturonic acid, pectin, soluble starch, dextrin, sucrose, maltose, cellobiose, D-glucuronic acid, D-glucose, D-mannose, D-fructose, D-galactose and L-arabinose. Isobutyric acid is required, and valeric acid is stimulatory for growth, but CO_2 is not required. The major fermentation products from glucose in axenic culture are formate, acetate and ethanol.

Ziolecki (1979) reported the isolation of large (0·7 × 12–25 μm) treponemes from the rumen of a cow fed on hay and concentrates. These strains fermented pectin, L-arabinose, inulin and sucrose, and one strain also fermented D-xylose. This is a much more restricted range of substrates than that of *T. bryantii*. The pectinolytic enzymes of these bacteria have been partially characterised and include pectinmethylesterase, endo-polygalacturonate lyase and (probably) a polygalacturonase (Wojciechowicz & Ziolecki, 1979).

Wolinella succinogenes

Wolinella succinogenes was originally named *Vibrio succinogenes* (Wolin et al., 1961). The cells are small, Gram-negative, curved rods, approximately 0·6 × 3 μm, found singly, in pairs and in spiral chains. The original isolates were from enrichment cultures designed for the isolation of methanogenic bacteria from the bovine rumen. Although it was not possible to enumerate these organisms in rumen samples, the description of their isolation suggests that they must have been present at least at 10^5/ml (Wolin et al., 1961). *V. succinogenes* was eventually renamed *Wolinella succinogenes* because the rumen organism (unlike the Vibrios) does not ferment sugars and is strictly anaerobic.

Sugars are not fermented (Table 5) but energy is conserved by oxidation–reduction reactions in which hydrogen or formate donate electrons to a range of electron acceptors, which can include fumarate, L-malate, L-aspartate, L-asparagine, nitrate or nitrous oxide. The major fermentation products are CO_2 and succinate, and H_2S is produced (Tanner & Socransky, 1984). Nitrous oxide is reduced to nitrogen, and nitrate to nitrite or ammonia but not to nitrogen. Succinate stimulates growth on nitrous oxide and the production of nitrogen (Yoshinari, 1980). The mechanism of energy conservation during fumarate reduction has been studied by Kröger & Winkler (1981). Cell pellets are pink due to the presence of cytochromes (Wolin et al., 1961), and menaquinones are present, serving in electron transfer from formate to fumarate. Kröger's group have established that sulphur can be used by *W. succinogenes* as a terminal electron acceptor with formate as electron donor (Macy et al., 1986). Growth was sustained only if precautions were taken to remove H_2S. Growth of *W. succinogenes* also occurred on H_2S and fumarate, with the formation of rhombic crystals of elemental sulphur.

Co-culture of *W. succinogenes* and *Ruminococcus albus* has been shown to result in the suppression of alcohol production in favour of acetate (Iannotti et al., 1973). *W. succinogenes* is capable of the reductive detoxification of ferulic acid. This process is accelerated by the presence of the fermentation products of *Ruminococcus albus*, which are assumed to act as electron donors (Ohmiya *et al.*, 1986). Utilisation of hydrogen might offer an enhanced role for *W. succinogenes* in the rumens of animals fed on diets supplemented with methane inhibitors.

Succinimonas amylolytica

Succinimonas amylolytica was first described by Bryant *et al.* (1958*a*) who isolated it from the rumens of cattle fed on mixtures of hay and grain. Like *Succinivibrio dextrinosolvens*, *Succinimonas amylolytica* is normally associated with starch digestion, although a wide range of glycoside hydrolases is produced (Williams *et al.*, 1984). The cells are Gram-negative straight rods or coccobacilli up to $1\cdot5 \times 3\cdot0$ μm in diameter, motile by means of a single polar flagellum. Fermentation characteristics are summarised in Table 5. Growth is stimulated by acetate. *S. amylolytica* does not produce indole (Bryant, 1984*b*), and is resistant to monensin and lasalocid (Table 4).

Succinivibrio dextrinosolvens

Succinivibrio dextrinosolvens was first described in detail by Bryant & Small (1956*b*) who isolated it from cattle fed on grain. Subsequently, its major ecological role in the rumen has been defined as the fermentation of dextrins in animals fed on starch (Bryant *et al.*, 1961; Bryant, 1984*c*). Bryant (1959) and Gomez-Alarcon *et al.* (1982) indicated that the 'Group B' organisms isolated from sheep by Wilson (1953) belong to this species. *S. dextrinosolvens* is probably ubiquitous in the rumen, and in addition to their starch digesting capacity, some strains possess enzymes which could enable products of plant cell wall breakdown to be used (Williams *et al.* 1984). The cells are Gram-negative, helically twisted, rods which are motile by means of a polar flagellum. Fermentation characteristics are summarised in Table 5, and the responses to monensin and lasalocid are shown in Table 4.

Succinivibrio dextrinosolvens does not produce acetoin or indole (Bryant, 1984*c*). Many strains are ureolytic (Wozny *et al.*, 1977). The composition of the fatty acids of *Succinivibrio* isolates studied by Miyagawa (1982) is summarised in Table 3.

Succinivibrio dextrinosolvens requires 1,4-naphthoquinone for growth in chemically defined media. Menadione or vitamin K_5 support only slow growth when substituted for 1,4-naphthoquinone (Gomez-Alarcon *et al.*, 1982). The type strain can utilise certain amino acids as nitrogen source in the absence of ammonia, but some other strains require ammonia for growth (Gomez-Alarcon *et al.*, 1982). Patterson & Hespell (1985) have studied the response of *S. dextrinosolvens* to changes in the ambient ammonia concentration. It was suggested that, when the concentration of ammonia is low, a high-affinity glutamine synthetase is the main route of ammonia incorporation; in the presence of excess ammonia a low-affinity glutamate dehydrogenase system is used.

Megasphaera elsdenii

Elsden and his colleagues (Elsden *et al.*, 1956) described bacteria similar to those isolated several years earlier (Elsden & Lewis, 1953) but lost before a complete description could be assembled. These were Gram-negative non-motile cocci, approximately 2·4 × 2·6 µm in diameter, which occurred in pairs and in chains of up to 20 cells. This organism (originally known as *Peptostreptococcus elsdenii*) is mainly found in the rumens of young animals (Hobson *et al.*, 1958) and in animals receiving high grain rations in which the fermentation of lactate assumes particular importance. Growth usually occurs on glucose, fructose and lactate and may occur on a number of other substrates (Table 6). The end-products of fermentation vary according to the substrate used; lactate is fermented mainly to butyrate, propionate (formed by the acrylate pathway), isobutyrate, valerate, CO_2 and some H_2, sometimes with traces of caproate. Glucose is fermented mainly to caproate and formate with some acetate, propionate, butyrate and valerate (Rogosa, 1984). A range of alcohols may also be produced (Table 6; Holdeman *et al.*, 1977). Acetate is stimulatory, and amino acids normally essential for growth (Hungate, 1966). However, it has proved possible to train strain B159 to grow without amino acids (Forsberg, 1978). Organic growth factor requirements can be met by yeast extract (Rogosa, 1984). *M. elsdenii* is resistant to avoparcin, monensin and lasalocid (Table 4). The conditions under which *M. elsdenii* was originally isolated (Elsden *et al.*, 1956; Hobson *et al.*, 1958) and the studies of Loesche (1969) suggest that it is not a particularly fastidious anaerobe.

The major phospholipids of *M. elsdenii* are serine- and ethanolamine-containing phosphoglycerides and their plasmalogen analogues. Major acyl-chain fatty acids in the phospholipids of *M. elsdenii* growing at 37°C are C_{19} cyclopropane chain, 16:0 and 18:1 moieties. The major alk-1-enyl chains that are present under similar conditions are 16:0, C_{17} and C_{19} cyclopropane chains and 18:1 moieties (Johnston & Goldfine, 1982).

The role of *M. elsdenii* in the fermentation of DL-lactate in the rumens of dairy cows was investigated by Counotte *et al.* (1981) using ^{13}C Fourier transform nuclear magnetic resonance. The contribution of *M. elsdenii* was influenced by the diet being fed and by the animal, but averaged 74 ± 13%. Since *M. elsdenii* is not subject to catabolite repression by glucose or maltose, its contribution to lactate catabolism is thought to increase after the feeding of soluble carbohydrates which repress lactate fermentation by *Selenomonas* and other lactate-using bacteria (Russell & Baldwin, 1978; Counotte *et al.*, 1983).

Megasphaera elsdenii is thought to play a major role in the production of branched-chain VFA in the rumen (Allison, 1978). The mechanism involves deamination and decarboxylation of amino acids. *M. elsdenii* catabolises amino acids, especially L-threonine and L-serine; Wallace (1986) concluded that this process may contribute to the maintenance energy requirement, but is mainly of importance for the production of NH_3 and branched-chain VFA which are growth factors for other bacteria (see Chapter 7).

The bacterium produces H_2 during fermentation, and the properties of the hydrogenase have been widely investigated (reviewed by Adams *et al.*, 1981). The hydrogenase has a molecular weight of 50 000 and ferredoxin serves as electron carrier. Twelve (non-haeme) Fe and 12 acid-labile sulphide groups are present per molecule (Adams *et al.*, 1981). When *M. elsdenii* is grown in iron-deficient media,

Table 6
Fermentation Characteristics of Some Rumen Cocci

	Ruminococcus albus	R. flavefaciens	R. bromii	Streptococcus bovis	Veillonella parvula	Megasphaera elsdenii
Cell wall Gram type (by EM)	+	+	+	+	−	−
Acid from:						
starch	−	−	+	+	−	−
cellulose	+	+	−	−	−	−
xylan	+	+	−	−	−	−
pectin	−	−	−	d	−	−
maltose	−	−	+	+	−	+
cellobiose	+	d	−	+	−	−
sucrose	d	d	−	+	−	d
D-xylose	d	d	−	d	−	−
L-arabinose	d	d	−	d	−	−
glucose	d	d	d	+	−	−
fructose	d	d	+	+	−	+
galactose	−	−	−	+	−	+
mannose	d	d	d	d	−	−
lactose	d	d	−	d	−	−
mannitol	d	−	−	d	−	d
glycerol	−	−	−	d	−	d
lactate	−	−	−	−	+	+
Aesculin hydrolysis	d	d	−	d	−	−
H$_2$S production	−	−	−	d	+	−
Nitrate reduction	−	−	−	−	+	−
Fermentation products:						
Major	A2	AS	A2	L	AP	CB45
Minor/some strains	FL	FL	LFPB	FA2	L	{ iVViBAS i4i5FP
Gas produced	H$_2$CO$_2$	H$_2$	H$_2$CO$_2$	CO$_2$	H$_2$CO$_2$	H$_2$CO$_2$

Sources and abbreviations as for Table 5; C = caproate; 2 = ethanol; 4 = butanol; 5 = pentanol; i4 = isobutanol; i5 = isopentanol. Note that products may vary according to the substrate and strain studied.

flavodoxin replaces ferredoxin as electron carrier for the hydrogenase. *M. elsdenii* flavodoxin (Moonen *et al.*, 1984) has a molecular mass of 15 000 and contains no metal ions but carries riboflavin-5'-monophosphate as a prosthetic group. Moonen & Müller (1984) concluded from proton NMR studies that *M. elsdenii* flavodoxin acts as a one-electron carrier, shuttling between semi-quinone and hydroquinone (more reduced) states.

During the fermentation of lactate by *M. elsdenii*, reducing equivalents are transferred from an FAD-containing D-lactate dehydrogenase to butyryl CoA dehydrogenase (BCD; Baldwin & Milligan, 1964; Brockman & Wood, 1975). It has been suggested that BCD activity may be regulated by the binding, or displacement, of a 'greening' ligand, CoA persulphide (Williamson *et al.*, 1982).

Megasphaera elsdenii possesses a thiaminase, but the substrate specificity of this enzyme is different from that of the enzyme in the rumen which is associated with the development of cerebro-cortical necrosis (Boyd, 1985; see also Chapter 14).

Veillonella parvula

Johns (1951) described a Gram-negative micrococcus isolated from the rumen of sheep. Originally named *V. gazogenes* (= *V. alcalescens* = *Micrococcus lactilytica*), these bacteria are now placed in the species *V. parvula* on the basis of DNA/DNA hybridisation studies (reviewed by Rogosa, 1984); however, the range in % homology values within this species is high (53–100%) and some further revision may prove appropriate. The cells are small (0·3–0·6 μm) and non-motile. *V. parvula* from the rumen ferments DL-lactate (Table 6), pyruvate, L-malate, fumarate and D-tartrate, but not sugars. The purines xanthine and hypoxanthine are also fermented (Whitely & Douglas, 1951). Succinic acid is decarboxylated to propionate and CO_2. Major products from lactate are acetate, propionate, CO_2 and H_2 (Hungate, 1966). Although the lactate-utilising capacity of *V. parvula* suggests a role in the rumen of animals fed on starchy rations (Hobson *et al.*, 1958), its contribution to the rumen fermentation generally seems minor compared to that of *M. elsdenii*.

Like *M. elsdenii*, *V. parvula* possesses phospholipids containing phosphatidyl-serine and -ethanolamine (Table 3) and plasmalogens. The acyl-chains of phospholipids in cells grown at 37°C were predominantly 17:1, 18:1 and 15:0 chains. The alk-1-enyl chains were composed mainly of 17:1, 15:0 and 17:0 moieties (Johnston & Goldfine, 1982).

Several aspects of the fermentative metabolism of *V. parvula* have been investigated in detail. The terminal step in succinate decarboxylation, the conversion of *S*-methylmalonyl CoA to CO_2 and propionyl CoA, is catalysed by methylmalonyl CoA decarboxylase, which has been shown to be membrane-bound (Hilpert & Dimroth, 1983). This enzyme acts as a sodium pump, exporting Na^+ ions and conserving the decarboxylation energy as a sodium gradient. Pestka & Delwiche (1983) postulated the presence in *V. parvula* of an alternative pathway for the formation of 3-phosphoglycerate from lactate via malate (by a malate-lactate transhydrogenase; Allen, 1983), glyoxylate and glycerate.

Streptococcus bovis

The cells of *Streptococcus bovis* are Gram-positive, non-motile, and ovoid to coccal (0·8–1·5 μm) in shape. Chains are sometimes formed and older cells may stain Gram-negative. Sometimes considered a facultative anaerobe, both strictly anaerobic and aerotolerant strains are found (Latham *et al.*, 1979). *S. bovis* is capable of very rapid growth; population doubling times from 21 minutes (Stewart, 1975) to between 24 and 27 minutes (Russell & Robinson, 1984) have been recorded. The major substrates fermented and products formed are listed in Table 6.

Analysis of the cell wall peptidoglycan of *S. bovis* by Latham *et al.* (1979) revealed, in addition to glutamic acid, the presence of lysine, alanine, threonine and serine, but no diaminopimelic acid (DAPA). However, Russell & Robinson (1984) reported that six strains of *S. bovis* which they examined contained DAPA. Carbon dioxide is required, or is stimulatory, for growth (Latham *et al.*, 1979), and NH_3 can serve as sole N source (Wolin *et al.*, 1959). Biotin is essential for growth and thiamine is stimulatory. Requirements for amino acids vary between strains, but arginine is generally most stimulatory (reviewed by Hungate, 1966). Early observations on the extracellular polysaccharides, carbohydrases and intracellular reserve polysaccharides are also reviewed by Hungate (1966). The strains of *S. bovis* so far examined vary in their response to antibiotics (Table 4).

Streptococcus bovis is of particular interest because of its role in the development of lactic acidosis in sheep and cattle fed on an excess of starch. The role of *S. bovis* in this condition, first identified by Hungate *et al.* (1952), has been reviewed by Hungate (1966) and Russell & Hino (1985). An elegant series of studies by Russell and his co-workers (Russell & Baldwin, 1978, 1979; Russell & Dombrowski, 1980; Russell *et al.*, 1981), together with observations on factors affecting the production of lactate by *S. bovis* (Russell & Hino, 1985), led to the presentation of a microbiological scenario for the development of lactic acidosis in sheep and cattle. The essential elements of this scenario (Russell & Hino, 1985; see also Chapters 6 and 14) are such that the production of lactate tends to be increased by the conditions that develop when large amounts of starch are fed and the rumen pH falls.

Streptococcus bovis belongs to Lancefield's antigenic group D (Hardie, 1986). An immunogenic glycan isolated from clinical strains, and presumably a feature of rumen isolates also, has been shown to consist of a tetraheteroglycan consisting of 6-deoxy L-talose, D-galactose, L-rhamnose and D-glucuronic acid (Pazur & Forsberg, 1978). These findings are consistent with the data of Hobson & Macpherson (1954) on the capsular polysaccharides of rumen amylolytic streptococci. *S. bovis* is susceptible to penicillin and possesses high-affinity penicillin-binding proteins (Williamson *et al.*, 1983).

Although principally of interest in relation to the feeding of starchy diets, *S. bovis* is able to grow on water-soluble cellodextrins derived from crystalline cellulose (Russell, 1985*b*). Thus *S. bovis* is able to survive in the rumens of animals fed on forage (Hungate, 1966). A pectin-degrading isolate of *S. bovis* from the bovine rumen was found by Wojciechowicz & Ziolecki (1984) to produce an endo-polygalacturonate lyase. This enzyme allows the organism to recover utilisable sugars (arabinose, xylose, galactose and rhamnose) which are associated with the pectic fraction of plant cell walls. The methyl esters of tetra- and tri-galacturonic acid which result from the enzyme action are not utilised by *S. bovis* but are presumably fermented by rumen microbes which possess pectinesterases (Wojciechowicz & Ziolecki, 1984).

Syntrophococcus sucromutans

Syntrophococcus sucromutans was first described by Krumholz & Bryant (1986a) who isolated this organism from a steer fed on a mixture of alfalfa hay and grain. The cells are non-motile Gram-negative cocci, 1·0–1·3 μm in diameter. Energy is obtained from electron donor/acceptor reactions involving (as donors) pyruvate, glucose, fructose, galactose, cellobiose, lactose, arabinose, maltose, ribose, xylose, salicin and aesculin. Electron acceptors are formate, caffeate, ferulate, syringate, 3,4,5-trimethoxybenzoate, vanillin and vanillate. In co-culture with *Methanobrevibacter smithii*, growth occurs on fructose. *S. sucromutans* requires 30% rumen fluid for optimal growth, though this requirement can be replaced by certain long-chain fatty acids (J. Dore & M. P. Bryant, personal communication). Krumholz & Bryant (1986a) considered that the main catabolic function of *S. sucromutans* in the rumen may be the use of methoxy groups of monobenzenoids as electron acceptors.

Ruminococcus

Sijpesteijn (1951) conferred the name *Ruminococcus flavefaciens* on Gram-positive, non-motile, cellulolytic streptococci with cells 0·8–0·9 μm in diameter, which occurred singly and in pairs and chains. A yellow pigment was produced, particularly during growth on cellulose. Later Hungate (1957) isolated from the rumen Gram-negative to Gram-variable, cellulolytic, non-motile cocci, with cells 0·8–2·0 μm in diameter and normally found as diplococci. This species, *R. albus*, is recognised together with *R. flavefaciens* as being among the most active bacteria involved in plant cell wall digestion in the rumen (Dehority & Scott, 1967; Stewart *et al.*, 1979; Morris & Van Gylswyk, 1980; Morris, 1984; Dehority, 1986; Bryant, 1986; Stewart, 1986; Chesson *et al.*, 1986; see also Chapter 8). *Ruminococcus* isolates having characteristics different from those of the two existing cellulolytic species have been found in most of the major studies (Hungate, 1957; Jarvis & Annison, 1967; Dehority, 1986). For example, the rumen contents of reindeer and musk oxen harbour strains of both *R. albus* and *R. flavefaciens*, accompanied by strains that produce equimolar amounts of ethanol, lactate and formate (Dehority, 1986). It was suggested by Dehority that these strains might form a biovar of one of the existing species comparable to isolates previously described as *R. flavefaciens* var. *lacticus* (Van Gylswyk & Roche, 1970). The gastrointestinal tract of guinea-pigs also harbours some atypical ruminococci (Dehority, 1977). Differences between certain strains of *R. albus* in SDS-PAGE patterns have also been reported (Stewart, 1986). There is a clear need for a re-evaluation of the taxonomic status of the cellulolytic ruminococci.

Ruminococcus flavefaciens

The morphological characteristics of *R. flavefaciens* are as described above. Almost all strains are cellulolytic, and additional fermentation characteristics are summarised in Table 6. Although most strains utilise relatively few substrates, Van der Linden *et al.* (1984) isolated an atypical strain that fermented 20 different substrates and produced propionate (Van der Toorn & Van Gylswyk, 1985). Trace amounts of branched-chain fatty acids are required for growth, and NH_3 is essential (Bryant, 1986).

The role of *R. flavefaciens* in plant cell wall breakdown, established in a series of studies on ruminococci and other rumen bacteria, has been further elucidated by scanning electron microscopy. Latham *et al.* (1978) showed that, when incubated with leaves of perennial ryegrass, *R. flavefaciens* mainly colonised the cut edges of the epidermis, schlerenchyma and phloem cells. Akin & Rigsby (1985) incubated *R. flavefaciens* with orchard grass and bermuda grass leaves and then studied the distribution and activity of the bacteria. The digestion of epidermis and parenchyma bundle-sheath cells was accomplished by attached bacteria. However, bacteria did not become attached either to the readily degraded mesophyll cells or to the indigestible xylem vessels.

Ruminococcus flavefaciens cellulase and xylanase have been partly characterised (Pettipher & Latham, 1979). The polysaccharidase activity has been shown to be strongly influenced by the growth substrate (Williams & Withers, 1982*b*). Cellodextrins are hydrolysed extracellularly prior to transport into the cell (Russell, 1985*b*).

When the pH of continuous cultures of *R. flavefaciens* is lowered, washout of the cells occurs at pH 6·1 or below (Russell & Dombrowski, 1980). Addition of readily fermentable carbohydrates to cellulose broths reduced cellulolysis as a consequence of the fall in pH during the fermentation (Hiltner & Dehority, 1983).

Cellulolysis and growth by *R. flavefaciens* was reduced by the plant phenolic acids *p*-coumaric and ferulic acid. When present at low (< 1 mM) concentrations, these acids were hydrogenated (and thereby detoxified) to phloretic and 3-methoxy-phloretic acids respectively. This detoxification could serve to protect other rumen microbial species (Chesson *et al.*, 1982).

Ruminococcus albus *and* Ruminococcus bromii

Ruminococcus albus typically ferments cellulose, cellobiose and glucose, and may ferment a range of other carbohydrates (Table 6) together with rhamnose, and salicin. Ammonia and one or more of the VFA isobutyric, isovaleric, 2-methylbutyric and *n*-valeric acid are required for growth; in addition, the type and other strains require 3-phenylpropanoic acid (PPA) (reviewed by Bryant, 1986).

Ruminococcus albus has attracted particular interest as a result of its production of high-molecular weight, cell-bound cellulases (Wood *et al.*, 1982; Wood & Wilson, 1984), cellobiosidase (Ohmiya *et al.*, 1982) and β-glucosidase (Ohmiya *et al.*, 1985). Cultivation of *R. albus* on different substrates influenced the production of polysaccharide depolymerases (Williams & Withers, 1982*b*). Greve *et al.* (1984) isolated from *R. albus* strain 8 an α-L-arabinofuranosidase which enhanced the activity of hemicellulase and pectinase against alfalfa cell walls. It has been shown that the PPA required for growth of some strains (Hungate & Stack, 1982) enhances the production of vesicular structures by the bacterium and increases the association of the cellulase enzyme with the cells in a high molecular-weight form (Stack & Hungate, 1984). This requirement for PPA is shared by type-strain 7 and strain 8 of *R. albus*, but not by *R. flavefaciens* or *B. fibrisolvens* (Stack & Cotta, 1986). In the presence of PPA, phenylacetic acid further enhances cellulose digestion by *R. albus* strain 8 (Stack et al., 1983).

Ruminococcus bromii has a significant role in the rumen and in the human large bowel. It is mainly implicated in starch digestion, and has many features in common with *R. albus* except for energy sources for growth (Bryant, 1986; Table 6).

Clostridium

A very large number of *Clostridium* species has been isolated from the rumen, and some representatives are listed in Table 1. Two of the species first isolated from the rumen, *C. longisporum* and *C. lochheadii*, were described by Hungate (1957). Both organisms were cellulolytic but, as cultures are not now available, the species do not appear in the approved list of bacterial names of Skerman et al. (1980). However, Sinha & Ranganathan (1983) recorded the isolation of bacteria similar to *C. lochheadii* and *C. longisporum* from the rumens of water buffaloes. Some of the fermentation characteristics of Hungate's isolates of *C. lochheadii* are summarised in Table 7.

Clostridium polysaccharolyticum
Van Gylswyk (1980) isolated *C. polysaccharolyticum* (originally identified as a *Eubacterium*) from the rumens of sheep fed on a mixture of maize stover and grain. The cells stain darkly Gram-negative (Van Gylswyk et al., 1980) and are motile by means of peritrichous flagella. Under some conditions, aseptate filaments up to 50 μm in length may be formed, but normally the cells are rods about 0·8 μm wide and 3–6 μm in length. Fermentation characteristics are summarised in Table 7.

Lactobacillus

Lactobacilli are particularly prominent members of the microflora of young ruminants (Bryant et al., 1958b; Jayne-Williams, 1979; see also Chapter 5); in animals fed on rations containing large amounts of readily fermentable carbohydrate they often proliferate in company with *Streptococcus* species, thus creating highly acidic conditions (reviewed by Hungate, 1966; Eadie & Mann, 1970; Allison et al., 1975; see also Chapter 14). Of the very large numbers of species isolated (some of which are listed in Table 1), two were first described following their isolation from the rumen. Both of these species require anaerobic conditions for growth, and meso-DAPA is present in the peptidoglycan (Sharpe et al., 1973).

Table 7
Fermentation Characteristics of Some Rumen Gram-positive Rods

	Lachnospira multiparus	Lactobacillus ruminis	Lactobacillus vitulinus	Clostridium polysaccharolyticum	Clostridium lochheadii	Eubacterium ruminantium	Eubacterium cellulosolvens
Acid from:							
starch	d			+	+	−	−
cellulose	−			+	+	−	+
xylan	−			+	−	d	d
pectin	+			+	−	d	d
maltose	d	+	+	+	+	d	+
cellobiose	+	+	+	+	+	+	+
sucrose	+	+	−	−	−	d	+
D-xylose	d	−	−	+	−	d	−
L-arabinose	−	−	−	−	−	d	−
glucose	+	+	+	+	+	+	+
fructose	d	+	+	d	d	−	d
galactose	d	+	+	−	−	−	−
mannose	d	−	−	−	−	d	+
lactose	d	d	d	−	−	d	+
mannitol	−	−	−	−	−	−	−
glycerol	−	−	−	−	−	−	−
lactate	−	−	−	+	−	−	−
Aesculin hydrolysis	+	+	+			d	d
H$_2$S production	−					−	−
Nitrate reduction	d			−	−	−	−
Fermentation products:							
Major	FAL	L	L	FBA	FBA2	FBL	L
Minor/some strains	S2			P2	L	AP	FASB
Gases produced	H$_2$CO$_2$			H$_2$	H$_2$CO$_2$	CO$_2$	H$_2$

Sources as for Table 5. Abbreviations as for Table 6. Lactobacilli from Sharpe *et al.* (1973). *C. polysaccharolyticum* from Van Gylswyk (1980).

Lactobacillus ruminis

Sharpe et al. (1973) described Gram-positive rods producing lactic acid (mainly L(+)) which they assigned to the new species *L. ruminis*, which was most closely related to strains of *L. plantarum*. Some of the sugars fermented by *L. ruminis* are shown in Table 7; in addition, sucrose, melibiose, raffinose, salicin and amygdalin are fermented. The cells are motile with peritrichous flagella.

Lactobacillus vitulinus

Lactobacillus vitulinus is a non-motile, Gram-positive rod which ferments a similar range of substrates to that of *L. ruminis* (Table 7), with the addition of lactose. Inulin, trehalose and sorbitol may be fermented. The cells are homofermentative, producing D(−)-lactic acid. *L. vitulinus*, as presently defined (Sharpe et al., 1973), appears actually to comprise two species that differ in some wall components, cell length, colony type, and ecology. Isolates consisting of short cells, which form smooth colonies, are found in both mature and young cattle (Bryant et al., 1958b). The longer type, originally known as anaerobic *L. lactis* (Mann & Oxford, 1954; Bryant et al., 1958b), has been found only in the rumens of animals 3–6 weeks of age.

Lachnospira multiparus

This pectin-degrading, Gram-positive curved rod, around 0·5 μm wide and 2–4 μm in length, was first described by Bryant & Small (1956b). *L. multiparus* has been detected in high numbers in the rumens of cattle fed on forage legumes (Bryant et al., 1960) which contain large amounts of pectic substances. Although the cells, which are motile (normally by means of one lateral flagellum), frequently stain Gram-negative, their wall ultrastructure is clearly Gram-positive (Cheng et al., 1979). The major substrates fermented and products formed are shown in Table 7. Growth requirements for B vitamins vary between strains; acetate stimulates growth, and NH_3, amino acids or peptides can serve as nitrogen sources (Bryant, 1984e). The fatty acid composition of some Japanese isolates is shown in Table 3, and the response to antibiotics in Table 4.

Lachnospira multiparus degrades plant pectic substances (Dehority, 1969; Gradel & Dehority, 1972). When incubated *in vitro* with surface-sterilised clover leaflets and grass leaves, *L. multiparus* penetrated the cut edges of the clover leaflets and caused extensive and rapid maceration of the tissues. Grass leaves were much less extensively attacked (Cheng et al., 1979). However, in contrast to the cellulolytic bacteria, *L. multiparus* causes only minor changes in weight of incubated plant material such as bermuda grass or orchard grass leaves (Akin & Rigsby, 1985) or barley straw (Stewart et al., 1979).

In a study of the pectinolytic enzymes of *L. multiparus*, Silley (1985) detected pectinesterase and pectinlyase but not polygalacturonase activity. The pectin lyase (PL) was able to macerate carrot tissue; PL production was maximal when the initial pH of the culture medium was adjusted to 6·1 (Silley, 1986).

Eubacterium

The first studies that recorded the presence of eubacteria in the rumen were those of Bryant & Burkey (1953a,b), Bryant et al. (1958c) and Bryant (1959). Apart from the two species described below, the type species, *E. limosum* has been isolated from the rumen (Table 1). Van Gylswyk & Van der Toorn (1985) have described two new species, *E. uniforme* and *E. xylanophilum*, which degraded xylan (but not cellulose) and were isolated from sheep fed on teff hay.

Eubacterium ruminantium

Bryant (1959) designated as *E. ruminantium* non-motile, short rods (0.4–$0.7\,\mu$m × 0.7–$1.5\,\mu$m) arranged mainly singly, in pairs, and in short chains. Young cultures stain Gram-positive, but older cultures destain readily. Major fermentation characteristics are shown in Table 7. Ammonia is the essential nitrogen source, and one or more of the VFA n-valerate, isovalerate, 2-methylbutyrate or isobutyrate are required for growth (Bryant & Robinson, 1962, 1963; reviewed by Moore & Holdeman-Moore, 1986).

Eubacterium cellulosolvens

Gram-positive, motile, cellulolytic rods, isolated from the rumen and named *Cillobacterium cellulosolvens*, were first described by Bryant et al. (1958c). Van Gylswyk & Hoffman (1970) later amended and extended the description of this species to include isolates obtained from sheep fed on supplemented teff hay diets. In some circumstances, *E. cellulosolvens* has been found to be among the most numerous cellulolytic bacteria in the rumen. Prins et al. (1972) estimated that this species contributed at least 50% of the total cellulolytic count in the rumen contents of cows fed on grass hay and concentrates.

The cells of *E. cellulosolvens* are typically 0.6–$0.8\,\mu$m × 1–$2\,\mu$m, and motile with a few peritrichous flagella; young cultures stain Gram-positive and contain many cells in chains. Older cells decolorise readily, and the chains break up (Prins et al., 1972). Some diagnostic features are shown in Table 7.

Eubacterium cellulosolvens has been shown to digest *Eragrostis tef* cell walls *in vitro*. Although less active than some ruminococci and *Bacteroides succinogenes*, this bacterium presumably has a significant role in plant cell wall digestion under some conditions (Morris & Van Gylswyk, 1980).

Eubacterium oxidoreducens

Eubacterium oxidoreducens was isolated by Krumholz & Bryant (1986b) from a hay-fed steer. The cells are Gram-positive, non-motile, curved rods around 0.5×1.5–$2.2\,\mu$m. *E. oxidoreducens* uses H_2 or formate as electron donor to degrade gallate, pyrogallol or phloroglucinol to acetate and butyrate, or quercetin to these plus 3,4-dihydroxyphenylacetate. Crotonate is fermented to acetate and butyrate and does not require the electron donors. This appears to be the most common rumen bacterium capable of degradation of gallate and pyrogallol to non-aromatics

(Krumholz & Bryant, 1986b; Krumholz et al., 1986). The pathway for anaerobic catabolism of gallate is described by Krumholz et al. (1987).

Mycoplasma

Two named species of anaerobic mycoplasmas have been isolated from the rumen. *Anaeroplasma* (formerly *Acholeplasma*) *bactoclasticum* (Robinson & Hungate, 1973; Robinson & Allison, 1975) and *A. abactoclasticum* (Robinson et al., 1975) both require sterol for growth, but differ in that *A. bactoclasticum* digests bacteria whereas *A. abactoclasticum* does not. These two species also differ in the range of substrates fermented and in principal fermentation products (Robinson, 1984). Characteristically, the named *Anaeroplasma* species show very low mol % $G+C$ contents of their DNA (29–34%); however, strains that do not require sterol and have around 40 mol % $G+C$ in their DNA have been reported (Robinson, 1984).

The large bacteria

Illustrated descriptions of some of the large bacteria that colonise the rumen have been presented by Moir & Masson (1952), Hungate (1966) and Ogimoto & Imai (1981). A list of some of the main types is given in Table 8. Relatively little is known of the ecological role and physiological properties of these bacteria. Kurihara et al. (1968) found that the total counts of *Oscillospira* and large oval forms tended to be lower when ciliate protozoa were introduced, suggesting that competitive or antagonistic relationships exist between these organisms. Clarke (1979) found *Lampropedia*, *Oscillospira* and oval bacteria colonising the intact cuticles of clover (*Trifolium repens*) leaflets in the rumen. These organisms did not apparently damage the cuticle, but may have been able to grow on soluble nutrients leaching from the plant material. The utilisation of soluble sugars is also thought to be the major role of the large bacterium Quin's Oval, which has been found to proliferate in the rumen when sugar-rich diets, including sugar-cane molasses, are fed (Vicini et al., 1987). Quin's Ovals are physiologically similar to the large selenomonads of sheep and are seldom seen in cattle. Phylogenetic studies are in progress. The large oval forms originally known as Eadie's Ovals have now been named *Magnoovum eadii* (Orpin, 1976) and grown in the laboratory in the presence of small bacteria. This Gram-negative bacterium has peritrichous flagella, and produces acid and gas from glucose, fructose, sucrose and lactose. Substrates that may be fermented include galactose, cellobiose, mannose, salicin and aesculin. The $G+C$ ratio of the DNA is 48%.

The rumen methanogens

The methanogens are very strictly anaerobic bacteria that require media poised at Eh values below -300 mV for growth. Oxygen is so toxic that very stringent conditions have to be maintained during cultivation (see Methods below).

The biology of the methanogens has been reviewed by Archer & Harris (1986)

Table 8
Large Bacteria of the Rumen

Organism	Morphology and comments
Oscillospira guillermondii	Motile rods, up to $5 \times 50\,\mu m$, Gram-negative with transverse septa; sometimes with sub-terminal circular 'spores' which have not been seen to germinate (Gibson, 1986)
Lampropedia	Gram-negative cocci, cells $1-2\,\mu m$, packed into sheets $20\,\mu m$ or more in diameter; probably related to the strict aerobe *L. hyalina* (Murray, 1984)
Sarcina bakeri	Gram-negative cocci $1-5\,\mu m$ in diameter aggregated in tetrads; assumed to attack starch grains (Moir & Masson, 1952; Canale-Parola, 1970)
Eadie's Ovals (E.O.)	Motile, Gram-negative ovals up to $20 \times 12\,\mu m$ in diameter (Eadie, 1962; Orpin, 1972); Orpin (1976) proposed the name *Magnoovum eadii* for this species, which has not been cultured axenically
Quin's Ovals	Similar to E.O. but typically smaller and more ellipsoidal, about $3 \times 7\,\mu m$ (Orpin, 1972)
Selenomonads	Gram-negative crescentic cells with a tuft of lateral flagella; most large forms now assumed to be related to *Selenomonas ruminantium*; some large selenomonads cultured axenically and named *S. ruminantium* subsp. *bryanti* by Prins (1971)
Rosettes	Clusters of Gram-negative rods, some with pointed ends; affiliation not known, but fresh isolates of *Bacteroides succinogenes* sometimes aggregate in this way, although the cells are smaller
Chains and filaments	Very long chains of cocci are illustrated in Hungate (1966); spiral forms similar to *Treponema* are often seen; it is not clear whether the recorded isolations of *Treponema* and *Leptospira* can account for these. Cellulolytic filaments were isolated by Leatherwood & Sharma (1972) and grown in axenic culture; they had some characteristics in common with *Butyrivibrio*, but differed in GC ratio

and Jones *et al.* (1987). Features that have been considered diagnostic for archaebacteria in general (Fox *et al.*, 1980) include (i) the cell wall structure (below), (ii) the composition of the membrane lipids which in general contain only small amounts of fatty acids and consist of C_{20}, C_{25} or C_{40} isoprenoid alcohols ether-linked to glycerol or to higher polyols (DeRosa *et al.*, 1986), (iii) the structure of the DNA-dependent RNA polymerase which has some features in common with the eukaryotic enzyme, and (iv) unusual modification patterns of transfer-RNA bases. Ribothymidine and 7-methylguanosine are absent, and the tRNA base sequence TψCG (ψ = pseudo-uridine) is replaced by one of four others in almost all

archaebacteria (Gupta & Woese, 1980). In addition, Luehrsen et al. (1985) have demonstrated the presence of non-ribosomal 7S RNA in representative archaebacteria, including many methanogens.

Of the rumen methanogens, *Methanobacterium formicicum* (Oppermann et al., 1957) and *Methanobrevibacter ruminantium* (Smith & Hungate, 1958) belong to the order Methanobacteriaceae, and *Methanomicrobium mobile* (Paynter & Hungate, 1968), *Methanosarcina barkeri* and *Ms. mazei* belong to two different families (the Methanomicrobiaceae and Methanoplanaceae respectively) of the order Methanomicrobiales. In the rumen, most methane is synthesised from H_2 and CO_2, but other potential substrates include formate, acetate, methylamine and methanol from the demethylation of plant polymers (see Chapters 1 and 11). The bacteria assumed to play the most significant role in ruminal methanogenesis are *Methanobrevibacter ruminantium* and *Methanosarcina* isolates similar to *Ms. barkeri*. However, new types continue to be isolated, and our perspectives on their relative importance will no doubt change. Lovley et al. (1984) isolated coccobacillary strains similar to *Methanobrevibacter* sp. that were capable, unlike *M. ruminantium*, of synthesising coenzyme M (2-mercaptoethanesulphonic acid), grew rapidly ($\mu = 0.24\,h^{-1}$) on H_2/CO_2, and showed a minimal concentration threshold for formate metabolism of less than 10 μM. Miller et al. (1986) isolated both coenzyme M-requiring and -non-requiring *Methanobrevibacter* strains in high numbers (10^8–10^9/ml) from the bovine rumen. Two of their strains required a fatty acid mixture for growth (isobutyric, isovaleric, 2-methylbutyric and valeric); growth of the remaining strains was stimulated by this mixture. None of the strains tested reacted with an antiserum against the type strain of *M. ruminantium*.

Methanobrevibacter ruminantium
First described by Smith & Hungate (1958) as a *Methanobacterium*, *M. ruminantium* consists of short, non-motile, Gram-positive coccobacilli about 0·7 μm wide and up to 1·8 μm in length. The major substrate is H_2/CO_2, but formate can be used when it is present at high concentrations above the rumen range (Hungate et al., 1970). Like the cell walls of other methanogens, those of *M. ruminantium* lack peptidoglycan. The major cell envelope component, pseudomurein, differs from eubacterial murein in several respects (reviewed by Archer & Harris, 1986; Jones et al., 1987). In particular, muramic acid is replaced by L-talosaminomuramic acid, and the amino acid sequences of the peptide crosslinks differ and contain principally L-amino acids. The glycan moieties of pseudomurein are thought to be β 1–3 linked. The absence of peptidoglycan makes these bacteria insensitive to antibiotics that inhibit its synthesis; even chloramphenicol, which inhibits protein synthesis in eubacteria, differs in its mode of action against some methanogens (reviewed by Harris et al., 1987).

Methanogens possess coenzymes that are common in other bacteria, including vitamin B_{12}-like cobamides (corrinoids). Krzycki & Zeikus (1980) and Shapiro (1982) have reported the corrinoid content of *M. ruminantium*. However, a special feature of methanogens is their possession of a number of coenzymes that are either unique or of limited distribution. The fluorescent flavin coenzyme F_{420} serves in *M.*

ruminantium as a cofactor for the hydrogenase, formate dehydrogenase, and NADP reductase (Tzeng *et al.*, 1975*a*), so that hydrogen and formate are equivalent sources of electrons for these bacteria (Tzeng *et al.*, 1975*b*). Initially discovered in methanogens, F_{420} has now been found in some actinomycetes and nocardioforms (Jones *et al.*, 1987). Coenzyme M, which is unique to methanogens, serves as methyl-group carrier, and it has been demonstrated with *M. thermoautotrophicum* that the methyl derivative is reduced at the terminal step in methane production (Gunsalus & Wolfe, 1980). The methyl CoM reductase of *M. ruminantium* has a prosthetic group, the yellow chromophore F_{430}, a nickel tetrapyrrole. Although CoM and F_{430} are associated, they do not appear to be linked covalently (Hüster *et al.*, 1985). Jones *et al.* (1985) suggested that methanofuran (MFR; formerly the CO_2-reduction factor) and tetrahydromethanopterin (H_4MPT) serve as carriers of C_1 groups in the reduction of CO_2. Formyl units are carried by MFR, and H_4MPT carries methenyl, methylene and methyl units.

The polar lipids of *M. ruminantium* consist of a mixture of $C_{20,20}$ and $C_{40,40}$ diethers; *Methanobrevibacter* species can be distinguished from other methanogens on the basis of their aminophospholipid composition (Grant *et al.*, 1985).

Methanosarcina barkeri
Most isolates of *Ms. barkeri* have been obtained from mud and anaerobic digesters, but Beijer (1952) reported the presence of acetate-utilising *Methanosarcina* in the rumen of a goat. When Rowe *et al.* (1979) fed sheep on a diet rich in molasses, the numbers of *Methanosarcina* in the rumen reached 10^9/ml and, although attempts to repeat this finding elsewhere were not successful, the occurrence of *Ms. barkeri* as a rumen methanogen has been confirmed. Patterson & Hespell (1979) detected *Methanosarcina* (on the basis of methylamine utilisation) at around 10^5-10^6/ml in the rumens of cattle fed on alfalfa grass, corn, and soybean meal, and Vicini *et al.* (1987) detected around 10^3 *Methanosarcina* per ml in the rumens of sheep fed on liquid molasses.

The cells are Gram-positive, non-motile spheres, 1·5–2·0 μm in diameter, which occur in packets or large clusters. Growth occurs in media containing methanol (or acetate), ammonia and sulphide (Bryant, 1974). Substrates for methane formation include H_2/CO_2, methanol, methylamines (formed in the rumen from the breakdown of choline; Patterson & Hespell, 1979) and acetate (Archer & Harris, 1986). Like other methanogens, the cell envelopes of *Ms. barkeri* lack peptidoglycan, but the major polymer differs from that present in *Methanobrevibacter*, consisting of D-glucuronic acid and *N*-acetylgalactosamine (Jones *et al.*, 1987). Analysis of the polar lipids of *Ms. barkeri* has revealed the presence of both $C_{20,20}$ and $C_{20,25}$ diethers, together with some unidentified components (Grant *et al.*, 1985).

Eikmanns & Thauer (1985) have suggested that a corrinoid enzyme is involved in the production of methane from acetate (but not from methanol or from H_2/CO_2) by *Ms. barkeri*. Coenzyme F_{420} is present in comparatively small amounts in *Ms. barkeri*, in which ferredoxin serves as an electron carrier (Jones *et al.*, 1987). Routes of coenzyme M synthesis, and the structure of the tetrahydromethanopterin (the yellow-fluorescent compound of Daniels & Zeikus, 1978) in *Ms. barkeri*, are described by White (1985) and Van Beelen et al. (1984). In contrast to methanogens that grow exclusively on H_2/CO_2, Ms. barkeri and other species that utilise methanol, methylamine or acetate possess membrane-bound cytochromes that function in the oxidation of methyl groups to CO_2 (Kühn & Gottschalk, 1983).

METHODS FOR THE ISOLATION, ENUMERATION AND CULTIVATION OF RUMEN BACTERIA

Comments here are limited to a small selection of particularly widely used procedures and are intended as a guide to the relevant literature which contains many specific details and suggestions that we have not attempted to cover. Useful reviews of the development and variety of anaerobic media and culture techniques used for rumen studies are presented by Kistner (1960), Hungate (1966, 1969), Hobson (1969), Leedle & Hespell (1980) and Leedle et al. (1982).

Rumen sampling and media preparation

For bacterial isolations or counts, rumen contents are normally obtained by suction, preferably after manually mixing the contents in the rumen (Leedle & Hespell, 1980). After blending under CO_2 for up to 1 minute in a laboratory homogeniser such as a Waring Blendor, the rumen contents are filtered through cheesecloth, and then may be further treated to release bacteria adherent to the rumen solids. Minato

Table 9
Media for Cultivation and Non-selective Isolation of Rumen Bacteria

	Ebroth[a]	RGCA[a]	Med10[b]	CC[c]	Med2[d]	98-5[e]
Clarified rumen fluid (ml)	30·0	30·0	—	16·0[f]	20·0	40·0
Peptone	0·05	—	—	—	—	—
Casitone	—	—	—	—	1·0	—
Trypticase	—	—	0·2	0·2	—	—
Yeast extract	0·05	—	0·05	0·05	0·25	—
Soluble starch	0·05	0·05	0·05	0·05	—	0·05
Sodium lactate (70% w/v)	—	—	—	—	1·0	—
Cellulose	—	—	—	0·05	—	—
Pectin	—	—	—	0·05	—	—
Xylan	—	—	—	0·05	—	—
Glycerin (ml)	—	—	—	0·05	—	—
Xylose	—	—	—	0·05	—	—
Glucose	0·05	0·0248	0·05	0·05	0·20	0·025
Maltose	0·05	—	—	0·05	0·20	—
Cellobiose	—	0·0248	0·05	0·05	0·20	0·025
$(NH_4)_2SO_4$	0·05	0·1	—	—	—	—
$NaHCO_3$	—	—	—	—	0·4	—
Na_2CO_3 soln. (8%) (ml)	—	—	5·0	5·0	—	5·0

Table 9 (*continued*)

	Ebroth[a]	RGCA[a]	Med10[b]	CC[c]	Med2[d]	98-5[e]
Mineral solutions:						
(Key no.) and vol. (ml)	(1)50	(1)50	(2)3·8 (3)3·8	(2)4·0 (4)4·0	(5)15 (6)15	(2)3·75 (7)3·75
VFA mix (ml)[g]	—	—	0·31	1·0	—	—
Haemin (mg)[h]	50	50	1·0	0·1	—	—
Vitamin K_1 (mg)	0·1	0·1	—	—	—	—
L-Cysteine HCl–H_2O[i]	0·05	0·05	0·025 ⎫	0·025 ⎫	0·05	0·025 ⎫
Na_2S–9H_2O[i]	—	—	0·025 ⎭	0·025 ⎭	—	0·025 ⎭
Rezazurin (mg)	0·1	0·1	0·1	0·1	0·1	0·1
Agar	—	2·0	2·0	2·0[j]	2·0	2·0
Dist. H_2O to (approx. final vol., ml)	100	100	100	100	100	100

Ingredients in g unless otherwise indicated.
Mineral solutions (g/l) dist. H_2O:
(1) $CaCl_2$ 0·2; $MgSO_4$ 0·2; K_2HPO_4 1·0; KH_2PO_4 1·0; $NaHCO_3$ 10·0; NaCl 2·0.
(2) K_2HPO_4 6·0.
(3) KH_2PO_4 6·0; $(NH_4)_2SO_4$ 6·0; NaCl 12·0; $MgSO_4$·7H_2O 2·5; $CaCl_2$·2H_2O 1·6.
(4) As (3) except $MgSO_4$·7H_2O 2·55; $CaCl_2$·2H_2O 1·69.
(5) K_2HPO_4 3·0.
(6) KH_2PO_4 3·0; $(NH_4)_2SO_4$ 6·0; NaCl 6·0; $MgSO_4$·7H_2O 0·6; $CaCl_2$ 0·6.
(7) KH_2PO_4 6·0; $(NH_4)_2SO_4$ 12·0; NaCl 12·0; $MgSO_4$ 1·2; $CaCl_2$ 1·2.
[a] Holdeman *et al.* (1977).
[b] Caldwell & Bryant (1966) cited in Holdeman *et al.* (1977).
[c] Leedle & Hespell (1980).
[d] Hobson (1969).
[e] Bryant & Robinson (1961), version used with 100% CO_2 and 40% clarified rumen fluid.
[f] Rumen fluid pre-incubated (see text).
[g] VFA mix: For M10, 17 ml acetic acid; 6 ml propionic; 4 ml *n*-butyric; and 1 ml each of *n*-valeric, isovaleric, isobutyric and 2-methylbutyric acid. For CC, the solution is made up to 100 ml with dist. H_2O after adjusting the pH to 7·5 with NaOH.
[h] Haemin prepared as a concentrated stock solution containing approx. 25% (M10) to 50% (CC) ethanol and 0·05M KOH (M10) or 0·025M NaOH (CC). For E broth and RGCA, dissolve 50 mg haemin in 1 ml NaOH, then make up to 100 ml with dist. H_2O.
[i] When added together, prepared as a single concentrated (1·25%, M10; 2·5% CC and 98-5) solution in dilute alkali, pH around 10 (Hespell & Bryant, 1981).
[j] Agar washed.

& Suto (1981) showed that rumen bacteria could be detached by treatment with 0·1% methylcellulose; alternatively, Dehority & Grubb (1980) found that the use of 0·1% Tween 80 and a 6–8 hour chilling period at 0°C increased total viable numbers of rumen bacteria recovered on roll-tubes. Samples are then diluted in a 10-fold dilution series in an anaerobic diluent which may contain either the same concentration of minerals as the medium (Table 9) supplemented with 0·05% cysteine HCl, 0·4% Na_2CO_3 and 0·0001% resazurin (Bryant & Burkey, 1953*a*), or a specialised buffer solution (S buffer; Leedle & Hespell, 1980) with 1·5 mM dithiothreitol as reducing agent.

Rumen fluid for use as a medium component may be prepared in different ways. Holdeman et al. (1977) prepared rumen fluid for E broths and RGCMA (Table 9) by squeezing whole rumen contents through two layers of cheesecloth, then autoclaving (121°C, 15 min) under CO_2 in sealed bottles. Prior to use, the liquid was clarified by centifuging at 10 000g for 20 minutes. Medium 98-5 can also be prepared with autoclaved rumen fluid clarified by centrifugation at low speeds (e.g. 1000g for 30 min; Henning & Van der Walt, 1978), but it may remain too cloudy to allow detection of very small colonies in roll-tubes. When lightly centrifuged rumen fluid is used, reduction of the volume of medium in the roll-tubes from 9 ml to 4 ml substantially eliminates the difficulty of seeing colonies against a background of cloudy rumen fluid (Grubb & Dehority, 1976). Dehority & Grubb (1976) devised a method to improve the selectivity of media by pre-incubation of rumen contents to remove fermentable substrates. The procedure was further refined by Leedle & Hespell (1980). To a flask containing a magnetic stirrer bar were added 300 ml of distilled water and 35 ml each of the mineral solutions 2 and 4 (Table 9). After bubbling with O_2-free CO_2 for 15 minutes, 2 ml of 2·5% (w/v) cysteine HCl were added, and bubbling with CO_2 was continued for a further 5 minutes. Strained, clarified rumen fluid (300 ml) was then added. The mixture was incubated in a sealed flask at 39°C for 3–5 days, with periodic adjustment of the pH to 6.8. After incubation, this rumen fluid was clarified by centrifugation (16 000g for 30 min, 4°C).

Anaerobic procedures for the preparation of media are illustrated by Hungate (1969), Bryant (1972) and Hespell & Bryant (1981). These accounts include details of tube furnaces (used to remove O_2 from CO_2 by passing the gas over hot copper turnings; Hungate, 1969), bent-needle gassing cannulas, and culture tubes with butyl-rubber bungs or septa (e.g. Hungate tubes, Bellco Glass Inc., Vineland, New Jersey, USA).

Bryant (1972) recommends the following procedure for media preparation. The heat-stable ingredients (excluding bicarbonate and reducing agents) are mixed and made up to volume (allowing for the later additions). The pH is adjusted to around 6·5 and the ingredients are boiled in a flask with at least 30% greater capacity than the final liquid volume. During the final stages of boiling, a stream of CO_2 (up to 800 ml/min) is passed into the flask to replace air. The flask is sealed with a black-rubber stopper which is wired in place. After autoclaving, the flask is cooled, opened under a stream of CO_2, and the heat-labile ingredients and (bi)carbonate and reducing agents are then added prior to dispensing the media into suitable sterile culture tubes under CO_2. If Na_2S is to be the reducing agent, the reducing solution is added to the tubes immediately before inoculation. For E broths and RGCA (Table 9), Holdeman et al. (1977) suggest that the ingredients (minus cysteine) should be mixed, boiled until the resazurin turns colourless, and then flushed with a stream of O_2-free CO_2 bubbling into the liquid. After cooling, the cysteine is added and the pH adjusted to between 7·0 and 7·1. After further bubbling with CO_2, O_2-free N_2 is used to exclude air while the medium is dispensed into tubes which are sealed and autoclaved using a tube press to retain the stoppers if necessary. This procedure is now used, with minor modification, at the Rowett Institute, substituting CO_2 for N_2.

Although resazurin is widely used as an oxidation–reduction indicator in culture media, it becomes oxidised at a relatively high Eh, indigo carmine (0·0005%) was suggested by Kistner (1960) to be more appropriate, though its colour change can be difficult to detect, and fresh solutions must be prepared regularly as it readily decomposes. Phenosafranin is an excellent indicator for extreme anaerobiosis but its oxidation–reduction potential is too low for most purposes (Costilow, 1981).

Table 9 summarises the composition of six widely used non-selective culture media for the enumeration, isolation and maintenance of rumen bacteria. Before preparation of these media, the original papers should be consulted for a detailed description of the routine employed. For counts and isolations, the media (with agar) are dispensed in 4–5 ml aliquots in roll-tubes (Hespell & Bryant, 1981). Roll-tubes are cooled to 47°C prior to inoculation with diluted rumen contents and rolling at low temperature to solidify the agar. For cultivation, liquid media are commonly tubed in 7·5–10 ml aliquots. From the many studies on the enumeration of rumen bacteria, it seems that medium 98-5 (Table 9), preferably amended to contain either 0·5% xylan or alfalfa fibre (Chung & Hungate, 1976), and the CC medium of Leedle & Hespell (1980) are likely to support the highest counts when inoculated with diluted rumen contents and incubated for up to 7 days at 39°C. Extension of the incubation period to 14 days may further increase the count (Bryant & Robinson, 1961). The use of a low-magnification stereoscopic microscope may aid the detection of colonies. Procedures for picking colonies from roll-tubes, using platinum–iridium needles or bent Pasteur pipettes, are described by Hungate (1969). Sub-cultured colonies often grow more successfully when stabbed into the base of slants, rather than into broths (Bryant, 1972).

In addition to non-selective media, many selective media have been described. Carbohydrate-specific subgroups have been analysed by Leedle *et al.* (1982) using media containing as major substrates either soluble starch, pectin, xylan, glucose or cellulose. Hobson (1969) described media for counting cellulolytic, amylolytic, proteolytic, lipolytic and methanogenic bacteria. Media or conditions which are selective for certain bacterial genera or species are listed in Table 10.

Cultivation and maintenance

Apart from the non-selective media used for enumeration and isolation of rumen bacteria, chemically defined or semi-defined media lacking rumen fluid have been devised that permit the growth of many species of rumen bacteria (Table 11). For fermentation studies, rather than simply for maintenance, sugars must not be autoclaved in culture media, since heating may cause interconversion of sugars (Meynell & Meynell, 1970) or, in the presence of amino acids, the formation of toxic Maillard products (Einarsson *et al.*, 1983).

Cultures of most rumen bacteria can be conveniently maintained in slants or 'sloppies' prepared from non-selective media such as RGCA, with 0·7–1·2% (w/v) agar. Cultures incubated at 39°C until growth is apparent, then stored at 4°C, usually remain viable for at least 1 month, though a decrease in the viable count is likely to occur (Teather, 1982). A particularly convenient procedure is to add

Table 10
Selective Media for Functional Groups of Species or Rumen Bacteria

Group or species	Reference and selective agent
Lipolytics, especially *Anaerovibrio*	Hobson & Mann (1961). Linseed oil Henderson (1973). Trilaurin
Cellulolytics, especially ruminococci *Butyrivibrio*, cellulolytic clostridia and *Eubacterium cellulosolvens*	Hungate (1950), Hungate (1957), Prins et al. (1972), Hungate (1966). Powdered (especially pebble-milled) cellulose
Xylanolytics, especially *Ruminococcus*, *Butyrivibrio* and eubacteria	Van der Linden et al. (1984). Xylan
Pectinolytics, especially *Lachnospira*, *Butyrivibrio* and *Bacteroides*	Dehority (1969). Pectin
Bacteroides succinogenes (and some *Ruminococcus flavefaciens*)	Stewart et al. (1981). Cotton-fibres enrichment procedure
B. succinogenes	Montgomery & Macy (1982) and Varel et al. (1984). Low-agar conc. with cellulose powder (non-rumen)
Selenomonas	Tiwari et al. (1969). Mannitol, pH, valerate
Butyrivibrio	Cheng et al. (1969). Rutin
Treponema bryantii	Stanton & Canale-Parola (1980). Rifampin
Syntrophococcus sucromutans	Krumholz & Bryant (1986a). Syringate
Eubacterium oxidoreducens	Krumholz & Bryant (1986b). Pyrogallol
Eubacterium limosum	Genthner et al. (1981). Methanol
Wolinella succinogenes	Wolin et al. (1961). Formate + fumarate
Methanogens	Lovley et al. (1984); Miller et al. (1986). H_2/CO_2; *Methanobrevibacter* sp. Patterson & Hespell (1979). *N*-Methylamines; *Methanosarcina* sp.

glycerol to a final concentration of 20% (v/v) to cultures which have been incubated for 12–24 hours at 39°C, then store at −20°C. These cultures remain viable for at least 1 year (Teather, 1982). A method for storage under liquid nitrogen (−196°C) in the presence of dimethyl sulphoxide is described by Hespell & Bryant (1981). Recovery of deep-frozen cultures may be most successful if the cultures are thawed rapidly by immersion in water at 32–35°C (Hespell & Bryant, 1981).

Several accounts have been published of successful freeze-drying of rumen bacteria; in some instances viability has been maintained for at least 5 years (White et al., 1974; Phillips et al., 1975).

Frequent sub-culture during the long-term maintenance of cultures is potentially mutagenic. Especially in view of the current interest in the genetics of rumen microbes, it is important to recognise that some genetic changes may also occur

Table 11
References to Media without Rumen Fluid for Cultivation of Rumen Bacteria

Butyrivibrio fibrisolvens	Roche *et al.* (1973); Schaefer *et al.* (1980)
Selenomonas ruminantium	Schaefer *et al.* (1980)
Anaerovibrio lipolytica	Prins *et al.* (1975)
Wolinella succinogenes	Kafkewitz (1975)
Succinimonas amylolytica	Bryant & Robinson (1962)
Succinivibrio dextrinosolvens	Gomez-Alarcon *et al.* (1982)
Treponema bryantii	Stanton & Canale-Parola (1980)
Bacteroides ruminicola	Schaefer *et al.* (1980)
Bacteroides succinogenes	Bryant *et al.* (1959); Schaefer *et al.* (1980)
Bacteroides amylophilus	Schaefer *et al.* (1980)
Megasphaera elsdenii	Schaefer *et al.* (1980); Rogosa (1984)
Veillonella parvula	Rogosa & Bishop (1964)
Eubacterium cellulosolvens	Van Gylswyk & Hoffman (1970); Prins *et al.* (1972)
Clostridium polysaccharolyticum	Van Gylswyk (1980)
Ruminococcus albus	Schaefer *et al.* (1980); Stack & Cotta (1986)
R. flavefaciens	Schaefer *et al.* (1980); Stack & Cotta (1986)
Streptococcus bovis	Prescott *et al.* (1957)
Lactobacillus ruminis *L. vitulinus*	Sharpe *et al.* (1973)
Methanobrevibacter ruminantium *Methanosarcina barkeri*	Balch *et al.* (1979)

during freezing, thawing and lyophilisation (Harrison & Pelczar, 1963; Calcott & Gargett, 1981).

Cultivation of methanogens

Balch & Wolfe (1976) and Balch *et al.* (1979) have described methods for the cultivation and maintenance of methanogens which employ serum tubes fitted with crimp-sealed black-rubber stoppers. Transfers are accomplished using syringes, and the tubes are pressurised to 2 atm with a mixture of 80% H_2/20% CO_2 to eliminate the need for replenishing the substrate as it is utilised. Although Zehnder & Wuhrman (1976) recommend titanium(III) citrate as a reducing agent in media for methanogens, this is not necessary for the rumen species. Because of its toxicity, this compound is not recommended in media for counting rumen bacteria (Wachenheim & Hespell, 1984). At high concentrations, agar may be toxic to methanogens, and Harris (1985) has recommended the use of an agar substitute, Gelrite (Kelco Div., Merck & Co. Inc., San Diego, California). Methods for the long-term storage of methanogens as cell suspensions in glycerol contained in double glass vials have been described by Winter (1983).

Anaerobic glove-box techniques

It has been possible for many years to grow anaerobic bacteria, including very fastidious methanogens, on Petri dishes in anaerobic glove-boxes (Aranki & Freter,

1972; Edwards & McBride, 1975). Special points for caution in the use of anaerobic chambers are discussed by Costilow (1981), and Leedle & Hespell (1980) describe the application of the glove-box technique to rumen microbiology. In particular, the use of Petri plates allows the application of replica-plating techniques to ecological studies on bacterial populations, and avoids exposure of bacteria to molten agar, which may depress the numbers of colonies detected (Leedle et al., 1982). Leedle & Hespell (1980) recommended equilibrating plastic Petri dishes inside glove-boxes for at least 24 hours before use, to ensure removal of traces of surface-adhering O_2. Media can either be introduced in bulk quantities (e.g. in crimp-topped bottles) and poured into plates within the glove-box, or (preferably) dispensed, outside the box, into anaerobic culture tubes, then introduced into the box and cooled to 50°C in a dry-block heater unit prior to pouring the plates.

The gas atmospheres used in anaerobic glove-boxes vary substantially from one laboratory to another. Since most of the media for cultivation of rumen bacteria were originally designed to be equilibrated with CO_2, it is important that the atmosphere contains sufficient CO_2 both to maintain the medium pH within the desired range and to provide CO_2 for those bacteria that utilise it. A gas mixture containing 95% CO_2/5% H_2 has been successfully employed. Alternatively, cultures can be incubated in gas-tight containers (anaerobic jars or modified pressure-cookers; Leedle & Hespell, 1980) flushed with CO_2 prior to incubation. Balch et al. (1979) describe the use of anaerobic glove-boxes for the cultivation of methanogens on plates incubated in cylinders pressurised (2 atm) with a gas mixture of 80% H_2/20% CO_2.

One particularly useful aspect of the use of anaerobic glove-boxes is that they facilitate the use of rapid identification systems such as API test strips (API System SA, La Balme Les Grottes, 38390 Montalien Vercieu, France; Flint & Stewart, 1987). In addition, the larger glove-boxes can accommodate centrifuges and other equipment, so that cell fractionations, enzyme purifications and other procedures can be performed anaerobically.

REFERENCES

Adams, M. W. W., Mortensen, L. E. & Chen, J. S. (1981). Hydrogenase. *Biochim. Biophys. Acta*, **594**, 105–76.

Akin, D. E. & Rigsby, L. L. (1985). Degradation of Bermuda and Orchard grass by species of rumen bacteria. *Appl. Environ. Microbiol.*, **50**, 825–30.

Allen, S. H. G. (1983). Lactate-oxaloacetate transhydrogenase from *Veillonella alcalescens*. *Meth. Enzymol.*, **89**, 367–76.

Allison, M. J. (1978). Production of branched-chain volatile fatty acids by certain anaerobic bacteria. *Appl. Environ. Microbiol.*, **35**, 872–7.

Allison, M. J., Robinson, I. M., Dougherty, R. W. & Bucklin, J. A. (1975). Grain overload in cattle and sheep: changes in microbial populations in the caecum and rumen. *Am. J. Vet. Res.*, **36**, 181–5.

Allison, M. J., Cook, H. M. & Jones, R. J. (1983). Detoxication of 3-hydroxy-4(1H)-pyridone, the goiterogenic of mimosine, by rumen bacteria from Hawaiian goats. In *Proc. 17th Rumen Function Conf. ARS/USDA*, p. 21.

Allison, M. J., Baetz, A. L. & Wiegel, J. (1984). Alternative pathways for biosynthesis of leucine and other amino acids in *Bacteroides ruminicola* and *Bacteroides fragilis*. *Appl. Environ. Microbiol.*, **48**, 1111–17.

Allison, M. J., Dawson, K. A., Mayberry, W. R. & Foss, J. G. (1985). *Oxalobacter formigenes* gen. nov. sp. nov.: oxalate degrading anaerobes that inhabit the gastrointestinal tract. *Arch. Microbiol.*, **141**, 1–7.

Aranki, A. & Freter, R. (1972). Use of anaerobic glove boxes for the cultivation of strictly anaerobic bacteria. *Am. J. Clin. Nutr.*, **25**, 1329–34.

Archer, D. B. & Harris, J. E. (1986). Methanogenic bacteria and methane production in various habitats. In *Anaerobic Bacteria in Habitats Other than Man*, ed. E. M. Barnes & G. C. Mead. Blackwell Scientific Publications, Oxford, pp. 185–223.

Balch, W. E. & Wolfe, R. S. (1976). New approach to the cultivation of methanogenic bacteria: 2-mercaptoethanesulphonic acid (HS-CoM)-dependent growth of *Methanobacterium ruminantium* in a pressurised atmosphere. *Appl. Environ. Microbiol.*, **32**, 781–91.

Balch, W. E., Fox, G. E., Magrum, L. J., Woese, C. R. & Wolfe, R. S. (1979). Methanogens: re-evaluation of a unique biological group. *Microbiol. Rev.*, **43**, 260–96.

Baldwin, R. L. & Milligan, L. P. (1964). Electron transport in *Peptostreptococcus elsdenii*. *Biochim. Biophys. Acta*, **92**, 421–32.

Begbie, R. & Stewart, C. S. (1984). Polyacrylamide gel electrophoresis of *Bacteroides succinogenes*. *Can. J. Microbiol.*, **30**, 863–6.

Beijer, W. H. (1952). Methane fermentation in the rumen of cattle. *Nature*, **170**, 576–7.

Boyd, J. W. (1985). Studies on thiaminase I activity in ruminant faeces and rumen bacteria. *J. Agric. Sci.*, **104**, 637–42.

Brockman, H. L. & Wood, W. A. (1975). Electron-transferring flavoprotein of *Peptostreptococcus elsdenii* that functions in the reduction of acrylyl-coenzyme A. *J. Bacteriol.*, **124**, 1447–53.

Bryant, M. P. (1952). The isolation and characteristics of a spirochaete from the bovine rumen, *J. Bacteriol.*, **64**, 325–35.

Bryant, M. P. (1959). Bacterial species of the rumen. *Bacteriol. Rev.*, **23**, 125–53.

Bryant, M. P. (1972). Commentary on the Hungate technique for cultivation of anaerobic bacteria. *Am. J. Clin. Nutr.*, **25**, 1324–8.

Bryant, M. P. (1974). Methane producing bacteria. In *Bergey's Manual of Determinative Bacteriology*, 8th edn, ed. R. E. Buchanan & N.E. Gibbons. Williams & Wilkins, Baltimore, pp. 472–7.

Bryant, M. P. (1984a). Butyrivibrio. In *Bergey's Manual of Systematic Bacteriology*, Vol. 1, ed. N. R. Krieg & J. G. Holt. Williams & Wilkins, Baltimore, pp. 641–3.

Bryant, M. P. (1984b). Succinimonas. In *Bergey's Manual of Systematic Bacteriology*, Vol. 1, ed. N. R. Krieg & J. G. Holt. Williams & Wilkins, Baltimore, pp. 643–4.

Bryant, M. P. (1984c). Succinivibrio. In *Bergey's Manual of Systematic Bacteriology*, Vol. 1, ed. N. R. Krieg & J. G. Holt. Williams & Wilkins, Baltimore, pp. 644–5.

Bryant, M. P. (1984d). Selenomonas. In *Bergey's Manual of Systematic Bacteriology*, Vol. 1, ed. N. R. Krieg & J. G. Holt. Williams & Wilkins, Baltimore, pp. 650–3.

Bryant, M. P. (1984e). Lachnospira. In *Bergey's Manual of Systematic Bacteriology*, Vol. 1, ed. N. R. Kreig & J. G. Holt. Williams & Wilkins, Baltimore, pp. 661–2.

Bryant, M. P. (1986). Ruminococcus. In *Bergey's Manual of Systematic Bacteriology*, Vol. 2, ed. P. H. A. Sneath. Williams & Wilkins, Baltimore, pp. 1093–7.

Bryant, M. P. & Burkey, L. A. (1953a). Cultural methods and some characteristics of some of the more numerous groups of bacteria in the bovine rumen. *J. Dairy Sci.*, **36**, 205–17.

Bryant, M. P. & Burkey, L. A. (1953b). Numbers and some predominant groups of bacteria in the rumen of cows fed different rations. *J. Dairy Sci.*, **36**, 218–24.

Bryant, M. P. & Robinson, I. M. (1961). An improved non-selective culture medium for ruminal bacteria and its use in determining the diurnal variation in numbers of bacteria in the rumen. *J. Dairy Sci.*, **44**, 1446–56.

Bryant, M. P. & Robinson, I. M. (1962). Some nutritional characteristics of predominant culturable ruminal bacteria. *J. Bacteriol.*, **82**, 605–14.

Bryant, M. P. & Robinson, I. M. (1963). Apparent incorporation of ammonia and amino acid carbon during growth of selected species of ruminal bacteria. *J. Dairy Sci.*, **46**, 150–4.

Bryant, M. P. & Robinson, I. M. (1968). Effect of diet, time after feeding and position sampled on numbers of viable bacteria in the bovine rumen. *J. Dairy Sci.*, **51**, 1950–5.

Bryant, M. P. & Small, N. (1956a). The anaerobic monotricous butyric acid producing curved rod shaped bacteria of the rumen. *J. Bacteriol.*, **72**, 16–21.

Bryant, M. P. & Small, N. (1956b). Characteristics of two new genera of anaerobic curved rods isolated from the rumen of cattle. *J. Bacteriol.*, **72**, 22–6.

Bryant, M. P., Robinson, I. M., Bouma, C. & Chu, H. (1958a). *Bacteroides ruminicola* n. sp. and the new genus and species *Succinimonas amylolytica*. Species of succinic-acid producing anaerobic bacteria of the bovine rumen, *J. Bacteriol.*, **76**, 15–23.

Bryant, M. P., Small, N., Bouma, C. & Robinson, I. M. (1958b). Studies on the composition of the ruminal flora and fauna of young calves. *J. Dairy Sci.*, **41**, 1747–67.

Bryant, M. P., Small, N., Bouma, C. & Robinson, I. M. (1958c). Characterisation of ruminal anaerobic cellulolytic cocci and *Cillobacterium cellulosolvens* n. sp. *J. Bacteriol.*, **76**, 529–37.

Bryant, M. P., Robinson, I. M. & Chu, H. (1959). Observations on *Bacteroides succinogenes*: a ruminal cellulolytic bacterium. *J. Dairy Sci.*, **42**, 1831–47.

Bryant, M. P., Barrentine, B. F., Sykes, J. F., Robinson, I. M., Shawver, C. B. & Williams, L. W. (1960). Predominant bacteria in the rumen of cattle on bloat-provoking ladino clover pasture. *J. Dairy Sci.*, **43**, 1435–44.

Bryant, M. P., Robinson, I. M. & Lindahl, I. L. (1961). A note on the flora and fauna in the rumen of steers fed a feed lot bloat-provoking ration and the effect of penicillin. *Appl. Microbiol.*, **9**, 511–15.

Calcott, P. H. & Gargett, A. M. (1981). Mutagenicity of freezing and thawing. *FEMS Microbiol. Lett.*, **10**, 151–5.

Caldwell, D. R. & Bryant, M. P. (1966). Medium without rumen fluid for non-selective enumeration and isolation of rumen bacteria. *Appl. Microbiol.*, **14**, 794–801.

Caldwell, D. R., Keeney, M., Barton, J. S. & Kelley, J. F. (1973). Sodium and other inorganic growth requirements of *Bacteroides amylophilus*. *J. Bacteriol.*, **114**, 782–9.

Canale-Parola, E. (1970). Biology of the sugar fermenting sarcinae. *Bacteriol. Rev.*, **34**, 84–97.

Chen, M. & Wolin, M. J. (1979). Effect of monensin and lasalocid-sodium on the growth of methanogenic and rumen saccharolytic bacteria. *Appl. Environ. Microbiol.*, **38**, 72–7.

Cheng, K.-J. & Costerton, J. W. (1977). Ultrastructure of *Butyrivibrio fibrisolvens*: a Gram-positive bacterium? *J. Bacteriol.*, **129**, 1506–12.

Cheng, K.-J., Jones, G. A., Simpson, F. J. & Bryant, M. P. (1969). Isolation and identification of anaerobic bacteria capable of rutin degradation. *Can. J. Microbiol.*, **15**, 1365–71.

Cheng, K.-J., Dinsdale, D. & Stewart, C. S. (1979). Maceration of clover and grass leaves by *Lachnospira multiparus*. *Appl. Environ, Microbiol.*, **38**, 723–9.

Cheng, K.-J., Stewart, C. S., Dinsdale, D. & Costerton, J. W. (1984). Electron microscopy of bacteria involved in the digestion of plant cell walls. *Anim. Feed Sci. Technol.*, **10**, 93–120.

Chesson, A., Stewart, C. S. & Wallace, R. J. (1982). Influence of plant phenolic acids on growth and cellulolytic activity of rumen bacteria. *Appl. Environ. Microbiol.*, **44**, 597–603.

Chesson, A., Stewart, C. S., Dalgarno, K. & King, T. P. (1986). Degradation of isolated grass mesophyll, epidermis and fibre cell walls in the rumen and by cellulolytic rumen bacteria in axenic culture. *J. Appl. Bacteriol.*, **60**, 327–16.

Chung, K.-T. & Hungate, R. E. (1976). Effect of alfalfa fiber substrate on culture counts of rumen bacteria. *Appl. Environ. Microbiol.*, **32**, 649–52.

Clarke, R. T. J. (1979). Niche in pasture-fed ruminants for the large rumen bacteria Oscillospira, Lampropedia and Quin's and Eadie's Ovals. *Appl. Environ. Microbiol.*, **37**, 654–7.

Cook, A. R. (1976). The elimination of urease activity in *Streptococcus faecium* as evidence for plasmid coded urease. *J. Gen. Microbiol.*, **92**, 49 58.
Costilow, R. N. (1981). Biophysical factors in growth. In *Manual of Methods for General Bacteriology*, ed. P. Gerhardt, R. G. E. Murray, R. N. Costilow, E. W. Nester, W. A. Wood, N. R. Krieg & G. B. Phillips. ASM, Washington, pp. 66–78.
Cotta, M. A. & Hespell, R. B. (1986). Proteolytic activity of the ruminal bacterium *Butyrivibrio fibrisolvens*. *Appl. Environ. Microbiol.*, **52**, 51–8.
Counotte, G. H. M., Prins, R. A., Janssen, R. H. A. M. & DeBie, M. J. A. (1981). Role of *Megasphaera elsdenii* in the fermentation of DL-[2-^{13}C]lactate in the rumen of dairy cattle. *Appl. Environ. Microbiol.*, **42**, 649–55.
Counotte, G. H. M., Lankhorst, A. & Prins, R. A. (1983). Role of DL-lactic acid as an intermediate in rumen metabolism of dairy cows. *J. Anim. Sci.*, **56**, 1222–35.
Daniels, L. & Zeikus, J. G. (1978). One-carbon metabolism in methanogenic bacteria: analysis of short-term fixation products of $^{14}CO_2$ and $^{14}CH_3OH$ incorporated into whole cells. *J. Bacteriol.*, **136**, 75–84.
Dawson, K. A., Preziosi, M. C. & Caldwell, D. R. (1979). Some effects of uncouplers and inhibitors on growth and electron transport in rumen bacteria. *J. Bacteriol.*, **139**, 384–92.
Dawson, K. A., Allison, M. J. & Hartman, P. A. (1980). Isolation and some characteristics of anaerobic oxalate degrading bacteria from the rumen. *Appl. Environ. Microbiol.*, **40**, 833–9.
Dehority, B. A. (1969) Pectin-fermenting bacteria isolated from the bovine rumen. *J. Bacteriol.*, **99**, 189–96.
Dehority, B. A. (1977). Cellulolytic cocci isolated from the cecum of guinea pigs. *Appl. Environ. Microbiol.*, **33**, 1278–83.
Dehority, B. A. (1986). Microbes in the foregut of arctic ruminants. In *Control of Digestion and Metabolism in Ruminants*, ed. L. P. Milligan, W. L. Grovum & A. Dobson. Reston, Prentice-Hall, Englewood Cliffs, New Jersey, pp. 307–25.
Dehority, B. A. & Grubb, J. A. (1976). Basal medium for the selective enumeration of rumen bacteria utilising specific energy sources. *Appl. Environ. Microbiol.*, **32**, 703–10.
Dehority, B. A. & Grubb, J. A. (1980). Effect of short-term chilling of rumen contents on viable bacterial numbers. *Appl. Environ. Microbiol.*, **39**, 376–81.
Dehority, B. A. & Scott, H. W. (1967). Extent of cellulose and hemicellulose digestion in various forages by pure cultures of rumen bacteria. *J. Dairy Sci.*, **50**, 1136–41.
Dennis, S. M., Nagaraja, T. G. & Bartley, E. E. (1981). Effects of lasalocid or monensin on lactate-producing or -using bacteria. *J. Anim. Sci.*, **52**, 418–26.
DeRosa, M., Gambacorta, A. & Gliozzi, A. (1986). Structure, biosynthesis and physicochemical properties of Archaebacterial lipids. *Microbiol. Rev.*, **50**, 70–80.
Eadie, J. M. (1962). The development of rumen microbial populations in lambs and calves under various conditions of management. *J. Gen. Microbiol.*, **29**, 563–78.
Eadie, J. M. & Mann, S. O. (1970). Development of the rumen microbial population: high starch diets and instability. In *Physiology of Digestion and Metabolism in the Ruminant*, ed. A. T. Phillipson. Oriel Press, Newcastle-upon-Tyne, pp. 335–47.
Edwards, T. & McBride, B. C. (1975). New method for the isolation and identification of methanogenic bacteria. *Appl. Microbiol.*, **29**, 540–5
Eikmanns, B., & Thauer, R. K. (1985). Evidence for the involvement and role of a corrinoid enzyme in methane formation from acetate in *Methanosarcina barkeri*. *Arch. Microbiol.*, **142**, 175–9.
Einarsson, H., Snygg, B. G. & Erikson, C. (1983). Inhibition of bacterial growth by Maillard reaction products. *J. Agric. Fd Chem.*, **31**, 1043–7.
Elsden, S. R. & Lewis, D. (1953). The production of fatty acids by a Gram-negative coccus. *Biochem. J.*, **55**, 183–9.
Elsden, S. R., Volcani, B. E., Gilchrist, F. M. C. & Lewis, D. (1956). Properties of a fatty acid forming organism from the rumen of sheep. *J. Bacteriol.*, **72**, 681–9.

Essers, L. (1982). Simple identification of anaerobic bacteria to genus level using typical antibiotic susceptibility patterns. *J. Appl. Bacteriol.*, **52**, 319–23.

Flint, H. J. & Stewart, C.S. (1987). Antibiotic resistance patterns and plasmids of ruminal strains of *Bacteroides ruminicola* and *Bacteroides multiacidus*. *Appl. Microbiol. Biotechnol.*, **26**, 450–5.

Flint, H. J., Duncan, S. H. & Stewart, C. S. (1987). Transmissible antibiotic resistance in strains of *Escherichia coli* isolated from the ovine rumen. *Lett. Appl. Microbiol.*, **5**, 47–9.

Fonty, G., Jouany, J.-P., Senaud, J., Gouet, P. & Grain, J. (1984). The evolution of microflora, microfauna and digestion in the rumen of lambs from birth to 4 months. *Can. J. Anim. Sci.*, **64** (Suppl.), 165–6.

Forsberg, C. W. (1978). Nutritional characteristics of *Megasphaera elsdenii*. *Can. J. Microbiol.*, **24**, 981–5.

Foster, T. J. (1983). Plasmid-determined resistance to antimicrobial drugs and toxic metal ions in bacteria. *Microbiol. Rev.*, **43**, 361–409.

Fox, G. E., Stackebrandt, E., Hespell, R. B., Gibson, J., Manilof, J., Dyer, T. A., Wolfe, R. S., Balch, W. E., Tanner, R. S., Magrum, L. J., Zableu, L. B., Blakemore, R., Gupta, R., Bonen, L., Lewis, B. J., Stahl, D. A., Luehrsen, K. R., Chen, K. N. & Woese, C. R. (1980). The phylogeny of prokaryotes. *Science*, **209**, 457–63.

Franklund, C. V. & Glass, T. L. (1987). Glucose uptake by the cellulolytic rumen anaerobe *Bacteroides succinogenes*. *J. Bacteriol.*, **169**, 500–6.

Fulghum, R. S. & Worthington, J. M. (1984). Superoxide dismutase in ruminal bacteria. *Appl. Environ. Microbiol.*, **48**, 675–7.

Gardner, R. M., Fuller, M. D. & Caldwell, D. R. (1983). Tetrapyrrole utilisation by protohaeme-synthesising anaerobes. *Curr. Microbiol.*, **9**, 59–61.

Genthner, B. R. S., Davis, C. L. & Bryant, M. P. (1981). Features of rumen and sewage sludge strains of *Eubacterium limosum*, a methanol and H_2–CO_2 using species. *Appl. Environ. Microbiol.*, **42**, 12–19.

Gibson, T. (1986). Oscillospira. In *Bergey's Manual of Systematic Bacteriology*, Vol. 2, ed. P. H. A. Sneath. Williams & Wilkins, Baltimore, p. 1207.

Gomez-Alarcon, R. A., O'Dowd, C., Leedle, J. A. Z. & Bryant, M. P. (1982). 1,4-Naphthoquinone and other nutrient requirements of *Succinivibrio dextrinosolvens*. *Appl. Environ. Microbiol.*, **44**, 346–50.

Gradel, C. M. & Dehority, B. A. (1972). Fermentation of isolated pectin and pectin from intact forages by pure cultures of rumen bacteria. *Appl. Microbiol.*, **23**, 332–40.

Graham, H., Aman, P., Theander, O., Kolankaya, N. & Stewart, C. S. (1985). Influence of heat sterilisation and ammoniation on straw composition and degradation by pure cultures of cellulolytic rumen bacteria. *Anim. Feed Sci. Technol.*, **12**, 195–203.

Grant, W. D., Pinch, G., Harris, J. E., DeRosa, M. & Gambacorta, A. (1985). Polar lipids in methanogen taxonomy. *J. Gen. Microbiol.*, **131**, 3277–86.

Greve, L. C., Labavitch, J. M. & Hungate, R. E. (1984). α-L-Arabinofuranosidase from *Ruminococcus albus* 8: purification and possible role in hydrolysis of alfalfa cell wall. *Appl. Environ. Microbiol.*, **47**, 1135–40.

Grubb, J. A. & Dehority, B. A. (1976). Variation in colony counts of total viable anaerobic rumen bacteria as influenced by media and cultural methods. *Appl. Environ. Microbiol.*, **31**, 262–7.

Gunsalus, R. P. & Wolfe, R. S. (1980). Methyl coenzyme M reductase from *Methanobacterium thermoautotrophicum*: resolution and properties of the components. *J. Biol. Chem.*, **255**, 1891–5.

Gupta, R. & Woese, C. R. (1980). Unusual modification patterns in the transfer ribonucleic acids of Archaebacteria. *Curr. Microbiol.*, **4**, 245–9.

Gutierrez, J. (1953). Numbers and characteristics of lactate-utilising organisms in the rumen of cattle. *J. Bacteriol.*, **66**, 123–8.

Halliwell, G. & Bryant, M. P. (1963). The cellulolytic activity of pure strains of bacteria from the rumen of cattle. *J. Gen. Microbiol.*, **32**, 441–8.

Hamlin, L. J. & Hungate, R. E. (1956). Culture and physiology of a starch-digesting bacterium (*Bacteroides amylophilus* n. sp.) from the bovine rumen. *J. Bacteriol.*, **72**, 548–54.
Harborth, P. B. & Hanert, H. H. (1982). Isolation of *Selenomonas ruminantium* from an aquatic ecosystem. *Arch. Microbiol.*, **132**, 135–40.
Hardie, J. M. (1986). Other *Streptococci*. In *Bergey's Manual of Systematic Bacteriology*, Vol. 2, ed. P. H. A. Sneath. Williams & Wilkins, Baltimore, pp. 1068–71.
Harris, J. E. (1985). Gelrite as an agar substitute for the cultivation of mesophilic *Methanobacterium* and *Methanobrevibacter* species. *Appl. Environ. Microbiol.*, **50**, 1107–9.
Harris, J. E., Evans, D. M., Knox, M. R. & Archer, D. B. (1987). Genetic approaches with methanogens important in mesophilic anaerobic digestion. In *Recent Advances in Anaerobic Bacteriology*, ed. S. P. Borriello & J. M. Hardie. Martinus Nijhof, Dordrecht, pp. 123–37.
Harrison, A. P. & Pelczar, M. J. (1963). Damage and survival of bacteria during freeze-drying and during storage over a 10-year period. *J. Gen. Microbiol.*, **30**, 395–400.
Hausinger, R. P. (1986). Purification of a nickel-containing urease from the rumen anaerobe *Selenomonas ruminantium*. *J. Biol. Chem.*, **261**, 7866–70.
Hazlewood, G. P. & Edwards, R. (1981). Proteolytic activities of a rumen bacterium *Bacteroides ruminicola* R8/4. *J. Gen. Microbiol.*, **125**, 11–15.
Heinrichova, K., Wojciechowicz, M. & Ziolecki, A. (1985). An exo-D-galacturonase of *Butyrivibrio fibrisolvens* from the bovine rumen. *J. Gen. Microbiol.*, **131**, 2053–8.
Henderson, C. (1973). An improved method for enumerating and isolating rumen lipolytic bacteria. *J. Appl. Bacteriol.*, **36**, 187–8.
Henderson, C. (1975). The isolation and characterisation of strains of lipolytic bacteria from the ovine rumen. *J. Appl. Bacteriol.*, **39**, 101–9.
Henderson, C. (1980). The influence of extracellular hydrogen on the metabolism of *Bacteroides ruminicola*, *Anaerovibrio lipolytica* and *Selenomonas ruminantium*. *J. Gen. Microbiol.*, **119**, 485–91.
Henderson, C. & Hodgkiss, W. (1973). An electron microscopic study of *Anaerovibrio lipolytica* (strain 5S) and its lipolytic enzyme. *J. Gen. Microbiol.*, **76**, 389–93.
Henderson, C., Hobson, P. N. & Summers, R. (1969). The production of amylase, protease and lipolytic enzymes by two species of anaerobic rumen bacteria. In *Continuous Cultivation of Microorganisms*. Academia, Prague, pp. 189–204.
Henning, P. A. & Van der Walt, A. E. (1978). Inclusion of xylan in a medium for the enumeration of total culturable rumen bacteria. *Appl. Environ. Microbiol.*, **35**, 1008–11.
Hespell, R. B. & Bryant, M. P. (1981). The genera *Butyrivibrio*, *Succinivibrio*, *Lachnospira* and *Selenomonas*. In *The Prokaryotes: A Handbook on Habitats, Isolation and Identification of Bacteria*, Vol. 2, ed. M. P. Starr, H. Stolp, H. G. Truper, A. Balows & H. G. Schlegel. Springer Verlag, Berlin, pp. 1479–94.
Hespell, R. B., Wolf, R. & Bothast, R. J. (1987). Fermentation of xylans by *Butyrivibrio fibrisolvens* and other ruminal bacteria. *Appl. Environ. Microbiol.*, **53**, 2849–53.
Hilpert, W. & Dimroth, P. (1983). Purification and characterisation of a new sodium-transport decarboxylase, methylmalonyl CoA decarboxylase from *Veillonella alcalescens*. *Eur. J. Biochem.*, **132**, 579–87.
Hiltner, P. & Dehority, B. A. (1983). Effect of soluble carbohydrates on digestion of cellulose by pure cultures of rumen bacteria. *Appl. Environ. Microbiol.*, **46**, 642–8.
Hobson, P. N. (1969). Rumen bacteria. In *Methods in Microbiology*, Vol. 3B, ed. J. R. Norris & D. W. Ribbons. Academic Press, London and New York, pp. 133–49.
Hobson, P. N. & Howard, B. H. (1969). Microbial transformations. In *Handbuch der Tierernährung*, Vol. I. Verlag Paul Parey, Hamburg, pp. 207–54.
Hobson, P. N. & Macpherson, M. J. (1954). Some serological and chemical studies on materials extracted from an amylolytic streptococcus from the rumen of sheep. *Biochem. J.*, **57**, 145–51.
Hobson, P. N. & Mann, S. O. (1957). Some studies on the identification of rumen bacteria with fluorescent antibodies. *J. Gen. Microbiol.*, **16**, 463–71.

Hobson, P. N. & Mann, S. O. (1961). The isolation of glycerol-fermenting and lipolytic bacteria from the rumen of the sheep. *J. Gen. Microbiol.*, **25**, 227–40.

Hobson, P. N., Mann, S. O. & Oxford, A. E. (1958). Some studies on the occurrence and properties of a large Gram-negative coccus from the rumen. *J. Gen. Microbiol.*, **19**, 462–72.

Hobson, P. N., Mann, S. O. & Summers, R. (1975/76a). Rumen microorganisms in red deer, hill sheep and reindeer in the Scottish highlands. *Proc. Roy. Soc. Edin.* (B), **75**, 171–80.

Hobson, P. N., Mann, S.O., Summers, R. & Staines, B. (1975/76b). Rumen function in red deer, hill sheep and reindeer in the Scottish highlands. *Proc. Roy. Soc. Edin.* (B), **75**, 181–98.

Holdeman, L. V., Cato, E. P. & Moore, W. E. C. (1977). *Anaerobe Laboratory Manual*. Virginia Polytechnic Institute, Blacksburg.

Holdeman, L. V., Kelley, R. W. & Moore, W. E. C. (1984). Bacteroides. In *Bergey's Manual of Systematic Bacteriology*, Vol. 1, ed. N. R. Krieg & J. G. Holt. Williams & Wilkins, Baltimore, pp. 604–31.

Howard, B. H., Jones, G. & Purdom, M. R. (1960). The pentosanases of some rumen bacteria. *Biochem. J.*, **74**, 173–80.

Hudman, J. F. (1984). Glucose-induced morphological variation in *Selenomonas ruminantium*. *FEMS Microbiol. Lett.*, **22**, 201–4.

Hudman, J. F. & Glenn, A. R. (1984). Selenite uptake and incorporation by *Selenomonas ruminantium*. *Arch. Microbiol.*, **140**, 252–6.

Hudman, J. F. & Glenn, A. R. (1985). Selenium uptake by *Butyrivibrio fibrisolvens* and *Bacteroides ruminicola*. *FEMS Microbiol. Lett.*, **27**, 215–20.

Hughes, P. E., Hunter, W. J. & Tove, S. B. (1982). Biohydrogenation of unsaturated fatty acids. Purification and properties of *cis*-9,*trans*-11-octadecadienoate reductase. *J. Biol. Chem.*, **257**, 3643–9.

Hungate, R. E. (1947). Studies on cellulose fermentation. III. The culture and isolation of cellulose-decomposing bacteria from the rumen of cattle. *J. Bacteriol.*, **53**, 631–45.

Hungate, R. E. (1950). The anaerobic mesophilic cellulolytic bacteria. *Bacteriol. Rev.*, **14**, 1–49.

Hungate, R. E. (1957). Microorganisms in the rumen of cattle fed a constant ration. *Can. J. Microbiol.*, **3**, 289–311.

Hungate, R. E. (1966). *The Rumen and Its Microbes*. Academic Press, New York and London.

Hungate, R. E. (1969). A roll tube method for cultivation of strict anaerobes. In *Methods in Microbiology*, Vol. 3B, ed. J. R. Norris & D. W. Ribbons. Academic Press, London and New York, pp. 117–32.

Hungate, R. E. & Stack, R. J. (1982). Phenylpropanoic acid: growth factor for *Ruminococcus albus*. *Appl. Environ. Microbiol.*, **44**, 79–83.

Hungate, R. E., Dougherty, R. W., Bryant, M. P. & Cello, R. M. (1952). Microbiological and physiological changes associated with acute indigestion in sheep. *Cornell Vet.*, **42**, 423–49.

Hungate, R. E., Smith, W., Bauchop, T., Yu, I. & Rabinowitz, J. C. (1970). Formate as an intermediate in the rumen fermentation. *J. Bacteriol.*, **102**, 389–97.

Hüster, R., Gilles, M.-H. & Thauer, R. K. (1985). Is coenzyme M bound to factor F430 in methanogenic bacteria? *Eur. J. Biochem.*, **148**, 107–11.

Iannotti, E. L., Kafkewitz, D., Wolin, M. J. & Bryant, M. P. (1973). Glucose fermentation products of *Ruminococcus albus* grown in continuous cultures with *Vibrio succinogenes*: changes caused by interspecies transfer of H_2. *J. Bacteriol.*, **114**, 1231–40.

Jarvis, B. D. W. & Annison, E. F. (1967). Isolation, classification and nutritional requirements of cellulolytic cocci in the sheep rumen. *J. Gen. Microbiol.*, **47**, 295–307.

Jayne-Williams, D. J. (1979). The bacterial flora of the rumen of healthy and bloating calves. *J. Appl. Bacteriol.*, **47**, 271–84.

Johns, A. T. (1951). Isolation of a bacterium, producing propionic acid, from the rumen of sheep. *J. Gen. Microbiol.*, **5**, 317–25.

Johnson, J. L. (1978). Taxonomy of the Bacteroides. 1. Deoxyribonucleic acid homologies among *Bacteroides fragilis* and other saccharolytic *Bacteroides* species. *Int. J. Syst. Bacteriol.*, **28**, 245–56.

Johnson, J. L. & Harich, B. (1986). Ribosomal ribonucleic acid homology among species of the genus *Bacteroides*. *Int. J. Syst. Bacteriol.*, **36**, 71–9.

Johnston, N. C. & Goldfine, H. (1982). Effect of growth temperature on fatty acid and alk-1-enyl group compositions of *Veillonella parvula* and *Megasphaera elsdenii*. *J. Bacteriol.*, **149**, 567–75.

Jones, R. J. (1985). Leucaena toxicity and the ruminal degradation of mimosine. In *Plant Toxicology*, ed. A. A. Seawright, M. P. Hegarty, L. F. James & R. F. Keeler. Queensland Poisonous Plants Committee, Yeerongpilly, pp. 111–19.

Jones, K. & Thomas, J. G. (1974). Nitrogen fixation by the rumen contents of sheep. *J. Gen. Microbiol.*, **85**, 97–101.

Jones, W. J., Donnelly, M. I. & Wolfe, R. S. (1985). Evidence of a common pathway of carbon dioxide reduction to methane in methanogens. *J. Bacteriol.*, **163**, 126–31.

Jones, W. J., Nagle, D. P. & Whitman, W. B. (1987). Methanogens and the diversity of Archaebacteria. *Microbiol. Rev.*, **51**, 135–77.

Kafkewitz, D. (1975). Improved growth media for *Vibrio succinogenes*. *Appl. Microbiol.*, **29**, 121–2.

Kamio, Y., Poso, H., Terawaki, Y. & Paulin, L. (1986). Cadaverine covalently linked to a peptidoglycan is an essential constituent of the peptidoglycan necessary for the normal growth in *Selenomonas ruminantium*. *J. Biol. Chem.*, **261**, 6585–9.

Kemp, P., White, R. W. & Lander, D. J. (1975). The hydrogenation of unsaturated fatty acids by five bacterial isolates from the sheep rumen, including a new species. *J. Gen. Microbiol.*, **90**, 100–14.

Kemp, P., Lander, D. J. & Gunstone, F. D. (1984a). The hydrogenation of some *cis*- and *trans*-octadecenoic acids to stearic acid by a rumen *Fusocillus* sp. *Br. J. Nutr.*, 165–70.

Kemp, P., Lander, D. J. & Holman, R. T. (1984b). The hydrogenation of the series of methylene-interrupted *cis*, *cis*-octadecenoic acids by pure cultures of six rumen bacteria. *Br. J. Nutr.*, **52**, 171–7.

Kepler, C. R., Hirons, K. P., McNeill, J. J. & Tove, S. B. (1966). Intermediates and products of the biohydrogenation of linoleic acid by *Butyrivibrio fibrisolvens*. *J. Biol. Chem.*, **241**, 1350–4.

Khambata, S. R. & Bhat, J. V. (1953). Studies on a new oxalate decomposing bacterium, *Pseudomonas oxalaticus*. *J. Bacteriol.*, **66**, 505–7.

Kistner, A. (1960). An improved method for viable counts of bacteria of the ovine rumen which ferment carbohydrates. *J. Gen. Microbiol.*, **23**, 565–76.

Kolankaya, N., Stewart, C. S., Duncan, S. H., Cheng, K.-J. & Costerton, J. W. (1985). The effect of ammonia treatment on the solubilisation of straw and the growth of cellulolytic rumen bacteria. *J. Appl. Bacteriol.*, **58**, 371–9.

Kröger, A. & Winkler, E. (1981). Phosphorylative fumarate reduction in *Vibrio succinogenes*: stoichiometry of ATP synthesis. *Arch. Microbiol.*, **129**, 100–4.

Krogh, N. (1960). Studies on alterations in the rumen fluid of sheep, especially concerning the microbial composition, when readily available carbohydrates are added to the food. II. Lactose. *Acta Vet. Scand.*, **1**, 383–410.

Krumholz, L. R. & Bryant, M. P. (1986a). *Syntrophococcus sucromutans* sp. nov. gen. nov. uses carbohydrates as electron donors and formate, methoxymonobenzenoids or *Methanobrevibacter* as electron acceptor systems. *Arch. Microbiol.*, **143**, 313–18.

Krumholz, L. R. & Bryant, M. P. (1986b). *Eubacterium oxidoreducens* sp. nov. requiring H_2 or formate to degrade gallate, pyrogallol, phloroglucinol and quercetin. *Arch. Microbiol.*, **144**, 8–14.

Krumholz, L. R., Crawford, R. L., Hemling, M. E. & Bryant, M. P. (1986). A rumen bacterium degrading quercitin and trihydroxybenzenoids with concurrent use of formate or H_2. In *Plant Flavonoids in Biology and Medicine: Biochemical, Pharmacological and Structure-Activity Relationships*, ed. E. Middleton. Liss, New York, pp. 211–14.

Krumholz, L. R., Crawford, R. L., Hemling, M. E. & Bryant, M. P. (1987). Metabolism of gallate and phloroglucinol in *Eubacterium oxidoreducens* via 3-hydroxy-5-oxohexanoate. *J. Bacteriol.*, **169**, 1886–90.

Krzycki, J. & Zeikus, J. G. (1980). Quantification of corrinoids in methanogenic bacteria. *Curr. Microbiol.*, **3**, 243–5.

Kühn, W. & Gottschalk, G. (1983). Characterisation of the cytochromes occurring in *Methanosarcina* species. *Eur. J. Biochem.*, **135**, 89–94.

Kunsman, J. E. & Caldwell, D. R. (1974). Comparison of the sphingolipid content of rumen *Bacteroides* species. *Appl. Microbiol.*, **28**, 1088–9.

Kurihara, Y., Eadie, J. M., Hobson, P. N. & Mann, S. O. (1968). Relationship between bacteria and ciliate protozoa in the sheep rumen. *J. Gen. Microbiol.*, **57**, 267–88.

Latham, M. J., Brooker, B. E., Pettipher, G. L. & Harris, P. J. (1978). Adhesion of *Bacteroides succinogenes* in pure culture and in the presence of *Ruminococcus flavefaciens* to cell walls in leaves of perennial ryegrass (L. perenne). *Appl. Environ. Microbiol.*, **35**, 1166–73.

Latham, M. J., Sharpe, E. & Weiss, N. (1979). Anaerobic cocci from the bovine alimentary tract, the amino acids of their cell wall peptidoglycans and those of various species of anaerobic *Streptococcus*. *J. Appl. Bacteriol.*, **47**, 209–21.

Leatherwood, J. M. & Sharma, M. P. (1972). Novel anaerobic cellulolytic bacterium. *J. Bacteriol.*, **110**, 751–3.

Leedle, J. A. Z. & Hespell, R. B. (1980). Differential carbohydrate media and anaerobic replica plating technique in delineating carbohydrate utilising subgroups in rumen bacterial populations. *Appl. Environ. Microbiol.*, **39**, 709–19.

Leedle, J. A. Z., Bryant, M. P. & Hespell, R. B. (1982). Diurnal variations in bacterial numbers and fluid parameters in ruminal contents of animals fed low- or high-forage diets. *Appl. Environ. Microbiol.*, **44**, 402–12.

Lewis, S. M. & Dehority, B. A. (1985). Microbiology and ration digestibility in the hind gut of the ovine. *Appl. Environ. Microbiol.*, **50**, 356–63.

Loesche, W. J. (1969). Oxygen sensitivity of various anaerobic bacteria. *Appl. Microbiol.*, **18**, 723–7.

Lovley, D. R., Greening, R. & Ferry, J. G. (1984). Rapidly growing rumen methanogenic organism that synthesises coenzyme M and has a high affinity for formate. *Appl. Environ. Microbiol.*, **48**, 81–7.

Luehresen, K. R., Nicholson, D. E. & Fox, G. E. (1985). Widespread distribution of a 7s RNA in Archaebacteria. *Curr. Microbiol.*, **12**, 69–72.

Mackie, R. I., Gilchrist, F. M., Robberts, A. M., Hannah, P. E. & Schwartz, H. M. (1978). Microbiological and chemical changes in the rumen during the stepwise adaptation of sheep to high concentrate diets. *J. Agric. Sci.*, **90**, 241–54.

Macy, J. M. & Probst, I. (1979). The biology of gastrointestinal bacteroides. *Ann. Rev. Microbiol.*, **33**, 561–94.

Macy, J. M., Addison, R. B. & Farrand, J. R. (1984). A new irregular coccus from the rumen of an oat-hay fed Jersey cow; effect on *in vitro* digestibility of corn stalks. *Arch. Microbiol.*, **140**, 66–73.

Macy, J. M., Schröder, I., Thauer, R. K. & Kröger, A. (1986). Growth (of) *Wolinella succinogenes* on H_2S plus fumarate and on formate plus sulphur as energy sources. *Arch. Microbiol.*, **144**, 147–50.

Maluzynska, G. M. & Janota-Bassalik, L. (1974). A cellulolytic rumen bacterium *Micromonospora ruminantium*. *J. Gen. Microbiol.*, **82**, 57–65.

Mann, S. O. & Ørskov, E. R. (1973). The effect of rumen and post-rumen feeding of carbohydrates on the caecal microflora of sheep. *J. Appl. Bacteriol.*, **36**, 475–84.

Mann, S. O. & Oxford, A. E. (1954). Studies of some presumptive lactobacilli isolated from the rumens of young calves. *J. Gen. Microbiol.*, **11**, 83–90.

Marounek, M. & Wallace, R. J. (1984). Influence of culture Eh on the growth and metabolism of the rumen bacteria *Selenomonas ruminantium*, *Bacteroides amylophilus*, *Bacteroides succinogenes* and *Streptococcus bovis* in batch culture. *J. Gen. Microbiol.*, **130**, 223–9.

Martin, S. A. & Russell, J. B. (1986). Phosphoenolpyruvate-dependent phosphorylation of hexoses by ruminal bacteria: evidence for the phosphotransferase transport system. *Appl. Environ. Microbiol.*, **52**, 1348–52.

McInerney, M. J., Bryant, M. P. & Pfennig, N. (1979). Anaerobic bacterium that degrades fatty acids in syntrophic association with methanogens. *Arch. Microbiol.*, **122**, 129–35.
Meynell, G. G. & Meynell, E. (1970). *Theory and Practice in Experimental Bacteriology*. CUP, Cambridge.
Miller, T. L. (1978). The pathway of formation of acetate and succinate from pyruvate by *Bacteroides succinogenes*. *Arch. Microbiol.*, **117**, 145–52.
Miller, T. L., Wolin, M. J., Zhao, H. & Bryant, M. P. (1986). Characteristics of methanogens isolated from bovine rumen. *Appl. Environ. Microbiol.*, **51**, 201–2.
Minato, H. & Suto, T. (1981). Technique for fractionation of bacteria in rumen microbial ecosystem. IV. Attachment of rumen bacteria to cellulose powder and elution of bacteria attached to it. *J. Gen. Appl. Microbiol.*, **27**, 21–31.
Miyagawa, E. (1982). Cellular fatty acid and fatty aldehyde composition of rumen bacteria. *J. Gen. Appl. Microbiol.*, **28**, 389–408.
Miyagawa, E., Azuma, R. & Suto, T. (1979). Cellular fatty acid composition in Gram-negative obligately anaerobic rods. *J. Gen. Appl. Microbiol.*, **25**, 41–51.
Moir, R. J. & Masson, M. (1952). An illustrated scheme for the microscopic identification of the rumen microorganisms of sheep. *J. Pathol. Bacteriol.*, **64**, 343–50.
Montgomery, L. & Macy, J. M. (1982). Characteristics of rat caecum cellulolytic bacteria. *Appl. Environ. Microbiol.*, **44**, 1435–43.
Moonen, C. T. W. & Müller, F. (1984). On the intermolecular transfer between different redox states of flavodoxin from *Megasphaera elsdenii*. *Eur. J. Biochem.*, **140**, 303–9.
Moonen, C. T. W., Scheek, R. M., Boelens, R. & Müller, F. (1984). The use of two-dimensional nuclear magnetic resonance spectroscopy and two-dimensional difference spectra in the elucidation of the active center of *Megasphaera elsdenii* flavodoxin. *Eur. J. Biochem.*, **141**, 323–30.
Moore, W. E. C. & Holdeman-Moore, L. V. (1986). Eubacterium. In *Bergey's Manual of Systematic Bacteriology*, Vol. 2, ed. P. H. A Sneath. Williams & Wilkins, Baltimore, pp. 1353–73.
Morris, E. J. (1984). Degradation of the intact plant cell wall of subtropical and tropical herbage by rumen bacteria. In *Herbivore Nutrition in the Subtropics and Tropics*, ed. F. M. C. Gilchrist & R. I. Mackie. The Science Press, South Africa, pp. 378–95.
Morris, E. J. & Van Gylswyk, N. P. (1980). Comparison of the action of rumen bacteria on cell walls of Eragrostis tef. *J. Agric. Sci.*, **95**, 313–23.
Mueller, R. E., Asplund, J. M. & Ianotti, E. L. (1984). Successive changes in the epimural bacterial community of young lambs as revealed by scanning electron microscopy. *Appl. Environ. Microbiol.*, **47**, 715–23.
Munn, E. A., Hazlewood, G. P. & Graham, M. (1983). Uptake and incorporation of the products of proteolysis by the rumen bacterium *Bacteroides ruminicola* R8/4. *Curr. Microbiol.*, **8**, 317–20.
Murray, R. G. E. (1984). Lampropedia. In *Bergey's Manual of Systematic Bacteriology*, Vol. 1, ed. N. R. Krieg & J. G. Holt. Williams & Wilkins, Baltimore, pp. 402–6.
Ogimoto, K. & Imai, S. (1981). *Atlas of Rumen Microbiology*. Japan Scientific Societies Press, Tokyo.
Ohmiya, K., Shimizu, M., Taya, M. & Shimizu, S. (1982). Purification and properties of cellobiosidase from *Ruminococcus albus*. *J. Bacteriol.*, **150**, 407–9.
Ohmiya, K., Shirai, M., Kurachi, Y. & Shimizu, S. (1985). Isolation and properties of β-glucosidase from *Ruminococcus albus*. *J. Bacteriol.*, **161**, 432–4.
Ohmiya, K., Takeuchi, M., Chen, W., Shimizu, S. & Kawakami, H. (1986). Anaerobic reduction of ferulic acid to dihydroferulic acid by *Wolinella succinogenes* from cow rumen. *Appl. Microbiol. Biotechnol.*, **23**, 274–9.
Oppermann, R. A., Nelson, W. O. & Brown, R. E. (1957). *In vitro* studies on methanogenic rumen bacteria. *J. Dairy Sci.*, **40**, 779–88.
Orpin, G. C. (1972). The culture of the rumen organism Eadie's oval *in vitro*. *J. Gen. Microbiol.*, **70**, 321–9.

Orpin, G. C. (1976). The characterisation of the rumen bacterium Eadie's oval *Magnoovum* gen. nov. *eadii* sp. nov. *Arch. Microbiol.*, **111**, 155–9.

Ørskov, E. R., Fraser, C., Mason, V. C. & Mann, S. O. (1970). Influence of starch digestion in the large intestine of sheep on caecal fermentation, caecal microflora and faecal nitrogen excretion. *Br. J. Nutr.*, **24**, 671–82.

Paster, B. J. & Canale-Parola, E. (1985). *Treponema saccharophilum* sp. nov., a large pectinolytic spirochaete from the bovine rumen. *Appl. Environ. Microbiol.*, **50**, 212–19.

Paster, B. J., Stackebrandt, E., Hespell, R. B., Hahn, C. M. & Woese, C. R. (1984). The phylogeny of the spirochaetes. *Syst. Appl. Microbiol.*, **5**, 337–51.

Paster, B. J., Ludwig, W., Weisburg, W. G., Stackebrandt, E., Hespell, R. B., Hahn, C. M., Reichenbach, H., Stetter, K. O. & Woese, C. R. (1985). A phylogenetic grouping of the Bacteroides, Cytophagas and certain Flavobacteria. *Syst. Appl. Microbiol.*, **6**, 34–42.

Patterson, J. A. & Hespell, R. B. (1979). Trimethylamine and methylamine as growth substrates for rumen bacteria and *Methanosarcina barkeri*. *Curr. Microbiol.*, **3**, 79–83.

Patterson, J. A. & Hespell, R. B. (1985). Glutamine synthetase activity in the ruminal bacterium *Succinivibrio dextrinosolvens*. *Appl. Environ. Microbiol.*, **50**, 1014–20.

Paynter, M. J. B. & Hungate, K. E. (1968). Characterisation of *Methanobacterium mobilis* sp. n., isolated from the bovine rumen. *J. Bacteriol.*, **95**, 1943–51.

Pazur, J. H. & Forsberg, L. S. (1978). Determination of the sugar sequences and the glycosidic band arrangements of immunogenic heteroglycans. *Carb. Res.*, **60**, 167–78.

Pestka, J. J. & Delwiche, E. A. (1983). An alternative pathway for 3-phosphoglycerate generation in *Veillonella*. *Can. J. Microbiol.*, **29**, 218–24.

Pettipher, G. L. & Latham, M. J. (1979). Characteristics of enzymes produced by *Ruminococcus flavefaciens* which degrade plant cell walls. *J. Gen. Microbiol.*, **110**, 21–7.

Phillips, B. A., Latham, M. J. & Sharpe, M. E. (1975). A method for freeze drying rumen bacteria and other strict anaerobes. *J. Appl. Bacteriol.*, **38**, 319–22.

Postgate, J. R. (1970). Nitrogen fixation by sporulating sulphate reducing bacteria including rumen strains. *J. Gen. Microbiol.*, **63**, 137–9.

Prescott, J. M., Ragland, R. S. & Stutts, A. L. (1957). Effects of carbon dioxide on the growth of *Streptococcus bovis* in the presence of various amino acids. *J. Bacteriol.*, **73**, 133–8.

Prins, R. A. (1971). Isolation, culture and fermentation characteristics of *Selenomonas ruminantium* var. *bryanti* var. n. from the rumen of sheep. *J. Bacteriol.*, **105**, 820–5.

Prins, R. A. (1984). Anaerovibrio. In *Bergey's Manual of Systematic Bacteriology*, Vol. 1, ed. N. R. Krieg & J. G. Holt. Williams & Wilkins, Baltimore, pp. 653–5.

Prins, R. A., van Vught, F., Hungate, R. E. & van Vorstenbosch, C. J. A. H. V. (1972). A comparison of strains of *Eubacterium cellulosolvens* from the rumen. *Ant. van Leeuwenhoek*, **38**, 153–61.

Prins, R. A., Lankhorst, A., van der Meer, P. & van Nevel, C. J. (1975). Some characteristics of *Anaerovibrio lipolytica*, a rumen lipolytic organism. *Ant. van Leeuwenhoek*, **41**, 1–11.

Reddy, C. A. & Bryant, M. P. (1977). Deoxyribonucleic acid base composition of certain species of the genus *Bacteroides*. *Can J. Microbiol.*, **23**, 1252–6.

Robinson, I. M. (1984). Anaeroplasma. In *Bergey's Manual of Systematic Bacteriology*, Vol. 1, ed. N. R. Krieg & J. G. Holt. Williams & Wilkins, Baltimore, pp. 787–90.

Robinson, I. M. & Allison, M. J. (1975). Transfer of *Acholeplasma bactoclasticum* Robinson and Hungate to the genus Anaeroplasma (*Anaeroplasma bactoclasticum* Robinson and Hungate comb. nov.): emended description of the species. *Int. J. Syst. Bacteriol.*, **25**, 182–6.

Robinson, I. M., Allison, M. J. & Hartman, P. A. (1975). *Anaeroplasma abactoclasticum* gen. gov. sp. nov: an obligately anaerobic mycoplasma from the rumen. *Int. J. Syst. Bacteriol.*, **25**, 173–81.

Robinson, I. M., Allison, M. J. & Bucklin, J. A. (1981). Characterisation of caecal bacteria of normal pigs. *Appl. Environ. Microbiol.*, **41**, 950–5.

Robinson, J. P. & Hungate, R. E. (1973). *Acholeplasma bactoclasticum* sp. nov., an anaerobic mycoplasma from the bovine rumen. *Int. J. Syst. Bacteriol.*, **23**, 171–81.

Roché, C., Albertyn, H., Van Gylswyk, N. O. & Kistner, A. (1973). The growth response of cellulolytic acetate-utilising and acetate-producing butyrivibrios to volatile fatty acids and other nutrients. *J. Gen. Microbiol.*, **78**, 253–60.

Rogosa, M. (1984). Anaerobic Gram-negative cocci. In *Bergey's Manual of Systematic Bacteriology*, Vol. 1, ed. N. R. Krieg & J. G. Holt. Williams & Wilkins, Baltimore, pp. 680–5.

Rogosa, M. & Bishop, F. S. (1964). The genus Veillonella. II. Nutritional Studies. *J. Bacteriol.*, **87**, 574–80.

Rowe, J. B., Loughnan, M. L., Nolan, J. V. & Leng, R. A. (1979). Secondary fermentation in the rumen of a sheep given a diet based on molasses. *Br. J. Nutr.*, **41**, 393–7.

Russell, J. B. (1983). Fermentation of peptides by *Bacteroides ruminicola* $B_1 4$. *Appl. Environ. Microbiol.*, **45**, 1566–74.

Russell, J. B. (1985a). Enrichment and isolation of rumen bacteria that reduce *trans*-aconitic acid to tricarballylic acid. *Appl. Environ. Microbiol.*, **49**, 120–6.

Russell, J. B. (1985b). Fermentation of cellodextrins by cellulolytic and non-cellulolytic rumen bacteria. *Appl. Environ. Microbiol.*, **49**, 572–6.

Russell, J. B. & Baldwin, R. A. (1978). Substrate preferences in rumen bacteria: evidence of catabolite regulatory mechanisms. *Appl. Environ. Microbiol.*, **36**, 319–29.

Russell, J. B. & Baldwin, R. L. (1979). Comparison of maintenance energy expenditure and growth yields among several rumen bacteria growth in continuous culture. *Appl. Environ. Microbiol.*, **37**, 537–43.

Russell, J. B. & Dombrowski, D. B. (1980). Effect of pH on the efficiency of growth by pure cultures of rumen bacteria in continuous culture. *Appl. Environ. Microbiol.*, **39**, 604–10.

Russell, J. B. & Hino, T. (1985). Regulation of lactate production in *Streptococcus bovis*: a spiralling effect that contributes to rumen acidosis. *J. Dairy Sci.*, **68**, 1712–21.

Russell, J. B. & Robinson, P. H. (1984). Composition and characteristics of strains of *Streptococcus bovis*. *J. Dairy Sci.*, **67**, 1525–31.

Russell, J. B., Cotta, M. A. & Dombrowski, D. B. (1981). Rumen bacterial composition in continuous culture: *Streptococcus bovis* versus *Megasphaera elsdenii*. *Appl. Environ. Microbiol.*, **41**, 1394–9.

Samah, O. A. & Wimpenny, J. W. T. (1982). Some effects of oxygen on the physiology of *Selenomonas ruminantium* WPL/151/1 grown in continuous culture. *J. Gen. Microbiol.*, **128**, 355–60.

Scardovi, V. (1981). The genus Bifidobacterium. In *The Prokaryotes: A Handbook on Habitats, Isolation and Identification of Bacteria*, Vol. 2, ed. M. P. Starr, H. Stolp, H. G. Truper, A. Balows & H. G. Schlegel. Springer Verlag, Berlin, pp. 1951–61.

Schaefer, D. M., Davis, C. L. & Bryant, M. P. (1980). Ammonia saturation constants for predominant species of rumen bacteria. *J. Dairy Sci.*, **63**, 1249–63.

Shah, H. N. & Collins, M. D. (1983). Genus *Bacteroides*: a chemotaxonomical perspective. *J. Appl. Bacteriol.*, **55**, 403–16.

Shane, B. S., Gouws, L. & Kistner, A. (1969). Cellulolytic bacteria occurring in the rumen of sheep conditioned to low protein Teff hay. *J. Gen. Microbiol.*, **55**, 445–57.

Shapiro, S. (1982). Do corrinoids function in the methanogenic dissimilation of methanol by *Methanosarcina barkeri*? *Can. J. Microbiol.*, **28**, 629–35.

Sharpe, M. E., Latham, M. J., Garvie, E. I., Zirngibl, J. & Kandler, O. (1973). Two new species of *Lactobacillus* isolated from the rumen, *Lactobacillus ruminis* sp. nov. and *Lactobacillus vitulinus* sp. nov. *J. Gen. Microbiol.*, **77**, 37–49.

Sharpe, M. E., Brock, J. M. & Phillips, B. K. (1975). Glycerol teichoic acid as an antigenic determinant in a Gram-negative bacterium *Butyrivibrio fibrisolvens*. *J. Gen. Microbiol.*, **88**, 355–63.

Scheifinger, C. C., Linehan, B. & Wolin, M. J. (1975). H_2 production by *Selenomonas ruminantium* in the absence and presence of methanogenic bacteria. *Appl. Microbiol.*, **29**, 480–3.

Sijpesteijn, A. K. (1951). On *Ruminococcus flavefaciens* a cellulose decomposing bacterium from the rumen of sheep and cattle. *J. Gen. Microbiol.*, **5**, 869–79.

Silley, P. (1985). A note on the pectinolytic enzymes of *Lachnospira multiparus*. *J. Appl. Bacteriol.*, **58**, 145–50.
Silley, P. (1986). The production and properties of a crude pectin lyase from *Lachnospira multiparus*. *Lett. Appl. Microbiol.*, **2**, 29–31.
Sinha, R. N. & Ranganathan, B. (1983). Cellulolytic bacteria in Buffalo rumen. *J. Appl. Bacteriol.*, **54**, 1–6.
Skerman, V. B. D., McGowan, V. & Sneath, P. H. A. (1980). Approved list of bacterial names. *Int. J. Syst. Bacteriol.*, **30**, 225–420.
Slyter, L. L., Kern, D. L. & Weaver, J. M. (1976). Effect of pH on ruminal lactic acid utilisation and accumulation *in vitro*. *J. Anim. Sci.*, **43**, 333–4.
Smibert, R. M. (1984). Treponema. In *Bergey's Manual of Systematic Bacteriology*, Vol. 1, ed. N. R. Krieg & J. G. Holt. Williams & Wilkins, Baltimore, pp. 49–57.
Smith, P. H. & Hungate, R. E. (1958). Isolation and characterisation of *Methanobacterium ruminantium* n. sp. *J. Bacteriol.*, **75**, 713–8.
Stack, R. J. & Cotta, M. A. (1986). Effect of 3-phenylpropanoic acid on growth of and cellulose utilisation by cellulolytic rumen bacteria. *Appl. Environ. Microbiol.*, **52**, 209–10.
Stack, R. J. & Hungate, R. E. (1984). Effect of 3-phenylpropanoic acid on capsule and cellulases of *Ruminococcus albus* 8. *Appl. Environ. Microbiol.*, **48**, 218–23.
Stack, R. J., Hungate, R. E. & Opsahl, W. P. (1983). Phenylacetic acid stimulation of cellulose digestion by *Ruminococcus albus* 8. *Appl. Environ. Microbiol.*, **46**, 539–44.
Stackebrandt, E. & Hippe, H. (1986). Transfer of *Bacteroides amylophilus* to a new genus *Ruminobacter* gen. nov. *Syst. Appl. Microbiol.*, **8**, 204–7.
Stackebrandt, E. & Woese, C. R. (1981). The evolution of prokaryokes. In *Molecular and Cellular Aspects of Microbial Evolution*, ed. M. J. Carlile, J. F. Collins & B. E. B. Moseley. CUP, Cambridge, pp. 1–31.
Stackebrandt, E., Pohla, H., Kroppenstedt, R., Hippe, H. & Woese, C. R. (1985). 16S rRNA analysis of *Sporomusa*, *Selenomonas* and *Megasphaera*: on the phylogenetic origin of Gram-positive Eubacteria. *Arch. Microbiol.*, **143**, 270–6.
Stanton, T. B. & Canale-Parola, E. (1979). Enumeration and selective isolation of rumen spirochaetes. *Appl. Environ. Microbiol.*, **38**, 965–73.
Stanton, T. B. & Canale-Parola, E. (1980). *Treponema bryantii* sp. nov., a rumen spirochaete that interacts with cellulolytic bacteria. *Arch. Microbiol.*, **127**, 145–56.
Stewart, C. S. (1975). Some effects of phosphate and volatile fatty acid salts on the growth of rumen bacteria. *J. Gen. Microbiol.*, **89**, 319–26.
Stewart, C. S. (1986). Rumen function with special reference to fibre digestion. In *Anaerobic Bacteria in Habitats Other than Man*, ed. E. M. Barnes & G. C. Mead. Blackwell Scientific Publishers, Oxford, pp. 263–86.
Stewart, C. S. & Duncan, S. H. (1985). The effect of avoparcin on cellulolytic bacteria of the ovine rumen. *J. Gen. Microbiol.*, **131**, 427–35.
Stewart, C. S., Dinsdale, D., Cheng, K.-J. & Paniagua, C. (1979). The digestion of straw in the rumen. In *Straw Decay and its Effect on Disposal and Utilisation*, ed. E. Grossbard. Wiley, Chichester, pp. 123–30.
Stewart, C. S., Paniagua, C., Dinsdale, D., Cheng, K.-J. & Garrow, S. H. (1981). Selective isolation and characteristics of *Bacteroides succinogenes* from the rumen of a cow. *Appl. Environ. Microbiol.*, **41**, 504–10.
Stewart, C. S., Crossley, M. V. & Garrow, S. H. (1983). The effect of avoparcin on laboratory cultures of rumen bacteria. *Eur. J. Appl. Microbiol. Biotechnol.*, **17**, 292–7.
Stewart, C. S., Fonty, G. & Gouet, P. (1988). The establishment of rumen microbial communities. *Anim. Feed Sci. Technol.*, **21**, (in press).
Tanner, A. C. R. & Socransky, S. S. (1984). Wolinella. In *Bergey's Manual of Systematic Bacteriology*, Vol. 1, ed. N. R. Krieg & J. G. Holt. Williams & Wilkins, Baltimore, pp. 646–50.
Teather, R. M. (1982). Maintenance of laboratory strains of obligately anaerobic rumen bacteria. *Appl. Environ. Microbiol.*, **44**, 499–501.

Thorley, C. M., Sharpe, M. E. & Bryant, M. P. (1968). Modification of the rumen bacterial flora by feeding cattle ground and pelleted roughage as determined with culture media with and without rumen fluid. *J. Dairy Sci.*, **51**, 1811–16.

Tiwari, A. D., Bryant, M. P. & Wolfe, R. S. (1969). Simple method for isolation of *Selenomonas ruminantium* and some nutritional characteristics of the species. *J. Dairy Sci.*, **52**, 2054–6.

Trovatelli, L. D. & Matteuzzi, D. (1976). Presence of bifidobacteria in the rumen of calves fed different rations. *Appl. Environ. Microbiol.*, **32**, 470–3.

Tzeng, S. F., Wolfe, R. S. & Bryant, M. P. (1975*a*). Factor 420-dependent pyridine nucleotide-linked hydrogenase system of *Methanobacterium ruminantium*. *J. Bacteriol.*, **121**, 184–91.

Tzeng, S. F., Bryant, M. P. & Wolfe, R. S. (1975*b*). Factor 420-dependent pyridine nucleotide-linked formate metabolism of *Methanobacterium ruminantium*, *J. Bacteriol.*, **121**, 192–6.

Ulyatt, M. J., Dellow, D. W., John, A., Reid, C. S. W. & Waghorn, G. C. (1986). Contribution of chewing during eating and rumination to the clearance of digesta from the ruminoreticulum. In *Control of Digestion and Metabolism in Ruminants*, ed. L. P. Milligan, W. L. Grovum & A. Dobson. Prentice-Hall, Englewood Cliffs, New Jersey, pp. 498–515.

Van Beelen, P., Van Neck, J. W., De Cock, R. M., Vogels, G. D., Guijt, W. & Haasnoot, A. G. (1984). 5,10-Methenyl-5,6,7,8-tetrahydromethanopterin, a one-carbon carrier in the process of methanogenesis. *Biochemistry*, **23**, 4448–54.

Van der Linden, Y., Van Gylswyk, N. O. & Schwartz, H. (1984). Influence of supplementation of corn stover with corn grain on the fibrolytic bacteria in the rumen of sheep and their relation to the intake and digestion of fiber. *J. Anim. Sci.*, **59**, 772–83.

Van der Toorn, J. J. T. K. & Van Gylswyk, N. O. (1985). Xylan-digesting bacteria from the rumen of sheep fed maize straw diets. *J. Gen. Microbiol.*, **131**, 2601–7.

Van Golde, L. M. G., Akkermans-Kruyswijk, J., Franklin-Klein, W., Lankhorst, A. & Prins, R. A. (1975). Accumulation of phosphatidylserine in strictly anaerobic lactate fermenting bacteria. *FEBS Lett.*, **53**, 57–60.

Van Gylswyk, N. O. (1980). *Fusobacterium polysaccharolyticum* sp. nov., a Gram-negative rod from the rumen that produces butyrate and ferments cellulose and starch. *J. Gen. Microbiol.*, **116**, 157–63.

Van Gylswyk, N. O. & Hoffman, J. S. L. (1970). Characteristics of cellulolytic Cillobacteria from the rumens of sheep fed teff (*Eragrostis tef*) hay diets. *J. Gen. Microbiol.*, **60**, 381–6.

Van Gylswyk, N. O. & Roche, C. E. G. (1970). Characteristics of *Ruminococcus* and cellulolytic *Butyrivibrio* species from the rumens of sheep fed differently substituted teff (*Eragrostis tef*) hay diets. *J. Gen. Microbiol.*, **64**, 11–17.

Van Gylswyk, N. O. & Van der Toorn, J. J. T. K. (1985). *Eubacterium uniforme* sp. nov. and *Eubacterium xylanophilum* sp. nov., fiber digesting bacteria from the rumina of sheep fed corn stover. *Int. J. Syst. Bacteriol.*, **35**, 323–6.

Van Gylswyk, N. O., Morris, E. J. & Els, H. J. (1980). Sporulation and cell wall structure of *Clostridium polysaccharolyticum* com. nov. (formerly *Eubacterium polysaccharolyticum*). *J. Gen. Microbiol.*, **121**, 491–3.

Varel, V. H., Fryda, S. J. & Robinson, I. M. (1984). Cellulolytic bacteria from pig large intestine. *Appl. Environ. Microbiol.*, **47**, 219–21.

Verkley, A. J., Ververgaert, P. H. J. T., Prins, R. A. & Van Golde, L. M. G. (1975). Lipid-phase transitions of the strictly anaerobic bacteria *Veillonella parvula* and *Anaerovibrio lipolytica*. *J. Bacteriol.*, **124**, 1522–8.

Vicini, J. L., Brulla, W. J., Davis, C. L. & Bryant, M. P. (1987). Quin's oval and other microbiota in the rumen of molasses-fed sheep. *Appl. Environ. Microbiol.*, **53**, 1273–6.

Wachenheim, D. E. & Hespell, R. B. (1984). Inhibitory effects of titanium(III) citrate on enumeration of bacteria from rumen contents. *Appl. Environ. Microbiol.*, **48**, 444–5.

Wallace, R. J. (1986). Catabolism of amino acids by *Megasphaera elsdenii* LC1. *Appl. Environ. Microbiol.*, **51**, 1141–3.

Wallace, R. J. & Brammall, M. L. (1985). The role of different species of bacteria in the hydrolysis of protein in the rumen. *J. Gen. Microbiol.*, **131**, 821–32.

Wallnöfer, P. & Baldwin, R. L. (1967). Pathway of propionate formation in *Bacteroides ruminicola*. *J. Bacteriol.*, **93**, 504–5.
Watanabe, T., Okuda, S.-I. & Takahashi, H. (1982). Physiological importance of even numbered fatty acids and aldehydes in plasmalogen phospholipids of *Selenomonas ruminantium*. *J. Gen. Appl. Microbiol.*, **28**, 23–33.
Watanabe, T., Okuda, S.-I. & Takahashi, H. (1984). Turn-over of phospholipids in *Selenomonas ruminantium*. *J. Biochem.* (Tokyo), **95**, 521–7.
Wegner, G. H. & Foster, E. M. (1963). Incorporation of isobutyrate and valerate into cellular plasmalogen by *Bacteroides succinogenes*. *J. Bacteriol.*, **85**, 53–61.
Wetzstein, H. G. & Gottschalk, G. (1985). A sodium-stimulated membrane-bound fumarate reductase system in *Bacteroides amylophilus*. *Arch. Microbiol.*, **143**, 157–62.
White, R. H. (1985). Biosynthesis of coenzyme M (2-mercaptoethane sulphonic acid). *Biochemistry*, **24**, 6487–93.
White, R. W., Mackenzie, A. R. & Bousefield, I. J. (1974). The successful freeze-drying and retention of biohydrogenation activity of bacteria isolated from the ovine rumen. *J. Appl. Bacteriol.*, **37**, vi.
Whitely, H. R. & Douglas, H. C. (1951). The fermentation of purines by *Micrococcus lactyliticus*. *J. Bacteriol.*, **61**, 605–16.
Williams, A. G. & Withers, S. E. (1982a). The production of plant cell wall polysaccharide-degrading enzymes by hemicellulolytic rumen bacterial isolates grown on a range of carbohydrate substrates. *J. Appl. Bacteriol.*, **52**, 377–87.
Williams, A. G. & Withers, S. E. (1982b). The effect of the carbohydrate growth substrate on the glycosidase activity of hemicellulose degrading rumen bacterial isolates. *J. Appl. Bacteriol.*, **52**, 389–401.
Williams, A. G. & Withers, S. E. (1983). *Bacillus* spp. in the rumen ecosystem. Hemicellulose depolymerases and glycoside hydrolases of *Bacillus* spp. and rumen isolates grown under anaerobic conditions. *J. Appl. Bacteriol.*, **55**, 283–92.
Williams, A. G. & Withers, S. E. (1985). Formation of polysaccharide depolymerase and glycoside hydrolase enzymes by *Bacteroides ruminicola* subsp. *ruminicola* grown in batch and continuous culture. *Curr. Microbiol.*, **12**, 79–84.
Williams, A. G., Withers, S. E. & Coleman, G. S. (1984). Glycoside hydrolases of rumen bacteria and protozoa. *Curr. Microbiol.*, **10**, 287–93.
Williamson, G., Engel, P., Mizzer, J. P., Thorpe, C. & Massey, V. (1982). Evidence that the greening ligand in native butyryl-CoA dehydrogenase is a CoA persulphide. *J. Biol. Chem.*, **257**, 4314–20.
Williamson, R., Calderwood, S. B., Maellering, R. C. & Tomasz, A. (1983). Studies on the mechanism of intrinsic resistance to β-lactam antibiotics in group D streptococci. *J. Gen. Microbiol.*, **129**, 813–22.
Wilson, S. N. (1983). Some carbohydrate fermenting organisms isolated from the rumen of the sheep. *J. Gen. Microbiol.*, **9**, i-ii.
Winter, J. (1983). Maintenance of stock cultures of methanogens in the laboratory. *Syst. Appl. Microbiol.*, **4**, 558–63.
Woese, C. R. & Fox, G. E. (1977). Phylogenetic structure of the prokaryotic domain: the primary kingdoms. *Proc. Nat. Acad. Sci. USA*, **74**, 5088–90.
Wojciechowicz, M. & Ziolecki, A. (1979). Pectinolytic enzymes of large rumen treponemes. *Appl. Environ. Microbiol.*, **37**, 136–42.
Wojciechowicz, M. & Ziolecki, A. (1984). A note on the pectinolytic enzyme of *Streptococcus bovis*. *J. Appl. Bacteriol.*, **56**, 515–18.
Wojciechowicz, M., Heinrichova, K. & Ziolecki, A. (1982). An exopectate lyase of *Butyrivibrio fibrisolvens* from the bovine rumen. *J. Gen. Microbiol.*, **128**, 2661–5.
Wolin, M. J., Manning, G. B. & Nelson, W. O. (1959). Ammonium salts as a sole source of nitrogen for the growth of *Streptococcus bovis*. *J. Bacteriol.*, **78**, 147–9.
Wolin, M. J., Wolin, E. A. & Jacobs, N. J. (1961). Cytochrome-producing anaerobic vibrio, *Vibrio succinogenes*, sp. n. *J. Bacteriol.*, **81**, 911–17.

Wood, T. M. & Wilson, C. A. (1984). Some properties of the endo-$(1\rightarrow 4)$-β-D-glucanase synthesised by the anaerobic cellulolytic rumen bacterium *Ruminococcus albus*. *Can. J. Microbiol.*, **30**, 316–21.

Wood, T. M., Wilson, C. A. & Stewart, C. S. (1982). Preparation of the cellulase from the cellulolytic anaerobic rumen bacterium *Ruminococcus albus* and its release from the bacterial cell wall. *Biochem. J.*, **105**, 129–37.

Wozny, M. A., Bryant, M. P., Holdeman, L. V. & Moore, W. E. C. (1977). Urease assay and urease producing species of anaerobes in the bovine rumen and human feces. *Appl. Environ. Microbiol.*, **33**, 1097–1104.

Wyburn, R. S. (1980). The mixing and propulsion of the stomach contents of ruminants. In *Digestive Physiology and Metabolism in Ruminants*, ed. Y. Ruckebusch & P. Thivend. MTP Press, Lancaster, pp. 35–51.

Yoshinari, T. (1980). N_2O reduction by *Vibrio succinogenes*. *Appl. Environ. Microbiol.*, **39**, 81–4.

Zehnder, A. J. B. & Wuhrman, K. (1976). Titanium (III) citrate as a non-toxic oxidation reduction buffering system for the culture of obligate anaerobes. *Science*, **194**, 1165–6.

Ziolecki, A. (1979). Isolation and characterisation of large treponemes from the bovine rumen. *Appl. Environ. Microbiol.*, **37**, 131–5.

3
The Rumen Protozoa

A. G. Williams
Hannah Research Institute, Ayr, UK

&

G. S. Coleman
AFRC Institute of Animal Physiology and Genetics Research, Cambridge, UK

In addition to the bacteria in the rumen there are many larger (5–250 μm long) organisms which at various times have been designated 'protozoa'. Of these the 'ovals' (Quin's and Eadie's) are now known to be large bacteria (Orpin, 1976) (see Chapter 2) and the 'flagellates' *Neocallimastix frontalis*, *Piromonas communis* and *Sphaeromonas communis* are the zoospores of phycomycete fungi (Orpin, 1977a,b). There are genuine flagellates in the rumen, e.g *Trichomonas* spp., *Monoceromonas* sp. and *Chilomastix* sp., but little is known about their metabolism (Jensen & Hammond, 1964). These organisms are described in Chapter 4. The largest, most obvious and most important protozoa are the ciliates, of which there are two groups, both in the subclass Trichostomatia. The so-called 'holotrich' protozoa belong to the order Vestibuliferida, and the entodiniomorphs to the order Entodiniomorphida, suborder Entodiniomorphina and family Ophryoscolecidae. As the properties and metabolism of these two protozoal groups are different they will be considered separately here.

THE ENTODINIOMORPHID PROTOZOA

Identification and classification

The identification of rumen ciliate protozoa is difficult because the investigator has access only to line drawings and sometimes photomicrographs with which to compare an unknown organism. Most of the many species described have never been cultured, and it is difficult to decide, especially with the entodinia, between intraspecies variation and interspecies differences, especially as the presence of spines can be dependent on growth conditions (see below). Many of the characters, such as size and spination, on which the original classifications were made are now believed to be variable and of poor taxonomic worth. The problems will be discussed under the individual genera. The difficulty of identification is compounded because, unless the rumen sample is taken after the host has been starved for at least 24 hours, the protozoa will be so full of starch that no internal detail will be visible.

Classification into genera

For detailed descriptions of rumen ciliates the reader is referred to Dogiel (1927), Kofoid & MacLennan (1930, 1932, 1933) and Ogimoto & Imai (1981), the last work containing many photomicrographs. Table 1 gives a summary of the characters used to identify the various genera. Those which, in the present authors' opinion, are most valuable are given in bold type. In the following descriptions the convention of Stein (1858) has been followed, whereby the side of the protozoon where the macronucleus is situated is called 'dorsal', with left, right and ventral following as in vertebrate morphology. However, Lubinsky (1957) disagreed with the convention and called the side with the macronucleus 'right' for *Entodinium* spp. and 'left' for the larger organisms; this orientation was used by Ogimoto & Imai (1981).

The genus Entodinium. The entodinia are the smallest and simplest of the entodiniomorphid protozoa and are the most difficult to identify and classify into species due to the existence of many which are morphologically similar. There is also a peculiar problem with *E. caudatum*, at least, as this produces its characteristic caudal spines only when grown (*in vivo* or *in vitro*) in the presence of *E. bursa*. It is also larger (65 μm long compared with 56 μm) when grown under these conditions (Coleman *et al.*, 1972; Coleman & Hall, 1984). Kofoid & MacLennan (1930) divided the entodinia into four groups and an unclassified section, on the basis of general morphology, and believed there was an evolution in each group from small to large and from no caudal spines to numerous spines. Latteur (1968) reclassified the entodinia into seven groups depending on the position of the contractile vacuole in relation to the macronucleus. The type species is *Entodinium bursa* (Stein, 1858), which is the largest and least typical of all the entodinia.

The genus Eodinium. This genus, the validity of which has been disputed by Latteur (1970) and by Hungate (1978), contains the small protozoa (32–60 μm long) with a dorsal band of cilia but no skeletal plate. Eodinia are comparatively rare, but are found more frequently in cattle than in sheep. The type species is *Eodinium lobatum* (Kofoid & MacLennan, 1932).

The genus Diplodinium. This genus includes the larger organisms of similar structure to the eodinia. Kofoid & MacLennan (1932) divided the diplodinia into four groups with 21 species including those which differed only in the number of caudal spines. More recently, Latteur (1970) has recognised 20 species and reverted to the practice of Dogiel (1927) of having different forms to describe organisms differing only in the number of caudal spines. The type species is *Diplodinium dentatum* (Stein).

The genus Eremoplastron. This genus, which was erected by Kofoid & MacLennan (1932), contains organisms with a dorsal band of cilia and one narrow

Table 1
Characteristics of Rumen Entodiniomorphid Protozoa

Genus	Dorsal cilia	Obvious skeletal plates	Macronucleus shape	Length (μm)	Width (μm)	Length/ Dorso-ventral diameter	Left-right diameter/ Dorso-ventral diameter	Operculum	Number of contractile vacuoles	Caudal spines	Member of population type
Entodinium	0	0	Various	22-95	11-68	1·0-2·0	0·7-1·0	None	1	0-6	All
Eodinium	**1 band ant. end**	0	Rod-shaped	**32-60**	20-40	1·25-2·0	0·85-0·94	Small	2	0-2	?
Diplodinium	1 band ant. end	0	Often bent rod	55-210	41-136	1·2-2·1	0·75-0·94	Wide	2	0-6	A and B
Eremoplastron	1 band ant. end	1 narrow	Often bent rod	45-500 but usually 70-100	21-260 but usually 40-50	1·3-2·0	0·75-0·95	Usually small	2	0-2	B
Eudiplodinium	1 band ant. end	1 narrow	**Hook shaped**	105-198	56-120	1·3-1·7	0·86-0·97	Inconspicuous	2	0	B
Ostracodinium	1 band ant. end	1 wide	Various	58-133	36-54	1·55-2·34	0·86-1·02	Variable	2-6	0-2 usually lobes	B(?A)
Polyplastron	1 band ant. end	**2 narrow**	Rod-shaped	123-205	98-123	1-7		**Large, protruding**	7-9	0	A
Diploplastron	1 band ant. end	2 narrow close at post. end	Rod-shaped	88-120	47-65	1-7		**Large, protruding**	2	0	A
Metadinium	1 band ant. end	2 narrow occ. fused	Rod-shaped 2-3 lobes	110-288	61-165	1·25-1·78	0·75-0·91	**Small**	2	0	B
Epidinium	**1 band behind ant. end**	3 variable width	Elongate	105-150	44-72	**1·63-2·86**	0·83-1·00	Small	0-5	0-5	B
Enoploplastron	1 band ant. end	3 narrow close tog. with window	Elongate	60-140	32-90	1·6-1·9		Small	2	0	A(?B)
Ophryoscolex	1 band round 3/4 of middle	3 variable width	Elongate	120-215	60-80	1·59-2·14	0·87-0·94	Wide, not protruding	9-15	0-?, usually array of spines	A
Epiplastron	1 band round 3/4 of middle	5 variable width	Elongate	90-140	41-60	2·55		Wide, not protruding	2	0	?
Elytroplastron	1 band ant. end	3 narrow (2 on right 1 on left)	Elongate	110-160	67-97	1·43-1·82	0·85-0·82	Wide, not protruding	4	0	A(?K)
Caloscolex	1 band round all middle	1 complex	Elongate	130-160	73-90	2·0			7	0-5	?
Opisthotrichum	1 band round 1/3 of middle	1 cylindrical	Elongate	60-80	21-28	2·2			2	1	?
Parentodinium	0	0	Round	26-39	14-21	1·85		None	1	0	B?

skeletal plate. These are the simplest organisms to have a skeletal plate, although the separation of *Eremoplastron* from *Eudiplodinium* has been disputed by Latteur (1966) and Hungate (1978). The type species is *Eremoplastron rostratum* (Fiorentini, 1889).

The genus Eudiplodinium. This genus contains organisms with dorsal cilia, one narrow skeletal plate, and a highly characteristic hook-shaped macronucleus with the micronucleus in the eye of the hook. On these criteria, *Eudiplodinium maggii* is the only common species and is the type species.

The genus Ostracodinium. The protozoa in this genus have dorsal cilia and a very distinctive broad skeletal plate which can extend up to half way round the organism. There are 15–20 species delineated on the body shape, number of contractile vacuoles, size and shape of the skeletal plate, and spination or the possession of caudal lobes. The authors have evidence that single organisms can give rise to protozoa with and without caudal lobes, but it is not known how variable the other characters are. The type species is *Ostracodinium mammosum* (Railliet, 1890) according to Kofoid & MacLennan (1932), and *O. dentatum* (Fiorentini, 1889) according to Hungate (1978).

The genus Polyplastron. This genus contains only six species, with two narrow parallel skeletal plates (which in rare species can be fused together) on the right side, and three, very difficult to see, small plates on the left. The commonest, and type, species is *P. multivesiculatum* (Dogiel & Fedorowa, 1925) which has nine contractile vacuoles and a protruding and characteristic operculum. *P. multivesiculatum* is easily confused with the slightly smaller *Diploplastron affine* with which it usually occurs in British sheep, and it is only easily distinguished by the configuration of the skeletal plates which in the latter converge towards the posterior end. The size of *P. multivesiculatum* is highly dependent on the growth conditions. It is 123 μm long when grown *in vitro* in the absence of other protozoa, 175 μm long when a member of a normal A-type population (see below) and 205 μm long when engulfing epidinia (Coleman *et al.*, 1972).

The genus Diploplastron. This genus, which has only one species, *Diploplastron affine* (Dogiel & Fedorowa, 1925), is very similar to *Polyplastron*, differing only in size and the configuration of the skeletal plates. Unfortunately there can be confusion over the name of this protozoon. It appears in the literature as *Metadinium affine* and as *Eudiplodinium affine*.

The genus Metadinium. This genus has been confused with *Polyplastron* in the past; it has a similar morphology, with two narrow skeletal plates beneath the right surface but no plates on the left side. *M. medium* and *M. ypsilon*, at least in the authors' experience, appear to be almost round when lying on their left side, whereas *P. multivesiculatum* is longer and thinner. This, together with the much less protruding operculum in *Metadinium* spp., enables them to be distinguished from

Polyplastron when both are full of starch. There are 10 species, of which *Metadinium medium* (Awerinzew & Mutafowa, 1914) is the type species.

The genus Epidinium. The members of this genus are characterised by their relatively long cylindrical bodies with a dorsal zone of cilia behind the adoral zone, giving the impression in live organisms of having a ciliated anterior 20% of the body. Epidinia tend to swim in a corkscrew manner. They have three skeletal plates, but the number can only be distinguished with difficulty. Many epidinia seen in the rumen have caudal spines, but these are probably different forms rather than different species as was suggested by Kofoid & MacLennan (1933). The spines tend to disappear on culture. It should be noted that the single-spined form has a small curved spine in sheep whereas in cattle the spine is much heavier and thicker and is straight. There are up to 18 different species or forms, the type species being *Epidinium ecaudatum* (Fiorentini, 1889).

The genus Enoploplastron. This genus is characterised by the presence of a dorsal zone of cilia at the anterior end of the organism together with three skeletal plates which are separated in the middle to form 'windows'. In *E. confluens* the plates are fused together. There are only four species, of which *Enoploplastron triloricatum* (Dogiel, 1925*b*) is the type species.

The genus Ophryoscolex. *Ophryoscolex* spp. are the most complex and distinctive of the genera found in domestic ruminants. The dorsal zone of cilia is displaced posteriorly about one-third of the length of the body and encircles three-quarters of its circumference. All common species possess a formidable array of caudal spines arranged in up to four circlets and sometimes surrounding a long slender main spine. Species and forms have been delineated on the basis of the number and distribution of the spines. There is also a spineless form, but this is of doubtful authenticity. Mah (1964) and the present authors (Coleman & Reynolds, 1982*a*) have observed that in culture the caudal spines decrease in number and can disappear completely. With *O. caudatus*, at least, the main caudal spine returns to its original size on inoculating cultured organisms into a defaunated rumen. The type species is *Ophryoscolex purkynjei* (Stein, 1858) although the dubious *O. inermis* may also have a claim to this position.

The genus Epiplastron. The protozoa in this genus are similar to the epidinia but have five skeletal plates. They have been found only in antelopes.

The genus Elytroplastron. There is only a single species, *E. bubali* (Dogiel, 1928), which is similar in appearance to *Metadinium* but is characterised by the presence of two long narrow skeletal plates on the right side and one similar plate on the left side.

The genera Caloscolex *and* Opisthotricum. These protozoa are not found in domestic ruminants, occurring in the camel and African antelope respectively. For descriptions see Dogiel (1927) and Table 1.

The genus Parentodinium. One species of these cycloposthid ciliates, *P. africanum*, has recently been found in the rumens of Brazilian cattle (Dehority, 1986*b*).

Evolution

There have been a number of attempts (Crawley, 1923; Dogiel, 1947; Lubinsky, 1957) to work out logical sequences for the evolution of ciliates based on the idea that structurally complex organisms evolved from structurally simple ones, i.e. *Ophryoscolex* developed from *Entodinium*. Lubinsky (1957) proposed that in the evolution of *Diplodinium* from *Entodinium* the nuclei and the contractile vacuole were displaced from the 'dorsal' to the left ventro-lateral side, and in the more complex ophryoscolecids to the ventral side. Evidence was also produced for a twisting of the whole body, including the skeletal plates, and it was suggested that this was associated with the way the ciliates rotate when swimming.

There is, however, an alternative way of looking at these ciliates apart from just morphology. Although *Entodinium* spp. are structurally simple and also biochemically simple, in that they do not possess cellulase or endopectate-lyase enzymes, they are a very successful group. They are the first species to become established in young ruminants (Eadie, 1962) and are the last to disappear and the first to reappear if the rumen pH drops and then recovers. In contrast, the so-called 'advanced' species, such as *Epidinium* and *Ophryoscolex*, establish late and can disappear from a normal rumen for no apparent reason (Coleman, personal observation). The survival value of caudal spines (except in cases such as *Entodinium caudatum*), the storage of amylopectin in skeletal plates rather than in free granules, and of an additional ciliary zone, in an environment that is full of food organisms and particles, is doubtful. The entodinia are also much more efficient at engulfing starch grains than the larger ciliates. It is therefore suggested that the more complex entodiniomorphid protozoa may be derived from other complex forms from elsewhere that already possessed cellulase and that they are still in the process of fully adapting to the rumen environment. The authors would agree with Dogiel (1947) that entodinia were historically the first group to become established in ruminants and have therefore had longest to adapt.

Structure and the uptake of particulate matter and soluble compounds

The entodiniomorphid protozoa, although able to take up soluble compounds, feed principally by the engulfment of particular matter. All species possess an extrudable U-shaped peristome on which is a band of cilia that runs round the outside and then down one side of the U-shaped oesophagus to the cytostome (Coleman & Hall, 1971). Particles are wafted into the oesophagus by ciliary action, and Furness & Butler (1983, 1985*a,b*) believe, on the basis of electron microscope studies, that in epidinia and eudiplodinia large engulfed particles are propelled towards the cytostome by a suction force derived from fibrillar contraction, although lamellar microtubules may be implicated in membrane and cytoplasmic movement. In the smaller entodinia, which are able to engulf only small particles, the necessary force may be produced by membrane movement and cytoplasmic flow across the cytopharyngeal microtubular ribbons. However, even among the larger ciliates there are feeding differences, and Grain & Senaud (1984) and Senaud *et al.* (1986) have evidence that epidinia attach themselves to plant fragments by means of 'pseudopods' derived from a non-ciliated part of the oesophagus wall. Epidinia, unlike the other rumen ciliates (Coleman & Laurie, 1974*b*), release enzymes into the medium and these are involved in the extracellular breakdown of plant tissues with the release of plant cells that are engulfed by the protozoa (Bauchop & Clarke, 1976; Bauchop, 1979). *Polyplastron multivesiculatum*, however, takes up plant fibres without prior attachment (Senaud *et al.*, 1986), although *Eremoplastron bovis* behaves similarly to the epidinia (Coleman, unpublished observations).

Entodinium caudatum, at least, takes up all particles that are small enough to be engulfed, whether they are digestible or not, voiding small, indigestible ones by the anus. Not all digestible fragments are taken up equally rapidly; starch grains are engulfed at a rate which completely fills the organism in a few seconds, whereas bacteria are taken up slowly over several hours (Coleman, 1964*b*; Coleman & Sandford, 1979*a*). It is noteworthy that *E. caudatum* even when completely filled with starch grains still takes in bacteria at over 35% of the maximum rate (Coleman, 1975*a*).

All the ciliates possess an anus (cytoproct) and rectum of varying structural complexity through which material is voided. Furness & Butler (1983, 1985*a,b*) have evidence that the rectum of the larger protozoa is more complex than that of the entodinia and that, in the former, myonemic elements may be involved in the forcible expulsion of material. However, despite assumptions to the contrary, it is not only obviously indigestible matter that is expelled but also starch grains and chloroplasts (Coleman & Hall, 1984).

All entodiniomorphid protozoa possess one or more contractile vacuoles, but little is known about their function except for the early work of MacLennan (1933) who described a pulsatory cycle in some detail. More recently, Furness & Butler (1985*a,b*) have provided information on the microstructure of the vacuoles and the surrounding cytoplasm.

These protozoa all take up soluble compounds, and approximately half of the cell, probably the ectoplasm, is freely permeable to low molecular weight compounds at room temperature. The permeability barrier could therefore be the boundary between the ectoplasm and endoplasm (Coleman, 1986*b*).

Cultivation

Although only about 20 of the 100-plus species of entodiniomorphid protozoa described in the literature have been grown *in vitro*, there is no reason why they should not all be culturable using the standard techniques. All the entodiniomorphs are strict anaerobes; the rationale of the methods used is the same as that of bacterial culture (Chapter 2) and is based on reproducing conditions in the rumen. These provide a warm, constant temperature environment, deficient in readily metabolisable compounds but rich in particulate matter such as bacteria, starch grains, chloroplasts and cellulose fibres. The culture media used are therefore O_2-free and of low redox potential, contain bacteria in limited numbers, and have the minimum amount (consistent with protozoal growth) of starch grains and/or dried grass added each day. Under these conditions most of the added food is rapidly engulfed by the protozoa and any excess is broken down only slowly, compared to free glucose, by the bacteria.

The mineral salts solutions used have all contained KH_2PO_4, K_2HPO_4 and NaCl in varying proportions (with or without sodium acetate), to which was added a reducing agent such as sulphide or cysteine, rumen fluid from which the protozoa had been removed (up to 10%) and, under some conditions, antibiotics. The media are usually deoxygenated by bubbling with 95% N_2 + 5% CO_2, but some species require 100% CO_2 in the gas phase. For these species, some of the potassium phosphate and sodium chloride is replaced by sodium hydrogen carbonate in order to keep the pH constant. For more information on the details of the methods used, the reader is referred to Hungate (1942, 1943), Coleman (1960, 1987b), Onodera & Henderson (1980) and Hino et al. (1973).

Various other additives have been used to stimulate protozoal growth *in vitro*. These include: cobalt salts or vitamin B_{12} (Kandatsu & Takahashi, 1956; Bonhomme et al., 1982a); heat-treated yeast, activated charcoal and β-sitosterol-treated starch (Onodera & Henderson, 1980); choline (Broad & Dawson, 1976); tryptophan, nicotinamide, stilboestrol and urea (Spisni et al., 1980); β-sitosterol and related sterols plus haemin (Coleman & Reynolds, 1982a); complex organic medium constituents plus sterols (Hino et al., 1973). In the authors' opinion, β-sitosterol or related sterols (normally found in grass), choline (normally associated with starch grains) and possibly haem and vitamin B_{12} are essential for the growth of rumen ciliates. The nature of the other nitrogenous compounds supplied by bacteria (living or dead) in the culture is unknown but, in view of the very limited biosynthetic ability demonstrated by these ciliates, it is likely that a large number of compounds are needed.

As all the entodiniomorphid protozoa contain bacteria in vesicles in their endoplasm they can be rendered bacteria-free only by treatment with antibiotics. There have been a number of successful attempts to produce axenic protozoa, but none of the protozoa has then grown and at best they have survived in decreasing numbers for up to 22 days (Coleman, 1962; Bonhomme et al., 1982b,c; Hino & Kametaka, 1977). *Entodinium caudatum* has, however, been grown for over 2 months in the presence of either *Escherichia coli* or *Streptococcus bovis* alone (Hino & Kametaka, 1977).

Entodiniomorphid protozoa do not grow well in conventional stirred-tank continuous-culture apparatus because they need a large dead space in which they can sequester and find food. Every successful apparatus has had a large hollow slowly moving piston, inside which were placed bags of nylon mesh that contained food materials such as crushed barley and hay. Population densities of 10^6 and 2×10^5/ml, in the bags and effluent respectively, have been obtained. For further details see Weller & Pilgrim (1974), Nakamura & Kurihara (1978), Abe & Kurihara (1984), Czerkawski & Breckenridge (1979), and Chapter 12.

Division and conjugation

Entodiniomorphid protozoa divide by binary fission. This has been observed by Warner (1962) to take, in entodinia, only 15 minutes from the first observed signs of division until its completion. During division the ciliary areas and the skeletal plates are retained by the anterior daughter, and new ciliary bands and skeletal plates form at about the midline (Noirot-Timothée, 1960). The time of protozoal division *in vivo* depends on a variety of factors, but the maximum times recorded are 6·8 hours (Potter & Dehority, 1973) and 5·5 hours (Warner, 1962) for entodinia and 14 hours (Senaud *et al.*, 1973) for the much larger *Polyplastron multivesiculatum*. Mean generation times *in vitro* vary from 6 hours with *Entodinium bursa* (Coleman *et al.*, 1977) to 38 hours with *Entodinium simplex* (Coleman, 1969*a*).

Conjugation, which is a form of sexual reproduction, occurs when two individuals join together by their anterior ends and exchange genetic material. It occurs infrequently and erratically in the rumen although, for no obvious reason, many conjugating individuals are occasionally seen in one animal (Noirot-Timothée, 1960; Warner, 1962). Conjugation has also been observed in cultures of *Entodinium caudatum* (Coleman, unpublished observation) and of *Ophryoscolex purkynjei* (Mah, 1962). For descriptions of the process see Dogiel (1925*a*) and Noirot-Timothée (1960). The importance of conjugation is not known but, as it has been observed in cultures that apparently grow indefinitely (>27 years) but not in those that die after 2–3 years, it may be a 'rejuvenating' process. (See Chapter 15 on survival of bacteria in gnotobiotic lambs.)

Uptake, digestion and metabolism of dietary components

Bacteria
Bacteria are probably the most important single source of nitrogenous compounds for protozoal growth although plant protein and free amino acids are also a valuable source for some species.

Although cultured *Entodinium caudatum* have taken up all bacteria tested, and take up individual bacteria from a mixed suspension in the proportion in which they are present (Coleman, 1964*b*), other protozoa selectively engulf or reject certain bacterial species. There is, however, no consistent pattern. Although *Selenomonas ruminantium* and *Butyrivibrio fibrisolvens* are almost always taken up at the same rate or faster than other bacteria, others such as *Klebsiella aerogenes* and *Proteus mirabilis*, which occur in protozoal cultures, appear to be either liked or disliked by many protozoal species. *Escherichia coli* and *Bacteroides ruminicola* are never taken up preferentially and are often not engulfed at all or taken up only slowly. For detailed information the reader is referred to Coleman (1986*b*).

The rate of uptake of bacteria has been compared (a) on the rate of uptake from an indefinitely dense suspension, which probably measures the rate at which a protozoon can pass bacteria down its oesophagus and form food vesicles, and (b) on the rate of clearance of bacteria from an infinitely dilute suspension, which measures the protozoon's ability to find and capture prey. The former is obviously more relevant to conditions in the rumen. Using these two criteria, it is apparent that *Entodinium caudatum* clears all *Proteus mirabilis* organisms from the medium at all suspension densities, i.e. if the bacterial density is doubled the rate of uptake is doubled. In contrast, the rate of uptake of the yeast *Saccharomyces fragilis* is almost independent of population density (Coleman, 1975a). The pattern of engulfment of most bacteria falls between these extremes. If one protozoon engulfs suspensions of different bacteria, the volume of medium cleared at infinitely dilute bacterial population density is almost independent of the bacterium preyed upon, suggesting that the protozoa can scavenge a certain volume of medium. For one species the volume is higher with cultured protozoa than with those grown in the rumen (Coleman, 1986b).

The rate of uptake of bacteria is relatively little affected by change in salt concentration from 60% to 150% of the optimum, but it is pH-sensitive with an optimum pH of 6·0 falling off to nothing at pH 5·0, 75% at pH 7·0 and 30% at pH 8·0 (Coleman & Sandford, 1979a). With some species, disruption of the bacteria prior to the incubation with the protozoa markedly increases the rate of uptake of bacterial carbon, but this is not universally true (Coleman & Laurie, 1974a).

Fate of engulfed bacteria. The limited evidence available suggests that in *Entodinium caudatum* different bacterial species are killed and digested at different rates and in different ways. *Escherichia coli* and *Klebsiella aerogenes* (lacking a polysaccharide capsule) are both killed rapidly, surviving for only a few minutes inside the protozoon, whereas *Proteus mirabilis* is comparatively resistant with 62% still being viable after 1 hour's continuous engulfment. Gram-negative bacteria such as *E. coli* are completely digested except for the lipopolysaccharide cell membrane. Gram-positive, lysozyme-sensitive bacteria, such as *Bacillus megaterium*, are digested very quickly, due to the rapid destruction of the cell wall, whereas bacteria such as *Staphylococcus aureus* and *Streptococcus faecalis*, the walls of which are comparatively resistant to lysozyme, lose their cell contents before there is extensive digestion of the wall (Coleman & Hall, 1972).

Some of the diaminopimelic acid in the cell walls is converted into lysine which is incorporated into protozoal protein (Denholm & Ling, 1984; Masson & Ling, 1986).

Bacterial protein and nucleic acid. On incubation of *Escherichia coli* labelled with one 14C amino acid with a suspension of an entodiniomorphid protozoon, some of the labelled amino acid is incorporated unchanged into protozoal protein, some may be incorporated as a related amino acid, and the remainder is released into the cell pool or the medium, often in an acetylated or formylated form. Some amino acids with hydrophobic side-chains are oxidatively deaminated (Coleman, 1967b).

Constituents of bacterial nucleic acid are incorporated into protozoal nucleic acid, the transfer taking place at the level of complexity of nucleotides (Coleman, 1968).

Release of bacterial constituents into the medium. The uptake, killing and digestion of bacteria by rumen ciliates and release of the digestion products into the surrounding medium is a very important part of the cycling of nitrogen in the rumen because any released amino acids are metabolised for growth by the remaining bacteria, often with the loss of ammonia. The rate of release of bacterial digestion products depends on several factors: (a) the bacterial species, with *Bacillus megaterium*, *Proteus mirabilis*, *Butyrivibrio fibrisolvens* and *Selenomonas ruminantium* usually being digested more rapidly than other species, with release of material (up to ten times that found in the protozoa) into the medium; (b) the energy status of the protozoa, with more bacterial carbon being released if the protozoa are energy-deficient; (c) the salt concentration, with more material being released at lower salt concentrations (Coleman, 1967*b*; Coleman & Sandford, 1979*a*).

Other protozoa
Polyplastron multivesiculatum. All strains of *P. multivesiculatum* engulf *Epidinium* spp. (at a rate of up to 10/day) although they will also take up *Diplodinium* spp., *Ostracodinium* spp. and *Eudiplodinium maggii*. Some strains will grow *in vitro* only in the presence of epidinia, but predation is always associated with an increase in size (Coleman *et al.*, 1972). The epidinia are engulfed posterior end first and initially are degraded only slowly until holes appear in the epiplasm; after this the epidinia disintegrate rapidly until all that remains is an almost empty vacuole containing fragments including heavily staining bodies that could have been pieces of epidinial cytoplasm (Coleman & Hall, 1978).

Entodinium bursa. In the authors' experience *Entodinium bursa* has an obligate requirement for the spineless form of *E. caudatum* when grown *in vitro*. Under these conditions 1·5–2·5 *E. caudatum* per hour are engulfed by each *E. bursa*, which grows and divides every 6 hours. After engulfment there is a slow attack on the pellicle (glycocalyx) followed by digestion of the underlying membranes and epiplasm. However, once a hole has been made, disintegration is rapid and all internal structure disappears leaving only bacteria, polysaccharide granules and pieces of pellicle in a large vesicle; the first two are utilised by the *E. bursa*. Amino acids from *E. caudatum* protein, and purines (at least) from its nucleic acid, are transferred intact into *E. bursa* protein and nucleic acid respectively, but appreciable amounts are lost into the medium, thus suggesting that *E. caudatum* is engulfed and digested more rapidly than the *E. bursa* can utilise the products (Coleman & Hall, 1984).

Protein
There is little information on the uptake of free protein, although Onodera & Kandatsu (1970) showed that insoluble casein is digested, presumably after engulfment. All the entodiniomorphid protozoa contain proteolytic enzymes, although there is considerable disagreement about the optimum pH for proteolysis which has been reported as 3·5 (for leaf fraction 1 protein and casein; Coleman, 1983), 6·5–7·0 (for casein; Abou Akkada & Howard, 1962) and 5·8 (for endogenous protein; Forsberg *et al.*, 1984). The activity is highest in *E. caudatum* and *E. simplex* and lowest in some cellulolytic species. All ciliates have a high concentration of leucine aminopeptidase (Forsberg *et al.*, 1984; Prins *et al.*, 1983).

Free amino acids
Amino acids, like all soluble compounds, are taken up by active and passive processes, the latter probably just being the trapping of medium containing the

solute in some part of the cell, possibly the contractile vacuole. The active process, which can be inhibited by suitable analogues, results in the incorporation of the amino acid, usually unchanged, into protozoal protein. The rates of uptake are lowest with *Entodinium* and *Epidinium* spp. and highest with *Eudiplodinium maggii* and *Polyplastron multivesiculatum* (Coleman, 1967a, 1972; Coleman & Laurie, 1974a, 1977; Owen & Coleman, 1977; Coleman & Sandford, 1979b; Coleman & Reynolds, 1982b). Amino acids which have been taken up can be released back into the medium and many are acetylated or formylated before release (Coleman, 1964a, 1967a).

Sources of amino acids for protozoal growth. As the protozoa cannot be grown axenically, it is impossible to determine directly the preferred sources of compounds, for example amino acids, which are required for growth. The best approximation is to measure the rate of incorporation into protozoal protein of amino acids from engulfed bacteria, from free amino acids at the concentration normally found in the rumen and, where appropriate, from other protozoa. It is then possible to calculate how long the amount of protozoal protein would take to double and to compare this with the mean generation time of the protozoon. Details of the calculations are given by Coleman (1986b) and these show that where another protozoon is engulfed this is an important source of amino acids, and this is reflected in the comparatively short generation time of *Entodinium bursa* and *Polyplastron multivesiculatum*. With *Entodinium caudatum, E. longinucleatum* and *Ophryoscolex caudatus* the uptake of bacteria is rapid enough to enable the protozoa to divide once a day, but with the others the uptake of bacteria and free amino acids is insufficiently rapid and it is suggested that protein associated with engulfed plant material might make up the shortfall. However, with some protozoal species the uptake of bacteria is highly dependent on population density, and a doubling of bacterial population density can almost double the rate of uptake. Except possibly with *Ophryoscolex caudatus*, there is no evidence for appreciable synthesis of amino acids from carbohydrate (e.g. Coleman, 1978).

Starch
All entodiniomorphid protozoa engulf starch grains which are fermented slowly to (principally) H_2, CO_2, acetic acid and butyric acid. Each *Entodinium caudatum* contains 18 ng starch when completely filled, and this disappears progressively over 42 hours. The highest amylase activities are in *Eremoplastron bovis, Diploplastron affine, Ophryoscolex caudatus* and *Polyplastron multivesiculatum*, and the lowest in *Ostracodinium obtusum bilobum, Entodinium caudatum* and *E. bursa* (Coleman, 1986c). Starch is digested to maltose and then glucose which is phosphorylated to glucose 6-phosphate; this is metabolised to produce energy, probably via classical glycolysis (Coleman, 1981), or is used in the ectoplasm (?) to form storage amylopectin (Eadie *et al.*, 1963; Wakita & Hoshino, 1980). In *Entodinium caudatum*, at least, the activities of the amylase and maltase are subject to negative feedback by the products of the reactions, and the concentrations of maltose and glucose inside the protozoa tend to remain almost constant despite a massive engulfment of starch.

Free ^{14}C-labelled sugars are also taken up, prolonging the life of starved protozoa, but there is no increase in cellular concentration in well fed cells, presumably due to inhibition of polysaccharide breakdown (Coleman, 1969b). In the rumen of sheep with natural A (entodinia, holotrichs, plus *Polyplastron multivesiculatum* and *Diploplastron affine*) or B (entodinia, holotrichs, plus epidinia, eudiplodinia, eremoplastron and ostracodinia) type protozoal populations, 33–57% of the total amylase is in the protozoal cytoplasm, whereas in a sheep containing only *Entodinium caudatum* only 2–15% is in the protozoal cytoplasm, most of the remainder being associated with the bacteria (Coleman, 1986c). (See also Chapter 5 for type-A and -B populations.)

Some of the glucose and maltose inside a protozoon is used by the intracellular bacteria as a source of energy and, in the case of *Klebsiella aerogenes* in cultured *Entodinium caudatum*, to produce a polysaccharide capsule which protects the bacteria against digestion by the protozoon's lytic enzymes (Coleman, 1969b, 1975b).

Cellulose
It was found many years ago that some of the larger entodiniomorphid protozoa engulf and digest cellulose and use the products for the synthesis of intracellular polysaccharide (Hungate, 1942, 1943). Although a soluble cellulase is present inside such protozoa, there is still no unambiguous evidence that it is of protozoal rather than bacterial origin. There is, however, indirect evidence, based on the effects of antibiotics and on the absence of bacteria attached to engulfed plant fibres, that at least some of the enzyme is produced by the protozoa (Coleman, 1978; Coleman & Hall, 1980) although Thines-Sempoux *et al.* (1980) believe, on the basis of studies with the electron microscope, that the enzyme is bacterial. All the rumen entodiniomorphid protozoa, except for *Entodinium* spp., contain cellulase, the highest specific activity being in *Eudiplodinium maggii*, *Epidinium ecaudatum caudatum* and *Ostracodinium obtusum bilobum* (Coleman, 1985a).

Using the amount of carboxymethylcellulase released from the various rumen fractions as parameter, the total cellulase in sheep containing only *Eudiplodinium maggii* is 2·9–4·2 times that in an animal containing no ciliate protozoa, and over 70% of the activity is associated with the protozoa. In a sheep containing the non-cellulolytic *Entodinium caudatum*, over 65% of the cellulase is bacterial (Coleman, 1986a). The rate of synthesis of intracellular amylopectin from cellulose by *Eudiplodinium maggii* is 2·6–4·5 ng per protozoon per hour (68–120 μg/mg protein), which in a rumen in which this is the only ciliate species corresponds to approximately 5 g per day (Coleman, 1987a).

Hemicellulases, etc.
The distribution of endopectate lyase, which breaks down polygalacturonic acids, hemicellulase B and xylanase, between the entodiniomorphid protozoa is similar to that of cellulase, with little activity in *Entodinium* spp. and appreciable amounts, except for endopectate lyase in *Polyplastron multivesiculatum*, in the larger organisms. All the ciliates, except the entodinia, contain the glycoside hydrolases

necessary for the breakdown of the primary degradation products to monosaccharides (Coleman et al., 1980; Williams & Coleman, 1985; Williams et al., 1984). However, the evidence that these protozoa can utilise anything except glucose and glucose polymers is poor, although they might use polygalacturonic acid and xylan to a very limited extent; fructose and fructosans are apparently never utilised (Coleman et al., 1980; Coleman, 1962).

It is of interest that the optimum pH of endopectate lyase is 8·5 (with 14% of maximum activity at pH 6·0) compared with optima of 5·0–7·5 for cellulase (Coleman, 1985a), 6·0 for amylase (Coleman, 1986c) and 3·5 for proteases (Coleman, 1983), and a probable vesicle pH of about 6.

Nucleic acid constituents

As mentioned above, bacteria are probably the principal sources of nucleotides for nucleic acid synthesis, although free purines, pyrimidines, phosphate and ribose are also taken up. There is some interconversion of the purines on incorporation into protozoal nucleic acid, the remainder being degraded via hypoxanthine and xanthine to form a compound in which C_8 becomes volatile. Uracil is taken up as both uracil and cytosine, and free uracil and thymine are both rapidly converted into their dihydro derivatives which are released into the medium. Free ribose is converted into glucose which is incorporated into protozoal polysaccharide (Coleman, 1968, 1969b; Coleman & Sandford, 1979a).

THE HOLOTRICH PROTOZOA

Holotrich ciliate protozoa occur widely in the reticulorumen of both domesticated and wild ruminants (e.g. Buisson, 1924; Jameson, 1925b; Van Hoven et al., 1979; Van Hoven, 1983; Dehority, 1986c). Their occurrence in the principal fermentative regions of the intestinal tracts of other members of the Artiodactyla (Buisson, 1923; Dogiel, 1928; Thurston & Grain, 1971; Kleynhans & Van Hoven, 1976), Perissodactyla (Hsiung, 1930; Grain, 1966) and Proboscidea (Latteur, 1967; Wolska, 1967b, 1968; Eloff & Van Hoven, 1980) has also been documented. The generic composition and overall size of the rumen holotrich population is influenced by a number of interacting factors, the more important of which are the type of host, its geographical location, the nature of the diet consumed, and the frequency of feeding. The three principal holotrich species in the rumen are *Isotricha intestinalis*, *Isotricha prostoma* and *Dasytricha ruminantium*; other genera that have been observed in the rumen environment, but not studied in detail, are listed in Table 2.

Holotrich numbers in the rumen are increased when the diet contains a source of readily available soluble carbohydrates, such as fresh temperate grasses (Clarke, 1965a) or sugar cane (Valdez et al., 1977). The number of holotrichs typically present in the rumens of domesticated animals ranges up to 10^5 per ml of rumen fluid, and on forage diets the holotrichs represent some 20% (12–40%) of the total ciliate population (Clarke, 1964). They have been observed more regularly in domesticated animals than in wild ruminants, and in Europe they occur principally in grazing

Table 2
Observations on the Occurrence of Holotrich Ciliates in Domesticated and Wild Hosts

Host	Dasytricha ruminantium	Isotricha prostoma	Isotricha intestinalis	Oligoisotricha bubali	Microcetus lappus	Buetschlia parva	Parabundleia ruminantium	Polymorphella bovis	Blepharoconus krugerensis	Blepharoprosthium parvum	Charonina ventriculi	Charonina equi	Parasitotricha sp.	Reference
Domesticated sheep (USA, Europe, Japan, China), *Ovis aries*	+	+	+			+								Jameson, 1925a; Hsiung, 1931; Dehority, 1970; Imai et al., 1978
North Ronaldshay sheep, *Ovis aries*	+	+	+											Eadie, 1957
Domesticated goat (India, Japan)	+	+	+											Das Gupta, 1935; Imai et al., 1978; Asada et al., 1980
Domesticated cattle (USA, Europe, Norway)	+	+	+	+		+					+			Dogiel, 1926; Wolska, 1967a; Dehority & Mattos, 1978; Dehority et al., 1983; Orpin & Mathiesen, 1986
Domesticated cattle (New Zealand)	+	+	+	+										Clarke, 1964, 1965a
Domesticated cattle (China, Japan, Korea)	+	+	+	+						+		+		Hsiung, 1932; Imai et al., 1978, 1982; Han. 1984
Bali cattle, *Bos javanicus domesticus*	+	+	+											Imai, 1985
Zebu cattle, *Bos indicus*	+	+	+	+			+	+			+			Kofoid & Maclennan, 1933; Imai & Ogimoto, 1983, 1984; Imai, 1984; Dehority, 1986b
Water buffalo, *Bubalus bubalis*	+	+	+	+		+					+			Michalowski, 1975; Dehority, 1979; Fujita et al., 1979; Imai, 1981, 1985; Imai & Ogimoto, 1984; Imai et al., 1981a,b; Ogimoto et al., 1983
Buffalo, *Bison bison bison*	+	+												Pearson, 1967; Giesecke, 1970
Musk-ox, *Ovibos moschatus*	+	+												Dehority, 1974, 1985
Gaur-ox, *Bos gaurus*	+	+	+											Kofoid & Christenson, 1933
Yak, *Poephagus poephagus*	+													Dogiel, 1934
Reindeer, *Rangifer tarandus*	+	+	+			+								Buisson, 1923; Dogiel, 1935; Giesecke, 1970; Dehority, 1975, 1986a
Caribou	+	+	+											Dehority, 1986a
Red deer, *Cervus elaphus*	+	+	+											Clarke, 1968; Giesecke, 1970
Mountain reedbuck, *Redunca fulvorufula fulvorufula*	+	+												Van Hoven, 1983
Sable antelope, *Hippotragus niger*	+	+												Van Hoven et al., 1979
Mouse deer, *Tragulus meminna*			+											Jameson, 1925b
Japanese serow, *Capricornus crispus*	+	+												Imai et al., 1981a
Camel	+	+												Buisson, 1923
Dromedary	+	+												Dogiel, 1928
Llama, *Auchenia lama*	+	+												Buisson, 1923
African elephant, *Loxodonta africana*													+	Eloff & Van Hoven, 1980
Giraffe													+	Noirot-Timothée, 1963; Kleynhans & Van Hoven, 1976

animals (Giesecke, 1970). However, in Africa the holotrichs tend to occur more regularly in browsers (Van Hoven, 1983). Surveys of the occurrence of the holotrichs in various hosts suggest that geographical variations in distributions also occur (Imai *et al.*, 1981*b*,*c*; Ogimoto *et al.*, 1983; Imai & Ogimoto, 1984; Dehority, 1986*a*).

The holotrich population is influenced by the diet and feeding practices of the host animal; these effects are discussed in Chapter 5 and have also been reviewed in detail elsewhere (Hungate, 1966; Warner, 1966; Clarke, 1977; Williams, 1986). In addition, the number of holotrichs in the rumen is not constant throughout the diurnal cycle in that the numbers of *Isotricha*, *Dasytricha* and *Buetschlia* spp. increase before feeding and decrease when feeding has ceased, although the timing of this decline has not been consistent in all studies (Williams, 1986). The variation in the numbers of *Charonina* spp. in the rumen during the diurnal cycle is different and resembles that of the entodiniomorphid ciliates. The marked decrease in holotrich numbers after feeding has been attributed to postfeed increases in rumen-outflow rates (Warner, 1966; Michalowski & Muszynski, 1978), protozoal settlement (Minor *et al.*, 1977; Valdex *et al.*, 1977), disintegration (Clarke, 1965*a*,*b*) and sequestration on food particles (Orpin & Letcher, 1978; Orpin, 1985) or the reticulum wall (Abe *et al.*, 1981). The contractions of the reticulum wall in anticipation of, and during, feeding, and the presence of glucose in the rumen immediately after feeding, have been shown to stimulate the migration of the holotrich protozoa into the rumen contents (Abe *et al.*, 1983; Murphy *et al.*, 1985). Recent studies by Leng *et al.* (1986) have confirmed that the holotrich ciliates are extensively retained within the rumen ecosystem and do not pass out of the rumen with the digesta.

Classification

The classification of the holotrichs is not consistent in the literature and different schemes have been published. Clarke (1964, 1977) and Hungate (1966, 1978) adopted the protozoal classification of Honigberg *et al.* (1964) in which the holotrich genera occur in the two orders Gymnostomatida and Trichostomatida of the subclass Holotrichia. The classification of the protozoa was subsequently revised by Levine *et al.* (1980) and this format was used by Ogimoto & Imai (1981). In this revision the rumen holotrichs were placed in two subclasses, Vestibuliferia and Gymnostomata, with the family Buetschliidae being located in the order Gymnostomatida whilst the Isotrichidae and Blepharocorythidae remained within the Trichostomatida. In a more recent reappraisal (Lee *et al.*, 1985) the Buetschliidae and Blepharocorythidae have been placed within the order Entodiniomorphida. The Isotrichidae belong to the Vestibuliferida; both orders are in the subclass Trichostomatia. These various classifications are summarised in Table 3.

Occurrence, morphology and ultrastructure

Isotrichidae
The holotrich genera occurring most frequently in the rumen are *Isotricha* and *Dasytricha*, and because of their size and motility they are easily seen during

Table 3
Classification of the Rumen Holotrich Ciliates

Scheme	Phylum	Subphylum	Class	Subclass	Order	Suborder	Family	Representative genera
1	Ciliophora		Ciliatea	Holotrichia	Trichostomatida		Isotrichidae	*Isotricha*
2	Ciliophora		Kinetofragminophorea	Vestibuliferia	Trichostomatida	Trichostomatina	Isotrichidae	*Dasytricha*
								Oligoisotricha
3	Ciliophora	Rhabdophora	Litostomatea	Trichostomatia	Vestibuliferida		Isotrichidae and Paraisotrichidae	*Paraisotricha*
1	Ciliophora		Ciliatea	Holotrichia	Trichostomatida		Blepharocorythidae	*Charonina*
2	Ciliophora		Kinetofragminophorea	Vestibuliferia	Trichostomatida	Blepharocorythina	Blepharocorythidae	
3	Ciliophora	Rhabdophora	Litostomatea	Trichostomatia	Entodiniomorphida	Blepharocorythina	Blepharocorythidae	
1	Ciliophora		Ciliatea	Holotrichia	Gymnostomatida	Rhabdophorina	Buetschliidae	*Buetschlia*
2	Ciliophora		Kinetofragminophorea	Gymnostomata	Prostomatida	Archistomatina	Buetschliidae	*Parabundleia*
3	Ciliophora	Rhabdophora	Litostomatea	Trichostomatia	Entodiniomorphida	Archistomatina	Buetschliidae	

1, Honigberg et al., 1964; 2, Levine et al., 1980; 3, Lee et al., 1985.

microscopic examination of rumen contents. The genera usually occur together, although there are reports of the presence of a single genus in, or the absence of both from, an individual host or group of animals. Dehority (1975) was unable to identify any host specificity for either genus. It would seem that the diet of the host animal is of major importance in holotrich establishment, although when present the protozoa are apparently able to modify the environment to aid their retention (Dehority & Purser, 1970; Dehority, 1978).

Isotricha. *Isotricha prostoma* Stein and *Isotricha intestinalis* Stein are elongated ovoid (ellipsoidal) organisms with a complete surface coverage of cilia arranged in rows parallel to the body axis. Although their sizes overlap, *I. prostoma* is the larger, ranging in length from 80 to 200 (av. 135) µm and in width from 50 to 120 (av. 70) µm; the corresponding figures for *I. intestinalis* are 90–200 (av. 110) µm and 45–150 (av. 60) µm. There are 6–12 contractile vacuoles present, although the number, size and distribution of food vacuoles is variable (Campbell, 1929). The macronucleus is rod-shaped and curved; its size (*c.* one-third of cell length and one-seventh of cell diameter), shape, and location near to the vestibulum are constant in non-dividing cells. The micronucleus is oval (3–6 µm in length) and is located in a depression on the ventral side of the macronucleus. Both nuclei are enclosed within a membrane and supported by a fibrillar nucleo-suspensory apparatus (Ten Kate, 1928; Campbell, 1929). The cytostome is an elliptical aperture covered by cilia that is situated at the end of the body which is to the rear whilst the protozoon is swimming. *I. intestinalis* closely resembles *I. prostoma* although the micronucleus may be more rounded or triangular (Bhatia, 1936; Ogimoto & Imai, 1981). The vestibulum is located ventrally in the species and is found one-third of the body length away from the posterior end of the cell (Grain, 1966).

Dasytricha. *Dasytricha ruminantium* Schuberg resembles *Isotricha* spp. in shape but is smaller (46–100 × 22–50 µm, av. 58 × 27 µm). Surface coverage with cilia is complete, although the rows of cilia are arranged obliquely to the body axis. The vestibulum is situated at the posterior end of the cell which, as with the isotrichs, is thus to the rear of swimming cells. There is a single contractile vacuole and the micronucleus is situated close to the elliptical macronucleus, the position of which can vary in the cell; a nucleo-suspensory apparatus has not been reported for *D. ruminantium* (Bhatia, 1936). Grain (1966) considered that the other rumen species of *Dasytricha* that had been described (Jirovec, 1933; Hukui, 1940) were, with the exception of *D. hukuokaensis* (Hukui & Nisida, 1954), not morphologically distinctive, and reassigned these species as *I. prostoma*, *I. intestinalis* or *D. ruminantium*. *Dasytricha hukuokaensis* is larger [120–182 (av. 151) × 68–122 (av. 95) µm] and has a lateral mouth and up to 11 contractile vacuoles. A spherical micronucleus lies in a depression of the ellipsoidal macronucleus. The protozoon is morphologically similar to *I. intestinalis* but lacks a nucleo-suspensory apparatus.

Oligoisotricha. The ciliate *Isotricha bubali* was first detected and described by Dogiel (1928) but was later assigned to the genus *Oligoisotricha* because of morphological differences from the genus *Isotricha* (Imai, 1981). *Oligoisotricha bubali* was initially observed in water buffalo, but has subsequently been found in a wider range of hosts (Table 2). It is ovoid in shape and relatively small, being 12–22 (av. 16) μm in length and 8–20 (av. 12) μm in width. The somatic ciliature is arranged in rows parallel to the body axis but is absent from the posterior one-sixth of the body surface. A single contractile vacuole is usually found at the posterior end of the cell, although this location may vary. The macronucleus is spherical or elliptical in shape and is inconsistently located within the endoplasm. The micronucleus is spherical and is positioned close to the anterior margin of the macronucleus. It has no nucleo-suspensory apparatus. Orpin & Mathiesen (1986) observed a small ciliate resembling *Oligoisotricha* sp. in the ruminal contents of Norwegian Red cattle. This isolate, like *O. bubali*, possessed a striated vestibulum and lacked cytopharyngeal rods, but differed in that its posterior end was smoothly rounded whereas that of *O. bubali* is concave. These differences suggest that this isolate may represent a previously undescribed species.

Microcetus. Another small ciliate was also present in some of the Norwegian cattle examined by Orpin & Mathiesen (1986). This protozoon differed from other small rumen ciliates, and the name *Microcetus lappus* was proposed. The cells are ovoid to elongate, being 18–29 μm long and 7·5–18 μm wide (av. 23·6 × 12·7 μm). The buccal cavity is apical and the cytoproct sub-terminal on the ventral surface or terminal. The somatic ciliature is absent from the cytoproct region and covers approximately 90% of the cell surface; the rows of cilia are inclined at an angle ($c.$ 20°) to the body axis. The buccal cilia occur on the dorsal side of the cavity. The macronucleus is spherical to ovoid and located close to the centre of the cell; a single contractile vacuole occurs close to the cytoproct. Two characteristic cytopharyngeal rods are present. The taxonomic position of the genus *Microcetus* is as yet unknown.

Microcetus lappus was not uniformly distributed in the cattle examined, but in two animals represented 12% and 15% of the total ciliate population. Orpin & Mathiesen (1986) believe that *M. lappus* may have been confused with *O. bubali* in earlier studies of protozoa in North American cattle (Dehority *et al.*, 1983); in this latter study the concentrations of this small ciliate ranged from up to 35% of the protozoal population in unweaned calves to as high as 72% in mature feedlot animals.

Ultrastructure. Detailed electron microscopic ultrastructural studies of the Isotrichidae have been undertaken infrequently and are restricted to preparations of *D. ruminantium* and the two species of *Isotricha* (Grain, 1966; Buckelew & Kontir, 1977; Stern *et al.*, 1977; Gerassimova, 1981; Ogimoto & Imai, 1981). Although the holotrichs are regarded as being less developed than the entodiniomorphid ciliates, morphological studies have revealed complex ultrastructural characteristics and surface structures. The endoplasm and ectoplasm are separated by a continuous double-layered fibrillar system (Noirot-Timothée, 1958; Vigues *et al.*, 1984). Certain of the protein components of the ecto-endoplasmic fibrillar boundary have been separated and shown to have molecular weights of 23 kd or more (Vigues & Groliere, 1985; Vigues et al., 1985).

Polysaccharide storage granules and the membrane-enveloped micro- and macro-nuclei are located in the endoplasm. Ingested food particles and bacteria occur in vacuoles that are distributed throughout the cytoplasm. Many electron-dense bodies are also present in the cytoplasm although, with the exception of the hydrogenosome, their functions have not been elucidated. *Dasytricha ruminantium* and *Isotricha* spp. both contain hydrogenosomes; these are microbody-like organelles which are approximately 500 nm in diameter and occur principally in the endoplasm close to the ecto-endoplasmic boundary (Yarlett et al., 1981, 1983a; Constantinescu & Dragos, 1984). Preliminary descriptions of mitochondria in these ciliates are erroneous (Grain, 1966; Gaumont & Grain, 1967); the organelles observed were, on the basis of morphology and location, hydrogenosomes. Many of the hydrolytic enzymes of *Dasytricha* and *Isotricha* are present in membrane-bound vesicles 100–800 nm in diameter (Yarlett et al., 1985; Williams et al., 1986) although their location in the cell ultrastructure has yet to be established.

Electron microscopic examination of the cell surface of *Isotricha intestinalis* showed that it is composed of longitudinal cytoplasmic ridges between the rows of cilia (Imai & Tsunoda, 1972; Orpin & Hall, 1983). On the dorsilateral surface the ridges are modified to form an attachment zone up to 35 μm long and extending up to 9 μm from the cell surface (Orpin & Hall, 1977, 1983). A similar attachment organelle has been observed on *I. prostoma* but not on *D. ruminantium* (Orpin & Letcher, 1978). This organelle enables the holotrich ciliates to attach to plant particles in the rumen (Orpin, 1985; Orpin & Letcher, 1978).

Buetschliidae

Several genera in the family Buetschliidae have been observed in the rumen. Their occurrence is, however, spasmodic and morphological descriptions will be restricted to genera represented most frequently.

Buetschlia. *Buetschlia* (syn. *Butschlia*) *parva* Schuberg has been observed in a range of hosts (Table 2). The population density of 10^3–10^4 cells/ml typically represented less than 1% of the total ciliate population (Dehority, 1970, 1974), although values of up to 2–3% have been reported in domestic cattle (Clarke, 1964).

This ciliate is described as ovoid (Ogimoto & Imai, 1981), although its shape and size may vary considerably. It is reported to range in length from 30 to 67 μm and in width from 20 to 48 μm (Schuberg, 1888; Clarke, 1964; Dehority, 1970). The macronucleus is elliptical; a single contractile vacuole is situated in the posterior portion of the cell and a concretion vacuole is located in the anterior part. The somatic ciliature is complete.

Other species (*B. neglecta*, *B. lanceolata* and *B. triciliata*) have been described in rumen contents (Becker & Talbott, 1927; Hsiung, 1932); *B. nana* and *B. omnivora* were described in the stomach contents of the camel (Dogiel, 1928).

Parabundleia. More recently, representatives of other genera in the family

Buetschliidae have been found in the rumen. *Parabundleia ruminantium* was detected in approximately 40% of the zebu cattle surveyed by Imai & Ogimoto (1983). This ciliate has an ovoid body with two anterior and one posterior ciliary zones. The ciliate is 37·5–50 (av. 42·5) μm long and 27·5–32·5 (av. 30·5) μm wide with a central elliptical macronucleus. The spherical micronucleus is sited close to the anterior margin of the macronucleus. The cell contains one posterior contractile vacuole and one anterior concretion vacuole.

Polymorphella. *Polymorphella bovis* was also present in the rumen of zebu cattle (Imai, 1984). This has a body that is generally ovoid, although the anterior one-third is tapered, varying in length from 26 to 37·5 (av. 34) μm and in width from 20 to 26 (av. 22) μm. Cilia are present in two zones; the larger zone is at the tapered anterior, whilst a smaller tuft is located close to the cytoproct. The macronucleus is in the centre of the body and two types of vacuole are again present. The contractile vacuole is in the posterior part of the cell and the concretion vacuole in the anterior portion.

Blepharoconus. *Blepharoconus krugerensis* was observed in the rumen contents of a single animal during a study of the rumen ciliates of Brazilian cattle (Dehority, 1986*b*). It had been observed previously only in the intestines of the elephant (Eloff & Van Hoven, 1980). The rumen isolates were more rounded, with body measurements of 30–65 (av. 46) × 21–60 (av. 35) μm, but in all other aspects were similar to the species description of Eloff & Van Hoven (1980). Other members of the Buetschliidae described by Hsiung (1932) and Jirovec (1933) are *Holophryozoon bovis*, *Pingius minutis* and *Blepharoprosthium parvum*. The latter author also observed *Butschliella bovis* in cattle rumen contents.

Blepharocorythidae

Jameson (1925*a*) described a small ciliate, *Charon ventriculi*, in rumen contents from cattle and sheep. Almost simultaneously this protozoon was described as *Blepharocorys bovis* by Dogiel (1926). Recognising the similarity, Dogiel (1934) proposed the name *Blepharocorys ventriculi*; in addition the generic name was changed to *Charonina* (syn. *Charonella*) (Strand, 1928). After a detailed study of the infraciliature and its morphogenesis, Wolska (1967*a*) concluded that the ciliate should be included in the more primitive genus *Charonina* Strand. Two other species included in the genus were *C. nuda* (Hsiung, 1932) which was detected in cattle, and *C. equi* in the horse colon (Hsiung, 1930); *C. equi* was subsequently detected in the bovine rumen (Clarke, 1964).

Charonina spp. are infrequently observed (Table 2), although when present they may be relatively numerous (Clarke, 1964; Dehority, 1979; Imai *et al.*, 1981*b*). The population size is influenced by the ration and has been reported to exceed 50% of the total rumen ciliate population (Syrjala *et al.*, 1976; Dehority & Mattos, 1978).

Charonina ventriculi is one of the smaller holotrichs, with a body size in the range 24–41 (av. 35) × 12–19 (av. 16·5) μm. The body is elongate, although one side is typically more convex. The somatic ciliature is restricted to an anterior zone and two tufts at the posterior extremity. The macronucleus is spherical to elliptical in shape and variable in location. A single, large contractile vacuole is located in the posterior part of the body.

Charonina equi is similar in size (30–48 × 10–14 µm, av. 39·5 × 12 µm) but has a shorter oesophagus and elongated macronucleus; the micronucleus is usually found some distance away from the macronucleus. The origin of the posterior ciliature is also different (Hsiung, 1930). *C. nuda* differs from *C. ventriculi* in that the posterior end of the body is not ciliated and the somatic ciliature is restricted to the anterior surface.

Paraisotrichidae
A small ciliate with many characteristics of the family Paraisotrichidae was observed in rumen contents of cattle from Brazil (Dehority, 1986b). The body was ellipsoidal, measuring 14–23 (av. 19) × 9–14 (av. 12) µm, and covered by long cilia that were arranged in eight spiral rows. An ovoid macronucleus was close to the centre of the body and one contractile vacuole was present in the posterior portion of the cell. Although an anterior concretion vacuole is characteristic of the family none was reported, and in this respect the protozoon resembled *Blepharoprosthium parva* (Hsiung, 1932). Superficially similar *Paraisotricha* spp. have been observed in the giraffe and horse (Hsiung, 1930; Kleynhans & Van Hoven, 1976).

Cultivation

Although, as previously described, some of the entodiniomorphid ciliates have been cultivated for extended periods *in vitro*, attempts to cultivate the holotrich ciliates have been less successful and only *Dasytricha* and *Isotricha* spp. have been grown *in vitro*. Both genera have been maintained as part of a mixed protozoal population in rumen simulation (continuous culture) systems of varying complexity (Abe & Kumeno, 1973; Czerkawski & Breckenridge, 1977; Michalowski, 1978; Nakamura & Kurihara, 1978). Protozoal retention can be improved by the provision of a solid matrix, and holotrich numbers approximating to 10^4 cells/ml have been obtained in such systems (Czerkawski & Breckenridge, 1979; Abe & Kurihara, 1984; see also Chapter 12).

The individual species have also been grown separately *in vitro* but have been maintained only for short periods of approximately 1–2 months. Their requirements in culture are similar to those detailed above in the section on the entodiniomorphid ciliates, and the media used typically contain complex extracts (microbial/plant) and rumen fluid, although Quin *et al.* (1962) have described a chemically defined medium suitable for holotrich cultivation *in vitro*. Supplementation of an anaerobic isotonic buffered salts solution with rumen fluid (Purser & Tompkin, 1965), protozoal extract (Clarke & Hungate, 1966; Kubo *et al.*, 1980), bacteria (Gutierrez & Hungate, 1957; Gutierrez, 1958), antibiotics (Purser & Weiser, 1963), soluble carbohydrates (Sugden & Oxford, 1952; Gutierrez, 1955; Bonhomme-Florentin & Hempel-Zawitkowska, 1977) or plant extract (Genskow *et al.*, 1969) will extend the cultivation period. However, cell division is impaired in cells containing excess storage polysaccharide.

Excessive polymer deposition and bacterial overgrowth are therefore limited in cultures by restricting the amount and frequency of sugar addition and by daily transfer of the cells to fresh medium (Clarke & Hungate, 1966; Kubo et al., 1980). It is not possible, however, to maintain single species in defined axenic culture for several generations, and in consequence metabolic studies on the holotrich ciliates have used cells isolated from conventionally faunated animals or those containing specifically manipulated protozoal populations (see Chapter 15). Cells isolated directly from the rumen have not been subjected to imposed nutritional constraints which may directly or indirectly influence the true metabolic capabilities of the organisms.

The rumen holotrichs *D. ruminantium* and *Isotricha* spp. may be readily recovered from mixed populations by differential filtration procedures using sintered glass filters (Williams & Harfoot, 1976) or, more effectively, defined-aperture textiles (Williams & Yarlett, 1982). The metabolic status of cells isolated by filtration is not affected, as preliminary incubation with glucose (Heald et al., 1952) and subsequent starvation periods (Heald & Oxford, 1953) are unnecessary. Contamination by free bacteria is minimised by thorough washing with sterile anaerobic buffer. Sedimentation procedures have also been used for holotrich isolation. Glucose was routinely added to the buffers by Oxford (1951) and Heald et al. (1952) to encourage polysaccharide deposition in the holotrichs and facilitate a more rapid sedimentation. *Isotricha* and *Dasytricha* spp. have been separated by differential sedimentation in buffer (Heald et al., 1952) or gradients containing rumen liquor (Gutierrez, 1955) or glycerol (Mould & Thomas, 1958). Holotrich separation by sedimentation is frequently incomplete, and cell preparations obtained in this manner are treated with antibiotics to reduce bacterial contamination and starved for 48 hours to deplete the amylopectin reserves. Species separation has also been achieved by adhesion to glass surfaces (Gutierrez, 1955) and sucrose-gradient centrifugation procedures (Mangan & West, 1977; Ogimoto & Imai 1981).

Uptake, digestion and metabolism of dietary components

Biochemical studies on the metabolic activities of the holotrich ciliates have been undertaken only with *Dasytricha ruminantium* and the two species of *Isotricha*. The reader is referred to the article by Williams (1986) for a detailed review of the subject.

Carbohydrate metabolism

Carbohydrate uptake. Although the ciliate protozoa may be responsible for 30–40% of microbial fibre degradation in the rumen (Demeyer, 1981), it is generally accepted that the holotrichs are most involved in the utilisation of non-structural polysaccharides and soluble sugars. The holotrich ciliates are able to utilise the monosaccharides fructose, glucose and galactose and certain soluble oligomers and polysaccharides containing one or more of these sugars. Fructose-containing carbohydrates are utilised most rapidly, whereas mannose and certain hexosamines are toxic. The rate of carbohydrate utilisation and proportions of the metabolites formed are substrate-dependent (Williams & Harfoot, 1976; Prins & Van Hoven, 1977; Van Hoven & Prins, 1977).

The range of carbohydrates, metabolised is genus-dependent. *Isotricha* spp., but not *D. ruminantium*, will ingest starch grains of a suitable size, whereas maltose, lactose and cellobiose are less effective substrates for *Isotricha* spp. than for *D. ruminantium* (Heald & Oxford, 1953; Gutierrez, 1955; Howard, 1959a; Yoshida & Katsuki, 1980). With the exception of the studies of Prins & Van Hoven (1977) on the carbohydrate metabolism of *I. prostoma*, all other work relating to the isotrichs has been based on studies using mixed preparations of *I. prostoma* and *I. intestinalis*; the range of carbohydrates utilised by *I. intestinalis* is not known.

Dasytricha ruminantium is not selective in its uptake of available carbohydrates and does not regulate sugar entry by preferential or sequential utilisation. *D. ruminantium* will take up approximately 175–250 pmol glucose/cell/hour *in vitro*. The mechanism of sugar uptake has not been elucidated but there is evidence to indicate that several sugars may be transported by the same route, as uptake rates are decreased in the presence of an alternative substrate (Williams, 1979a). Galactose is, however, taken up more rapidly by *D. ruminantium* in the presence of a readily assimilated substrate, indicating that an energy-dependent mechanism is involved.

The holotrichs are attracted by chemotaxis to sources of soluble sugars (Orpin & Letcher, 1978) and non-selectively assimilate them as they become available in the immediate environment of the protozoa. The rate at which sugars are taken up is affected by the nature and concentration of the sugar and by the environmental pH and temperature (Williams & Harfoot, 1976). The rate of glucose uptake by *D. ruminantium* is relatively constant throughout the diurnal cycle of the rumen and not obviously affected by amylopectin deposition within the cell. Glucose uptake and metabolism is inhibited by lactic acid (Williams & Morrison, 1982) but stimulated by molybdenum (Zenbayashi *et al.*, 1969).

Carbohydrate degradation. The holotrich ciliates are able to colonise plant tissues and maintain a close proximity to carbohydrate sources (Orpin & Letcher, 1978; Bauchop, 1980; Orpin, 1985). The rumen protozoal population contains a wide range of polysaccharide-degrading enzymes (Williams & Strachan, 1984), and it is apparent that the holotrichs are able to degrade plant structural and storage polysaccharides.

The principal plant reserve polysaccharides are starch and inulin. Plant fructosans are rapidly fermented, although *Dasytricha* is the most active genus (Thomas, 1960; Williams & Harfoot, 1976). Holotrich invertase is non-specific and degrades both sucrose and fructosan. *Isotricha* and *Dasytricha* spp. also have amylolytic enzymes (Mould & Thomas, 1958; Williams *et al.*, 1984), although starch ingestion has not been demonstrated frequently. The amylases of the two genera are both inhibited by short-chain oligomeric degradation products but have different pH optima at pH 5 and 6 (*D. ruminantium*) or 4·8 (*Isotricha*).

The holotrichs also have a limited ability to degrade plant cell wall structural polysaccharides. Cell-free extracts of *D. ruminantium* and *Isotricha* spp. will release reducing sugars from carboxymethylcellulose (CMC) (Williams *et al.*, 1986) and although metabolite formation is not enhanced by CMC it does cause an initial stimulation of gas formation (Williams & Harfoot, 1976; Van Hoven & Prins, 1977). These activities are, however, low and may arise from non-specific carbohydrase activity.

Although the information, thus far, on holotrich cellulase is inconclusive, it is apparent that both genera have hemicellulolytic and pectolytic activities. Ryegrass hemicellulose and oat-spelt xylan are degraded to alcohol-soluble oligomers, and the component monosaccharides by endo-acting β-D-xylanase(s), β-D-xylosidase and α-L-arabinofuranosidase (Williams *et al.*, 1984; Williams & Coleman, 1985). Holotrich cell extracts possess both pectin esterase and polygalacturonase activities, and products typical of both lyase and hydrolase action have been detected (Abou Akkada & Howard, 1961; Coleman *et al.*, 1980). Methanol is a product of pectin hydrolysis by both genera; polygalacturonic acid, oligo-uronides and galacturonic acid, unlike pectin, are not fermented by the holotrichs. Various neutral sugars are covalently linked to pectin and it is their release by glycosidases and their subsequent fermentation that is likely to stimulate gas formation. The depolymerisation products of arabinoxylan and pectin are not metabolised further (this explains the failure of manometric techniques to detect hemicellulolytic action). However, although the products are not utilised, these degradative activities may enable the protozoa to weaken the plant cell wall sufficiently enough to release the more readily fermentable, intracellular, soluble carbohydrates. Pectolytic and certain carbohydrase enzymes have been detected extracellularly (Abou Akkada & Howard, 1961; Williams, 1979*b*). The secretion of extracellular enzymes through the attachment organelle would serve to localise the activities at the point of attachment to the plant cell wall, and the released cell contents could be taken directly into the cell through the organelle, as proposed by Orpin & Letcher (1978).

The holotrich ciliates also contain a wide range of carbohydrase (glycoside hydrolase) enzymes. The presence and certain characteristics of invertase, α-glucosidase (maltase) and β-glucosidase (cellobiase, laminaribiase) in cell-free extracts of both genera were determined with disaccharide substrates (Christie & Porteous, 1957; Carnie & Porteous, 1959; Howard, 1959*b*; Thomas, 1960; Bailey & Howard, 1963). Later surveys using natural and chromogenic substrates have confirmed that both genera have activity against glycosides of pentoses, hexoses, hexosamines and hexuronic acids (Williams, 1979*b*; Yarlett *et al.*, 1981, 1983*a*; Williams *et al.*, 1984). Although there are differences in the range of disaccharides fermented by *Dasytricha* and *Isotricha* spp., the enzyme profiles of both genera are the same. Most of the polysaccharide-degrading enzyme activity is located in membrane-bound subcellular organelles (Yarlett *et al.*, 1981, 1983*a*; Williams *et al.*, 1986); the glucosidase- and acid phosphatase-containing organelle of *D. ruminantium* has a mean equilibrium density of $1\cdot13$ g/ml (Yarlett *et al.*, 1985).

Products of carbohydrate fermentation. The principal metabolites formed during carbohydrate fermentation by *Isotricha* are *Dasytricha* spp. and D- and L-lactic acid, butyric acid, acetic acid, hydrogen, carbon dioxide and storage polysaccharide.

Under certain conditions formic acid and propionic acid are also detected. The range of metabolites formed is the same with all substrates; however, the proportions of the products formed and their synthesis rates are influenced by the carbohydrate available, its initial concentration and the environmental conditions (Williams & Harfoot, 1976; Prins & Van Hoven, 1977; Van Hoven & Prins, 1977). Metabolic interaction with other microbial groups (e.g. methanogens; Hillman, Lloyd & Williams, personal communication) and the presence of oxygen (Prins & Prast, 1973; Yarlett et al., 1983b; Hillman et al., 1985b) can affect the metabolites produced.

The reserve polysaccharide is a branched homoglucan, similar to amylopectin (Forsyth & Hirst, 1953), that is formed when sugars are readily available. A wide range of carbohydrates is converted into the polymer, and its deposition has been used as the criterion for substrate utilisation (Oxford, 1951; Masson & Oxford, 1951). The rate of deposition is affected by both the substrate and its concentration. The maximum rates of deposition by *D. ruminantium* and *I. prostoma* are 130 and 1300 pmol of amylopectin-hexose/cell/hour from glucose but only 7 and 120 pmol/cell/hour from galactose (Prins & Van Hoven, 1977; Van Hoven & Prins, 1977). Polymer deposition is affected by temperature (Eadie & Oxford, 1955). Amylopectin formation by the phosphorylase mechanism (Mould & Thomas, 1958) is an energy-consuming process.

Simultaneous amylopectin synthesis and utilisation can occur in the presence of available sugars. However, there is a threshold concentration for each sugar (e.g. glucose 20 mM) at which amylopectin breakdown is prevented (Prins & Van Hoven, 1977). The protozoa control the amount of amylopectin deposited, and the observed lysis of holotrichs in the presence of excess substrate is not, as first proposed, due to an overaccumulation of polysaccharide (Sugden & Oxford, 1952; Oxford, 1955). The lysis occurs as the cells do not control sugar entry (Williams, 1979a) and its metabolism causes a detrimental intracellular build-up of acidic fermentation products, notably lactic acid (Van Hoven & Prins, 1977).

Approximately equal proportions of acetic acid and butyric acid are produced during the endogenous fermentation of reserve amylopectin. Racemic lactic acid is a minor product of *I. prostoma*, but the amount of L-lactic acid formed by *D. ruminantium* is more variable. Formic and propionic acids are not produced endogenously by either genus, although approximately equimolar amounts of carbon dioxide and hydrogen are released (Williams & Harfoot, 1976; Prins & Van Hoven, 1977; Van Hoven & Prins, 1977).

Intermediary metabolism of carbohydrates. The principal pathways of product formation in the holotrichs are summarised in Fig. 1. The enzymes involved are located in either the cytosol or a microbody-like organelle, the hydrogenosome (Yarlett et al., 1981, 1982, 1983a, 1985). The pathways in *D. ruminantium* and *Isotricha* spp. appear to be the same. Sugars are metabolised to pyruvate by the Embden–Meyerhof–Parnas pathway and the protozoa obtain at least 4 ATP for each glucose moiety converted into acetate; ATP is also generated in the formation of butyrate (Yarlett et al., 1985). Rate-limiting enzymes in glycolysis and the

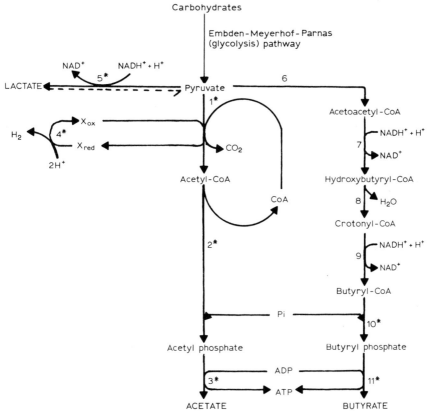

Fig. 1. Proposed pathways leading to metabolite formation in *Dasytricha ruminantium* and *Isotricha* spp. The enzymes indicated are: 1, pyruvate synthase; 2, phosphoacetyltransferase; 3, acetate kinase; 4, hydrogenase; 5, lactate dehydrogenase; 6, acetyl-CoA:acetyl-CoA acetyltransferase; 7, 3-hydroxybutyryl-CoA dehydrogenase; 8, 3-hydroxyacyl-CoA hydrolase; 9, 3-hydroxyacyl-CoA reductase; 10, phosphobutyryl transferase; 11, butyrate kinase. Some, or all, of the activity of enzymes marked * is present in the hydrogenosomal fraction; the other enzymes are located in the cell cytosol.

formation of butyrate by *D. ruminantium* have been identified (Yarlett *et al.*, 1985) and factors regulating the activity of lactate dehydrogenase in *I. prostoma* described (Counotte, 1979; Counotte *et al.*, 1980). However, our understanding of the metabolic pathways and their control in the holotrich ciliates is far from complete.

Nitrogen metabolism

Ingestion of bacteria. Both *Isotricha* and *Dasytricha* spp. will ingest rumen and non-rumen bacteria, although both genera exhibit some selectivity in the range of bacteria ingested (Sugden & Oxford, 1952; Gutierrez & Hungate, 1957; Gutierrez, 1958; Wallis & Coleman, 1967). Ingested bacteria have been observed in digestive vacuoles within the ciliates. Following engulfment, the bacterial cells rapidly lose viability and are extensively degraded. Unchanged bacterial amino acids are incorporated directly into the protozoal protein (Wallis & Coleman, 1967). The holotrich ciliates obtain some of their nitrogen from the ingestion and digestion of bacteria, but they are also able to take up amino acids from their surroundings (Wallis & Coleman, 1967; Harmeyer, 1971a). The amino acids taken up are assimilated directly and not interconverted or degraded, although the holotrichs do excrete nitrogenous material (Heald & Oxford, 1953; Harmeyer, 1971b).

Amino acid synthesis. In addition to utilising preformed amino acids released by the proteolysis of ingested plant material and bacteria, the holotrichs are also able to synthesise amino acids *de novo*. Chesters (1968) was able to demonstrate protein synthesis by polysomes prepared from a rumen protozoal preparation containing over 80% holotrichs. Incorporation of ^{14}C into the cellular protein fraction of both genera has been demonstrated using ^{14}C-labelled sodium carbonate, sodium acetate, glucose or galactose as a precursor (Harmeyer, 1965; Harmeyer & Hekimoglu, 1968; Williams & Harfoot, 1976; Williams, 1979a).

Lipid metabolism
It is probable that the holotrichs are able to satisfy their lipid requirements by the utilisation of ingested plant and bacterial lipids. They are, however, capable of the *de novo* synthesis of fatty acids and phospholipids using monosaccharides or acetate as precursors (Harmeyer & Hekimoglu, 1968; Emmanuel, 1974; Demeyer *et al.*, 1978). Direct incorporation of radioactivity from sugars is low (Williams & Harfoot, 1976; Williams, 1979a) and principally into the glycerol moiety of the lipid. Long-chain fatty acids are incorporated directly into phospholipids and sterol esters (Demeyer *et al.*, 1978; Girard & Hawke, 1978), whilst glycerol enters both phosphoglycerides and plasmalogens following conversion into *sn*-glycerol-3-phosphate (Prins & Van Golde, 1976). The lipid component of the holotrichs is approximately 70% phospholipid; the principal fatty acid in all lipid classes is palmitic acid. The major sterol, cholestanol, is formed by the modification of dietary sterols (Katz & Keeney, 1967; Williams & Dinusson, 1973).

The long-chain fatty acids taken up by the holotrichs are usually incorporated into the free fatty-acid pool and lipid components of the cell, there being little evidence of catabolic activity (Gutierrez *et al.*, 1962; Williams *et al.*, 1963; Demeyer *et al.*, 1978; Girard & Hawke, 1978). However, some short-chain fatty acids (C_2 to C_6), tributyrin and the methyl derivatives of C_{12} and C_{14} medium-chain fatty acids have been observed to stimulate manometrically determined fermentative activity of *I. prostoma* and *I. intestinalis* (Gutierrez et al., 1962; Williams *et al.*, 1963).

The involvement of the holotrich ciliates in biohydrogenation has not been conclusively established. In three separate studies (Chalupa & Kutches, 1968; Abaza *et al.*, 1975; Girard & Hawke, 1978) it was concluded that the holotrichs were either unable, or only had limited ability, to biohydrogenate in the absence of rumen bacteria. However, both *I. prostoma* (Gutierrez *et al.*, 1962) and *I. intestinalis* (Williams *et al.*, 1963) have been shown to hydrogenate oleic acid to stearic acid, and Demeyer *et al.* (1978) observed that *Isotricha* spp. were able to hydrogenate exogenously supplied linoleic acid before it was incorporated into cellular lipids. The ability of holotrich ciliates to desaturate fatty acids is similarly equivocal (Emmanuel, 1974; Abaza et al., 1975).

Metabolic effects of oxygen
The holotrich ciliates are regarded as being obligately anaerobic. Oxygen, however, is present in the rumen ecosystem (Scott et al., 1983; Hillman et al., 1985a), and it is a common observation that the holotrichs survive longer than other rumen ciliates when some oxygen is present. Although *D. ruminantium* has limited viability in the presence of air, it will survive prolonged periods under lower oxygen tensions. Under such conditions *Dasytricha* and *Isotricha* spp. have K_m values for O_2 in the range 0·3–2·5 μM. These values are similar to those of non-rumen aerobic (*Tetrahymena pyriformis*) and aerotolerant (*Tritrichomonas foetus*) protozoa (Lloyd et al., 1982; Yarlett et al., 1982; Hillman et al., 1985b). *Isotricha* and *Dasytricha* spp. may therefore be regarded as aerotolerant or oxyduric, rather than obligate, anaerobes. These holotrichs do not have mitochondria and lack detectable cytochromes. They do, however, possess hydrogenosomes, and these organelles will consume oxygen, although the intermediary carriers and terminal oxidase involved have still to be characterised. Hydrogenosomes are also present in some rumen entodiniomorphid ciliates (Yarlett et al., 1984) and exhibit respiratory activity (Snyers et al., 1982). Other enzyme systems present in the holotrichs that are involved in oxygen scavenging include NADH oxidase, NADH peroxidase and catalase (Prins & Prast, 1973; Yarlett et al., 1981, 1983a). The metabolic reduction of oxygen has both a protective and an energy-generating function.

The holotrichs sequester on the reticulum wall and may have an important function in oxygen consumption in the rumen (Williams, 1986). However, oxygen affects the proportions of the metabolites formed by the holotrichs. Hydrogen formation is reversibly inhibited at physiological oxygen concentrations, and acetate formation is increased in the presence of oxygen with a corresponding fall in butyrate production (Prins & Prast, 1973; Yarlett et al., 1983b; Hillman et al., 1985b). The metabolic contribution of the holotrichs to the rumen ecosystem is undoubtedly influenced by the environmental levels of oxygen.

THE IMPORTANCE OF PROTOZOA TO THE RUMINANT

Becker et al. (1929) concluded that the rumen ciliates were commensals that had no obvious benefit to the host. Although there is now considerably more information available on the protozoa, there is still no consensus on the value of the protozoa to the host ruminant. The role of the ciliates has been discussed in earlier reviews (Abou Akkada, 1965; Hungate, 1966; Coleman, 1979, 1980; Williams, 1986), and their nutritional effects on the host have recently been considered in more detail by Veira (1986). Certain of the data on the effects of the ciliates are summarised in Table 4; the objective of this section is to identify some of the areas in which the protozoa have

Table 4
Survey of Some Reported Effects of the Absence of Rumen Ciliate Protozoa

Animal or metabolic parameter	Reported change in characteristic when ciliate protozoa absent[a]		
	Increase	Decrease	No effect
1. Effect on rumen environment			
Rumen volume	27 65	84	
Retention time	27 65	47 84	
Bacterial population	10 25 26 32 47 49 52 53 62 65 70 71 72		
ATP levels		63 92	
Ammonia concentration		1 9 13 14 20 25 28 32 34 35 39 40 41 45 47 48 49 52 53 55 57 58 62 66 84 87 88 89 90	
Volatile fatty acid concentration	25 29 76 87 88	1 9 12 14 39 40 43 45 47 52 53 57 62 66 73 85 90 95 97	
Acetic acid (molar proportion)	1 14 32 72 95	12 20 29 31 58 76 85 93 94 97	
Propionic acid (molar proportion)	18 20 29 31 34 47 48 53 58 76 84 85 93 94 97	1 12 13 14 22 25 32 35 57 72 87 95	
Butyric acid (molar proportion)	1 12 13 22 35 39 49 70 72 93	14 18 20 28 29 31 34 48 58 62 84 85 94 97	
Formic acid concentration		1	
Lactic acid concentration	11 12 22 47 60 62 79 96	31	
Bicarbonate concentration	72		
Rumen pH	14 97	18 25 29 45 87 88	
2. Effects on blood components			
Blood haemoglobin levels	93	2 80	25
Plasma levels of:			
urea	2 80	67 89 93	
oleic acid	56	49 56	
linoleic acid	49 56		
linolenic acid	33 49 93		
amino acids	2 34 75 80 93	2 80	
glucose		35	
volatile fatty acids		73	25
bicarbonate/CO_2			72
copper	37		

insulin		34	
albumin			73
β-globulin			73
γ-globulin			
3. Effect on ruminal metabolism			
Organic matter digestibility	rumen	50	2 22 43 47 48 49 50 55 67 78 83 84 87 88 89
	intestinal	48 55 84	
ADF breakdown			21 40 43 46 48 67 89
Cellulose breakdown			2 16 18 38 39 42 53	4 51
Starch breakdown			17 39 60 87 88 89
ME supply			71
Methanogenesis			20 35 50 85 94
Biohydrogenation				19
Formation of choline-containing and other specific phospholipids			19 61
Ruminal nitrogen digestibility				23 25 41 58
Proteolytic activity			74 81 82	44
Urea utilisation		79	
Nitrate/nitrite reduction			96
Lysine synthesis			64
Selenomethionine metabolism				30
Efficiency of bacterial protein synthesis		20 21 48 54 59 77 83 84 86 91	
Nitrogen flow to duodenum		44 48 54 59 70 71 77 83 84 87 88 89 90 91	
Zn, Mn, Cu, Fe flow to lower tract				36
4. Effect on host ruminant			
Food conversion efficiency		5 6 7 8 22	66 69
Live weight gain		3 5 6 7 8 22 24	1 9 14 37 66 68 69	5 25 95
Wool growth		6 7 8	
Quantity of carcass fat			87
Susceptibility to, and severity of, bloat			15	69
Incidence of scours			68
Hepatic copper levels			37
Physical condition			68	4

[a] The references are given in Appendix 1.

been shown to influence the nutrition of the host. Various aspects of the role of the protozoa are also considered in Chapters 13–15.

Defaunation has been widely used to evaluate the role of the protozoa in the nutrition and productivity of the ruminant (see Chapter 15). The removal of the protozoa is achieved by dosing the rumen with an antiprotozoal compound. Ideally these agents should have no adverse effects upon the host or the other rumen microorganisms; in reality no such agent has been discovered. The defaunating agents used include acids, copper sulphate and surface-active chemicals (Becker, 1929; Wright & Curtis, 1975; Orpin, 1977c; Burggraaf & Leng, 1980). The rumen can also be defaunated by a washing procedure (Jouany & Senaud, 1979). The most effective means of obtaining ciliate-free animals is by rearing the ruminant in isolation from birth so that contact with, and inoculation from, faunated animals is prevented (Eadie, 1962).

Host digestion of protozoal cell constituents

The importance of the cellular protein, lipid and carbohydrate fractions of the protozoa to the host is dependent upon the quantity and availability of the biomass reaching the lower intestinal tract. Because most diets support a large protozoal biomass in the rumen, it was always assumed that the ciliates were an important source of nutrients to the host. In particular, the protozoal proteins, although of similar biological value to bacterial protein, are readily digested (McNaught *et al.*, 1954; Bergen *et al.*, 1968) and the lipid fraction contains proportionally more unsaturated fatty acids (Katz & Keeney, 1967). However, studies on the dynamics of protozoa in the rumen (Leng, 1982; Leng *et al.*, 1981, 1986) and comparative measurements of bacterial and protozoal nitrogen reaching the lower tract (e.g. Weller & Pilgrim, 1974; Harrison *et al.*, 1979; Michalowski *et al.*, 1986) have confirmed that the ciliates are selectively retained within the rumen. The value and availability of protozoal cell components to the host must therefore be reconsidered.

Protozoa and rumen metabolism

The rumen ciliates are able to transform the principal dietary components consumed by the animal into a variety of metabolites that are utilised by the host ruminant. In addition to these direct activities a protozoal presence modifies the rumen environment with resultant indirect effects on the rate, extent and location of the digestive processes (see Chapter 15).

The presence of protozoa in the rumen has been shown to influence the dry matter content of rumen digesta and its retention time, the rumen volume, the numbers and type of rumen bacteria present, the overall concentration and proportion of volatile fatty acids, the environmental pH and the concentration of ammonia (Table 4). These modulations of the ruminal ecosystem will have an effect on the bacterial activities. The protozoa also contribute directly to the digestive processes, and the digestion of organic matter in the rumen is reduced in their absence. In defaunated animals an increased amount of protein is available for intestinal digestion;

however, the amount of starch and plant fibre digested in the small intestines and hind gut is also greater, with differing implications for animal productivity. The amount of choline available for absorption from the lower tract of defaunated animals is markedly reduced (Neil et al., 1979).

Approximately one-quarter to one-third of fibre degradation in the rumen is protozoal (Demeyer, 1981; Orpin, 1984). The protozoa also stimulate bacterial cellulolysis (Yoder et al., 1966), and their released enzymes may also be of importance in the ruminal degradation of plant constituents (Coleman, 1985b). Lactic acid, which has been shown to depress fibre degradation (Fay & Ovejero, 1986), is actively taken up by the entodiniomorphid ciliates (Newbold et al., 1987), and the activities of xylanolytic enzymes are increased following refaunation (Williams et al., 1987). Bacterial ingestion and proteolysis by the protozoa, however, decreases the efficiency of nitrogen utilisation by the host (Leng & Nolan, 1984; Demeyer & Van Nevel, 1986). Protozoal proteolysis and the selective retention of the ciliates in the rumen reduces the microbial nitrogen yield (net microbial synthesis) in the rumen, and thus the protozoa directly influence the proportion of protein and energy in the absorbed nutrients.

The total concentration and proportions of the acidic fermentation products are different in faunated and defaunated animals (Table 4). Methane formation is sometimes lower in defaunated animals, and this may be due to the shift in metabolite profile and hydrogen availability (Eadie & Gill, 1971; Whitelaw et al., 1984). The protozoa also consume oxygen (Williams, 1986), and oxygen levels increase transiently in the immediate post-feed period in defaunated animals (Hillman et al., 1985a); the higher oxygen concentrations would also diminish methanogenesis (Scott et al., 1983). The volatile fatty acids are utilised by the host, and hence alterations in the proportions formed may be reflected in the metabolic capabilities of the host (e.g. milk fat synthesis). As with the rumen metabolites, the levels of metabolites in the blood are different in faunated and defaunated animals (Table 4).

The changes in the proportions of the principal rumen metabolites are not consistent in the studies reported (Table 4). The nature of the diet and the concomitant changes in the bacterial population will also influence product formation. However, the rumen fermentation is more stable when an active ciliate population is present, as the protozoa exert some control over the rate of acid formation. The rumen pH is different in faunated and conventional animals receiving the same diet (Table 4), and these differences may exert some effect on ruminal digestion rates. However, a more important consequence of the protozoa is their ability to ingest starch and soluble sugars, so preventing the alternative rapid bacterial fermentation to lactic acid (Mackie et al., 1978). The subsequent fermentation of the ingested starch and storage polysaccharide stabilises the rate of VFA formation and allows their synthesis over a longer period. In addition, the rumen ciliates are more effective than the bacteria in taking up lactic acid (Chamberlain et al., 1983; Newbold et al., 1986, 1987) and they have an important role in regulating rumen lactate metabolism and in preventing the onset of lactic acid acidosis (see Chapter 14).

The protozoa also have consequences in the aetiology of other ruminant disorders (see also Chapter 14). Pounden & Hibbs (1950) observed a lower incidence of scours in faunated animals. The absorption and retention of copper is higher in defaunated animals, and an outbreak of chronic copper toxicity has occurred in animals maintained without ciliate protozoa (Ivan et al., 1986). The protozoa are also more active than the bacteria in reducing nitrates and nitrites (Yoshida et al., 1982) and in degrading certain mycotoxins (Kiessling et al., 1984). Conversely, protozoa appear to be important contributors to legume bloat in ruminants (Clarke & Reid, 1974). The protozoal contribution to bloat relates to their gas production and enhancement of foam stability. The cell contents of the ciliates produce stable foams and contribute substantially to ruminal foams (Clarke, 1965c; Jones & Lyttleton, 1972). Additionally, the ciliates ingest plant chloroplasts which contain glycolipids that have antifoaming properties (Mangan, 1959). The severity of the bloat can be reduced by treatments that reduce the size of the protozoal population (Clarke et al., 1969; Clarke & Reid, 1974; Katz et al., 1986). Defaunated animals may however succumb to bloat, and it would therefore appear that, although the protozoa are not necessary for bloat to occur, both the incidence and severity are greater when protozoa are present.

It is apparent that, although the protozoa are not essential for the growth and development of the ruminant, their presence can have pronounced effects. Improvements in animal productivity have been achieved with both faunated and defaunated animals (Table 4). The advantages to be gained by manipulating the protozoal population will depend upon the age of the animal, the type of diet available, and other pertinent economic and management factors (see Chapter 13).

REFERENCES

Abaza, M. A., Abou Akkada, A. R. & El Shazly, K. (1975). Effect of rumen protozoa on dietary lipid in sheep. *J. Agric. Sci.*, **85**, 135–43.

Abe, M. & Kumeno, F. (1973). *In vitro* simulation of rumen fermentation: apparatus and effects of dilution rate and continuous dialysis on fermentation and protozoal population. *J. Anim. Sci.*, **36**, 941–8.

Abe, M. & Kurihara, Y. (1984). Long-term cultivation of certain rumen protozoa in a continuous fermentation system supplemented with sponge materials. *J. Appl. Bacteriol.*, **56**, 201–13.

Abe, M., Iriki, T., Tobe, N. & Shibui, H. (1981). Sequestration of holotrich protozoa in the reticulo-rumen of cattle. *Appl. Environ. Microbiol.*, **41**, 758–65.

Abe, M., Suzuki, Y., Okano, H. & Iriki, T. (1983). Specific difference in fluctuation pattern of holotrich concentration in the rumen of cattle, goat and sheep. *Jap. J. Zootech. Sci.*, **54**, 457–62.

Abou Akkada, A. R. (1965). The metabolism of ciliate protozoa in relation to rumen function. In *Physiology of Digestion in the Ruminant*, ed. B. S. Dougherty, R. S. Allen, W. Burroughs, N. L. Jacobson & A. D. McGilliard. Butterworths, Washington, pp. 335–45.

Abou Akkada, A. R. & Howard, B. H. (1961). The biochemistry of rumen protozoa. 4. Decomposition of pectic substances. *Biochem. J.*, **78**, 512–17.

Abou Akkada, A. R. & Howard, B. H. (1962). The biochemistry of rumen protozoa. 5. The nitrogen metabolism of *Entodinium*. *Biochem. J.*, **83**, 313–20.

Asada, T., Konno, T. & Katsuki, T. (1980). Studies on the contents of the rumen in Shiba goats. 1. Comparison of the contents of the rumen between Shiba goats and Saanen goats under the same feeding conditions. *Bull. Nippon Vet. Zootech. Coll.*, **29**, 69–73.
Awerinzew, S. & Mutafowa, R. (1914). Material zur Kenntnis der Infusorien aus dem Magen der Wiederkauer. *Arch. Protistenk.*, **38**, 109–18.
Bailey, R. W. & Howard, B. H. (1963). The biochemistry of rumen protozoa. 6. The maltases of *Dasytricha ruminantium*, *Epidinium ecaudatum* (Crawley) and *Entodinium caudatum*. *Biochem. J.*, **86**, 446–52.
Bauchop, T. (1979). The rumen ciliate *Epidinium* in primary degradation of plant tissue. *Appl. Environ. Microbiol.*, **37**, 1217–23.
Bauchop, T. (1980). Scanning electron microscopy in the study of microbial digestion of plant fragments in the gut. In *Contemporary Microbial Ecology*, ed. D. C. Ellwood, J. N. Hedger, M. J. Latham, J. M. Lynch & J. H. Slater. Academic Press, New York, pp. 305–26.
Bauchop, T. & Clarke, R. T. J. (1976). Attachment of the ciliate *Epidinium* Crawley to plant fragments in the sheep rumen. *Appl. Environ. Microbiol.*, **32**, 417–22.
Becker, E. R. (1929). Methods for rendering the rumen and reticulum of ruminants free from their normal infusorian fauna. *Proc. Nat. Acad. Sci.*, **15**, 435–40.
Becker, E. R. & Talbott, M. (1927). The protozoan fauna of the rumen and reticulum of American cattle. *Iowa State Coll. J. Sci.*, **1**, 345–71.
Becker, E. R., Schulz, J. A. & Emmerson, M. A. (1929). Experiments on the physiological relationships between the stomach infusoria of ruminants and their hosts, with a bibliography. *Iowa State Coll. J. Sci.*, **4**, 215–51.
Bergen, W. G., Purser, D. B. & Cline, J. H. (1968). Determination of limiting amino acids of rumen isolated microbial proteins fed to rats. *J. Dairy Sci.*, **51**, 1698–1700.
Bhatia, B. L. (1936). *The Fauna of British India including Ceylon and Burma*, ed. R. B. S. Sewell. Taylor and Francis, London, pp. 156–62.
Bonhomme, A., Durand, M. & Quintana, C. (1982*a*). Influence du cobalt et de la vitamine B_{12} sur la croissance et la survie des ciliés du rumen *in vitro*, en fonction de la population bactérienne. *Reprod. Nutr. Develop.*, **22**, 107–22.
Bonhomme, A., Fonty, G. & Senaud, J. (1982*b*). Essai d'obtention et de survie Entodiniomorphes du rumen en culture axeniques. *Ann. Microbiol.* (Inst. Pasteur), **133B**, 335–41.
Bonhomme, A., Fonty, G. & Senaud, J. (1982*c*). Obtention de *Polyplastron multivesiculatum* (cilié Entodiniomorphe du rumen) en condition axenique. *J. Protozool.*, **29**, 231–3.
Bonhomme-Florentin, A. & Hempel-Zawitkowska, J. (1977). Comportment des populations ciliaires du rumen en culture dans le milieux saccharose et saccharose-uree. *Ann. Nutr. Alim.*, **31**, 331–48.
Broad, T. E. & Dawson, R. M. C. (1976). Role of choline in the nutrition of the rumen protozoon *Entodinium caudatum*. *J. Gen. Microbiol.*, **92**, 391–7.
Buckelew, T. P. & Kontir, D. M. (1977). The ultrastructure of two cow rumen ciliates, the holotrich, *Isotricha* spp. and the spirotrich *Epidinium ecaudatum*. *Proc. Pennsylvania Acad. Sci.*, **51**, 95.
Buisson, J. (1923). Sur quelques infusoires nouveaux ou peu connus parasites des mannifères. *Ann. Parasitol.*, **1**, 209–46.
Buisson, J. (1924). Quelques infusoires parasites d'antilopes africaines. *Ann. Parasitol.*, **2**, 155–60.
Burggraaf, W. & Leng, R. A. (1980). Antiprotozoal effects of surfactant detergents in the rumen of sheep. *N.Z J. Agric. Res.*, **23**, 287–91.
Campbell, A. S. (1929). The structure of *Isotricha prostoma*. *Arch. Protistenk.*, **66**, 331–8.
Carnie, J. A. & Porteous, J. W. (1959). The kinetic properties of invertases obtained from different genera of sheep rumen holotrich protozoa. *Biochem. J.*, **73**, 47P–48P.
Chalupa, W. & Kutches, A. J. (1968). Biohydrogenation of linoleic-1-^{14}C acid by rumen protozoa. *J. Anim. Sci.*, **27**, 1502–8.
Chamberlain, D. G., Thomas, P. C. & Anderson, F. J. (1983). Volatile fatty acid proportions

and lactic acid metabolism in the rumen in sheep and cattle receiving silage diets. *J. Agric. Sci.*, **101**, 47–58.

Chesters, J. K. (1968). Cell-free protein synthesis by rumen protozoa. *J. Protozool.*, **15**, 509–12.

Christie, A. O. & Porteous, J. W. (1957). An invertase from the holotrich protozoa of sheep rumen liquor. *Biochem. J.*, **67**, 19P.

Clarke, R. T. J. (1964). Ciliates of the rumen of domestic cattle (*Bos taurus* L). *NZ J. Agric. Res.*, **7**, 248–57.

Clarke, R. T. J. (1965a). Diurnal variation in the numbers of rumen ciliate protozoa in cattle. *NZ J. Agric. Res.*, **8**, 1–9.

Clarke, R. T. J. (1965b). Quantitative studies of digestion in the reticulorumen. 3. Fluctuations in the numbers of rumen protozoa and their possible role in bloat. *Proc. NZ Soc. Anim. Prod.*, **25**, 96–103.

Clarke, R. T. J. (1965c). Role of rumen ciliates in bloat in cattle. *Nature* (London), **205**, 95–6.

Clarke, R. T. J. (1968). The ophryoscolecid ciliates of red deer (*Cervus elaphus* L) in New Zealand. *NZ J. Sci.*, **11**, 686–92.

Clarke, R. T. J. (1977). Protozoa in the rumen ecosystem. In *Microbial Ecology of the Gut*, ed. R. T. J. Clarke & T. Bauchop. Academic Press, London, pp. 251–75.

Clarke, R. T. J. & Hungate, R. E. (1966). Culture of the rumen holotrich ciliate *Dasytricha ruminantium* Schuberg. *Appl. Microbiol.*, **14**, 340–5.

Clarke, R. T. J. & Reid, C. S. W. (1974). Foamy bloat of cattle: a review. *J. Dairy Sci.*, **57**, 753–85.

Clarke, R. T. J., Reid, C. S. W. & Young, P. W. (1969). Bloat in cattle. 32. Attempts to prevent legume bloat in dry and lactating cows by partial or complete elimination of the rumen holotrich protozoa with dimetridazole. *NZ J. Agric. Res.*, **12**, 446–66.

Coleman, G. S. (1960). The cultivation of sheep rumen oligotrich protozoa *in vitro*. *J. Gen. Microbiol.*, **22**, 555–63.

Coleman, G. S. (1962). The preparation and survival of almost bacteria-free suspensions of *Entodinium caudatum*. *J. Gen. Microbiol.*, **28**, 271–81.

Coleman, G. S. (1964a). The metabolism of ^{14}C-glycine and ^{14}C-bicarbonate by washed suspensions of the rumen ciliate *Entodinium caudatum*. *J. Gen. Microbiol.*, **35**, 91–103.

Coleman, G. S. (1964b). The metabolism of *Escherichia coli* and other bacteria by *Entodinium caudatum*. *J. Gen. Microbiol.*, **37**, 209–23.

Coleman, G. S. (1967a). The metabolism of free amino acids by washed suspensions of the rumen ciliate *Entodinium caudatum*. *J. Gen. Microbiol.*, **47**, 433–47.

Coleman, G. S. (1967b). The metabolism of the amino acids of *Escherichia coli* and other bacteria by the rumen ciliate *Entodinium caudatum*. *J. Gen. Microbiol.*, **47**, 449–64.

Coleman, G. S. (1968). The metabolism of bacterial nucleic acid and of free components of nucleic acid by the rumen ciliate *Entodinium caudatum*. *J. Gen. Microbiol.*, **54**, 83–96.

Coleman, G. S. (1969a). The cultivation of the rumen ciliate *Entodinium simplex*. *J. Gen. Microbiol.*, **57**, 81–90.

Coleman, G. S. (1969b). The metabolism of starch, maltose, glucose and some other sugars by the rumen ciliate *Entodinium caudatum*. *J. Gen. Microbiol.*, **57**, 303–32.

Coleman, G. S. (1972). The metabolism of starch, glucose, amino acids, purines, pyrimidines and bacteria by the rumen ciliate *Entodinium simplex*. *J. Gen. Microbiol.*, **71**, 117–31.

Coleman, G. S. (1975a). In *Digestion and Metabolism in the Ruminant*, ed. I. W. Mcdonald & A. C. I. Warner. University of New England Publishing Unit, Armidale, Australia, pp. 149–64.

Coleman, G. S. (1975b). In *Symbiosis*, ed. D. H. Jennings & D. L. Lee. *Symp. Soc. Exp. Biol.*, **29**, 533–58.

Coleman, G. S. (1978). The metabolism of cellulose, glucose and starch by the rumen ciliate protozoon *Eudiplodinium maggii*. *J. Gen. Microbiol.*, **107**, 359–66.

Coleman, G. S. (1979). The role of rumen protozoa in the metabolism of ruminants given tropical feeds. *Trop. Anim. Prod.*, **4**, 199–213.

Coleman, G. S. (1980). Rumen ciliate protozoa. *Adv. Parasitol.*, **18**, 121–73.

Coleman, G. S. (1981). Alternate metabolic pathways in protozoan energy metabolism (Workshop 1, EMOP3). *Parasitol.*, **82**, 5–6.
Coleman, G. S. (1983). Hydrolysis of fraction 1 leaf protein and casein by rumen entodiniomorphid protozoa. *J. Appl. Bacteriol.*, **55**, 111–18.
Coleman, G. S. (1985a). The cellulase content of 15 species of entodiniomorphid protozoa, mixed bacteria and plant debris isolated from the ovine rumen. *J. Agric. Sci.*, **104**, 349–60.
Coleman, G. S. (1985b). Possible causes of the high death rate of ciliate protozoa in the rumen. *J. Agric. Sci.*, **105**, 39–43.
Coleman, G. S. (1986a). The distribution of carboxymethyl cellulase between fractions taken from the rumens of sheep containing no protozoa or one of five different protozoal populations. *J. Agric. Sci.*, **106**, 121–7.
Coleman, G. S. (1986b). The metabolism of rumen ciliate protozoa. *FEMS Microbiol. Rev.*, **39**, 321–44.
Coleman, G. S. (1986c). The amylase activity of 14 species of entodiniomorphid protozoa and the distribution of amylase in rumen digesta fractions of sheep containing no protozoa or one of seven different protozoal populations. *J. Agric. Sci.*, **107**, 709–21.
Coleman, G. S. (1987a). The rate of synthesis of amylopectin from cellulose and of the degradation of intracellular 'starch' by rumen ciliate protozoa. *J. Protozool. Abstracts*, **33**, Abstract No. 101.
Coleman, G. S. (1987b). Rumen entodiniomorphid protozoa. In *In vitro Methods for Parasite Cultivation*, ed. A. E. R. Taylor & J. R. Baker. Academic Press, London, pp. 29–51.
Coleman, G. S. & Hall, F. J. (1971). A study in the light and electron microscope of the extruded peristome and related structures of the rumen ciliate *Entodinium caudatum*. *Tissue and Cell*, **3**, 371–80.
Coleman, G. S. & Hall, F. J. (1972). Fine structural studies on the digestion of bacterial species in the rumen ciliate, *Entodinium caudatum*. *Tissue and Cell*, **4**, 37–48.
Coleman, G. S. & Hall, F. J. (1978). Digestion of *Epidinium ecaudatum caudatum* by the rumen ciliate *Polyplastron multivesiculatum* as shown by studies in the electron microscope. *Soc. Gen. Microbiol. Q.*, **6**, 29–30.
Coleman, G. S. & Hall, F. J. (1980). The digestion of grass particles by rumen cellulolytic protozoa. *Soc. Gen. Microbiol. Q.*, **7**, 87.
Coleman, G. S. & Hall, F. J. (1984). The uptake and utilization of *Entodinium caudatum*, bacteria, free amino acids and glucose by the rumen ciliate *Entodinium bursa*. *J. Appl. Bacteriol.*, **56**, 283–94.
Coleman, G. S. & Laurie, J. I. (1974a). The metabolism of starch, glucose, amino acids, purines, pyrimidines and bacteria by three *Epidinium* spp. isolated from the rumen. *J. Gen. Microbiol.*, **85**, 244–56.
Coleman, G. S. & Laurie, J. I. (1974b). The utilization of *Bacillus megaterium* and the release of a lytic enzyme by three *Epidinium* spp. isolated from the rumen. *J. Gen. Microbiol.*, **85**, 257–64.
Coleman, G. S. & Laurie, J. I. (1977). The metabolism of starch, glucose, amino acids, purines, pyrimidines and bacteria by the rumen ciliate *Polyplastron multivesiculatum*. *J. Gen. Microbiol.*, **95**, 29–37.
Coleman, G. S. & Reynolds, D. J. (1982a). The effect of sterols and haemin on the growth of the rumen ciliate *Ophryoscolex caudatus* and some other Entodiniomorphid protozoa. *J. Appl. Bacteriol.*, **52**, 129–34.
Coleman, G. S. & Reynolds, D. J. (1982b). The uptake of bacteria and amino acids by *Ophryoscolex caudatus*, *Diploplastron affine* and some other rumen Entodiniomorphid protozoa. *J. Appl. Bacteriol.*, **52**, 135–44.
Coleman, G. S. & Sandford, D. C. (1979a). The engulfment and digestion of mixed rumen bacteria and individual bacterial species by single and mixed species of rumen ciliate protozoa grown *in vivo*. *J. Agric. Sci.*, **92**, 729–42.
Coleman, G. S. & Sandford, D. C. (1979b). The uptake and utilization of bacteria, amino acids

and nucleic acid components by the rumen ciliate *Eudiplodinium maggii*. *J. Appl. Bacteriol.*, **47**, 409–19.

Coleman, G. S., Davies, J. I. & Cash, M. A. (1972). The cultivation of the rumen ciliates *Epidinium ecaudatum caudatum* and *Polyplastron multivesiculatum in vitro*. *J. Gen. Microbiol.*, **73**, 509–21.

Coleman, G. S., Laurie, J. I. & Bailey, J. E. (1977). The cultivation of the rumen ciliate *Entodinium bursa* in the presence of *Entodinium caudatum*. *J. Gen. Microbiol.*, **101**, 253–8.

Coleman, G. S., Sandford, D. C. & Beahon, S. (1980). The degradation of polygalacturonic acid by rumen ciliate protozoa. *J. Gen. Microbiol.*, **120**, 295–300.

Constantinescu, E. & Dragos, N. (1984). Unele aspecte de structura electronomicroscopica a infurzorilor ruminali de la ovine. *Bull. Inst. Agron. Cluj-Napoca Ser. Zooteh. Med. Vet.*, **33**, 17–18.

Counotte, G. H. M. (1979). Kinetic parameters of lactate dehydrogenases of some rumen bacterial species, the anaerobic rumen ciliate *Isotricha prostoma* and mixed rumen microorganisms. *Antonie van Leeuwenhoek J. Microbiol. Serol.*, **45**, 614.

Counotte, G. H. M., De Groot, M. & Prins, R. A. (1980). Kinetic parameters of lactate dehydrogenases of some rumen bacterial species, the anaerobic ciliate *Isotricha prostoma* and mixed rumen microorganisms. *Antonie van Leeuwenhoek J. Microbiol. Serol.*, **46**, 363–81.

Crawley, H. (1923). Evolution in the family Ophryoscolecidae. *Proc. Acad. Nat. Sci. Philad.*, **75**, 393–412.

Czerkawski, J. W. & Breckenridge, G. (1977). Design and development of a long-term rumen simulation technique (Rusitec). *Br. J. Nutr.*, **38**, 371–84.

Czerkawski, J. W. & Breckenridge, G. (1979). Experiments with the long-term rumen simulation technique (Rusitec); use of soluble food and an inert matrix. *Br. J. Nutr.*, **42**, 229–45.

Das Gupta, M. (1935). Preliminary observations on the protozoan fauna of the rumen of the Indian goat, *Capra hircus* Linn. *Arch. Protistenk.*, **85**, 153–72.

Dehority, B. A. (1970). Occurrence of the ciliate protozoa *Butschlia parva* Schuberg in the rumen of the ovine. *Appl. Microbiol.*, **19**, 179–81.

Dehority, B. A. (1974). Rumen ciliate fauna of Alaskan moose (*Alces americana*), musk ox (*Ovibos moschatus*) and Dall mountain sheep (*Ovis dalli*). *J. Protozool.*, **21**, 26–32.

Dehority, B. A. (1975). Rumen ciliate protozoa of Alaskan reindeer and caribou (*Rangifer tarandus* L). In *Proc. 1st Int. Reindeer and Caribou Symp.*, Fairbanks, Alaska (Biological papers of the University of Alaska, Special Report No. 1), pp. 228–40.

Dehority, B. A. (1978). Specificity of rumen ciliate protozoa in cattle and sheep. *J. Protozool.*, **25**, 509–13.

Dehority, B. A. (1979). Ciliate protozoa in the rumen of Brazilian water buffalo, *Bubalus bubalis* Linnaeus. *J. Protozool.*, **26**, 536–44.

Dehority, B. A. (1985). Rumen ciliates of Musk-oxen (*Ovibos moschatus* Z.) from the Canadian Arctic. *J. Protozool.*, **32**, 246–50.

Dehority, B. A. (1986a). Microbes in the foregut of arctic ruminants. In *Control of Digestion and Metabolism in Ruminants*, ed. L. P. Milligan, W. L. Grovum & A. Dobson. Reston Books, Prentice-Hall, Englewood Cliffs, New Jersey, pp. 307–25.

Dehority, B. A. (1986b). Rumen ciliate fauna of some Brazilian cattle: occurrence of several ciliates new to the rumen, including the cycloposthid *Parentodinium africanum*. *J. Protozool.*, **33**, 416–21.

Dehority, B. A. (1986c). Protozoa of the digestive tract of herbivorous mammals. *Insert Sci. Applic.*, **7**, 279–96.

Dehority, B. A. & Mattos, W. R. S. (1978). Diurnal changes and effect of ration on concentrations of the rumen ciliate *Charon ventriculi*. *Appl. Environ. Microbiol.*, **36**, 953–8.

Dehority, B. A. & Purser, D. B. (1970). Factors affecting the establishment and numbers of holotrich protozoa in the ovine rumen. *J. Anim. Sci.*, **30**, 445–9.

Dehority, B. A., Damron, W. S. & McLaren, J. B. (1983). Occurrence of the rumen ciliate *Oligoisotricha bubali* in domestic cattle (*Bos taurus*). *Appl. Environ. Microbiol.*, **45**, 1394–7.
Demeyer, D. I. (1981). Rumen microbes and digestion of plant cell walls. *Agric. Environ.*, **6**, 295–337.
Demeyer, D. I. & Van Nevel, C. (1986). Influence of substrate and microbial interaction on efficiency of rumen microbial growth. *Reprod. Nutr. Develop.*, **26**, 161–79.
Demeyer, D. I., Henderson, C. & Prins, R. A. (1978). Relative significance of exogenous and *de novo* synthesized fatty acids in the formation of rumen microbial lipids in vitro. *Appl. Environ. Microbiol.*, **35**, 24–31.
Denholm, A. M. & Ling, J. R. (1984). *In vitro* metabolism of bacterial cell wall by rumen protozoa. *Can J. Anim. Sci.*, **64**(Suppl.), 18–19.
Dogiel, V. A. (1925a). Die Geschlechtsprozesse bei Infusorien (speziell bei den Ophryoscoleciden); neue Tatsachen und theoretische Erwagungen. *Arch. Protistenk.*, **50**, 283–442.
Dogiel, V. A. (1925b). Nouveux infusoires de la familie des Ophryoscolecides parasites d'antilopes africaines *Ann. Parasitol.*, **3**, 116–42.
Dogiel, V. A. (1926). Une nouvelle espece du genre Blepharocorys, *B. bovis* n.sp. habitant l'estomac du boeuf. *Ann. Parasitol.*, **4**, 61–4.
Dogiel, V. A. (1927). Monographie der Familie Ophryoscolecidae. *Arch. Protistenk.*, **59**, 1–288.
Dogiel, V. A. (1928). La faune d'infusoires habitant l'estomac du buffle et du dromadaire. *Ann. Parasitol.*, **6**, 323–38.
Dogiel, V. A. (1934). Angaben uber die Ophryoscolecidae des wildschafes aus kamtschatka, des Elches und des Yaks, nebst deren zoogeographischen verwertung. *Arch. Protistenk.*, **82**, 290–7.
Dogiel, V. A. (1935). Eine notiz uber die infusoires des renntiermagens. *Trans. Arctic Inst. Leningrad*, **24**, 144–8.
Dogiel, V. A. (1947). The phylogeny of the stomach-infusorians of ruminants in the light of palaentological and parasitological data. *Quart J. Microsc. Sci.* (Sect. 3), **88**, 337–43.
Dogiel, V. A. & Fedorowa, T. (1925). Ueber den Bau und die Funktion des inneren Skeletts des Ophryoscoleciden. *Zool. Anz.*, **62**, 97–107.
Eadie, J. M. (1957). The mid-winter rumen micro-fauna of the seaweed-eating sheep of North Ronaldshay. *Proc. Roy. Soc. Edinburgh, B*, **66**, 276–87.
Eadie, J. M. (1962). The development of rumen microbial populations in lambs and calves under various conditions of management. *J. Gen. Microbiol.*, **29**, 563–78.
Eadie, J. M. & Gill, J. C. (1971). The effect of the absence of rumen ciliate protozoa on growing lambs fed on a roughage-concentrate diet. *Br. J. Nutr.*, **26**, 155–67.
Eadie, J. M. & Oxford, A. E. (1955). Factors involved in the production of a novel kind of derangement of storage mechanism in living holotrich ciliate protozoa from sheep rumen. *J. Gen. Microbiol.*, **12**, 298–310.
Eadie, J. M., Manners, D. J. & Stark, J. R. (1963). The molecular structure of a reserve polysaccharide from *Entodinium caudatum*. *Biochem. J.*, **89**, 91P.
Eloff, A. K. & Van Hoven, W. (1980). Intestinal protozoa of the African elephant *Loxodonta africana* (Blumenbach). *S. Afr. J. Zool.*, **15**, 84–90.
Emmanuel, B. (1974). On the origin of rumen protozoan fatty acids. *Biochim. Biophys. Acta*, **337**, 404–13.
Fay, J. P. & Ovejero, F. M. A. (1986). Effect of lactate on the *in vitro* digestion of *Agropyron elongatum* by rumen microorganisms. *Anim. Feed Sci. Technol.*, **16**, 161–7.
Fiorentini, A. (1889). *Intorno ai Protisti Stomaco dei Bovini*. frat. Fusi, Pavia, Italy.
Forsberg, C. W., Lovelock, L. K., Krumholtz, L. & Buchanan-Smith, J. G. (1984). Protease activities of rumen protozoa. *Appl. Environ. Microbiol.*, **47**, 101–10.
Forsyth, G. & Hirst, E. L. (1953). Protozoal polysaccharides. Structure of the polysaccharide produced by the holotrich ciliates present in the sheep's rumen. *J. Chem. Soc.*, 2132–5.

Fujita, J., Imai, S. & Ogimoto, K. (1979). Bacterial flora, protozoal fauna and volatile fatty acids in the rumen of the water buffalo in Taiwan. *Jap. J. Zootech. Sci.*, **50**, 850–4.

Furness, D. N. & Butler, R. D. (1983). The cytology of sheep rumen ciliates. I. Ultrastructure of *Epidinium caudatum* Crawley. *J. Protozool.*, **30**, 676–87.

Furness, D. N. & Butler, R. D. (1985a). The cytology of sheep rumen ciliates. II. Ultrastructure of *Eudiplodinium maggii*. *J. Protozool.*, **32**, 205–14.

Furness, D. N. & Butler, R. D. (1985b). The cytology of sheep rumen ciliates. III. Ultrastructure of the genus *Entodinium* (Stein). *J. Protozool.*, **32**, 699–707.

Gaumont, R. & Grain, J. (1967). L'anaerobiose et les mitochondries chez les protozoaires du tube digestif. *Ann. Univ. Arers Rheims*, **5**, 174–6.

Genskow, R. D., Lynch, D. L. & Hess, E. A. (1969). Evaluation of the contribution of forage material to the *in vitro* culture of *Isotricha*. *Trans. Illinois State Acad. Sci.*, **62**, 80–4.

Gerassimova, Z. P. (1981). The ultrastructure and systematic position of the rumen ciliate *Isotricha intestinalis* Stein. *Tsitologiya*, **23**, 861–6.

Giesecke, D. (1970). Comparative microbiology of the alimentary tract. In *Physiology of Digestion and Metabolism in the Ruminant*, ed. A. T. Phillipson, E. F. Annison, D. G. Armstrong, C. C. Balch, R. S. Comline, R. N. Hardy, P. N. Hobson & R. D. Keynes. Oriel Press, Newcastle-upon-Tyne, pp. 306–18.

Girard, V. & Hawke, J. C. (1978). The role of holotrichs in the metabolism of dietary linoleic acid in the rumen. *Biochim. Biophys. Acta*, **528**, 17–27.

Grain, J. (1966). Etude cytologique de quelques ciliés holotrichs endocommensaux des ruminants et des equides (Parts 1 & 2). *Protistologica*, **2**, 5–141.

Grain, J. & Senaud, J. (1984). New data on the degradation of fresh lucerne fragments by the rumen ciliate *Epidinium ecaudatum*: attachment, ingestion and digestion. *Can. J. Anim. Sci.*, **64** (Suppl.), 26.

Gutierrez, J. (1955). Experiments on the culture and physiology of holotrichs from the bovine rumen. *Biochem. J.*, **60**, 516–22.

Gutierrez, J. (1958). Observations on bacterial feeding by the rumen ciliate *Isotricha prostoma*. *J. Protozool.*, **5**, 122–6.

Gutierrez, J. & Hungate, R. E. (1957). Inter-relationship between certain bacteria and the rumen ciliate *Dasytricha ruminantium*. *Science*, **126**, 511.

Gutierrez, J., Williams, P. P., Davis, R. E. & Warwick, E. J. (1962). Lipid metabolism of rumen ciliates and bacteria. 1. Uptake of fatty acids by *Isotricha prostoma* and *Entodinium simplex*. *Appl. Microbiol.*, **10**, 548–51.

Han, S. S. (1984). Rumen ciliate protozoal fauna of the native cattle in Korea. *Jap. J. Zootech. Sci.*, **55**, 279–86.

Harmeyer, J. (1965). Fixation of carbon dioxide in amino acids by isolated rumen protozoa (*Isotricha prostoma* and *I. intestinalis*). *Zentralbl. Veterinaermed.*, **A12**, 10–17.

Harmeyer, J. (1971a). Amino acid metabolism of isolated rumen protozoa (*Isotricha prostoma* and *Isotricha intestinalis*). 1. Catabolism of amino acids. *Z. Tierphysiol. Tierernähr. Futtermittelkd*, **28**, 65–75.

Harmeyer, J. (1971b). Amino acid metabolism of isolated rumen protozoa (*Isotricha prostoma* and *Isotricha intestinalis*). 2. Excretion of amino acids. *Z. Tierphysiol. Tierernähr. Futtermittelkd*, **28**, 75–85.

Harmeyer, J. & Hekimoglu, H. (1968). Acetatinkorporation durch isolierte pansenprotozoenarten. *Zentralbl. Veterinaermed.*, **A15**, 242–54.

Harrison, D. G., Beever, D. E. & Osbourn, D. F. (1979). The contribution of protozoa to the protein entering the duodenum of sheep. *Br. J. Nutr.*, **41**, 521–7.

Heald, P. J. & Oxford, A. E. (1953). Fermentation of soluble sugars by anaerobic holotrich ciliate protozoa of the genera *Isotricha* and *Dasytricha*. *Biochem. J.*, **53**, 506–12.

Heald, P. J., Oxford, A. E. & Sugden, B. (1952). A convenient method for preparing massive suspensions of virtually bacteria-free ciliate protozoa of the genera *Isotricha* and *Dasytricha* for manometric studies. *Nature* (London), **169**, 1055–6.

Hillman, K., Lloyd, D. & Williams, A. G. (1985a). Use of a portable quadrupole mass spectrometer for the measurement of dissolved gas concentrations in ovine rumen liquor in situ. *Curr. Microbiol.*, **12**, 335–40.

Hillman, K., Lloyd, D., Scott, R. I. & Williams, A. G. (1985b). The effects of oxygen on hydrogen production by rumen holotrich protozoa as determined by membrane-inlet mass spectrometry. In *Microbial Gas Metabolism: Mechanistic, Metabolic and Biotechnological Aspects*, ed. R. K. Poole & C. S. Dow. Academic Press, London, pp. 271–7.

Hino, T. & Kametaka, M. (1977). Gnotobiotic and axenic cultures of a rumen protozoon, *Entodinium caudatum*. *J. Gen. Appl. Microbiol.*, **23**, 37–48.

Hino, T., Kametaka, M. & Kandatsu, M. (1973). The cultivation of rumen oligotrich protozoa. III. White clover factors which stimulate the growth of entodinia. *J. Gen. Appl. Microbiol.*, **19**, 397–413.

Honigberg, B. M., Balamuth, W., Bovee, E.C., Corliss, J. O., Gojdics, M., Hall, R. P., Kudo, R. R., Levine, N. D., Loeblich, A. R., Weiser, J. & Wenrich, D. H. (1964). A revised classification of the phylum protozoa. *J. Protozool.*, **11**, 7–20.

Howard, B. H. (1959a). The biochemistry of rumen protozoa. 1. Carbohydrate fermentation by *Dasytricha* and *Isotricha*. *Biochem. J.*, **71**, 671–5.

Howard, B. H. (1959b). The biochemistry of rumen protozoa. 2. Some carbohydrases in cell-free extracts of *Dasytricha* and *Isotricha*. *Biochem. J.*, **71**, 675–80.

Hsiung, T. S. (1930). Some new ciliates from the large intestine of the horse. *Trans. Amer. Microsc. Soc.*, **49**, 34–41.

Hsiung, T. S. (1931). The protozoan fauna of the rumen of the Chinese sheep. *Bull. Fan Memorial Inst. Biol.*, **2**, 29–41.

Hsiung, T. S. (1932). A general survey of the protozoan fauna of the rumen of the Chinese cattle. *Bull. Fan Memorial Inst. Biol.*, **3**, 87–104.

Hukui, T. (1940). Untersuchungen uber die drei neuen und viersehn bekannten ciliaten von *Bos taurus* var. *domesticus* Gmelin in West Japan. *J. Sci. Hiroshima Univ.*, **B7**, 169–83.

Hukui, T. & Nisida, K. (1954). On *Dasytricha hukuokaensis* n. sp. *Zool. Mag.*, **63**, 367–9.

Hungate, R. E. (1942). The culture of *Eudiplodinium neglectum* with experiments on the digestion of cellulose. *Biol. Bull. Mar. Biol. Lab. Woods Hole*, **83**, 303–19.

Hungate, R. E. (1943). Further experiments on cellulose digestion by protozoa in the rumen of cattle. *Biol. Bull. Mar. Biol. Lab. Woods Hole*, **84**, 157–63.

Hungate, R. E. (1966). *The Rumen and Its Microbes*. Academic Press, New York.

Hungate, R. E. (1978). The rumen protozoa. In *Parasitic Protozoa*, Vol. 2, ed. J. P. Kreier. Academic Press, New York, pp. 655–95.

Imai, S. (1981). Four new rumen ciliates *Entodinium ogimotoi* sp. n., *E. bubalum* sp. n., and *E. tsunodai* sp. n. and *Oligoisotricha bubali* (Dogiel, 1928) n. combi. *Jap. J. Vet. Sci.*, **43**, 201–9.

Imai, S. (1984). New rumen ciliates *Polymorphella bovis* sp. n. and *Entodinium longinucleatum* forms *spinolobum* f.n., from the zebu cattle in Thailand. *Jap. J. Vet. Sci.*, **46**, 391–5.

Imai, S. (1985). Rumen ciliate protozoal faunae of Bali cattle (*Bos javanicus domesticus*) and water buffalo (*Bubalus bubalis*) in Indonesia with the description of a new species *Entodinium javanicum* sp. nov. *Zool. Sci.*, **2**, 591–600.

Imai, S. & Ogimoto, K. (1983). *Parabundleia ruminantium* gen. n. sp. n., *Diplodinium mahidoli* sp. n. with two formae, and *Entodinium parvum* forma *monospinosum* forma n. from the zebu cattle (*Bos indicus* L., 1758) in Thailand. *Jap. J. Vet. Sci.*, **45**, 585–91.

Imai, S. & Ogimoto, K. (1984). Rumen ciliate protozoal fauna and bacterial flora of the zebu cattle (*Bos indicus*) and the water buffalo (*Bubalus bubalis*) in Thailand. *Jap. J. Zootech. Sci.*, **55**, 576–83.

Imai, S. & Tsunoda, K. (1972). Scanning electron microscopic observations on the surface structures of ciliated protozoa in sheep rumen. *Nat. Inst. Anim. Health. Quart.*, **12**, 74–88.

Imai, S., Katsuno, M. & Ogimoto, K. (1978). Distribution of rumen ciliate protozoa in cattle, sheep and goat and experimental transfaunation of them. *Jap. J. Zootech. Sci.*, **49**, 494–505.

Imai, S., Abe, M. & Ogimoto, K. (1981a). Ciliate protozoa from the rumen of the Japanese serow, *Capricornus crispus* (Temminck). *Jap. J. Vet. Sci.*, **43**, 359–67.

Imai, S., Chang, C.-H., Wang, J.-S., Ogimoto, K. & Fujita, J. (1981b). Rumen ciliate protozoal fauna of the water buffalo (*Bubalus bubalis*) in Taiwan. *Bull. Nippon Vet. Zootech. Coll.*, **29**, 77–81.

Imai, S., Ogimoto, K. & Fujita, J. (1981c). Rumen ciliate protozoal fauna of water buffalo, *Bubalus bubalis* (Linnaeus), in Okinawa, Japan. *Bull. Nippon Vet. Zootech. Coll.*, **29**, 82–5.

Imai, S., Shimizu, M., Kinoshita, M., Toguchi, M., Ishi, T. & Fujita, J. (1982). Rumen ciliate protozoal fauna and composition of the cattle in Japan. *Bull. Nippon Vet. Zootech. Coll.*, **31**, 70–4.

Ivan, M., Veira, D. M. & Kelleher, C. A. (1986). The alleviation of chronic copper toxicity in sheep by ciliate protozoa. *Br. J. Nutr.*, **55**, 361–7.

Jameson, A. P. (1925a). A new ciliate *Charon ventriculi* n.g., n. sp., from the stomach of ruminants. *Parasitol.*, **17**, 403–5.

Jameson, A. P. (1925b). A note on the ciliates from the stomach of the mouse deer (*Tragulus meminna* Milne-Edwards) with the description of *Entodinium ovalis* n. sp. *Parasitol.*, **17**, 406–9.

Jensen, E. A. & Hammond, D. M. (1964). A morphological study of Trichomonads and related flagellates from the bovine digestive tract. *J. Protozool.*, **11**, 386–94.

Jirovec, O. (1933). Beobachtungen uber die fauna des rinderpansens. *Z. Parasitenkunde*, **5**, 584–91.

Jones, W. T. & Lyttleton, J. W. (1972). Bloat in cattle. 37. The foaming properties of bovine salivary secretions and protozoal proteins. *NZ J. Agric. Res.*, **15**, 506–11.

Jouany, J. P. & Senaud, J. (1979). Defaunation du rumen de mouton. *Ann. Biol. Anim. Biochim. Biophys.*, **19**, 619–24.

Kandatsu, M. & Takahashi, N. (1956). Studies on reticulo-rumen digestion. Part 4. On the artificial culture of some *Entodinia*. III. *J. Agric. Chem. Soc., Japan*, **30**, 96–102.

Katz, I. & Keeney, M. (1967). The lipids of some rumen holotrich protozoa. *Biochim. Biophys. Acta*, **144**, 102–12.

Katz, I., Nagaraja, T. G. & Fina, L. R. (1986). Ruminal changes in monensin- and lasalocid-fed cattle grazing bloat-provocative alfalfa pasture. *J. Anim. Sci.*, **63**, 1246–57.

Kiessling, K.-H., Pettersson, H., Sandholm, K. & Olsen, M. (1984). Metabolism of aflatoxin, ochratoxin, zearalenone, and three trichothecenes by intact rumen fluid, rumen protozoa and rumen bacteria. *Appl. Environ. Microbiol.*, **47**, 1070–3.

Kleynhans, C. J. & Van Hoven, W. (1976). Rumen protozoa of the giraffe with a description of two new species. *E. Afr. Wildl. J.*, **14**, 203–14.

Kofoid, C. A. & Christenson, J. F. (1933). Ciliates from *Bos gaurus* H. Smith. *Univ. Calif. Publ. Zool.*, **39**, 341–91.

Kofoid, C. A. & MacLennan, R. F. (1930). Ciliates from *Bos indicus* Linn. I. The genus *Entodinium* (Stein). *Univ. Calif. Publ. Zool.*, **33**, 471–544.

Kofoid, C. A. & MacLennan, R. F. (1932). Ciliates from *Bos indicus* Linn. II. A revision of *Diplodinium* Schuberg. *Univ. Calif. Publ. Zool.*, **37**, 53–152.

Kofoid, C. A. & MacLennan, R. F. (1933). Ciliates from *Bos indicus* Linn. 3. *Epidinium* Crawley, *Epiplastron* gen. nov., and *Ophryoscolex* Stein. *Univ. Calif. Publ. Zool.*, **39**, 1–33.

Kubo, T., Kobayashi, K., Ishigami, H., Sekija, A. & Mashiyama, F. (1980). The cultivation of the rumen ciliate *Dasytricha ruminantium*. *Bull. Coll. Agric. Utsunomiya Univ.*, **11**, 9–16.

Latteur, B. (1966). Contribution a la systematique de la Famille des Ophyroscolescidae. Stein. *Ann. Soc. Roy. Zool., Belg.*, **96**, 117–44.

Latteur, B. (1967). *Helicozoster indicus* n. gen., n. sp., ciliate holotriche du caecum de l'elephant des Indes. *Acta Zool. Pathol. Antverpiensia*, **43**, 93–106.

Latteur, B. (1968). Revision systematique de la Famille des *Ophryoscolescidae* Stein, 1898: sous-Famille des *Entodiniinae* Lubinsky, 1957, Genre *Entodinium* Stein, 1858. *Ann. Soc. Roy. Zool., Belg.*, **98**, 1–41.

Latteur, B. (1970). Revision systematique de la Famille des *Ophryoscolescidae* Stein, 1858:

sous-Famille des *Diplodiniinae* Lubinsky, 1957, Genre *Diplodinium* (Schuberg, 1888) sensu novo. *Ann. Soc. Roy. Zool., Belg.*, **100**, 275–312.

Lee, J. J., Hutner, S. H. & Bovee, E. C. (1985). In *An Illustrated Guide to the Protozoa*. Society of Protozoologists, Lawrence, Kansas.

Leng, R. A. (1982). Dynamics of protozoa in the rumen of sheep. *Br. J. Nutr.*, **48**, 399–415.

Leng, R. A. & Nolan, J. V. (1984). Nitrogen metabolism in the rumen. *J. Dairy Sci.*, **67**, 1072–89.

Leng, R. A., Gill, M., Kempton, T. J., Rowe, J. B., Nolan, J. V., Stachiw, S. J. & Preston, T. R. (1981). Kinetics of large ciliate protozoa in the rumen of cattle given sugar cane diets. *Br. J. Nutr.*, **46**, 371–84.

Leng, R. A., Dellow, D. & Waghorn, G. (1986). Dynamics of large ciliate protozoa in the rumen of cattle fed on diets of freshly cut grass. *Br. J. Nutr.*, **56**, 455–62.

Levine, N. D., Corliss, J. O., Cox, F. E. G., Deroux, G., Grain, J., Honigberg, B. M., Leedale, G. F., Loeblich, A. R., Lom, J., Lynn, D., Merinfeld, E. G., Page, F. C., Poljansky, G., Sprague, V., Vavra, J. & Wallace, F. G. (1980). A newly revised classification of the protozoa. *J. Protozool.*, **27**, 37–58.

Lloyd, D., Williams, J., Yarlett, N. & Williams, A. G. (1982). Oxygen affinities of the hydrogenosome-containing protozoa *Tritrichomonas foetus* and *Dasytricha ruminantium* and two aerobic protozoa determined by bacterial bioluminescence. *J. Gen. Microbiol.*, **128**, 1019–22.

Lubinsky, G. (1957). Studies on the evolution of the Ophryoscolecidae (Ciliata: Oligotricha). III. Phylogeny of the ophryoscolecidae based on their comparative morphology. *Can. J. Zool.*, **35**, 141–59.

Mackie, R. I., Gilchrist, F. M. C., Robberts, A. M., Hannah, P. E. & Schwartz, H. M. (1978). Microbiological and chemical changes in the rumen during the stepwise adaptation of sheep to high concentrate diets. *J. Agric. Sci.*, **90**, 241–54.

MacLennan, R. F. (1933). A pulsatory cycle of the contractile vacuole in the Ophryoscolecidae ciliates from the stomach of cattle. *Univ. Calif. Publ. Zool.*, **39**, 205–49.

Mah, R. A. (1962). Experiments on the culture and physiology of *Ophryoscolex purkynei*. PhD thesis, University of California, Davis. [Quoted by Hungate, R. E. (1966). In *The Rumen and Its Microbes*. Academic Press, New York.]

Mah, R. A. (1964). Factors influencing the *in vitro* culture of the rumen ciliate *Ophryoscolex purkynei* (Stein). *J. Protozool.*, **11**, 546–52.

Mangan, J. L. (1959). Bloat in cattle. 11. The foaming properties of proteins, saponins and rumen liquor. *NZ J. Agric. Res.*, **2**, 47–61.

Mangan, J. L. & West, J. (1977). Ruminal digestion of chloroplasts and the protection of protein by glutaraldehyde treatment. *J. Agric. Sci.*, **89**, 3–15.

Masson, H. A. & Ling, J. R. (1986). The *in vitro* metabolism of free and bacterially-bound 2,2'-diaminopimelic acid by rumen microorganisms. *J. Appl. Bacteriol.*, **60**, 341–9.

Masson, F. M. & Oxford, A. E. (1951). The action of the ciliates of the sheep's rumen upon various water-soluble carbohydrates including polysaccharides. *J. Gen. Microbiol.*, **5**, 664–72.

McNaught, M. L., Owen, E. C., Henry, K. M. & Kon, S. K. (1954). The utilization of non-protein nitrogen in the bovine rumen. 8. The nutritive value of the proteins of preparations of dried rumen bacteria, rumen protozoa and bakers yeast for rats. *Biochem. J.*, **56**, 151–6.

Michalowski, T. (1975). Effect of different diets on the diurnal concentrations of ciliate protozoa in the rumen of water buffalo. *J. Agric. Sci.*, **85**, 145–50.

Michalowski, T. (1978). A simple system for continuous culture of rumen ciliates. *Bull. Acad. Pol. Sci., Ser. Sci. Biol.*, **27**, 581–3.

Michalowski, T. & Muszynski, P. (1978). Diurnal variations in number of ciliate protozoa in the rumen of sheep fed once and twice daily. *J. Agric. Sci.*, **90**, 1–5.

Michalowski, T., Harmeyer, J. & Breves, G. (1986). The passage of protozoa from the reticulo-rumen through the omasum of sheep. *Br. J. Nutr.*, **56**, 625–34.

Minor, S., Macleod, N. A. & Preston, T. R. (1977). Effect of sampling by fistula or at slaughter on estimation of rumen protozoa. *Trop. Anim. Prod.*, **2**, 62–7.

Mould, D. L. & Thomas, G. J. (1958). The enzymic degradation of starch by holotrich protozoa from sheep rumen. *Biochem. J.*, **69**, 327–37.

Murphy, M. R., Drone, P. E. & Woodford, S. T. (1985). Factors stimulating migration of holotrich protozoa into the rumen. *Appl. Environ. Microbiol.*, **49**, 1329–31.

Nakamura, F. & Kurihara, Y. (1978). Maintenance of a certain rumen protozoal population in a continuous *in vitro* fermentation system. *Appl. Environ. Microbiol.*, **35**, 500–6.

Neil, A. R., Grime, D. W., Snoswell, A. M., Northrop, A. J., Lindsay, D. B. & Dawson, R. M. C. (1979). The low availability of dietary choline for the nutrition of the sheep. *Biochem. J.*, **180**, 559–65.

Newbold, C. J., Chamberlain, D. G. & Williams, A. G. (1986). The effects of defaunation on the metabolism of lactic acid in the rumen. *J. Sci. Food Agric.*, **37**, 1083–90.

Newbold, C. J., Williams, A. G. & Chamberlain, D. G. (1987). The *in vitro* metabolism of DL-lactic acid by rumen microorganisms. *J. Sci. Food Agric.*, **38**, 9–18.

Noirot-Timothée, C. (1958). L'ultrastructure de la limite ectoplasme-endoplasme et des fibres formant le caryophore chez les ciliés du genre *Isotricha* Stein (Holotriches Trichostomes). *Compt. Rend. Acad. Sci.*, **247**, 692–5.

Noirot-Timothée, C. (1960). Etude d'une familie des cilies: les 'Ophyroscolecidae': structures et ultrastructures. *Ann. Sci. Nat. Zool. Biol.*, Ser. 12, **2**, 527–718.

Noirot-Timothée, C. (1963). Sur les ciliés du rumen de *Giraffa camelopardalis* L. *Compt. Rend. Acad. Sci.*, **256**, 5400–1.

Ogimoto, K. & Imai, S. (1981). *Atlas of Rumen Microbiology*. Japan Scientific Societies Press, Tokyo.

Ogimoto, K., Imai, S., Asada, T. & Fujita, J. (1983). Bacterial flora, protozoal fauna and volatile fatty acids in the rumen of the water buffalo (*Bubalus bubalis*) in tropical Asia. *S. Afr. J. Anim. Sci.*, **13**, 59–61.

Onodera, R. & Henderson, C. (1980). Growth factors of bacterial origin for the culture of the rumen oligotrich protozoon, *Entodinium caudatum*. *J. Appl. Bacteriol.*, **48**, 125–34.

Onodera, R. & Kandatsu, M. (1970). Amino acids and protein metabolism of rumen ciliate protozoa. 4. Metabolism of casein. *Jap. J. Zootech. Sci.*, **41**, 307–13.

Orpin, C. G. (1976). The characterization of the rumen bacterium Eadie's oval, *Magnoovum* gen. nov. *eadii* sp. nov. *Arch. Microbiol.*, **111**, 155–9.

Orpin, C. G. (1977a). The rumen flagellate *Piromonas communis*: its life-history and invasion of plant material in the rumen. *J. Gen. Microbiol.*, **99**, 107–17.

Orpin, C. G. (1977b). The occurrence of chitin in the cell walls of the rumen organisms *Neocallimastix frontalis*, *Piromonas communis* and *Sphaeromonas communis*. *J. Gen. Microbiol.*, **99**, 214–8.

Orpin, C. G. (1977c). Studies on the defaunation of the ovine rumen using dioctyl sodium sulphosuccinate. *J. Appl. Bacteriol.*, **43**, 309–18.

Orpin, C. G. (1984). The role of ciliate protozoa and fungi in the rumen digestion of plant cell walls. *Anim. Feed Sci. Technol.*, **10**, 121–43.

Orpin, C. G. (1985). Association of rumen ciliate populations with plant particles *in vitro*. *Microbial Ecol.*, **11**, 59–69.

Orpin, C. G. & Hall, F. J. (1977). Attachment of the rumen holotrich protozoon *Isotricha intestinalis* to grass particles. *Proc. Soc. Gen. Microbiol.*, **4**, 82–3.

Orpin, C. G. & Hall, F. J. (1983). Surface structures of the rumen holotrich protozoon *Isotricha intestinalis* with particular reference to the attachment zone. *Curr. Microbiol.*, **8**, 321–5.

Orpin, C. G. & Letcher, A. J. (1978). Some factors controlling the attachment of the rumen holotrich protozoa *Isotricha intestinalis* and *I. prostoma* to plant particles *in vitro*. *J. Gen. Microbiol.*, **106**, 33–40.

Orpin, C. G. & Mathiesen, S. D. (1986). *Microcetus lappus* gen. nov., sp. nov.: new species of ciliated protozoon from the bovine rumen. *Appl. Environ. Microbiol.*, **52**, 527–30.

Owen, R. W. & Coleman, G. S. (1977). The uptake and utilization of bacteria, amino acids and carbohydrate by the rumen ciliate *Entodinium longinucleatum* in relation to the sources of amino acids for protein synthesis. *J. Appl. Bacteriol.*, **43**, 67–74.

Oxford, A. E. (1951). The conversion of certain soluble sugars to a glucosan by holotrich ciliates in the rumen of sheep. *J. Gen. Microbiol.*, **5**, 83–90.

Oxford, A. E. (1955). The rumen ciliate protozoa: their chemical composition, metabolism, requirements for maintenance and culture and physiological significance for the host. *Exp. Parasitol.*, **4**, 569–605.

Pearson, H. A. (1967). Rumen microorganisms in buffalo from southern Utah. *Appl. Microbiol.*, **15**, 1450–1.

Potter, E. L. & Dehority, B. A. (1973). Effects of changes in feed level, starvation and level of feed after starvation upon the concentration of rumen protozoa in the ovine. *Appl. Microbiol.*, **26**, 692–8.

Pounden, W. D. & Hibbs, J. W. (1950). The development of calves raised without protozoa and certain other characteristic rumen microorganisms. *J. Dairy Sci.*, **33**, 639–44.

Prins, R. A. & Prast, E. R. (1973). Oxidation of NADH in a coupled oxidase-peroxidase reaction and its significance for the fermentation in rumen protozoa of the genus *Isotricha. J. Protozool.*, **20**, 471–7.

Prins, R. A. & Van Golde, L. M. G. (1976). Entrance of glycerol into plasmalogens of some strictly anaerobic bacteria and protozoa. *FEBS Lett.*, **63**, 107–11.

Prins, R. A. & Van Hoven, W. (1977). Carbohydrate fermentation by the rumen ciliate *Isotricha prostoma. Protistologica*, **13**, 549–56.

Prins, R. A., van Rheenen, D. L. & van't Klooster, A. T. (1983). Characterization of microbial proteolytic enzymes in the rumen. *Antonie van Leeuwenhoek J. Microbiol. Serol.*, **49**, 585–95.

Purser, D. B. & Tompkin, R. B. (1965). The influence of rumen fluid source upon viable bacterial counts and the cultivation of oligotrich and holotrich rumen protozoa. *Life Sci.*, **4**, 1493–1501.

Purser, D. B. & Weiser, H. H. (1963). Influence of time of addition of antibiotic on the *in vitro* life of rumen holotrich protozoa. *Nature* (London), **200**, 290.

Quin, L. Y., Burroughs, W. & Christiansen, W. C. (1962). Continuous culture of ruminal microorganisms in chemically defined medium. 2. Culture medium studies. *Appl. Microbiol.*, **10**, 583–92.

Railliet, A. (1890). Protozoaires. In *Nouveau Dictionaire de Medicine, de Chirurgie et d'Hygiene Veterinaires*, Vol. 18, ed. H. Bouley, A. Sansen, L. Trasbot & E. Nocard. Paris, pp. 288–321.

Schuberg, A. (1888). Die protozoen des wiederkauermagens. *Zool. Jahrb. Abt. Syst. Oekol. Geograph. Tiere*, **3**, 365–418.

Scott, R. I., Yarlett, N., Hillman, K., Williams, T. N., Williams, A. G. & Lloyd, D. (1983). The presence of oxygen in rumen liquor and its effects on methanogenesis. *J. Appl. Bacteriol.*, **55**, 143–9.

Senaud, J., Jouany, J.-P., Grain, J. & de Puytorac, P. (1973). Dynamique d'une population de *Polyplastron multivesiculatum* (Cilie, Oligotriche) en equilibre dans le rumen de mouton. *Comp. Rend. Acad. Sci., Paris*, **277**, 197–200.

Senaud, J., Bohatier, J. & Grain, J. (1986). Comparisons du comportement alimentaire de 2 protozoaires ciliés du rumen, *Epidinium caudatum, Polyplastron multivesiculatum*; mécanismes d'ingestion et de digestion. *Reprod. Nutr. Develop.*, **26**(1B), 287–9.

Snyers, L., Hellings, P., Bovy-Kesler, C. & Thines-Sempoux, D. (1982). Occurrence of hydrogenosomes in the rumen ciliates *Ophryoscolecidae. FEBS Lett.*, **137**, 35–9.

Spisni, D., Frateschi, T. L., Mariani, A. P., Martelli, F., Preziuso, F. & Sighiere, C. (1980). Miglioramento dei methodi di coltura *in vitro* dei protozoi del rumine e determinazione del loro contenuto di lisina. Nota 1. Prove *in vitro* sui protozoi del rumine con aggiunta ai normali terreni di coutura di alcuni substrati organici. *Ann. Fac. Med. Vet. Pisa Univ. Studi Pisa*, **33**, 49–60.

Stein, F. (1858). Ueber mehere neue im Pansen der Wiederkauer lebende Infusionsthiere. *Abh. Kais. Bohm. Ges. Wiss.*, **10**, 69–70.

Stern, M. D., Hoover, W. H. & Leonard, J. B. (1977). Ultrastructure of rumen holotrichs by electron microscopy. *J. Dairy Sci.*, **60**, 911–18.

Strand, E. (1928). Miscelanea nomenclatorica et paleontologica. I und II. *Arch. Naturgesch.*, **92**, 30–75.

Sugden, B. & Oxford, A. E. (1952). Some cultural studies with holotrich ciliate protozoa of the sheep's rumen. *J. Gen. Microbiol.*, **7**, 145–53.

Syrjala, L., Saloniemi, H. & Laalhati, L. (1976). Composition and volume of the rumen microbiota of sheep fed grass silage with different sucrose, starch and cellulose supplements. *J. Sci. Agric. Soc. Finland*, **48**, 138–53.

Ten Kate, C. G. B. (1928). Das fibrillensystem der Isotrichen. *Arch. Protistenk.*, **62**, 328–354.

Thines-Sempoux, D., Delfosse-Debusscher, J. & Latteur, B. (1980). Mechanism of cellulose digestion by the rumen ciliates. *Arch. Int. Physiol. Biochim.*, **88**, 102B–104B.

Thomas, G. J. (1960). Metabolism of the soluble carbohydrates of the grasses in the rumen of sheep. *J. Agric. Sci.*, **54**, 360–72.

Thurston, J. P. & Grain, J. (1971). Holotrich ciliates from the stomach of *Hippopotamus amphibius* with descriptions of two new genera and four new species. *J. Protozool.*, **18**, 133–41.

Valdez, R. E., Alvarez, F. J., Ferreiro, H. M., Guerra, F., Lopez, J., Priego, A., Blackburn, T. H., Leng, R. A. & Preston, T. R. (1977). Rumen function in cattle given sugar cane. *Trop. Anim. Prod.*, **2**, 260–72.

Van Hoven, W. (1983). Rumen ciliates with descriptions of two new species from the three African reedbuck species. *J. Protozool.*, **30**, 688–91.

Van Hoven, W. & Prins, R. A. (1977). Carbohydrate fermentation by the rumen ciliate *Dasytricha ruminantium*. *Protistologica*, **13**, 599–606.

Van Hoven, W., Hamilton-Attwell, V. L. & Grobler, J. H. (1979). Rumen ciliate protozoa of the sable antelope *Hippotragus niger*. *S. Afr. J. Zool.*, **14**, 37–42.

Veira, D. M. (1986). The role of ciliate protozoa in nutrition of the ruminant. *J. Anim. Sci.*, **63**, 1547–60.

Vigues, B. & Groliere, C.-A. (1985). Evidence for Ca^{2+}-binding protein associated to non-actin microfilamentous systems in two ciliated protozoans. *Exp. Cell Res.*, **159**, 366–76.

Vigues, B., Metenier, G. & Groliere, C.-A. (1984). Biochemical and immunological characterization of the microfibrillar ecto-endoplasmic boundary in the ciliate *Isotricha prostoma*. *Biol. Cell*, **51**, 67–78.

Vigues, B., Metenier, G., Groliere, C.-A., Grain, J. & Senaud, J. (1985). Biochemical study of proteins of cortical cytoskeleton in the ciliate *Isotricha prostoma*. *J. Protozool.*, **32**, 38–44.

Wakita, M. & Hoshino, S. (1980). Physiochemical properties of a reserve polysaccharide from sheep rumen ciliate genus *Entodinium*. *Comp. Biochem. Physiol.*, **65B**, 571–4.

Wallis, O. C. & Coleman, G. S. (1967). Incorporation of ^{14}C-labelled components of *Escherichia coli* and of amino acids by *Isotricha intestinalis* and *Isotricha prostoma* from the sheep rumen. *J. Gen. Microbiol.*, **49**, 315–23.

Warner, A. C. I. (1962). Some factors influencing the rumen microbial population. *J. Gen. Microbiol.*, **28**, 129–46.

Warner, A. C. I. (1966). Diurnal changes in the concentrations of microorganisms in the rumens of sheep fed limited diets once daily. *J. Gen. Microbiol.*, **45**, 213–35.

Weller, R. A. & Pilgrim, A. F. (1974). Passage of protozoa and volatile fatty acids from the rumen of a sheep and from a continuous *in vitro* fermentation system. *Br. J. Nutr.*, **32**, 341–51.

Whitelaw, F. G., Eadie, J. M., Bruce, L. A. & Shand, W. J. (1984). Methane formation in faunated and ciliate-free cattle and its relationship with rumen volatile fatty acid proportions. *Br. J. Nutr.*, **52**, 261–75.

Williams, A. G. (1979a). The selectivity of carbohydrate assimilation by the anaerobic rumen ciliate *Dasytricha ruminantium*. *J. Appl. Bacteriol.*, **47**, 511–20.

Williams, A. G. (1979b). Exocellular carbohydrase formation by rumen holotrich ciliates. *J. Protozool.*, **26**, 665-72.
Williams, A. G. (1986). Rumen holotrich ciliate protozoa. *Microbiol. Rev.*, **50**, 25-49.
Williams, A. G. & Coleman, G. S. (1985). Hemicellulose-degrading enzymes in rumen ciliate protozoa. *Curr. Microbiol.*, **12**, 85-90.
Williams, P. P. & Dinusson, W. E. (1973). Amino acid and fatty acid composition of bovine ruminal bacteria and protozoa. *J. Anim. Sci.*, **36**, 151-5.
Williams, A. G. & Harfoot, C. G. (1976). Factors affecting the uptake and metabolism of soluble carbohydrates by the rumen ciliate *Dasytricha ruminantium* isolated from ovine rumen contents by filtration. *J. Gen. Microbiol.*, **96**, 125-36.
Williams, A. G. & Morrison, I. M. (1982). Studies on the production of saccharinic acids by the alkaline treatment of young grasses and their effectiveness as substrates for mixed rumen microorganisms *in vitro*. *J. Sci. Food Agric.*, **33**, 21-9.
Williams, A. G. & Strachan, N. H. (1984). The distribution of polysaccharide-degrading enzymes in the bovine rumen digesta ecosystem. *Curr. Microbiol.*, **10**, 215-20.
Williams, A. G. & Yarlett, N. (1982). An improved technique for the isolation of holotrich protozoa from rumen contents by differential filtration with defined aperture textiles. *J. Appl. Bacteriol.*, **52**, 267-70.
Williams, P. P., Gutierrez, J. & Davis, R. E. (1963). Lipid metabolism of rumen ciliates and bacteria. 2. Uptake of fatty acids and lipid analysis of *Isotricha intestinalis* and rumen bacteria with further information on *Entodinium simplex*. *Appl. Microbiol.*, **11**, 260-4.
Williams, A. G., Withers, S. E. & Coleman, G. S. (1984). Glycoside hydrolases of rumen bacteria and protozoa. *Curr. Microbiol.*, **10**, 287-94.
Williams, A. G., Ellis, A. B. & Coleman, G. S. (1986). Subcellular distribution of polysaccharide depolymerase and glycoside hydrolase enzymes in rumen ciliate protozoa. *Curr. Microbiol.*, **13**, 139-47.
Williams, A. G., Strachan, N. H. & Ellis, A. B. (1987). Factors affecting the xylanolytic enzyme activity of microbial populations recovered from the liquid and solid phases of the rumen ecosystem. *Proceedings IV Int. Symp. Microbial. Ecol.*, (in press).
Wolska, M. (1967a). Study on the family *Blepharocorythidae* Hsiung. 2. *Charonina ventriculi* (Jameson). *Acta Protozool.*, **4**, 279-83.
Wolska, M. (1967b). Study on the family *Blepharocorythidae* Hsiung. 3. *Raabena bella* gen. n. sp. n. from the intestine of the Indian elephant. *Acta Protozool.*, **4**, 285-90.
Wolska, M. (1968). Study on the family *Blepharocorythidae* Hsiung. 4. *Pararaabena dentata* gen. n. sp. n. from the intestine of Indian elephant. *Acta Protozool.*, **5**, 219-24.
Wright, D. E. & Curtis, M. W. (1976). Bloat in cattle. 42. The action of surface active chemicals on ciliated protozoa. *NZ J. Agric. Res.*, **19**, 19-23.
Yarlett, N., Hann, A. C., Lloyd, D. & Williams, A. G. (1981). Hydrogenosomes in the rumen protozoon *Dasytricha ruminantium* Schuberg. *Biochem. J.*, **200**, 365-72.
Yarlett, N., Lloyd, D. & Williams, A. G. (1982). Respiration of the rumen ciliate *Dasytricha ruminantium* Schuberg. *Biochem. J.*, **206**, 259-66.
Yarlett, N., Hann, A. C., Lloyd, D. & Williams, A. G. (1983a). Hydrogenosomes in a mixed isolate of *Isotricha prostoma* and *Isotricha intestinalis* from ovine rumen contents. *Comp. Biochem. Physiol.*, **74B**, 357-64.
Yarlett, N., Scott, R. I., Williams, A. G. & Lloyd, D. (1983b). A note on the effects of oxygen on hydrogen production by the rumen protozoon *Dasytricha ruminantium* Schuberg. *J. Appl. Bacteriol.*, **55**, 359-61.
Yarlett, N., Coleman, G. S., Williams, A. G. & Lloyd, D. (1984). Hydrogenosomes in known species of rumen entodiniomorphid protozoa. *FEMS Microbiol. Lett.*, **21**, 15-19.
Yarlett, N., Lloyd, D. & Williams, A. G. (1985). Butyrate formation from glucose by the rumen protozoon *Dasytricha ruminantium*. *Biochem. J.*, **228**, 187-92.
Yoder, R. D., Trenkle, A. & Burroughs, W. (1966). Influence of rumen protozoa and bacteria upon cellulose digestion *in vitro*. *J. Anim. Sci.*, **25**, 609-12.

Yoshida, M. & Katsuki, T. (1980). Ingestion and digestion of carbohydrates by rumen ciliate protozoa in vitro. *Bull. Nippon Vet. Zootech. Coll.*, **29**, 63–7.

Yoshida, J., Nakamura, Y. & Nakamura, R. (1982). Effect of protozoal fraction and lactate on nitrate metabolism of microorganisms in sheep rumen. *Japan. J. Zootech. Sci.*, **53**, 677–85.

Zenbayashi, M., Kawashima, R. & Uesaka, S. (1969). Studies on importance of trace elements in farm animal feeding. 41. Effect of molybdenum on carbohydrate fermentation by rumen protozoa. *Bull. Res. Inst. Food Sci. Kyoto Univ.*, **32**, 32–8.

APPENDIX 1: REFERENCES FOR TABLE 4

1. Abou Akkada, A. R. & El Shazly, K. (1964). Effect of absence of ciliate protozoa from the rumen on microbial activity and growth of lambs. *Appl. Microbiol.*, **12**, 384–90.
2. Abou Akkada, A. R. & El Shazly, K. (1965). Effect of presence or absence of rumen ciliate protozoa on some blood components, nitrogen retention and digestibility of food constituents in lambs. *J. Agric. Sci.*, **64**, 251–5.
3. Becker, E. R. & Everett, R. C. (1930). Comparative growths of normal and infusoria-free lambs. *Am. J. Hyg.*, **11**, 362–70.
4. Becker, E. R., Schulz, J. A. & Emmerson, M. A. (1929). Experiments on the physiological relationships between the stomach infusoria of ruminants and their hosts with a bibliography. *Iowa State Coll. J. Sci.*, **4**, 215–51.
5. Bird, S. H. & Leng, R. A. (1978). The effects of defaunation of the rumen on the growth of cattle on low-protein high-energy diets. *Br. J. Nutr.*, **40**, 163–7.
6. Bird, S. H. & Leng, R. A. (1983). The influence of the absence of rumen protozoa on ruminant production. In *Recent Advances in Animal Nutrition in Australia*, eds. D. J. Farrell & P. Vohra. University of New England Publishing Unit, Armidale, Australia, pp. 110–18.
7. Bird, S. H. & Leng, R. A. (1984). Further studies on the effects of the presence or absence of protozoa in the rumen on live-weight gain and wool growth of sheep. *Br. J. Nutr.*, **52**, 607–11.
8. Bird, S. H., Hill, M. K. & Leng, R. A. (1979). The effects of defaunation of the rumen on the growth of lambs on low-protein high-energy diets. *Br. J. Nutr.*, **42**, 81–7.
9. Borhami, B. E. A., El Shazly, K., Abou Akkada, A. R. & Ahmed, A. I. (1967). Effect of early establishment of ciliate protozoa in the rumen on microbial activity and growth of early weaned buffalo calves. *J. Dairy Sci.*, **50**, 1654–60.
10. Bryant, M. P. & Small, N. (1960). Observations on the ruminal microorganisms of isolated and inoculated calves. *J. Dairy Sci.*, **43**, 654–67.
11. Chamberlain, D. G., Thomas, P. C. & Anderson, F. J. (1981). Lactic acid metabolism in the rumen of animals given silage diets. In *Silage Production and Utilisation* (Proc. 6th Silage Conf., Queen's University, Belfast), pp. 25–6.
12. Chamberlain, D. G., Thomas, P. C. & Anderson, F. J. (1983). Volatile fatty acid proportions and lactic acid metabolism in the rumen in sheep and cattle receiving silage diets. *J. Agric. Sci.*, **101**, 47–58.
13. Chamberlain, D. G., Thomas, P. C., Wilson, W., Newbold, C. J. & MacDonald, J. C. (1985). The effect of carbohydrate supplements on ruminal concentrations of ammonia in animals given diets of grass silage. *J. Agric. Sci.*, **104**, 331–40.
14. Christiansen, W. C., Kawashima, R. & Burroughs, W. (1965). Influence of protozoa upon rumen acid production and liveweight gains in lambs. *J. Anim. Sci.*, **24**, 730–4.
15. Clarke, R. T. J., Reid, C. S. W. & Young, P. W. (1969). Bloat in cattle. 32. Attempts to prevent legume bloat in dry and lactating cows by partial or complete elimination of the rumen holotrich protozoa with dimetridazole. *NZ J. Agric. Res.*, **12**, 446–66.
16. Coleman, G. S. (1986). The distribution of carboxymethylcellulase between fractions

taken from the rumen of sheep containing no protozoa or one of five different protozoal populations. *J. Agric. Sci.*, **106**, 121–7.
17. Coleman, G. S. (1986). The amylase activity of 14 species of entodiniomorphid protozoa and the distribution of amylase in rumen digesta fractions of sheep containing no protozoa or one of seven different protozoal populations. *J. Agric. Sci.*, **107**, 709–21.
18. Conrad, H. R., Hibbs, J. W. & Frank, N. (1958). High roughage systems for raising calves based on early development of rumen function. 9. Effects of rumen inoculations and chlortetracycline on rumen function of calves fed high roughage pellets. *J. Dairy Sci.*, **41**, 1248–61.
19. Dawson, R. M. C. & Kemp, P. (1969). The effect of defaunation on the phospholipids and on the hydrogenation of unsaturated fatty acids in the rumen. *Biochem. J.*, **115**, 351–2.
20. Demeyer, D. I. & Van Nevel, C. J. (1979). Effect of defaunation on the metabolism of rumen microorganisms. *Br. J. Nutr.*, **42**, 515–24.
21. Demeyer, D. I. & Van Nevel, C. J. (1986). Influence of substrate and microbial interaction on efficiency of rumen microbial growth. *Reprod. Nutr. Develop.*, **26**, 161–79.
22. Demeyer, D. I., Van Nevel, C. J. & Van de Voorde, G. (1982). The effect of defaunation on the growth of lambs fed three urea-containing diets. *Arch. Tierernährung*, **32**, 595–604.
23. Demeyer, D. I., Todorov, N., Van Nevel, C. J. & Vets, J. (1982). α,ε-Diaminopimelic acid (DAPA) in a sheep rumen infused with a synthetic diet of sugars and urea: evidence for degradation of bacteria. *Z. Tierphysiol. Tierernähr. Futtermittelkd*, **48**, 21–32.
24. Demeyer, D. I., Van Nevel, C. J., Teller, E. & Godeau, J. M. (1986). Manipulation of rumen digestion in relation to the level of production in ruminants. *Arch. Anim. Nutr.* (Berlin), **36**, 132–43.
25. Eadie, J. M. & Gill, J. C. (1971). The effect of the absence of rumen ciliate protozoa on growing lambs fed on a roughage-concentrate diet. *Br. J. Nutr.*, **26**, 155–67.
26. Eadie, J. M. & Hobson, P. N. (1962). Effect of the presence or absence of rumen ciliate protozoa on the total rumen bacterial count in lambs. *Nature*, **193**, 503–4.
27. Faichney, G. J. & Griffiths, D. A. (1978). Behaviour of solute and particle markers in the stomach of sheep given a concentrate diet. *Br. J. Nutr.*, **40**, 71–82.
28. Fonty, G., Jouany, J.-P., Thivend, P., Gouet, P. & Senaud, J. (1983). A descriptive study of digestion in meroxenic lambs according to the nature and complexity of the microflora. *Reprod. Nutr. Develop.*, **23**, 857–73.
29. Grummer, R. R., Staples, C. R. & Davis, L. (1983). Effect of defaunation on ruminal volatile fatty acids and pH of steers fed a diet high in dried whole whey. *J. Dairy Sci.*, **66**, 1738–41.
30. Hidiroglou, M. & Jenkins, K. J. (1974). Influence de la defaunation sur l'utilisation de la selenomethionine chez la mouton. *Ann. Biol. Anim. Biochem. Biophys.*, **14**, 157–65.
31. Hinkson, R. S., Stern, M. D. & Gray, H. G. (1976). Effect of defaunation and faunation on acid concentrations in the rumen of the mature bovine. *Sci. Biol. J.*, **2**, 116–23.
32. Itabashi, H. & Katada, A. (1976). Nutritional significance of rumen ciliate protozoa in cattle. 1. Effect of protozoa on microbial population and rumen metabolism. *Bull. Tohoku Nat. Agric. Exp. Stn* (*Morioka*), **52**, 161–8.
33. Itabashi, H. & Katada, A. (1976). Nutritional significance of rumen ciliate protozoa in cattle. 2. Effects of protozoa on amino acid concentrations in some rumen fractions and blood plasma. *Bull. Tohoku Nat. Agric. Exp. Stn* (*Morioka*), **52**, 169–76.
34. Itabashi, H., Kobayashi, T., Morii, R. & Okamoto, S. (1982). Effects of ciliate protozoa on the concentration of ruminal and duodenal volatile fatty acids and plasma glucose and insulin after feeding. *Bull. Nat. Inst. Anim. Ind.* (Japan), **39**, 21–32.
35. Itabashi, H., Kobayashi, T. & Matsumoto, M. (1984). The effects of rumen ciliate protozoa on energy metabolism and some constituents in rumen fluid and blood plasma of goats. *Jap. J. Zootech. Sci.*, **55**, 248–56.
36. Ivan, M. & Veira, D. M. (1982). Duodenal flow and soluble proportions of zinc, manganese, copper and iron in the rumen fluid and duodenal digesta of faunated and defaunated sheep. *Can. J. Anim. Sci.*, **62**, 979–82.

37. Ivan, M., Veira, D. M. & Kelleher, C. A. (1986). The alleviation of chronic copper toxicity in sheep by ciliate protozoa. *Br. J. Nutr.*, **55**, 361–7.
38. Jouany, J.-P. & Senaud, J. (1979). Role of rumen protozoa in the digestion of food cellulosic materials. *Ann. Rech. Vet.*, **10**, 261–3.
39. Jouany, J.-P. & Senaud, J. (1982). Influence des ciliés du rumen sur la digestion de different glucides chez le mouton. 1. Utilisation des glucides parietaux (cellulose et hemicelluloses) et de l'amidon. *Reprod. Nutr. Develop.*, **22**, 735–52.
40. Jouany, J.-P. & Senaud, J. (1983). Influence des ciliés du rumen sur l'utilisation digestive de different regimes riches en glucides solubles et sur les produits terminaux formes dans le rumen. 2. Régimes contenant de l'inuline, du saccharose et du lactose. *Reprod. Nutr. Develop.*, **23**, 607–23.
41. Jouany, J.-P. & Thivend, P. (1983). Influence of protozoa on nitrogen digestion in the ruminant. In *Proc. 4th International Symp. Protein Metabolism and Nutrition* (INRA Coll. No. 16), ed. R. Pion, M. Arnal & D. Bonin. INRA Publications, Paris, Vol. II, pp. 287–90.
42. Jouany, J.-P., Senaud, J., Groliere, C. A., Thivend, P. & Grain, J. (1981). Influence de traitment par la soude et de l'inoculation du cilie *Polyplastron multivesiculatum* dans le rumen sur la digestion d'un régime riche en glucides parietaux. *Reprod. Nutr. Develop.*, **21**, 866.
43. Jouany, J.-P., Zainab, B., Senaud, J., Groliere, C. A., Grain, J. & Thivend, P. (1981). Role of the rumen ciliate protozoa *Polyplastron multivesiculatum, Entodinium* sp. and *Isotricha prostoma* in the digestion of a mixed diet in sheep. *Reprod. Nutr. Develop.*, **21**, 871–84.
44. Jouany, J.-P., Demeyer, D. & Grain, J. (1987). Effects of defaunating the rumen. In *Proceedings of OECD Workshop on Microbial and Multipurpose Utilisation of Lignocellulose* (in press).
45. Kahlon, T. S., Ranhotra, G. S. & Langar, P. N. (1970). Partial defaunation of the rumen and its influence on the biochemical activity. 1. Effect of partial defaunation on the intraruminal environment of the buffalo and zebu. *Ind. J. Anim. Sci.*, **40**, 593–9.
46. Kayouli, C., Demeyer, D. & Van Nevel, C. (1982). La défaunation du rumen: technique pour ameliorer la production des ruminants avec aliments tropicaux. In *Proc. Int. Colloq. Tropical Animal Production for Benefit of Man*, Prince Leopold Inst. Trop. Med., Antwerp, Belgium, pp. 302–8.
47. Kayouli, C., Demeyer, D. I., Van Nevel, C. J. & Dendooven, R. (1983/84). Effect of defaunation on straw digestion in sacco and on particle retention in the rumen. *Anim. Feed Sci. Technol.*, **10**, 165–72.
48. Kayouli, C., Van Nevel, C. J., Dendooven, R. & Demeyer, D. I. (1986). Effect of defaunation and refaunation of the rumen on rumen fermentation and N-flow in the duodenum of sheep. *Arch. Anim. Nutr.* (Berlin), **36**, 827–37.
49. Klopfenstein, T. J., Purser, D. B. & Tyznik, W. J. (1966). Effects of defaunation on feed digestibility, rumen metabolism and blood metabolites. *J. Anim. Sci.*, **25**, 765–73.
50. Kreuzer, M., Kirchgessner, M. & Muller, H. L. (1986). Effect of defaunation on the loss of energy in wethers fed different quantities of cellulose and normal or steam flaked maize starch. *Anim. Feed Sci. Technol.*, **16**, 233–41.
51. Kumar, O. M. & Raghavan, G. V. (1978). Digestibility of nutrients and concentration of rumen metabolites in faunated and defaunated lambs. *Ind. Vet. J.*, **55**, 775–80.
52. Kurihara, Y., Eadie, J. M., Hobson, P. N. & Mann, S. O. (1968). Relationship between bacteria and ciliate protozoa in the sheep rumen. *J. Gen. Microbiol.*, **51**, 267–88.
53. Kurihara, Y., Takechi, T. & Shibata, F. (1978). Relationship between bacteria and ciliate protozoa in the rumen of sheep fed a purified diet. *J. Agric. Sci.*, **90**, 373–81.
54. Leng, R. A. & Nolan, J. V. (1984). Nitrogen metabolism in the rumen. *J. Dairy Sci.*, **67**, 1072–89.
55. Lindsay, J. R. & Hogan, J. P. (1972). Digestion of two legumes and rumen bacterial growth in defaunated sheep. *Aust. J.Agric. Res.*, **23**, 321–30.
56. Lough, A. K. (1968). Component fatty acids of plasma lipids of lambs with and without rumen ciliate protozoa. *Proc. Nutr. Soc.*, **27**, 30A–31A.

57. Luther, R., Trenkle, A. & Burroughs, W. (1966). Influence of rumen protozoa on volatile acid production and ration digestibility in lambs. *J. Anim. Sci.*, **25**, 1116–22.
58. Males, J. R. & Purser, D. B. (1970). Relationship between rumen ammonia levels and the microbial population and volatile fatty acid proportions in faunated and defaunated sheep. *Appl. Microbiol.*, **19**, 485–90.
59. Meyer, J. H. F., van der Walt, S. I. & Schwartz, H. M. (1986). The influence of diet and protozoal numbers on the breakdown and synthesis of protein in the rumen of sheep. *J. Anim. Sci.*, **62**, 509–20.
60. Nagaraja, J. G., Dennis, S. M., Galitzer, S. J. & Harmon, D. L. (1986). Effect of lasalocid, monensin and thiopeptin on lactate production from *in vitro* rumen fermentation of starch. *Can. J. Anim. Sci.*, **66**, 129–39.
61. Neil, A. R., Grime, D. W., Snoswell, A. M., Northrop, A. J., Lindsay, D. B. & Dawson, R. M. C. (1979). The low availability of dietary choline for the nutrition of the sheep. *Biochem. J.*, **180**, 559–65.
62. Newbold, C. J., Chamberlain, D. G. & Williams, A. G. (1986). The effects of defaunation on the metabolism of lactic acid in the rumen. *J. Sci. Food Agric.*, **37**, 1083–90.
63. Nuzback, D. E., Bartley, E. E., Dennis, S. M., Nagaraja, T. G., Galitzer, S. J. & Dayton, A. D. (1983). Relation of rumen ATP concentration to bacterial and protozoal numbers. *Appl. Environ. Microbiol.*, **46**, 533–8.
64. Onodera, R. (1986). Contribution of protozoa to lysine synthesis in the *in vitro* rumen microbial ecosystem. *Appl. Environ. Microbiol.*, **51**, 1350–1.
65. Orpin, C. G. & Letcher, A. J. (1983/84). Effect of absence of ciliate protozoa on rumen fluid volume, flow rate and bacterial populations in sheep. *Anim. Feed Sci. Technol.*, **10**, 145–53.
66. Osman, H., Abou Akkada, A. R. & Agabawi, K. A. (1970). Influence of rumen ciliate protozoa on conversion of food and growth rate in early weaned zebu calves. *Anim. Prod.*, **12**, 267–71.
67. Perkins, J. L. & Luther, R. M. (1967). Effects of defaunation and energy level on ration utilisation in sheep. *J. Anim. Sci.*, **26**, 928.
68. Pounden, W. D. & Hibbs, J. W. (1950). The development of calves raised without protozoa and certain other characteristic rumen microorganisms. *J. Dairy Sci.*, **33**, 639–44.
69. Ramaprasad, J. & Raghavan, G. V. (1981). Note on the growth rate and body composition of faunated and defaunated lambs. *Ind. J. Anim. Sci.*, **51**, 570–2.
70. Rowe, J. B., Davies, A. & Broome, A. W. J. (1981). Quantitative effects of defaunation on rumen fermentation and digestion in sheep. *Proc. Nutr. Soc.*, **40**, 49A.
71. Rowe, J. B., Davies, A. & Broome, A. W. J. (1983). Changing rumen fermentation by chemical means. In *Recent Advances in Animal Nutrition in Australia*, ed. D. J. Farrell & P. Vohra. University of New England Publishing Unit. Armidale, Australia, pp. 102–9.
72. Rowe, J. B., Davies, A. & Broome, A. W. J. (1985). Quantitative effects of defaunation on rumen fermentation and digestion in sheep. *Br. J. Nutr.*, **54**, 105–19.
73. Sedloev, N. (1970). Effect of defaunation on volatile fatty acid levels, cellogel electrophoretic serum, protein spectrum and rumen movement of sheep. *Nauch. Tr. Vissh. Veterinarnomed. Inst., Sofia*, **22**, 301–6.
74. Stinchi, S., Itoh, T., Abe, M. & Kandatsu, M. (1986). Effect of rumen ciliate protozoa on the proteolytic activity of cell free rumen liquid. *Jap. J. Zootech. Sci.*, **57**, 89–96.
75. Singh, S. & Makkar, G. S. (1976). Effect of partial defaunation on the utilisation of carbohydrates in buffalo and zebu cattle. *Ind. J. Anim. Sci.*, **46**, 463–7.
76. Stern, M. D. & Hinkson, R. S. (1974). Effect of defaunation and faunation on intraruminal factors. *J. Anim. Sci.*, **39**, 253.
77. Sutton, J. D., Knight, R., McAllan, A. B. & Smith, R. N. (1983). Digestion and synthesis in the rumen of sheep given diets supplemented with free and protected oils. *Br. J. Nutr.*, **49**, 419–32.
78. Szuecs, E., Regius, A., Weber, A. & Szoelloesi, T. (1982). Effect of various nutrient supply

and copper supplementation on the performance and carcass value of growing-finishing bulls. *Z. Tierphysiol. Tierernähr. Futtermittelkd*, **47**, 276–88.
79. Takahashi, F. & Kametaka, M. (1976). The effect of the absence of rumen ciliate protozoa on the utilisation of urea by goats. *Jap. J. Zootech. Sci.*, **47**, 192–6.
80. Thamman, O. P., Ranhotra, G. S. & Langar, P. N. (1971). Partial defaunation of the rumen and its influence on the biochemical activity. 2. Effect of partial defaunation on some blood components of buffalo and zebu. *Ind. J. Anim. Sci.*, **41**, 73–7.
81. Ushida, K. & Jouany, J.-P. (1985). Effect of protozoa on rumen protein degradation in sheep. *Reprod. Nutr. Develop.*, **25**, 1075–81.
82. Ushida, K. & Jouany, J.-P. (1986). Influence des protozoaires sur la dégradation des protéines mesurée *in vitro* et *in sacco*. *Reprod. Nutr. Develop.*, **26**, 293–4.
83. Ushida, K., Jouany, J.-P., Lassalas, B. & Thivend, P. (1984). Protozoal contribution to nitrogen digestion in sheep. *Can. J. Anim. Sci.*, **64** (Suppl.), 20–1.
84. Ushida, K., Jouany, J.-P. & Thivend, P. (1986). Role of rumen protozoa in nitrogen digestion in sheep given two isonitrogenous diets. *Br. J. Nutr.*, **56**, 407–19.
85. Ushida, K., Miyazaki, A. & Kawashima, R. (1986). Effect of defaunation on ruminal gas and VFA production *in vitro*. *Jap. J. Zootech. Sci.*, **57**, 71–7.
86. Van Nevel, C. J. & Demeyer, D. I. (1981). Effect of methane inhibitors on the metabolism of rumen microbes *in vitro*. *Arch. Tierernährung*, **31**, 141–51.
87. Van Nevel, C. J., Demeyer, D. I. & Van de Voorde, G. (1985). Effect of defaunating the rumen on growth and carcass composition of lambs. *Arch. Tierernährung*, **35**, 331–7.
88. Veira, D. M., Ivan, M. & Jui, P. Y. (1981). The effects of partial defaunation on rumen metabolism in sheep. *Can. J. Anim. Sci.*, **61**, 1086–7.
89. Veira, D. M., Ivan, M. & Jui, P. Y. (1983). Rumen ciliate protozoa: effects on digestion in the stomach of sheep. *J. Dairy Sci.*, **66**, 1015–22.
90. Veira, D. M., Ivan, M. & Jui, P. Y. (1984). The effect of ciliate protozoa on the flow of amino acids from the stomach of sheep. *Can. J. Anim. Sci.*, **64** (Suppl.), 22–3.
91. Verite, R., Durand, M. & Jouany, J.-P. (1986). Influence des facteurs alimentaires sur la protéosynthèse microbienne dans le rumen. *Reprod. Nutr. Develop.*, **26**, 181–201.
92. Wallace, R. J. & West, A. A. (1982). Adenosine-5'-triphosphate and adenylate energy charge in sheep digesta. *J. Agric. Sci.*, **98**, 523–8.
93. Whitelaw, F. G., Eadie, J. M., Mann, S. O. & Reid, R. S. (1972). Some effects of rumen ciliate protozoa in cattle given restricted amounts of a barley diet. *Br. J. Nutr.*, **27**, 425–37.
94. Whitelaw, F. G., Eadie, J. M., Bruce, L. A. & Shand, W. J. (1984). Methane formation in faunated and ciliate-free cattle and its relationship with rumen volatile fatty acid proportions. *Br. J. Nutr.*, **52**, 261–75.
95. Williams, P. P. & Dinusson, D. E. (1973). Ruminal volatile fatty acid concentrations and weight gains of calves reared with and without ruminal ciliated protozoa. *J. Anim. Sci.*, **36**, 588–91.
96. Yoshida, J., Nakamura, Y. & Nakamura, R. (1982). Effects of protozoal fraction and lactate on nitrate metabolism of microorganisms in sheep rumen. *Jap. J. Zootech. Sci.*, **53**, 677–85.
97. Youssef, F. G. & Allen, D. M. (1968). Part played by ciliate protozoa in rumen function. *Nature*, **217**, 777–8.

4
The Rumen Anaerobic Fungi

C. G. Orpin

AFRC Institute of Animal Physiology and Genetics Research, Cambridge, UK
and
Department of Arctic Biology, Institute of Medical Biology, University of Tromsø, Norway

&

K. N. Joblin

Biotechnology Division, Department of Scientific and Industrial Research, Palmerston North, New Zealand

Yeasts and aerobic fungi have long been known to be normal inhabitants of the rumen (Clarke & DiMenna, 1961; Sivers, 1962*a,b*; Lund, 1974) but most species isolated are considered to be transient and non-functional, entering the rumen with the feed. Some aerobic fungi are capable of growth under anaerobic conditions, and Brewer *et al.* (1972) concluded that *Aspergillus fumigatus, Mucor rouxii* and *Sporormia minima*, which are implicated as causative agents in ovine ill-thrift, could survive in the rumen. Two other groups of fungi are now known to occur, one group parasitic upon ciliate protozoa, the other saprophytic on plant tissues.

The parasitic fungi were found by Lubinsky (1955*a,b*). Two distinct species were identified parasitising Ophryoscolecid ciliates, *Sphaerita hoari* parasitising principally *Eremoplastron bovis*, and *Sagittospora cameroni* parasitising *Diplodinium minor* and *Eudiplodinium maggii*. Both were tentatively identified as chytrid fungi despite the fact that the spores were apparently non-flagellated and similar to microsporidian spores. Each species of parasite consists of a plasmodial thallus, usually situated in the endoplasm near the macronucleus of the host. On maturity the thallus develops into a sporangium in which large numbers of spores develop. Little is known about these species, but it is assumed that infection occurs after ingestion of a spore released from an infected and possibly dying ciliate. *Sphaerita hoari* may do little harm to its host since lightly infected ciliates continue to divide, although heavily infected cells degenerate. In contrast, *Sagittospora cameroni* invariably kills its host. The significance of these, apparently uncommon, parasites remains to be elucidated.

Anaerobic fungi saprophytic on ruminal digesta have only recently been discovered in the rumen (Orpin, 1975, 1976, 1977*b*; Bauchop, 1979), although the zoospores of the species had been recognised many years previously by pioneering microbiologists when small flagellated microorganisms were observed in rumen liquor. These organisms, smaller than the readily identified ciliate protozoa, were all believed to be flagellate protozoa, and descriptions were published by Liebetanz (1910) and Braune (1913) of several distinct morphological types of flagellates occurring in domestic ruminants. Little further attention was paid to rumen flagellates until Jensen & Hammond (1964) cultured a number of flagellates, identified correctly as flagellate protozoa on morphological criteria, from the bovine rumen. Some flagellates, notably *Callimastix frontalis*, *Piromonas communis* and *Sphaeromonas communis* (Braune, 1913; Liebetanz, 1910), apparently were not cultivated by these authors.

Warner (1966) measured diurnal fluctuations in the population densities of rumen microorganisms in sheep, and observed that motile polyflagellated cells ('polymastigates'), similar to *Callimastix frontalis*, showed a significant increase in population density 1 hour after feeding the animal. The reasons for this fluctuation were examined by Orpin (1974) who found that it was considerably greater than reported by Warner, with the peak population density of polyflagellated cells occurring 15–30 minutes after the animal had eaten. Further studies showed that migration from the rumen epithelium was unlikely, and that the population density increase could be duplicated *in vitro* by treating large digesta fragments, centrifuged from the rumen fluid, with an aqueous extract of oats. It was concluded that either the flagellates sequestered on or within the plant fragments, or a multiple reproductive phase of the organism was implicated. At the time, both possibilities were novel for rumen microorganisms which, to maintain a constant population density from day to day, need to divide only about 1–2 times per day, depending on the rumen dilution rate.

Orpin (1975) first succeeded in culturing one of these polyflagellated organisms under anaerobic conditions by initially following the culture methods of Jensen & Hammond (1964) employed for the flagellated protozoa. Light microscopy of growing cultures revealed that the highly motile flagellates were liberated from sporangia borne on structures resembling the vegetative stage of certain aquatic phycomycete fungi. The flagellates were identified as *Callimastix frontalis* (Braune, 1913), which had been transferred by Vavra & Joyon (1966) to the 'zooflagellate' genus, *Neocallimastix*. Thus the major stages in the life cycle of *Neocallimastix* sp. were discovered. Later, other species, *Sphaeromonas communis* (Orpin, 1976), *Piromonas communis* (Orpin, 1977*b*), *Neocallimastix patriciarum* (Orpin & Munn, 1986) and unidentified, morphologically distinct strains (Orpin, C. G. & Kemp, P., unpublished), were cultured and shown to have similar life cycles. All grew only anaerobically and were subsequently found to be cellulolytic. It is highly probable that more species will be discovered in domestic and wild herbivores.

The fungal nature of these organisms was strongly indicated by the morphology of the vegetative stage, and the life cycles were similar to known species of aquatic phycomycetes (Sparrow, 1960), particularly the chytridiomycetes. The demonstration that the cell walls of these species contained chitin (Orpin, 1977*b*) and chitin synthetase (Brownlee, 1986*a*) proved that these organisms were true fungi, despite being strict anaerobes. The flagellates should therefore be termed, more correctly, 'zoospores'. Until these fungi had been discovered, fungi were regarded as being either aerobes or facultative anaerobes.

The distribution of these strictly anaerobic fungi appears to be limited to the gut of herbivores. Joblin has extensively sampled terrestrial and aquatic environments for anaerobic fungi of the type found in herbivores, but without success. Strictly anaerobic fungi have now been found in the gut of the foregut-fermenters sheep (Orpin, 1975, 1976, 1977b), cattle, deer, impala, kangaroo (Bauchop, 1980, 1983), goat (Orpin, C. G. & Joblin, K., unpublished), reindeer (Orpin et al., 1985), musk ox (Orpin, C. G. & Mathiesen, S. D., unpublished) and camel (Joblin, K., unpublished), and in the hindgut-fermenters horse (Orpin, 1981), elephant (Bauchop, 1983) and rhinoceros (Joblin, K., unpublished). It seems likely that these fungi are a normal component of the fermentative microflora in herbivores on a highly fibrous diet. In the case of the giant panda, an animal that has a high intake of plant fibre but is not a true herbivore, no anaerobic fungi could be isolated from fresh faeces (Joblin, K., unpublished). Lowe et al. (1987d) isolated anaerobic fungi from the oesophagus and faeces of sheep, so transfer between animals may be mediated by either route.

GENERA AND SPECIES

A variety of different morphological types (Fig. 1) have been identified, but only one genus, *Neocallimastix* (Heath et al., 1983), has been legitimately described. *Piromonas* and *Sphaeromonas* have not so far been legitimately described. Orpin (1976, 1977b) cultured two species with zoospores morphologically similar to flagellates, called *Sphaeromonas communis* and *Piromonas communis* by Liebetanz (1910), which were described further by Braune (1913), and these names were retained. Now that it is known that these organisms are zoospores of true fungi, it is necessary to legitimise the nomenclature according to mycological procedures. Judging from the wide range of sporangial morphology seen during electron microscopic examination of plant fragments from large herbivores (Bauchop, 1980) it is likely that a number of genera will eventually be described. The life cycles of species so far studied are similar, and consist of an alternation between a zoospore stage and a vegetative stage carrying a fruiting body (sporangium) (Fig. 2). All representatives of these genera are monocentric and holocarpic with endogenous development; their vegetative stages are saprophytic on plant tissues. The zoospores are actively motile, somewhat variable in shape, with the flagellum or flagellar apparatus posteriorly orientated, and are capable of amoeboid movement. The zoospores locate and germinate on freshly ingested plant tissues (Orpin, 1977a,b) where they develop into the vegetative stage, often in high numbers (Bauchop, 1979, 1980, 1981). The rhizoidal system of the fungal thallus normally penetrates the plant tissue, with the sporangium developing on the exterior surface of the plant fragment (Fig. 3). Flagellates are released from mature sporangia to infect other plant fragments in the rumen; in animals fed once per day this occurs soon after the animal is fed (Orpin, 1975, 1976, 1977b).

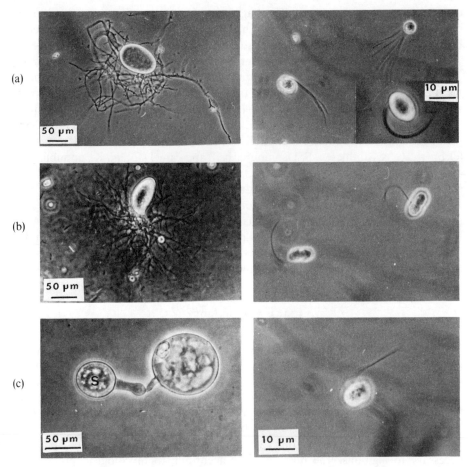

Fig. 1. Light micrographs of vegetative (left) and zoospore stages (right) of rumen fungi cultured *in vitro*. Glutaraldehyde-fixed preparations, except the inset photograph of *N. patriciarum* zoospore. (a) *Neocallimastix patriciarum*; inset, living zoospore showing the flagella united to form a single locomotory organelle. (b) *Piromonas communis*; the vegetative stage is in culture with a methanogenic bacterium. (c) *Sphaeromonas communis*; note the limited rhizoidal system and associated spherical body; S = sporangium.

TAXONOMY

The taxonomic status of rumen fungi is not yet clear. The classification of the chytrids is based upon the ultrastructure of the mature zoospore. *Neocallimastix* spp. zoospores carry up to 17 flagella; as the existing fungal taxonomic system at the time of the isolation of *Neocallimastix* accommodated only zoospores with either 1 or 2 flagella, the taxonomic position of *Neocallimastix* was in some doubt. Heath *et al.* (1983) later assigned *N. frontalis* to the Spizellomycetales (Chytridiomycetes) after examining the fine-structure of the zoospore, particularly the kinetosomes

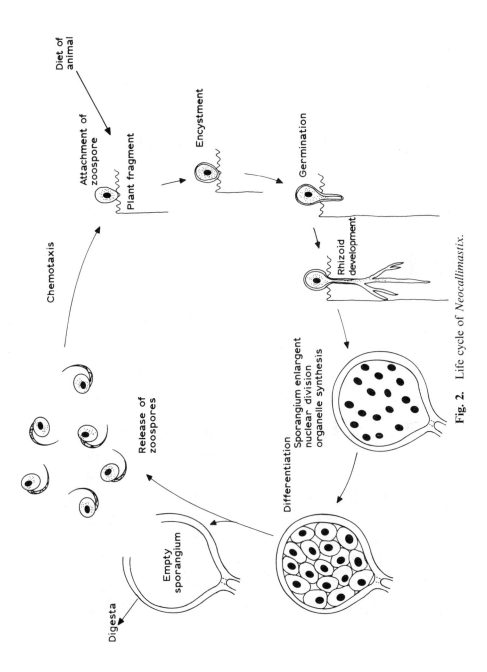

Fig. 2. Life cycle of *Neocallimastix*.

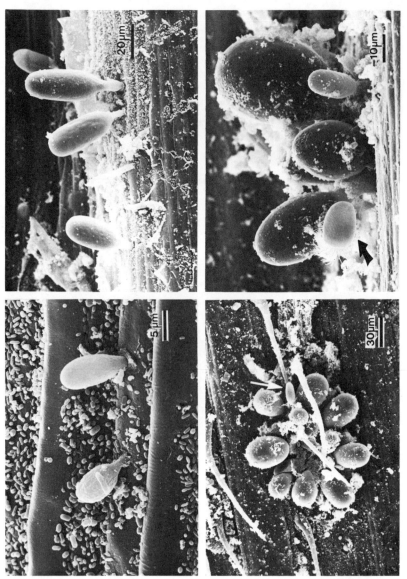

Fig. 3. Scanning electron micrographs of sporangia of rumen fungi growing together with bacteria and protozoa (arrowed) on ryegrass stems from rumen digesta of a sheep fed on meadow hay.

which have been used in assessing relationships between zoosporic fungi (Barr, 1981). The inclusion of *Piromonas* and *Sphaeromonas* in the Chytridiomycetes is supported by the structure of the kinetosomes and accessory structures (Munn *et al.*, 1988). There does not, however, appear to be the non-functional kinetosome which is normally characteristic of the Chytridiomycetes. Additionally, a key feature of the Spizellomycetales, the association of the nucleus via a beak-like projection with the kinetosomes, does not appear to be present (Munn *et al.*, 1987), although Heath *et al.* (1983) did find some evidence for it. An examination of the fine-structure of zoospores of *Neocallimastix* spp., *Piromonas* and *Sphaeromonas* has demonstrated a number of common structural features which suggests that they may represent a new, distinct, taxon (Munn *et al.*, 1987, 1988). These workers have suggested that 16

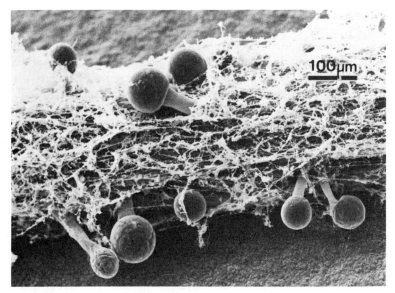

Fig. 4. Scanning electron micrograph of a plant fibre colonised by an axenic culture of a *Neocallimastix* sp. isolated from a camel.

fine structural features that rumen fungi have in common, collectively define the group. Some of the common features (two distinctive surface layers, the organisation of the ribosomes, the structures associated with the kinetosomes in the zoospores, and the presence of hydrogenosomes) were suggested individually to distinguish these obligate anaerobes from the aerobic and facultatively anaerobic fungi.

In the case of the mature fungal thallus, the size and branching of the rhizoids and the size and shape of the sporangium may differ between different species (Joblin, 1981, Figs 1, 3 and 4). So far, little attention has been paid to the use of rhizoidal structural features or gross morphology in classification.

FINE STRUCTURE

The most obvious difference between the zoospores of different genera of rumen fungi is that those of *Neocallimastix* are multiflagellate (Orpin & Munn, 1986) whilst *Piromonas* and *Sphaeromonas* are uniflagellate. In all genera the surface of the cell is coated with a characteristic layer of particles (Munn *et al.*, 1988) and the surface layer of the flagellum is composed of parallel fibrils. No non-functional kinetosomes or kinetosomal props are present. Zoospores of all genera contain two types of microtubule-associated vesicular entities, amorphous globules identified by Yarlett *et al.* (1986*a*) as hydrogenosomes, associated with the kinetosomes (Munn *et al.*, 1981), and small coated vesicle-like structures (Munn *et al.*, 1988).

Ribosome particles occur in two distinct forms, ribosome helices and ribosome granules (Munn *et al.*, 1981). Other inclusions such as crystallites and glycogen granules are common. In zoospores of *N. frontalis*, the larger intracellular structures, excluding the nucleus, are associated with the end of the cell close to the kinetosomes, whereas in *N. patriciarum* they are (except the hydrogenosomes) distributed throughout the cytoplasm. Hydrogenosomes and crystallites occur throughout the rhizoids and undifferentiated sporangia. The DNA from an unidentified rumen fungus (Brownlee, 1986*b*) was found to be unusually rich in adenosine and thymidine. Further work will be necessary to determine whether the GC content of the DNA can be used as a diagnostic character for these fungi. Mitosis, and the fate of the kinetosomes during encystment, has been studied by Heath & Bauchop (1985) and Heath *et al.* (1986).

Neocallimastix (Heath *et al.*, 1983)

This genus embraces the polyflagellated species *N. frontalis* and *N. patriciarum*. Vegetative growth consists of a single sporangium borne on a much branched rhizoidal system (Fig. 1). The formal description of the genus (Heath *et al.*, 1983) is based on the ultrastructure of the zoospore of *N. frontalis*, particularly the kinetosomes, since kinetosome structure forms the basis of current taxonomy of zoosporic fungi (Barr, 1981).

The original description of the genus *Neocallimastix* was emended by Orpin & Munn (1986) to include *N. patriciarum*. Zoospores of *N. patriciarum* differ from those of *N. frontalis* in not showing the strong bipartite distribution of intracellular organelles or consistently waisted appearance, and in having more flagella (9–17) than *N. frontalis* (7–10). At the fine-structure level, *N. patriciarum* lacks the positive interaction of microtubules of the fan-shaped array radiating from the kinetosomes with the nucleus, and the beak-shaped nuclear projection of *frontalis*. At a nutritional level, *patriciarum* requires D-biotin (Orpin & Munn, 1986) whereas *frontalis* apparently does not, since the method of media preparation used by Heath *et al.* (1983) would destroy D-biotin.

The zoospores of species in this genus are broadly ellipsoidal; see Fig. 1(a). Organelles are clustered, and either have a bipartite distribution (*frontalis*) or are distributed throughout the cytoplasm (*partriciarum*). The nucleus is approximately central, and the 7–17 flagella are inserted mainly in two adjacent and approximately parallel rows. Prior to encystment of the zoospore, the flagella become detached from the cell. Shedding or withdrawal of fungal zoospore cilia or flagella normally results in the kinetosomes remaining within the cell, but this does not occur with *Neocallimastix* (Heath et al., 1986) where the kinetosomes remain attached to the shed flagella.

Piromonas

The vegetative stage (Fig. 1(b)) of *P. communis* is smaller than that of *Neocallimastix*, with more highly branched rhizoids. The zoospore has a single flagellum, and swims with a rather fast, jerky movement. The flagellum is posteriorly orientated during swimming. Orpin, C. G. & Kemp, P. (unpublished) isolated a very similar species, but with larger zoospores than those of *communis*, which swam with a slower rolling movement and had vegetative growth similar to *Neocallimastix*.

Sphaeromonas

The vegetative stage of *Sphaeromonas* (Orpin, 1976) can be differentiated from that of the other known genera by the absence of a developed rhizoidal system (Fig. 1(c)), and it consists of a single sporangium borne on a short thick thallus on which are carried a number (up to 7, but normally 1–4) of spherical bodies of unknown function. These spherical bodies possibly make up a bulbous type of rhizoid. The zoospore has a single flagellum.

LIFE CYCLES *IN VIVO*

In animals fed once per day, zoosporogenesis and liberation of zoospores from the vegetative stages occurs soon after the animal is fed. With *Neocallimastix* a peak in the population density of zoospores is found 15–30 minutes after feeding (Orpin, 1974, 1975), but with *Sphaeromonas* and *Piromonas* the peak occurs about 1 hour after feeding (Orpin, 1976, 1977b). Zoosporogenesis in *Neocallimastix* (and possibly the other species) was inducible by a water-soluble component of the diet (Warner, 1966; Orpin, 1974, 1977d). Subsequently haemes were shown to be active in inducing zoosporogenesis (Orpin & Greenwood, 1986a). Haemes in chemically oxidised or reduced form are present in all living plant tissues as enzyme prosthetic groups, and oxidised forms occur in dead plant-tissues. Thus, whenever the animals eat, haemes enter the rumen.

Free zoospores of *Neocallimastix* showed a chemotactic response to soluble carbohydrates (Orpin & Bountiff, 1978). Soluble carbohydrates present in freshly ingested plant tissues diffuse into the rumen liquor, and *Neocallimastix* zoospores probably locate these plant fragments by migrating up soluble-carbohydrate gradients to damaged surface regions and plant stomata. Four chemoreceptors were identified, sensing glucose, fructose, mannitol and mannose, respectively, and low concentrations of mixtures of glucose, sucrose and fructose resulted in a synergistic response. This synergism could be responsible for specific tissue location by the zoospores. Chemotactic response was very sensitive, being elicited by as little as 1 μm sucrose. After the plant tissue has been located by the zoospores, attachment, encystment and germination occurs, followed by penetration of the plant tissues by the fungal rhizoidal system and subsequent growth of the sporangium. From studies *in vitro*, it appears that the zoospore, still bearing flagella, adopts an amoeboid stage and may move across the plant surface, presumably to locate exactly the right site for encystment.

The life cycle (Fig. 2), from studies *in vivo* and *in vitro*, of *Neocallimastix* spp. lasts about 24–32 hours (Joblin, 1981; Bauchop, 1983; Lowe *et al.*, 1987a), although zoosporogenesis of young sporangia may occur under appropriate conditions as early as 8 hours after encystment (Orpin, 1977d). In continuously fed animals the life cycle may therefore be only 8 hours.

ESTIMATION OF POPULATION DENSITIES AND BIOMASS

The zoospore population density of the sheep rumen as measured by different methods (Orpin, 1974; Joblin, 1981; Stewart *et al.*, 1985) is in the range 10^3–10^5 ml^{-1}. Population densities of zoospores may be determined by direct microscopic observation of 10- to 100-fold dilutions of samples of filtered rumen fluid stained with Lugol's iodine as used for protozoa (Coleman, 1978). The zoospores stain yellowish brown, and remain refractile under phase-contrast illumination. Difficulties are, however, encountered because the zoospores are often associated with the particulate material in the sample. Joblin (1981) published a cultural method of enumerating zoospores by dilution in an agar-containing medium, followed by incubation and counting of fungal colonies. This procedure has usually employed cellobiose as substrate, but there is some evidence that the use of glucose may give a better estimation of the zoospore population (Joblin, 1981). If single colonies, selected at random, are removed in agar cores and incubated separately in liquid media and identified, it is possible to assign population densities to each species present.

There is as yet no reliable method for the determination of the biomass of anaerobic fungi in the rumen. Chitin was used as a marker by Orpin (1981) who estimated the fungal biomass to be up to 8% of the total biomass. The method, however, assumes an identical and constant chitin content for each fungus irrespective of age and species, and is susceptible to error due to the long enzyme assay required to measure chitin in the sample. In addition, most ruminant foods contain a variable background level of chitin due to contamination of the food by aerobic fungi prior to consumption by the animal. Zoospore counts cannot be used to estimate biomass for a number of reasons, but chiefly because of a diurnal fluctuation in population density (Orpin, 1975) and the difficulties in differentiating between species. It was, however, suggested by Kemp *et al.* (1984) that tetrahymanol, a triterpenol present in *Neocallimastix* and *Piromonas* and rare in (or absent from) the diet, could possibly be used as a biomass marker.

ISOLATION AND CULTURE

Rumen anaerobic fungi can be isolated directly from rumen fluid using the roll-bottle method of Joblin (1981). In this procedure, a dilution series of molten agar medium is inoculated with filtered rumen contents and the medium is solidified. Although cellobiose has been the most commonly used substrate, other soluble carbohydrates can serve equally well. An insoluble substrate, cellulose powder, has been used for isolation by taking advantage of the cellulose-clearing action of the colonies to locate fungi (Joblin, K., et al., unpublished). The addition of antibiotics such as penicillin and streptomycin is essential to suppress the growth of bacteria which would otherwise outgrow developing fungal colonies. After incubation at 39°C for 48 hours, single colonies of fungi (Fig. 5) are visible to the naked eye, but they are best examined using a low-power inverted or dissecting microscope, when it is often possible to discriminate between colonies of different size and appearance (Joblin, 1981).

Individual colonies removed anaerobically from the agar using a sterile Pasteur pipette are incubated at 39°C in cellobiose-containing liquid medium. Free-swimming zoospores can be detected in the liquid medium within 48 hours when cultures are examined under an inverted microscope.

Other methods of isolation employ the overlaying of sloppy agar media with filtered rumen fluid, followed by a period of incubation during which the zoospores migrate downwards into the agar to grow into vegetative stages (Orpin, 1975) from which they can subsequently be isolated in pure culture, and the use of large, infected plant-tissue particles to inoculate semi-solid media (Orpin, 1977b) from which to isolate single strains.

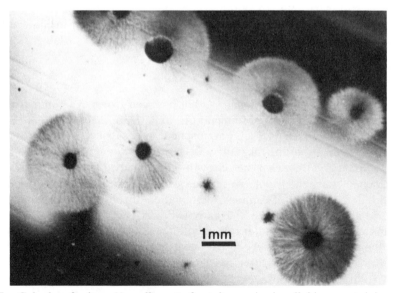

Fig. 5. Colonies of a sheep *Neocallimastix frontalis* growing in cellobiose-containing agar in a roll-rube.

Methanogenic bacteria sometimes contaminate fresh isolates of rumen anaerobic fungi (Bauchop & Mountfort, 1981), but they can be removed by treatment with chloramphenicol.

When rumen fungi are cultured axenically on plant fibre, rhizoid tissue may form a mat on the surface of the fibre (Fig. 4). This is in contrast to fungi growing in the rumen where fungal rhizoids are located beneath the surface of the plant tissue.

Culture media

The most commonly employed, undefined culture media used for growing anaerobic rumen fungi are similar to that of Orpin (1975), and consist of centrifuged rumen fluid, tryptone, yeast extract, a carbon source, a carbon dioxide–bicarbonate buffer at pH 6·7–6·9, L-cysteine as reducing agent, and vitamins. This medium may be solidified with 0·8–1·5% agar for use in isolating clones from single colonies (Joblin, 1981). If bacterial contamination of cultures occurs, antibiotics such as ampicillin, streptomycin and chloramphenicol may be incorporated into the medium.

Some strains of the polyflagellated species have been grown in defined media, for example *N. patriciarum* (Orpin & Greenwood, 1986b) and unidentified *Neocallimastix* spp. (Lowe et al., 1985).

Culture media incorporating plant tissues as carbon source are valuable for ensuring that the organisms do not lose their ability to ferment plant structural carbohydrates *in vitro* (Orpin & Letcher, 1979). Cultures in the liquid media of Orpin (1975) gave high cell yields but needed subculturing after 24–48 hours to maintain viability, whereas cultures on plant tissues may be viable for up to 7 days, thus simplifying routine maintenance. Maintenance of cultures for periods of several months without subculturing has been achieved with *Neocallimastix* and *Piromonas* spp. by the storage at 39°C of cultures on plant tissues embedded in agar (Joblin, 1981). Long-term storage has been achieved by controlled freezing in liquid nitrogen or at $-70°C$ using 5% dimethyl sulphoxide as cryoprotectant (Yarlett et al., 1986b).

Nutrition

Only one species, *Neocallimastix patriciarum*, has been grown in a minimal medium (Orpin & Greenwood, 1986b). Minimal nutritional requirements for growth of *N. patriciarum* on cellobiose in a CO_2 atmosphere were satisfied by the provision of sources of haem, D-biotin, thiamin or its precursors, ammonium ions, and of reduced sulphur and trace elements. Growth was stimulated by amino acids, straight and branched short-chain volatile fatty acids, low concentrations of long-chain fatty acids and a number of vitamins. Germination of zoospores was stimulated by acetic acid and soluble, fermentable carbohydrates.

Haems appear to play a major role in both the nutrition and zoosporogenesis of rumen fungi (Orpin, 1986a). Although haems probably occur in rumen contents during the entire day, the addition of haems to rumen contents of animals fed once per day resulted in zoosporogenesis of *Neocallimastix* sp., but this could not be satisfactorily repeated *in vitro* with pure cultures, suggesting that the control of zoosporogenesis and zoospore release may be more complex than suggested by the work *in vivo*. Haems may act synergistically with other dietary components to stimulate zoosporogenesis. Because of the low redox potential of the rumen, dietary haems in an oxidised state would be rapidly reduced, and it is likely that the reduced rather than the oxidised haem is functional *in vivo*. Zoosporogenesis in monoflagellated species may also be triggered by similar compounds (Orpin, 1977d).

INTERMEDIARY METABOLISM

In *Neocallimastix patriciarum* the Embden–Meyerhof pathway appears to be the major glycolytic pathway used in the catabolism of carbohydrates (Yarlett *et al.*, 1986a). This is supported by the distribution of $^{14}CO_2$ produced by cultures fermenting glucose which had been labelled in C-1, C-2 and C-6 positions (Orpin, C. G., unpublished).

Glycolysis generates phosphoenolpyruvate which is converted via oxaloacetate into malate. Hydrogenosomal enzymes generate energy from the oxidation of malate to pyruvate which is coupled, via NADPH:ferridoxin oxidoreductase, ferridoxin and hydrogenase, to production of molecular hydrogen. Some of the pyruvate generated is converted into acetyl-CoA via pyruvate:ferridoxin oxidoreductase, and subsequently into acetate. Cytosolic lactate dehydrogenase is responsible for lactate generation and, when cultures are grown under a N_2 rather than under a CO_2 atmosphere, low levels of alcohol dehydrogenase convert acetaldehyde into ethanol (Yarlet *et al.*, 1986a). Carbon dioxide appears to suppress hydrogenosomal acetate generation as well as cytosolic ethanol production, perhaps by end-product inhibition of the pyruvate:ferridoxin oxidoreductase and pyruvate decarboxylase reactions, both of which generate CO_2. One strain of *Neocallimastix* was found to require 7% CO_2 in the gas phase and could not be grown under a 100% N_2 atmosphere (Joblin, K., unpublished).

FERMENTATION OF CARBOHYDRATES

It is clear that the rumen fungi so far examined produce a wide range of enzymes that can digest the major structural carbohydrates of plant cell walls (Pearce & Bauchop, 1985; Wood *et al.*, 1986; Williams & Orpin, 1987a; Lowe *et al.*, 1987b,c), hydrolyse a range of glycosidic linkages (Williams & Orpin, 1987b), and enable the fungi to grow on a number of polysaccharides (Orpin & Letcher, 1979; Mountfort & Asher, 1983). Many of the polysaccharide-hydrolysing enzymes are produced by the vegetative stage and the zoospores of the fungi, and are also present in the extracellular culture medium (Williams & Orpin, 1987a,b; Lowe *et al.*, 1987b). The particular activity depends upon the growth substrate. Thus *Neocallimastix frontalis* and *N. patriciarum* will utilise cellulose, xylan, starch and hemicelluloses (Orpin & Letcher, 1979; Bauchop, 1983; Orpin & Munn, 1986), as will *P. communis* and another monoflagellated species (Williams O Orpin, 1987a). No isolates have yet been found to grow on pectin, but very low levels of pectin-hydrolysing enzymes may be produced in culture media (Williams & Orpin, 1987a), and pectin is lost from plant fragments digested by isolates of rumen fungi (Orpin, 1983/84).

The major products of cellobiose fermentation by all species are lactate, acetate, hydrogen and carbon dioxide, with traces of formate and ethanol (Orpin & Munn, 1986) being produced by *N. patriciarum*. However, *N. frontalis* produces significant quantities of formate and ethanol from cellulose (Bauchop & Mountfort, 1981), as do two *Piromonas* species (Joblin, K. et al., unpublished). Monensin, an ionophore which inhibits hydrogen- and formate-producing bacteria (Chen & Wolin, 1979) also has an inhibitory effect on *N. frontalis* (Stewart et al., 1987).

A wide range of mono- and di-saccharides, including glucose, cellobiose, fructose, maltose, sucrose and xylose, support growth of all species examined (Orpin, 1975, 1976, 1977b; Bauchop, 1980; Mountfort & Asher, 1983), but strain differences are evident. Chemotaxis of zoospores of *N. patriciarum* to soluble carbohydrates was, however, elicited to compounds such as mannose, sorbose, sorbitol and fucose which were not utilised as carbon sources (Orpin & Bountiff, 1978). Cellobiose and sucrose were utilised preferentially over fructose and glucose respectively (Mountfort & Asher, 1983). In *N. frontalis*, catabolite repression by glucose was observed in the presence of fructose, xylose and cellulose (Mountfort & Asher, 1983, 1985).

The cellulose enzymes necessary for solubilising both the amorphous and highly ordered cellulose present in plant fibre are produced by rumen fungi. High levels of endo-1,4-β-glucanase (CMCase) were released into culture supernatants by *N. frontalis* (Mountfort & Asher, 1985; Wood et al., 1986) and by other species (Williams & Orpin, 1987a). Exoglucanase active against microcrystalline cellulose (Avicel) was also detected, but at lower levels than CMCase (Mountfort & Asher, 1985). To date, the fungi are the only rumen microorganisms that can solubilise the most resistant form of cellulose known, cotton fibre. An *N. frontalis* strain produced extracellular exoglucanase(s) capable of solubilising highly resistant cotton fibre as well as Avicel (Wood et al., 1986). Exoglucanase(s) with the capacity to degrade the crystalline cellulose in cotton fibre were also produced by *Piromonas* species (Wood, T. M. & Joblin, K., unpublished). Production of these very active exoglucanases by *N. frontalis* was markedly increased when the fungus was co-cultured with the methanogen *Methanobrevibacter smithii* (Wood et al., 1986).

Partial hydrolysis of cellulose would yield cellodextrins, and culture supernatants of *N. frontalis* have been shown to contain a cellodextrinase with a greater specific activity than CMCase. The combination of extracellular exocellulase, endocellulase and cellodextrinase thus has the capacity to convert cellulose into cellobiose which can be fermented by the organisms.

CHEMICAL COMPOSITION

The lipids of *P. communis* and *Neocallimastix* sp. have been found to be quite similar (Kemp et al., 1984; Body & Bauchop, 1985). The major phospholipids were phosphatidylethanolamine, phosphatidylcholine and phosphatidylinositol. No sphingolipids, glycolipids, plasmalogens or phosphonyl lipids were detected. Free fatty acids, triacylglycerols, 1,2-diacylglycerols and some 1,3-diacylglycerol were identified, as well as small amounts of squalene and a triterpenol which was probably tetrahymanol. No sterols were detected by Kemp et al. (1984), and it is likely that squalene and tetrahymanol replace sterols in membranes. Body & Bauchop (1985) have suggested that sterols may be present in a strain of Neocallimastix sp. However, due to the anaerobic growth of the organism and the necessity for molecular oxygen in steroid synthesis, the latter is unlikely in these fungi. Growth of fungi was not inhibited by amphotericin B or nystatin (Lowe et al., 1987a) which normally inhibit sterol-synthesising fungi. Of the fatty acids, about half were straight-chain even-numbered 12–24 carbon saturated acids; the remainder were even-numbered 16–24 carbon mono-unsaturated acids, with the double bond in the 9ω position, except in the 16 carbon acid where it was probably 7ω. Evidence was presented (Kemp et al., 1984) which suggested that oleate was synthesised by desaturation of stearate. The authors ruled out the possibility of oxygen being involved in the reaction by the vigorous exclusion of air from the system, the use of inhibitors, and the lack of spectroscopic evidence for cytochromes which might couple with other electron acceptors. Thus an anaerobic Δ^9 desaturase could be operative in these organisms. Body & Bauchop (1985) also suggest that an anaerobic pathway for monoenoic fatty acid biosynthesis occurs in Neocallimastix sp.

Fatty acid synthesis took place from both ^{14}C-acetate and ^{14}C-glucose (Kemp et al., 1984); long-chain fatty acids from the culture medium could be incorporated into complex lipids. ^{14}C-labelled choline, ethanolamine and serine were incorporated directly into complex lipids but, whilst ethanolamine was incorporated into phosphatidylcholine as well as phosphatidylethanolamine, serine was only incorporated as either phosphatidylethanolamine or phosphatidylserine.

The cell walls of the vegetative stages contain chitin (Orpin, 1977c), the proportion of chitin varying with species from at least 8% to 22%. Chitin synthetase has been located in the rhizoids (Brownlee, 1986a), and nikkomycin (an inhibitor of chitin synthetase) inhibited the growth of Neocallimastix sp. (Lowe et al., 1987a).

The free- and protein-amino acids of Neocallimastix and Piromonas communis have been determined by Kemp et al. (1985). N. patriciarum and P. communis contained, respectively, 24% and 30% protein by dry weight. About half of the proteins examined by HPLC were in the 200 000 molecular weight range and 40% in the 50 000 range; the native dissociated proteins were predominantly in the range from 25 000 to 50 000. The amino acid profiles of the two fungi were similar and compared favourably with casein and lucerne Fraction 1 protein. These organisms could therefore be a valuable source of amino acids to the host animal if they were digested in the lower digestive tract; crude protein values (N $\times$ 6·25) were 44% for N. patriciarum and 42% for P. communis. N. frontalis is, however, proteolytic (Wallace & Joblin, 1985), and this may offset any nutritional advantage resulting from its digestion by the animal. The proteolytic activity associated with the solid fraction of digesta in the rumen is considerably increased by the presence of Neocallimastix frontalis (Wallace & Munro, 1986).

Although the fate of the vegetative structure remaining after zoosporogenesis is not known, autolysis apparently occurs *in vitro* to some extent, indicating that the spent growth could itself contribute to rumen fermentation (Lowe *et al.*, 1987*a*).

ATTACK ON PLANT TISSUES

As shown previously, most of the major polysaccharides of plants are fermented by rumen fungi, and fungi are capable of solubilising a high proportion of the dry weight of plant fragments (Akin *et al.*, 1983; Orpin, 1983/84; Gordon & Ashes, 1984; Joblin, K. *et al.*, unpublished). In pure cultures *in vitro*, about 40–45% of the dry weight of wheat straw fragments was lost in 4 days. Whilst about 50% of the cellulose and hemicellulose was digested, about 16% of the lignin in the plant fragments was also lost (Orpin, 1983/84). The loss of lignin probably represents solubilisation rather than digestion, for rumen anaerobic fungi cannot ferment simple phenolic acids (Orpin, 1983/84), and free phenolic acids inhibit plant fibre digestion *in vitro* (Akin & Rigsby, 1985).

Rumen fungi have the ability to gain access to plant polysaccharides not available to cellulolytic bacteria. When barley straw was challenged by cultures of *Neocallimastix* or *Piromonas*, or the cellulolytic rumen bacterium *Ruminococcus albus*, the fungi solubilised 30–40% of the fragments, whereas *R. albus* solubilised only 8% (Joblin, K. *et al.*, unpublished). Thus fungi may be less dependent than bacteria on the comminution of plant fragments by the animal.

The means whereby anaerobic fungi penetrate plant cell walls, by mechanical force, enzymic action or a combination of the two (Kolattukudy, 1985), is not known. Rumen fungi are proteolytic, unlike the rumen cellulolytic bacteria (Wallace & Joblin, 1985), and their proteolytic action probably facilitates the penetration of the proteinaceous layer by the rhizoids; this layer prevents cellulolytic bacteria from gaining access to the secondary cell wall (Engels & Brice, 1985).

Fungi have been found attached in large numbers to refractory plant fragments removed from the rumens of domestic herbivores (Orpin, 1977*a*; Bauchop, 1980; Akin *et al.*, 1983), particularly with animals fed on a high-fibre diet (Bauchop, 1986). Fungi can be detected on fragments 2–3 hours after the fragments have been ingested (Bauchop, 1981). Initial colonisation occurs at sites of tissue damage (Bauchop, 1980) and at stomata (Orpin, 1977*a*; Akin *et al.*, 1983). In colonised plant fragments, lignified cell walls are preferentially colonised, although non-lignified cell walls are also attacked. After fungal attack the tensile strength of fragments is substantially lowered (Akin *et al.*, 1983). This weakening of tissues might substantially increase digestion by rumen bacteria by making more sites available for bacterial colonisation. The ability of rumen fungi to reduce the size of plant fragments (Orpin, 1983/84) suggests that fungal activity could affect the residence time of particles in the rumen, for only the smallest plant particles pass out of the rumen (Ullyat *et al.*, 1985).

Because of the above observations, the concept of a significant role for rumen fungi in fibre digestion seems to be compelling. Nevertheless, the net effect of the

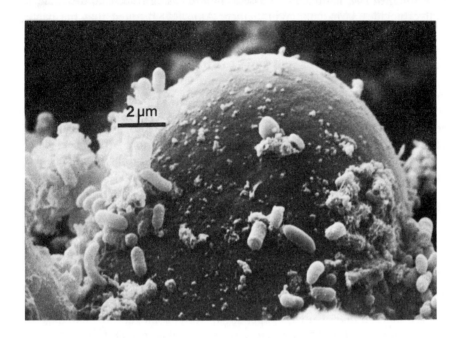

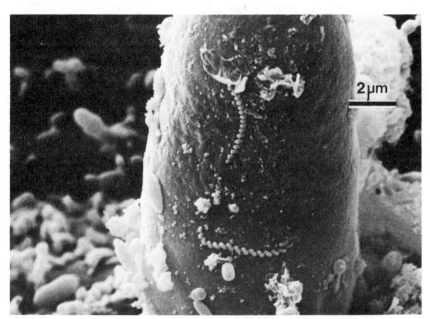

Fig. 6. Scanning electron micrograph showing bacteria attached to sporangia of rumen fungi growing on plant particles from the ruminal digesta of a sheep.

rumen fungi *in vivo* is difficult to assess. Indeed, using rumen fluid treated with selective biocides as inocula, Windham & Akin (1984) found that the bacterial population was more effective than the fungal population at degrading resistant grass fibre. This study suggested that the number of sporangia found on plant fragments does not indicate a substantial role for fungi in fibre digestion. Moreover, the authors noted that the most fibrous substrate did not support the greatest population of fungi.

The apparently conflicting conclusions on the significance of fungi in fibre digestion may be related to the choice of animal diets in the respective studies. Diet composition influences the populations and proportions of rumen bacteria (Hungate, 1966; see also Chapter 5), and not only dietary fibre content but also interactions within the microbial community are likely to affect fungal development.

In the rumen, bacteria are often found attached to fungal sporangia (Fig. 6), and growth of rumen fungi in association with bacteria has been shown to affect the activity of some of the fungal enzymes. *Neocallimastix* and *Piromonas* species form stable associations with methanogenic bacteria (Bauchop & Mountfort, 1981; Mountfort *et al.*, 1982; Wood *et al.*, 1986; Joblin, K. *et al.*, unpublished), and this has previously been mentioned as increasing cellulase activity. Such co-cultures degrade paper at a faster rate and to a greater extent than do fungal monocultures (Bauchop & Mountfort, 1981). With lignocellulosic substrates, this synergism is also apparent, but is less marked than that with paper as substrate. Some fungus–methanogen co-cultures solubilised straw fragments to a greater extent than did fungal monocultures and showed a shorter lag before digestion. However, the rate of straw solubilisation by a *Piromonas* sp. was not increased in the presence of a methanogen (Joblin, K. *et al.*, unpublished).

Little is known about the interaction of rumen fungi with other types of bacteria, but one report notes that when the lactate-utilising bacteria *Megasphaera elsdenii* and *Veillonella alcalescens* were co-cultured with a strain of *N. frontalis* the cellulolytic activity of the fungus was always enhanced, whereas culturing the fungus with rumen cellulolytic bacteria gave two distinct types of response. With *Ruminococcus albus* and *Ruminococcus flavefaciens* fungal digestion of straw was inhibited, whereas with *Bacteroides succinogenes* straw digestion was increased (Richardson *et al.*, 1986).

CONCLUSIONS

The recent discovery of fungi able to compete successfully in the anaerobic environment of the rumen offers a new dimension not only to the rumen microbial ecosystem but also to microbiology itself. However, relatively little is known about the physiology of the members of this novel group of anaerobes, and further studies are needed to increase our knowledge of their properties and potential. Studies of fungi growing under controlled conditions should improve our understanding of fungal interactions with lignocellulosic materials, and with other microorganisms involved in rumen function. Rumen fungi are not essential to rumen function *per se*, as they are present only in very low numbers in animals on a low-fibre diet, but it seems probable that fungi are important in the digestion of poor quality forage. In domestic animals under some dietary conditions manipulation of rumen fungal populations may improve fibre digestion (Gordon, 1986). An assessment of their contribution to fibre digestion in ruminants awaits a method for accurately quantifying fungal tissue in the solid phase of digesta, and comparative studies on animals with and without fungi in their rumen microflora.

ACKNOWLEDGEMENT

The preparation of samples for electron microscopy and of electron micrographs by D. H. Hopcroft and R. J. Bennett is gratefully acknowledged.

REFERENCES

Akin, D. E. & Rigsby, L. L. (1985). Influence of phenolic acids on rumen fungi. *Agron. J.*, **77**, 180–2.

Akin, D. E., Gordon, G. L. R. & Hogan, J. P. (1983). Rumen bacterial and fungal degradation of *Digitaria pentzii* grown with and without sulphur. *Appl. Env. Microbiol.*, **46**, 738–48.

Barr, D. J. S. (1981). The phylogenetic and taxonomic implications of flagellar rootlet morphology among zoosporic fungi. *Biosystems*, **14**, 359–70.

Bauchop, T. (1979). Rumen anaerobic fungi of cattle and sheep. *Appl. Environ. Microbiol.*, **38**, 148–58.

Bauchop, T. (1980). Scanning electron microscopy in the study of microbial digestion of plant fragments in the gut. In *Contemporary Microbial Ecology*, ed. D. C. Ellwood, J. N. Hedger, M. J. Latham, J. M. Lynch & J. H. Slater. Academic Press, London, pp. 305–26.

Bauchop, T. (1981). The anaerobic fungi in rumen fibre digestion. *Agric. Environ.*, **6**, 338–48.

Bauchop, T. (1983). The gut anaerobic fungi: colonizers of dietary fibre. In *Fibre in Human and Animal Nutrition*, ed. G. Wallace & L. Bell. Royal Society of New Zealand, Wellington, pp. 143–8.

Bauchop, T. (1986). Rumen anaerobic fungi and the utilization of fibrous feeds. *Rev. Rural Sci.*, **6**, 118–23.

Bauchop, T. & Mountfort, D. O. (1981). Cellulose fermentation by a rumen anaerobic fungus in both the absence and presence of rumen methanogens. *Appl. Environ. Microbiol.*, **42**, 1103–10.

Body, D. R. & Bauchop, T. (1985). Lipid composition of an obligatory anaerobic fungus *Neocallimastix frontalis* isolated from a bovine rumen. *Can. J. Microbiol.*, **31**, 463–6.

Braune, R. (1913). Untersuchungen uber die in Wiederkauermagen vorkommenden Protozoen. *Arch. Protist.*, **32**, 111–70.

Brewer, D., Duncan, J. M., Safe, S. & Taylor, A. Ovine ill-thrift in Nova Scotia. 4. The survival at low oxygen partial pressure of fungi isolated from the contents of the ovine rumen. *Can. J. Microbiol.*, **18**, 1119–28.

Brownlee, A. G. (1986*a*). Properties of chitin synthetase from a rumen anaerobic fungus. *Abstracts of XIV International Congress of Microbiology*, P.M4-3.

Brownlee, A. G. (1986*b*). Genome organization in an anaerobic fungus from sheep rumen. *Abstracts of XIV International Congress of Microbiology*, P.G1-19.

Chen, M. & Wolin, M. J. (1979). Effect of monensin and lasalocid-sodium on the growth of methanogenic and rumen saccharolytic bacteria. *Appl. Environ. Microbiol.*, **38**, 72–7.

Clarke, R. T. J. & DiMenna, M. E. (1961). Yeasts from the bovine rumen. *J. Gen. Microbiol.*, 113–17.
Coleman, G. S. (1978). Rumen entodiniomorphid protozoa. In *Methods of Cultivating Parasites* in vitro, ed. A. E. R. Taylor & J. R. Baker. Academic Press, London.
Engels, F. M. & Brice, R. E. (1985). A barrier covering lignified cell walls of barley straw that resists access by rumen micro-organisms. *Curr. Microbiol.*, **12**, 217–24.
Gordon, G. L. R. (1986). The potential for manipulation of rumen fungi. *Rev. Rural Sci.*, **6**, 124–8.
Gordon, G. L. R. & Ashes, J. R. (1984). *In vitro* digestion of wheat straw by different rumen anaerobic fungi. *Can. J. Anim. Sci.*, **64** (suppl.), 156–7.
Heath, I. B. & Bauchop, T. (1985). Mitosis and the phylogeny of the genus *Neocallimastix*. *Can. J. Bot.*, **63**, 1595–1604.
Heath, I. B., Bauchop, T. & Skipp, R. A. (1983). Assignment of the rumen anaerobe *Neocallimastix frontalis* to the Spizellomycetales (Chytridiomycetes) on the basis of its polyflagellate zoospore ultrastructure. *Can. J. Bot.*, **61**, 295–307.
Heath, I. B., Kaminsky, J. & Bauchop, T. (1986). Basal body loss during fungal zoospore encystment: evidence against centriole autonomy. *J. Cell. Sci.*, **83**, 135–40.
Hungate, R. E. (1966). *The Rumen and its Microbes*. Academic Press, New York.
Jensen, E. H. C. & Hammond, D. M. (1964). A morphological study of Trichomonads and related flagellates from the bovine digestive tract. *J. Protozool.*, **11**, 386–94.
Joblin, K. N. (1981). Isolation, enumeration and maintenance of rumen anaerobic fungi in roll tubes. *Appl. Environ. Microbiol.*, **42**, 1119–22.
Kemp, P., Lander, D. & Orpin, C. G. (1984). The lipids of the anaerobic rumen fungus *Piromonas communis*. *J. Gen. Microbiol.*, **130**, 27–37.
Kemp, P., Jordan, D. J. & Orpin, C. G. (1985). The free and protein-amino acids of the rumen phycomycete fungi *Neocallimastix frontalis* and *Piromonas communis*. *J. Agric. Sci.*, **105**, 523–6.
Kolattukudy, P. E. (1985). Enzymatic penetration of the plant cuticle by fungal pathogens. *Ann. Rev. Phytopath.*, **23**, 223–50.
Liebetanz, E. (1910). Die parasitischen Protozoen der Wiederkauermagens. *Arch. Protist.*, **19**, 19.
Lowe, S. E., Theodorou, M. K., Trinci, A. P. J. & Hespell, R. B. (1985). Growth of anaerobic rumen fungi on defined and semi-defined media lacking rumen fluid. *J. Gen. Microbiol.*, **131**, 2225–9.
Lowe, S. E., Griffiths, G. G., Milne, A., Theodorou, M. K. & Trinci, A. P. J. (1987a). The lifecycle and growth kinetics of an anaerobic rumen fungus. *J. Gen. Microbiol.*, **133**, 1751–8.
Lowe, S. E., Theodoru, M. K. & Trinci, A. P. J. (1987b). Cellulases and xylanase of an anaerobic rumen fungus grown on wheat straw, wheat straw holocellulose, cellulose and xylan. *Appl. Environ. Microbiol.*, **53**, 1216–23.
Lowe, S. E., Theodorou, M. K. & Trinci, A. P. J. (1987c). Growth and fermentation of an anaerobic rumen fungus on various carbon sources and effect of temperature on development. *Appl. Environ. Microbiol.*, **53**, 1210–15.
Lowe, S. E., Theodorou, M. K. & Trinci, A. P. J. (1987d). Isolation of anaerobic fungi from saliva and faeces of sheep. *J. Gen. Microbiol.*, **133**, 1829–934.
Lubinsky, G. (1955a). On some parasites of parasitic protozoa. 1. *Sphaerita hoari* sp.n.—a chytrid parasitizing *Eremoplastron bovis*. *Can. J. Microbiol.*, **1**, 440–50.
Lubinsky, G. (1955b). On some parasites of parasitic protozoa. II. *Sagittospora cameroni* gen. n., sp. n.—a phycomycete parasitizing Ophryoscolecidae. *Can. J. Microbiol.*, **1**, 675–84.
Lund, A. (1974). Yeasts and moulds in the bovine rumen. *J. Gen. Microbiol.*, **81**, 453–62.
Mountfort, D. O. & Asher, R. A. (1983). Role of catabolite regulatory mechanisms in control of carbohydrate utilization by the rumen anaerobic fungus *Neocallimastix frontalis*. *Appl. Environ. Microbiol.*, **46**, 1331–8.
Mountfort, D. O. & Asher, R. A. (1985). Production and regulation of cellulase by two strains of the rumen anaerobic fungus *Neocallimastix frontalis*. *Appl. Environ. Microbiol.*, **49**, 1314–22.

Mountfort, D. O., Asher, R. A. & Bauchop, T. (1982). Fermentation of cellulose to methane and carbon dioxide by a rumen anaerobic fungus in a triculture with *Methanobrevibacter* sp. strain RA1 and *Methanoscarcina barkeri*. *Appl. Environ. Microbiol.*, **44**, 128–34.

Munn, E. A., Orpin, C. G. & Hall, F. J. (1981). Ultrastructural studies of the free zoospore of the rumen phycomycete *Neocallimastix frontalis*. *J. Gen. Microbiol.*, **125**, 311–23.

Munn, E. A., Orpin, C. G. & Greenwood, C. A. (1987). Organization of the kinetosomes and associated structures of the zoospores of the rumen chytridiomycete *Neocallimastix*. *Can. J. Bot.*, **65**, 456–65.

Munn, E. A., Orpin, C. G. & Greenwood, C. A. (1988). The relationship of the obligate anaerobic fungi of the rumen. *Biosystems*, in press.

Orpin, C. G. (1974). The rumen flagellate *Callimastix frontalis*: does sequestration occur? *J. Gen. Microbiol.*, **84**, 395–8.

Orpin, C. G. (1975). Studies on the rumen flagellate *Neocallimastix frontalis*. *J. Gen. Microbiol.*, **91**, 249–62.

Orpin, C. G. (1976). Studies on the rumen flagellate *Sphaeromonas communis*. *J. Gen. Microbiol.*, **94**, 270–80.

Orpin, C. G. (1977a). Invasion of plant tissue in the rumen by the flagellate *Neocallimastix frontalis*. *J. Gen. Microbiol.*, **98**, 423–30.

Orpin, C. G. (1977b). The rumen flagellate *Piromonas communis*: its life-history and invasion of plant material in the rumen. *J. Gen. Microbiol.*, **99**, 107–17.

Orpin, C. G. (1977c). The occurrence of chitin in the cell walls of the rumen organisms *Neocallimastix frontalis*, *Piromonas communis* and *Sphaeromonas communis*. *J. Gen. Microbiol.*, **99**, 215–18.

Orpin, C. G. (1977d). On the induction of zoosporogenesis in the rumen phycomycetes *Neocallimastix frontalis*, *Piromonas communis* and *Sphaeromonas communis*. *J. Gen. Microbiol.*, **101**, 181–9.

Orpin, C. G. (1981). Isolation of cellulolytic phycomycete fungi from the caecum of the horse. *J. Gen. Microbiol.*, **123**, 187–96.

Orpin, C. G. (1983/84). The role of ciliate protozoa and fungi in the rumen digestion of plant cell walls. *Anim. Feed Sci. Technol.*, **10**, 121–43.

Orpin, C. G. & Bountiff, L. (1978). Zoospore chemotaxis in the rumen phycomycete *Neocallimastix frontalis*. *J. Gen. Microbiol.*, **104**, 113–22.

Orpin, C. G. & Greenwood, Y. (1986a). Effects of haems and related compounds on growth and zoosporogenesis of the rumen phycomycete *Neocallimastix frontalis* H8. *J. Gen. Microbiol.*, **132**, 2179–85.

Orpin, C. G. & Greenwood, Y. (1986b). Nutrition and germination requirements of the rumen phycomycete *Neocallimastix patriciarum*. *Trans. Br. Mycol. Soc.*, **86**(1), 103–9.

Orpin, C. G. & Letcher, A. J. (1979). Utilization of cellulose, starch, xylan and other hemicelluloses for growth by the rumen phycomycete *Neocallimastix frontalis*. *Curr. Microbiol.*, **3**, 121–4.

Orpin, C. G. & Munn, E. A. (1986). *Neocallimastix patriciarum*: new member of the Neocallimasticaceae inhabiting the sheep rumen. *Trans. Br. Mycol. Soc.*, **86**(1), 178–81.

Orpin, C. G., Mathiesen, S. D., Greenwood, Y. & Blix, A. S. (1985). Seasonal changes in the ruminal microflora of the High-Arctic Svalbard reindeer (*Rangifer tarandus platyrhynchus*). *Appl. Environ. Microbiol.*, **50**, 144–51.

Pearce, P. D. & Bauchop, T. (1985). Glycosidases of the anaerobic rumen fungus *Neocallimastix frontalis* grown on cellulosic substrates. *Appl. Environ. Microbiol.*, **49**, 1265–9.

Richardson, A. J., Stewart, C. S., Campbell, G. P., Wilson, A. B. & Joblin, K. N. (1986). Influence of co-culture with rumen bacteria on the lignocellulolytic activity of phycomycetous fungi from the rumen. *Abstracts of XIV International Congress of Microbiology*, PG2-24, 233.

Sivers, V. S. (1962a). Fungi of the order Mucorales in the rumen of cattle. *Mikrobiol. Zh.* (Kyyiv), **24**, 14–19.

Sivers, V. S. (1962b). Quantity of microscopic fungi in the rumen of cattle during winter and summer kelp. *Mikrobiol. Zh.* (Kyyiv), **24**, 51–8.

Sparrow, F. K., Jr (1960). *Aquatic Phycomycetes*, 2nd edn. University of Michigan Press, Ann Arbor.

Stewart, C. S., Duncan, S. H. & Joblin, K. N. (1985). The use of tritiated cellulose for the rapid enumeration of cellulolytic anaerobes. *Lett. Appl. Microbiol.*, **1**, 45–99.

Stewart, C. S., Duncan, S. H. & Joblin, K. N. (1987). Antibiotic manipulation of the rumen microflora: the effects of avoparcin and monensin on the release of tritrium from labelled cellulose by *Bacteroides succinogenes* and the rumen fungus *Neocallimastix frontalis*. In *Recent Advances in Anaerobic Bacteriology*, ed. S. P. Borriello & J. M. Hardie. Martinus Nijhoff, Lancaster, pp. 108–17.

Ullyat, M. J., Dellow, D. W., John, A., Reid, C. S. W. & Waghorn, G. C. (1985). The contribution of chewing during eating and rumination to the clearance of digesta from the reticulo-rumen. In *Control of Digestion and Metabolism in Ruminants*, ed. L. P. Milligan & W. L. Grovum. Prentice-Hall, New Jersey, USA, pp. 498–515.

Vavra, J. & Joyon, L. (1966). Etude sur la morphologie, le cycle evolutif et la position systematique de *Callimastix cyclopsis* Weissenberg 1912. *Protistologica*, **2**, 5–16.

Wallace, R. J. & Joblin, K. N. (1985). Proteolytic activity of a rumen anaerobic fungus. *FEMS Microbiol. Lett.*, **29**, 19–25.

Wallace, R. J. & Munro, C. A. (1986). Influence of the rumen anaerobic fungus *Neocallimastix frontalis* on the proteolytic activity of a defined mixture of rumen bacteria growing on a solid substrate. *Lett. Appl. Microbiol.*, **3**, 23–6.

Warner, A. C. S. (1966). Diurnal changes in the concentrations of microorganisms in the rumens of sheep fed limited diets once daily. *J. Gen. Microbiol.*, **45**, 213–35.

Williams, A. L. & Orpin, C. G. (1987a). Polysaccharide degrading enzymes formed by three species of rumen fungi grown on a range of carbohydrate substrates. *Can. J. Microbiol.*, **33**, 418–26.

Williams, A. L. & Orpin, C. G. (1987b). Glycoside hydrolase enzymes present in the zoospore and vegetative growth stages of the rumen fungi *Neocallimastix patriciarum*, *Piromonas communis* and an unidentified isolate, grown on a range of carbohydrates. *Can. J. Microbiol.*, **33**, 427–34.

Windham, W. R. & Akin, D. E. (1984). Rumen fungi and forage fibre degradation. *Appl. Environ. Microbiol.*, **48**, 473–6.

Wood, T. M., Wilson, C. A., McCrae, S. I. & Joblin, K. N. (1986). A highly active extracellular cellulase from the anaerobic rumen fungus *Neocallimastix frontalis*. *FEMS Microbiol. Lett.*, **34**, 37–40.

Yarlett, N., Orpin, C. G., Munn, E. A., Yarlett, N. C. & Greenwood, C. A. (1986a). Evidence for hydrogenosomes in the rumen fungus, *Neocallimastix patriciarum*. *Biochem. J.*, **236**, 729–39.

Yarlett, N. C., Yarlett, N., Orpin, C. G. & Lloyd, D. (1986b). Cryopreservation of the anaerobic rumen fungus *Neocallimastix patriciarum*. *Lett. Appl. Microbiol.*, **3**, 1–3.

5
Development of, and Natural Fluctuations in, Rumen Microbial Populations

B. A. Dehority

Department of Animal Sciences, Ohio State University, Wooster, Ohio, USA

&

C. G. Orpin

AFRC Institute of Animal Physiology and Genetics Research, Cambridge, UK
and
Department of Arctic Biology, University of Tromsø, Norway

Establishment of microbes in the young ruminant is primarily dependent upon its exposure to an adult animal and having a rumen environment compatible with growth of the microorganisms. The major portion of the rumen microbial population consists of strictly anaerobic bacteria and ciliate protozoa, which appear to account for most of the fermentative activity in this organ. Microbial numbers and the composition of the population are affected by a number of factors, of which diet is probably one of the most important. Smaller numbers of facultatively anaerobic bacteria (which may be of importance in very young ruminants), aerobic bacteria, flagellate protozoa, fungi and mycoplasmas are also present; however, their contribution to the overall fermentation is considered minimal. The only exception might be in the case of the anaerobic fungi, whose ecological niche in the rumen has been covered in Chapter 4.

Early studies on development and fluctuation of rumen microbial populations have been reviewed by Hungate (1966). Improved methodology and a growing interest in the overall rumen fermentation have stimulated research in this area, providing a better understanding of these parameters. The present chapter will focus chiefly on the ciliate protozoa and anaerobic bacteria, describing enumeration and counting procedures, inoculation, factors affecting the population, comparisons between domestic and wild ruminants, and interrelationships between the bacterial and protozoal populations.

Most of the quantitative information on microbial numbers will be presented on the basis of concentration or population density, i.e. number of cells per ml or g of rumen contents. Total numbers are reported in a few studies (e.g. Table 3) and generally represent the product of concentration and rumen volume or total contents weight.

RUMEN CILIATE PROTOZOA

Counting procedures

Concentrations of rumen ciliates are generally determined by counting fixed cells in a volumetric chamber deep enough to accommodate the largest of the protozoa. Although several different sizes and types of counting chambers have been used (Boyne et al., 1957; Warner, 1962a), the Sedgwick–Rafter cell is commercially available and, with proper precautions in filling, appears to be quite suitable for routine use (Purser & Moir, 1959). The cell measures exactly 1 ml ($50 \times 20 \times 1$ mm) and, using 30% glycerol as a diluent, all of the protozoa in a 1 ml volume settle to the bottom of the chamber within 5–10 minutes. Using a measured eyepiece grid, a specific number of fields are then counted at a magnification of $\times 100$.

A somewhat different technique has been used by Coleman (1958), primarily for counting protozoa cultured *in vitro*. A sample of the culture is diluted with 0.02M iodine in KI solution, and all of the protozoa in 0.1 ml of the mixture are counted microscopically.

Most studies reported in the literature have counted protozoan numbers in strained rumen fluid (Warner, 1962a; Clarke, 1964; Nakamura & Kanegasaki, 1969; Abe et al., 1973; Dennis et al., 1983). Dehority (1984) found that concentrations of rumen protozoa can be significantly lower in rumen fluid than in whole rumen contents, by a factor depending on the time of sampling and method used to separate the fluid fraction. Differences were more pronounced in samples taken after feeding, and it was suggested that the increased percentage of solids acted as a filter mat retaining more of the protozoa. Of particular interest was the fact that generic composition was markedly affected at all sampling times and under all conditions of separation. The percentage of *Entodinium* increased while percentages of the larger protozoa, *Diplodinium* and *Ophryoscolex*, decreased in the fluid fraction. *Isotricha* and *Dasytricha* concentrations were relatively unaffected.

Species identification

Identification of the protozoa is based entirely on microscopic observation, and identification and classification have been described in Chapter 3.

Faunation

Young animals
Since no resistant forms or cysts of rumen protozoa have been detected in feed, water or faeces, direct contact with a faunated animal appears to be the only route for transmission of these rumen organisms (Becker & Hsiung, 1929; Becker, 1932). Several studies have experimentally demonstrated that calves and sheep isolated from other ruminants at birth do not become faunated (Bryant et al., 1958; Bryant & Small, 1960; Eadie, 1962a; Dehority, 1978). Infection can occur by several means, some more subtle than others. When the mother grooms a young animal the

protozoa present in her mouth from rumination can be transferred to the offspring. Salivation on the feed, or even pasture, by a faunated animal, followed by ingestion of the contaminated material by a young animal, also leads to faunation. Observations by Eadie (1962a) also suggest that some of the smaller protozoa can be transmitted short distances between animals in airborne water droplets.

Assuming that the young animal is housed in contact with other ruminants and routinely exposed to infection as described above, then establishment of a protozoan fauna is dependent upon suitable environmental conditions within the rumen. Rumen pH is supposedly quite acid in young animals, presumably due to the rapid fermentation of soluble sugars in their diet of milk or starter ration (Becker, 1932). Bryant et al. (1958) noted a sequence of establishment of protozoa in calves, *Entodinium* first, followed by *Diplodinium* and then the holotrichs. In a later study with both sheep and calves, Eadie (1962a) found that *Entodinium* became established at a pH a little above 6·0, and *Diplodinium* and the holotrichs did not develop until the pH reached 6·5 or above.

Fonty et al. (1984) studied the establishment of protozoa in 15 lambs reared with their dams in the flock. The lambs were weaned at 6 weeks, but had access to meadow grass hay and concentrate from birth. *Entodinium* appeared in 15–20 days, *Polyplastron*, *Eudiplodinium* and *Epidinium* in 20–25 days and *Isotricha* by 50 days. The total population increased until the lambs were 2 months old ($5.7 \pm 3.6 \times 10^5$ protozoa/ml). These data would support the earlier studies of Bryant et al. (1958) and Eadie (1962a). In 60% of these lambs, protozoa disappeared, or their numbers decreased markedly, for a period of 1–4 weeks in the third month, indicating a somewhat unstable environment. Four other lambs were fed only cow's milk until 4 months old, but had contact with conventional lambs. *Entodinium* and *Polyplastron* became established in these animals, but at low population densities (5×10^4 and 5×10^2 protozoa/ml, respectively). At 3 months old, *Polyplastron* disappeared from all four lambs.

Data on rumen pH in very young animals are limited, with values reported from calves generally pH 6·0 or higher from 5 days onward (Bryant & Small, 1960; Williams & Dinusson, 1972). However, even in calves inoculated at 1 week old, protozoa did not become established until the animals were 3–6 weeks old (Bryant & Small, 1960). Feeding concentrates to young animals can lower rumen pH, or at least cause fluctuations to as low as pH 5·0, which presumably prevents protozoan establishment (Eadie et al., 1959, 1967; Eadie, 1962a). Purser & Moir (1959) have demonstrated that minimum rumen pH is a major factor in controlling concentrations of the protozoa. However, several other factors might influence the time at which protozoa can establish, i.e. the oesophageal-groove reflex in suckling animals should limit the amount of readily fermentable milk in the rumen, and the rate of passage of primarily fluid material from the rumen could result in protozoal washout. It is of interest that Eadie (1962a) noted a high rumen pH in calves fed milk by bottle and an earlier development of ciliates.

Protozoa-free ruminants
Williams & Dinusson (1972) inoculated rumen ciliate protozoa-free calves, 60–285

days of age, with 2–50 cells of either entodiniomorphs or holotrichs. Establishment was successful and the ciliate population densities obtained in 5–7 weeks were equal to those found 1–2 weeks after inoculation with 10 000–94 000 cells. About 16 weeks were required to establish similar protozoan population densities when a protozoa-free calf was allowed to run with a mature goat. These authors were able to establish holotrichs in calves free of entodiniomorphs, and suggested that the sequence of establishment noted earlier may be a result of the numbers of the particular genera normally occurring in rumen contents, i.e. entodinia occurring in the highest density. Unfortunately they did not report rumen pH values which would have allowed a more critical appraisal of those factors influencing the sequence of establishment of the different genera.

Six protozoa-free sheep which had been obtained by hysterectomy and reared in isolation for a year were inoculated with 300 ml of rumen fluid from a steer fed on alfalfa hay (Dehority, 1978). In three of the sheep also fed on alfalfa hay, the approximate maximum numbers of total protozoa per ml were attained by day 7 after inoculation. All genera, except *Isotricha* in one animal, became established in roughly the same proportions as in the inoculum. Total numbers per ml were higher at 7 days in the other three sheep fed on a concentrate diet, but numbers tended to fluctuate and the fauna was much less diverse. Protozoa-free ruminants are discussed further in Chapters 13 and 15.

Factors influencing population size and composition

Diurnal variations
Marked diurnal variations have been noted in the concentration of rumen protozoa. Purser & Moir (1959) first reported a distinct diurnal cycle for *Entodinium* in sheep. Numbers decreased for 6–8 hours after feeding and then gradually rose to prefeeding levels by 20–24 hours. Subsequent work by Purser (1961) established that a diurnal cycle also existed for the holotrichs; however, it differed from the cycle for *Entodinium*. Peak concentrations occurred at feeding time (animals fed once daily) and then numbers gradually diminished until 20 hours after feeding when a rapid increase occurred up to feeding time. These concentration cycles were confirmed by Warner (1962*b*).

In subsequent work, Warner (1966*a,b,c*) studied diurnal changes in protozoan concentrations in sheep fed a limited diet once daily, fed to appetite in pens or pasture, and fed a limited diet every three hours. His results for animals fed once daily were in agreement with previous data and expanded the observed diurnal cycle for *Entodinium* to include almost all the entodiniomorphs. Diurnal changes in protozoan concentrations for sheep fed to appetite were similar to those in animals fed once daily (Warner, 1966*b*). From monitoring of time spent eating, it appeared that almost all the daily intake was consumed during one major period. The studies on sheep given small amounts of feed every three hours suggested a three-hour cycle of concentration changes (Warner, 1966*c*); however, a gradual decline in concentration occurred over time, presumably from the repeated sampling.

Warner's (1966*a*) conclusions, based upon measurements of dilution rate, were that the diurnal fluctuation in entodiniomorph concentrations was the end result of changes in dilution rate associated with eating (increased saliva flow and drinking) and changes in protozoan growth rate in response to incoming nutrients. This explanation appears to be in agreement with the results obtained with cattle by Clarke (1965) who found that total numbers of entodiniomorphs in the rumen-reticulum did not decrease after feeding.

Purser (1961) and Warner (1966a) both observed a marked increase in holotrich concentrations in sheep just prior to feeding. For lack of a better explanation, Warner (1966a) suggested a very rapid multiplication of the holotrichs within a 4–8 hour period around feeding time, with no further divisions for 16–18 hours. However, the numbers of dividing cells observed during the time of rapid increase in numbers did not substantiate this explanation (Warner, 1966a; Michalowski, 1977; Dehority & Mattos, 1978).

Somewhat in contrast to the previous data on holotrichs, Clarke (1965) did not observe a rise in total holotrich numbers in cattle until feeding time, with numbers peaking in the first several hours after feeding. Similar cycles for holotrich concentrations in cattle were later noted by Abe *et al.* (1981) and Murphy *et al.* (1985). Visual and microscopic observations of the inner walls of the rumen and reticulum by Abe *et al.* (1981) suggested that the holotrichs sequester on the reticulum wall a few hours after feeding, and migrate into the rumen again at the next feeding. On the basis of the chemotaxis of *Isotricha* to soluble carbohydrates previously demonstrated by Orpin & Letcher (1978), the above authors proposed that this migration at feeding could be a chemotactic response to soluble sugars in the incoming feed. Their studies also suggested that the quantity of feed and act of ingesting feed could be additional stimuli for migration. Murphy *et al.* (1985) were able to show that glucose solution infused into the reticulum stimulated migration of the holotrich protozoa into the rumen, whereas water, artificial saliva, NaCl or starch solutions had no effect. In addition, they found that bypassing the act of feed ingestion by placing chopped straw directly into the rumen also elicited migration of the holotrichs. The rapid decrease in holotrich numbers after feeding would appear to be the result of their return to the reticulum wall; however, the factors controlling the sequestration remain to be studied.

In general, the prefeeding rise in holotrich concentrations has primarily been observed in sheep, while increases in cattle occur immediately after feeding. However, it should be noted that Michalowski (1975) did observe a prefeeding increase in holotrich concentrations in water buffalo housed in a zoo. Abe *et al.* (1981), reporting on some preliminary experiments in goats, found an appreciable increase in holotrich numbers before feeding. Any explanation for the rise in holotrich concentrations based on composition of feed or act of feeding would apply only to increases after feeding. Data presented by Warner (1966a) and Dehority (1970) indicate that, although holotrich concentrations begin to rise before feeding, there is an additional increase after feeding, presumably a chemotactic response to incoming feed. Other unknown factors, unique to the small ruminant, could be responsible for the prefeeding increase in holotrich concentrations.

The diurnal curves for protozoan concentrations presented by Michalowski

(1977), Warner (1966a,b,c) and others indicate that percentage generic distribution varies considerably during a 24 hour period. Little information is available on species distribution during this same time period; however, data presented by Clarke (1965) and Dehority (1970) suggest that the proportion of *Dasytricha ruminantium* to *Isotricha* (two species) changes with time after feeding.

Diet effects

The influence of diet on protozoan concentrations in sheep has been studied by Nakamura & Kanegasaki (1969) and by Grubb & Dehority (1975). In the study by Nakamura & Kanegasaki (1969), sheep were changed from a diet of 1500 g orchard grass hay plus 600 g of concentrates (28·5% concentrate) per day to 1500 g orchard grass hay per day. The rations were fed in equal portions twice a day. Protozoan concentrations were in the range $7-12 \times 10^5$ per ml on the hay-concentrate diet, and $2-4 \times 10^5$ per ml on hay alone. Grubb & Dehority (1975) abruptly changed their sheep from an all-roughage diet to a 60% corn–40% roughage diet, with 800 g of diet being fed once daily. Concentrations ranged between 4 and 6×10^5 protozoa per ml on 100% orchard grass hay, rose markedly during the five days following the diet change, and then stabilized between 10 and 18×10^5 protozoa per ml. Although there were differences in the experimental design of these two studies, they both used similar types of diets and, when the amount of available energy in the ration increased, protozoan concentrations increased. Similar increases in protozoan concentrations have been observed in cattle and water buffalo when concentrates were added to the diet (Abe *et al.*, 1973; Michalowski, 1975; Dehority & Mattos, 1978; Dennis *et al.*, 1983).

As the percentage of concentrates in the diet increases to 60% or more, there is generally a corresponding decrease in minimum rumen pH values (Briggs *et al.*, 1957; Abe *et al.*, 1973; Mackie *et al.*, 1978; Wedekind *et al.*, 1986). This can result in a decrease in protozoan concentrations, a shift towards *Entodinium* species and, in some cases, complete disappearance of the protozoa (Latham *et al.*, 1971; Vance *et al.*, 1972; Abe *et al.*, 1973; Schwartz & Gilchrist, 1975; Mackie *et al.*, 1978). The type of grain also influences rumen pH and protozoan concentrations (Slyter *et al.*, 1970). It would appear that rations containing about 40–50% roughage will support maximal protozoan numbers with a diverse fauna containing species of most of the genera.

Experiments by Nakamura & Kurihara (1978) and Czerkawski & Breckenridge (1979a), using continuous *in vitro* fermentation systems, have demonstrated the importance of solid digesta in the maintenance of protozoan numbers. The protozoa apparently sequester in the solid digesta (see Chapter 12) and concentrations in the effluent are only 10–20% of those associated with the particulate matter. An inert solid matrix (wood shavings or predigested hay residue) plus a balanced soluble substrate did not provide adequate conditions for maintenance of the protozoa (Czerkawski & Breckenridge, 1979b). Regular addition of solid digestible hay was necessary to simulate a normal rumen fermentation.

As discussed earlier, straining the rumen contents can markedly affect generic composition, particularly numbers of *Entodinium* (Dehority, 1984). In several studies where distribution has been determined in whole rumen contents, proportions of *Entodinium* have ranged from about 90–98% on concentrate-type diets to 40–90% on hay or pasture diets (Michalowski, 1975; Puch, 1977; Dehority, 1978, 1979). The majority of the remaining ciliates were from genera in the subfamily Diplodininae and, as would be expected, constituted about 2–10% on concentrates to 10–55% of the population on hay or pasture. One exception would be the high incidence of *Epidinium* (20–25%) which has been observed in New Zealand cattle grazing on fresh red-clover (Clarke, 1964) and caribou on native pastures (Dehority, 1986b).

Level of intake
When sheep were fed a pelleted high-concentrate ration to appetite, rumen protozoa were eliminated in most cases or reduced to a very low concentration (Christiansen *et al.*, 1964). However, relatively high concentrations of protozoa were obtained when the same ration constituted only two-thirds of the total feed. When various physical forms of a ration were fed, protozoan numbers were inversely related to particle size and rate of passage of feed through the rumen.

Warner (1962b) fed the same diet to two sheep at levels ranging from 300 to 1200 g per day. Some decrease in protozoan concentrations was observed at the 300 g intake level; however, he concluded that the level of a given diet above a certain minimum has little effect on protozoan concentrations. These observations were later substantiated in a more comprehensive study by Potter & Dehority (1973). Their data indicated that energy may be the important factor controlling protozoan concentrations at low intake levels, whereas feed passage rate becomes the controlling factor at higher intakes.

Dearth *et al.* (1974) fed the same diet to sheep at either 1·0 or 1·8 times their daily maintenance energy requirement, and found that protozoan numbers were significantly decreased at the 1·8 times maintenance intake. The concentration decrease occurred primarily in the genera *Dasytricha*, *Entodinium* and *Ophryoscolex*.

Dehority (1978) fed 3 sheep on 800 g of a roughage diet and 3 sheep on 1400 g of a concentrate diet. Mean protozoan concentrations were $38·9 \times 10^4$ per ml for the roughage fed animals and $118·4 \times 10^4$ per ml for the concentrate fed animals. Average liquid rumen volumes were 6·37 and 2·57 litres for the roughage and concentrate sheep, respectively, while fluid turnover rates were similar on both diets. The differences in protozoan concentrations and rumen volumes were both significant. Although the amount of dry matter in the rumen contents may vary slightly (3–5%) between roughage and concentrate feeds, multiplication of volume by concentration should give an estimate of total protozoa in the rumen. The resulting values, $2·43 \times 10^9$ and $2·98 \times 10^9$ protozoa in the rumens of the roughage and concentrate fed animals were not significantly different. Thus rumen volume, as influenced by level and type of diet, can be of major importance when evaluating protozoan populations.

Frequency of feeding
The effects of multiple feedings upon rumen protozoan concentrations were first demonstrated by Moir & Somers (1956). In a Latin-square design experiment with

sheep, they found that feeding the same quantity of feed 4 times daily instead of once a day resulted in a doubling of protozoan concentrations. If the same quantity of diet was fed twice daily, protozoan concentrations were intermediate but still significantly higher than those of the once-a-day feeding.

The most plausible explanation for this increase in numbers would be that multiple feedings prevent the drastic fluctuations in rumen pH which can be inhibitory to protozoa. For example, when a given level of concentrates was fed to cows twice a day, rumen pH ranged from about 5·85 to 6·65; however, when it was fed 6 times daily, rumen pH fluctuated only between 6·15 and 6·4 (Kaufmann et al., 1980). In a recent study, Bragg et al. (1986) determined the diurnal pattern for rumen pH and protozoan concentrations of steers fed on corn silage concentrate diets (40:60) either 2 or 8 times a day. Using a Latin-square design, minimum pH values when the animals were fed twice daily were 5·45 compared to 5·8 when they were fed 8 times. This was reflected in slightly higher protozoan concentrations and considerably less fluctuation over the day in those animals fed 8 times daily.

Clarke et al. (1982) fed two levels of chaffed alfalfa hay, either hourly or once a day, to 32 sheep. The sheep were slaughtered at the end of the experiment to measure weight of rumen contents. In general, protozoan concentrations were highest in those animals fed the high level of hay at hourly intervals. However, calculated total protozoan dry matters varied as much as from 14 to 70 g in two sheep in the same group. The authors concluded that there was marked variability in numbers, sizes and masses of ciliate protozoa in the rumens of individual sheep fed the same diet. These data would suggest that an experimental design like the Latin square, where each animal is on all treatments, is almost essential in these types of studies. Obviously this would prevent collection of any data requiring samples which could only be obtained by slaughter.

Seasonal differences

For animals grazing native pastures, seasonal changes can cause marked variation in protozoan numbers. The two principal seasonal changes, i.e. hot to cold or wet to dry, both inhibit or slow down plant growth and result in a decrease in energy available to the animal. Pearson (1965, 1969) observed a marked decrease in rumen protozoan concentrations in mule deer from Utah and in white-tailed deer from Texas during the winter months. Similar winter decreases in numbers occurred in red deer and sheep in the Scottish Highlands (Hobson et al, 1976). Westerling (1970) has reported a 45% decrease in protozoan numbers in Finnish reindeer between August and November, the latter samples taken about 2 weeks after snow cover. *Entodinium anteronucleatum* appeared to be the only species which increased in concentration during winter when the reindeer fed mainly on lichen (Westerling, 1970).

Protozoan concentrations in zebu cattle doubled between the dry (5.9×10^4/ml) and wet (12×10^4/ml) seasons in Senegal (Bonhomme-Florentin et al., 1978). Entodinia comprised 89% of the total ciliates in the dry season, with 5·7% *Diplodinium* and 4·2% holotrichs. During the wet season, the genus *Epidinium* accounted for 6% of the population and the holotrichs 7%. *Entodinium* ranged from

35% to 85%, with the remainder of the ciliates in the genus *Diplodinium*. Van Hoven (1978) followed protozoan numbers in the blesbok over an 18-month period. Concentrations of protozoa were highest during the wet season. Changes in *Entodinium* and *Opisthotrichum* concentrations were similar: high in summer (wet season) and low in winter (dry season). *Diplodinium* species showed the opposite trend.

A compilation of protozoan concentrations to illustrate the effects of the feeding variables discussed above is extremely difficult. Variations in type of feed, diet composition, intake level, time of feeding, time of sampling, number of feedings per day, and season and animal species make comparisons somewhat questionable. In general, from the references cited, protozoan concentrations in domestic ruminants fed on mostly roughage diets or pasture range from about 10 to 50×10^4 per ml. Values for animals fed concentrate type rations are usually $50-150 \times 10^4$ protozoa per ml; however, concentrations up to 300×10^4 per ml are occasionally reported.

Differences within and between domestic ruminant species

Protozoan concentrations, occurrence of species, and number of species can vary markedly between animals within a single ruminant species, as well as between different ruminant species. An obvious factor contributing to these differences would be geographical location, which probably reflects differences in diet, origin of the animals and their possible isolation from other ruminants. Specificity of the rumen ciliates as well as potential antagonism between species could also contribute to the variations observed.

Geographic difference

Clarke (1964) did not observe the genus *Ophryoscolex* in New Zealand cattle and sheep; also, although its absence is not known for certain, this genus is not mentioned in a number of papers from Australia. A report from Finland (Westerling, 1969), on samples obtained from 24 cattle and 8 sheep, does not include the genus *Ophryoscolex*, nor was it observed in sheep or cattle from Japan (Nakamura & Kanegasaki, 1969; Imai *et al*, 1982). In samples collected by Dehority from Brazil and Peru, *Ophryoscolex* was not observed in the rumen contents of *Bos indicus*, *Bos taurus*, sheep, goats, water buffalo or alpaca. However, this genus was found in several cattle samples obtained from Chile.

A very common protozoan in the USA, Europe and Japan, *Polyplastron multivesiculatum*, was not observed in New Zealand (Clarke, 1964), or in rumen contents of *Bos indicus* (humped cattle) and *Bos gaurus* (gaur) from India and Ceylon (Kofoid & MacLennan, 1932; Kofoid & Christenson, 1934). Occurrence of this species in Brazil was limited to very low numbers in only one animal (*Bos taurus*). It was also absent from the Peruvian material, but present in Chilean cattle. It seems unusual that *Polyplastron* would be absent from so many geographic areas; however, this may reflect the source of importation of ruminants and their subsequent isolation in certain areas, i.e. New Zealand and Brazil.

In other studies, Abou Akkada & el-Shazly (1964) and Naga *et al*. (1969) observed the protozoan population in both cow and buffalo calves and in sheep from various regions in Egypt, and suggested that the genus *Epidinium* is absent from all ruminants in Egyptian territory.

Protozoan concentrations and number of species per animal, for water buffalo and cattle located in various geographical locations, are presented in Table 1. Considerable variation can be seen, both within and between ruminant species.

Table 1
Comparison of Protozoan Concentrations and Number of Species per Animal between Water Buffalo and Cattle Located in Different Geographical Areas

Host and location	No. of animals	Total protozoa $\times 10^4$/ml	Number of species	Ref.[a]
Water buffalo				
Indonesia	17	1·5 (0·1–31·6)[b]	12·9 (8–20)	1
Thailand	10	0·7 (0·2–2·0)	9·4 (2–17)	2
Taiwan	29	8·9 (0·5–316·2)	11·5 (3–25)	3
Philippines	2	4·7 (1·5–7·8)	8·0 (7–9)	4
Brazil	4	22·9 (16·6–35·8)	29·0 (22–35)	5
Okinawa	5	37·4 (27·0–49·5)	9·6[c]	6
Zebu cattle				
Thailand	46	7·1 (0·6–31·6)	26·1 (14–39)	2
Philippines	4	15·8 (13·2–18·1)	20·0 (18–22)	4
Senegal	24	9·0 (3·6–31·0)	13·2 (8–18)	7
Sri Lanka	20	2·9 (0·1–31·6)	18·4 (6–29)	8
Brazil	4	26·4 (9·0–51·2)	30·2 (22–36)	9
Cattle				
Japan	69	13·5 (0·5–3981)	10·3 (4–25)	10

[a] References: (1) Imai, 1985; (2) Imai & Ogimoto, 1984; (3) Imai *et al.*, 1981*a*; (4) Shimizu *et al.*, 1983; (5) Dehority, 1979; (6) Imai *et al.*, 1981*b*; (7) Bonhomme-Florentin *et al.*, 1978; (8) Imai, 1986; (9) Dehority, 1986*a*; (10) Imai *et al.*, 1982.
[b] Mean and range.
[c] Range not reported.

Between ruminant species, concentrations and number of species per animal tend to be higher in zebu cattle, while within ruminant species the animals located in Brazil show higher values. Unfortunately not enough quantitative data, particularly with regard to number of protozoan species, are available for sheep and other domestic ruminants to include in such a comparison.

Important information not included in Table 1 is the occurrence of the different species. Presentation of species compositions requires considerable space, and the reader is referred to the references in Table 1, Hungate (1966) and Ogimoto & Imai (1981) for more information. Imai (1985) has attempted to make such comparisons; e.g. he found that 80·8% of the protozoan species in Thailand water buffalo were common with the species detected in Indonesian water buffalo. In contrast, only 48·8% of the species were common between Brazilian and Indonesian water buffalo.

He discusses these differences in relation to geographical location and origin of the animals.

Since publication of the species compilations cited above, several new genera and unusual species of rumen protozoa have been reported (see also Chapter 3). *Parabundleia ruminantium, Polymorphella bovis* and *Blepharoconus krugerensis* (all in the family Buetschlidae) were observed in zebu cattle (Imai & Ogimoto, 1983; Imai, 1984; Dehority, 1986a). *Microcetus lappus* has been found in rumen contents of Norwegian Red cattle (Orpin & Mathiesen, 1986). A single species of Cycloposthiidae, *Parentodinium africanum*, has been identified in rumen contents of zebu cattle in Brazil (Dehority, 1986a), and this appears to be the first reported occurrence of a cycloposthid in the rumen habitat.

Specificity of rumen ciliates

Two types of specificity might be postulated for the rumen protozoa, i.e. host specificity and protozoan specificity. In host specificity, the animal itself by some unknown 'physiological factors' could influence the genera and species which establish in its rumen. These factors would include the type and amount of feed consumed, rate of consumption of feed, and saliva production, which in turn would influence rumen pH, rate and type of fermentation, osmolality, and turnover times of fluid and particulate matter. Dogiel (1927) once considered 10 species of rumen protozoa to be specific for the reindeer. However, 9 of these species have subsequently been observed in other animal hosts with a wide geographical distribution (Dehority, 1986b). On the other hand, the occurrence of these species is sporadic and they do not occur as a group, suggesting the existence of a specific rangifer-type fauna in feral reindeer and caribou. Additional evidence for host specificity would be that two genera of Ophryoscolecidae, *Epiplastron* and *Opisthotrichum*, have only been observed in several species of African antelope (Dogiel, 1932; Kofoid & MacLennan, 1933; Van Hoven, 1975). Experiments involving inoculation of defaunated cattle and sheep with African antelope rumen contents would add much to our knowledge concerning host specificity of these two genera.

The genus *Caloscolex*, family Ophryoscolecidae, is specific to the camel (Dogiel, 1926). Although the camel is not a true ruminant, its forestomach is inhabited by rumen-type ciliates.

Protozoan specificity can be defined on the basis of 'races' existing within a given species, e.g. a race of cattle *Epidinium ecaudatum* which would differ from a race of sheep *E. ecaudatum*. Bovine-rumen ciliates were successfully established in protozoa-free or 'almost protozoa-free' goats by Becker & Hsiung (1929) and Dogiel & Winogradowa-Fedorowa (1930). However, not all species became established, which suggested to the authors that a slight degree of specificity existed between domesticated ruminants. Naga *et al.* (1969) reached a similar conclusion in experiments involving inoculation of newly born cow and buffalo calves with rumen contents from adult cows, buffaloes and sheep.

This question of protozoan specificity was investigated by Dehority (1978) who inoculated 6 protozoa-free sheep with rumen contents from a steer fed on alfalfa hay. All 24 species of protozoa in the steer inoculum became established in 3 of the sheep which were fed on the same alfalfa hay diet. In contrast, only 9 species established in the 3 remaining sheep which were fed on a concentrate type diet. It was concluded that the diet of the donor and recipient could be a major factor in determining the fauna established by cross-inoculation. This appeared to be a very feasible explanation for the slight specificity noted in the earlier studies, where diet differences existed between donor and recipient.

Antagonism between species

The type of fauna established in a given animal can be influenced by antagonistic relationships which appear to exist between certain rumen ciliate species. Eadie (1962a,b) first observed that two general types of ciliate populations seemed to occur in domestic ruminants. Type A contained entodinia and holotrich species, *Diplodinium (Diploplastron) affine* and *Ophryoscolex tricoronatus*, plus *Polyplastron multivesiculatum* as the predominant large entodiniomorph. Type B contained entodinia, holotrichs, *Diplodinium* and *Ostracodinium* species, together with a large entodiniomorph, either *Eudiplodinium* or *Epidinium*, or both. In general the type A fauna was predominant, and inoculation of an animal containing a type B rumen fauna with type A rumen fluid always resulted in an irreversible change to the type A population. Subsequent studies (Eadie, 1967) indicated that the type B organisms disappeared because of predation by *Polyplastron multivesiculatum*. Eadie found that *Polyplastron* will eliminate *Epidinium* and *Eudiplodinium maggii* plus many other *Diplodinium* and *Ostracodinium* species from a protozoal population. This predation by *Polyplastron* differs from accidental predation as discussed by Lubinsky (1957a), in that it leads to the complete removal of the prey species from the population.

In these same studies, Eadie also observed that there appears to be some type of antagonism between *Ophryoscolex* and *Epidinium*. When both genera are present in the same fauna, *Ophryoscolex* slowly disappears. *Ophryoscolex* was also difficult to establish in young sheep or goats, and its presence appeared to be variable in older ruminants.

As mentioned in Chapter 3, Coleman et al. (1977) found that *Entodinium bursa* (*E. vorax*) grown *in vitro* had an obligate requirement for the spineless form of *E. caudatum*. Attempts to replace *E. caudatum* with *E. simplex*, *E. longinucleatum*, mixed entodinia from the rumen, or killed cells of *E. caudatum* were all unsuccessful, so a specific antagonism between protozoan populations is again shown.

Protozoa in wild ruminants

Although studies of wild ruminants are somewhat limited, they can provide insight into the ecology of rumen protozoa. Early studies by Eberlein (1895) and Kopperi (1926, 1929) on the rumen ciliate fauna of reindeer housed in zoos clearly established the need for sampling wild ruminants in their native habitat. Dogiel (1947) and Lubinsky (1957b,c,d) discussed phylogeny of the rumen ciliates, and in general concluded that specific genera and species evolved after divergence of the animal families (e.g. *Opisthotrichum* in African antelope). Isolation of the animals and effect of available feedstuffs are probably the major contributing factors to the fauna occurring in various wild ruminants.

Protozoan concentrations were determined in 6 species of South African wild ruminants by Giesecke & Van Gylswyk (1975). The buffalo, blue wildebeest and gemsbok, classified as grazers on the basis that grasses comprised 85–100% of their diet, had concentrations of 169, 398 and 311 × 10^3 protozoa/g of rumen contents, respectively. In contrast, protozoan concentrations in the browsers, impala, springbok and kudu, were 400, 579 and 1059 × 10^3/g, respectively. Rumen contents and rumen water, as a percentage of body weight, were lower in the browsers, indicating a smaller rumen volume. Presumably this was a reflection of the higher nutrient density of feed consumed by the browsers.

Ogimoto & Imai (1981) have recently compiled a table listing the distribution of rumen ciliate genera in various hosts. There are obvious differences between the various animal hosts; however, the more subtle species differences are not shown. Space requirements prohibit the presentation of species distribution in the various wild ruminants, but several examples can be cited. *Entodinium alces* was first described from rumen contents of Alaskan moose (Dehority, 1974) and subsequently found to occur in South African giraffe and springbok (Kleynhans & Van Hoven, 1976; Wilkinson & Van Hoven, 1976). Occurrence of this species in such widely separated geographical areas with completely different climates and vegetation is quite perplexing. It will be of interest to determine whether any other hosts for this species can be found.

The fauna of Dall sheep in Alaska was found to differ markedly from that of other wild arctic ruminants (Dehority, 1974) but was fairly similar to the fauna observed in a Sierra Nevada bighorn sheep from California (Bush & Kofoid, 1948). Until 1948 only one species of *Polyplastron* had been described, *P. multivesiculatum*, which occurred in domestic cattle and sheep around the world (Dogiel, 1927). Bush & Kofoid (1948) described a new species, *P. californiense*, from the Sierra Nevada bighorn sheep. Subsequently, Lubinsky (1958) described *P. arcticum* from Canadian reindeer, and Dehority (1974) described *P. alaskum* from Alaskan Dall sheep. From these observations one might speculate that the other species evolved from *P. multivesiculatum* after the specific animal hosts became isolated. The natural habitats of these particular hosts would be somewhat similar, providing rather harsh nutritional and environmental pressures. This in turn, over a long period of time, could have been selective for *Polyplastron* species which are better adapted to these conditions.

RUMEN FLAGELLATE PROTOZOA

In his studies on diurnal changes in the concentrations of rumen microorganisms in sheep fed once daily, Warner (1966a) observed that a flagellate protozoon, presumably *Neocallimastix frontalis*, increased in numbers dramatically after feeding. Population density then decreased rapidly, reaching a low point about 10 hours after feeding, and remained low until the next feeding. To accomplish such marked concentration changes, at least 2–5 divisions of *Neocallimastix* would have to occur within 1 hour after feeding, followed by rapid death and lysis. Warner suggested that the flagellates probably sequestered on the rumen wall, returned to the rumen contents by a chemotactic response to feeding, and then migrated back to the wall in response to another chemical stimulus. Orpin (1974) later investigated this question and did not observe any flagellates in washings of the rumen wall or in rumen fluid immediately adjacent to the wall. Population density could be increased *in vivo* by adding an extract of oats to the rumens of sheep which had not been fed for 24 hours. By centrifugation techniques he determined that *Neocallimastix* was associated with the larger feed particles, and he suggested that the organisms sequestered there, or their life cycle involved a multiple reproduction phase which was associated with the large particle fraction of rumen fluid.

In subsequent studies, Orpin (1975) found that the rapid increase in *Neocallimastix* shortly after feeding resulted from stimulation of a reproductive body on a vegetative phase of the organism, which differentiated and liberated the flagellates. The stimulant was a component of the host's diet, later identified as haeme or haeme-containing compounds (Orpin, 1978). The flagellates liberated *in vivo* lose motility within 1 hour and develop into the vegetative phase, thus explaining the rapid decrease in concentration. The author also noted that the vegetative stage was morphologically similar to certain species of aquatic phycomycete fungi. Investigation of the other rumen flagellates revealed that two additional organisms, *Sphaeromonas communis* and *Piromonas communis*, also had similar life cycles and their vegetative stages resembled phycomycete fungi (Orpin, 1976, 1977*a*). For these 3 organisms, maximum production of flagellates occurred at pH 6·5 and 39°C, in the presence of CO_2 and absence of O_2, all of which are normal rumen conditions.

Since fungi are the only non-photosynthetic microorganisms which contain chitin or cellulose in their vegetative cell walls, Orpin (1977*b*) analysed cultures of the three rumen organisms *N. frontalis*, *S. communis* and *P. communis* for these compounds. He found that the vegetative cell walls of all three organisms contained chitin, confirming that they are true fungi despite their ability to grow under low redox potential and in the absence of O_2. The flagellates released from the vegetative stage (sporangia) would thus be correctly called zoospores. The rumen fungi have been discussed in detail in Chapter 4.

At the present time, five species of true flagellate protozoa from the rumen have been described. They are *Monocercomonas ruminantium*, *Monocercomonoides caprae*, *Chilomastix caprae*, *Tetratrichomonas buttreyi* and *Pentatrichomonas hominis*, all in the class Zoomastigophorea, subphylum Mastigophora, phylum Sarcomastigophora (Ogimoto & Imai, 1981). The rumen flagellates are rather small, ranging from 4 to 15 µm in length. The body is elliptical to piriform with 3–5 anterior flagella and a nucleus located at the anterior end of the body. A posterior flagellum is present in one species, *Pentatrichomonas hominis*.

Jensen & Hammond (1964) successfully cultivated clone cultures of *Pentatrichomonas hominis* from the rumen, and found a mean generation time of about 5·2 hours for this species. It was also noted that the number of anterior flagella in *P. hominis* varied from 3 to 5 in clone cultures. *Tetratrichomonas buttreyi* strains were much more difficult to maintain in culture, while *Monocercomonas ruminantium* could not be cultured for more than a few days.

Reported concentrations of the rumen flagellates are generally low, $<10^5$/ml, with fluctuations over the feeding period and considerable variation between animals (Eadie, 1962*a*; Warner, 1962*b*, 1966*a,b,c*). In addition, these values are probably inflated by inclusion of the recently identified fungal zoospores in the count. Since the flagellates occur in low numbers and probably utilize only soluble substrates, their contribution to the overall rumen fermentation is considered to be minor (Clarke, 1977). Flagellate numbers are reported to be higher in young animals before ciliates become established (Eadie, 1962*a*) or after ciliates have been removed from older animals (Hungate, 1978). However, further studies are needed to differentiate between the true flagellates and fungal zoospores.

Transmission of the flagellates probably occurs by similar routes to those of the ciliates, i.e. by direct contact or in aerosols. Early observations indicated that the flagellates were transmitted much more easily and over greater distances than the ciliates; however, the organisms mentioned were generally those now identified as fungal zoospores (Becker, 1929; Becker & Hsiung, 1929; Eadie, 1962*a*).

RUMEN BACTERIA

Enumeration procedures

In general, two methods have been used to estimate bacterial numbers in the rumen: direct counts and culture or viable counts. The reader is referred to Hungate (1966) for a comprehensive discussion on the relative merits and disadvantages of these two methods, as well as a comparison of the rumen bacterial numbers obtained with each. Sampling procedures, use of whole rumen contents instead of rumen fluid, and various culture techniques have been reviewed by Hungate (1966) and Ogimoto & Imai (1981). The following section will attempt to describe briefly the newer methodologies which have been proposed for enumeration of rumen bacteria. Classification and properties of the individual rumen bacteria are discussed in Chapter 2 and some more details of media and methods are given there.

Bacterial concentrations
Caldwell & Bryant (1966) developed a medium without rumen fluid which gave colony counts similar to those obtained in a habitat-simulating medium containing rumen fluid. Essentially the media were identical except that rumen fluid was replaced by haemin, trypticase, yeast extract and a mixture of volatile fatty acids. These same two media were compared in a later study in England, and the medium without rumen fluid was found to give significantly higher colony counts (Thorley *et al.*, 1968). Grubb & Dehority (1976) proposed a procedure for total counts using 4 ml per tube of standard 40% rumen fluid in agar medium containing 0·1% total added

carbohydrate (0·025% each of glucose and cellobiose and 0·05% soluble starch). Total colony counts of rumen contents from sheep consuming four different types of rations were significantly higher with their medium and procedures than were those obtained with a medium without rumen fluid. Counts were also higher than those obtained with the rumen fluid medium of Caldwell & Bryant (1966); however, this difference was only significant with rumen contents from animals fed on high-concentrate diets.

Specific carbohydrate-utilizing subgroups
Dehority & Grubb (1976) developed a 40% rumen-fluid basal medium for the selective enumeration of rumen bacteria utilizing specific energy sources. Essentially their procedure included a 7-day preincubation of the medium (minus carbohydrates, Na_2CO_3 and cysteine) and the addition of xylose as a fourth substrate (with glucose, cellobiose and starch) when determining total counts. With no added carbohydrates this preincubated medium supported only 10% or less of the viable counts measured in the presence of added carbohydrates, but without preincubation the value was 80%. Thus, for selective enumeration of bacteria utilizing carbohydrate energy sources, preincubation of the medium is essential in reducing the growth of bacteria growing on carbon sources contributed by the rumen fluid.

Chung & Hungate (1976) found that inclusion of alkali-treated alfalfa fibre as a substrate in their 40% rumen-fluid medium resulted in increased colony counts. Henning & Van der Walt (1978) subsequently found that xylan could replace the fibre fraction, and they concluded that to obtain the highest possible colony counts some source of pentose is needed in the medium. They suggested xylan rather than xylose since not all bacteria which degrade xylan can utilize xylose (Dehority, 1973).

Leedle & Hespell (1980) incorporated preincubated rumen fluid into a new medium which they used both in roll-tubes and on agar plates (prepared and incubated in an anaerobic glove-box). Using either their new medium or several previously described media, they found colony counts to be higher on plates than in roll-tubes. Replica plating methods were used with the anaerobic plates, allowing differentiation of the population into specific carbohydrate-utilizing subgroups.

Prior to development of selective media, organisms utilizing specific carbohydrates had to be isolated from high dilutions and identified by appropriate methods. This was very time-consuming, which limited the number of animals to be studied, and could be biased unless sufficient bacterial colonies were picked completely at random. These selective techniques appear to offer a major advance in methodology for characterizing rumen microbial populations.

Specific media
Two media relatively specific for individual bacterial species have also been reported. Using a low pH medium (5·9–6·1) with mannitol as the only energy source, Tiwari *et al.* (1969) were able selectively to culture *Selenomonas ruminantium* from rumen contents. Omission of branched-chain volatile acids and haeme also contributed to the specificity of the medium. Iverson & Millis (1977) monitored *Streptococcus bovis* concentrations in rumen fluid by including antibiotics (nalidixic

acid and colistin sulphate) in an agar plate medium. Using potato starch as a substrate, and counting only colonies surrounded by a halo (indicating starch hydrolysis), the procedure was very specific for the enumeration of *S. bovis*.

Inoculation and establishment of rumen bacteria

Inoculation of young ruminants with normal or typical rumen bacteria is undoubtedly dependent upon transmission from the dam by routes similar to those described earlier for faunation. However, limited information would suggest that the rumen bacteria are more resistant to environmental stresses and can survive under conditions where the protozoa are killed (Bryant *et al.*, 1958; Bryant & Small, 1960). Mann (1963) found three genera of normal anaerobic rumen bacteria in air samples collected within a cow barn. The extent to which authentic rumen bacteria can be transmitted over extended distances through air, water or by carriers (such as clothing of the animal caretaker) remains to be determined.

Young amimals
Early studies by Bryant *et al.* (1958) and Bryant & Small (1960) showed that total anaerobic bacterial counts were higher in isolated calves than in normal or inoculated calves, up to 17–18 weeks of age. Predominant bacteria in both isolated and inoculated calves 1–3 weeks old differed from those found in mature cattle. Many bacteria similar to those occurring in mature animals were present at 6 weeks, and by 9–17 weeks the predominant bacteria were those typical of mature cattle.

Williams & Dinusson (1972) transferred 31 calves into isolation stalls within 24–48 hours after birth. Total anaerobic bacterial numbers were quite high at 1 week old (11×10^9/g), decreased during weeks 2 and 3, returned to their original level by week 5 and remained at this level until at least week 34. In contrast, total numbers of aerobic and facultatively anaerobic bacteria were very high at week 1 (90×10^7/g), decreased by half during week 2, fell to 2.5×10^7/g at week 5, and after 34 weeks were 1.0×10^7. Based on differential counts (anaerobic haemolytics, lactate-fermenters, and lipolytics, plus aerobically cultured aciduric-acidophiles, coliforms, haemolytics, lactobacilli and non-lactose-fermenters) they noted that the different bacterial subgroups shifted to levels comparable to those in mature ruminants as the calves matured. Cellulolytic bacteria (1×10^4/g) were present in calf rumen contents at 5–7 days old, with numbers increasing to 2×10^6/g by 2 weeks and remaining at a similar level until at least 34 weeks.

Fonty *et al.* (1984) found that a strictly anaerobic microflora was established in flock-reared lambs within 2 days after birth, in numbers almost comparable to those in adult sheep. However, the predominant genera observed in lambs under 10 days old were different from those in adult ruminants. Numbers of facultative anaerobes were also quite high at day 2, but then decreased steadily until the animals were at least 4 months old. Cellulolytic bacteria had become established in many animals by 4 days, preceding the intake of solid food.

The bacterial populations of sheep taken by hysterectomy and reared in strict isolation on sterilized diets were studied by Males (1973). In general, dry matter digestion was 2–10% less in isolated sheep than in normally reared sheep, whereas cellulose digestion was decreased by 15–40%. Bacterial populations in the rumens of the isolated sheep were atypical, containing a high proportion of aerobic and facultatively anaerobic bacteria along with a high proportion of the less-common obligate anaerobes usually found in the rumen. These differences persisted for the 6 months duration of the study. It would appear that the rumen is a very suitable habitat for microbial growth, and in the absence of normal rumen bacteria this niche is filled by available organisms which can adapt to this environment. The development of defined bacterial populations in gnotobiotic lambs is described in Chapter 15.

Factors affecting the bacterial population

Diurnal changes

Using direct counting procedures, Warner (1966a) found that, for sheep fed once daily, the concentration of total rumen bacteria decreased from 1 to 4 hours after feeding, increased slowly to a maximum between 12 and 20 hours, and then gradually decreased till the next feeding. Concentrations of 3 morphologically distinct bacterial groups (selenomonads, Eadie's ovals and peptostreptococci) were also determined, and in general they followed similar growth patterns. It was concluded that these concentration patterns reflect an initial dilution by feed, water and saliva, an increase in growth rate in response to incoming nutrients which exceeds the dilution rate, and finally a depletion of nutrients with a corresponding decrease in growth rate until it is less than the dilution rate. It is of interest that this pattern was very similar to the changes in viable bacterial concentrations observed earlier by Bryant & Robinson (1961).

The concentrations of rumen bacteria in sheep fed to appetite in pens or pastures followed similar growth patterns to those in animals fed once daily (Warner, 1966b). This appeared to be the result of the animals consuming most of their daily intake in one period of continuous eating. In contrast, rumen bacterial concentrations were fairly stable with time in sheep fed on a limited ration every 3 hours (Warner, 1966c).

Bryant & Robinson (1968) investigated the effects of diet and sampling site within the rumen on bacterial concentrations in cattle at various times after feeding. Four diets were used (chopped hay, pelleted hay, hay-grain and silage) and the animals were fed at 12 hour intervals. For all diets, concentrations were lowest at 1 hour after feeding, and increased significantly between 1 and 2.5 and 2.5 and 5.5 hours. Concentrations did not differ between 5.5 and 10 hours. Bacterial concentrations were highest in samples taken from the dorsal rumen, with lower (but similar) numbers occurring in samples taken from the ventral rumen and reticulum.

Leedle et al. (1982) used direct counts and viable counts (determined with agar plates) to study the diurnal variations in bacterial numbers in cattle fed on maintenance levels of high-forage or high-concentrate diets once daily. Direct and viable counts decreased after feeding; the lowest values were observed at 2 hours and 4 hours after feeding with the high-concentrate and high-forage diets, respectively. Concentrations then increased steadily, reaching their highest values at 16 hours.

These data would support the previous observations of Warner (1966a) with sheep. The lowest viable proportion of the direct-count bacterial populations occurred at 2 hours after feeding (14·6% and 14·1%), while the highest values (48·6% and 73·5% on the high-forage and high-concentrate diets, respectively) were found at 16 hours. Both Bryant & Burkey (1953) and Maki & Foster (1957) had previously observed a higher percentage of viable bacteria when animals were fed on concentrate diets. Because of the marked differences in bacterial viability with time after feeding, the magnitudes of concentration changes are greater with viable counts than with direct counts. The authors suggested that, in addition to dilution and increased passage from the rumen, the loss of viable bacteria after feeding may be due to rapidly changing rumen conditions such as osmotic shock effects, temperature changes, pH changes, entrance of oxygen, and attachment of organisms to incoming feed particles.

Diet

A number of recent studies can be found in the literature which compare total viable bacterial concentrations in different animals fed on either high-forage or high-concentrate diets. In general, bacterial concentrations are higher in those animals receiving a high-concentrate diet (Caldwell & Bryant, 1966; Hungate, 1966; Grubb & Dehority, 1976; Dehority & Grubb, 1980; Leedle & Hespell, 1980). However, there are also several reports in which numbers are equal or higher in animals fed on high-roughage diets (Bryant & Robinson, 1968; Latham et al., 1971; Dehority & Grubb, 1977; Van der Linden et al., 1984; Leedle et al., 1986). Differences between such factors as percentage of concentrate in the diet, feeding frequency, feeding level and sampling time, and individual animal variation, all appear to influence bacterial concentrations and in turn make comparisons difficult. The data compiled in Table 2 present bacterial concentrations found in the same animals fed on high-forage or high-concentrate diets. These data indicate that bacterial concentrations do tend to

Table 2
Comparison of Rumen Bacterial Concentrations Measured in the Same Animals Fed on High-forage or High-concentrate Diets

Animal species	No. of animals	Sampling time (h after feeding)	Total bacteria $\times 10^9$/ml or g		Ref.[a]
			High-forage	High-concentrate	
Cattle	1	4	2·4	11·0	1
Cattle	2	16	11·0	18·6	2
Cattle	3	4–5	0·30	0·30–0·51[b]	3
Sheep	3	0	5·6	21·0	4
Sheep	4	2	2·6	8·5	5

[a] (1) Bryant & Robinson, 1961; (2) Leedle et al., 1982; (3) Latham et al., 1971; (4) Grubb & Dehority, 1975; (5) Mackie & Gilchrist, 1979.
[b] Lower count (0.30×10^9) obtained when feeding 80% rolled barley; higher count when feeding 80% flaked corn.

Table 3
Effect of Diet on Rumen Volumes, Bacterial Concentrations and Total Bacterial Numbers in Sheep[a]

Sheep	Measurement	Diet[b]	
		Orchardgrass hay	60% corn–40% orchardgrass hay
1	Rumen volume (litres)	6·45[e]	4·78[f]
	Bacterial concentrations ($\times 10^8$)/g	38·2[c]	60·2[d]
	Total bacteria ($\times 10^{12}$)	24·6	28·8
2	Rumen volume (litres)	5·96	5·72
	Bacterial concentrations ($\times 10^8$)/g	57·1[c]	255·9[d]
	Total bacteria ($\times 10^{12}$)	34·0[e]	146·4[f]
3	Rumen volume (litres)	5·13[e]	4·54[f]
	Bacterial concentrations ($\times 10^8$)/g	74·1[c]	314·0[d]
	Total bacteria ($\times 10^{12}$)	38·0[c]	142·5[d]

[a] Data from Grubb & Dehority (1975).
[b] 800 g fed once daily at 9 am.
[c,d] Means within a row followed by different superscripts are significantly different at $P < 0.01$.
[e,f] Means significantly different at $P < 0.05$.

increase with an increased intake of available energy. However, rumen volume can also be influenced by the type of ration. Data for individual sheep from the study by Grubb & Dehority (1975) are shown in Table 3. Although bacterial concentrations on high-roughage or high-concentrate diets were different in all three animals, adjusting for rumen volume eliminated the difference between diets in sheep 1.

Thorley et al. (1968) compared bacterial concentrations in two cows fed *ad libitum* on either long grass or the same grass ground and pelleted. Mean colony counts were significantly higher when the animals were given ground grass (15.7×10^9/g) rather than long grass (10.5×10^9/g). However, such factors as pH and rate of fluid and particulate matter turnover could have affected these values.

Level and frequency of feeding
Studies which focus on either of these two parameters specifically are quite limited. Moir & Somers (1957) found that rumen bacterial concentrations were similar in sheep fed 1, 2, or 4 times daily. In a later study by Warner (1966c), bacterial numbers did not show much fluctuation with time in animals fed every 3 hours.

Using a poor quality teff hay diet, Gilchrist & Kistner (1962) fed 3 sheep on 1200, 600 and 300 g/day for three consecutive periods of 60, 109 and 42 days. They found no differences in viable counts of bacteria fermenting cellulose, starch, glucose, xylose or lactate between feed intake levels. Warner (1962b) reached a similar conclusion when comparing total bacterial concentrations in sheep fed with a similar range of intakes.

In a more controlled study, Dearth et al. (1974) determined total bacterial concentrations in animals fed on the same diet at maintenance or 1·8 × maintenance level. Using 8 sheep in a 4 × 4 Latin square design, bacterial concentrations were significantly increased by the higher feed intake.

Animals fed on pasture or range
Bryant et al. (1960) concluded that cattle fed on ladino clover pasture had higher bacterial numbers than animals fed on dry roughage or silage. However, no obvious differences were found by Puch (1977) when comparing bacterial numbers in cattle which were fed on pasture and then, later, on the same kind of hay as the pasture they had grazed. The animals were allowed to eat the hay *ad libitum*.

Rumen bacterial counts were made on grazing Red deer, hill sheep and reindeer in the Scottish Highlands by Hobson et al. (1976). For the different seasons over several years, counts were lowest in winter and highest during the summer. The authors suggested that this reflected the low nitrogen content of winter herbage. A study on Svalbard reindeer by Orpin et al. (1985) indicated that rumen bacterial concentrations during the summer months were similar to those found in domestic ruminants consuming high-concentrate diets ($20·9 \times 10^9/g$). However, bacterial concentrations decreased to less than 20% of this value during the winter. As might be expected, this difference appeared to be related to nutritional quality, and availability, of food.

Giesecke & Van Gylswyk (1975) measured viable bacterial numbers in rumen contents from six species of South African wild ruminants. Three species were classified as grazers and three as browsers; however, no obvious relationship between animal feeding type and bacterial concentrations was noted. Viable counts, taken during the dry season, ranged from $13·3 \times 10^8/g$ in buffalo to $46·6 \times 10^8/g$ in springbok.

Antibiotics
The carboxylic polyether ionophore antibiotics have received considerable attention in recent years because of their ability to improve efficiency of production of growing ruminants. One action of these compounds appears to be inhibition of the Gram-positive rumen bacteria, with a corresponding enrichment of the Gram-negative population (Brulla & Bryant, 1980; Dennis et al., 1981) (see Chapter 2). As a result, propionate production is increased, methane production is depressed, and protein degradation is decreased in the rumen (Van Nevel & Demeyer, 1977). Chen & Wolin (1979) have suggested that the ionophore selects for succinate-forming *Bacteroides* and for *Selenomonas ruminantium*, a propionate producer that decarboxylates succinate to propionate. Selection against the Gram-positive hydrogen- and formate-producing organisms could account for the decreased production of methane. Feeding the ionophores monensin or salinomycin was also found to increase the percentage of resistant rumen bacteria (Dawson & Boling, 1983; Olumeyan et al., 1986). The principal mode of action of the ionophores appears to be a disruption of primary membrane transport in cells, both bacterial and animal (Bergen & Bates, 1984; Schelling, 1984) (see Chapter 2).

Distribution of species or specific carbohydrate-fermenting groups

Most of the early studies on changes in the distribution of rumen bacterial species were based on the isolation of organisms from roll-tubes containing high dilutions of rumen contents. As mentioned earlier, this procedure is very laborious and time-consuming, and some questions might be raised about the probability of picking colonies from high dilutions that are truly representative of the whole population. Development of satisfactory selective media and use of replica plating techniques have allowed examination of the broader changes in specific carbohydrate-fermenting groups. Because of their relative simplicity, these latter techniques can be used to monitor changes associated with time after feeding, and with diet, as well as differences between animals.

Diurnal changes

Using the anaerobic replica plating technique, Leedle et al. (1982) studied diurnal variations of specific carbohydrate-utilizing subgroups in rumen bacterial populations. Two cattle, first fed on a high-forage diet and then changed to a high-concentrate diet, were fed once daily, and rumen samples for determination of bacterial numbers and subgroups were taken 1 hour before feeding and 2, 4, 8, 12 and 16 hours after feeding. On both diets, concentrations of total viable bacteria and those subgroups utilizing glucose, soluble starch, pectin, xylan-xylose or cellulose were lowest at 2 or 4 hours after feeding. Peak concentrations on the high-forage diet occurred at 16 hours for total viable count and subgroups utilizing xylan-xylose or cellulose. The other subgroups reached their highest levels at 12 hours. In contrast, all of these groups peaked at 16 hours when the animals were fed on the high-concentrate diet. In general, the major carbohydrate-utilizing subgroups at all sampling times were those using glucose and soluble starch. Xylan-xylose, pectin and cellulose subgroups comprised about one-half, one-third and one-tenth of the population, respectively, when the high-forage diet was fed. The proportion of all of these latter subgroups decreased when the animals were fed on the high-concentrate diet.

Diet effects

Although information in this area is limited, distributions have been reported in studies comparing roughage form, i.e. long or ground and pelleted (Thorley et al., 1968), type and level of grain and diethylstilboestrol (Slyter et al., 1970), and purified diets with or without starch and different nitrogen sources (Slyter et al., 1971). Three studies have been conducted comparing high-roughage with high-concentrate diets, and the data are shown in Table 4 (Caldwell & Bryant, 1966; Latham et al., 1971, 1972). For purposes of combining the data from the different studies, organisms have been grouped at the generic level. In general, *Butyrivibrio* spp. were the predominant organisms on high-roughage diets, while the selenomonads, peptostreptococci, streptococci and lactobacilli predominated when high-concentrate diets were fed. Caldwell & Bryant (1966) further differentiated *Bacteroides* into *B. ruminicola*, which predominated with the high-roughage diet, and *B. amylophilus* which predominated with the high-concentrate diet.

Table 4
Distribution of Major Bacterial Genera in Rumen Contents of Cattle Fed on High-roughage or High-concentrate Diets[a]

Presumptively identified genera	Percentage of total isolates					
	Caldwell & Bryant (1966)		Latham et al. (1971)		Latham et al. (1972)	
	High-roughage (HR)	High-concentrate (HC)	HR	HC	HR	HC
Butyrivibrio	23	9	38	10	25	2
Selenomonas	0	16	5	12	7	19
Bacteroides	22	26	8	9	11	4
Ruminicoccus	0	1	4	4	11	10
Peptostreptococcus	0	0	0	8	6	14
Streptococcus	0	0	3	6	4	8
Lactobacillus[b]	0	4	2	17	6	20

[a] All isolations were made from Medium 10 of Caldwell & Bryant (1966).
[b] Organisms in *Bifidobacterium* were included in this genus by Latham *et al.* (1971, 1972).

Mackie & Gilchrist (1979) estimated the changes in amylolytic and lactate-utilizing bacteria during stepwise adaptation of 4 sheep to a high-concentrate diet. The percentages of amylolytic and lactate-utilizing bacteria increased from 1·6% and 0·2% on the low (10%) concentrate diet to 21·2% and 22·3% on the high (71%) concentrate feed. *Lactobacillus, Butyrivibrio* and *Eubacterium* were the principal genera of amylolytic bacteria with the high-concentrate diet, while *Anaerovibrio* and *Propionibacterium* became the predominant lactate-utilizers. A somewhat similar study was reported by Roxas (1980), in which the numbers of amylolytic and lactate-utilizing bacteria were determined in sheep during abrupt changes from roughage to high-concentrate diets. Percentages of amylolytic and lactate-utilizing bacteria were 50% and 25% during roughage feeding and 85% and 32% when feeding high-concentrate. The predominant amylolytic species isolated when feeding high-concentrate were *Butyrivibrio, Selenomonas, Bacteroides* and *Streptococcus*, while *Selenomonas, Propionibacterium* and *Anaerovibrio* were the predominant lactate-utilizers. These results differ from those of Mackie & Gilchrist (1979) in that percentages of both amylolytic and lactate-utilizing bacteria were much higher and, except for *Butyrivibrio*, different genera of bacteria predominated in the high-concentrate rumen population.

Van der Linden *et al.* (1984) observed no significant changes in the concentrations of cellulolytic or xylanolytic bacteria in sheep fed on diets containing from 0 to 39% grain. The predominant cellulolytic bacterium resembled *Ruminococcus albus*, and *Butyrivibrio fibrisolvens*-like rods were the major hemicellulolytic organisms. The percentage of *R. albus* decreased with grain supplementation due to increases in concentrations of the other species (*R. flavefaciens*-like cocci and *B. fibrisolvens*-like rods); however, proportions of the hemicellulolytics did not change. *Bacteroides succinogenes*, an extremely active rumen cellulolytic species, was not isolated in this study; however, recent studies have shown that this species cannot produce clearings in 2% agar medium, the criterion for cellulolysis used by Van der Linden *et al.* (Stewart *et al.*, 1981; Macy *et al.*, 1982).

Variation between ruminant species
Using selective media, Dehority & Grubb (1976) and Leedle & Hespell (1980) estimated the proportions of specific carbohydrate-utilizing bacteria in cattle and sheep fed on high-forage and high-concentrate diets (Table 5). Dehority & Grubb

Table 5
Percentages of Specific Carbohydrate-utilizing Subgroups in Rumen Contents of Cattle and Sheep Fed on High-forage or High-concentrate Rations

Carbohydrate source in medium	% of total population			
	High-forage		High-concentrate	
	Cattle[a]	Sheep[b]	Cattle[a]	Sheep[b]
Starch	64	61	81	96
Xylan[c]	52	64	47	61
Pectin	29	59	50	60
Glycerol	18	43	60	34
Mannitol	5	32	51	37

[a] Leedle & Hespell (1980).
[b] Dehority & Grubb (1976).
[c] Xylose was included in this medium by Leedle & Hespell (1980).

(1976) obtained samples just before feeding from sheep fed once daily. In contrast, Leedle & Hespell (1980) obtained samples from cattle fed twice a day about 3–4 hours after the morning feeding. The two studies gave fairly similar results for the starch-utilizing (amylolytic) and xylan-utilizing (hemicellulolytic) subgroups; however, marked differences were observed with the other carbohydrate subgroups. It is not known whether these discrepancies are the result of diet differences, animal species, sampling times or methods.

Bacterial species in wild ruminants

Most studies on rumen bacteria in wild ruminants have been limited to microscopic observations on morphology and the Gram stain. In general, bacterial populations in red deer, reindeer, mule deer, white-tailed deer and buffalo appeared similar to those described in domestic ruminants (Pearson, 1965, 1967, 1969; Hobson *et al.*, 1976). In other studies, Hungate *et al.* (1959) have reported colonies resembling

Bacteroides succinogenes and *Butyrivibrio* in rumen contents from eland, kongoni and camel.

Information obtained by isolation and subsequent identification of bacterial species from wild ruminants is extremely limited. Studies have been conducted with rumen contents from elk (McBee *et al.*, 1969), Alaskan semi-domestic reindeer (Dehority, 1975), Svalbard reindeer (Orpin *et al.*, 1985) and Alaskan moose (Dehority, 1986*b*). It is unusual that almost all the animal species studied are in the family Cervidae (deer and allied animals), primarily those residing in Arctic environments. *Bacteroides* spp. were predominant in the elk, followed by *Butyrivibrio* spp. and *Ruminococcus* spp. *Butyrivibrio* spp. predominated in both species of reindeer, with *Streptococcus bovis* and *Selenomonas ruminantium* also occurring in significant percentages. The Svalbard reindeer also contained fairly high numbers of *Bacteroides* spp., which were not found in Alaskan reindeer. The predominant species in moose was *Streptococcus bovis* followed by *Butyrivibrio* spp. and *Lachnospira multiparus*. Based on these limited data, it would appear that *Butyrivibrio* spp., *S. bovis* and *S. ruminantium* are the principal bacterial species occurring in Arctic ruminants. Aside from the fact that a much higher percentage of the *Butyrivibrio* spp. isolated from Svalbard reindeer were cellulolytic, these organisms appear to be similar in most characteristics studied to the same species isolated from domestic ruminants. Proteolytic and ureolytic strains were also common amongst the Svalbard reindeer isolates.

INTERRELATIONSHIPS BETWEEN RUMEN BACTERIAL AND PROTOZOAL POPULATIONS

Some aspects of the roles of the bacterial and protozoal populations in the overall rumen fermentation have been discussed in Chapters 2 and 3 and will be further considered in later chapters. In the literature most attention has been given to the question of whether the protozoa are essential to the ruminant animal, probably because it is possible to establish and study protozoa-free or defaunated ruminants (see also Chapters 13 and 15). Beginning with the experiments of Becker *et al.* (1930), it has been established that the rumen protozoa are not essential to the host. In a recent review, Veira (1986) has summarized and evaluated most of the experiments comparing faunated and defaunated animals, and he concluded that the major nutritional effect of rumen protozoa is upon the protein/energy ratio of nutrients available for absorption in the small intestine. This would offer a possible explanation for the major areas in which the rumen protozoa appear to be involved: affecting animal growth rate, feed intake and feed digestibility, the effects varying with diet and physiological age; exerting a levelling or buffering effect on the rumen, resulting in a more stable rumen fermentation; and influencing the quality and quantity of protein passing down the digestive tract.

In general, the metabolic capabilities of the bacteria and protozoa appear to be similar (Prins, 1977; Coleman, 1980; Chapters 2 and 3). Bacterial concentrations have been found to be much higher in defaunated than faunated animals, and to decrease markedly following faunation (Eadie & Hobson, 1962; Kurihara et al., 1968; Eadie & Gill, 1971). Thus it would appear that, if protozoa are absent, the fermentation of feedstuffs is taken over by an increased bacterial population. Orpin & Letcher (1984) observed a 327% increase in rumen fluid bacterial concentrations after defaunation, which because of volume changes represented a 480% increase in total rumen bacterial numbers. Total microbial protein concentrations in rumen samples (liquid–small particle phase) were measured by Teather et al. (1984). They found that the variation among animals was less for total microbial protein concentrations than for either the bacterial or protozoal protein levels. A highly significant negative correlation was found between concentrations of the bacterial and protozoal proteins. In 22 lactating cows, the ratio of bacterial to protozoal protein concentrations ranged from 0·08 to >1000.

In summary, different levels of energy intake would be expected to support different amounts of total microbial protoplasm. The proportions of bacteria and protozoa contributing to this total, as well as the different species involved, could be influenced by a variety of factors, most of which have been enumerated and discussed earlier in this chapter.

REFERENCES

Abe, M., Shibui, H., Iriki, T. & Kumeno, F. (1973). Relation between diet and protozoal populations in the rumen. *Br. J. Nutr.*, **29**, 197–202.

Abe, M., Iriki, T., Tobe, N. & Shibui, H. (1981). Sequestration of holotrich protozoa in the reticulo-rumen of cattle. *Appl. Environ. Microbiol.*, **41**, 758–65.

Abou Akkada, A. R. & el-Shazly, K. (1964). Effect of absence of ciliate protozoa from the rumen on microbial activity and growth of lambs. *Appl. Microbiol.*, **12**, 384–90.

Becker, E. R. (1929). Methods of rendering the rumen and reticulum of ruminants free from their normal infusorian fauna. *Proc. Nat. Acad. Sci. USA*, **15**, 435–8.

Becker, E. R. (1932). The present status of problems relating to the ciliates of ruminants and equidae. *Quart. Rev. Biol.*, **7**, 282–97.

Becker, E. R. & Hsiung, T. S. (1929). The method by which ruminants acquire their fauna of infusoria, and remarks concerning experiments on the host-specificity of these protozoa. *Proc. Nat. Acad. Sci. USA*, **15**, 684–90.

Becker, E. R., Schulz, J. A. & Emmerson, M. A. (1930). Experiments on the physiological relationships between the stomach infusoria of ruminants and their hosts, with a bibliography. *Iowa St. J. Sci.*, **4**, 215–51.

Bergen, W. G. & Bates, D. B. (1984). Ionophores: their effect on production efficiency and mode of action. *J. Anim. Sci.*, **58**, 1465–83.

Bonhomme-Florentin, A., Blancou, J. & Latteur, B. (1978). Étude des variations saisonnières de la microfauna du rumen de zebus. *Protistologica*, **14**, 282–9.

Boyne, A. W., Eadie, J. M. & Raitt, K. (1957). The development and testing of a method of counting rumen ciliate protozoa. *J. Gen. Microbiol.*, **17**, 414–23.

Bragg, D. St. A., Murphy, M. R. & Davis, C. L. (1986). Effect of source of carbohydrate and frequency of feeding on rumen parameters in dairy steers. *J. Dairy Sci.*, **69**, 392–402.

Briggs, P. K., Hogan, J. P. & Reid, R. L. (1957). The effect of volatile fatty acids, lactic acid, and ammonia on rumen pH in sheep. *Aust. J. Agric. Res.*, **8**, 674–90.

Brulla, W. J. & Bryant, M. P. (1980). Monensin induced changes in the major species of the rumen bacterial population. *Abstr. Ann. Meeting Am. Soc. Microbiol.*, p. 104.

Bryant, M. P. & Burkey, L. A. (1953). Numbers and some predominant groups of bacteria in the rumen of cows fed different rations. *J. Dairy Sci.*, **36**, 218–24.
Bryant, M. P. & Robinson, I. M. (1961). An improved nonselective culture medium for ruminal bacteria and its use in determining diurnal variation in numbers of bacteria in the rumen. *J. Dairy Sci.*, **44**, 1446–56.
Bryant, M. P. & Robinson, I. M. (1968). Effects of diet, time after feeding, and position sampled on numbers of viable bacteria in the bovine rumen. *J. Dairy Sci.*, **51**, 1950–5.
Bryant, M. P. & Small, N. (1960). Observations on the ruminal microorganisms of isolated and inoculated calves. *J. Dairy Sci.*, **43**, 654–67.
Bryant, M. P., Small, N., Bouma, C. & Robinson, I. (1958). Studies on the composition of the ruminal flora and fauna of young calves. *J. Dairy Sci.*, **41**, 1747–67.
Bryant, M. P., Barrentine, B. F., Sykes, J. F., Robinson, I. M., Shawver, C. V. & Williams, L. W. (1960). Predominant bacteria in the rumen of cattle on bloat-provoking ladino clover pasture. *J. Dairy Sci.*, **43**, 1435–44.
Bush, M. & Kofoid, C. A. (1948). Ciliates from the Sierra Nevada bighorn *Ovis canadensis sierrae* Grinnel. *Univ. Calif. Publ. Zool.*, **53**, 237–62.
Caldwell, D. R. & Bryant, M. P. (1966). Medium without rumen fluid for nonselective enumeration and isolation of rumen bacteria. *Appl. Microbiol.*, **14**, 794–801.
Chen, M. & Wolin, M. J. (1979). Effect of monensin and lasalocid-sodium on the growth of methanogenic and rumen saccharolytic bacteria. *Appl. Environ. Microbiol.*, **38**, 72–7.
Christiansen, W. C., Woods, W. & Burroughs, W. (1964). Ration characteristics influencing rumen protozoal populations. *J. Anim. Sci.*, **23**, 984–8.
Chung, K.-T. & Hungate, R. E. (1976). Effect of alfalfa fiber substrate on culture counts of rumen bacteria. *Appl. Environ. Microbiol.*, **32**, 649–52.
Clarke, R. T. J. (1964). Ciliates of the rumen of domestic cattle (*Bos taurus* L.). *NZ J. Agric. Res.*, **7**, 248–57.
Clarke, R. T. J. (1965). Diurnal variation in the numbers of rumen ciliate protozoa in cattle. *NZ J. Agric. Res.*, **18**, 1–9.
Clarke, R. T. J. (1977). Protozoa in the rumen ecosystem. In *Microbial Ecology of the Gut*, ed. R. T. J. Clarke & T. Bauchop. Academic Press, London, pp. 251–75.
Clarke, R. T. J., Ulyatt, M. J. & John, A. (1982). Variation in numbers and mass of ciliate protozoa in the rumens of sheep fed chaffed alfalfa (*Medicago sativa*). *Appl. Environ. Microbiol.*, **43**, 1201–4.
Coleman, G. S. (1958). Maintenance of oligotrich protozoa from the sheep rumen *in vitro*. *Nature*, **182**, 1104–5.
Coleman, G. S. (1980). Rumen ciliate protozoa. In *Advances in Parasitology*, ed. W. H. R. Lumsden, R. Muller & J. R. Baker. Academic Press, New York, pp. 121–73.
Coleman, G. S., Laurie, J. I. & Bailey, J. E. (1977). The cultivation of the rumen ciliate *Entodinium bursa* in the presence of *Entodinium caudatum*. *J. Gen. Microbiol.*, **101**, 253–8.
Czerkawski, J. W. & Breckenridge, G. (1979a). Experiments with the long-term rumen simulation technique (Rusitec); response to supplementation of basal rations. *Br. J. Nutr.*, **42**, 217–28.
Czerkawski, J. W. & Breckenridge, G. (1979b). Experiments with the long-term rumen simulation technique (Rusitec); use of soluble food and an inert solid matrix. *Br. J. Nutr.*, **42**, 229–45.
Dawson, K. A. & Boling, J. A. (1983). Monensin-resistant bacteria in the rumens of calves on monensin-containing and unmedicated diets. *Appl. Environ. Microbiol.*, **46**, 160–4.
Dearth, R. N., Dehority, B. A. & Potter, E. L. (1974). Rumen microbial numbers in lambs as affected by level of feed intake and dietary diethylstilbesterol. *J. Anim. Sci.*, **38**, 991–6.
Dehority, B. A. (1970). Occurrence of the ciliate protozoa *Bütschlia parva* Schuberg in the rumen of the ovine. *Appl. Microbiol.*, **19**, 179–81.
Dehority, B. A. (1973). Hemicellulose degradation by rumen bacteria. *Fed. Proc.*, **32**, 1819–25.
Dehority, B. A. (1974). Rumen ciliate fauna of Alaskan moose (*Alces americana*), musk-ox (*Ovibos moschatus*) and Dall mountain sheep (*Ovis dalli*). *J. Protozool.*, **21**, 26–32.

Dehority, B. A. (1975). Characterization studies on rumen bacteria isolated from Alaskan reindeer (*Rangifer tarandus* L.). In *Proceedings of the 1st International Reindeer and Caribou Symposium*, Fairbanks, Alaska (Biological Papers of the University of Alaska, Special Report No. 1), pp. 228–40.

Dehority, B. A. (1978). Specificity of rumen ciliate protozoa in cattle and sheep. *J. Protozool.*, **25**, 509–13.

Dehority, B. A. (1979). Ciliate protozoa in the rumen of Brazilian water buffalo, *Bubalus bubalis* Linnaeus. *J. Protozool.*, **26**, 536–44.

Dehority, B. A. (1984). Evaluation of subsampling and fixation procedures used for counting rumen protozoa. *Appl. Environ. Microbiol.*, **48**, 182–5.

Dehority, B. A. (1986a). Rumen ciliate fauna of some Brazilian cattle: occurrence of several ciliates new to the rumen, including the cycloposthid *Parentodinium africanum*. *J. Protozool.*, **33**, 416–21.

Dehority, B. A. (1986b). Microbes in the foregut of arctic ruminants. In *Control of Digestion and Metabolism in Ruminants*, ed. L. P. Milligan, W. L. Grovum & A. Dobson. Prentice-Hall, Englewood Cliffs, New Jersey, pp. 307–25.

Dehority, B. A. & Grubb, J. A. (1976). Basal medium for the selective enumeration of rumen bacteria utilizing specific energy sources. *Appl. Environ. Microbiol.*, **32**, 703–10.

Dehority, B. A. & Grubb, J. A. (1977). Glucose-1-phosphate as a selective substrate for enumeration of *Bacteroides* species in the rumen. *Appl. Environ. Microbiol.*, **33**, 998–1001.

Dehority, B. A. & Grubb, J. A. (1980). Effect of short term chilling of rumen contents on viable bacterial numbers. *Appl. Environ. Microbiol.*, **39**, 376–81.

Dehority, B. A. & Mattos, W. R. S. (1978). Diurnal changes and effect of ration on concentrations of the rumen ciliate *Charon ventriculi*. *Appl. Environ. Microbiol.*, **36**, 953–8.

Dennis, S. M., Nagaraja, T. G. & Bartley, E. E. (1981). Effects of lasalocid or monensin on lactate-producing or -using rumen bacteria. *J. Anim. Sci.*, **52**, 418–26.

Dennis, S. M., Arambel, M. J., Bartley, E. E. & Dayton, A. D. (1983). Effect of energy concentration and source of nitrogen on numbers and types of rumen protozoa. *J. Dairy Sci.*, **66**, 1248–54.

Dogiel, V. A. (1926). Sur quelques infusoires nouveaux habitant l'estomac du dromadaire (*Camelus dromedarius*). *Ann. Parasit.*, **4**, 241–71.

Dogiel, V. A. (1927). Monographie der familie Ophryoscolecidae. *Arch. Protistenk.*, **59**, 1–288.

Dogiel, V. A. (1932). Beschreibung einiger neuer vertreter der familie Ophryoscolecidae aus afrikanischen antilopen nebst revision der infusorienfauna afrikanischer wiederkäuer. *Arch. Protistenk.*, **77**, 92–107.

Dogiel, V. A. (1947). The phylogeny of the stomach-infusorians of ruminants in the light of palaeontological and parasitological data. *Quart. J. Microsc. Sci.*, **88**, 337–43.

Dogiel, V. A. & Winogradowa-Federowa, T. (1930). Experimentelle untersuchungen zur biologie der infusiorien des wiederkäuermagens. *Wiss. Arch. Landwirtsch.*, B, **3**, 172–88.

Eadie, J. M. (1962a). The development of rumen microbial populations in lambs and calves under various conditions of management. *J. Gen. Microbiol.*, **29**, 563–78.

Eadie, J. M. (1962b). Inter-relationships between certain rumen ciliate protozoa. *J. Gen. Microbiol.*, **29**, 579–88.

Eadie, J. M. (1967). Studies on the ecology of certain rumen ciliate protozoa. *J. Gen. Microbiol.*, **49**, 175–94.

Eadie, J. M. & Gill, J. C. (1971). The effect of the absence of rumen ciliate protozoa on growing lambs fed on a roughage-concentrate diet. *Br. J. Nutr.*, **26**, 155–67.

Eadie, J. M. & Hobson, P. N. (1962). Effect of the presence or absence of rumen ciliate protozoa on the total rumen bacterial count in lambs. *Nature*, **193**, 503–5.

Eadie, J. M., Hobson, P. N. & Mann, S. O. (1959). A relationship between some bacteria, protozoa and diet in early weaned calves. *Nature*, **183**, 624–5.

Eadie, J. M., Hobson, P. N. & Mann, S. O. (1967). A note on some comparisons between the rumen content of barley-fed steers and that of young calves also fed on a high concentrate ration. *Anim. Prod.*, **9**, 247–50.

Eberlein, R. (1895). Über die in wiederkäuermagen vorkommenden ciliaten infusorien. *Zeitschr. Wiss. Zool.*, **59**, 233–303.
Fonty, G., Jouany, J. P., Senaud, J., Gouet, Ph. & Grain, J. (1984). The evolution of microflora, microfauna and digestion in the rumen of lambs from birth to 4 months. *Can. J. Anim. Sci.*, **64** (Suppl.), 165–6.
Giesecke, D. & Van Gylswyk, N. O. (1975). A study of feeding types and certain rumen functions in six species of South African wild ruminants. *J. Agric. Sci. (Camb.)*, **85**, 75–83.
Gilchrist, F. M. C. & Kistner, A. (1962). Bacteria of the ovine rumen. I. The composition of the population on a diet of poor teff hay. *J. Agric. Sci.*, **59**, 77–83.
Grubb, J. A. & Dehority, B. A. (1975). Effects of an abrupt change in ration from all roughage to high concentrate upon rumen microbial numbers in sheep. *Appl. Microbiol.*, **30**, 404–12.
Grubb, J. A. & Dehority, B. A. (1976). Variation in colony counts of total viable anaerobic rumen bacteria as influenced by media and cultural methods. *Appl. Environ. Microbiol.*, **31**, 262–7.
Henning, P. A. & Van der Walt, A. E. (1978). Inclusion of xylan in a medium for the enumeration of total culturable rumen bacteria. *Appl. Environ. Microbiol.*, **35**, 1008–11.
Hobson, P. N., Mann, S. O. & Summers, R. (1976). Rumen micro-organisms in red deer, hill sheep and reindeer in the Scottish highlands. *Proc. Roy. Soc. Edinburgh, B*, **75**, 171–80.
Hungate, R. E. (1966). *The Rumen and Its Microbes*. Academic Press, New York.
Hungate, R. E. (1978). The rumen protozoa. In *Parasitic Protozoa*, ed. J. P. Kreier, Vol. II. Academic Press, New York, pp. 655–95.
Hungate, R. E., Phillips, G. D., McGregor, A., Hungate, D. P. & Buechner, H. K. (1959). Microbial fermentation in certain mammals. *Science*, **130**, 1192–4.
Imai, S. (1984). New rumen ciliates, *Polymorphella bovis* sp. n. and *Entodinium longinucleatum* forma *spinolobum* f.n., from the zebu cattle in Thailand. *Jap. J. Vet. Sci.*, **46**, 391–5.
Imai, S. (1985). Rumen ciliate protozoal faunae of Bali cattle (*Bosjavanicus domesticus*) and water buffalo (*Bubalus bubalis*) in Indonesia, with the description of a new species, *Entodinium javanicum* sp. nov. *Zool. Sci.*, **2**, 591–600.
Imai, S. (1986). Rumen ciliate protozoal fauna of zebu cattle (*Bos taurus indicus*) in Sri Lanka, with the description of a new species, *Diplodinium sinhalicum* sp. nov. *Zool. Sci.*, **3**, 699–706.
Imai, S. & Ogimoto, K. (1983). *Parabundleia ruminantium* gen. n. sp. n., *Diplodinium mahidoli* sp. n. with two formae, and *Entodinium parvum* forma *monospinosum* forma n. from the zebu cattle (*Bos indicus* L., 1758) in Thailand. *Jap. J. Vet. Sci.*, **45**, 585–91.
Imai, S. & Ogimoto, K. (1984). Rumen ciliate protozoal fauna and bacterial flora of the zebu cattle (*Bos indicus*) and the water buffalo (*Bubalus bubalis*) in Thailand. *Jap. J. Zootech. Sci.*, **55**, 576–83.
Imai, S., Chang, C.-H., Wang, J.-S., Ogimoto, K. & Fujita, J. (1981a). Rumen ciliate protozoal fauna of the water buffalo (*Bubalus bubalis*) in Taiwan. *Bull. Nippon Vet. Zootech. Coll.*, **29**, 77–81.
Imai, S., Ogimoto, K. & Fujita, J. (1981b). Rumen ciliate protozoal fauna of water buffalo, *Bubalus bubalis* (Linnaeus), in Okinawa, Japan. *Bull. Nippon Vet. Zootech. Coll.*, **29**, 82–5.
Imai, S., Shimizu, M., Kinoshita, M., Toguchi, M., Ishii, T. & Fujita, J. (1982). Rumen ciliate protozoal fauna and composition of the cattle in Japan. *Bull. Nippon Vet. Zootech. Coll.*, **31**, 70–4.
Iverson, W. G. & Millis, N. F. (1977). Succession of *Streptococcus bovis* strains with differing bacteriophage sensitivities in the rumens of two fistulated sheep. *Appl. Environ. Microbiol.*, **33**, 810–13.
Jensen, E. A. & Hammond, D. M. (1964). A morphological study of trichomonads and related flagellates from the bovine digestive tract. *J. Protozool.*, **11**, 386–94.
Kaufmann, W., Hagemeister, H. & Dirksen, G. (1980). Adaptation to changes in dietary composition, level and frequency of feeding. In *Digestive Physiology and Metabolism in Ruminants*, ed. Y. Ruckebusch and P. Thivend. MTP Press, Lancaster, pp. 587–602.

Kleynhans, C. J. & Van Hoven, W. (1976). Rumen protozoa of the giraffe with a description of two new species. *E. Afr. Wildl. J.*, **14**, 203–14.
Kofoid, C. A. & Christenson, J. F. (1934). Ciliates from *Bos gaurus* H. Smith. *Univ. Calif. Publ. Zool.*, **30**, 341–91.
Kofoid, C. A. & MacLennan, R. F. (1932). Ciliates from *Bos indicus* Linn., II. A revision of *Diplodinium* Schuberg. *Univ. Calif. Publ. Zool.*, **37**, 53–152.
Kofoid, C. A. & MacLennan, R. F. (1933). Ciliates from *Bos indicus* Linn., III. *Epidinium* Crawley, *Epiplastron* gen. nov. and *Ophryoscolex* Stein. *Univ. Calif. Publ. Zool.*, **39**, 1–34.
Kopperi, A. J. (1926). Märehlijöitten rapamahan infusoreista (Zusammenfassung: Die mageninfusorien der wiederkäuer). *Ann. Soc. Zool.-Bot. Vanamo*, **4**, 225–38.
Kopperi, A. J. (1929). Über eine diplodinium-art aus dem magen des renntieres. *Ann. Soc. Zool.-Bot. Vanamo*, **8**, 27–33.
Kurihara, Y., Eadie, J. M., Hobson, P. N. & Mann, S. O. (1968). Relationship between bacteria and ciliate protozoa in the sheep rumen. *J. Gen. Microbiol.*, **51**, 267–88.
Latham, M. J., Sharpe, M. E. & Sutton, J. D. (1971). The microbial flora of the rumen of cows fed hay and high cereal rations and its relationship to the rumen fermentation. *J. Appl. Bacteriol.*, **34**, 425–34.
Latham, M. J., Storry, J. E. & Sharpe, M. E. (1972). Effect of low-roughage diets on the microflora and lipid metabolism in the rumen. *Appl. Microbiol.*, **24**, 871–7.
Leedle, J. A. Z. & Hespell, R. B. (1980). Differential carbohydrate media and anaerobic replica plating techniques in delineating carbohydrate-utilizing subgroups in rumen bacterial populations. *Appl. Environ. Microbiol.*, **39**, 709–19.
Leedle, J. A. Z., Bryant, M. P. & Hespell, R. B. (1982). Diurnal variations in bacterial numbers and fluid parameters in ruminal contents of animals fed low- or high-forage diets. *Appl. Environ. Microbiol.*, **44**, 402–12.
Leedle, J. A. Z., Barsuhn, K. & Hespell, R. B. (1986). Postprandial trends in estimated ruminal digesta polysaccharides and their relation to changes in bacterial groups and ruminal fluid characteristics. *J. Anim. Sci.*, **62**, 789–803.
Lubinsky, G. (1957*a*). Note on the phylogenetic significance of predatory habits in the Ophryoscolecidae (Ciliate: Oligotricha). *Can. J. Zool.*, **35**, 579–80.
Lubinsky, G. (1957*b*). Studies on the evolution of the Ophryoscolecidae (Ciliata: Oligotricha). I. A new species of Entodinium with 'caudatum', 'loboso-spinosum', and 'dubardi' forms, and some evolutionary trends in the genus Entodinium. *Can. J. Zool.*, **35**, 111–33.
Lubinsky, G. (1957*c*). Studies on the evolution of the Ophryoscolecidae (Ciliata: Oligotricha). II. On the origin of the higher Ophryoscolecidae. *Can. J. Zool.*, **35**, 135–40.
Lubinsky, G. (1957*d*). Studies on the evolution of the Ophryoscolecidae (Ciliata: Oligotricha). III. Phylogeny of the Ophryoscolecidae based on their comparative morphology. *Can. J. Zool.*, **35**, 141–59.
Lubinsky, G. (1958). Ophryoscolecidae (Ciliate: Entodinomorphida) of reindeer (*Rangifer tarandus* L.) from the Canadian arctic, II. Diplodiniinae. *Can. J. Zool.*, **36**, 937–59.
Mackie, R. I. & Gilchrist, F. M. C. (1979). Changes in lactate-producing and lactate-utilizing bacteria in relation to pH in the rumen of sheep during stepwise adaptation to a high-concentrate diet. *Appl. Environ. Microbiol.*, **38**, 422–30.
Mackie, R. I., Gilchrist, F. M. C., Robberts, A. M., Hannah, P. E. & Schwartz, H. M. (1978). Microbiological and chemical changes in the rumen during the stepwise adaptation of sheep to high concentrate diets. *J. Agric. Sci.*, **90**, 241–54.
Macy, J. M., Garrand, J. R. & Montgomery, L. (1982). Celluloytic and non cellulolytic bacteria in rat gastrointestinal tracts. *Appl. Environ. Microbiol.*, **44**, 1428–34.
Maki, L. R. & Foster, E. M. (1957). Effect of roughage in the bovine ration on types of bacteria in the rumen. *J. Dairy Sci.*, **40**, 905–13.
Males, J. R. (1973). Ration digestibility, rumen bacteria and several rumen parameters in sheep born and reared in isolation. PhD dissertation, Ohio State University, Columbus.
Mann, S. O. (1963). Some observations on the airborne dissemination of rumen bacteria. *J. Gen. Microbiol.*, **33**, IX.

McBee, R. H., Johnson, J. L. & Bryant, M. P. (1969). Ruminal microorganisms from elk. *J. Wildl. Mgmt*, **33**, 181–6.

Michalowski, T. (1975). Effect of different diets on the diurnal concentrations of ciliate protozoa in the rumen of water buffalo. *J. Agric. Sci.*, **85**, 145–50.

Michalowski, T. (1977). Diurnal changes in concentration of rumen ciliates and in occurrence of dividing forms in water buffalo (*Bubalus bubalus*) fed once daily. *Appl. Environ. Microbiol.*, **33**, 802–4.

Moir, R. J. & Somers, M. (1956). A factor influencing the protozoal population in sheep. *Nature*, **178**, 1472.

Moir, R. J. & Somers, M. (1957). Ruminal flora studies. VIII. The influence of rate and method of feeding a ration upon its digestibility, upon ruminal function, and upon the ruminal population. *Aust. J. Agric. Res.*, **8**, 253–65.

Murphy, M. R., Drone, P. E., Jr & Woodford, S. T. (1985). Factors stimulating migration of holotrich protozoa into the rumen. *Appl. Environ. Microbiol.*, **49**, 1329–31.

Naga, M. A., Abou Akkada, A. R. & el-Shazly, K. (1969). Establishment of rumen ciliate protozoa in cow and water buffalo (*Bos bubalus* L.) calves under late and early weaning systems. *J. Dairy Sci.*, **52**, 110–12.

Nakamura, K. & Kanegasaki, S. (1969). Densities of ruminal protozoa of sheep established under different dietary conditions. *J. Dairy Sci.*, **52**, 250–5.

Nakamura, F. & Kurihara, Y. (1978). Maintenance of a certain rumen protozoal population in a continuous *in vitro* fermentation system. *Appl. Environ. Microbiol.*, **35**, 500–6.

Ogimoto, K. & Imai, S. (1981). *Atlas of Rumen Microbiology*. Japan Scientific Societies Press, Tokyo.

Olumeyan, D. B., Nagaraja, T. G., Miller, G. W., Frey, R. A. & Boyer, J. E. (1986). Rumen microbial changes in cattle fed diets with or without salinomycin. *Appl. Environ. Microbiol.*, **51**, 340–5.

Orpin, C. G. (1974). The rumen flagellate *Callimastix frontalis*: does sequestration occur? *J. Gen. Microbiol.*, **84**, 395–8.

Orpin, C. G. (1975). Studies on the rumen flagellate *Neocallimastix frontalis*. *J. Gen. Microbiol.*, **91**, 249–62.

Orpin, C. G. (1976). Studies on the rumen flagellate *Sphaeromonas communis*. *J. Gen. Microbiol.*, **94**, 270–80.

Orpin, C. G. (1977*a*). The rumen flagellate *Piromonas communis*: its life-history and invasion of plant material in the rumen. *J. Gen. Microbiol.*, **99**, 107–17.

Orpin, C. G. (1977*b*). The occurrence of chitin in the cell walls of the rumen organisms *Neocallimastix frontalis*, *Piromonas communis* and *Sphaeromonas communis*. *J. Gen. Microbiol.*, **99**, 215–18.

Orpin, C. G. (1978). Induction of zoosporogenesis in the rumen phycomcyete *Neocallimastix frontalis* in rumen fluid after the addition of haem-containing compounds. *Proc. Soc. Gen. Microbiol.*, **5**, 46–7.

Orpin, C. G. & Letcher, A. J. (1978). Some factors controlling the attachment of the rumen holotrich protozoa *Isotricha intestinalis* and *Isotricha prostoma* to plant particles *in vitro*. *J. Gen. Microbiol.*, **106**, 33–40.

Orpin, C. G. & Letcher, A. J. (1984). Effect of absence of ciliate protozoa on rumen fluid volume, flow rate and bacterial populations in sheep. *Anim. Feed Sci. Tech.*, **10**, 145–53.

Orpin, C. G. & Mathiesen, S. D. (1986). *Microcetus lappus* gen. nov., sp. nov.: new species of ciliated protozoon from the bovine rumen. *Appl. Environ. Microbiol.*, **52**, 527–30.

Orpin, C. G., Mathiesen, S. D., Greenwood, Y. & Blix, A. S. (1985). Seasonal changes in the ruminal microflora of the high-arctic Svalbard reindeer (*Rangifer tarandus platyrhynchus*). *Appl. Environ. Microbiol.*, **50**, 144–51.

Pearson, H. A. (1965). Rumen organisms in white-tailed deer from south Texas. *J. Wildl. Mgmt*, **29**, 493–6.

Pearson, H. A. (1967). Rumen microorganisms in buffalo from southern Utah. *Appl. Microbiol.*, **15**, 1450–1.

Pearson, H. A. (1969). Rumen microbial ecology in mule deer. *Appl. Microbiol.*, **17**, 819–24.
Potter, E. L. & Dehority, B. A. (1973). Effects of changes in feed level, starvation, and level of feed after starvation upon the concentration of rumen protozoa in the ovine. *Appl. Microbiol.*, **26**, 692–8.
Prins, R. A. (1977). Biochemical activities of gut micro-organisms. In *Microbial Ecology of the Gut*, ed. R. T. J. Clarke & T. Bauchop. Academic Press, New York, pp. 73–183.
Puch, H. C. (1977). MS thesis, Ohio State University, Columbus.
Purser, D. B. (1961). A diurnal cycle for holotrich protozoa of the rumen. *Nature*, **190**, 831–2.
Purser, D. B. & Moir, R. J. (1959). Ruminal flora studies in the sheep. IX. The effect of pH on the ciliate population of the rumen *in vivo*. *Aust. J. Agric. Res.*, **10**, 555–64.
Roxas, D. B. (1980). Effects of abrupt changes in the ration on rumen microflora of sheep. PhD dissertation, Ohio State University, Columbus.
Schelling, G. T. (1984). Monensin mode of action in the rumen. *J. Anim. Sci.*, **58**, 1518–27.
Schwartz, H. M. & Gilchrist, F. M. C. (1975). Microbial interactions with the diet and the host animal. In *Digestion and Metabolism in the Rumen*, ed. I. W. McDonald & A. C. I. Warner. University of New England Publishing Unit, Armidale, pp. 165–79.
Shimizu, M., Kinoshita, M., Fujita, J. & Imai, S. (1983). Rumen ciliate protozoal fauna and composition of the zebu cattle, *Bos indicus*, and water buffalo, *Bubalus bubalis*, in Philippines. *Bull. Nippon Vet. Zootech. Coll.*, **32**, 83–8.
Slyter, L. L., Oltjen, R. R., Kern, D. L. & Blank, F. C. (1970). Influence of type and level of grain and diethylstilbesterol on the rumen microbial populations of steers fed all-concentrate diets. *J. Anim. Sci.*, **31**, 996–1002.
Slyter, L. L., Kern, D. L., Weaver, J. M., Oltjen, R. R. & Wilson, R. L. (1971). Influence of starch and nitrogen sources on ruminal microorganisms of steers fed high fiber purified diets. *J. Nutr.*, **101**, 847–54.
Stewart, C. S., Paniagua, C., Dinsdale, D., Cheng, K.-J. & Garrow, S. H. (1981). Selective isolation and characteristics of *Bacteroides succinogenes* from the rumen of a cow. *Appl. Environ. Microbiol.*, **41**, 504–10.
Teather, R. M., Mahadevan, S., Erfle, J. D. & Sauer, F. D. (1984). Negative correlation between protozoal and bacterial levels in rumen samples and its relation to the determination of dietary effects on the rumen microbial population. *Appl. Environ. Microbiol.*, **47**, 566–70.
Thorley, C. M., Sharpe, M. E. & Bryant, M. P. (1968). Modification of the rumen bacterial flora by feeding cattle ground and pelleted roughage as determined with culture media with and without rumen fluid. *J. Dairy Sci.*, **51**, 1811–16.
Tiwari, A. D., Bryant, M. P. & Wolfe, R. S. (1969). Simple method for isolation of *Selenomonas ruminantium* and some nutritional characteristics of the species. *J. Dairy Sci.*, **52**, 2054–6.
Van der Linden, Y., Van Gylswyk, N. O. & Schwartz, H. M. (1984). Influence of supplementation of corn stover with corn grain on the fibrolytic bacteria in the rumen of sheep and their relation to the intake and digestion of fiber. *J. Anim. Sci.*, **59**, 772–83.
Van Hoven, W. (1975). Rumen ciliates of the tsessebe (*Damaliscus lunatus lunatus*) in South Africa. *J. Protozool.*, **22**, 457–62.
Van Hoven, W. (1978). Development and seasonal changes in the rumen protozoan population in young blesbok (*Damaliscus dorcas phillipsi* Harper 1939). *S. Afr. J. Wildl. Res.*, **8**, 127–30.
Van Nevel, C. J. & Demeyer, D. I. (1977). Effect of monensin on rumen metabolism *in vitro*. *Appl. Environ. Microbiol.*, **34**, 251–7.
Vance, R. D., Preston, R. L., Klosterman, E. W. & Cahill, V. R. (1972). Utilization of whole shelled and crimped corn grain with varying proportions of corn silage by growing-finishing steers. *J. Anim. Sci.*, **35**, 598–605.
Veira, D. M. (1986). The role of ciliate protozoa in nutrition of the ruminant. *J. Anim. Sci.*, **63**, 1547–60.
Warner, A. C. I. (1962a). Enumeration of rumen micro-organisms. *J. Gen. Microbiol.*, **28**, 119–28.

Warner, A. C. I. (1962b). Some factors influencing the rumen microbial population. *J. Gen. Microbiol.*, **28**, 129–46.

Warner, A. C. I. (1966a). Diurnal changes in the concentrations of microorganisms in the rumens of sheep fed limited diets once daily. *J. Gen. Microbiol.*, **45**, 213–35.

Warner, A. C. I. (1966b). Periodic changes in the concentrations of microorganisms in the rumen of a sheep fed a limited ration every three hours. *J. Gen. Microbiol.*, **45**, 237–42.

Warner, A. C. I. (1966c). Diurnal changes in the concentrations of microorganisms in the rumens of sheep fed to appetite in pens or pasture. *J. Gen. Microbiol.*, **45**, 243–51.

Wedekind, K. J., Muntifering, R. B. & Barker, K. B. (1986). Effects of diet concentrate level and sodium bicarbonate on site and extent of forage fiber digestion in the gastrointestinal tract of wethers. *J. Anim. Sci.*, **62**, 1388–95.

Westerling, B. (1969). Våmciliatfaunan hos tamboskap i finska Lappland, med speciell hänsyn til arter ansedda som specifika för ren. *Nord. Vet. Med.*, **21**, 14–19.

Westerling, B. (1970). Rumen ciliate fauna of semi-domestic reindeer (*Rangifer tarandus* L.) in Finland: composition, volume and some seasonal variations. *Acta Zool. Fennica*, **127**, 1–76.

Wilkinson, R. C. & Van Hoven, W. (1976). Rumen protozoa of the giraffe with a description of two new species. *E. Afr. Wildl. J.*, **14**, 203–14.

Williams, P. P. & Dinusson, W. E. (1972). Composition of the ruminal flora and establishment of ruminal ciliated protozoal species in isolated calves. *J. Anim. Sci.*, **34**, 469–74.

6
Energy Yielding and Consuming Reactions

J. B. Russell

Agricultural Research Service, US Department of Agriculture and Department of Animal Science, Cornell University, Ithaca, New York, USA

&

R. J. Wallace

Rowett Research Institute, Aberdeen, UK

The work of biological growth depends on the transfer of energy from catabolic (yielding) to anabolic (consuming) reactions (Fig. 1). Classically the two reactions have been depicted to be connected by ATP, in the sense that the energy released by, for example, the glycolytic breakdown of sugars is conserved in the form of ATP, which then provides the energy necessary for the biosynthesis of cell material. It is now clear that most organisms, including anaerobes, can conserve energy, either in the form of a transmembrane electrochemical gradient of protons or in protonmotive force (Mitchell, 1961; Thauer et al., 1977; Dawes, 1986). However, the protonmotive force (PMF) drives few biosynthetic processes directly, and the transfer of energy from the PMF to ATP via a membrane-bound reversible ATPase is a vital link between catabolic and anabolic reactions.

The efficiencies of ATP production, transfer and utilization during growth are important factors influencing the survival of microorganisms in an environment such as the rumen, where energy sources are only occasionally abundant. Equally important are the strategies that rumen organisms have developed for coping with the intervening periods of starvation. At these times ability to store energy is crucial, not only to maintain viability but to ensure that the organism can respond rapidly and effectively to the subsequent influx of fermentable energy sources.

Our knowledge of the energy metabolism of rumen microorganisms is generally sketchy, and it is frequently necessary to make inferences from similar or related microorganisms. This extrapolation has enhanced our understanding of rumen microbial metabolism, and has fortunately only seldom been misleading.

ENERGY-YIELDING REACTIONS

ATP

The role of phosphate esters in alcoholic fermentation of sugars was recognized by Harden & Young in 1906, but it was not until the early 1940s that the significance of

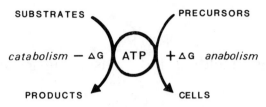

Fig. 1. A schematic representation of anabolism and catabolism.

phosphate esters was more fully appreciated. Lipmann (1941) used the term 'energy rich' to describe ATP and other phosphorylated intermediates; with time, phosphate-bond formation and breakage were recognized as means of energy exchange. However, the 'energy rich' concept is not entirely correct. Nicholls (1982) recently noted: 'It is frequently, and misleadingly, supposed that the phosphate anhydride bonds of ATP are "high energy" bonds which are capable of storing energy and driving reactions in otherwise unfavourable directions. However, it should be clear that it is the extent to which the observed mass action ratio is displaced from equilibrium which defines the capacity of the reactants to do work, rather than the attribute of a single component'. The re-definition of 'bioenergetics' as 'chemical equilibria' provides a central theme for understanding and unifying the once diverse concepts of electron transport, ATP formation, and active transport processes.

Glycolysis

The Embden–Meyerhof–Parnas (EMP) pathway was first discovered in muscle, but it is the most common pathway of hexose metabolism in both aerobic and anaerobic microorganisms (Gottschalk, 1979). After glucose is phosphorylated at carbon-6 by an ATP-dependent kinase or the phosphotransferase system of transport, the phosphorylated derivative is isomerized and phosphorylated again at the carbon-1 position by another ATP-dependent kinase. These initial steps in glucose oxidation require an initial input of ATP rather than its release, but the derivatives are less permeable to the cell membrane and thus more easily 'trapped' by the cell. Aldolase cleaves a carbon–carbon bond and the resulting triose phosphates are oxidized by two dehydrogenases and dephosphorylated by two kinases which produce additional ATP.

The ATP yield of the EMP pathway has traditionally been assumed to be 2 ATP but in *Propionibacterium shermanii* phosphofructokinase is replaced by a pyrophosphate-6-phosphofructokinase (O'Brien *et al.*, 1975). This enzyme uses pyrophosphate as the phosphoryl donor and does not require a direct input of ATP. The substitution of pyrophosphate for ATP means that phosphate-bond energy is still used; the free-energy changes of ATP hydrolysis to ADP and of pyrophosphate cleavage are 7·6 and 5·2 kcal/mol respectively (Thauer *et al.*, 1977). Pyrophosphate hydrolysis is otherwise thought to 'help assure the completeness of certain biosynthetic reactions' like fatty acid and nucleoside synthesis (Lehninger, 1975).

This aspect of glycolytic ATP formation has not been investigated in rumen bacteria.

Many bacteria metabolize hexoses via alternative catabolic routes employing the Entner–Doudoroff, pentose, and pentose phosphoketolase pathways. Each of these pathways has the enzyme glucose 6-phosphate dehydrogenase. Baldwin et al. (1963) studied the ruminal metabolism of glucose labelled (^{14}C) in the 1, 2, and 6 positions and found that the labelling pattern in acetate was consistent with the operation of the EMP pathway. However, they 'did not attempt to obtain complete recoveries of ^{14}C, nor did they attempt to determine the amount of unlabelled acetate arising from the endogenous substrates present in the incubation media' (Baldwin, 1965). Later studies indicated that the amount of $^{14}CO_2$ from [1-^{14}C]glucose was consistently greater than the amounts produced from [2-^{14}C]- or [6-^{14}C]-glucose. These results indicated that small amounts of glucose may have been decarboxylated by glucose 6-phosphate dehydrogenase, but more than 90% of the glucose appeared to be degraded by the EMP pathway (Wallnofer et al., 1966). Later studies with pure cultures supported the assumption that the EMP pathway is the primary pathway of hexose metabolism in the rumen (Mountfort & Roberton, 1978; Miller & Wolin, 1979). The use of the EMP pathway is advantageous to anaerobic bacteria because it maximizes the yield of ATP.

Pentose metabolism

Feedstuff polysaccharides are primarily composed of hexose or hexose derivatives (pectin), but hemicellulose may contain varying amounts of pentose. Pentose fermentation can proceed by the transketolase and transaldolase reactions of the pentose cycle or by a phosphorolytic cleavage (phosphoketolase). The ATP yield of the pentose cycle is greater than that obtained by the pathway involving phosphoketolase, and studies indicated that 75% of labelled xylan was fermented by the pentose pathway while 25% was fermented by phosphoketolase (Wallnofer et al., 1966).

Pyruvate metabolism

Pyruvate is the central intermediary metabolite in the rumen. It is formed as a result of the catabolism of sugars and is the branch point where pathways diverge to form fermentation products. During glycolysis, NAD has been converted into NADH, and it is essential that pyruvate metabolism results in the re-oxidation of this NADH so that fermentation can continue. The simplest way of re-oxidizing NADH is by the formation of lactate, as occurs in muscle, or by ethanol production as in yeast.

Many species of rumen bacteria produce lactate in pure culture, but the concentration of lactate *in vivo* is usually less than 1 mM (Counotte et al., 1983; Mackie et al., 1984). Lactate can be used by a secondary, lactate-fermenting population consisting of *Veillonella alcalescens*, *Megasphaera elsdenii* and *Selenomonas ruminantium*, but the turnover of lactate in the rumen is usually small

(Mackie et al., 1984). The lack of lactate in the rumen is also due to the thermodynamic and regulatory factors governing interspecies H_2 transfer (see below). Similarly, ethanol is formed in pure cultures as an alternative electron-sink product only when the accumulation of H_2 prevents further formation of the gas.

Rumen bacteria have evolved fermentation pathways that enable them to oxidize NADH, and the main products are volatile fatty acids (VFA; principally acetic, propionic and butyric), carbon dioxide and methane. A typical stoichiometry for hexose fermentation might be (Wolin, 1979)

$$57\cdot5 C_6H_{12}O_6 \rightarrow 65 HAc + 20 HPr + 15 HBu + 35 CH_4 + 60 CO_2 + 25 H_2O$$

but these proportions change according to diet. The proportion of propionate is higher with concentrate diets, because starch-fermenters tend to be propionate- and succinate-producers. However, acetate is always the most abundant product.

Aerobically, pyruvate is oxidized by an NAD-linked pyruvate dehydrogenase to form acetyl-CoA but, when *Escherichia coli* and other enterobacteria are grown anaerobically, pyruvate is oxidized by pyruvate-formate lyase. Because formate is produced by many rumen bacteria, it was originally thought that the coliform type of pyruvate oxidation was important in the rumen. However, Miller & Wolin (1979) found that ^{14}C from formate was not exchanged into the carboxyl group of pyruvate, whereas CO_2 exchange, a reaction typical of pyruvate oxidoreductases, did occur. The pyruvate oxidoreductases of intestinal anaerobes reduced flavin mononucleotide, flavin adenine dinucleotide, and viologen dyes (Miller & Wolin, 1979). However, the electron acceptor in *Bacteroides succinogenes* is more likely to be a flavoprotein (Miller, 1978), while in *Ruminococcus albus* ferredoxin is the probable acceptor (Glass et al., 1977). Electrons linked to ferredoxin are at a very high potential and are thus more easily oxidized by hydrogenases. Hydrogen formation by ruminococci, *M. elsdenii* and *V. alcalescens* is consistent with ferredoxin-linked pyruvate oxidoreduction. The prevalence of pyruvate oxidoreductase rather than pyruvate-formate lyase in intestinal bacteria led Miller & Wolin (1979) to speculate that these bacteria produce formate by the direct reduction of CO_2. Carbon dioxide and H_2 can also be derived from formate (formate-hydrogen lyase), and this cleavage reaction contributed to the confusion over whether rumen bacteria had a pyruvate-formate lyase or a pyruvate-ferredoxin oxidoreductase.

Acetyl-CoA resulting from either method of pyruvate oxidoreduction and decarboxylation is converted into acetate by a reversible phosphotransacetylase reaction which conserves much of the free energy as acetyl-phosphate. Acetyl-phosphate in turn is acted on by acetate kinase which yields another ATP from substrate-level phosphorylation (Table 1). In rumen holotrich protozoa the conversion of pyruvate into acetate and hydrogen occurs in hydrogenosomes, membranous organelles analogous to mitochondria (Yarlett et al., 1981, 1982, 1983). Hydrogenosomes consume oxygen and thereby protect associated O_2-sensitive pyruvate synthase and hydrogenase (Yarlett et al., 1981), which probably contributes to the aerotolerance of these organisms (Williams, 1986) (see also Chapter 3).

The main butyrate-producing bacterium in the rumen is *Butyrivibrio fibrisolvens*. It uses thiolase, β-hydroxybutyryl-CoA dehydrogenase, crotonase, crotonyl-CoA reductase, phosphate butyryltransferase and finally butyrate kinase, to condense two molecules of acetyl-CoA. Acetoacetyl-CoA is then converted into butyryl-CoA by a series of reactions similar to a reversal of β-oxidation (Miller & Jenesel, 1979). Thus two pairs of reducing equivalents are oxidized and 1 ATP is formed by the conversion of acetoacetyl-CoA in butyrate via butyryl-CoA and/or butyryl phosphate (Table 1). Flavoproteins rather than free cofactors were implicated in the

Table 1
Enzymatic Reactions Producing ATP ($\sim$P) or Reducing Equivalents (2H) and the Balance of these Reactions in Various Fermentations[a]

Enzyme	Final product					
	Lactate	Acetate	Propionate[b]	Butyrate	Ethanol	Valerate
Glucokinase	−1	−1	−1	−1	−1	−1
Phosphofructokinase	−1	−1	−1	−1	−1	−1
Glycerate kinase	2	2	2	2	2	2
Pyruvate kinase	2	2	2	2	2	2
Acetate kinase	—	2	—	—	—	—
Fumarate reductase[c]	—	—	2	—	—	—
Butyrate kinase	—	—	—	1	—	—
Total ($\sim$P)	2	4	4	3	2	2
Glyceraldehyde 3-phosphate dehydrogenase	2	2	2	2	2	2
Lactate dehydrogenase	−2	—	—	—	—	—
Pyruvate oxidoreductase	—	2	—	2	2	1
Alcohol dehydrogenase	—	—	—	—	−4	—
Malate dehydrogenase	—	—	−2	—	—	−1
Fumarate reductase	—	—	−2	—	—	−1
β-Hydroxybutyrate dehydrogenase	—	—	—	−1	—	—
Butyryl-CoA dehydrogenase	—	—	—	−1	—	—
β-Hydroxyvalerate dehydrogenase	—	—	—	—	—	−1
Valeryl-CoA dehydrogenase	—	—	—	—	—	−1
Total (2H)	0	4	−2	2	0	−1

[a] From 1 molecule of hexose via Embden–Meyerhof–Parnas pathway.
[b] The randomizing pathway employing succinate as an intermediate. If the non-randomizing pathway via acrylyl-CoA reductase were used, the (2H) balance would be the same, but the $\sim$P is thought to be only 2.
[c] Assumes an ATP-linked fumarate reductase reaction; *Megasphaera elsdenii*, the predominant organism making valerate, does not have this enzyme since it uses the acrylate pathway to make propionyl-CoA.

reduction of crotonyl-CoA (Miller & Jenesel, 1979). An essentially similar pathway appears to operate in the holotrich protozoa (Williams, 1986).

Propionate is synthesized by two different pathways in the rumen, and these pathways can be readily differentiated on the basis of ^{14}C-labelling studies (Baldwin, 1965). When glucose is labelled in the second position and metabolized by the EMP pathway, the label is found in the 2-position of pyruvate. If the pyruvate (or phosphoenolpyruvate) is metabolized by the 'randomizing pathway' the label is seen in both the second and third positions of propionate because the intermediates, fumarate and succinate, are symmetrical molecules. When [2-^{14}C]glucose is metabolized by the direct reductive pathway involving acrylyl-CoA, the molecules are not symmetrical, and the label shows up only in the second position of propionate (Baldwin et al., 1963).

Propionate production via the randomizing pathway involves either (i) carboxylation of pyruvate or phosphoenolpyruvate to form oxaloacetate, the oxaloacetate being reduced to malate by malic dehydrogenase, or (ii) the reductive carboxylation of pyruvate directly to malate by the malic enzyme. The former mechanism was indicated in *Bacteroides ruminicola* and *Bacteroides (Ruminobacter) amylophilus*, while the malic enzyme was found in the ruminococci, *B. succinogenes* and *S. ruminantium* (Joyner & Baldwin, 1966). Both enzymes occurred in *B. fibrisolvens*. Malate is then dehydrated by fumarase, with fumarate being hydrogenated by a fumarate reductase. The final stages of propionate metabolism entail formation of succinyl-CoA, a vitamin B_{12}-dependent mutase reaction, a decarboxylation, and hydrolysis of the CoA ester.

The energetics of the randomizing pathway were at first difficult because there were no obvious sites of ATP formation in the conversion of either pyruvate or phosphoenolpyruvate into propionate. Rather, the carboxylation step and formation of succinyl-CoA might have caused a loss of ATP. The synthesis of oxaloacetate can, however, be catalysed by a biotin-dependent transcarboxylation reaction that conserves the energy of succinyl-CoA decarboxylation. The hydrolysis of the propionyl-CoA ester does not lead to a phosphate intermediate or to ATP formation by a kinase, but the free energy is conserved by a CoA transferase that drives the synthesis of succinyl-CoA. It was generally assumed that strict anaerobes did not possess electron transport systems capable of driving ATP formation, but the discovery of cytochromes in *B. ruminicola* challenged this assumption (White et al., 1962). The requirement of bacteroides species for haemin to synthesize cytochromes (Caldwell et al., 1965), and the influence of haemin on growth yields and succinate production, suggested that fumarate reduction might be linked to ATP formation (Macy et al., 1975). Recent work supports the possibility of ATP synthesis by a fumarate reductase step in *Vibrio succinogenes* (Kroger & Winkler, 1981). Since the transcarboxylation and CoA transferase reactions do not require ATP, and since fumarate reduction causes ATP formation, the ATP yield can be as great for propionate formation as for acetate (Table 1).

In the direct reductive pathway of propionic acid fermentation, lactate (or pyruvate) is converted into a CoA ester which is dehydrated, and subsequently reduced by a flavoprotein (Baldwin, 1965; Brockman & Wood, 1975). This pathway does not include a carboxylation or decarboxylation step. The reduction of acrylyl-CoA involves an electron transport chain employing ferredoxin and an 'electron-transferring flavoprotein' (Brockman & Wood, 1975), but ATP synthesis has not been demonstrated.

Among the organisms most prevalent in rumen propionogenesis, *S. ruminantium* uses the randomizing pathway (Paynter & Elsden, 1970) and *M. elsdenii* the acrylate pathway (Ladd & Walker, 1965). The former is thought to be most important *in vivo* although, when concentrates are fed, the increased numbers of *M. elsdenii* may augment the amount of propionate produced via acrylate (Baldwin, 1965). Propionate is also formed by the fermentation of lactate, and the labelling pattern is similar to the pattern described above. *M. elsdenii*, which is the only rumen lactate fermenter to use the non-randomizing pathway, was estimated by this method to metabolize an average of 74% of the lactate produced in the rumen (Counotte et al., 1981, 1983).

Higher volatile fatty acids (valeric, caproic etc.) are formed by the condensation of acetyl-CoA and/or propionyl-CoA. The reversal of β-oxidation which leads to their formation represents another means of reducing-equivalent disposal. As in butyrate synthesis, the ATP yield is decreased because the free energy of the CoA ester is used for carbon–carbon bond formation rather than phosphate ester and ATP formation (Table 1).

Methane

Methane production in the rumen has a profound effect on fermentation end products, as well as on the ATP yield. In pure culture, rumen bacteria sometimes produce ethanol and lactate, but these products are rarely observed *in vivo* unless very large amounts of readily fermentable carbohydrates have been consumed. If methanogens are present, reduced nucleotides can be re-oxidized by hydrogenase activity rather than by alcohol- or lactate-dehydrogenase. The thermodynamics of hydrogenase activity are unfavourable if hydrogen accumulates, but the methanogens keep the partial pressure of hydrogen low enough to pull the reaction towards hydrogen formation. Examples of this type of interspecies H_2 transfer are given in Chapter 11. In terms of energy production, more of the carbon can proceed through acetate kinase, and the ATP yield of the H_2-producer increases (Table 1).

Methanogenesis involves the uptake of hydrogen and the stepwise reduction of carbon dioxide. Formate can also serve as a substrate for rumen methanogenesis, but Hungate has indicated that most of the ruminal formate is converted into hydrogen and carbon dioxide (formate-hydrogen lyase) prior to methanogenesis (Hungate et al., 1970). This conclusion was based on the observation that the K_m for formate was much higher than the concentration in rumen fluid, and that the rate of methane production was more closely correlated with the concentration of dissolved hydrogen than with formate (Hungate, 1967). In sewage digesters and many other anaerobic habitats, volatile fatty acids can serve as substrates for methanogens. Organisms using these substrates grow slowly and appear to be washed out of the rumen. As a result, little acetate is converted into methane except under exceptional dietary conditions (Opperman et al., 1961; Rowe et al., 1979). Short alcohols arising from pectin breakdown (Czerkawski & Breckenridge, 1972), methylamine and trimethylamine (Neill et al., 1978) can also be converted into methane, but H_2 and CO_2 are the primary substrates for methanogenesis in the rumen. Conversely, most of the H_2 production arising from carbohydrate fermentation ends up in methane. H_2 can be used as a substrate in fumarate reduction but the K_m of the methanogens for hydrogen (1 μM) is much less (Hungate et al., 1970) than the K_m of hydrogenases from B. ruminicola, Anaerovibrio lipolytica and S. ruminantium (4·5, 14, and 44 μM, respectively; Henderson, 1980).

Our understanding of methane metabolism has increased significantly in the last 15 years, but specific aspects of the energy transduction still remain elusive. The various steps of carbon dioxide reduction have standard free-energy changes that are generally low or even positive, and only the last step, methanol reduction, has a free-energy change ($-26·9$ kcal) that would be sufficient to drive ATP formation. However, the free-energy change of the overall process is very negative ($-32·4$ kcal) and sufficient for ATP synthesis (most probably through electron transport phosphorylation).

A variety of electron carriers have been isolated from methanogens, but the sequence of electron transfer has not been thoroughly elucidated. In some cases, the function of a carrier also remains uncertain. Coenzyme F_{420} accepts electrons from hydrogenase as well as formate, NADP, 2-oxoglutarate, carbon monoxide, and pyruvate dehydrogenases, but it may be more important in biosynthesis than methanogenesis (Large, 1983). Coenzyme M (2-mercaptoethanesulphonic acid) is required by certain strains of Methanobrevibacter ruminantium, and it is produced by other, non-requiring strains (Lovley et al., 1984). Methyl-coenzyme M reductase is involved in the final stage of methane formation, as well as the assimilation of hydrogen. One of the disturbing aspects of the study of methanogenesis has been that ATP is required, not produced, in coenzyme M reactions, but the ATP requirement may be peculiar to highly purified systems in vitro. Other factors involved in methanogenesis include component B, factors F_{342}, F_{430}, a yellow fluorescent compound, methanopterin, carbon dioxide reduction factor, and $c_1 - x - t$ (Large, 1983).

Amino acids

Ammonia is usually a net product of rumen fermentation, and under natural feeding conditions much of this nitrogen compound arises from the fermentation of amino acids. Pathways of amino acid fermentation in rumen microorganisms have not been studied in any detail (Allison, 1970), but studies with non-rumen bacteria may serve as models. Clostridia ferment alanine, glutamate, histidine, aspartate, glycine, serine, cystine and tryptophan to acetate or butyrate via pyruvate, while threonine, homoserine, homocysteine and methionine are fermented to propionate via 2-oxobutyrate. Arginine and lysine are degraded to aminovaleric acid (Gottschalk, 1979). The branched-chain amino acids phenylalanine and tyrosine are oxidatively deaminated, and decarboxylated, by rumen bacteria (Allison & Bryant, 1963;

Allison, 1978; Hungate & Stack, 1982). The corresponding branched-chain or phenolic acid is esterified by coenzyme A, and this ester indirectly provides the free energy for ATP formation.

The importance of the redox state to amino acid fermentation was first recognized by Stickland in the 1930s (Nisman, 1954). He and later workers showed that highly reduced amino acids (alanine, leucine, isoleucine, valine) could be fermented by clostridia only if oxidized amino acids (glycine, proline, arginine, tryptophan) served as electron acceptors in closely coupled reactions. Although it has been suggested that such Stickland-type reactions might be important in the fermentation of amino acids by rumen microorganisms, results with pairs of oxidized and reduced amino acids were inconclusive (Lewis & Emery, 1962; Van den Hende et al., 1963). Arsenate, an inhibitor of the reduction step in Stickland reactions (Nisman, 1954), decreased amino acid deamination by mixed rumen bacteria (Broderick & Balthrop, 1979), but it was not clear if this compound was having a specific effect on deamination.

In the rumen ecosystem, however, methanogenesis is a primary means of reducing-equivalent disposal, and inhibition of methane production by carbon monoxide, a hydrogenase inhibitor, caused a decrease in ammonia production (Russell & Jeraci, 1984). The decline in ammonia was associated with a large decrease in branched-chain volatile fatty acid production and little change in that of straight-chain acids (Fig. 2). Since branched-chain volatile fatty acids are normally derived from branched-chain amino acids, it appeared that the inhibition of

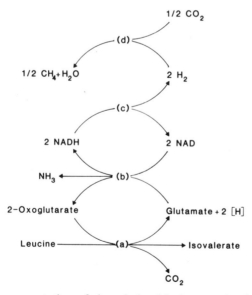

Fig. 2. Schematic representation of the relationship between leucine fermentation and methane production: (a) transamination and decarboxylation; (b) deamination and hydrogenation; (c) hydrogenase activity; (d) methanogenesis. (Taken with permission from Hino & Russell, 1985.)

methane production caused a selective decrease in branched-chain amino acid fermentation (Hino & Russell, 1985).

Amino acid fermentation is regulated by the availability of carbohydrate, and the accumulation of ammonia can be great if the solubility and availability of protein exceeds that of the carbohydrate fraction (Russell *et al.*, 1983). Most rumen bacteria are unable to grow on amino acids or amino acid sources alone (Bladen *et al.*, 1961; Russell *et al.*, 1983; Wallace, 1986). Since the deaminase activity of cell-free extracts has been found to be greater than that of whole cells (Hino & Russell, 1985), the failure of rumen bacteria to grow on peptides or amino acids is probably related to inefficient uptake systems for amino acid sources. *B. ruminicola* and *M. elsdenii*, two of the best deaminating bacteria, were only able to deaminate amino acids at rates fast enough to meet part of their maintenance requirements (Russell, 1983; Wallace, 1986). Since they were unable to meet their maintenance requirement, growth was impossible.

ENERGY-CONSUMING REACTIONS

Transport

Feedstuffs are primarily composed of large, relatively insoluble, and sometimes complex carbohydrate polymers which must be degraded by extracellular enzymes before their components can be transported by rumen microorganisms. Microbial competition for these low molecular weight products is very intense, and their extracellular concentration is usually very low (Russell, 1984). Because rapid and efficient bacterial growth requires much higher concentrations of sugars within the cell than those found outside, high-affinity transport mechanisms are needed, and transport of energy sources against a concentration gradient means that these mechanisms must be energy-linked. In protozoa the situation is somewhat different, because they are able to engulf particles prior to digestion. Hydrolytic products may be retained in a fashion similar to that in intestinal digestion, and the concentration gradient is apt to be more favourable.

There are three main categories of membrane-transport systems in bacteria (Hengge & Boos, 1983). The first is ion gradient-linked active transport, of which the best known example is the lac permease of *E. coli*. Here protons, or occasionally other cations, are co-transported with the solute into the cell, and the energy necessary to drive accumulation is therefore derived from the protonmotive force or other electrochemical gradient across the membrane (Fig. 3). A second group of transport mechanisms are the so-called 'shock-sensitive' systems, in which the substrate binds to a specific periplasmic binding protein before being transported across the cell membrane at the expense of high-energy phosphate (ATP, acetyl phosphate etc.). Examples include arabinose, galactose and maltose translocation in *E. coli*. With both of these transport systems, the substrate enters the cytoplasm unmodified, so it must be phosphorylated before glycolysis. In the third major type of transport, group translocation, the substrate is chemically modified as it passes through the membrane (Postma & Lengeler, 1985). The glucose phosphotransferase system (PTS) of *E. coli* phosphorylates glucose to glucose 6-phosphate during transport using phosphoenolpyruvate (PEP). The PTS is energetically more economical than the other systems, because it presents sugar phosphates rather than free sugars to glycolysis (Fig. 3). The glucose PTS has an additional benefit in many organisms because it regulates transport of other sugars, and thereby enables bacteria to exhibit substrate preferences (Saier, 1977; Postma, 1986).

The PTS type of group translocation is found in a wide variety of bacteria, most notably anaerobes and facultative anaerobes. In this system, PEP is dephosphorylated to pyruvate and the phosphate is donated to a soluble enzyme (enzyme I). The phosphate is then transferred to a histidine-containing protein (HPr) and in some cases to still another protein (enzyme III). This cascade of phosphorylation and dephosphorylation leads to enzyme II, which is a substrate-specific protein in the cell membrane. The sugar or sugar alcohol is then phosphorylated by enzyme II as it passes across the cell membrane (Fig. 3).

Until recently the mechanisms of sugar transport in rumen bacteria had not been

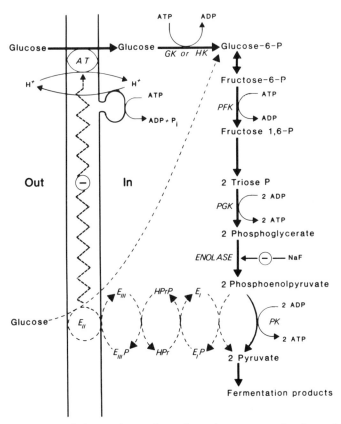

Fig. 3. The transport of glucose into a bacterium via protonmotive force driven active transport (AT) or the phosphoenolpyruvate dependent phosphotransferase system (PTS).

studied, but there is now evidence for the presence of a glucose PTS in *M. elsdenii*, *Streptococcus bovis* and *S. ruminantium* (Dills *et al.*, 1981; Martin & Russell, 1986). PEP-dependent phosphorylation of glucose was low for *B. fibrisolvens*, *B. succinogenes* and *B. ruminicola*, indicating that other types of transport must have been more important than the PTS. Franklund & Glass (1987) indicated that *B. succinogenes* had a PMF-linked active transport system for glucose.

Disaccharides entering the cell by active transport must be degraded to monosaccharides, and these reactions can be catalysed by hydrolases or phosphorylases. When the disaccharide is hydrolysed, the free-energy of the bond is dissipated as heat and free sugars are released. Both sugars must then be phosphorylated by a kinase. If disaccharides are cleaved by a phosphorylase, some of the free-energy from the bond is conserved and one of the sugars is phosphorylated. The phosphorylation usually occurs at the carbon-1 position but the phosphate can be transferred from the 1 to the 6 position by a mutase. Since one of the sugars is phosphorylated by the cleavage step, only half as much ATP is needed in a kinase reaction.

Cellobiose phosphorylase was first recognized in *Clostridium thermocellum* (Sih & McBee, 1955), but it was detected thereafter in *R. flavefaciens* (Ayers, 1958). Bailey (1963) reported the presence of a sucrose phosphorylase in *S. bovis*. We were unable to detect any such activity in the JBI strain of *S. bovis*, but this strain did have an inducible maltose phosphorylase. Since maltose (not maltose phosphate arising from the PTS) is the substrate for this enzyme, a maltose phosphorylase would be of little value unless maltose was transported into the cell by active transport, and *S. bovis* had an active-transport system for maltose as well as a maltose-specific PTS (Martin & Russell, 1987).

Growth yields of rumen bacteria

When rumen bacteria are compared with other bacteria fermenting sugars anaerobically, the growth yields of the rumen bacteria appear to be anomalously high (Bauchop & Elsden, 1960; Stouthamer, 1969, 1979). For example, *E. coli* has an anaerobic Y_{gluc} of 26 g mol^{-1} (Stouthamer, 1969) and an aerobic Y_{gluc} of 83 g mol^{-1} (Shiloach & Bauer, 1975). The Y_{gluc} values for *S. ruminantium* and *S. bovis* for homolactic fermentation are similar to the anaerobic yield of *E. coli*, but yields from pure cultures carrying out mixed fermentations cover the range 29–100 g mol^{-1} of hexose (Table 2). In other words, yields observed with rumen bacteria can be as great as those observed with an actively respiring aerobic organism. Mixed cultures of rumen bacteria *in vitro* give yields towards the upper limit of the range found with pure cultures even though growth conditions are not always optimal and some of the bacteria are consumed by protozoa (Czerkawski, 1978).

Since the overall mean value for pure cultures of rumen bacteria is 57 g mol^{-1} hexose, it is probable that reactions other than substrate-level phosphorylation (SLP) contribute substantially to ATP synthesis. The contribution by electron transport-linked phosphorylation (ETP) was investigated by Dawson *et al.* (1979), and they suggested that bacteria with high yields also possessed electron transport

Table 2
Growth Yields of Rumen Microorganisms

Organism	Energy source	Y_{substr}[a] ($g\,mol^{-1}$)	Y_{ATP} ($g\,mol^{-1}$)	Reference
A. lipolytica	Glycerol	20	10	Hobson & Summers (1967)
	Fructose	60	15	Hobson & Summers (1967)
	Fructose	59	16[b]	Henderson (1980)
B. ruminicola	Glucose	66	17[b]	Howlett et al. (1976)
	Glucose	88	15	Dawson et al. (1979)
	Glucose	82	21[b]	Russell & Baldwin (1979)
	Xylose	62	20[b]	Turner & Roberton (1979)
	Arabinose	68	19[b]	Turner & Roberton (1979)
	Arabinose	36	16[b]	Turner & Roberton (1979)
	Glucose	46	20[b]	Henderson (1980)
	Glucose	95	[c]	Russell (1983)
B. succinogenes	Glucose	64	15	Dawson et al. (1979)
	Glucose	42	[c]	Marounek & Wallace (1984)
B. fibrisolvens	Glucose	72	[c]	Russell & Baldwin (1979)
	Glucose	62	15	Dawson et al. (1979)
M. elsdenii	Glucose	73	[c]	Russell & Baldwin (1979)
	Glucose	51	21[b]	Henderson (1980)
	Lactate	11	11[b]	Henderson (1980)
R. albus	Cellobiose	102	11	Hungate (1963)
R. amylophilus	Maltose	160	20	Hobson & Summers (1967)
	Maltose	101	[c]	Jenkinson & Woodbine (1979)
	Maltose	76	13[b]	Marounek & Wallace (1984)
R. flavefaciens	Cellobiose	92	16	Pettipher & Latham (1979)
	Glucose	29	13[b]	Hopgood & Walker (1967)
S. ruminantium	Glucose	62	15[b]	Hobson & Summers (1972)
	Glucose	29	14	Dawson et al. (1979)
	Glucose	100	25[b]	Russell & Baldwin (1979)
	Glucose	49	16[b]	Henderson (1980)
	Glucose	30	19[b]	Marounek & Wallace (1984)
S. bovis	Glucose	57	[c]	Russell & Baldwin (1979)
	Glucose	36	18	Hayashi & Kozaki (1980)
	Glucose	25	13[b]	Marounek & Wallace (1984)
Mixed bacteria in vitro	Glucose	84	27[b]	Isaacson et al. (1975)
	[Hexose]	95[b]	26[b]	Demeyer & Van Nevel (1979)
Mixed population in vivo	[Hexose]	39[d]	11[b]	Hobson & Wallace (1982), from Czerkawski (1978)
	[Hexose]	[c]	21	Kennedy & Milligan (1978)

[a] Highest measured yield.
[b] Estimate calculated by the present authors, assuming 2 ATP derived from all fermentation acid products except lactate (1 ATP) and C_3 acids (3), i.e. assuming a stoichiometry of 1 ATP/fumarate reduced.
[c] Insufficient information to give a reliable estimate.
[d] Average of 75 estimations.

carriers. High yields were depressed by uncouplers and inhibitors of electron transport, but the lowest yields were not. Many propionate- and succinate-producing species possess electron transport carriers, but the mechanism of energy transduction is not fully understood (reviewed by Hobson & Wallace, 1982; Erfle et al., 1986). While ETP must occur in some of these species, a reversible ATPase that converts ADP into ATP in response to a transmembrane flux of protons has never been demonstrated. Even the assumption that 1 ATP is produced by ETP in fumarate reduction may be an underestimate. Depending on the concentrations of substrates and products, the free-energy change could be great enough to synthesize 2 ATP (Thauer et al., 1977) (Table 1). In *Bacteroides (Ruminobacter) amylophilus*, where electron transport carriers are not apparent, oxidation–reduction-linked proton translocation is used to maintain $\Delta\psi$ across the membrane (Wetzstein et al., 1987). The possibility that proton-linked end-product efflux contributes to energy conservation in rumen organisms has also been the subject of speculation (Hobson & Wallace, 1982; Konings et al., 1986), but there was little evidence for such a mechanism in *B. amylophilus* (Wetzstein et al., 1987) or *S. ruminantium* (Wallace & Robertson, 1985). While some of the variation in Y_{ATP} may be due to methods of measurement, the likely existence of as yet unknown mechanisms for energy conservation, and hence a greater yield of ATP equivalents, means that the true Y_{ATP} could be much closer to the original value of 10.5 g mol^{-1} (Bauchop & Elsden, 1960).

Although Y_{ATP} was regarded for some time as a biological constant, later work has indicated that it actually depends on cell composition, substrate and growth conditions (Pirt, 1965, 1982; Stouthamer, 1973; Stouthamer & Bettenhaussen, 1973; Tempest & Neijssel, 1984). Despite this variation, it has been difficult to reconcile measured values of Y_{ATP} with estimates based on the ATP requirements of all known biosynthetic reactions (Stouthamer, 1973). The Y_{ATP} calculated from measured yields is usually much less than would be expected from the theoretical biosynthetic summation approach (approximately 32 g mol^{-1}). This discrepancy occurs with virtually all bacteria and probably means that we have underestimated the amounts of ATP devoted to membrane energization or futile cycles (Stouthamer, 1979). The discrepancy seemed less for rumen bacteria (Harrison & McAllan, 1980), for which experimentally derived Y_{ATP} (Table 2) were in some cases closer to the theoretical value (Hespell & Bryant, 1979), but this may have been due to the exclusion from the calculations of ETP or other modes of ATP production described previously.

Variables such as changes in the principal energy source, the availability of pre-formed monomers, changes in cell composition and changes in growth rate would all be expected to cause variations in $Y_{substrate}$ and Y_{ATP}. Since carbohydrates are the primary carbon and energy sources in the rumen, major variations in Y_{ATP} would not be expected to arise from carbon flowing into the metabolic pathways at different points (Stouthamer, 1973). The main exception to this is the case of the methanogens, where Y_{ATP} would be expected to be exceptionally low because of the energy necessary to fix CO_2 (Stouthamer, 1979). Although the availability of pre-formed monomers for biosynthesis would make a large change in Y_{ATP} for methanogens, the theoretical difference in Y_{ATP} values is relatively small (approximately 10%) for organisms growing on hexoses (Stouthamer, 1973; Hespell & Bryant, 1979) and synthesizing all cell material or using pre-formed monomers. However, despite this prediction, experiments *in vivo* (Hume, 1970) and *in vitro* (Russell & Sniffen, 1984) have indicated that amino acid sources could increase bacterial yields by 28% and 19% respectively. In these cases, amino acids appeared to alter the overall efficiency of energy utilization (see 'energy spilling', below).

Changes in growth rate and cell composition often have a much greater influence than growth substrates on Y_{ATP} for rumen bacteria. Since polysaccharides are only one-third as expensive to make, in terms of ATP, as corresponding quantities of protein or nucleic acids (Stouthamer, 1973), cells containing 40% polysaccharide (conditions of energy excess) would have a Y_{ATP} that is approximately one-third greater than cells without reserve carbohydrate. The high growth yields of rumen bacteria can sometimes be attributed to their high content of storage polysaccharides (Czerkawski, 1978; Hespell & Bryant, 1979). Cell composition also varies according to the growth rate of the cells, and therefore affects Y_{ATP}. For example, the carbohydrate content of *S. ruminantium* increased from 16% to 32% as the growth rate (μ) increased from 0·035 to 0·38 h^{-1} (Wallace, 1980).

Based on work with *E. coli* (Maaloe & Kjeldgaard, 1966), the RNA content of rumen bacteria would also be expected to change with growth rate. Although no evidence of such changes was found with *B. ruminicola* (Russell, 1983), Bates *et al.* (1985) found that the RNA/protein ratio of 6 rumen bacteria increased from about 0·3 at $\mu = 0·1$ h^{-1} to 0·9 at $\mu = 1·0$ h^{-1}. However, the main impact of growth rate on Y_{ATP} is not via changes in cell composition, but through its effect on the proportion of energy that is channelled to maintenance energy.

Maintenance energy

With the advent of continuous culture techniques in the late 1940s, it became apparent that cell yield could vary with growth rate. Since cell yields were generally lower at slow growth rates, the idea was introduced that energy could be used for growth-related functions and for functions not directly related to growth (Herbert *et al.*, 1956). These latter 'maintenance functions' have not been precisely defined (Pirt, 1965) but ion balance across the cell membrane, motility, and turnover of macromolecules are probably most important (Fig. 4).

In the 1960s Marr *et al.* (1963) and Pirt (1965) described maintenance energy using double reciprocal plots of growth rate and cell mass or yield, respectively. In the mathematical derivation by Marr *et al.*, maintenance is defined as a fractional rate which is used to predict the number of additional cells that would be obtained if there were no maintenance requirement. The derivation by Pirt also related maintenance to the difference between the theoretical maximum growth yield (Y_G—no maintenance) and the observed yield (Y). The Pirt equation is $1/Y = m/\mu + 1/Y_G$. Algebraic rearrangements of this relationship employing the specific rate of substrate utilization (q) versus growth rate (μ) have likewise been used to estimate a maintenance coefficient (m) (Neijssel & Tempest, 1976; Tempest & Neijssel, 1984). In all these transformations, maintenance has been assumed to be a time-dependent function that is proportional to cell mass and independent of growth rate (Pirt, 1982).

Since maintenance is generally defined in terms of yield or the specific rate of energy-source utilization, it is apparent that any factor affecting the production of bacterial biomass will affect the estimate of maintenance. When Pirt (1965) plotted yield data for *S. ruminantium*, the relationship was not linear, and he concluded that

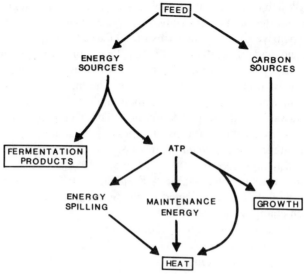

Fig. 4. A schematic diagram showing the fermentation of feed by rumen microorganisms and the partition of the feed energy among the various products.

'it seems doubtful whether the rapid fall in yield of the selenomonad can be attributed to the maintenance energy requirement'. Subsequently Scheifinger et al. (1975) showed that *S. ruminantium* changed from an acetate and propionate fermentation to lactate at a growth rate of approximately $0.2\,h^{-1}$. Since the switch in the fermentation products would affect the amount of ATP available for growth, it is not surprising that the estimate of maintenance was not constant. As mentioned earlier, bacteria sometimes accumulate large amounts of polysaccharide when nutrients other than energy-source limit growth. Since polysaccharide requires less ATP for its synthesis than other cell components, the yield increases. Polysaccharide accumulation is usually correlated with an increase in the theoretical maximum growth yield, but it could also affect (increase) the estimate of maintenance if polysaccharide increased at particular (rapid) growth rates.

The maintenance coefficient of mixed rumen bacteria in continuous cultures was estimated at 0.26 mmol glucose (g dry wt)$^{-1}$ h^{-1} (Isaacson et al., 1975), which is somewhat lower than the maintenance coefficient of *E. coli*: 0.39 mmol glucose (g dry wt)$^{-1}$ h^{-1} (Marr et al., 1963). Individual species of rumen bacteria had maintenance coefficients that appeared to be of a similar magnitude, although changes in fermentation products made estimation difficult in some cases (Russell & Baldwin, 1979). *Bacteroides amylophilus* is exceptional in having a maintenance coefficient about ten-fold higher (Jenkinson & Woodbine, 1979), which may be a consequence of the fragility of its cell envelope, and consequent lysis at low growth rates. Maintenance coefficients *in vivo* are much more difficult to estimate, but the indications are that the maintenance requirement of the mixed population is quite similar to values determined *in vitro* (Harrison & McAllan, 1980). However, even these low maintenance values have a major influence on microbial productivity at the low dilution rates pertaining *in vivo*. At a dilution rate of 0·06 h^{-1}, well within the range of values found in the rumen, 32% of the energy generated by bacteria is dissipated by maintenance (Harrison & McAllan, 1980).

In continuous culture, growth rate is proportional to dilution rate, and maintenance makes up a smaller fraction of the total energy utilization at faster growth rates. While the relationship between growth rate and dilution rate is probably not as absolute *in vivo*, one might expect higher bacterial yields at higher dilution rates. Under conditions of cold stress, the effective dilution rate in the rumen increased from 0·068 to 0·115 h^{-1}, and this was accompanied by a 19% increase in Y_{ATP} (Kennedy & Milligan, 1978). Unfortunately, decreased digestibility of the diet negated the beneficial effect of the enhanced yield.

Energy-spilling reactions

Experiments by Neijssel & Tempest (1976) showed that nitrogen-, sulphur- and phosphate-limited cultures of *Klebsiella aerogenes* consumed energy sources at a rapid rate even though the rate of biomass formation was slow. Consequently, the molar growth yields were low, which prompted Neijsel & Tempest to question the mathematical derivations of maintenance and postulate that bacteria must have mechanisms to hydrolyse ATP, if energy sources are in excess. Since then, the terms 'overflow metabolism', 'slip reactions', 'energy-spilling reactions' and 'futile cycles' have all been used in explanations of this uncoupling of catabolism and anabolism (Fig. 4).

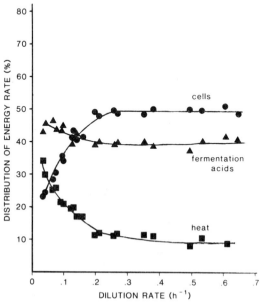

Fig. 5. Distribution of energy in the fermentation of glucose by *Bacteroides ruminicola* grown in continuous-culture. (Taken with permission from Russell, 1986.)

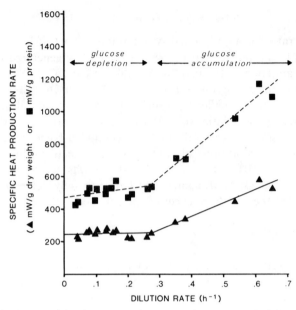

Fig. 6. Specific heat production rate of *Bacteroides ruminicola* in continuous-culture. (Taken with permission from Russell, 1986.)

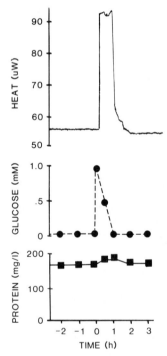

Fig. 7. Effect of a pulse dose of glucose (time 0) on the heat production of *Bacteroides ruminicola* which had been growing in a glucose-limited chemostat at a dilution rate of $(0.167\,h^{-1})$. (Taken with permission from Russell, 1986.)

Since maintenance is, by definition, the amount of energy that cannot be used for growth, it is apparent that maintenance must contribute to heat production. When *B. ruminicola* was grown in continuous culture, the heat of fermentation ranged from 34% to 8% of the energy provided, and the production of cells was inversely related to heat production (Fig. 5). The specific rate of heat production remained relatively constant so long as the glucose was completely depleted; because the heat of growth was insignificant relative to the heat of maintenance, total heat production provided a reasonable estimate of maintenance (Fig. 6). However, at higher dilution rates glucose eventually accumulated in the chemostat vessel and there was a marked increase in specific heat-production. Experiments with pulse doses of glucose likewise indicated that glucose accumulation caused a large increase in heat production even though there was at the time little change in cell protein or mass (Fig. 7).

ENERGY REGULATION

Enzyme activity

Because soluble energy sources are often limiting in the rumen, and because the yield of ATP is generally low in anaerobic habitats, one might expect rumen bacteria to select fermentation pathways that maximize the amount of ATP per unit of substrate and use this ATP conservatively. This does not always happen. *S. bovis* and *S. ruminantium* are fast-growing rumen organisms (Russell & Baldwin, 1978), and their numbers increase dramatically after animals have consumed large amounts of cereal grain or grasses containing an abundance of sugar. At such times, both of these organisms produce lactate even though the ATP yield is then low relative to the amount of hexose fermented (Scheifinger *et al.*, 1975; Russell & Baldwin, 1979). As pointed out by Hungate (1979), the ability of these species to outgrow the acetate producers that obtain more ATP per unit of hexose is due to the speed of their metabolism. By growing very fast they are able to generate more ATP per unit of time than the more slowly growing acetate producers. When energy sources become limiting and rapid growth rates are not possible, they switch to pathways yielding greater ATP per hexose and once again try to maximize ATP per unit time.

In *S. ruminantium* the change in fermentation products between lactate and acetate plus propionate is due to the homotropic activation of the lactate dehydrogenase (LDH) by its substrate, pyruvate (Wallace, 1978). When growth rates are low, pyruvate concentrations within the cell are low, the LDH is not active, and pyruvate from the EMP pathway proceeds towards acetate and propionate. If hexose is abundant and growth rate is rapid, pyruvate builds up, the LDH is able to produce lactate, and the 'bottle-neck' in acetate and propionate production is by-passed. If *S. ruminantium* cultures were limited by nitrogen, lactate was the primary product even at slow growth rates, because glucose, and thus conceivably intracellular pyruvate, was never limiting.

The LDH of *S. bovis* (Wolin, 1964) and *B. fibrisolvens* (Van Gylswyk, 1977) are activated by fructose 1,6-bisphosphate (FBP). The increased production of lactate by *S. bovis* at rapid growth rates and neutral pH was associated with an increased concentration of FBP within the cells. However, large amounts of lactate were produced at low pH even though growth rate and FBP were low (Russell & Hino, 1985). As extracellular pH declines (usually due to lactate accumulation), intracellular pH also declines. At low intracellular pH (5·4) the LDH has a much higher maximum velocity and a lower requirement for FBP (Russell & Hino, 1985). Alternative pathways via pyruvate-formate lyase would be inhibited by the low intracellular pH (Takahashi *et al.*, 1982). The organism's LDH continues to function, but the organism is unable to switch to an acetate, formate, and ethanol fermentation, so the ATP formation per unit time is low under these conditions (Russell *et al.*, 1981). However, *S. bovis* is more resistant to low pH than most of the other bacteria (Russell & Dombrowski, 1980; Therion *et al.*, 1982), and thus is not faced with as much competition for hexose.

Neither pyruvate nor FBP was found to be of major importance in a study of the kinetic properties of LDH extracted from the mixed rumen population (Counotte *et al.*, 1980). ATP, which inhibited the LDH of *S. ruminantium* (Wallace, 1978), and the ratio of NADH to NAD were found to be of greater importance. A similar regulatory pattern was found with the rumen ciliate *Isotricha prostoma* (Counotte *et al.*, 1980).

Futile cycles

Energy-sufficient cultures of rumen bacteria consumed more energy source and produced more heat than energy-limited cultures (Russell, 1986). The mechanisms involved in this energy spilling need to be elaborated further. Because the heat production increased so quickly after a pulse dose of glucose, and decreased with the same rapidity when the glucose was depleted (Fig. 7), changes arising from protein synthesis are probably not responsible. According to the chemiosmotic model (Mitchell, 1961), energy is expended to translocate protons outwardly through the membrane to establish a proton-motive force. Ion recycling could provide a rapid mechanism for ATP hydrolysis and added heat production. The advantage of added heat production to rumen bacteria is not yet clear. Westerhoff *et al.* (1983) theorized that microbial growth is analogous to free enthalpy changes in nonequilibrium thermodynamics: 'some thermodynamic efficiency may be sacrificed to make the process run faster'. Tempest (1967) noted that fast growing bacteria contain greater concentrations of potassium, magnesium and ammonium ion than the same bacteria growing slowly. It is conceivable that rumen bacteria use the proton-motive force to transport more of these ions in preparation for an eventual increase in growth rate. If some of the ions continue to leak back outside, there would be a mechanism for sustained heat production. A less direct advantage from the waste of energy as heat would be the denial of energy source to competing organisms.

The 100 mM concentration of VFA in rumen fluid poses an interesting challenge to metabolic regulation in rumen bacteria. The intracellular pH of neutrophilic organisms like *E. coli* is strictly regulated (Booth, 1985). However, the ability of VFA to traverse membranes in their uncharged form is well known (Kell *et al.*, 1981), and this activity dissipates the ΔpH component of PMF, particularly at lower pH where more of the acids are in the permeant, un-ionized form; in effect, the VFA act as uncouplers. The growth of *E. coli* was inhibited by rumen VFA concentrations at pH 6·0, but it was less affected at pH 7·0 (Wolin, 1969). *E. coli* in fact dies rapidly in rumen liquor from concentrate-fed sheep, whereas the death rate is much less at the higher pH of roughage-fed animals (Wallace, unpublished observation). Fermentative bacteria in general (Booth, 1985; Kell *et al.*, 1981), and rumen bacteria in particular (Russell & Hino, 1985; Wetzstein *et al.*, 1987; Wallace & Robertson, unpublished results), regulate intracellular pH very poorly and do not always maintain pH gradients typical of other bacteria (Booth, 1985). This variation has been interpreted as being a response to the prevailing high concentrations of membrane-permeable VFA produced by anaerobic metabolism (Kell *et al.*, 1981).

The high CO_2 concentrations that prevail in the rumen may also be involved in energy dissipation. At high CO_2 concentrations, *K. aerogenes* grew less efficiently (Teixeira de Mattos *et al.*, 1984), and it was suggested that ATP hydrolysis occurred via the futile cycle of PEP carboxylase and PEP carboxykinase. It remains to be seen if rumen organisms respond in a similar fashion.

Monensin is often used as an additive in the beef cattle industry because it inhibits bacteria that produce hydrogen, a precursor of methane (Chen & Wolin, 1979) (see also Chapter 2). In spite of its widespread application, until recently little was known about its biochemical mechanism of action. EDTA treatments are routinely used to increase the permeability of the Gram-negative outer membrane to ionophores (Booth *et al.*, 1979), and Gram-positive bacteria, like *S. bovis*, which lack an outer membrane are much more sensitive to monensin (Muir & Barreto, 1979). When monensin (5 mg l^{-1}) was added to actively growing *S. bovis* cultures there was an almost immediate inhibition of growth but glucose utilization and lactate production continued for another 8 hours (Russell, 1987). Monensin caused a decrease in intracellular pH, a decrease in intracellular potassium, and an increase in intracellular sodium (Fig. 8). The net exchange of potassium for sodium and protons was driven by the difference in concentration of potassium and sodium across the cell membrane (accumulation ratios of 68 and 2·7 respectively). Since *S. bovis* is able to tolerate an internal pH as low as 5·4 (Russell & Hino, 1985), it appeared that growth inhibition resulted from the utilization of ATP in expelling the excess of protons from the cell. Previous work by Dawson & Boling (1984) indicated that increased potassium decreased the sensitivity of some bacterial species to monensin, and these observations are consistent with the model described above.

Catabolite regulatory mechanisms

Catabolite repression and PTS repression allow microorganisms to utilize some energy sources in preference to, or to the exclusion of, others, and to control the level of expression of different enzymes. Monod studied catabolite repression in the late

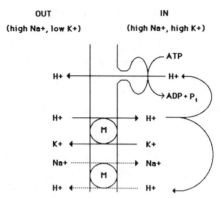

Fig. 8. A schematic diagram showing a proposed mechanism of monensin action in Gram-positive rumen bacteria. (Taken with permission from Russell, 1987.)

1940s and referred to it as the 'glucose effect' (Monod, 1947). Magasanik (1961) coined the term 'catabolite repression' to describe the phenomenon when it was discovered that substrates other than glucose could produce the same effect. The mechanism was not well understood until cyclic 3′,5′-AMP (cAMP) was discovered in bacteria and it was shown that the depletion of glucose or other preferred substrates caused an increase in cAMP; cAMP could bind a small regulatory protein designated as CAP (or sometimes CRP), and the cAMP–CAP complex could bind to DNA at a site adjacent to the promoter and facilitate the transcription of genes needed for the utilization of less-preferred substrates (Postma, 1986). Cellular concentrations of cAMP are regulated by adenyl cyclase activity, which in turn is thought to be activated by the phosphorylated form of PTS enzyme III for glucose. Since enzyme III is dephosphorylated in the presence of glucose, cAMP levels are low and the synthesis of enzymes such as the *lac* operon of *E. coli* is repressed. Cyclic AMP is not totally responsible for catabolite repression, but the other factors responsible have not yet been clearly identified (Postma, 1986).

Catabolite repression provides a means of regulating energy source utilization at the level of protein synthesis, but preferred substrates have sometimes caused an immediate cessation of non-preferred substrate utilization (McGinnis & Paigen, 1969). An immediate cessation could not easily be explained by an inhibition of protein synthesis because protein already present would have to be diluted out by successive divisions of cells. Subsequent work showed that preferred energy sources could inhibit the uptake of non-preferred energy sources, and that this 'exclusion of inducer' occurred at the level of the cell membrane. The PTS is a regulatory system as well as a transport mechanism in most bacteria (Saier, 1977; Postma & Roseman, 1976), and the uptake of sugars by the PTS often inhibits the uptake of sugars by active transport permeases (Fig. 3). When sugars are transported by the PTS, enzyme III is largely dephosphorylated and in this state it can inhibit active transport. PTS sugars generally inhibit non-PTS (actively transported) sugars, but some PTS sugars can inhibit the utilization of other PTS sugars as well. The latter form of regulation is probably due to a competition for the phosphorylated HPr.

Enzymes III and II are sugar-specific and these components can have different affinities for the non specific HPr (Saier, 1977; Postma & Lengeler, 1985).

A variety of rumen bacteria (Russell & Baldwin, 1978) and the rumen fungus *Neocallimastix frontalis* (Mountfort & Asher, 1983) were found to have preferences for particular energy sources and to use some of them to the exclusion of others. Because these microorganisms often preferred different substrates they were able to occupy separate niches. *S. bovis* used glucose and sucrose in preference to maltose and cellobiose, and the preferred substrates caused an almost immediate cessation of non-preferred substrate utilization (Russell & Baldwin, 1978). All four of these sugars were transported by the PTS (Martin & Russell, 1986), but only the glucose PTS was constitutive (Martin & Russell, 1987). The maltose-, sucrose- and cellobiose-PTS were inducible. When maltose-grown cells were incubated with [^{14}C]maltose, a ten-fold excess of glucose caused a 90% inhibition of PEP-dependent phosphorylation of [^{14}C]maltose. However, if glucose-grown cells were incubated with [^{14}C]glucose, a ten-fold excess of maltose caused less than 5% inhibition of [^{14}C]glucose phosphorylation. Since a significant inhibition was not observed in both cases, it is unlikely that both sugars were transported by the same enzyme II. The inhibition of maltose transport by glucose was most easily explained by the competition of separate enzymes II for phosphorylated HPr. In this case, glucose enzyme II must have had a higher affinity for phosphorylated HPr than maltose enzyme II. Similar effects were also noted between maltose and sucrose, but the kinetics were somewhat different (Martin & Russell, 1987). Maltose utilization was immediately inhibited by glucose addition, but the inhibition caused by sucrose did not occur immediately (Russell & Baldwin, 1978). This lag was consistent with PTS regulation by intracellular sugar phosphates (Saier, 1985). After sucrose addition, sucrose phosphate resulting from the action of sucrose enzyme II would eventually accumulate and inhibit maltose enzyme II.

S. bovis had a maltose PTS as well as an active-transport system for maltose (Martin & Russell, 1987), and the reasons for the apparent redundancy of these maltose transport mechanisms were at first not known. However, this duality of transport may allow *S. bovis* to adapt to greater ranges in extracellular pH. The PTS of oral streptococci had a high affinity at neutral pH values, but it showed sensitivity to decline in pH (Hamilton & Ellwood, 1978). Because low pH repressed the PTS while the glycolytic activity of the intact cells remained high, active transport must have been more resistant to low pH.

Since *E. coli* (and presumably other bacteria) has more than 1000 different intracellular proteins (O'Farrell, 1975), it might be thought that the synthesis of a single enzyme would represent a minor load on the overall energetics of bacterial protein synthesis. Such an assumption does not consider the fact that the transcription and translation rate of different genes can vary widely. For instance, in *E. coli*, β-galactosidase can account for more than 4% of the total cell protein (Novick, 1961). Thus it is no small wonder that most bacteria regulate enzyme synthesis as a means of both conserving energy and adapting to a changing environment.

Since *E. coli*, the organism most used as a model for microbial metabolism, does not synthesize extracellular enzymes, less is known about the regulation of extracellular enzymes than intracellular enzymes. Many extracellular substrates are too large to enter the cell, and the organism must have some means of detecting the presence of these molecules. It has been proposed that this problem is overcome by a low constitutive rate of enzyme secretion and a significant induction when the right substrate is present (Priest, 1984). Little is known about induction and repression mechanisms in rumen bacteria. Pettipher & Latham (1979) grew *R. flavefaciens* in cellobiose-limited chemostats and concluded that the carboxymethylcellulase and xylanase formed were constitutive. Since neither cellulose nor xylan was added, it is not known whether additional activity could have been induced.

Survival during starvation

The rumen has frequently been cited as an example of a naturally occurring continuous-culture. There is a clear difference between the conditions prevailing in the rumen and those found in a chemostat, however, because in a meal-fed ruminant the energy supply to the microorganisms consists of a short period of plenty followed by a much longer period of starvation. Only in the grazing animal would steady-state conditions be remotely approached.

Rumen bacteria therefore prepare for starvation by laying down vast amounts of cytoplasmic polysaccharide storage granules. Carbohydrate contents of rumen bacteria as high as 75% of the dry weight of the organism have been reported (Stewart *et al.*, 1981), and electron micrographs of rumen bacteria *in vivo* invariably show abundant cytoplasmic glycogen-like granules. Cell carbohydrate, and possibly nucleic acids, protein and other cell polymers, may provide energy during starvation (Allison, 1978; Wallace, 1980; Mink & Hespell, 1981*a,b*; Wachenheim & Hespell, 1985). Different species lose viability at different rates under starvation conditions. Surprisingly, *S. ruminantium* survived poorly, half of the cells dying within 2·5 hours (Mink & Hespell, 1981*a*). *R. flavefaciens* (Wachenheim & Hespell, 1985) and *M. elsdenii* (Mink & Hespell, 1981*b*) had half-lives of up to 12 and 13 hours respectively. Since the number of microorganisms dying in the rumen should be related to retention time, increased yields at high dilution rates (Kennedy & Milligan, 1978) might actually be due to a decrease in microbial death, not just a decrease in the proportion of energy expended on maintenance.

CONCLUSIONS

During rumen fermentation, feedstuffs are converted into short-chain acids, ammonia, methane, carbon dioxide, cell material, and heat. Animal performance is dependent on the balance of these products, and this balance is ultimately controlled by the types and activities of microorganisms in the rumen. The acids are used by the animal as an energy source, while the microbes serve as an important source of amino acids for protein synthesis. Ammonia, methane and heat, by contrast, represent a loss of either nitrogen or energy to the animal. Microbial energy transformations affect all aspects of microbial ecology, and hence ruminant nutrition. They determine not only the quantity and composition of the fermentation acids, but the amount of biomass present. Since the rates of feedstuff fermentation and degradation depend on the cell density, and since the animal is dependent on microbial protein as an amino acid source, the energy metabolism of rumen microorganisms affects both the energy and nitrogen status of the animal.

Ruminant nutritionists have been successful in reducing methane production and protein fermentation, but these reductions have sometimes had a deleterious effect on other ruminal parameters. Ionophores which decrease methane can also decrease the yield of ATP, increase energy spilling, and decrease the efficiency of microbial protein synthesis. Low solubility proteins are often fermented at a slower rate, and the decrease in ammonia can have a positive impact on nitrogen retention. However, if the protein is overprotected, the rumen microbes can be starved for amino acid nitrogen and grow less efficiently.

Optimization of the rumen fermentation will necessitate a more thorough analysis of rumen microbial physiology as well as of the properties of the diet. While quick answers and simplistic approaches are at least initially attractive, fundamental and mechanistic strategies are almost always more rewarding. Studies of rumen microbiology have to a large degree been limited to the organismic level of organization. We are only beginning to understand the various reactions actually responsible for the success or failure of microorganisms in the rumen.

REFERENCES

Allison, M. J. (1970). Nitrogen metabolism of ruminal micro-organisms. In *Physiology of Digestion and Metabolism in the Ruminant*, ed. A. T. Phillipson. Oriel Press, Newcastle-upon-Tyne, pp. 456–73.

Allison, M. J. (1978). Production of branched-chain volatile fatty acids by certain anaerobic bacteria. *Appl. Environ. Microbiol.*, **35**, 872–7.

Allison, M. J. & Bryant, M. P. (1963). Biosynthesis of branched-chain amino acids from branched-chain fatty acids by rumen bacteria. *Arch. Biochem. Biophys.*, **101**, 269–77.

Ayers, W. A. (1958). Phosphorylation of cellobiose and glucose by *Ruminococcus flavefaciens*. *J. Bacteriol.*, **76**, 515–7.

Bailey, R. W. (1963). The intracellular galactosidase of a rumen strain of *Streptococcus bovis*. *Biochem. J.*, **86**, 509–14.

Baldwin, R. L. (1965). Pathways of carbohydrate metabolism in the rumen. In *Physiology of Digestion in the Ruminant*, ed. R. W. Dougherty. Butterworth, Washington, DC, pp. 379–89.

Baldwin, R. L., Wood, W. A. & Emery, R. S. (1963). Conversion of glucose-^{14}C to propionate by the rumen microbiota. *J. Bacteriol.*, **85**, 1346–9.

Bates, D. B., Gillett, J. A., Barao, S. A. & Bergen, W. G. (1985). The effect of specific growth rate and stage of growth on nucleic acid-protein values of pure cultures and mixed ruminal bacteria. *J. Anim. Sci.*, **61**, 713–24.

Bauchop, T. & Elsden, S. R. (1960). The growth of microorganisms in relation to their energy supply. *J. Gen. Microbiol.*, **23**, 457–69.

Bladen, H. A., Bryant, M. P. & Doetsch, R. N. (1961). A study of bacterial species from the rumen which produce ammonia from protein hydrolyzate. *Appl. Microbiol.*, **9**, 175–80.

Booth, I. R. (1985). Regulation of cytoplasmic pH in bacteria. *Microbiol. Rev.*, **49**, 359–78.

Booth, I. R., Mitchell, W. J. & Hamilton, W. A. (1979). Quantitative analysis of proton-linked transport systems. *Biochem. J.*, **182**, 687–96.

Brockman, H. L. & Wood, W. A. (1975). Electron-transferring flavoprotein of *Peptostreptococcus elsdenii* that function in the reduction of acrylyl-coenzyme A. *J. Bacteriol.*, **123**, 1447–53.

Broderick, G. A. & Balthrop, J. E. (1979). Chemical inhibition of amino acid deamination by ruminal microbes *in vitro*. *J. Anim. Sci.*, **49**, 1101–11.

Caldwell, D. R., White, D. C., Bryant, M. P. & Doetsch, R. N. (1965). Specificity of the heme requirement for growth of *Bacteroides ruminicola*. *J. Bacteriol.*, **90**, 1645–54.

Chen, M. & Wolin, M. J. (1979). Effect of monensin and lasalocid-sodium on the growth of methanogenic and rumen saccharolytic bacteria. *Appl. Environ. Microbiol.*, **38**, 72–7.

Caunotte, G. H. M., deGroot, M. & Prins, R. A. (1980). Kinetic parameters of lactate dehydrogenase of some rumen bacterial species, the anaerobic ciliate *Isotricha prostoma* and mixed rumen microorganisms. *Ant. van Leeuwenhoek*, **46**, 363–81.

Counotte, G. H. M., Prins, R. A., Jansen, R. H. A. M. & de Bie, M. J. A. (1981). Role of *Megasphaera elsdenii* in the fermentation of DL-[2-^{13}C]lactate in the rumen of dairy cattle. *Appl. Environ. Microbiol.*, **42**, 649–55.

Counotte, G. H. M., Lankhorst, A. & Prins, R. A. (1983). Role of DL-lactic acid as an intermediate in rumen metabolism of dairy cows. *J. Anim. Sci.*, **56**, 1222–35.

Czerkawski, J. W. (1978). Reassessment of efficiency of synthesis of microbial matter in the rumen. *J. Dairy Sci.*, **61**, 1261–73.

Czerkawski, J. W. & Breckenridge, G. (1972). Fermentation of various glycolytic intermediates and other compounds by rumen microorganisms with particular reference to methane production. *Br. J. Nutr.*, **27**, 131–46.

Dawes, E. A. (1986). *Microbial Energetics*. Blackie, London.

Dawson, K. A. & Boling, J. A. (1984). Factors affecting resistance of monensin-resistant and sensitive strains of *Bacteroides ruminicola*. *Can. J. Anim. Sci.*, **64** (Suppl.), 132–3.

Dawson, K. A., Preziosi, M. C. & Caldwell, D. R. (1979). Some effects of uncouplers and inhibitors on growth and electron transport in rumen bacteria. *J. Bacteriol.*, **139**, 384–92.

Demeyer, D. I. & Van Nevel, C. J. (1979). Effect of defaunation on the metabolism of rumen microorganisms. *Br. J. Nutr.*, **42**, 515–24.

Dills, S. S., Lee, C. A. & Saier, M. H. (1981). Phosphoenolpyruvate-dependent sugar phosphotransferase activity in *Megasphaera elsdenii*. *Can. J. Microbiol.*, **27**, 949–52.

Erfle, J. D., Sauer, F. D. & Mahadevan, S. (1986). Energy metabolism in rumen microbes. In *Control of Digestion and Metabolism in Ruminants*, ed. L. P. Milligan, W. L. Grovum & A. Dobson. Prentice-Hall, Englewood Cliffs, New Jersey, pp. 81–99.

Franklund, C. V. & Glass, T. L. (1987). Glucose uptake by the cellulolytic rumen anaerobe *Bacteroides succinogenes*. *J. Bacteriol.*, **169**, 500–6.

Glass, T. L., Bryant, M. P. & Wolin, M. J. (1977). Partial purification of ferredoxin from *Ruminococcus albus* and its role in pyruvate metabolism and reduction of nicotinamide adenine dinucleotide by H_2. *J. Bacteriol.*, **131**, 463–72.

Gottschalk, G. (1979). In *Bacterial Metabolism*. Springer Verlag, New York, pp. 15, 212–19.

Hamilton, I. R. & Ellwood, D. C. (1978). Effects of fluoride on carbohydrate metabolism by washed cells of *Streptococcus mutans* grown at various pH values in a chemostat. *Infect. and Immunity*, **19**, 434–42.

Harden, A. & Young, W. J. (1906). The alcoholic fermentation of yeast juice. II. The coferment of yeast juice. *Proc. Roy. Soc.*, **B78**, 369–75.

Harrison, D. G. & McAllan, A. B. (1980). Factors affecting microbial growth yields in the reticulo-rumen. In *Digestive Physiology and Metabolism in Ruminants*, ed. Y. Ruckebusch & P. Thivend. MTP Press, Lancaster, pp. 205–26.

Hayashi, T. & Kozaki, M. (1980). Growth yield of an orange-colored *Streptococcus bovis* No. 148. *J. Gen. Appl. Microbiol.*, **26**, 245–53.

Henderson, C. (1980). The influence of extracellular hydrogen on the metabolism of

Bacteroides ruminicola, Anaerovibrio lipolytica and *Selenomonas ruminantium. J. Gen. Microbiol.*, **119**, 485–91.

Hengge, R. & Boos, W. (1983). Maltose and lactose transport in *Escherichia coli*: examples of two different types of concentrative transport systems. *Biochim. Biophys. Acta*, **737**, 443–78.

Herbert, D., Elsworth, R. & Telling, R. O. (1956). The continuous culture of bacteria: a theoretical and experimental study. *J. Gen. Microbiol.*, **25**, 227–38.

Hespell, R. B. & Bryant, M. P. (1979). Efficiency of rumen microbial growth: influence of some theoretical and experimental factors on Y_{ATP}. *J. Dairy Sci.*, **49**, 1640–59.

Hino, T. & Russell, J. B. (1985). The effect of reducing equivalent disposal and NADH/NAD on the deamination of amino acids by intact and cell-free extracts of rumen microorganisms. *Appl. Environ. Microbiol.*, **50**, 1368–74.

Hobson, P. N. & Summers, R. (1967). The continuous culture of anaerobic bacteria. *J. Gen. Microbiol.*, **47**, 53–65.

Hobson, P. N. & Summers, R. (1972). ATP pool and growth yield in *Selenomonas ruminantium. J. Gen. Microbiol.*, **70**, 351–60.

Hobson, P. N. & Wallace, R. J. (1982). Microbial ecology and activities in the rumen: part II. *Crit. Rev. Microbiol.*, **9**, 253–320.

Hopgood, M. F. & Walker, D. J. (1967). Succinic acid production by rumen bacteria. I. Isolation and metabolism of *Ruminococcus flavefaciens. Aust. J. Biol. Sci.*, **20**, 165–82.

Howlett, M. R., Mountfort, D. O., Turner, K. W. & Robertson, A. M. (1976). Metabolism and growth yields in *Bacteroides ruminicola* strain B_14. *Appl. Environ. Microbiol.*, **32**, 274–83.

Hume, I. D. (1970). Synthesis of microbial protein in the rumen. III. Effect of dietary protein. *Aust. J. Agric. Res.*, **21**, 305–14.

Hungate, R. E. (1963). Polysaccharide storage and growth efficiency in *Ruminococcus albus. J. Bacteriol.*, **86**, 848–54.

Hungate, R. E. (1967). Hydrogen as an intermediate in the rumen fermentation. *Arch. Mikrobiol.*, **59**, 158–64.

Hungate, R. E. (1979). Evolution of a microbial ecologist. *Ann. Rev. Microbiol.*, **33**, 1–20.

Hungate, R. E. & Stack, R. J. (1982). Phenylpropanoic acid: growth factor for *Ruminococcus albus. Appl. Environ. Microbiol.*, **44**, 79–83.

Hungate, R. E., Smith, W., Bauchop, T., Yu, I. & Rabinowitz, J. C. (1970). Formate as an intermediate in the bovine rumen fermentation. *J. Bacteriol.*, **102**, 389–97.

Isaacson, H. R., Hinds, F. C., Bryant, M. P. & Owens, F. N. (1975). Efficiency of energy utilisation by mixed rumen bacteria in continuous culture. *J. Dairy Sci.*, **58**, 1645–59.

Jenkinson, H. F. & Woodbine, M. (1979). Growth and energy production in *Bacteroides amylophilus. Arch. Microbiol.*, **120**, 275–81.

Joyner, A. E. & Baldwin, R. L. (1966). Enzymatic studies of pure cultures of rumen microorganisms. *J. Bacteriol.*, **92**, 1321–30.

Kell, D. B., Peck, M. W., Rodger, G. & Morris, J. G. (1981). On the permeability to weak acids and bases of the cytoplasmic membrane of *Clostridium pasteurianum. Biochem. Biophys. Res. Commun.*, **99**, 81–8.

Kennedy, P. M. & Milligan, L. P. (1978). Effects of cold exposure on digestion, microbial synthesis and nitrogen transformations in sheep. *Br. J. Nutr.*, **39**, 105–17.

Konings, W. N., Otto, R. & ten Brink, B. (1986). Energy transduction and solute transport in streptococci. In *Control of Digestion and Metabolism in Ruminants*, ed. L. P. Milligan, W. L. Grovum & A. Dobson. Prentice-Hall, Englewood Cliffs, New Jersey, pp. 100–21.

Kroger, A. & Winkler, E. (1981). Phosphorylative fumarate reduction in *Vibrio succinogenes* stoichiometry of ATP synthesis. *Arch. Microbiol.*, **129**, 100–4.

Ladd, J. N. & Walker, D. J. (1965). Fermentation of lactic acid by the rumen microorganism *Peptostreptococcus elsdenii. Ann. N. Y. Acad. Sci.*, **119**, 1038–45.

Large, P. J. (1983). *Methylotrophy and Methanogenesis*. Amer. Soc. Microbiol., Washington, DC, pp. 11–24.

Lehninger, A. L. (1975). In *Biochemistry*. Worth Publishers, New York, p. 411.

Lewis, T. R. & Emery, R. S. (1962). Relative deamination rates of amino acids by rumen microorganisms. *J. Dairy Sci.*, **45**, 765–8.

Lipmann, F. (1941). Metabolic generation and utilization of phosphate bond energy. *Adv. Enzymol.*, **1**, 99–162.

Lovley, D. R., Greening, R. C. & Ferry, J. G. (1984). Rapidly growing rumen methanogenic organism that synthesizes coenzyme and has a high affinity for formate. *Appl. Environ. Microbiol.*, **48**, 81–7.

Maaloe, O. & Kjeldgaard, N. O. (1966). *Control of Macromolecular Synthesis*. Benjamin, New York.

Mackie, R. I., Gilchrist, F. M. C. & Heath, S. (1984). An *in vivo* study of ruminal microorganisms influencing lactate turnover and its contribution to volatile fatty acid production. *J. Agric. Sci.* (Camb.), **103**, 37–51.

Macy, J., Probst, J. & Gottschalk, G. (1975). Evidence for cytochrome involvement in fumarate reduction and adenosine 5′-triphosphate synthesis by *Bacteroides fragilis* grown in the presence of hemin. *J. Bacteriol.*, **123**, 436–42.

Magasanik, B. (1961). Catabolite repression. *Cold Spring Harbor Symp. Quant. Biol.*, **26**, 249–56.

Marounek, M. & Wallace, R. J. (1984). Influence of culture E_h on the growth and metabolism of the rumen bacteria *Selenomonas ruminantium*, *Bacteroides amylophilus*, *Bacteroides succinogenes* and *Streptococcus bovis* in batch culture. *J. Gen. Microbiol.*, **130**, 223–9.

Marr, A. G., Nilson, E. H. & Clark, D. J. (1963). The maintenance requirements of *Escherichia coli*. *Ann. N. Y. Acad. Sci.*, **102**, 536–48.

Martin, S. A. & Russell, J. B. (1986). Phosphoenolpyruvate-dependent phosphorylation of hexoses by ruminal bacteria: evidence for the phosphotransferase transport system. *Appl. Environ. Microbiol.*, **52**, 1348–52.

Martin, S. A. & Russell, J. B. (1987). Transport and phosphorylation of disaccharides by the ruminal bacterium *Streptococcus bovis* JB1. *Applied and Environ. Microbiol.*, **53**, 2388–93.

McGinnis, J. F. & Paigen, K. (1969). Catabolite inhibition: a general phenomenon in the control of carbohydrate utilization. *J. Bacteriol.*, **100**, 902–13.

Miller, T. L. (1978). The pathway of formation of acetate and succinate from pyruvate by *Bacteroides succinogenes*. *Arch. Microbiol.*, **117**, 145–52.

Miller, T. L. & Jenesel, S. E. (1979). Enzymology of butyrate formation by *Butyrivibrio fibrisolvens*. *J. Bacteriol.*, **138**, 99–104.

Miller, T. L. & Wolin, M. J. (1979). Fermentations by saccharolytic intestinal bacteria. *Am. J. Clin. Nutr.*, **32**, 164–72.

Mink, R. M. & Hespell, R. B. (1981*a*). Long-term nutrient starvation of continuously cultured (glucose-limited) *Selenomonas ruminantium*. *J. Bacteriol.*, **148**, 541–50.

Mink, R. M. & Hespell, R. B. (1981*b*). Survival of *Megasphaera elsdenii* during starvation. *Curr. Microbiol.*, **5**, 51–6.

Mitchell, P. (1961). Coupling of phosphorylation to electron and hydrogen transfer by a chemiosmotic type of mechanism. *Nature* (London), **191**, 144–7, 423–7.

Monod, J. (1947). The phenomenon of enzymatic adaptation. *Growth*, **11**, 223–89.

Mountfort, D. O. & Asher, R. A. (1983). Role of catabolite regulatory mechanisms in control of carbohydrate utilization by the rumen anaerobic fungus *Neocallimastix frontalis*. *Appl. Environ. Microbiol.*, **46**, 1331–8.

Mountfort, D. O. & Roberton, A. M. (1978). Origins of fermentation products formed during growth of *Bacteroides ruminicola* on glucose. *J. Gen. Microbiol.*, **106**, 353–60.

Muir, L. A. & Barreto, A. (1979). Sensitivity of *Streptococcus bovis* to various antibiotics. *J. Anim. Sci.*, **48**, 468–73.

Neijssel, O. M. & Tempest, D. W. (1976). Bioenergetic aspects of aerobic growth of *Klebsiella aerogenes* NCTC 418 in carbon-limited and carbon-sufficient culture. *Arch. Microbiol.*, **107**, 215–21.

Neill, A. R., Grime, D. W. & Dawson, R. M. C. (1978). Conversion of chotine methyl groups through trimethylamine into methane in the rumen. *Biochem. J.*, **170**, 529–35.

Nicholls, D. G. (1982). In *Bioenergetics: An Introduction to the Chemiosmotic Theory*. Academic Press, London, p. 48.
Nisman, B. (1954). The Stickland reaction. *Bacteriol. Rev.*, **18**, 16–42.
Novick, A. (1961). Bacteria with high levels of specific enzymes. In *Growth in Living Systems*, ed. M. X. Zarrow. Basic Books, New York, pp. 93–106.
O'Brien, W. E., Bowien, S. & Wood, H. G. (1975). Isolation and characterization of a pyrophosphate-dependent phosphofructokinase from *Propionibacterium shermanii*. *J. Biol. Chem.*, **250**, 8690–5.
O'Farrell, P. H. (1975). High resolution two-dimensional electrophoresis of proteins. *J. Biol. Chem.*, **250**, 4007–21.
Opperman, R. A., Nelson, W. O. & Brown, R. E. (1961). *In vivo* studies of methanogenesis in the bovine rumen: dissimilation of acetate. *J. Gen. Microbiol.*, **25**, 103–11.
Paynter, M. J. B. & Elsden, S. R. (1970). Mechanism of propionate formation by *Selemononas ruminantium*, a rumen bacterium. *J. Gen. Microbiol.*, **61**, 1–7.
Pettipher, G. L. & Latham, M. J. (1979). Production of enzymes degrading plant cell walls and fermentation of cellobiose by *Ruminococcus flavefaciens* in bath and continuous culture. *J. Gen. Microbiol.*, **110**, 29–38.
Pirt, S. J. (1965). The maintenance energy of bacteria in growing cultures. *Proc. Roy. Soc.* (London), *B***163**, 224–31.
Pirt, S. J. (1982). Maintenance energy: a general model for energy-limited and energy-sufficient growth. *Arch. Microbiol.*, **133**, 300–2.
Postma, P. W. (1986). Catabolite repression and related processes. In *Regulation of Gene Expression—25 Years On* (Soc. Gen. Microbiol. Symposium No. 39), ed. I. R. Booth and C. F. Higgins. Cambridge University Press, pp. 21–9.
Postma, P. W. & Lengeler, J. W. (1985). Phosphoenol pyruvate: carbohydrate phosphotransferase system of bacteria. *Microbiol. Rev.*, **49**, 232–69.
Postma, P. W. & Roseman, S. (1976). The bacterial phosphoenolpyruvate: sugar phosphotransferase system. *Biochem. Biophys. Acta*, **457**, 213–57.
Priest, F. G. (1984). *Extracellular Enzymes*. American Society for Microbiology, Washington, DC, pp. 17–31.
Rowe, J. B., Loughnan, M. L., Nolan, J. V. & Leng, R. A. (1979). Secondary fermentation in the rumen of a sheep given a diet based on molasses. *Br. J. Nutr.*, **41**, 393–7.
Russell, J. B. (1983). Fermentation of peptides by *Bacteroides ruminicola* $B_1 4$. *Appl. Environ. Microbiol.*, **45**, 1566–74.
Russell, J. B. (1984). Factors influencing competition and composition of the rumen bacterial flora. In *Herbivore Nutrition in the Subtropics and Tropics*, ed. F. M. C. Gilchrist & R. I. Mackie. Science Press, Craighall, South Africa, pp. 313–45.
Russell, J. B. (1986). Heat production by ruminal bacteria in continuous culture and its relationship to maintenance energy. *J. Bacteriol.*, **168**, 694–701.
Russell, J. B. (1987). A proposed model of monensin action in inhibiting rumen bacterial growth: effects on ion flux and protonmotive force. *J. Anim. Sci.*, **64**, 1519–25.
Russell, J. B. & Baldwin, R. L. (1978). Substrate preferences in rumen bacteria: evidence of catabolite regulatory mechanisms. *Appl. Environ. Microbiol.*, **36**, 319–29.
Russell, J. B. & Baldwin, R. L. (1979). Comparison of maintenance energy expenditures and growth yields among several rumen bacteria grown on continuous culture. *Appl. Environ. Microbiol.*, **37**, 537–43.
Russell, J. B. & Dombrowski, D. B. (1980). Effect of pH on the efficiency of growth by pure cultures of rumen bacteria in continuous culture. *Appl. Environ. Microbiol.*, **39**, 604–10.
Russell, J. B. & Hino, T (1985). Regulation of lactate production in *Streptococcus bovis*: a spiraling effect that leads to rumen acidosis. *J. Dairy Sci.*, **68**, 1712–21.
Russell, J. B. & Jeraci, J. L. (1984). Effect of carbon monoxide on fermentation of fiber, starch, and amino acids by mixed rumen microorganisms *in vitro*. *Appl. Environ. Microbiol.*, **48**, 211–17.

Russell, J. B. & Sniffen, C. J. (1984). Effect of carbon-4 and carbon-5 volatile fatty acids on growth of mixed rumen bacteria *in vitro*. *J. Dairy Sci.*, **67**, 987–94.

Russell, R. B., Cotta, M. A. & Dombrowski, D. B. (1981). Rumen bacterial competition in continuous culture: *Streptocococcus bovis* versus *Megasphaera elsdenii*. *Appl. Environ. Microbiol.*, **41**, 1394–9.

Russell, J. B., Sniffen, C. F. & Van Soest, P. J. (1983). Effect of carbohydrate limitation on degradation and utilization of casein by mixed rumen bacteria. *J. Dairy Sci.*, **66**, 763–75.

Saier, M. H. (1977). Bacterial phosphoenolpyruvate: sugar phosphotransferase systems: functional and evolutionary interrelationships. *Bacteriol. Rev.*, **41**, 856–71.

Saier, M. H. (1985). *Mechanisms and Regulation of Carbohydrate in Bacteria*. Academic Press, pp. 14–48.

Scheifinger, C. C., Latham, M. J. & Wolin, M. J. (1975). Relationship of lactate dehydrogenase specificity and growth rate to lactate metabolism by *Selenomonas ruminantium*. *Appl. Microbiol.*, **30**, 916–21.

Shiloach, J. & Bauer, S. (1975). High-yield growth of *E. coli* at different temperatures in a bench scale fermentor. *Biotechnol. Bioeng.*, **17**, 227–39.

Sih, C. J. & McBee, R. H. (1955). A cellobiose phosphorylase in *Clostridium thermocellum*. *Proc. Montana Acad. Sci.*, **15**, 21–2.

Stewart, C. S., Paniagua, C., Dinsdale, D., Cheng, K.-J. & Garrow, S. H. (1981). Selective isolation and characteristics of *Bacteroides succinogenes* from the rumen of a cow. *Appl. Environ. Microbiol.*, **41**, 504–10.

Stouthamer, A. H. (1969). Determination and significance of molar growth yields. In *Methods in Microbiology*, Vol. 1, ed. J. R. Norris & D. W. Ribbons. Academic Press, London, pp. 629–63.

Stouthamer, A. H. (1973). A theoretical study on the amount of ATP required for synthesis of microbial cell material. *Ant. van Leeuwenhoek*, **39**, 545–65.

Stouthamer, A. H. (1979). The search for correlation between theoretical and experimental growth yields. In *International Review of Biochemistry and Microbial Biochemistry*, Vol. 21, ed. J. R. Quayle. University Park Press, Baltimore, pp. 1–47.

Stouthamer, A. H. & Bettenhaussen, C. (1973). Utilization of energy for growth and maintenance in continuous and batch cultures of microorganisms. *Biochim. Biophys. Acta*, **301**, 53–70.

Takahashi, S., Abbe, K. & Yamada, T. (1982). Purification of pyruvate-formate lyase from *Streptococcus mutans* and its regulatory properties. *J. Bacteriol.*, **149**, 1034–42.

Teixeira de Mattos, M. J., Plomp, P. J. A. M., Neijssel, O. M. & Tempest, D. W. (1984). Influence of metabolic end-products on the growth efficiency of *Klebsiella aerogenes* in anaerobic chemostat culture. *Ant. van Leeuwenhoek*, **50**, 461–72.

Tempest, D. W. (1967). Quantitative relationships between inorganic ion and anionic polymers in growing bacteria. In *Microbial Growth* (Soc. Gen. Microbiol. Symp. No. 19), pp. 87–111.

Tempest, D. W. & Neijssel, O. M. (1984). The status of Y_{ATP} and maintenance energy as biologically interpretable phenomena. *Ann. Rev. Microbiol.*, **38**, 459–86.

Thauer, R. K., Jungermann, K. & Decker, K. (1977). Energy conservation in chemotrophic anaerobic bacteria. *Bacteriol. Rev.*, **41**, 100–80.

Therion, J. J., Kistner, A. & Kornelius, J. H. (1982). Effect of pH on growth rates of rumen amylolytic and lactilytic bacteria. *Appl. Environ. Microbiol.*, **44**, 428–34.

Turner, K. W. & Roberton, A. M. (1979). Xylose, arabinose, rhamnose fermentation by *Bacteroides ruminicola*. *Appl. Env. Microbiol.*, **38**, 7–12.

Van den Hende, C., Oyaert, W. & Bouckaert, J. H. (1963). Metabolism of glycine, alanine, valine, leucine and isoleucine by rumen bacteria. *Res. Vet. Sci.*, **4**, 382–9.

Van Gylswyk, N. O. (1977). Activation of NAD-dependent lactate dehydrogenase in *Butyrivibrio fibrisolvens* by fructose 1,6-diphosphate. *J. Gen. Microbiol.*, **99**, 441–3.

Wachenheim, D. E. & Hespell, R. B. (1985). Responses of *Ruminococcus flavefaciens*, a ruminal cellulolytic species, to nutrient starvation. *Appl. Environ. Microbiol.*, **50**, 1361–7.

Wallace, R. J. (1978). Control of lactate production by *Selenomonas ruminantium*: homotrophic activation of lactate dehydrogenase by pyruvate. *J. Gen. Microbiol.*, **107**, 45–52.

Wallace, R. J. (1980). Cytoplasmic reserve polysaccharide of *Selenomonas ruminantium*. *Appl. Environ. Microbiol.*, **39**, 630–4.

Wallace, R. J. (1986). Catabolism of amino acids by *Megasphaera elsdeni* LC1. *Appl. Environ. Microbiol.*, **51**, 1141–3.

Wallace, R. J. & Robertson, J. D. (1985). Lactate efflux from *Selenomonas ruminantium*. In *Proceedings of 18th Conference on Rumen Function*, Chicago, 1985, p. 18.

Wallnofer, P., Baldwin, R. L. & Stagno, E. (1966). Conversion of ^{14}C-labeled substrates to volatile fatty acids by the rumen microbiota. *Appl. Microbiol.*, **14**, 1004–10.

Westerhoff, H. V., Hellingwerf, K. J. & Van Dam, K. (1983). Thermodynamic efficiency of microbial growth is low but optimal for maximal growth rate. *Proc. Nat. Acad. Sci.*, **80**, 305–9.

Wetzstein, H. G., McCarthy, J. E. G. & Gottschalk, G. (1987). The membrane potential in a cytochrome-deficient species of *Bacteroides*: its magnitude and mode of generation. *J. Gen. Microbiol.*, **133**, 73–83.

White, D. C., Bryant, M. P. & Caldwell, D. R. (1962). Cytochrome-linked fermentation in *Bacteroides ruminicola*. *J. Bacteriol.*, **84**, 822–8.

Williams, A. G. (1986). Rumen holotrich ciliate protozoa. *Microbiol. Rev.*, **50**, 25–49.

Wolin, M. J. (1964). Fructose-1,6-diphosphate requirement of streptococcal lactic dehydrogenases. *Science*, **146**, 775–7.

Wolin, M. J. (1969). Volatile fatty acids and the inhibition of *Escherichia coli* growth by rumen fluid. *Appl. Microbiol.*, **17**, 83–7.

Wolin, M. J. (1979). The rumen fermentation: a model for microbial interactions in anaerobic ecosystems. *Adv. Micro. Ecol.*, **3**, 49–77.

Yarlett, N., Hann, A. C., Lloyd, D. & Williams, A. G. (1981). Hydrogenosomes in the rumen protozoon *Dasytricha ruminantium* Schuberg. *Biochem. J.*, **200**, 365–72.

Yarlett, N., Lloyd, D. & Williams, A. G. (1982). Respiration of the rumen ciliate *Dasytricha ruminantium* Schuberg. *Biochem. J.*, **206**, 259–66.

Yarlett, N., Hann, A. C., Lloyd, D. & Williams, A. G. (1983). Hydrogenosomes in a mixed isolate of *Isotricha prostoma* and *Isotricha intestinalis* from ovine rumen contents. *Comp. Biochem. Physiol.*, **B74**, 357–64.

7
Metabolism of Nitrogen-Containing Compounds

R. J. Wallace

Rowett Research Institute, Aberdeen, UK

&

M. A. Cotta

US Department of Agriculture, Peoria, Illinois, USA

The way in which the rumen has evolved as their first digestive organ potentially affords ruminants an efficiency of protein nutrition that is not available to monogastric herbivores. Protein is synthesised in the gut in the form of rumen microorganisms. The necessary energy is derived from plant polysaccharides such as cellulose, and the nitrogen is derived mainly from ammonia in the rumen. The energy and nitrogen sources can therefore be substrates of little value to most monogastrics. Even more important, however, is the direct availability of that microbial protein for digestion and absorption by the host animal. Indeed, microbial protein is generally the ruminant's principal source of amino acids.

The benefit actually derived from the rumen fermentation, in terms of protein nutrition, varies according to the diet. Virtanen (1966) elegantly demonstrated that cows can maintain unimpaired milk production on diets lacking protein and with cellulose as principal carbon source and urea as the main nitrogenous nutrient. Yet, especially with intensive production systems, the nitrogen metabolism of rumen microorganisms is usually regarded as being inefficient. Dietary protein is broken down much too rapidly relative to the breakdown of the energy-containing plant fibre, excessive ammonia production results, and the biological value of the dietary protein is severely reduced.

The objectives of research in this field have therefore been twofold. One is to capitalise on the microbial capacity to form protein from ammonia by feeding non-protein nitrogen (NPN). The other is to minimise protein breakdown in the rumen, and thereby to increase the 'bypass' dietary protein (i.e. protein unaffected by passage through the rumen) reaching the lower tract. The contents of this chapter describe the basic microbiological research that has been done in this general area.

FLOW OF NITROGENOUS COMPOUNDS THROUGH THE RUMEN

It is customary for ruminant nutritionists to determine the N content of feeds by a Kjeldahl procedure, then to multiply that value by 6·25 and call it the 'crude protein' content of the diet (Ørskov, 1982). This is a useful device for evaluating diets, but clearly oversimplifies the very complex nitrogenous components of plant materials. Titles such as 'Urea as a Protein Supplement' (Briggs, 1967) might well puzzle the uninitiated.

The plant materials that comprise the bulk of ruminant feeds are composed of a vast array of nitrogenous compounds. Protein is almost always the most abundant, but nucleic acids will always occur in association with protein, and substantial amounts of nitrate and ammonia may be present, depending on the diet. Plant materials may be supplemented with NPN in the form of compounds such as urea.

Other nitrogenous inputs to the rumen fermentation are salivary mucoprotein and urea; the latter enters both in saliva and by diffusion across the rumen wall. Gaseous nitrogen also enters the rumen, but nitrogen fixation is almost certainly not of quantitative significance.

The interconversions that occur in the rumen are illustrated for an example of sheep receiving a lucerne-chaff diet (Fig. 1). This diagram ignores the distinction between crude protein and true protein or other materials, and markedly different diets, e.g. straws or cereals, might give substantially different patterns, but the following general observations apply to most situations:

All interconversions are by microbial enzymes.
Microbial protein is the most abundant form of protein N leaving the rumen.
Endogenous urea will only be of quantitative significance for microbial protein synthesis when dietary protein is low.
Ammonia is the major product of catabolism, and also the main substrate for microbial protein synthesis.
Ammonia overflow leads to inefficient N retention.

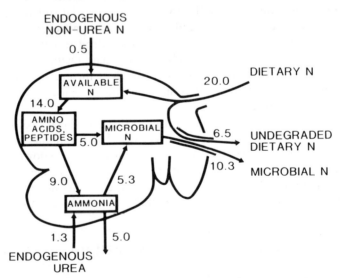

Fig. 1. Flow of N (g day^{-1}) in the rumen of sheep receiving a lucerne-chaff diet (from Nolan, 1975).

BREAKDOWN OF DIETARY NITROGENOUS COMPOUNDS

Protein degradation

Dietary factors

Protein is the most abundant source of N in most ruminant diets. Feed protein is usually hydrolysed rapidly in the rumen, although the precise rate and extent of breakdown depend on a number of factors, which ultimately determine the nutritive value of the protein. Early work produced the general rule that protein breakdown in the rumen is proportional to solubility (Chalmers & Synge, 1954; Henderickx & Martin, 1963; Henderickx, 1976), but subsequent research has shown that other properties are also important (Kaufmann & Lupping, 1982). For example, some (pure) soluble proteins are broken down more slowly than insoluble proteins, depending on the degree of secondary and tertiary structure, and cleavage of disulphide bonds enhances the breakdown of albumin and similarly heavily crosslinked molecules (Mahadevan et al., 1980; Nugent et al., 1983; Wallace & Kopecny, 1983). Conversely, the introduction of artificial cross-links into proteins inhibits their hydrolysis (Friedman & Broderick, 1977; Wallace, 1983). Heating and formaldehyde treatments, affecting both solubility and cross-linking, have been used to protect proteins from rumen degradation and thereby provide bypass protein to the lower tract (Kaufmann & Lupping, 1982). However, these treatments may impair the subsequent availability of some amino acids, notably lysine, cysteine and tyrosine (Ashes et al., 1984). Gentler methods, such as coating protein supplements with an undegradable shell of heated, dried blood, may be preferable (Ørskov et al., 1980). The proteins in some plants often have some degree of natural protection, depending on the matrix in which they are located. The hydrolysis of non-protein polymers, such as polysaccharides, may limit the access of proteolytic organisms to their substrate (Ganev et al., 1979; Siddons & Paradine, 1981).

The nature of the diet has a major influence on the proteolytic activity of rumen contents. Fresh herbage promotes an activity up to nine times higher than that found with dry rations, the higher soluble-protein content of the herbage enriching for proteolytic bacteria (Nugent & Mangan, 1981; Hazlewood et al., 1983; Nugent et al., 1983). Cereal diets also yield higher activities than do dry forage diets, probably because proteolytic rumen microorganisms tend to be amylolytic rather than cellulolytic (Siddons & Paradine, 1981). The nature of the protein substrate is another factor which affects proteolytic activity. Hydrolysis of leaf Fraction I protein was stimulated, relative to casein, by fresh fodder, in which it would be abundant, more than by a diet consisting of hay plus concentrates (Hazlewood et al., 1983; Nugent et al., 1983). In contrast, when albumin replaced casein as an experimental protein supplement to a sheep diet, the rate of breakdown of albumin relative to casein was unchanged, despite a modified proteolytic flora (Wallace et al., 1987a). Furthermore, the proteolytic activity of the rumen contents was hardly changed. The effect of different dietary proteins on ruminal proteolysis is not yet clear, and probably varies from protein to protein and with the other constituents of the diet.

Cilate protozoa

The rumen ciliates have for many years been known to be proteolytic. Holotrichs undergo rapid endogenous protein breakdown in the absence of an exogenous source of nitrogen (Heald & Oxford, 1953), and ingest and digest rumen bacteria (Coleman, 1980) (see Chapter 3). Many entodiniomorphs, including *Entodinium caudatum* (Abou Akkada & Howard, 1962; Onodera & Kandatsu, 1970), *Eudiplodinium medium* (Naga & el-Shazly, 1968) and *Ophryoscolex* spp. (Williams *et al.*, 1961; Mah & Hungate, 1965), are also proteolytic. Cell-free extracts of fourteen individual entodiniomorphid species hydrolysed leaf Fraction I protein, with *E. caudatum* and *E. simplex* having the highest activity and the cellulolytic species the lowest (Coleman, 1983). The optimum pH of proteolysis in cell extracts of *E. caudatum* and *Eudiplodinium maggii* was 3·2 with Fraction I protein or casein as substrate (Coleman, 1983). However, maximal breakdown of casein was also reported to occur at pH 6·5–7·0 with extracts of *E. caudatum* (Abou Akkada & Howard, 1962). In view of its stimulation by cysteine (Abou Akkada & Howard, 1962) and inhibition by N-ethylmaleimide and leupeptin (Coleman, 1983), the main activity of *E. caudatum* was apparently a thiol (cysteine) protease, although partial inhibition by pepstatin indicated the presence of some carboxyl (aspartic) type of enzyme (Coleman, 1983). Trypsin-like specificity was also detected (Abou Akkada & Howard, 1962).

Mixed intact and autolysed protozoa had fairly similar types of activity to those observed with individual species. Cysteine proteases were most prevalent and some aspartic protease was also present (Forsberg *et al.*, 1984). Mixed protozoa had a much higher aminopeptidase activity than rumen bacteria and a slightly higher trypsin-like activity (Forsberg *et al.*, 1984), which is consistent with these activities being lower in the rumen of ciliate-free sheep than in faunated animals (Wallace *et al.*, 1987b). Maximum proteolytic activity of intact mixed protozoa occurred at pH 5–9, and of sonicated organisms at pH 5·8 (Forsberg *et al.*, 1984).

The rumen ciliates are of relatively little importance in the breakdown of soluble proteins in rumen contents. Brock *et al.* (1982) estimated that they were, on a specific activity basis, one-tenth as active in azocasein breakdown as rumen bacteria. A similar conclusion was made by Nugent & Mangan (1981), as the hydrolytic products of Fraction I protein appeared first in the bacteria, then only subsequently became associated with protozoa as predation occurred. The proteolytic activity of rumen fluid, with ^{14}C-labelled casein as substrate, was little changed by faunation (Wallace *et al.*, 1987b), again consistent with a minor role for protozoa in soluble protein breakdown. Similar conclusions were reached by Hino & Russell (1987) using inhibitors of protozoal growth. The role of the protozoa is instead likely to be in the hydrolysis of particulate proteins of an appropriate particle size. Chloroplasts, for example, are avidly engulfed by protozoa (Mangan & West, 1977), and stained particles of casein have a similar fate (Abou Akkada, 1965), with both types of particle being rapidly degraded intracellularly. Naga & el-Shazly (1968) demonstrated that the precise size of protein particles was important to *Eudiplodinium medium*. If particles were too small, their rate of hydrolysis decreased.

Even more significant is the predatory activity of protozoa against rumen bacteria (see Chapter 3), which is of enormous significance to bacterial protein turnover in the rumen. In the absence of protozoa, bacterial protein turnover varied from 0·3% to 2·7% h^{-1} depending on species; in their presence, the rates increased to 2·4–37·0% h^{-1} (Wallace & McPherson, 1987). Clearly therefore the main role of the protozoa is not in the hydrolysis of exogenous soluble protein, but in metabolising bacterial protein.

Bacteria

No common species of rumen bacteria use protein as a major energy source, but between 30% and 50% of the bacteria isolated from rumen fluid have proteolytic activity towards extracellular protein (Fulghum & Moore, 1963; Prins *et al.*, 1983). Representatives of most species have some activity, with the possible exception of the main cellulolytic bacteria, *Bacteroides succinogenes*, *Ruminococcus flavefaciens* and *R. albus*. *Bacteroides* (*Ruminobacter*) *amylophilus* is one of the most active proteolytic species isolated; since it is amylolytic it would be expected to be of particular importance in starchy diets (Blackburn & Hobson, 1962; Abou Akkada & Blackburn, 1963). Proteolytic *Butyrivibrio fibrisolvens* can be the predominant proteolytic organism isolated from some animals (Blackburn & Hobson, 1962; Fulghum & Moore, 1963; Hazlewood *et al.*, 1983; Wallace *et al.*, 1987*a*), and is probably enriched for when more resistant types of protein are present in the diet (Wallace *et al.*, 1987*a*). Probably the most numerous proteolytic bacterium is *Bacteroides ruminicola*, proteolytic strains of which occur on both all-roughage and mixed roughage–concentrate diets (Blackburn & Hobson, 1962; Abou Akkada & Blackburn, 1963; Hazlewood & Nugent, 1978; Wallace & Brammall, 1985). However, proteolytic strains of *B. ruminicola* cannot be isolated from all animals (Hazlewood *et al.*, 1983). Bacteria with higher activity have also been isolated. These are often atypical organisms, like *Clostridium* (Hungate, 1966) or *Fusobacterium* (Wallace & Brammall, 1985). Other proteolytic bacteria more typical of predominant rumen species include species of the genera *Eubacterium*, *Lachnospira*, *Selenomonas* and *Succinivibrio* (Blackburn & Hobson, 1962; Abou Akkada & Blackburn, 1963; Fulghum & Moore, 1963; Wallace & Brammall, 1985). Proteolytic Gram-positive cocci were noted in early isolation experiments (Appleby, 1955; Blackburn & Hobson, 1960*a*, 1962), but *Streptococcus bovis* was not recognised as a potentially major proteolytic organism until quite recently (Russell *et al.*, 1981; Hazlewood *et al.*, 1983; Wallace & Brammall, 1985). Other proteolytic, facultatively anaerobic bacteria isolated in earlier studies are probably of little importance in the mainly strictly anaerobic flora.

Most attention has been focused on the three species considered to be the major proteolytic organisms, namely *B. amylophilus*, *B. fibrisolvens* and *B. ruminicola*. *B. amylophilus* at first sight appears to be peculiar, because it does not require peptides or amino acids for growth (Hobson *et al.*, 1968; Jenkinson *et al.*, 1979). Even if these are available, ammonia remains the principal nitrogen source, with only a small proportion of cell N derived from labelled protein, peptides or amino acids (Hobson *et al.*, 1968; Hullah & Blackburn, 1971). Furthermore, protease is produced in medium devoid of proteins or amino acids (Blackburn, 1968a). The retention of such an apparently gratuitous protease activity in a competitive ecosystem like the rumen prompted Cotta & Hespell (1986a) to suggest that its function was not a nutritional one, but rather to break down structural protein within cereal particles, thereby exposing starch granules to amylolytic attack. In contrast, both *B. ruminicola* and *B. fibrisolvens* can grow in a medium containing protein as the sole source of N (Hazlewood & Nugent, 1978; Wallace & Brammall, 1985; Cotta & Hespell, 1986b).

The production of protease activity seems to be only loosely regulated in these bacteria. Blackburn (1968a) concluded that the composition of the growth medium in batch cultures had little influence on the expression of protease activity by *B. amylophilus*, whereas growth rate had a small effect, with more activity tending to be produced at lower growth rates (Henderson *et al.*, 1969). In contrast, the activity produced by *B. fibrisolvens* with different carbon sources increased as the growth rate elicited by each compound increased, the activity at $0.62\,h^{-1}$ on fructose as carbon source being double that at $0.22\,h^{-1}$ on xylose (Cotta & Hespell, 1986b). High concentrations ($20\,g\,l^{-1}$) of amino acids or peptides in the medium severely repressed activity, while amino acids at $1\,g\,l^{-1}$ caused threefold stimulation relative to that expressed with casein (Cotta & Hespell, 1986b). Similar studies of the regulation of protease activity in *B. ruminicola* have not been done, although Hazlewood *et al.* (1981) found that the activity expressed depended on the protein used as growth substrate. Only 37% of the proteolytic activity found in cultures containing Fraction I occurred when albumin was the substrate. The protease activity of other rumen bacteria was affected little by different sources of N in the medium (Wallace & Brammall, 1985).

Fractionation of rumen fluid has shown clearly that most proteolytic activity is cell-associated (Blackburn & Hobson, 1960b; Nugent & Mangan, 1981; Brock *et al.*, 1982; Kopecny & Wallace, 1982; Prins *et al.*, 1983). Indeed the soluble activity that can be found in the cell-free supernatant fluid may have been largely displaced from the main site of enzyme activity, namely exocellular polysaccharide capsular material (Kopecny & Wallace, 1982). Cell-associated proteases can be liberated by gentle shaking, treatment in a Waring Blendor or extraction with Triton X-100 (Kopecny & Wallace, 1982; Prins *et al.*, 1983). This cell-surface location results in a mechanism of proteolysis whereby the substrate protein adsorbs rapidly and irreversibly to rumen bacteria as an integral part of the process (Nugent & Mangan, 1981; Wallace, 1985a). Within individual species, protease activity is substantially cell-bound in *B. amylophilus* (Blackburn, 1968a; Blackburn & Hullah, 1974), *B. ruminicola* (Hazlewood *et al.*, 1981; Wallace & Brammall, 1985), *Eubacterium* sp. (Wallace & Brammall, 1985) and some low-activity isolates of *B. fibrisolvens*, *S. ruminantium* and *S. bovis* (Wallace & Brammall, 1985). The activities of both *B. ruminicola* and *B. amylophilus* remain almost entirely cell-associated during growth, and are released into the medium largely as the result of autolysis in stationary phase (Lesk & Blackburn, 1971; Hazlewood *et al.*, 1981). In contrast, high-activity strains of *B. fibrisolvens* produce an activity that is always extracellular (Wallace & Brammall, 1985; Cotta & Hespell, 1986b).

Proteolytic activity has a broad pH optimum around approximately pH 5·5–7·0 in rumen fluid (Blackburn & Hobson, 1960a), mixed rumen bacteria (Kopecny & Wallace, 1982) and extracted bacterial enzymes (Kopecny & Wallace, 1982). *B. fibrisolvens* protease has a similar pH optimum (Cotta & Hespell, 1986b), while *B. amylophilus*, pH range 5·5–9·5 for the soluble enzyme (Blackburn, 1986b), and 4·5–12·0 for the cell-associated enzyme (Lesk & Blackburn, 1971), and *B. ruminicola*, pH 5·9–8·2 (Hazlewood et al., 1981), were active at the same and higher pH values.

It can be concluded from the effects of protease inhibitors that the predominant type of protease present in rumen contents (Wallace, 1984), mixed rumen bacteria (Brock et al., 1982; Kopecny & Wallace, 1982; Wallace & Brammall, 1985) and extracted capsular enzymes (Kopecny & Wallace, 1982; Prins et al., 1983) is a cysteine-protease type, sensitive to p-chloromercuribenzoate (PCMB). Other types of activity are also present, but are more variable. These include phenylmethylsulphonyl fluoride (PMSF)-sensitive serine protease (present at 0–41% of total activity), metalloprotease (9–30%) and aspartic protease (2–15%) (Brock et al., 1982; Kopecny & Wallace, 1982; Prins et al., 1983; Wallace, 1984; Wallace & Brammall, 1985). Among the proteolytic species that have been examined, *B. ruminicola* produces an activity most similar to that of rumen contents. PCMB was highly inhibitory (56–89%), and serine-protease inhibitors affected the activity by 21–43% (Hazlewood & Edwards, 1981; Wallace & Brammall, 1985). *B. ruminicola* was more sensitive to EDTA than mixed rumen bacteria (Hazlewood & Edwards, 1981; Wallace & Brammall, 1985), but the effects of chelators may be complex and they do not always indicate a metalloprotease (Hazlewood & Edwards, 1981). The same is, of course, true of the effects of EDTA on rumen contents. *B. amylophilus* had, in contrast, a predominantly serine-protease activity (Wallace & Brammall, 1985), as did most strains of *B. fibrisolvens* (Wallace & Brammall, 1985; Cotta & Hespell, 1986b). Indeed, all ten bands of activity in supernatant fluids from *B. fibrisolvens* were inhibited by PMSF (Strydom et al., 1986). A low-activity strain of *B. fibrisolvens* had an activity that was inhibited only 12% (cf. >90% for other strains) by PMSF and 93% by PCMB (Wallace & Brammall, 1985). *S. bovis* and a proteolytic *Eubacterium* also had PMSF-sensitive activity, whereas the low activity of *S. ruminantium* was insensitive to PMSF, but inhibited by PCMB and to a lesser extent by EDTA (Wallace & Brammall, 1985).

There is some evidence that the type of protease present is influenced by animal diet (Prins et al., 1983; Wallace & Brammall, 1985). Indeed, a cow on a hay diet had no serine protease in its rumen bacterial fraction (Prins et al., 1983). Clearly, since the microbial flora will alter with diet, such changes would be expected, but much more work is required to correlate, for example, the increased numbers of proteolytic *Butyrivibrio* species under some circumstances (Hazlewood et al., 1983; Wallace et al., 1987a) with proteases present in rumen contents. Also required is more work on the distribution of activity between free-swimming and particle-bound microorganisms. Brock et al. (1982) reported that 75% of the proteolytic activity of whole rumen contents was particle-associated. Wallace (unpublished results) noted that the proteolytic activity of rumen solids had, on a wet-weight basis, from 1·5 to 4 times the activity of the filtered rumen fluid with animals on a mixed diet, and from

2·5 to 6 times the activity with animals on a hay diet. The proportion of activity associated with solids therefore depends on the concentration of particles in the rumen suspension. If particles constituted about 3% of the wet weight, as they do on a hay diet, less than 20% of the activity would be particle-associated. Interestingly, the particle-associated activity is quite similar to that of the free-swimming population. PCMB inhibits most of the activity, with PMSF inhibiting 30% of the activity in sheep fed on hay plus concentrate and 39% in animals fed on hay alone (Wallace, 1985b). Metalloprotease is much more variable (see p. 226). The flora associated with the rumen wall (see Chapter 12) also has protease activity, of a slightly different type from that of rumen contents (Wallace, 1984). The bacteria have a very high specific activity, in keeping with their evident active digestion of epithelial tissue, but the numbers of these organisms are low relative to total rumen contents, and their contribution to proteolysis will be limited to their effect at the rumen wall (Dinsdale et al., 1980; Wallace, 1984).

The proteases of rumen bacteria have several different kinds of specificity. Among mixed rumen bacteria, work with synthetic substrates has shown mainly trypsin-like and leucine aminopeptidase activities, as judged by the hydrolysis of benzoyl-arginine p-nitroanilide (BAPNA) and leucine p-nitroanilide (LNA) respectively (Brock et al., 1982; Prins et al., 1983; Wallace & Kopecny, 1983) and by the inhibition of proteolysis by the trypsin substrate analogues N-tosyl-L-lysine chloromethyl ketone (TLCK; Brock et al., 1982; Wallace & Kopecny, 1983) and N-tosyl-L-lysine chloromethane (Prins et al., 1983). Prins et al. (1983) and Wallace & Kopecny (1983), using synthetic chymotrypsin substrates, found that chymotrypsin-like activity was present, but low, in rumen bacteria and extracted bacterial cell-envelope enzymes. Chymotrypsin inhibitors had little effect (Prins et al., 1983; Wallace & Brammall, 1985). In contrast, Brock et al. (1982) obtained 21% inhibition of bacterial proteolysis by N-tosyl-L-phenylalanine chloromethyl ketone (TPCK), and a substantial activity against the chymotrypsin substrate N-3-(carboxypropionyl)-L-phenylalanine p-nitroanilide. The latter activity was, however, measured with sonicated bacteria and therefore included intracellular activity not usually involved in the breakdown of exogenous protein. Carboxypeptidase activity was of a similar magnitude to chymotrypsin-like activity (Prins et al., 1983; Wallace & Kopecny, 1983), i.e. of less importance than the two main activities of trypsin- and leucine aminopeptidase-like specificity.

None of the proteolytic bacteria isolated from the rumen has a particularly distinctive activity, except possibly S. bovis, which has an exceptionally high leucine aminopeptidase activity (Wallace & Brammall, 1985). The cell-associated activity of B. amylophilus (Blackburn, 1968b) and the soluble protease released in late stationary phase (Lesk & Blackburn, 1971) were active against trypsin substrates and inhibited by trypsin substrate analogues. Some aminopeptidase (Blackburn, 1968b) activity was also observed. B. ruminicola has no activity against BAPNA, although it was inhibited slightly by TLCK (Wallace & Brammall, 1985), leupeptin and soybean trypsin inhibitor (Hazlewood & Edwards, 1981), indicating some trypsin-like activity. Chymostatin was more inhibitory, giving 38–63% inhibition

(Hazlewood & Edwards, 1981), so the specificity may be more of a chymotrypsin type. High-activity *Butyrivibrio* spp. also did not hydrolyse BAPNA (Wallace & Brammall, 1985) and were not significantly inhibited by TLCK or TPCK (Wallace & Brammall, 1985; Cotta & Hespell, 1986*b*) or trypsin inhibitor I-S (Cotta & Hespell, 1986*b*). A low-activity isolate differed in that TLCK and TPCK caused 32% and 28% inhibition (Wallace & Brammall, 1985). The low activity of *S. ruminantium* was significantly inhibited by TPCK, indicative of chymotrypsin-like activity, and that of a *Eubacterium* sp. was active against LNA, indicating a leucine aminopeptidase activity.

What bacterial species is therefore most important in protein digestion in the rumen? *B. ruminicola* has properties most similar to those of rumen contents, is widely distributed, and probably is in general the predominant proteolytic organism. *B. amylophilus* and *B. fibrisolvens* have characteristics that are not consistent with the major part of rumen activity. Nevertheless, there is no question that these two organisms will be important under some dietary circumstances. The lower-activity isolates will also be important, depending on their numbers, and *S. bovis* has a special role in its aminopeptidase activity. These conclusions must be carefully qualified, however, in the light of two different sorts of experimental observation.

One is that, not surprisingly, the bacteria can interact synergistically with each other in the degradation of protein. Cooperativity of this kind has been observed between *B. ruminicola* and *S. ruminantium*, *S. bovis* and *S. ruminantium*, and some other pairs of species (Wallace, 1985*c*). The interaction between *S. bovis* and *S. ruminantium* enabled rapid growth on a medium containing casein as sole N source, in which either species inoculated alone grew poorly. Presumably more complex mixtures of species would give even better cooperativity, depending on the type of proteolytic specificity possessed by each and required in the breakdown of the protein. Consequently, it is not possible to predict how combinations of species will interact, so conclusions should not be based solely on pure-culture work. Apparently unimportant organisms with low activities of an atypical type, could conceivably catalyse a cascade of different types of hydrolysis carried out by other species, and therefore occupy a pivotal position in the rumen impossible to predict from their solitary behaviour in pure culture.

The second qualification is that the types of proteolytic enzymes actually found in rumen contents seem to be many and varied. The research carried out with rumen contents and inhibitors or synthetic substrates, as described above, indicated that the properties of rumen contents could be quite different with different diets and even with samples taken from different animals on the same diet. Direct analysis by a zymogram technique of the enzymes present in the rumens of different sheep on the same and different diets has illustrated just how variable the proteolytic population can be (Fig. 2). Sometimes the same band of protease activity can be detected in different samples. However, the major bands of activity are rarely the same. Hence, there is an intrinsic variability that probably reflects the many different species possessing proteolytic activity that can from time to time dominate the flora.

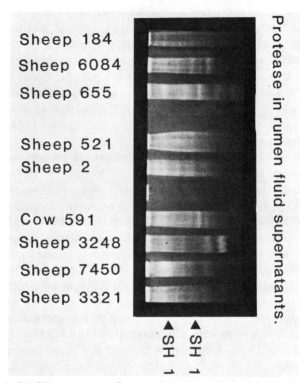

Fig. 2. Gelatin-PAGE zymogram of extracellular protease activities in different samples of rumen fluid. Arrows refer to the location of protease bands from *B. fibrisolvens* SH1, of approximate M 83,500 and 133,000. Sheep 184, 521 and 2 received a hay diet; sheep 6084, 3248, 7450 and 3321 were fed on dried grass cubes; sheep 655 and cow 591 received mixed diets consisting of 2:1 hay + cereals concentrate and equal parts of hay, grass cubes and concentrate respectively.

Fungi

The strictly anaerobic phycomycete *Neocallimastix frontalis* has a high protease activity (Wallace & Joblin, 1985). From its inhibition by EDTA and 1,10-phenanthroline (the former inhibition can be reversed by the addition of divalent cations, particularly Zn^{2+}), and its sensitivity to TLCK, the activity was concluded to be a zinc metalloprotease with a trypsin-like specificity. Other fungal isolates of different morphology also had metalloprotease activity (Wallace, unpublished observations). The significance of this activity to proteolysis *in vivo* has not been established. Because fungal biomass is associated primarily with insoluble plant fibres (see Chapter 4), it might be expected that fibres would be the site at which evidence of protease activity would be found. Experiments *in vitro*, in which the proteolytic activity associated with solids increased tenfold when *N. frontalis* was added to a consortium of rumen bacteria digesting a grass/fish-meal pelleted diet (Wallace & Munro, 1986), are consistent with this view. However, similar *in vivo* evidence has been more difficult to obtain. Solids-associated activity is highly variable, as has already been described, and is usually sensitive to EDTA and 1,10-phenanthroline when it is exceptionally high (Wallace, 1985b), but no experiments have yet been done to correlate EDTA-sensitive protease in rumen contents with fungal numbers or biomass.

Peptide metabolism

Hydrolysis of proteins by rumen microbial enzymes releases oligopeptides, which are then broken down in turn to smaller peptides and, finally, to amino acids. Whether or not peptides accumulate in rumen fluid depends on the nature of the protein. Casein hydrolysis by mixed rumen microorganisms *in vitro* is very rapid, and peptides do accumulate (Russell *et al.*, 1983). Nugent & Mangan (1981) did not detect peptide accumulation following the hydrolysis of ^{14}C-Fraction I protein *in vivo*, but Sniffen *et al.* (1985) reported significant peptide accumulation from the breakdown of soybean meal in the bovine rumen. Less accumulated when the soybean meal was heat-treated, but the quantity of peptides flowing from the rumen was still significant. Wallace & Broderick (unpublished results) found that peptides accumulated in the sheep rumen when casein was used as a protein supplement, but were not observed when the supplement was the more slowly degraded egg albumin. Thus proteins that are rapidly hydrolysed in the rumen could lead to the build-up of peptides in rumen fluid, and the flow of intact peptides from the rumen could be significant. However, for many proteins the rate of utilization of peptides probably exceeds their rate of production. Calculations based on the rate of uptake of di- and tri-peptides and on the rate of proteolysis support this view (Broderick *et al.*, 1988).

Peptides are utilised more rapidly by mixed rumen microorganisms, and incorporated into cellular materials more efficiently, than are the corresponding free amino acids (Wright, 1967; Prins, 1977; Prins *et al.*, 1979). The reason for the preference for peptides is not known but, as with *E. coli*, it may be energetically more efficient to transport peptides rather than free amino acids into bacteria (Payne, 1983). Small peptides are converted mainly into VFA, while a higher proportion of larger molecules tends to be incorporated into bacterial protein (Wright & Hungate, 1967). The rate of uptake of amino acid residues from alanine oligopeptides by mixed rumen organisms was highest with $(ala)_3$, followed by $(ala)_4$, $(ala)_5$ and then $(ala)_2$ (Broderick *et al.*, 1988). The fate of the alanine was not determined. The composition of tripeptides caused their rate of uptake to vary from 0·5 [ala(gly)$_2$] to 1·6 [(ala)$_3$] nmol min^{-1} (mg cells)$^{-1}$ (dry weight). Dipeptides were less variable (0·7–0·9 nmol min^{-1} mg^{-1}, dry weight), except for Gly-Pro which was taken up at 0·4 nmol min^{-1} mg^{-1} (dry weight). However, much more work is required to determine how larger oligopeptides of differing composition are utilised, and how peptide metabolism is influenced by factors such as diet.

Tryptic peptides prepared from *Chlorella* protein were metabolised almost exclusively by cell-associated microbial activity (Wright & Hungate, 1967). Removal of (ala)$_3$ and higher peptides was catalysed by a similar fraction (Broderick *et al.*, 1987). However, 43% of dipeptidase activity was present in the extracellular fluid (Broderick *et al.*, 1988). Tripeptide, but not dipeptide, uptake was significantly higher in faunated than in ciliate-free sheep (Wallace et al., 1987b), so it seems likely that protozoa could be as important as bacteria in this metabolic activity.

Which species of protozoa might be most active in peptide metabolism can, in the light of present knowledge, only be implied from the stimulation of growth or production of amino acids that occurs in the presence of protein. Presumably both protease and peptidase activities are necessary for this to occur. *Eremoplastron bovis* (Coleman & Sandford, 1979a), *Epidinium ecaudatum caudatum* (Coleman & Laurie, 1974), *Diploplastron affine, Enoploplastron triloricatum, Ostracodinium obtusum bilobum, Diplodinium affine* and *Diplodinium anisacantham* (Coleman & Reynolds, 1982) use plant protein for growth, and virtually all entodiniomorphs and holotrichs digest engulfed bacterial protein (Coleman, 1980). The protozoa making the greatest contribution are probably, due to their generally higher numbers, the small entodiniomorphs (Coleman, 1980; Coleman, 1983; Wallace & McPherson, 1987).

Bladen et al. (1961) screened rumen contents for bacteria that produced ammonia from Trypticase, and found significant activity in 28% of isolates, half of which were *B. ruminicola*. *Selenomonas* and *Butyrivibrio* had ammonia-producing strains, and *M. elsdenii* was also implicated. There does not, however, seem to have been a systematic search for bacteria that convert peptides into amino acids rather than ammonia. This may be why the very active peptidolytic activity of *S. bovis* (Russell & Robinson, 1984; Broderick, Roubal, McKain & Wallace, unpublished results) has not been generally recognised.

Most pure-culture work on bacterial peptide metabolism has been done with *B. ruminicola*. Peptides did not support the growth of this organism in the absence of an energy source, and even when glucose was present the energy derived from the subsequent deamination was relatively small (Russell, 1983). Peptides such as the octapeptides oxytocin and vasopressin, or enzyme-hydrolysed casein, were however able to replace ammonia as the main nitrogen source for growth, whereas small peptides of less than four residues, free amino acids or some other small M_r compounds could not (Pittman & Bryant, 1964). The oligopeptides are hydrolysed on entering the cells and simply provide intracellular amino acids for growth (Pittman et al., 1967). Amino acid transport systems do occur in *B. ruminicola*, but they appear to be inhibited by a substance, possibly acetate, in the medium (Stevenson, 1979). It is therefore likely that, given the presence of high concentrations of VFA in the rumen, *B. ruminicola* will utilise mainly peptides and ammonia *in vivo*. It is not known if the same applies to the many other species of rumen bacteria that grow better in the presence of enzyme-hydrolysed casein.

Amino acid metabolism

Metabolism of the amino acids themselves is the next stage in the metabolism of most of the constituents of dietary protein. As our knowledge of the importance of peptide metabolism in microbial growth and ammonia production improves, it becomes clearer that studying the metabolism of extracellular free amino acids, particularly single amino acids, may be misleading if we are interested in amino acids that originate from dietary protein. The rates of entry of peptide-bound amino acids into cells may be quite different from the rates of entry of the corresponding free acids (Prins *et al.*, 1979). Furthermore, experiments with high concentrations of amino acids in incubations *in vitro*, such as the 10 mM used by Lewis & Emery (1962), may be misleading because concentrations *in vivo* are much less.

There is little free amino acid in rumen fluid, and what there is occurs for the most part intracellularly (Wright & Hungate, 1967; Wallace, 1979). Glutamate, for example, is present almost totally as intracellular pools (Wright & Hungate, 1967; Wallace, 1979). Even 1 hour after feeding, when there is a major increase in the α-amino N, the free amino acid content of rumen fluid is quite low (Leibholz, 1969). The extent of accumulation varies with diet, with the highest concentrations observed after feeding alfalfa hay (Leibholz, 1969). Nugent & Mangan (1981) also found the overflow of free amino acids during the hydrolysis of Fraction I protein to be low.

The observed rates of amino acid deamination at so-called 'physiological' concentrations are indeed rapid. Of the amino acids essential to the animal, lysine, phenylalanine, leucine and isoleucine are broken down at 0.2–0.3 mmol h^{-1}, while arginine and threonine are more labile (0.5–0.9 mmol h^{-1}) and valine and methionine are most stable (0.10–0.14 mmol h^{-1}) (Chalupa, 1976). Experiments with non-essential amino acids suggest that they are metabolised at least as rapidly as essential amino acids (Broderick & Balthrop, 1979). Hino & Russell (1985) compared the deaminase activity of intact microorganisms and cell extracts, and concluded that the high capacity for deamination implied that rate of uptake of peptides or amino acids into cells might limit the rate of ammonia production.

The fate of amino acids in the rumen is predominantly to be broken down rather than to be assimilated directly into microbial protein (Al-Rabbat *et al.*, 1971a; Mathison & Milligan, 1971; Chalupa, 1976). Most microbial protein N is derived from the rumen NH_3 pool (Pilgrim *et al.*, 1970; Al-Rabbat *et al.*, 1971a; Nolan *et al.*, 1976), but the proportion can vary from 42% to 100% depending on the availability of energy (Al-Rabbat *et al.*, 1971b).

Ammonia is, of course, the main product of deamination, and the remaining carbon skeletons give rise to a variety of volatile fatty acid products, as reviewed by Blackburn (1965) and Allison (1970). Presumably the purpose of this metabolism must be to provide energy for the microorganisms, otherwise there would be no advantage in their having the activity. There is some evidence of Stickland-like reactions in rumen fluid, whereby pairs of amino acids are metabolised and provide energy via coupled oxidation and reduction (Barker, 1981), because sodium arsenite caused a 70% inhibition of the metabolism of the amino acids in an acid hydrolysate of casein (Broderick & Balthrop, 1979). However, other experiments with pairs of amino acids and rumen microorganisms were inconclusive as to whether these react in the rumen (Lewis & Emery, 1962; Van den Hende *et al.*, 1963). Energy is presumably also produced from the metabolic sequence leading to VFA production from branched-chain amino acids (Harwood & Canale-Parola, 1981).

The disposal of reducing equivalents can prove to be problematic in anaerobic ecosystems. In the rumen, disposal is usually achieved by methanogenesis. When methane formation was inhibited by carbon monoxide, which inhibits bacterial hydrogenases, the fermentation stoichiometry switched to a higher propionate production (Russell & Jeraci, 1984). A more surprising secondary effect was that ammonia production declined, primarily as a consequence of inhibition of branched-chain amino acid fermentation (Russell & Jeraci, 1984; Russell & Martin, 1984). In cell extracts of rumen bacteria, the $NADH/NAD^+$ ratio was an important effector of branched-chain amino acid fermentation, with NAD^+ being essential as an electron acceptor (Hino & Russell, 1985). Thus, when hydrogenase was inhibited by carbon monoxide, the $NADH/NAD^+$ ratio increased and amino acid deamination declined (Hino & Russell, 1985). The feed ionophores, monensin and lasalocid, also inhibited both methanogenesis and deamination, but by a mechanism that could not be explained solely by their effect on hydrogenase activity (Russell & Martin, 1984).

Despite the energy potentially available from the deamination of amino acids, neither mixed rumen bacteria (Russell *et al.*, 1983) nor pure cultures of *B. ruminicola* (Russell, 1983) or *M. elsdenii* (Russell & Baldwin, 1979; Wallace, 1986*a*) gave much higher yields as a result of amino acid fermentation. It was suggested from the pure-culture work that the energy produced would contribute to the maintenance energy of the bacteria. The importance of deamination within the rumen ecosystem may therefore be principally in the provision of branched-chain VFA, which enhance the growth of many species of rumen bacteria, particularly of the predominant cellulolytic species (see Chapter 2 and below). These acids were produced by *M. elsdenii* during stationary phase, in low-energy conditions, rather than during growth (Allison, 1978), again consistent with their role in maintenance.

Ciliate protozoa seem to have a significant role in deamination. It has been known for a long time that most species of protozoa produce ammonia from protein or amino acids (Abou Akkada, 1965; Allison, 1970; Coleman, 1980; Williams, 1986), but it was only recently that the specific activities of mixed protozoa were found to be approximately three times those of bacteria (Hino & Russell, 1985). This may explain why ammonia concentrations in faunated sheep can be about twice those in ciliate-free animals (Eadie & Gill, 1971) and why deaminase activities were higher in faunated sheep, particularly when only small entodinia were present (Wallace *et al.*, 1987*b*). Products other than ammonia arising from protozoal breakdown of amino acids include 2-oxobutyric and 2-aminobutyric acids, from threonine and methionine (Onodera & Migita, 1985), pipecolic acid from lysine (Onodera & Kandatsu, 1972), δ-aminovaleric acid from proline (Onodera *et al.*, 1983) and methionine sulphoxide (Onodera & Takei, 1986). The main products of amino acid catabolism by protozoa are, however, similar to those of the bacteria, i.e. short- and branched-chain VFA (Allison, 1970; Coleman, 1980). Exogenously supplied amino acids were not catabolised by *Isotricha* spp. (Wallis & Coleman, 1967; Harmeyer, 1971), but deamination of amino acids has been observed with *E. caudatum* (Coleman, 1967; Wakita & Hoshino, 1975), which released pipecolic acid (Onodera & Kandatsu, 1969), and with *E. ecaudatum caudatum* (Coleman & Laurie, 1974).

Rumen bacteria which produce significant quantities of ammonia from amino acids include *B. ruminicola*, *S. ruminantium*, *M. elsdenii* and some strains of *Butyrivibrio* (Bladen *et al.*, 1961). Based on a decreased α-amino N concentration in spent growth medium, *Streptococcus* and *Eubacterium* spp. were considered to have important activities as well (Scheifinger *et al.*, 1976). *B. ruminicola* is probably most important in the rumen, due to its generally high activity and the numbers present (Bladen *et al.*, 1961). Different bacterial species utilise different spectra of amino acids during growth (Scheifinger *et al.*, 1976), which may cause different patterns of amino acid deamination in different animals and in animals on different diets, depending on the bacterial populations present. Some organisms, such as species of *Megasphaera, Streptococcus* and *Eubacterium*, removed a substantial proportion of all amino acids present in a mixture in the medium, while *Butyrivibrio* and *Selenomonas* were more selective. Methionine was actually produced by 3 out of 7 isolates tested (Scheifinger *et al.*, 1976). As is found with the mixed population, the pattern of amino acid utilisation by individual species is somewhat different when amino acids are present in peptides rather than in the free form (Cotta & Russell, 1982; Wallace, 1986a). The degradation products of amino acids in pure cultures are principally short- and branched-chain VFA (Blackburn, 1965; Allison, 1970). However, the product of tryptophan breakdown in the rumen, skatole (3-methylindole), is of interest as it causes bovine pulmonary emphysema (Carlson & Dickinson, 1978). The skatole-producing organism is a *Lactobacillus* sp. which produces skatole by decarboxylation of indoleacetic acid (Yokoyama *et al.*, 1977; Yokoyama & Carlson, 1981). Presumably because lactobacilli are sensitive to this ionophore, monensin prevents the disease (Hammond *et al.*, 1978).

The inhibition of amino acid degradation is an obvious objective for manipulation (see also Chapter 13). Even if the amino acids were not to pass undegraded from the rumen, if they were to be incorporated directly into microbial protein rather than be degraded to ammonia and then re-synthesised, the energy cost of re-synthesis would be saved. As has already been mentioned, ionophores achieve partial inhibition, but this is not their primary effect. Diaryliodonium compounds were intended specifically as inhibitors of amino acid degradation (Chalupa, 1977, 1980) and were found to be successful in improving N retention *in vivo* (Chalupa *et al.*, 1976). The organism most sensitive to diphenyliodonium chloride was *B. ruminicola* (Wallace, 1986b), and the improved N retention due to inhibition of *B. ruminicola* is consistent with the proposed central role of this organism in amino acid catabolism. Few chemicals will be entirely specific, however. Effects other than those on amino acid metabolism were seen with diaryliodonium compounds, including inhibited methanogenesis (Chalupa, 1980). Deamination can be inhibited by hydrazine and similar compounds, but their toxicity prohibits their use *in vivo* (Broderick & Balthrop, 1979).

Breakdown of urea

Urea is broken down extremely rapidly in the rumen, releasing ammonia. This activity, combined with rumen microbial protein synthesis from ammonia, enables ruminants to utilise urea entering the rumen either with the feed (Virtanen, 1966; Roffler & Satter, 1975), or in endogenous salivary secretion, or by diffusion across the rumen wall (Kennedy & Milligan, 1980). The enzyme mechanism is a simple hydrolysis by urease, which can be inhibited *in vitro* by acetohydroxamic acid (Jones, 1968; Brent *et al.*, 1971; Cook, 1976; Makkar *et al.*, 1981). The rumen enzyme is probably similar to jackbean urease in its Ni content, judging by the stimulation of rumen urease activity by dietary Ni (Spears *et al.*, 1977; Spears & Hatfield, 1978). Urease is associated with the particulate microbial fraction of rumen fluid, and is mainly of bacterial origin (Gibbons & McCarthy, 1957; Jones *et al.*, 1964; Mahadevan *et al.*, 1976). Urea is not hydrolysed in the absence of the microbial population, when its concentration in the rumen is the same as that in blood (Cheng & Wallace, 1979).

Precisely which bacterial species are most important in the hydrolysis of urea *in vivo* is not known, but the issue has always provoked interesting speculation and discussion, which can be read in more detail elsewhere (Jones, 1967; John *et al.*, 1974; Wozny *et al.*, 1977; Hobson & Wallace, 1982). Some studies found strictly anaerobic bacteria which hydrolysed urea to be elusive (Jones *et al.*, 1964; Cook, 1976), but isolates from the genera *Lactobacillus*, *Peptostreptococcus*, *Propionibacterium*, *Bacteroides*, *Ruminococcus*, *Butyrivibrio*, *Treponema*, *Selenomonas*, *Bifidobacterium* and *Succinivibrio* have been obtained (Gibbons & Doetsch, 1959; Slyter *et al.*, 1968; John *et al.*, 1974; Wozny *et al.*, 1977). Ureolytic, facultatively anaerobic bacteria are more readily isolated from rumen contents. These have generally been present in smaller populations than the strict anaerobes, and include species of *Streptococcus*, *Staphylococcus*, *Micrococcus*, *Propionibacterium* and *Corynebacterium* (Cook, 1976; Cheng *et al.*, 1979; Wallace *et al.*, 1979; Cheng & Costerton, 1980). Their similarity to the ureolytic population of the rumen wall prompted suggestions that these latter bacteria might be the more significant urease producers *in vivo* (Cheng *et al.*, 1979; Cheng & Costerton, 1980). Certainly, when the epithelial cells to which they are attached slough into rumen contents, wall bacteria confer a urease activity on the fluid sufficient to account for that found in normal rumen contents (Wallace *et al.*, 1979). The ecological argument is basically between a numerous population of strict anaerobes, mostly with a low urease activity, and a much smaller population of atypical, facultative anaerobes with characteristically high specific activity. There are weighty arguments in favour of each, and indeed the two populations may be equally important (Hobson & Wallace, 1982). One might speculate that, due to location, those organisms residing on the rumen wall may be most important in the hydrolysis of urea transferred across the rumen wall, while the organisms in the fluid may be important in the hydrolysis of dietary and salivary urea. Experiments related to this topic and using gnotobiotic lambs are described in Chapter 15.

The urease activity present in rumen contents has been partially purified, and appears to be associated with a single polypeptide, of smaller molecular weight than jackbean urease (Mahadevan *et al.*, 1976). The urease of *S. ruminantium* strain D, which has been studied in some detail (John *et al.*, 1974; Wozny *et al.*, 1977; Smith & Bryant, 1979), differs from the enzyme purified from rumen contents in its specific activity, which is 20- to 30-fold higher than rumen urease, and its molecular weight, which is 3-fold greater (Hausinger, 1986). This should not be taken as evidence dismissing the other anaerobes (this strain was fairly atypical in any case in several of its properties) until similar work is done with other facultatively and strictly anaerobic urease producers.

Urease is one of the most variable enzyme activities in rumen contents. The effects of dietary Ni have been mentioned, but many other factors influence its activity. Ammonia may repress activity (John et al., 1974; Cook, 1976; Wozny et al., 1977; Cheng & Wallace, 1979), and urea is likely to be an inducer (Czerkawski & Breckenridge, 1982), but other regulatory factors are ill-understood, except in *S. ruminantium* (John et al., 1974; Smith & Bryant, 1979). It is of interest to understand urease regulation, because the principal disadvantage of urea as a source of NPN is that it is broken down too rapidly, resulting in ammonia overflow and inefficient N retention.

Nucleic acids

Nucleic acids, despite their comprising 5·2–9·5% of the total N in grasses and hay (Smith & McAllan, 1970; Coelho da Silva et al., 1972), have received much less attention than proteins with regard to their breakdown by rumen microorganisms. DNA and RNA are rapidly hydrolysed in the rumen whether added as pure compounds or as plant material (Smith & McAllan, 1970; McAllan & Smith, 1973*a*). Transient products formed are a mixture of nucleotides, nucleosides and bases (McAllan & Smith, 1973*a*). McAllan & Smith (1973*b*) investigated the breakdown of nucleic acids in rumen fluid *in vitro*, and found that purine nucleotides formed hypoxanthine and xanthine, while pyrimidine nucleotides formed uracil and thymine. Cytosine was deaminated to uracil. Although these products were formed *in vitro*, no such accumulation was seen *in vivo*.

The microbial ecology of nucleic acid metabolism is poorly understood. For once, more is known about protozoal than bacterial metabolism, from the work of Coleman and his colleagues, but even then our knowledge is patchy. It is not known, for example, if protozoa are more important than the bacterial flora in this activity. Coleman (1968, 1980) showed that *E. caudatum* assimilated bacterial nucleic acid components intact, and took up nucleotides from the medium. Several species, including *E. caudatum* (Coleman, 1968), *E. ecaudatum caudatum* (Coleman & Laurie, 1974), *P. multivesiculatum* (Coleman & Laurie, 1977) and *E. maggii* (Coleman & Sandford, 1979*b*), convert adenine and guanine into hypoxanthine and xanthine, metabolise pyrimidines, and incorporate exogenous bases. The principal function of these protozoal activities is thought to be the utilisation of bacterial nucleic acids (Coleman, 1980). Presumably the bacteria also degrade and incorporate nucleic acids, but derived mainly from the ruminant's food. A strain of *B. fibrisolvens* was found to be highly nucleolytic (Orpin, personal communication).

Other nitrogenous compounds in the diet

The other naturally occurring N-containing compounds in the diet that have received some attention are nitrate, ethanolamine and choline, and the possibilities that N_2 can be utilised from ingested air and that novel non-protein N sources can be developed have also been investigated.

Nitrate is quite abundant in some plant materials, and conceivably could not only provide N for microbial protein synthesis but also provide a terminal electron acceptor for anaerobic respiration, thus enhancing energy production. Evidence for either of these possibilities is scant, however. Some strains of *S. ruminantium* use nitrate as a nitrogen source (John *et al.*, 1974), and presumably other rumen bacteria do so too. Nitrate metabolism *in vivo* can cause nitrite poisoning in the host animal if nitrite is not reduced to ammonium quickly enough (Lewis, 1951; Holtenius, 1957). Nitrate reduction was stimulated *in vitro* by the addition of H_2 or glucose as electron donors, and nitrite accumulated (Jones, 1972). Only formate was a good enough electron acceptor to prevent nitrite accumulation. Denitrification was considered to be unlikely in the rumen (Jones, 1972). Nothing is known of the organisms that are principally responsible for the different steps of nitrate reduction *in vivo*.

Choline is essential for the growth of the protozoon *E. caudatum*, and is rapidly incorporated into phospholipid (Broad & Dawson, 1975). Ethanolaminine is less rapidly used and cannot replace choline (Bygrave & Dawson, 1976). Indeed, choline incorporation is a good index of protozoal activity in the mixed rumen population (Newbold, personal communication). However, its main fate in the mixed population is to be converted into trimethylamine, which in turn is converted into methane (Itabashi & Kandatsu, 1978; Neill *et al.*, 1978) by the methanogen *Methanosarcina barkeri* (Patterson & Hespell, 1979).

The rate of N_2 fixation in the rumen is insignificant compared to the total N entering the rumen (Moisio *et al.*, 1969; Hobson *et al.*, 1973; Jones & Thomas, 1974). Even with daily inoculation of a N_2-fixing *Bacillus macerans* into sheep on a 10% molasses diet, N_2 fixation in the rumen amounted to only 0·75 g per day (Jones & Thomas, 1974).

Cheap, synthetic N-containing compounds that can yield ammonia by rumen microbial hydrolysis could be an attractive alternative to urea, particularly if they released ammonia at a rate balanced with the rate of energy production, unlike urea which is broken down too quickly. The microbial population would likely have to adapt to these new substrates (Nikolic *et al.*, 1980), but compounds like biuret, creatine, ammoniated molasses and glycosyl ureas could be useful (Schwartz, 1967; Chalupa, 1972). Little is known of the microbiology of their breakdown.

ANABOLIC NITROGEN METABOLISM

Ammonia assimilation

Mixed rumen microorganisms

Ammonia is the most important source of N for protein synthesis in the rumen. Its concentration in the rumen fluctuates markedly, from less than 1 mM observed in some animals on extremely low protein roughages to perhaps 40 mM, transiently after feeding, in animals receiving rapidly degraded protein or urea. There are several different enzymic mechanisms for ammonia uptake into amino acids, each with a different affinity for its substrate. The most important mechanisms, and perhaps organisms as well, therefore probably vary as the ruminal NH_3 concentration changes. Because it is the central pathway for protein synthesis, the mechanism of ammonia uptake is of great interest to microbiologists and nutritionists dealing with ruminants. A recent review (Hespell, 1984) dealt comprehensively with the microbiology and biochemistry of ammonia assimilation by rumen bacteria.

The first step in ammonia uptake into a cell is its transport across the cell membrane. Nothing is known about this mechanism in rumen microorganisms, but it is improbable that the rate of ammonia transport limits the rate of ammonia assimilation into amino acids. What might be more important, at least in the interpretation of enzyme kinetics measurements, is whether ammonia is accumulated within cells. Without this information, predicting the likely enzyme involved from the K_m (ammonia) of different enzymes is hazardous.

The highest-affinity enzyme system for ammonia assimilation is the glutamine synthetase–glutamate synthase (GS-GOGAT) couple (Brown et al., 1974). Ammonia is first incorporated into the amide group of glutamine, using glutamate as substrate, and ATP is hydrolysed. The amide $-NH_2$ is then transferred to α-oxoglutarate to form 2 molecules of glutamate. GOGAT has been demonstrated to be present in rumen microorganisms under conditions of low ammonia (Erfle et al., 1977), but it is not significant at higher ammonia concentrations (Wallace, 1979; Lenartova et al., 1985). The low K_m (ammonia) of GS (1·8 mM; Woolfolk et al., 1966) is consistent with an effective scavenging role at low rumen ammonia concentrations. Because ATP is required, however, this system might be expected to be a handicap to organisms at higher ammonia concentrations where ATP-independent enzymes can function. For this reason, the GS-GOGAT system is only expressed by *Klebsiella aerogenes*, for example, under conditions of ammonia limitation (Brown et al., 1974). When fermentation *in vitro* was artificially ammonia-limited, GS activity increased tenfold (Erfle et al., 1977). Rumen ammonia concentrations *in vivo* would seldom be so low as to require the high affinity of the GS-GOGAT mechanism for effective assimilation.

Lower-affinity, higher K_m, systems present in rumen microorganisms include NADP-glutamate dehydrogenase (NADP-GDH; K_m for ammonia, 1·8–3·1 mM; Wallace, 1979), NAD-GDH (20–33 mM; Erfle et al., 1977; Wallace, 1979) and alanine dehydrogenase (70 mM; Wallace, 1979). Other possibilities that have been investigated have been asparagine synthetase replacing GS in a coupled system analogous to GS-GOGAT (Erfle et al., 1977), NADP-alanine dehydrogenase and aspartate dehydrogenase (Wallace, 1979) and carbamyl phosphokinase (Chalupa et al., 1970), but these activities were negligible or of very high K_m (ammonia).

NAD-linked GDH is the highest-activity ammonia-assimilating enzyme present, not only in rumen contents (Hoshino et al., 1966; Palmquist & Baldwin, 1966; Chalupa et al., 1970; Wallace, 1979; Bhatia et al., 1980; Lenartova et al., 1985) but in rumen mucosa (Hoshino et al., 1966; Chalupa et al., 1970) and in bacteria attached to the rumen wall (Lenartova et al., 1985). Sometimes this activity is much higher than that of the NADP-linked enzyme (Chalupa et al., 1970; Wallace, 1979), but activity varies with diet, and the two activities can be quite similar (Erfle et al., 1977; Lenartova et al., 1985). Under the latter circumstances, the higher K_m (ammonia) of NAD-GDH would restrict its activity to one primarily of glutamate catabolism.

Various aminotransferase (transaminase) activities are present in rumen contents, which transfer the trapped $-NH_2$ from the primary amino acid throughout the amino acid pool. The most commonly found are glutamate-pyruvate and glutamate-oxaloacetate aminotransferases (Chalupa et al., 1970; Bhatia et al., 1979; Wallace, 1979; Lenartova et al., 1985), but many others exist to disperse the bound ammonia (Bhatia et al., 1969).

A question then arises as to which of these mechanisms actually provides the main route of ammonia assimilation. From the enzymic observations, glutamate would be expected to be the first amino acid into which ammonia would be assimilated, and this coincides with glutamate usually being the most abundant amino acid in the free amino acid pool (Wright & Hungate, 1967; Shimbayashi et al., 1975; Erfle et al., 1977; Wallace, 1979; Blake et al., 1983). However, alanine was surprisingly prominent in these pools and often exceeded glutamate, particularly under conditions of high ammonia concentration. Clearly this need not mean that alanine is the primary product of ammonia assimilation. Imbalances in rates of formation by transaminases and rates of utilisation in protein synthesis could easily result in alanine accumulation. However, the earlier indications of Shimbayashi et al. (1975) that alanine was a primary product were confirmed by Blake et al. (1983). [^{15}N]Ammonium chloride enriched alanine more than glutamate or other amino acids in the microbial pool after only 2 minutes (Blake et al., 1983). The role of alanine and alanine dehydrogenase in ammonia uptake, rather than simply overproduction of alanine for some other physiological purpose, is still in need of clarification (Blake et al., 1983). Past experience would suggest that the correct electron donor has not been found (Hespell, 1984). Furthermore, given the high K_m (ammonia) and low activity of alanine dehydrogenase that has so far been measured, the possible energy-linked accumulation of ammonia within cells, as occurs with *E. coli* (Stevenson & Silver, 1977) and *Clostridium pasteurianum* (Kleiner & Fitzke, 1979), then becomes critical to the efficient assimilation of ammonia into alanine.

The different enzymic mechanisms for ammonia uptake probably reflect different niches that the microorganisms can occupy. For example, the maximum rate of degradation of barley in the rumen occurs at an ammonia concentration of 9 mM or above (Mehrez et al., 1977; Wallace, 1979; Odle et al., 1985), higher than that required for corn (Slyter et al., 1979; Nikolic & Filipovic, 1981; Odle et al., 1985), and much higher than the K_s (ammonia) of predominant rumen bacteria (less than 50 μM; Schaefer et al., 1980). It has been suggested that this apparent increase in excess ammonia in total rumen contents may be necessary for sufficient penetration of ammonia to the site of digestion of a feed component. Ammonia could easily be limiting within that microenvironment, depending on the nature of the diet. The microenvironment probably varies enormously from one plant material to another, explaining the different effects of gross ammonia concentration on the rate at which different feeds are fermented. Furthermore, it would also explain why rumen bacteria retain the ability to form the GS-GOGAT system when ammonia becomes limiting (Erfle et al., 1977; Hespell, 1984).

Individual species
Ciliate protozoa are known to synthesise some of their protein *de novo* (reviewed by Coleman, 1980; Williams, 1986), but to our knowledge the extent to which ammonia

is necessary for protein synthesis, and the assimilation mechanism, have not been described. In contrast, the fact that the anaerobic fungus *Neocallimastix frontalis* grew in a defined medium without pre-formed amino acids (Lowe *et al.*, 1985) suggested that these organisms may depend heavily on ammonia for protein synthesis *in vivo*. Again, information on the enzymic mechanism of uptake is lacking.

Most species of rumen bacteria can use ammonia as their main source of nitrogen for growth (Bryant, 1974), and indeed do so under conditions normally prevailing in the rumen (Nolan, 1975). Ammonia is in fact essential for the growth of many species (Allison, 1969, 1970; Bryant, 1974). The enzymology of ammonia uptake has been studied in some of these species.

Ammonia-limited growth of *B. amylophilus* resulted in a repression of the main glutamate dehydrogenase activity, NADP-GDH, and a stimulation of GS (Jenkinson *et al.*, 1979). GOGAT activity was not detected (Jenkinson *et al.*, 1979), implying that the high-affinity GS-GOGAT couple could not function. Several mechanisms for ammonia assimilation are therefore possible. The first is that NADP-GDH, with its K_m (ammonia) of 1·0–1·7 mM, is the principal ammonia uptake enzyme. In that case, the very low ammonia saturation constant (6–13 μM; Schaefer *et al.*, 1980) for the whole organism would have to be explained by an active accumulation of ammonia intracellularly. Alternatively, GS may indeed be the first uptake enzyme, with the secondary aminotransferase activity being different from GOGAT. As in the mixed population, alanine was prominent in the intracellular free amino acid pools, yet alanine dehydrogenase was low (Jenkinson *et al.*, 1979).

With *Selenomonas ruminantium*, GS activity was again stimulated in ammonia-limited cultures (Smith *et al.*, 1980, 1981; Hespell, 1984), particularly at high growth rate (Hespell, 1984). Unlike *B. amylophilus*, a low GOGAT activity was found in *S. ruminantium* (Smith *et al.*, 1981), so the GS-GOGAT couple would be expected to function at low ammonia concentrations. Surprisingly, GOGAT was not induced by ammonia limitation (Smith *et al.*, 1981). NADP-GDH was higher in glucose-limited (ammonia-sufficient) cultures, suggesting that it was the route used under these conditions (Smith *et al.*, 1980; Hespell, 1984). Hespell (1984) calculated that 50% of the difference in growth yields obtained in glucose- and ammonia-limited chemostats could be accounted for by a switch from GDH to the ATP-consuming GS-GOGAT route of ammonia uptake.

Other species have been less well studied. *S. bovis* possesses GS (Griffith & Carlsson, 1974) and NADP-GDH (Burchall *et al.*, 1964; Griffith & Carlsson, 1974), but not GOGAT (Griffith & Carlsson, 1974). NADP-GDH was much higher in ammonia-limited cells, and it was concluded that this was the only pathway of ammonia assimilation in *S. bovis* (Griffith & Carlsson, 1974). The NADP-GDH of *R. flavefaciens* was stimulated in a similar way (Pettipher & Latham, 1979). *S. dextrinosolvens* has both NADP-GDH and GS activities (Patterson & Hespell, 1985).

In all of these pure cultures, the mechanism of ammonia uptake has been implied from the enzymes present, and by the way that they are regulated. The work of

Smith et al. (1980), Hespell (1984) and others has shown that the conditions for measurement of the enzymes really must be investigated in detail before such a conclusion can be drawn. In fact, the only sure way of establishing the true mechanism would be to use a ^{13}N or ^{15}N technique using labelled ammonia.

Amino acid biosynthesis

Once nitrogen has been fixed into an appropriate compound, such as glutamic acid, transfer of nitrogen to specific carbon skeletons is required. As many rumen microorganisms are able to grow in the absence of pre-formed amino acids, synthesis of amino acid carbon skeletons from available energy and carbon sources must occur. Again, our information is greatest for amino acid biosynthesis by rumen bacteria. There is evidence for *de novo* synthesis of amino acids in protozoa but, since protozoa are effective utilisers of bacterial protein, synthesis is likely to be of minor importance (reviewed by Coleman, 1980; Williams, 1986). Rumen fungi are able to grow in media lacking pre-formed amino acids and therefore must be able to synthesise needed amino acids, but information on amino acid biosynthesis is lacking (Lowe et al., 1985). Because of this, the discussion presented here will centre on amino acid biosynthesis by rumen bacteria.

The biosynthetic pathways for the production of amino acids by bacteria and fungi have been reviewed by Umbarger (1978), and the reader is referred to this work for a more detailed discussion of the synthesis of individual amino acids. In brief, amino acids can be divided into groups or families based on the source of carbon used for their synthesis. These are: the glutamate family—glutamate, glutamine, proline, arginine; the serine family—serine, glycine, cysteine; the aspartate family—aspartate, asparagine, lysine, methionine, threonine, isoleucine; the pyruvate family—alanine, isoleucine, leucine, valine; the aromatic family—phenylalanine, tyrosine, tryptophan; and histidine. While absolute proof of the synthesis of all amino acids by the same pathways in rumen bacteria is lacking, evidence for a number of these exists and radioactive tracer experiments on amino acid synthesis by mixed rumen bacteria yielded amino acids with labelling patterns consistent with the pathways described by Umbarger (Sauer et al., 1975). As illustrated in the section on ammonia assimilation, glutamic acid occupies a central role in the nitrogen metabolism of organisms, and thus the generation of α-oxoglutarate is of great importance to nitrogen metabolism in rumen bacteria. Since these bacteria are anaerobes, and lack a functional tricarboxylic acid (TCA) cycle, α-oxoglutarate is not produced as a normal intermediate of energy metabolism as in aerobic organisms. Synthesis of α-oxoglutarate by rumen bacteria has been examined using both mixed and pure cultures of bacteria. Milligan (1970) showed that rumen contents incubated with $NaH^{14}CO_3$ produced glutamate labelled in the C-1, C-2 and C-5 positions. This labelling pattern suggested that α-oxoglutarate was synthesised by both forward and reverse TCA cycle activity. That is, α-oxoglutarate is formed by reductive carboxylation of succinic acid for reverse TCA cycle, as against condensation of oxaloacetic acid and acetyl-CoA to form citrate and subsequent forward TCA activities. The specific activity of labelled glutamate carbons indicated that the reverse TCA cycle route of α-oxoglutarate synthesis was the predominant pathway employed by rumen microorganisms. Later experiments by Sauer et al. (1975), using a mixed rumen microbial population maintained in continuous culture, confirmed the presence of both modes of α-oxoglutarate synthesis. In their experiments, however, forward TCA cycle function was concluded to be the major pathway of α-oxoglutarate generation. Representatives of individual species of rumen bacteria can be cited for both mechanisms of α-oxoglutarate synthesis.

Megasphaera elsdenii produces α-oxoglutarate by the forward TCA method of synthesis. This was first suggested by specific labelling of glutamate carbons when this organism was provided with [1-^{14}C]lactate as the growth substrate (Somerville & Peel, 1967). Later, Somerville (1968) demonstrated that *M. elsdenii* produces the required enzymic machinery for this synthesis. Allison & Robinson (1970) showed that *B. ruminicola* forms α-oxoglutarate by reductive carboxylation of succinate and described some characteristics of the α-oxoglutarate synthase reaction. Subsequently, strains of *S. ruminantium*, *Veillonella alcalescens*, and other gastrointestinal tract *Bacteroides* spp. were shown to synthesise α-oxoglutarate by reductive carboxylation of succinate (Allison et al., 1979). Conversion of [^{14}C]succinate into [^{14}C]glutamate could not be demonstrated for *R. flavefaciens*, *Methanobrevibacter ruminantium*, *S. bovis*, *B. fibrisolvens* and *Succinivibrio dextrinosolvens*, but these strains failed to take up exogenous [^{14}C]succinate, and this evidence does not preclude existence of this pathway for α-oxoglutarate synthesis.

Direct demonstration of the production of the other important carbon skeletons for amino acid biosynthesis is variable. Pyruvate is generated in the energy metabolism of the majority of rumen bacteria and can also be produced by the reductive carboxylation of acetate (Allison, 1969; Prins, 1977). Serine is produced from phosphoglyceric acid (a glycolytic intermediate) by conversion of this compound into phosphohydroxypyruvate then phosphoserine and serine (Somerville, 1968; Sauer et al., 1975). Aspartate is the transamination product of oxaloacetic acid which is produced in the energy metabolism of most succinic acid-producing rumen bacteria (Joyner & Baldwin, 1966; Prins, 1977). Oxaloacetic acid is also generated in both of the pathways of α-oxoglutarate synthesis discussed previously. Evidence for the biosynthetic pathways resulting in the formation of the aromatic amino acids and histidine is more indirect. The results of the radioactive tracer studies of Sauer et al. (1975) are consistent with the synthesis of histidine from phosphoribosyl pyrophosphate and aromatic amino acids via the shikimic acid pathway, although the results suggest that this may not be the major route of aromatic amino acid synthesis used by rumen bacteria.

While a number of rumen bacteria can form all the needed carbon skeletons for amino acid synthesis, many organisms have the ability to utilise products formed by other organisms as intermediates in the synthesis of amino acids. In fact, in many cases bacteria have an absolute nutritional requirement for such intermediates. Probably the most widely recognised example of this phenomenon is the branched-chain volatile fatty acid requirement of the predominant cellulolytic rumen bacteria.

Bacteroides succinogenes, R. flavefaciens and R. albus require one or more of the branched-chain fatty acids isobutyric acid, 2-methylbutyric acid and isovaleric acid for the synthesis of valine, isoleucine and leucine respectively (Bryant & Robinson, 1962; Allison & Bryant, 1963; Dehority et al., 1967). Conversion of the branched-chain fatty acid into the appropriate amino acid is the result of reductive carboxylation and transamination of the fatty acid. Other organic acids that can be converted into amino acids in this manner include phenylacetate and indoleacetic acid, for the synthesis of phenylalanine and tryptophan respectively (Allison, 1965; Allison & Robinson, 1967). It is interesting that organisms such as B. ruminicola and M. elsdenii, which produce the branched-chain fatty acids, can also utilise these for the synthesis of amino acids (Allison & Peel, 1971; Allison, 1978). Allison et al. (1984) found that, when B. ruminicola was grown in media lacking isovaleric acid, carbon from glucose was used for the synthesis of leucine. However, when isovaleric acid was added to the medium, the conversion of [^{14}C]glucose into leucine was markedly inhibited. They also found that synthesis of phenylalanine or isoleucine from glucose was reduced when phenylacetate or 2-methylbutyrate were provided in the medium. The authors concluded that this organism has the ability to regulate these pathways of amino acid biosynthesis, and will use pre-formed intermediates for synthesis of these amino acids in preference to *de novo* synthesis. Furthermore, since these intermediates are generally present in rumen fluid, their reductive carboxylation is likely to be the predominant pathway for synthesis of the related amino acids in the rumen.

REFERENCES

Abou Akkada, A. R. (1965). The metabolism of ciliate protozoa in relation to rumen function. In *Physiology of Digestion in the Ruminant*, ed. R. W. Dougherty, R. S. Allen, W. Burroughs, N. L. Jacobson & A. D. McGilliard. Butterworths, London, pp. 335–45.

Abou Akkaa, A. R. & Blackburn, T. H. (1963). Some observations on the nitrogen metabolism of rumen proteolytic bacteria. *J. Gen. Microbiol.*, **31**, 461–9.

Abou Akkada, A. R. & Howard, B. H. (1962). The biochemistry of rumen protozoa 5. The nitrogen metabolism of *Entodinium*. *Biochem. J.*, **82**, 313–20.

Allison, M. J. (1965). Phenylalanine biosynthesis from phenylacetic acid by anaerobic bacteria from the rumen. *Biochem. Biophys. Res. Comm.*, **18**, 30–5.

Allison, M. J. (1969). Biosynthesis of amino acids by ruminal microorganisms. *J. Anim. Sci.*, **29**, 797–807.

Allison, M. J. (1970). Nitrogen metabolism of ruminal micro-organisms. In *Physiology of Digestion and Metabolism in the Ruminant*, ed. A. T. Phillipson. Oriel Press, Newcastle-upon-Tyne, pp. 456–73.

Allison, M. J. (1978). Production of branched-chain volatile fatty acids by certain anaerobic bacteria. *Appl. Environ. Microbiol.*, **35**, 872–7.

Allison, M. J. & Bryant, M. P. (1963). Biosynthesis of branched-chain amino acids from branched-chain fatty acids by rumen bacteria. *Arch. Biochem. Biophys.*, **101**, 269–77.

Allison, M. J. & Peel, J. L. (1971). The biosynthesis of valine from isobutyrate by *Peptostreptococcus elsdenii* and *Bacteroides ruminicola*. *Biochem. J.*, **121**, 431–7.

Allison, M. J. & Robinson, I. M. (1967). Tryptophan biosynthesis from indole-3-acetic acid by anaerobic bacteria from the rumen. *Biochem. J.*, **102**, 36–7.

Allison, M. J. & Robinson, I. M. (1970). Biosynthesis of α-ketoglutarate by the reductive carboxylation of succinate in *Bacteroides ruminicola*. *J. Bacteriol.*, **104**, 50–6.

Allison, M. J., Robinson, I. M. & Baetz, A. L. (1979). Synthesis of α-ketoglutarate by reductive carboxylation of succinate in *Veillonella, Selenomonas*, and *Bacteroides* sp. *J. Bacteriol.*, **140**, 980–6.
Allison, M. J., Baetz, A. L. & Wiegel, J. (1984). Alternative pathways for biosynthesis of leucine and other amino acids in *Bacteroides ruminicola* and *Bacteroides fragilis*. *Appl. Environ. Microbiol.*, **48**, 1111–17.
Al-Rabbat, M. F., Baldwin, R. L. & Weir, W. C. (1971a). *In vitro* nitrogen tracer technique for some kinetic measurements of ruminal ammonia. *J. Dairy Sci.*, **54**, 1150–61.
Al-Rabbat, M. F., Baldwin, R. L. & Weir, W. C. (1971b). Microbial growth dependence on ammonia nitrogen in the bovine rumen: a quantitative study. *J. Dairy Sci.*, **54**, 1162–72.
Appleby, J. C. (1955). The isolation and classification of proteolytic bacteria from the rumen of the sheep. *J. Gen. Microbiol.*, **12**, 526–33.
Ashes, J. R., Mangan, J. L. & Sidhu, G. S. (1984). Nutritional availability of amino acids from protein cross-linked to protect against degradation in the rumen. *Br. J. Nutr.*, **52**, 239–47.
Barker, H. A. (1981). Amino acid degradation by anaerobic bacteria. *Ann. Rev. Biochem.*, **50**, 23–40.
Bhatia, S K., Pradhau, K. & Singh, R. (1979). Microbial transaminase activities and the relationship with bovine rumen metabolites. *J. Dairy Sci.*, **62**, 441–6.
Bhatia, S. K., Pradhau, K. & Singh, R. (1980). Ammonia anabolising enzymes in cattle and buffalo fed varied non-protein nitrogen and carbohydrates. *J. Dairy Sci.*, **63**, 1104–8.
Blackburn, T. H. (1965). Nitrogen metabolism in the rumen. In *Physiology of Digestion in the Ruminant*, ed. R. W. Dougherty, R. S. Allen, W. Burroughs, N. L. Jacobson & A. D. McGilliard. Butterworths, London, pp. 322–34.
Blackburn, T. H. (1968a). Protease production by *Bacteroides amylophilus* strain H18. *J. Gen. Microbiol.*, **53**, 27–36.
Blackburn, T. H. (1968b). The protease liberated from *Bacteroides amylophilus* strain H18 by mechanical disintegration. *J. Gen. Microbiol.*, **53**, 37–51.
Blackburn, T. H. & Hobson, P. N. (1960a). Isolation of proteolytic bacteria from the sheep rumen. *J. Gen. Microbiol.*, **22**, 282–9.
Blackburn, T. H. & Hobson, P. N. (1960b). Proteolysis in the sheep rumen by whole and fractionated rumen contents. *J. Gen. Microbiol.*, **22**, 272–81.
Blackburn, T. H. & Hobson, P. N. (1962). Further studies on the isolation of proteolytic bacteria from the sheep rumen. *J. Gen. Microbiol.*, **29**, 69–81.
Blackburn, T. H. & Hullah, W. A. (1974). The cell-bound protease of *Bacteroides amylophilus* H18. *Can. J. Microbiol.*, **20**, 435–41.
Bladen, H. A., Bryant, M. P. & Doetsch, R. N. (1961). A study of bacterial species from the rumen which produce ammonia from protein hydrolysate. *Appl. Microbiol.*, **9**, 175–80.
Blake, J. S., Salter, D. N. & Smith, R. H. (1983). Incorporation of nitrogen into rumen bacterial fractions of steers given protein- and urea-containing diets. Ammonia assimilation into intracellular bacterial amino acids. *Br. J. Nutr.*, **50**, 769–82.
Brent, B. E., Adepoju, A. & Portela, F. (1971). *In vitro* inhibition of rumen urease with acetohydroxamic acid. *J. Amin. Sci.*, **32**, 794–8.
Briggs, M. H. (ed.) (1967). *Urea as a Protein Supplement*. Pergamon, London.
Broad, T. E. & Dawson, R. M. C. (1975). Phospholipid biosynthesis in the anaerobic protozoan *Entodinium caudatum*. *Biochem. J.*, **146**, 317–28.
Brock, F. M., Forsberg, C. W. & Buchanan-Smith, J. G. (1982). Proteolytic activity of rumen microorganisms and effects of proteinase inhibitors. *Appl. Environ. Microbiol.*, **44**, 561–9.
Broderick, G. A. & Balthrop, J. E. (1979). Chemical inhibition of amino acid deamination by ruminal microbes *in vitro*. *J. Anim. Sci.*, **49**, 1101–11.
Broderick, G. A., Wallace, R. J. & Brammall, M. L. (1988). Uptake of small neutral peptides by mixed rumen microorganisms *in vitro*. *J. Sci. Food. Agric.*, **42**, 109–18.
Brown, C. M., MacDonald-Brown, D. S. & Meers, J. L. (1974). Physiological aspects of microbial inorganic nitrogen metabolism. *Adv. Microb. Physiol.*, **11**, 1–52.

Bryant, M. P. (1974). Nutritional features and ecology of predominant anaerobic bacteria of the intestinal tract. *Am. J. Clin. Nutr.*, **27**, 1313–19.
Bryant, M. P. & Robinson, I. M. (1962). Some nutritional characteristics of predominant culturable ruminal bacteria. *J. Bacteriol.*, **84**, 605–14.
Burchall, J. J., Niederman, R. A. & Wolin, M. J. (1964). Amino group formation and glutamate synthesis in *Streptococcus bovis*. *J. Bacteriol.*, **88**, 1038–44.
Bygrave, F. L. & Dawson, R. M. C. (1976). Phosphatidylcholine biosynthesis and choline transport in the anaerobic protozoan *Entodinium caudatum*. *Biochem. J.*, **160**, 481–90.
Carlson, J. R. & Dickinson, E. O. (1978). Tryptophan-induced pulmonary edema and emphysema in ruminants. In *Effects of Poisonous Plants on Livestock*. Academic Press, New York, pp. 261–72.
Chalmers, M. I. & Synge, R. L. M. (1954). The digestion of protein and nitrogenous compounds in ruminants. *Adv. Prot. Chem.*, **9**, 93–120.
Chalupa, W. (1972). Metabolic aspects of non-protein nitrogen utilization in ruminant animals. *Fed. Proc.*, **31**, 1152–64.
Chalupa, W. (1976). Degradation of amino acids by the mixed rumen microbial population. *J. Anim. Sci.*, **43**, 828–34.
Chalupa, W. (1977). Manipulating rumen fermentation. *J. Anim. Sci*, **45**, 585–99.
Chalupa, W. (1980). Chemical control of rumen microbial metabolism. In *Digestive Physiology and Metabolism in Ruminants*, ed. Y. Ruckebusch & P. Thivend. MTP Press. Lancaster, pp. 325–47.
Chalupa, W., Clark, J., Opliger, P. & Lavker, R. (1970). Ammonia metabolism in rumen bacteria and mucosa from sheep fed soy protein or urea. *J. Nutr.*, **100**, 161–9.
Chalupa, W., Patterson, J. A., Chow, A. W. & Parish, R. C. (1976). Deaminase inhibitor effects on N-utilization. *J. Anim. Sci.*, **43**, 316–17.
Cheng, K.-J. & Costerton, J. W. (1980). Adherent rumen bacteria—their role in the digestion of plant material, urea and epithelial cells. In *Digestive Physiology and Metabolism in Ruminants*, ed. Y. Ruckebusch & P. Thivend. MTP Press, Lancaster, pp. 227–50.
Cheng, K.-J. & Wallace, R. J. (1979). The mechanism of passage of endogenous urea through the rumen wall and the role of ureolytic epithelial bacteria in the urea flux. *Br. J. Nutr.*, **42**, 553–7.
Cheng, K.-J., McCowan, R. P. & Costerton, J. W. (1979). Adherent epithelial bacteria in ruminants and their roles in digestive tract function. *Am. J. Clin. Nutr.*, **32**, 139–48.
Coelho da Silva, J. F., Seeley, R. C., Beever, D. E. & Prescott, J. H. D. (1972). The effect in sheep of physical form and stage of growth on the sites of digestion of a dried grass. *Br. J. Nutr.*, **28**, 357–71.
Coleman, G. S. (1967). The metabolism of free amino acids by washed suspensions of the rumen ciliate *Entodinium caudatum*. *J. Gen. Microbiol.*, **47**, 433–47.
Coleman, G. S. (1968). The metabolism of bacterial nucleic acid and of free components of nucleic acid by the rumen ciliate *Entodinium caudatum*. *J. Gen. Microbiol.*, **54**, 83–96.
Coleman, G. S. (1980). Rumen ciliate protozoa. *Adv. Parasitol.*, **18**, 121–73.
Coleman, G. S. (1983). Hydrolysis of Fraction I leaf protein and casein by rumen entodiniomorphid protozoa. *J. Appl. Bacteriol.*, **55**, 111–18.
Coleman, G. S. & Laurie, J. I. (1974). The metabolism of starch, glucose, amino acids, purines, pyrimidines and bacteria by three *Epidinium* spp. isolated from the rumen. *J. Gen. Microbiol.*, **85**, 244–56.
Coleman, G. S. & Laurie, J. I. (1977). The metabolism of starch, glucose, amino acids, purines, pyrimidines and bacteria by the rumen ciliate *Polyplastron multivesiculatum*. *J. Gen. Microbiol.*, **98**, 29–37.
Coleman, G. S. & Reynolds, D. J. (1982). The uptake of bacteria and amino acids by *Ophryoscolex caudatus*, *Diploplastron affine* and some other rumen entodiniomorphid protozoa. *J. Appl. Bacteriol.*, **52**, 135–44.
Coleman, G. S. & Sandford, D. C. (1979a). The engulfment and digestion of mixed rumen bacteria and individual bacterial species by single and mixed species of rumen ciliate protozoa grown *in vivo*. *J. Agric. Sci.* (Camb.), **92**, 729–42.

Coleman, G. S. & Sandford, D. C. (1979b). The uptake and utilization of bacteria, amino acids, and nucleic acid components by the rumen ciliate *Eudiplodinium maggii*. *J. Appl. Bacteriol.*, **47**, 409–19.

Cook, A. R. (1976). Urease activity in the rumen of sheep and the isolation of ureolytic bacteria. *J. Gen. Microbiol.*, **92**, 32–48.

Cotta, M. A. & Hespell, R. B. (1986a). Protein and amino acid metabolism of rumen bacteria. In *Control of Digestion and Metabolism in Ruminants*, ed. L. P. Milligan, W. L. Grovum & A. Dobson. Prentice-Hall, Englewood Cliffs, New Jersey, pp. 122–36.

Cotta, M. A. & Hespell, R. B. (1986b). Proteolytic activity of the the ruminal bacterium *Butyrivibrio fibrisolvens*. *Appl. Environ. Microbiol.*, **52**, 51–8.

Cotta, M. A. & Russell, J. B. (1982). Effect of peptides and amino acids on efficiency of rumen bacterial protein synthesis in continuous culture. *J. Dairy Sci.*, **65**, 226–34.

Czerkawski, J. W. & Breckenridge, G. (1982). Distribution and changes in urease (EC 3.5.1.5) activity in Rumen Simulation Technique (Rusitec). *Br. J. Nutr.*, **47**, 331–48.

Dehority, B. A., Scott, H. W. & Kowaluk, P. (1967). Volatile fatty acid requirements of cellulolytic rumen bacteria. *J. Bacteriol.*, **94**, 537–43.

Dinsdale, D., Cheng, K.-J., Wallace, R. J. & Goodlad, R. A. (1980). Digestion of epithelial tissue of the rumen wall by adherent bacteria in infused and conventionally fed sheep. *Appl. Environ. Microbiol.*, **39**, 1059–66.

Eadie, J. M. & Gill, J. C. (1971). The effect of the absence of rumen ciliate protozoa on growing lambs fed on a roughage-concentrate diet. *Br. J. Nutr.*, **26**, 155–67.

Erfle, J. D., Sauer, F. D. & Mahadevan, S. (1977). The effect of ammonia concentration on activity of enzymes of ammonia assimilation and on synthesis of amino acids by mixed rumen bacteria in continuous culture. *J. Dairy Sci.*, **60**, 1064–72.

Forsberg, C. W., Lovelock, L. K. A., Krumholz, L. & Buchanan-Smith, J. G. (1984). Protease activities of rumen protozoa. *Appl. Environ. Microbiol.*, **47**, 101–10.

Friedman, M. & Broderick, G. A. (1977). Protected proteins in ruminant nutrition. *In vitro* evaluation of casein derivatives. In *Protein Crosslinking: Nutritional and Medical Consequences* (Advances in Experimental Medicine and Biology, Vol. 86B), ed. M. Friedman. Plenum Press, New York, pp. 545–58.

Fulghum, R. S. & Moore, W. E. C. (1963). Isolation, enumeration, and the characteristics of proteolytic ruminal bacteria. *J. Bacteriol.*, **85**, 808–15.

Ganev, G., Ørskov, E. R. & Smart, R. (1979). The effect of roughage or concentrate feeding and rumen retention time on total degradation of protein in the rumen. *J. Agric. Sci.* (Camb.), **93**, 651–6.

Gibbons, R. J. & Doetsch, R. N. (1959). Physiological study of an obligately anaerobic ureolytic bacterium. *J. Bacteriol.*, **77**, 417–28.

Gibbons, R. J. & McCarthy, R. D. (1957). Obligately anaerobic urea-hydrolyzing bacteria in the bovine rumen. *Univ. Maryland Agric. Exp. Stn Misc. Publ.*, **291**, 12–16.

Griffith, C. J. & Carlsson, J. (1974). Mechanism of ammonia assimilation in *Streptococci*. *J. Gen. Microbiol.*, **82**, 253–60.

Hammond, A C., Carlson, J. R. & Breeze, R. G. (1978). Monensin and the prevention of tryptophan-induced acute bovine pulmonary edema and emphysema. *Science*, **201**, 153–5.

Harmeyer, J. (1971). Amino acid metabolism of isolated rumen protozoa (*Isotricha prostoma* and *Isotricha intestinalis*). 1. Catabolism of amino acids. *Z. Tierphysiol. Tierernähr. Futtermittelkd*, **28**, 65–75.

Harwood, C. S. & Canale-Parola, E. (1981). Adenosine 5′-triphosphate-yielding pathways of branched-chain amino acid fermentation by marine spirochete. *J. Bacteriol.*, **148**, 117–23.

Hausinger, R. P. (1986). Purification of nickel-containing urease from the rumen anaerobe *Selenomonas ruminantium*. *J. Biol. Chem.*, **261**, 7866–70.

Hazlewood, G. P. & Edwards, R. (1981). Proteolytic activities of a rumen bacterium, *Bacteroides ruminicola* R8/4. *J. Gen. Microbiol.*, **125**, 11–15.

Hazlewood, G. P. & Nugent, J. H. A. (1978). Leaf Fraction I protein as a nitrogen source for the growth of a proteolytic rumen bacterium. *J. Gen. Microbiol.*, **106**, 369–71.

Hazlewood, G. P., Jones, G. A. & Mangan, J. L. (1981). Hydrolysis of leaf fraction 1 protein by the proteolytic rumen bacterium *Bacteroides ruminicola* R8/4. *J. Gen. Microbiol.*, **123**, 223–32.

Hazlewood, G. P., Orpin, C. G., Greenwood, Y. & Black, M. E. (1983). Isolation of proteolytic rumen bacteria by use of selective medium containing leaf Fraction I protein (riboluse bisphosphate carboxylase). *Appl. Environ. Microbiol.*, **45**, 1780–4.

Heald, P. J. & Oxford, A. E. (1953). Fermentation of soluble sugars by anaerobic holotrich ciliate protozoa of the genera *Isotricha* and *Dasytricha*. *Biochem. J.*, **53**, 506–12.

Henderickx, H. K. (1976). Quantitative aspects of the use of non-protein nitrogen in ruminant feeding. *Cuban J. Agric. Sci.*, **10**, 1–18.

Henderickx, H. & Martin, J. (1963). *In vitro* study of the nitrogen metabolism in the rumen. *Compt. Rend. de Recherches, IRSIA, IWONL*, **31**, 9–66.

Henderson, C., Hobson, P. N. & Summers, R. (1969). The production of amylase, protease and lipolytic enzymes by two species of anaerobic rumen bacteria. In *Continuous Culture of Microorganisms* (Proceedings of 4th Symposium). Academia, Prague, pp. 189–204.

Hespell, R. B. (1984). Influence of ammonia assimilation pathways and survival strategy on ruminal microbial growth. In *Herbivore Nutrition in the Subtropics and Tropics*, ed. F. M. C. Gilchrist & R. I. Mackie. Science Press, South Africa, pp. 346–58.

Hino, T. & Russell, J. B. (1985). Effect of reducing-equivalent disposal and NADH/NAD on deamination of amino acids by intact rumen microorganisms and their cell extracts. *Appl. Environ. Microbiol.*, **50**, 1368–74.

Hino, T. & Russell, J. B. (1987). Relative contributions of ruminal bacteria and protozoa to the degradation of protein *in vitro*. *J. Anim. Sci.*, **64**, 261–70.

Hobson, P. N. & Wallace, R. J. (1982). Microbial ecology and activities in the rumen: part II. *CRC Crit. Rev. Microbiol.*, **9**, 253–320.

Hobson, P. N., McDougall, E. I. & Summers, R. (1968). The nitrogen sources of *Bacteroides amylophilus*. *J. Gen. Microbiol.*, **50**, i.

Hobson, P. N., Summers, R., Postgate, J. R. & Ware, D. A. (1973). Nitrogen fixation in the rumen of a living sheep. *J. Gen. Microbiol.*, **77**, 225–6.

Holtenius, P. (1957). Nitrite poisoning in sheep, with special reference to the detoxification of nitrite in the rumen. *Acta Agric. Scand.*, **7**, 113–63.

Hoshino, S., Skatsuhara, K. & Morimotu, K. (1966). Ammonia metabolism in ruminants. *J. Dairy Sci.*, **49**, 1523–8.

Hullah, W. A. & Blackburn, T. H. (1971). Uptake and incorporation of amino acids and peptides by *Bacteroides amylophilus*. *Appl. Microbiol.*, **21**, 187–91.

Hungate, R. E. (1966). *The Rumen and its Microbes*. Academic Press, London.

Itabashi, H. & Kandatsu, M. (1978). Formation of methylamine by rumen microorganisms. *Jap. J. Zootech. Sci.*, **49**, 110–18.

Jenkinson, H. F., Buttery, P. J. & Lewis, D. (1979). Assimilation of ammonia by *Bacteroides amylophilus* in chemostat cultures. *J. Gen. Microbiol.*, **113**, 305–13.

John, A., Isaacson, H. R. & Bryant, M. P. (1974). Isolation and characteristics of a ureolytic strain of *Selenomonas ruminantium*. *J. Dairy Sci.*, **57**, 1003–14.

Jones, G. A. (1967). Ureolytic rumen bacteria. In *Urea as a Protein Supplement*, ed. M. H. Briggs. Pergamon, London, pp. 111–24.

Jones, G. A. (1968). Influence of acetohydroxamic acid on some activities *in vitro* of the mixed rumen biota. *Can. J. Microbiol.*, **14**, 409–16.

Jones, G. A. (1972). Dissimilatory metabolism of nitrate by the rumen microbiota. *Can. J. Microbiol.*, **18**, 1783–7.

Jones, G. A., MacLeod, R. A. & Blackwood, A. C. (1964). Ureolytic rumen bacteria. I. Characteristics of the microflora from a urea-fed sheep. *Can. J. Microbiol.*, **10**, 371–8.

Jones, K. & Thomas, J. G. (1974). Nitrogen fixation by rumen contents of sheep. *J. Gen. Microbiol.*, **85**, 97–101.

Joyner, A. E., Jr. & Baldwin, R. L. (1966). Enzymatic studies of pure cultures of rumen microorganisms. *J. Bacteriol.*, **92**, 1321–30.

Kaufmann, W. & Lupping, W. (1982). Protected proteins and protected amino acids for ruminants. In *Protein Contribution of Feedstuffs for Ruminants*, ed. E. L. Miller, I. H. Pike & A. J. H. Van Es. Butterworths, London, pp. 36–75.

Kennedy, P. M. & Milligan, L. P. (1980). The degradation and utilization of endogenous urea in the gastrointestinal tract of ruminants: a review. *Can. J. Anim. Sci.*, **60**, 205–21.

Kleiner, D. & Fitzke, E. (1979). Evidence for ammonia translocation by *Clostridium pasteurianum*. *Biochem. Biophys. Res. Comm.*, **86**, 211–17.

Kopecny, J. & Wallace, R. J. (1982). Cellular location and some properties of proteolytic enzymes of rumen bacteria. *Appl. Environ. Microbiol.*, **43**, 1026–33.

Leibholz, J. (1969). Effect of diet on the concentration of free amino acids, ammonia and urea in the rumen liquor and blood plasma of the sheep. *J. Anim. Sci.*, **29**, 628–33.

Lenartova, V., Holovska, K., Havassy, I., Javorsky, P. & Rybosova, E. (1985). Ammonia-utilizing enzymes of adherent bacteria in the sheep's rumen. *Physiol. Bohemoslov.*, **34**, 512–17.

Lesk, E. M. & Blackburn, T. H. (1971). Purification of *Bacteroides amylophilus* protease. *J. Bacteriol.*, **106**, 394–402.

Lewis, D. (1951). The metabolism of nitrate and nitrite in the sheep. *Biochem. J.*, **48**, 175–80.

Lewis, T. R. & Emery, R. S. (1962). Relative deamination rates of amino acids by rumen microorganisms. *J. Dairy Sci.*, **45**, 765–8.

Lowe, S. E., Theodorou, M. K., Trinci, A. P. J. & Hespell, R. B. (1985). Growth of anaerobic rumen fungi on defined and semi-defined media lacking rumen fluid. *J. Gen. Microbiol.*, **131**, 2225–9.

Mah, R. A. & Hungate, R. E. (1965). Physiological studies of the rumen ciliate *Ophryoscolex purkynei* Stein. *J. Protozool.*, **12**, 131–6.

Mahadevan, S., Sauer, F. & Erfle, J. D. (1976). Studies on bovine rumen bacterial urease. *J. Anim. Sci.*, **42**, 745–53.

Mahadevan, S., Erfle, J. D. & Sauer, F. D. (1980). Degradation of soluble and insoluble proteins by *Bacteroides amylophilus* protease and by rumen microorganisms. *J. Anim. Sci.*, **50**, 723–8.

Makkar, H. P. S., Sharma, O. P., Dawra, R. K. & Negi, S. S. (1981). Effect of acetohydroxamic acid on rumen urease activity *in vitro*. *J. Dairy Sci.*, **64**, 643–8.

Mangan, J. L. & West, J. (1977). Ruminal digestion of chloroplasts and the protection of protein by glutaraldehyde treatment. *J. Agric. Sci.* (Camb.), **89**, 3–15.

Mathison, G. W. & Milligan, L. P. (1971). Nitrogen metabolism in sheep. *Br. J. Nutr.*, **25**, 351–66.

McAllan, A. B. & Smith, R. H. (1973a). Degradation of nucleic acids in the rumen. *Br. J. Nutr.*, **29**, 331–45.

McAllan, A. B. & Smith, R. H. (1973b). Degradation of nucleic acid derivatives by rumen bacteria *in vitro*. *Br. J. Nutr.*, **29**, 467–74.

Mehrez, A. Z., Ørskov, E. R. & McDonald, I. (1977). Rates of rumen fermentation in relation to ammonia concentration. *Br. J. Nutr.*, **38**, 437–43.

Milligan, L. P. (1970). Carbon dioxide fixing pathways of glutamic acid synthesis in the rumen. *Can. J. Biochem.*, **48**, 463–8.

Moisio, R., Kreula, M. & Virtanen, A. E. (1969). Experiments on nitrogen fixation in cow's rumen. *Suomen Kemistilehti*, **B42**, 432–3.

Naga, M. A. & el-Shazly, K. (1968). The metabolic characterization of the ciliate protozoon *Eudiplodinium medium* from the rumen of buffalo. *J. Gen. Microbiol.*, **53**, 305–15.

Neill, A. R., Grime, D. W. & Dawson, R. M. C. (1978). Conversion of choline methyl groups through trimethylamine to methane in the rumen. *Biochem. J.*, **170**, 529–35.

Nikolic, J. A. & Filipovic, R. (1981). Degradation of maize protein in rumen contents. Influence of ammonia concentration. *Br. J. Nutr.*, **45**, 111–16.

Nikolic, J. A., Pavlicevic, A., Zeremski, D. & Negovanovic, D. (1980). Adaptation to diets containing significant amounts of non-protein nitrogen. In *Digestive Physiology and Metabolism in Ruminants*, ed. Y. Ruckebusch & P. Thivend. MTP Press, Lancaster, pp. 603–20.

Nolan, J. V. (1975). Quantitative models of nitrogen metabolism in sheep. In *Digestion and Metabolism in the Ruminant*, ed. I. W. McDonald & A. C. I. Warner. University of New England Publishing Unit, Armidale, Australia, pp. 416–31.

Nolan, J. V., Norten, B. W. & Leng, R. A. (1976). Further studies on the dynamics of nitrogen metabolism in sheep. *Br. J. Nutr.*, **35**, 127–47.

Nugent, J. H. A. & Mangan, J. L. (1981). Characteristics of the rumen proteolysis of fraction I (18S) leaf protein from lucerne (*Medicago sativa* L.). *Br. J. Nutr.*, **46**, 39–58.

Nugent, J. H. A., Jones, W. T., Jordan, D. J. & Mangan, J. L. (1983). Rates of proteolysis in the rumen of the soluble proteins casein, Fraction I (18S) leaf protein, bovine serum albumin and bovine submaxillary mucoprotein. *Br. J. Nutr.*, **50**, 357–68.

Odle, J., Schaefer, D. M. & Costerton, J. W. (1985). Influence of rumen ammonia concentration on *in situ* fractional degradation rates of barley and corn. In *Report on 18th Conference on Rumen Function*, Chicago, p. 32.

Onodera, R. & Kandatsu, M. (1969). Occurrence of L-(−)-pipecolic acid in the culture medium of rumen ciliate protozoa. *Agric. Biol. Chem.*, **33**, 113–15.

Onodera, R. & Kandatsu, M. (1970). Amino acid and protein metabolism of rumen ciliate protozoa. *Jap. J. Zootech. Sci.*, **41**, 307–13.

Onodera, R. & Kandatsu, M. (1972). Conversion of lysine to pipecolic acid by rumen ciliate protozoa. *Agric. Biol. Chem.*, **36**, 1989–95.

Onodera, R. & Migita, R. (1985). Metabolism of threonine, methionine, and related compounds in mixed rumen ciliate protozoa. *J. Protozool.*, **32**, 326–30.

Onodera, R. & Takei, K. (1986). Methionine sulfoxide in the incubation medium of mixed rumen ciliate protozoa. *Agric. Biol. Chem.*, **50**, 767–9.

Onodera, R., Yamaguchi, Y. & Morimoto, S. (1983). Metabolism of arginine, citrulline, ornithine and proline by starved rumen ciliate protozoa. *Agric. Biol. Chem.*, **47**, 821–8.

Ørskov, E. R. (1982). *Protein Nutrition in Ruminants*. Academic Press, London.

Ørskov, E. R., Mills, C. F. & Robinson, J. J. (1980). The use of whole blood for the protection of organic materials from degradation in the rumen. *Proc. Nutr. Soc.*, **39**, 60A.

Palmquist, D. L. & Baldwin, R. L. (1966). Enzymic techniques for the study of pathways of carbohydrate utilization in the rumen. *Appl. Microbiol.*, **14**, 60–9.

Patterson, J. A. & Hespell, R. B. (1979). Trimethylamine and methylamine as growth substrates for rumen bacteria and *Methanosarcina barkeri*. *Curr. Microbiol.*, **3**, 79–83.

Patterson, J. A. & Hespell, R. B. (1985). Glutamine synthetase activity in the ruminal bacterium *Succinivibrio dextrinosolvens*. *Appl. Environ. Microbiol.*, **50**, 1014–20.

Payne, J. W. (1983). Peptide transport in bacteria: methods, mutants and energy coupling. *Biochem. Soc. Trans.*, **11**, 794–8.

Pettipher, G. L. & Latham, M. J. (1979). Production of enzymes degrading plant cell walls and fermentation of cellobiose by *Ruminococcus flavefaciens* in batch and continuous culture. *J. Gen. Microbiol.*, **110**, 29–38.

Pilgrim, A. F., Gray, F. V., Weller, R. A. & Belling, G. B. (1970). Synthesis of microbial protein from ammonia in the sheep's rumen and the proportion of dietary nitrogen converted into microbial N. *Br. J. Nutr.*, **24**, 589–98.

Pittman, K. A. & Bryant, M. P. (1964). Peptides and other nitrogen sources for growth of *Bacteroides ruminicola*. *J. Bacteriol.*, **88**, 401–10.

Pittman, K. A., Lakshmanan, S. & Bryant, M. P. (1967). Oligopeptide uptake by *Bacteroides ruminicola*. *J. Bacteriol.*, **93**, 1499–1508.

Prins, R. A. (1977). Biochemical activities of gut microorganisms. In *Microbial Ecology of the Gut*, ed. R. T. J. Clarke & T. Bauchop. Academic Press, London, pp. 73–183.

Prins, R. A., Van Hal-Van Gestel, J. C. & Counotte, G. H. M. (1979). Degradation of amino acids and peptides by mixed rumen microorganisms. *Z. Tierphysiol. Tierernähr. Futtermittelkde*, **42**, 333–9.

Prins, R. A., van Rheenen, D. L. & van't Klooster, A. T. (1983). Characterization of microbial proteolytic enzymes in the rumen. *Ant. van Leeuwen.*, **49**, 585–95.

Roffler, R. E. & Satter, L. D. (1975). Relationship between ruminal ammonia and nonprotein nitrogen utilization by ruminants. II. Application of published evidence to the development of a theoretical model for predicting nonprotein nitrogen utilization. *J. Dairy Sci.*, **58**, 1889–98.

Russell, J. B. (1983). Fermentation of peptides by *Bacteroides ruminicola* B_14. *Appl. Environ. Microbiol.*, **45**, 1566–74.

Russell, J. B. & Baldwin, R. L. (1979). Comparison of maintenance energy expenditures and growth yields among several rumen bacteria grown in continuous culture. *Appl. Environ. Microbiol.*, **37**, 537–43.

Russell, J. B. & Jeraci, J. L. (1984). Effect of carbon monoxide on fermentation of fiber, starch and amino acids by mixed rumen microorganisms *in vitro*. *Appl. Environ. Microbiol.*, **48**, 211–17.

Russell, J. B. & Martin, S. A. (1984). Effects of various methane inhibitors on the fermentation of amino acids by mixed rumen microorganisms *in vitro*. *J. Anim. Sci.*, **59**, 1329–38.

Russell, J. B. & Robinson, P. H. (1984). Composition and characteristics of strains of *Streptococcus bovis*. *J. Dairy Sci.*, **67**, 1525–31.

Russell, J. B., Bottje, W. G. & Cotta, M. A. (1981). Degradation of protein by mixed cultures of rumen bacteria: identification of *Streptococcus bovis* as an actively proteolytic rumen bacterium. *J. Anim. Sci.*, **53**, 242–52.

Russell, J. B., Sniffen, C. J. & Van Soest, P. J. (1983). Effect of carbohydrate limitation on degradation and utilization of casein by mixed rumen bacteria. *J. Dairy Sci.*, **66**, 763–75.

Sauer, F. D., Erfle, J. D. & Mahadevan, S. (1975). Amino acid biosynthesis in mixed rumen cultures. *Biochem. J.*, **150**, 357–72.

Schaefer, D. M., Davis, C. L. & Bryant, M. P. (1980). Ammonia saturation constants for predominant species of rumen bacteria. *J. Dairy Sci.*, **63**, 1248–63.

Scheifinger, C., Russell, N. & Chalupa, W. (1976). Degradation of amino acids by pure cultures of rumen bacteria. *J. Anim. Sci.*, **43**, 821–7.

Schwartz, H. M. (1967). The rumen metabolism of non-protein nitrogen. In *Urea as a Protein Supplement*, ed. M. H. Briggs. Pergamon, London, pp. 95–109.

Shimbayashi, K., Obara, Y. & Yonemura, T. (1975). Changes of free amino acids during rumen fermentation and incorporation of urea ^{15}N into microorganisms *in vitro*. *Jap. J. Zootech. Sci.*, **46**, 243–50.

Siddons, R. C. & Paradine, J. (1981). Effect of diet on protein degrading activity in the sheep rumen. *J. Sci. Food Agric.*, **32**, 973–81.

Slyter, L. L., Oltjen, R. R., Kern, D. L. & Weaver, J. M. (1968). Microbial species including ureolytic bacteria from the rumen of cattle fed purified diets. *J. Nutr.*, **94**, 185–92.

Slyter, L. L., Satter, L. D. & Dinius, D. A. (1979). Effect of ruminal ammonia concentration on nitrogen utilization by steers. *J. Anim. Sci.*, **48**, 906–12.

Smith, C. J. & Bryant, M. P. (1979). Introduction to metabolic activities of intestinal bacteria. *Am. J. Clin. Nutr.*, **32**, 149–57.

Smith, C. J., Hespell, R. B. & Bryant, M. P. Ammonia assimilation and glutamate formation in the anaerobe *Selenomonas ruminantium*. *J. Bacteriol.*, **141**, 593–602.

Smith, C. J., Hespell, R. B. & Bryant, M. P. (1981). Regulation of urease and ammonia assimilatory enzymes in *Selenomonas ruminantium*. *Appl. Environ. Microbiol.*, **42**, 89–96.

Smith, R. H. & McAllan, A. B. (1970). Nucleic acid metabolism in the ruminant. 2. Formation of microbial nucleic acids in the rumen in relation to the digestion of food nitrogen, and the fate of dietary nucleic acids. *Br. J. Nutr.*, **24**, 545–56.

Sniffen, C. J., Chen, G. & Russell, J. B. (1985). Concentration and flow of peptides from the rumen of cattle fed different amounts and types of protein. In *Report on XVIII Conference on Rumen Function*, Chicago, p. 28.

Somerville, H. J. (1968). Enzymatic studies on the biosynthesis of amino acids from lactate by *Peptostreptococcus elsdenii*. *Biochem. J.*, **108**, 107–19.

Somerville, H. J. & Peel, J. L. (1967). Tracer studies on the biosynthesis of amino acids from lactate by *Peptostreptococcus elsdenii*. *Biochem. J.*, **105**, 229–310.

Spears, J. W. & Hatfield, E. E. (1978). Nickel for ruminants. I. Influence of dietary nickel on ruminal urease activity. *J. Anim. Sci.*, **47**, 1345–50.

Spears, J. W., Smith, C. J. & Hatfield, E. E. (1977). Rumen bacterial urease requirement for nickel. *J. Dairy Sci.*, **60**, 1073–6.

Stevenson, R. M. W. (1979). Amino acid uptake systems in *Bacteroides ruminicola*. *Can. J. Microbiol.*, **25**, 1161–8.

Stevenson, R. & Silver, S. (1977). Methylammonium uptake by *Escherichia coli*: evidence for a bacterial NH_4^+ transport system. *Biochem. Biophys. Res. Comm.*, **75**, 1133–9.

Strydom, E., Mackie, R. I. & Woods, D. R. (1986). Detection and characterization of extracellular proteases in *Butyrivibrio fibrisolvens* H17c. *Appl. Microbiol. Biotechnol.*, **24**, 214–17.

Umbarger, H. E. (1978). Amino acid biosynthesis and its regulation. *Ann. Rev. Biochem.*, **47**, 533–606.

Van den Hende, C., Oyaert, W. & Bouchaert, J. H. (1963). Metabolism of glycine, alanine, valine, leucine and isoleucine by rumen bacteria. *Res. Vet. Sci.*, **4**, 382–9,.

Virtanen, A. I. (1966). Milk production of cows on protein-free feed. *Science*, **153**, 1603–14.

Wakita, M. & Hoshino, S. (1975). A branched chain amino acid aminotransferase from the rumen ciliate genus *Entodinium*. *J. Protozool.*, **22**, 281–5.

Wallace, R. J. (1979). Effect of ammonia concentration on the composition, hydrolytic activity and nitrogen metabolism of the microbial flora of the rumen. *J. Appl. Bacteriol.*, **47**, 443–55.

Wallace, R. J. (1983). Hydrolysis of ^{14}C-labelled proteins by rumen microorganisms and by proteolytic enzymes prepared from rumen bacteria. *Br. J. Nutr.*, **50**, 345–55.

Wallace, R. J. (1984). A comparison of the ureolytic and proteolytic activities of rumen bacteria from lambs fed conventionally and by intragastric infusion. *Can. J. Anim. Sci.*, **64** (Suppl.), 140–1.

Wallace, R. J. (1985a). Adsorption of soluble proteins to rumen bacteria and the role of adsorption in proteolysis. *Br. J. Nutr.*, **53**, 399–408.

Wallace, R. J. (1985b). Proteolytic activity of large particulate material in the rumen. In *Abstracts of XIII International Congress of Nutrition*, Brighton, UK, p. 10.

Wallace, R. J. (1985c). Synergism between different species of proteolytic rumen bacteria. *Curr. Microbiol.*, **12**, 59–64.

Wallace, R. J. (1986a). Catabolism of amino acids by *Megasphaera elsdenii* LC1. *Appl. Environ. Microbiol.*, **51**, 1141–3.

Wallace, R. J. (1986b). Rumen microbial metabolism and its manipulation. In *Proceedings of XIII International Congress of Nutrition*, ed. T. G. Taylor & N. K. Jenkins. John Libbey, London, pp. 215–20.

Wallace, R. J. & Brammall, M. L. (1985). The role of different species of bacteria in the hydrolysis of protein in the rumen. *J. Gen. Microbiol.*, **131**, 821–32.

Wallace, R. J. & Joblin, K. N. (1985). Proteolytic activity of rumen anaerobic fungus. *FEMS Microbiol. Lett.*, **29**, 19–25.

Wallace, R. J. & Kopecny, J. (1983). Breakdown of diazotized proteins and synthetic substrates by rumen bacterial proteases. *Appl. Environ. Microbiol.*, **45**, 212–17.

Wallace, R. J. & McPherson, C. A. (1987). Factors affecting the breakdown of bacterial protein in rumen fluid. *Br. J. Nutr.*, **58**, 313–23.

Wallace, R. J. & Munro, C. A. (1986). Influence of the rumen anaerobic fungus *Neocallimastix frontalis* on the proteolytic activity of a defined mixture of rumen bacteria growing on a solid substrate. *Lett. Appl. Microbiol.*, **3**, 23–6.

Wallace, R. J., Cheng, K.-J., Dinsdale, D. & Ørskov, E. R. (1979). An independent microbial flora of the epithelium and its role in the ecomicrobiology of the rumen. *Nature* (Lond.), **279**, 424–6.

Wallace, R. J., Broderick, G. A. & Brammall, M. L. (1987a). Protein degradation by ruminal microorganisms from sheep fed dietary supplements of urea, casein or albumin. *Appl. Environ. Microbiol.*, **53**, 751–3.

Wallace, R. J., Broderick, G. A. & Brammall, M. L. (1987b). Microbial protein and peptide metabolism in rumen fluid from faunated and ciliate-free sheep. *Br. J. Nutr.*, **58**, 87–93.

Wallis, O. C. & Coleman, G. S. (1967). Incorporation of ^{14}C-labelled components of *Escherichia coli* and of amino acids by *Isotricha intestinalis* and *Isotricha prostoma* from the sheep rumen. *J. Gen. Microbiol.*, **49**, 315–23.

Williams, A. G. (1986). Rumen holotrich ciliate protozoa. *Microbiol. Rev.*, **50**, 25–49.

Williams, P. P., Davis, R. E., Doetsch, R. N. & Gutierrez, J. (1961). Physiological studies of the rumen ciliate *Ophryoscolex caudatus* Eberlein. *Appl. Microbiol.*, **9**, 405–9.

Woolfolk, C. A., Shapiro, B. & Stadtman, E. R. (1966). Regulation of glutamine synthetase. 1. Purification and properties of glutamine synthetase from *Escherichia coli*. *Arch. Biochem. Biophys.*, **116**, 177–92.

Wozny, M. A., Bryant, M. P., Holdeman, L. V. & Moore, W. E. C. (1977). Urease assay and urease-producing species of anaerobes in the bovine rumen and human feces. *Appl. Environ. Microbiol.*, **33**, 1097–104.

Wright, D. E. (1967). Metabolism of peptides by rumen microorganisms. *Appl. Microbiol.*, **15**, 547–50.

Wright, D. E. & Hungate, R. E. (1967). Amino acid concentrations in rumen fluid. *Appl. Microbiol.*, **15**, 148–51.

Yokoyama, M. T. & Carlson, J. R. (1981). Production of skatole and *para*-cresol by a rumen *Lactobacillus* sp. *Appl. Environ. Microbiol.*, **41**, 71–6.

Yokoyama, M. T., Carlson, J. R. & Holdeman, L. V. (1977). Isolation and characteristics of a skatole-producing *Lactobacillus* sp. from the bovine rumen. *Appl. Environ. Microbiol.*, **34**, 837–42.

8
Polysaccharide Degradation by Rumen Microorganisms

A. Chesson
Rowett Research Institute, Aberdeen, UK

&

C. W. Forsberg
University of Guelph, Ontario, Canada

Most polysaccharides entering the rumen can be considered as belonging to one of two general types: plant storage polysaccharides such as starch and the fructosans, or the structural polysaccharides which compose the greater part of all plant cell walls and which are loosely considered to form the fibrous component of animal feedstuffs. Additional, but limited, amounts of polymeric carbohydrate may also be ingested by the animal in the form of lower molecular weight oligosaccharides, as the sugar moiety of various glyco-conjugates and as a component of animal and fish by-products.

Storage polysaccharides function as food reserves and as such must be readily mobilised when required by the plant. As a consequence they are easily degraded by plant hydrolytic activities and, similarly, are susceptible to attack by enzymes secreted by rumen microorganisms. Any limitation to starch degradation usually results from extraneous factors, such as the presence of intact cell walls restricting enzyme access, or the presence of plant tannins, and only more rarely from the structure and organisation of starch itself. In distinct contrast, the structural polysaccharides have a skeletal function in the living plant and are, by their very nature and organisation within the cell wall, far more resistant to microbial attack. However, it is the ability to utilise such materials as an energy source that provides ruminants with their particular ecological niche, and the ability of rumen microorganisms to degrade plant polysaccharides efficiently is of paramount importance once the animal no longer depends on a milk diet.

STRUCTURE AND DIGESTION OF PLANT STORAGE POLYSACCHARIDES

Starch structure

Although starch is generally thought of as a single polysaccharide, it is in fact a composite of two structurally distinct α-linked polymers of glucose: amylose and amylopectin (Whistler & Daniel, 1984). Amylose is the simpler of the two, consisting of an essentially linear chain of α-1–4 linked sugar residues with a chain length of several hundred units. In solution, amylose assumes a helical structure with approximately six glucose residues forming each turn. It is this helical structure and its ability to form inclusion complexes which accounts for the well known colour reaction formed between starch and iodine. Amyloses are degraded by randomly acting α-amylases (1,4-α-D-glucan glucanohydrolase, EC 3.2.1.1) which bring about a rapid reduction in viscosity and the release of low molecular weight oligosaccharides (maltodextrins) and by β-amylases (1,4-α-D-glucan maltohydrolase, EC 3.2.1.1) which remove successive maltose units from the non-reducing end of the chain. It was the action of β-amylase which first suggested that the amylose chain was not as homogeneous as it first appeared. Treatment with this enzyme produced only 70% degradation (instead of the 100% expected) indicating that there were chemically distinct regions of the chain which formed a barrier to further attack by the enzyme. Subsequent work has demonstrated the presence of very occasional side-chains linked α-1–6 to in-chain glucose units (Hizukuri et al., 1981); the presence of such branch points inhibits the action of the β-amylase.

Amylopectin is similarly composed of α-1–4 linked glucose residues but, unlike amylose, it is a considerably larger molecule with a far more extensively branched structure. On average, one in every 20–25 glucose units carries a glucan side-chain linked through position 6. Various model structures have been proposed which attempt to satisfy the observed physical and chemical properties of amylopectin. Virtually all models envisage the presence of three distinct types of 1–4 linked glucan chains:

(a) A chains: linked to the rest of the molecule by the reducing group only
(b) B chains: similarly linked through the terminal reducing group to the rest of the molecule, but also bearing side-chains linked to the 6 position of one or more residues
(c) C chains: as B above, but carrying a free reducing end

However, the models differ with respect to the arrangement proposed for the three chain types. To some extent these difficulties have been resolved by a redetermination of the ratio of A to B chains in amylopectin from waxy maize (Manners & Matheson, 1981) which gave values of 1:1, consistent with the currently favoured 'cluster model' first proposed by French (1973, 1984) and which is shown in Fig. 1.

Approximately 50% of amylopectin can be degraded to maltose by the action of β-amylase, leaving a residue known as a 'β-limit dextrin' which is protected from further attack by the 6-linked branch points (Fig. 1). Complete hydrolysis of amylopectin (and amylose) requires the action of enzymes capable of cleaving α-1–6 linkages, notably glucoamylase (1,4-α-D-glucan glucohydrolase, EC 3.2.1.3) and a series of enzymes variously known as pullulanases, dextrinases and 'debranching enzymes', but now described as α-dextrin 6-glucanohydrolase (EC 3.2.1.41) and isoamylase (glycogen 6-glucohydrolase, EC 3.2.1.68). Maltose and low molecular weight maltodextrins are degraded to glucose by α-glucosidase (α-D-glucoside glucohydrolase, EC 3.2.1.20).

Amylose and amylopectin are packaged in the plant in the form of discrete granules which may vary in size from the short-lived particles of less than 1 μm found within chloroplasts (leaf starch), to the far larger granules found in stems, seeds, roots and tubers and used by the plant for longer-term storage (French, 1984;

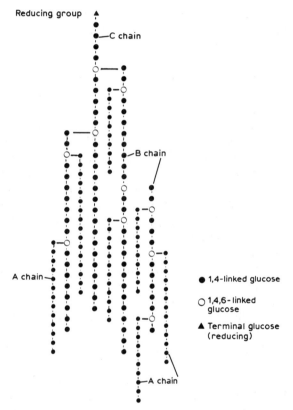

Fig. 1. The amylopectin 'cluster' model; a small section only of the whole amylopectin molecule is shown.

Fitt & Snyder, 1984). The amylose content of native starch is usually in the range 25–27% of dry matter, although starches exist with a far higher (high-amylose maize, 40–80% amylose) and lower ('waxy' cereal starches) amylose content. Granules often have an external morphology characteristic of the plant or plant part from which they are derived, and a well defined, semi-crystalline, internal structure. The very low level of hydration associated with native starch ensures its insolubility, although this insolubility does not significantly affect the ability of plants or microorganisms to metabolise starch.

Starch metabolism

The ability to utilise starch as a carbon source is widespread amongst strains of the bacteria, protozoa and fungi which form the rumen microbial population (see also Chapters 2–4). Principal amongst the amylolytic rumen organisms are the bacteria *Bacteroides amylophilus*, *Streptococcus bovis*, *Succinimonas amylolytica*, many strains of *Selenomonas ruminantium*, *Butyrivibrio fibrisolvens* and *Eubacterium ruminantium*, and *Clostridium* spp. (Russell & Hespell, 1981; Marounek & Bartos, 1986), virtually all of the larger entodiniomorph protozoa, and the chytrid fungus *Neocallimastix frontalis* (Orpin & Letcher, 1979; Pearce & Bauchop, 1985). As it has proved possible to establish a defined bacterial flora in gnotobiotic lambs fed starchy diets which supports growth and normal rumen function (Hobson et al., 1981; see Chapter 15), it is probable that the major bacteria involved in starch degradation have now been identified. It is also evident that amylolytic protozoa and fungi and are not essential elements in starch utilisation. However, it has been suggested that engulfing of starch granules by protozoa limits the amount of starch available for the rapid bacterial fermentation and so helps to prevent a detrimental lowering of rumen pH (Mackie et al., 1978; see also Chapters 3 and 14).

Despite the importance of starch in production diets for ruminants, the mechanism of starch degradation by rumen bacteria has been little studied. Amylases from three strains of rumen bacteria—*S. bovis* (Walker, 1965), *Clostridium butyricum* (Walker & Hope, 1964) and *B. amylophilus* (McWethy & Hartman, 1977)—have been isolated and characterised. In each case the enzymes proved to be α-amylases with properties similar to the α-amylases isolated from mammalian or other microbial sources. However, the enzymes examined were extracellular and represented only a fraction of the total amylolytic activity of the cells. As might be expected of essentially aquatic microorganisms, much of the amylolytic activity produced was cell-bound and this enzyme fraction has not been studied extensively. α-D-Glucosidase activity appears ubiquitous amongst rumen microorganisms and was present in all the bacterial and protozoal strains examined by Williams and his colleagues (Williams et al., 1984).

The ability of the larger rumen ciliates to engulf and subsequently to metabolise starch granules is well known (see Chapter 3). Amongst the holotrichs this ability is limited to *Isotricha* (Prins & Van Hoven, 1977), but amongst the entodiniomorphs it is virtually universal. Thus the major sources of carbon for energy and growth for all Entodinia and Diplodinia are starch grains and soluble sugars. The major route of starch fermentation involves hydrolysis of starch to glucose, often with a partial release of glucose and maltose into the medium, followed by phosphorylation of glucose to glucose-6-phosphate (Coleman & Laurie, 1976). Amylase activity has been demonstrated in most protozoa examined (Coleman, 1980). However, intracellular amylase activity cannot be taken as evidence of the ability to utilise starch. Virtually all protozoan species synthesise amylopectin as a storage polysaccharide, and the same activities would be involved in its mobilisation as in feed-starch utilisation. Starch phosphorylase (1,4-α-D-glucan: orthophosphate α-D-glucosyltransferase, EC 2.4.1.1), although often detected in protozoa and bacteria, does not seem to contribute significantly to starch catabolism. An exception to this general rule is *Polyplastron multivesiculatum*, in which ingested starch undergoes phosphorolysis to glucose-1-phosphate which is, in turn, isomerised to glucose-6-phosphate (Coleman & Laurie, 1977).

Fructosan structure and metabolism

Polymers of fructose occur in two forms, the inulins which serve as reserve carbohydrates in tubers of the Compositae family and in which the fructofuranosyl units are linked β-2–1, and the levans in which the units are linked β-2–6. The latter are commonly found in the vegetative parts of temperate grasses where they can reach levels of 1–5% of total dry matter (Smith, 1973) and as soluble components of cereal grains (Åman & Hesselman, 1984; Dahlqvist & Nilsson, 1984). In both inulin and levan the polymer chain is terminated by a glucopyranosyl residue linked α-1–2 at the non-reducing end, forming sucrose as the terminal disaccharide (Tomasic *et al.*, 1978).

Fructosans are rapidly and completely fermented in the rumen (Kosaric *et al.*, 1984) by organisms which include the holotrich (Williams, 1986) and entodiniomorph (Coleman, 1980) protozoa. Studies with a holotrich invertase indicated that this enzyme alone was able fully to degrade grass fructosans by sequentially removing a terminal fructose unit (Thomas, 1960). Although inulinase (2,1-β-D-fructan fructanohydrolase, EC 3.2.1.7) and levanase (2,6-β-D-fructan fructanohydrolase, EC 3.2.1.64), enzymes with an endo mode of action against fructosans, have been described, their specific presence in rumen microorganisms has not been demonstrated. Many rumen bacteria are able to utilise inulin (Hungate, 1966) and it is to be assumed that they can metabolise the levans with equal or greater facility.

STRUCTURE OF PLANT CELL WALLS

It is customary for animal nutritionists to refer to the poorly degraded fraction of a plant feedstuff as its 'fibre' content. However, 'fibre', except when used in the special sense of textile fibre, is not a recognisable anatomical or biochemical entity within the plant. Rather it is the insoluble residue prepared from plant material, the composition of which depends upon the reagents and conditions used for its extraction (Chesson, 1986). 'Fibre' analyses were developed as a purely pragmatic response to the need for chemical data with which to predict animal performance. Although 'fibre' can be equated with the plant cell wall to a greater or lesser extent depending on the method of analysis chosen, it has no sound basis in biology. Animals consume intact plant cell walls, not 'fibre', and these and their constituent polysaccharides form the substrates for the rumen microflora.

Polysaccharides of the plant cell wall

Hydrolysis of plant cell walls invariably yields the same 7 neutral and 2 acidic monosaccharides in amounts which account for 70–90% of the total dry matter

Table 1
Water-soluble Carbohydrate (WSC), Starch and Non-starch Polysaccharide (NSP) Content of Seeds, Vegetative Parts and Storage Organs of Monocotyledonous and Dicotyledonous Plant Feeds

Plant part	Sample	WSC	Starch	NSP	% Dry matter Monosaccharide residues[a] contributing to NSP						
					Rha	Fuc	Ara	Xyl	Man	Gal	Glc	Uronic acid[b]
Monocotyledon												
Seed	Barley grain	4·7	51·3	19·7	0	0·2	2·9	6·5	0·2	0·3	9·2	0·4
Vegetation	Ryegrass	9·0	0·9	56·3	0·2	0·5	3·7	14·3	0·3	1·4	33·1	2·8
Dicotyledon												
Seed	Faba bean	3·3	23·3	15·9	0·1	0·2	2·2	1·1	trace	0·6	8·7	2·9
Vegetation	Lucerne	6·6	2·6	42·6	0·5	0·4	2·5	5·8	1·0	1·8	23·1	7·5
Storage organ	Kale	31·3	0·8	35·5	0·6	0·4	4·2	2·8	0·8	2·9	13·3	10·5

[a] Rha, rhamnose; Fuc, fucose; Ara, arabinose; Xyl, xylose; Man, mannose; Gal, galactose; Glc, glucose.
[b] Uronic acid: sum of galacturonic acid and glucuronic acid or its 4-O-methyl derivative.

present. However, the proportion of the different monosaccharides recovered differs considerably, depending on the phylogenic origin of the plant and the plant part examined (Table 1). In particular, there exists a fundamental difference in the structure of cell walls from monocotyledonous (monocot) and dicotyledonous (dicot) plants.

Monosaccharide residues in plants are interconnected by a single type of linkage, the glycosidic bond, to give a series of linear or branched polysaccharides containing one (homopolysaccharides) or more (heteropolysaccharides) types of sugar residue. With only nine monosaccharide precursors and a single bond type, it might be supposed that the potential complexity of polysaccharide structure is limited. In fact there are twenty ways of linking together two identical hexose sugars, depending on the carbon atoms involved and the configuration of the glycosidic bond.

The polysaccharides of varying composition that may be extracted from intact cell walls are considered conventionally to belong to one of three groups: cellulose, hemicellulose and the pectic substances.

Pectic substances

This complex of polysaccharides is based on chains of α-1–4 linked galacturonic acid units in which the carboxylic acid groups are variably esterified with methanol and the uronic acid residues variably substituted at carbon-2 with acetyl groups. Rhamnose units are found throughout the chain, linked 1–2 to adjacent uronic acid residues (rhamnogalacturonan) (Stephen, 1983). The presence of 2-linked units was said to induce kinks in the otherwise linear chain; however, X-ray diffraction studies have failed to support this view (Atkins *et al.*, 1979). Many of the in-chain rhamnose units are also linked through carbon-4 to other polysaccharides composed of galactose and arabinose residues. Several types of galactan/arabinogalactan have been described (Clarke *et al.*, 1979; Aspinall, 1981; Kato & Nevins, 1984). Pectic substances are found in all plant feedstuffs but occur in far higher proportions in dicot cell walls (Table 1).

Hemicellulose

The distinction between the pectic and hemicellulosic polymers is predominantly a historical one which arose from a method commonly employed to fractionate plant cell walls by sequential extraction with reagent of increasing strength. Hemicellulosic polymers are those polymers not extracted by hot water or chelating agents, but which are soluble in alkali. The predominant alkali-soluble polysaccharides of primary (young) monocot cell walls are (4-O-methyl) glucuronoarabinoxylans and lesser amounts of mixed-linked glucan (Wilkie, 1979, 1983). The former are based on β-1–4 linked chains of xylopyranosyl units carrying side-chains of single glucuronic acid units and short branched chains of α-1–3 and 1–5 linked arabinofuranose residues. The latter is a homopolymer of glucose in which approximately 70% of units are linked β-1–4 and the remainder β-1–3. Xyloglucans replace arabinoxylans as the major hemicellulosic polymer of the primary dicot cell wall. Here a β-1–4 linked glucan chain is substituted at intervals through carbon-6 with single xylopyranose units which may be further linked to terminal galactose or arabinose residues (Burke et al., 1974; Stephen, 1983). Substantial amounts of additional hemicellulose are laid down during the secondary thickening of plant cell walls. In both monocots and dicots this predominantly takes the form of glucuronoarabinoxylans, but these characteristically show a lower level of substitution with glucuronic acid and arabinose residues than the comparable polymer formed by the primary monocot wall (Chesson et al., 1985). Glucuronoarabinoxylans from both primary and secondary walls also are invariably substituted with ester-linked non-carbohydrate compounds, notably with acetyl groups and phenolic acid residues.

Cellulose

The unique properties of cellulose are conferred by its secondary, rather than its primary, structure. The linear chains of β-1–4 linked glucose units aggregate to form microscopically visible fibrils in which the individual glucan chains are extensively cross-linked by hydrogen-bonding. The degree of order found within and between fibrils varies from regions in which the glucan chains are held firmly in parallel, and where X-ray diffraction studies indicate a high degree of crystallinity, to regions in which this order is somewhat reduced (amorphous regions). Microfibrils are insoluble, hydrophobic in nature, and show considerably more resistance to chemical or enzymatic attack than the glucan chains from which they are formed (Krassig, 1985). Resistance is directly related to the degree of order within the molecule, with celluloses with a high crystallinity index showing the lowest rates of degradation when incubated in the rumen (Chesson, 1981).

Organisation of the cell wall

The amounts and types of polymers present in examples of monocot and dicot primary cell walls are shown in Table 2. Cellulose is the single major component of both cell wall types and is the only polymer which can be distinguished by microscopic observation. In the primary cell wall, a mesh of apparently randomly orientated cellulose microfibrils can be seen embedded in an amorphous matrix of the remaining cell wall components. Evidence to date suggests that the matrix polysaccharides are all glycosidically linked to form what is, in effect, a single macromolecule. A variety of model structures have been proposed for dicot cell walls, showing how the various polymers are interlinked (Monro et al., 1976), the most widely accepted being that of Albersheim et al. (1973). No such models have been published for the monocot cell wall. Work subsequent to the publication of the original Albersheim model for the structure of the wall of sycamore callus cells has shown that the individual polymers might not be as homogeneous as first thought (Lau et al., 1985) and that forms of bonding other than glycosidic and hydrogen-bonding may be important. It is now recognised, for example, that divalent cations and diferulic acid (Hartley & Jones, 1976; Markwalder & Neukom, 1976) can form bridges between adjacent matrix polysaccharides and may be of considerable importance in stabilising the structure of some cell walls.

The region where adjacent cell walls make contact can be distinguished microscopically from the remaining cell wall. Commonly referred to as the middle

Table 2
Polymer Composition of Primary Monocotyledonous and Dicotyledonous Cell Walls

Wall component	% Cell wall	
	Monocot[a]	Dicot[b]
Rhamnogalacturonan	4	16
Arabinan	—	10
(Arabino)galactan	4	10
Xyloglucan	11	21
Arabinoxylan	21	—
Mixed-linked glucan	3	—
Cellulose	46	23
Protein	7	10

[a] Mesophyll cell wall of perennial ryegrass (Chesson et al., 1985).
[b] Cell walls from sycamore callus culture (Albersheim et al., 1973).

lamella, this section of the wall appears free from cellulose microfibrils but is correspondingly rich in the polymers found in the cell wall matrix. In dicot tissue, pectic substances form a substantial part of the middle lamella, and in consequence primary dicot tissues are susceptible to maceration (cell separation and loss of tissue cohesion) by pectinolytic microorganisms (Chesson, 1980). This susceptibility may have practical consequences, as the rapid maceration of some ingested forage legumes may lead to bloating (Howarth et al., 1978, 1986; see also Chapter 14). Pectic substances are found in far lower amounts in most monocot tissues, particularly Gramineae (Table 2), which are resistant to maceration by pectic enzymes alone (Ishii, 1984).

Cells from the meristematic regions of the plant differentiate to produce either metabolically active cells in which the cell wall is little changed, or cells in which the cell wall undergoes substantial secondary thickening. During the process of secondary thickening, additional cellulose is laid down in layers, with the fibrils in each layer showing a more clearly defined orientation than in the primary wall. Hemicellulosic polymers, notably xylans characterised by a low level of substitution with arabinofuranosyl and glucuronosyl side-chains, are formed simultaneously and laid down with the cellulose. In virtually all cases of secondary thickening, the polyphenolic polymer lignin is also deposited in both primary and secondary layers, leading to the death of the cell and the loss of cell contents.

Plant tissues are thus invariably heterogeneous, consisting of several cell types each of whose cell walls can show considerable differences in chemical composition and physical properties (Akin, 1979; Gordon et al., 1985) and in resistance to attack by rumen microorganisms (Akin, 1986, 1988; Chesson et al., 1986; Grenet & Demarquilly, 1987).

DIGESTION OF PLANT CELL WALLS BY RUMEN MICROORGANISMS

As Van Soest (1981) has pointed out, it is difficult for an aquatic microorganism, such as is found in the rumen, to ensure its food supply and to avoid being passively removed by the normal flow of the liquid medium. Organisms which utilise plant structural polysaccharides as their major energy source can achieve some dominance over their substrate by associating closely with plant particles entering the rumen and, at the same time, by so doing extend their residence time in the rumen to that of the least digestible part of the animal's diet. The ability of cell wall degraders to adhere to plant material is of primary importance and appears an essential first stage in the digestive process.

Adhesion by rumen microorganisms

Bacteria, fungi and protozoa colonise practically all plant materials that enter the rumen, with the exception of intact, outer plant surfaces which reportedly are not colonised by any microbe (Bauchop, 1980). The major route of invasion appears to be via epidermal lesions. Colonisation by entry through stomatae is comparatively unimportant with stem fragments of legumes and grasses, but it can be of greater importance for colonisation of leaves (Cheng et al., 1983/4).

Bacteria become attached to cell walls as the initial step in the degradation process (Akin, 1986; Cheng et al., 1983/4). The major species that attach are the cellulolytic bacteria *Ruminococcus albus*, *R. flavefaciens* and *Bacteroides succinogenes*, which are commonly found on cut edges of cell walls and damaged areas of cell surfaces (Akin, 1986; Latham et al., 1978a,b). The *Ruminococcus* species appear to be loosely associated with cell walls, while *B. succinogenes* exhibits a tight adhesion, frequently conforming to the surface of the material being digested (Figs 2 and 3) (Forsberg et al., 1981; Cheng et al., 1983/4). Readily digestible parenchyma

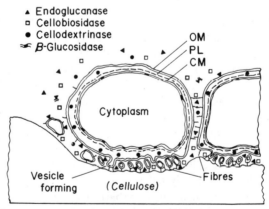

Fig. 2. Sketch of a transverse section of *Bacteroides succinogenes* growing on cellulose, showing the locations of the enzymes of the cellulase complex (OM, outer membrane; PL, plasmalemma; CM, cytoplasmic membrane).

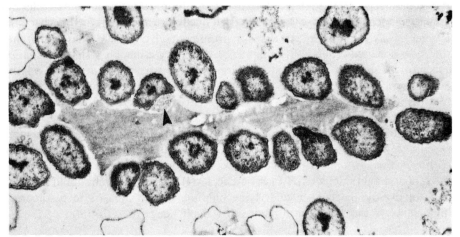

Fig. 3. Morphology of *Bacteroides succinogenes* cells growing on filter paper cellulose in pure culture; note the tight adhesion of the cells to the cellulose and the vesicles formed between one cell and the cellulose (arrow). Kindly provided by T. S. Beveridge.

cell walls are heavily colonised by bacteria and rapidly digested, whereas the thick, recalcitrant, walls of the vascular and sclerenchymal tissue of the same plant are sparsely colonised (Cheng et al., 1983/4). Partial removal of lignin from cell walls by treatment with alkali reportedly increased tenfold the numbers of attached rumen bacteria (Latham et al., 1979).

Rumen fungi and protozoa colonise plant fragments and degrade them to differing degrees (Akin, 1986). The processes involved in the attachment of these organisms are poorly understood. At the macro level a distinct preference is shown by anaerobic fungal zoospores for stomata and for damaged areas of plant particles from which soluble sugars diffuse and to which the zoospores demonstrate a chemotaxic response. Four distinct chemoreceptors have been identified in zoospores of *Neocallimastix frontalis* (Orpin & Bountiff, 1978). Zoospores swim to acceptable sites for colonisation on plant materials, encyst, and invade the tissue via thallus formation and rhizoids (Bauchop, 1981) (see Chapter 4).

The mechanism of attachment of different rumen microorganisms to cellulose, and indeed other complex fibrous materials, may involve specific binding by: cell surface-associated enzymes; adhesins (molecules on the microbial cell surface that bind to receptors on the plant material); or, possibly, non-specific ionic interaction (Forsberg, 1986). Bacterial attachment has received the most study, and it has been observed, for example, that many rumen bacteria attach to crystalline cellulose powder, and factors influencing such an attachment have been examined by Minato & Suto (1981). In the case of the adherent cellulolytic bacteria, addition of methylcellulose causes their detachment, *B. succinogenes* being released almost completely (Kudo et al., 1987). Methylcellulose also inhibited bacterial attachment and growth on cellulose, but not growth on glucose or cellobiose. Since methylcellulose inhibited cellulolytic activity of cell-free preparations by less than

20% it has been suggested that cellulases play a minor role in attachment of cells, but that attachment to cellulose was necessary for cellulose hydrolysis by these bacteria. *Ruminococcus flavefaciens* possesses a prominent glycoprotein coat and it has surface-associated cellulases (Latham et al., 1978a); presumably a similar situation exists for *B. succinogenes*. Although these data are very informative, the relative importance of surface-associated cellulases and adhesins in attachment to cellulose cannot be conclusively resolved at the present time because of a lack of basic knowledge about both the cellulases and the cell surface architecture.

An ionic interaction is quite different, and it could account for much of the non-specific binding of non-cellulolytic cells to cellulose. For example, even the faecal pathogen *Escherichia coli* suspended in water binds to cellulose in the presence of calcium or magnesium ions (Hwa & Ferenc, 1984). Future research should be aimed at resolving the mechanism underlying binding, since it appears to be of prime importance in the digestion of plant cell walls (Cheng et al., 1983/4).

Glycanases of rumen microorganisms

Very small proportions, often less than 15%, of cellulase, hemicellulase and glycosidase activities are found in the cell-free culture fluid of rumen digesta (Williams & Strachan, 1984). Bacteria with high glycosidase activity form a subpopulation loosely bound to plant materials, while the cellulolytic and hemicellulolytic bacteria are more firmly bound (Williams & Strachan, 1984).

Mechanism of cellulose digestion

Cellulases of the aerobic fungi. The mechanism of cellulose hydrolysis by rumen microorganisms has received rather little attention compared to that by aerobic cellulolytic fungi. However, the fungal cellulase system (Wood, 1985) serves as a useful model on which to base a consideration of rumen cellulases. The aerobic fungi excrete cellulase enzymes in an active form. The cellulase complex usually consists of three major types of enzymes which function synergistically in the hydrolysis of crystalline cellulose. These are: endo-1,4-β-glucanase (endo-1,4-β-D-glucan 4-glucanohydrolase, endoglucanase, endocellulase EC 3.2.1.4); cellobiohydrolase (1,4-β-D-glucan cellobiohydrolase, exoglucanase, exocellulase EC 3.2.1.91); β-glucosidase (cellobiase EC 3.2.1.21). The endoglucanase attacks carboxymethylcellulose or phosphoric acid-swollen cellulose in a random fashion that results in a rapid decrease in chain length and gives rise to cello-oligosaccharides, but it is extremely inefficient at the hydrolysis of crystalline cellulose. The cellobiohydrolase degrades cellulose by splitting off cellobiose units from the non-reducing end of the chain. Substituted celluloses (e.g. carboxymethylcellulose) are not attacked, but 'acid-swollen', partially degraded, cellulose and water-soluble cello-oligosaccharides are readily degraded, and Avicel (a highly crystalline cellulose with a low degree of polymerisation) is degraded by some cellobiohydrolase preparations (Berghem & Pettersson, 1973). The β-glucosidases hydrolyse cellobiose, and soluble cello-oligosaccharides with a low degree of polymerisation, to glucose, but cellulose is not degraded.

The currently accepted model for the enzymic hydrolysis of cellulose envisages a mechanism in which endoglucanase initiates the attack and the newly formed, nonreducing, chain ends are then degraded by the endwise-acting cellobiohydrolase. End-product inhibition due to an accumulation of cellobiose is prevented by action of the β-glucosidase. The significant feature of the cellulase system of the aerobic fungi is the synergism observed between the endoglucanase and cellobiohydrolase which brings about the solubilisation of crystalline cellulose (Wood & McCrae, 1979). Reese et al. (1951) suggested that a non-hydrolytic chain-separating enzyme might be involved. Studies by Griffin et al. (1984) with cellulase preparations from *Trichoderma reesei* support this hypothesis, although a component with this property has never been isolated.

Cellulases of rumen bacteria. The major cellulolytic bacteria in the rumen include *R. albus*, *R. flavefaciens* and *B. succinogenes* (Van der Linden et al., 1984; Cheng et al., 1983/4). Studies by Leatherwood (1965) involving the use of antibodies against extracellular material in culture fluid from *R. albus* strain 7 revealed that this bacterium produced about 20% of the endoglucanase in rumen fluid. Some 58% of the cellulase in *R. albus* was released from the cells (Leatherwood, 1965). The cellulase in *R. albus* strain RAM was active over a pH range of 6·0–6·8, with an optimum temperature of 45°C (Smith et al., 1973). Yu & Hungate (1979) isolated 4 cellulases, ranging in molecular weight from 39 000 to greater than 6×10^5, from *R. albus* strain 6 grown in rumen fluid-based medium. Wood et al. (1982) and Wood & Wilson (1984) isolated from strain SY3 a single, low molecular weight (30 000), extracellular endoglucanase and a cell wall-bound, aggregated, form with a high molecular weight (1.5×10^6). The molecular weight of the enzyme was dependent upon the conditions of bacterial cultivation. When cells were grown without rumen fluid, or once they reached the stationary phase of growth on cellulose, mainly the low molecular weight enzyme was present in the culture, whereas rumen fluid or cellobiose cultures contained mainly the high molecular weight aggregate. Neither cellulase had significant hydrolytic activity on crystalline cellulose (Wood et al., 1982). Stack & Hungate (1984) discovered that 3-phenylpropionic acid in rumen fluid was the active compound which, upon inclusion in the growth medium in the place of rumen fluid, caused *R. albus* to grow faster and to synthesise a very active, cell-associated, high molecular weight cellulase. The 3-phenylpropionic acid had no effect on *R. flavefaciens* or *B. fibrisolvens* (Stack & Cotta, 1986).

An extracellular cellobiosidase (Ohmiya et al., 1982) and a membrane-bound β-glucosidase (Ohmiya et al., 1985) have been isolated from *R. albus*. The cellobiosidase, which cleaved *p*-nitrophenylcellobioside to give cellobiose and *p*-nitrophenol, was synthesised and released into the culture fluid after growth had ceased. The native enzyme was a dimer with a subunit molecular weight of 100 000. It had low activity against cellulose and no function has been attributed to it. The β-glucosidase was loosely associated with the cell surface and was extracted from cells by potassium phosphate. The molecular weight of the enzyme was 82 000. Because of the membrane location of the β-glucosidase, it was thought to enhance uptake of cellobiose.

A non-hydrolytic affinity factor was postulated as necessary for the adsorption of the *R. albus* cellulase to cellulose (Leatherwood, 1969), although this factor was not detected by Wood *et al.* (1982).

Cellulase was synthesised by *R. albus* when cells were grown on ball-milled cellulose and cellobiose, but little or no activity was produced by cells grown with polygalacturonic acid or beet araban (Greve *et al.*, 1984*b*), which clearly shows that the enzyme(s) are subject to either induction or repression. No data are available about the mechanism involved.

Characteristics of the cellulase enzymes of *R. flavefaciens* have been studied by Pettipher & Latham (1979*a,b*). The pH optimum was between 6·4 and 6·6 and the temperature optimum between 39 and 45 °C. The molecular weight of the cell-free enzyme from cultures grown in a defined medium (containing phenylacetic acid) ranged from $2·5 \times 10^4$ to greater than 3×10^6. Cellulase activity was primarily cell-associated during exponential growth, but cell-free enzyme accumulated during the stationary phase. Endoglucanase (or CM-cellulase) was produced in both ammonium-limited and cellobiose-limited continuous cultures, which suggests that production of this enzyme is constitutive in *R. flavefaciens*. Culture conditions did not seem to affect greatly the amount of cell-associated endoglucanase (in whole cells 69% was not accessible to substrate), while the cell-free cellulase was higher in cellobiose-limited cultures and in cultures grown at lower dilution rates. Changes in viscosity and production of reducing sugars from carboxymethylcellulose, and the production of cellobiose from milled filter paper and Avicel cellulose, suggested that the cellulase(s) of *R. flavefaciens* included a divalent cation-requiring endoglucanase(s) and a cellobiohydrolase-like enzyme(s). Although not elaborated upon, Pettipher & Latham (1979*a*) did observe swelling factor activity (Nisizawa *et al.*, 1966) and short-fibre-forming activity (Halliwell & Riaz, 1970).

Bacteroides succinogenes is one of the rumen bacteria most active in growth-related degradation of recalcitrant forms of cellulose such as cotton fibres and cellulose powder, although cell-free culture fluid and cell-extracts had very low hydrolytic activity against either filter paper or other forms of crystalline cellulose (Groleau & Forsberg, 1981; Halliwell & Bryant, 1963). *B. succinogenes* produced high levels of endoglucanase and β-glucosidase (Groleau & Forsberg, 1981). More than 60% of the endoglucanase activity from stationary-phase cultures grown on cellulose was released from the cells (Table 3; Forsberg *et al.*, 1981). The extracellular endoglucanase was present in a low molecular weight form associated with sizes ranging from 45 000 daltons (28–38%) to large non-sedimentable aggregates with molecular weights in excess of $4·0 \times 10^6$ (9–13%), and with sedimentable membrane fragments (50–62%) (Groleau & Forsberg, 1983). An endoglucanase was isolated from the low molecular weight fraction which had a molecular weight of 64 400 with a pH optimum of 7·0 and a temperature optimum of 39°C (McGavin & Forsberg, 1987). It cleaved acid-swollen cellulose to give cellotriose and cellobiose as the major hydrolytic products.

An extracellular, chloride-stimulated, cellobiosidase (Huang & Forsberg, 1987*a*) and a periplasmic cellodextrinase have also been detected in *B. succinogenes* (Huang

Table 3
Distribution of Hydrolytic Enzyme Activities Between the Cells and the Cell-free Culture Fluid of a *Bacteroides succinogenes* Culture Grown with Cellulose as the Carbon Source

Enzyme	Activity ($\mu mol/min/mg$ protein)	
	Cells	Fluid
Endoglucanase	0·298 (39)	0·642 (61)
Cellobiase	0·027 (88)	0·005 (12)
Cellobiosidase	0·016 (62)	0·012 (39)
Xylanase	0·076 (18)	0·476 (82)
Cellulase	0·002 (24)	0·010 (76)

The data, with the exception of those for cellobiosidase, are from 48 hour batch cultures (Forsberg *et al.*, 1981). The cellobiosidase activities are from a chemostat culture with cellulose as the carbon source and a dilution rate of $0.025\,h^{-1}$ (Huang & Forsberg, 1987).
Percent total enzyme activity is shown within parentheses.

& Forsberg, 1987b) and purified to homogeneity for detailed characterisation. The function of the extracellular cellobiosidase has not yet been well characterised, although it did cleave cello-oligosaccharides from C_3 to C_6 and slowly hydrolysed acid-swollen cellulose, to give primarily cellobiose. The periplasmic cellodextrinase also cleaved cell-oligosaccharides (C_3 to C_6) to give glucose plus cellobiose, but it lacked activity on acid-swollen cellulose. It presumably functions in the hydrolysis of cellodextrins, which have entered the periplasm through pores in the cell wall, to give glucose and cellobiose which can be readily transported into the cell. Locations of the cellulase enzymes of *B. succinogenes* are shown in Fig. 2.

Cells grown on cellulose produced 7–8 times more endoglucanase activity than either cellobiose-grown or glucose-grown cells (Groleau & Forsberg, 1981), which is consistent with the suggestion that the endoglucanase is subject to a type of catabolite repression. The role of catabolite repression in regulation of endoglucanase activity is further supported by the apparent catabolite repression of a *cel* gene from *B. succinogenes* expressed in *Escherichia coli* (Taylor *et al.*, 1987). This observation is particularly interesting since cAMP, the mediator of catabolite repression documented in many aerobic bacteria, may not be present in some anaerobes (Setlow & Sacks, 1983).

Many non-cellulolytic bacteria, including strains of *S. ruminantium*, *B. ruminicola* and *S. bovis*, were able to grow at the expense of cellobiose and the higher molecular weight cellodextrins. In most cases growth rates were similar on both substrates. Russell (1985) suggested that the ability to utilise the degradation products of cellulolysis may explain why high numbers of non-cellulolytic organisms are found in the rumens of animals fed on 'high-fibre' diets (see Chapter 11).

Hemicellulases of rumen bacteria

Hemicellulose, which represents some 37–48% of plant cell walls, is a very complex material (Chesson et al., 1986) with xylan normally present as one of the major polymers, as has been described earlier. Some major structural features of xylan and the sites of enzyme action are summarised by Fig. 4. Table 4 lists the distribution and specific activities of some of the more prominent hemicellulase and glycosidase enzymes of selected rumen bacteria (Williams & Withers, 1982a,b). Detailed information on the glycanases and glycosidases is provided in other papers by Williams and his co-workers. Xylanases are more widely distributed than are cellulases among rumen bacteria. Microorganisms often produce more than one

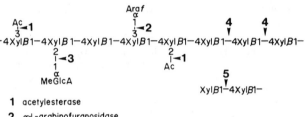

1 acetylesterase
2 α-L-arabinofuranosidase
3 α-glucuronidase
4 endo-1,4-β-xylanase
5 β-xylosidase

Fig. 4. A generalised structure of plant glucuronoarabinoxylan showing the sites of action of the major enzymes responsible for the cleavage of glycosidic links between sugar residues and the ester linkage to acetyl (Ac) groups (Xyl, xylopyranosyl residue; Ara*f*, arabinofuranosyl residue; MeGlcA, 4-*O*-methyl-glucuronosyl residue).

xylanase, and *B. succinogenes*, for example, produces at least two (Forsberg, McGavin & Huang, personal communication). β-Xylosidase (1,4-β-D-xylan xylanohydrolase, EC 3.2.1.37), β-glucosidase and α-L-arabinofuranosidase (α-L-arabinofuranoside arabinofuranohydrolase, EC 3.2.1.55) are essential for the complete degradation of oligomeric fragments arising from the hydrolysis effected by the polysaccharide-degrading enzymes, and they are widely distributed among the rumen bacteria (Williams et al., 1984).

Interactions of the hemicellulolytic glycanase and glycosidase activities of rumen bacteria are on the whole rather poorly characterised, although Greve et al. (1984b) have found that *R. albus* possesses a complement of fibrolytic glycanases and glycosidases able to digest alfalfa cell wall polymers and producing a mixture of monosaccharides. They subsequently purified an α-arabinofuranosidase and a β-1,4-xylanase (1,4-β-D-xylan xylanohydrolase, EC 3.2.1.8) from *R. albus* and demonstrated that the two enzymes co-operatively released reducing material from alfalfa cell walls at 5·2 times the rate of the enzymes acting separately (Greve et al., 1984b). Similar co-operative action had been demonstrated earlier for the identical enzymes from the rumen protozoon *Epidinium ecaudatum* (Bailey & Gaillard, 1965).

Acetyl substitutions have been reported to inhibit the digestion of isolated plant

Table 4
Effect of Carbon Source on the Hemicellulolytic and Pectolytic Activities of Some Fibre-digesting Rumen Bacteria[a]

Culture	Activities (nmol/h/mg protein)					
	Xylanase	Hemicellulase	α-L-Arabino-furanosidase	β-D-Glucuronidase	Esterase	Pectinase
Bacteroides ruminicola brevis GA3	3 maltose ND glucose	41 arabinan 5 glucose	— —	7 xylose ND glucose	186 arabinoxylan 34 glucose	11 arabinan ND arabinose
B. ruminicola ruminicola H2b	26 maltose ND cellobiose	45 arabinan ND cellobiose	9 × 10⁴ arabinan 9 glucose	22 xylan ND arabinoxylan	171 xylose 92 cellobiose	58 arabinogalactan 3 glucose
Bacteroides succinogenes S85	27 xylan ND glucose	50 maltose 1 arabinoxylan	— —	6 xylan ND cellobiose	293 lactose 134 glucose	4 xylose ND cellobiose
Butyrivibrio fibrisolvens H10b	48 glucose 10 arabinose	51 arabinoxylan 3 xylose	77 arabinoxylan 9 glucose	932 xylan 11 arabinoxylan	207 xylose 75 arabinoxylan	150 arabinan 2 arabinogalactan
Ruminococcus albus RUM5	65 xylose ND glucose	42 xylose ND glucose	— —	741 maltose 37 glucose	232 arabinan 121 arabinose	18 glucose ND xylose
Ruminococcus flavefaciens 123	23 arabinan ND glucose	38 arabinan ND glucose	— —	138 maltose 2 arabinogalactan	381 xylan 293 lactose	35 cellobiose ND glucose

[a] Each bacterium was grown with 10 different carbon sources. Growth was reported on some substrates on which it would not be expected; this may have been due to impurities in the substrates. The cells from each were harvested and disrupted, and assayed for the various enzymic activities. The carbon source giving the highest enzymic activity is listed in the top of each pair of rows beside the specific activity, while the carbon source giving the lowest specific activity is listed on the bottom row beside the specific activity. The substrates for the enzyme assays included larch-wood xylan, hemicellulose B from delignified *Lolium multiflorum* cell walls, *p*-nitrophenyl-L-α-arabinofuranoside, *p*-nitrophenyl-β-D-glucuronide, *p*-nitrophenyl acetate, and pectin. Data modified from Williams & Withers (1982a,b).
ND not detected. — not tested.

polysaccharides by xylanases, but this does not appear to occur in the rumen (Chesson, 1981). Acetylesterase (or acetyl xylan esterase) is present in a number of aerobic fungi, and this enzyme also cleaved p-nitrophenyl acetate, but with lower activity (Biely et al., 1985). Since a number of rumen bacteria possess esterases that cleave the synthetic substrate p-nitrophenyl acetate (Table 4), these enzymes may have the capability to function co-operatively with xylanases in the degradation of plant polymers, as was demonstrated for the aerobic fungi (Biely et al., 1986). The role of α-glucuronidase in the hydrolysis of hemicellulose has not been studied, although it has been reported that the α-glucuronidase from the fungus *Agaricus bisporus* acts synergistically with xylanases and liberates 4-O-methylglucuronic acid from methylglucuronic acid-substituted xylo-oligomers (Puls et al., 1987). Phenolic compounds, including the *trans* isomers of p-coumaric and ferulic acids, are covalently bound to cell wall polysaccharides of graminaceous plants, probably through ester linkages as feruloylated arabinoxylans and coumaroylated arabinoxylans (Mueller-Harvey et al., 1986). Alkali treatment increases the rate of degradation of forage by ruminants through the cleavage of ester linkages. The possibility exists that some rumen microorganisms may possess esterases able to cleave the bond, but for some reason they do not play a dominant role in the hydrolytic process within the rumen.

The glycanases and glycosidases with roles in hemicellulose digestion are subject to carbon source-dependent regulation (Table 4; Williams & Withers, 1982b; Greve, 1984). The xylanase from *R. albus* is produced when cells are grown on filter paper but not when they are grown on cellobiose (Greve et al., 1984a), and that from *R. flavefaciens* is produced when cells grow on either filter paper or cellobiose (Pettipher & Latham, 1979a,b). The xylanase from *B. succinogenes* is produced by cells grown on filter paper and on Avicel (Forsberg et al., 1981; Forsberg, unpublished), but not by cells grown on glucose (Table 4; Williams & Withers, 1982b). These data would suggest that xylanases from *R. albus* and *B. succinogenes* are subject to repression. It has been observed that a cloned xylanase gene from *B. succinogenes* expressed in *E. coli* is expressed constitutively (Sipat et al., 1987). This would appear to be inconsistent with the results reported by Williams & Withers (1982b) mentioned above. However, since *B. succinogenes* produces more than one xylanase activity, it may be that the cloned xylanase gene is either differently regulated in *B. succinogenes* or is expressed at a very low level.

Glycanases and glycosidases of rumen protozoa
Glycanases and glycosidases are widely distributed among the rumen ciliate protozoa (Table 5; see also Chapter 3). The least widely distributed glycanase is cellulase, and the highest activities have been found in *Epidinium ecaudatum caudatum*, *Eremoplastron bovis*, *Ostracodinium obtusum bilobum* and *Eudiplodinium maggi*, with little or no activity in five *Entodinium* species tested (Coleman, 1985). The component activities of the cellulase complex, including the endoglucanase, cellobiosidase and β-glucosidase activities, were more widely distributed. Protozoa have been trained to grow on grass as a carbohydrate source, although whether they actually used cellulose could not be ascertained because grass contains a range of

Table 5

Occurrence and Specific Activities of Glycanase and Glycosidase Enzymes in Entodiniomorphid and Holotrich Rumen Ciliate Protozoa

Entodiniomorphid or holotrich rumen ciliate protozoon		Activity (nmol/min/mg protein)											%[a]	
		α-L-Arabinofuranosidase	β-D-Galacturonidase	β-D-Glucuronidase	α-D-Galactosidase	β-D-Galactosidase	β-D-Xylosidase	Xylanase	Hemicellulase	β-D-Glucosidase	β-D-Cellobiosidase	Polygalacturonidase	Endoglucanase	Cellulase
Epidinium ecaudatum caudatum	C	38.8	7.7	12.1	26.4	23.6	8.2	NA	NA	2.3	NA	NA	41	NA
	R	66.6	6.9	7.8	59.3	51.3	33.6	42.2	57.8	10.5	43.2	6.6	100	100
Ophryoscolex caudatus	C	31.4	16.5	5.8	18.5	23.3	12.7	NA	NA	10.9	18.7	NA	72	NA
	R	24.1	3.4	13.6	30.0	17.5	17.2	61.7	67.3	4.2	NA	NA	116	80
Large (80 μm) Entodiniomorphid fraction	R	22.6	9.1	1.5	16.4	12.1	19.0	NA	NA	22.7	NA	NA	NA	NA
Polyplastron multivesiculatum	C	34.8	13.8	0.67	38.8	5.3	24.7	30.0	37.8	24.0	2.2	NA	55	NA
	R	61.5	30.4	3.7	28.0	15.7	45.4	59.2	60.8	44.4	34.3	6.3	NA	NA
Diploplastron affine	C	72.5	15.5	0.97	25.8	18.7	14.2	95.2	NA	29.7	33.4	NA	54	NA
	R	45.9	6.6	0.81	23.0	22.3	11.0	77.6	101.8	17.0	2.1	3.8	31	36
Eremoplastron bilobum	C	20.4	6.1	0.30	12.3	0.52	6.8	38.5	NA	11.6	6.8	NA	32	NA
Eremoplastron bovis	C	52.0	11.5	1.0	37.2	5.7	18.1	NA	NA	39.6	13.6	NA	16	NA
	R	60.1	14.3	2.2	38.8	13.8	47.2	51.2	53.8	56.0	33.5	6.4	43	122
Ostracodinium obtusum	R	51.8	17.0	1.6	23.6	12.8	29.0	62.3	82.5	43.9	24.4	6.0	89	92
Eudiplodinium maggii	R	48.1	18.1	2.5	23.8	10.0	38.5	34.7	35.5	40.4	33.6	4.3	64	79
Mixed Entodinium spp.	R	13.3	0.77	0.70	22.2	11.3	15.7	10.7	27.8	19.5	4.6	NA	NA	NA
Entodinium simplex	C	4.4	0.37	0.40	9.7	2.0	3.8	1.1	3.0	9.2	2.2	NA	0	0
Entodinium bursa	C	3.2	0.49	0.28	7.6	1.4	1.9	2.0	NA	27.1	0.68	NA	NA	NA
Entodinium caudatum	C	1.3	0.16	0.48	11.0	1.0	0.9	1.6	4.5	41.0	5.3	NA	NA	NA
Mixed Isotricha spp.	R	4.6	1.3	0.65	5.5	2.0	7.8	16.3	18.3	32.3	10.3	NA	2	NA
Dasytricha ruminantium	R	1.7	0.23	0.59	3.8	0.59	6.9	2.3	13.0	138.5	25.7	NA	NA	NA

The data are from Williams et al. (1984), with the exception of the hemicellulase B and xylase data which are from Williams & Coleman (1985), the cellulase data from Coleman (1985), and the polygalacturonidase data from Williams et al. (1986). NA, not assayed; C, cultured in vitro; R, isolated from rumen. Other glycosidases widely distributed among rumen protozoa include β-D-fucosidase, α-D-glucosidase, β-D-mannosidase, α-D-xylosidase and N-acetyl-β-D-glucosaminidase.

[a] Results expressed as activities/mg protein relative to that of Epidinium ecaudatum caudatum taken as 100%.

other polysaccharides (Coleman et al., 1976). Coleman (1985) found no consistent effect of growth conditions on cellulase activity, although protozoa grown in vitro on grass alone tended to exhibit higher levels of cellulase activity than those grown on starch and grass.

Hemicellulase activity was first detected in the rumen ciliate protozoon *Epidinium caudatum* (Bailey et al., 1962). Since then hemicellulase and xylanase activities have been detected in most protozoa tested (Williams & Coleman, 1985; Table 5). The specific activities were higher in the cellulolytic entodiniomorph ciliates than in other protozoa.

A wide range of glycoside hydrolases are synthesised by rumen ciliate protozoa (Williams et al., 1984). The glycoside hydrolases involved in the degradation of plant cell wall structural polysaccharides, including α-L-arabinofuranosidase, β-D-galacturonidase, β-D-glucuronidase, β-D-glucosidase and β-D-cellobiosidase, were higher in the cellulolytic entodiniomorphid ciliates.

The only reported separation of a rumen protozoal glycanase or glycosidase enzyme was the separation of arabinofuranosidase from xylanase activity in a cytoplasmic extract prepared from *Epidinium caudatum* Crawley (Bailey & Gaillard, 1965). This enabled the authors to demonstrate that hydrolysis of arabinoxylan involved the prior removal of single-unit arabinose side-chains before extensive hydrolysis of the xylan backbone could occur.

The glycanase and glycosidase activities of ciliate protozoa were found to be quite low in comparison with those of active bacterial species. Since protozoa used for these trials were either purified from a monoculture in a sheep or grown in vitro in the presence of bacteria, there was the possibility of surface or intracellular contamination of the protozoal preparation by bacteria able to produce glycanase and glycosidase activities. To circumvent the problem of bacterial contamination, the protozoa were gently lysed by methods which did not disrupt bacteria, and then centrifuged to sediment bacterial cells and any intact protozoa (Coleman, 1985).

Of the broad range of glycosidases possessed, certain of these enzymes (e.g. uronidase, mannosidase, xylosidase and arabinofuranosidase) reportedly generate monomers which are not utilised by the protozoa.

Glycanases and glycosidases of rumen anaerobic fungi

The anaerobic rumen fungi (see Chapter 4) secrete a wide range of glycanases and glycosidases during growth, and the activity of each is dependent upon the source of carbohydrate for growth (Mountfort & Asher, 1985; Pearce & Bauchop, 1985; Williams & Orpin, 1987a,b). The activities of some of these enzymes are shown in Table 6. *Neocallimastix frontalis* has been reported to produce a highly active, extracellular, cellulase, several-fold higher in activity than that of the aerobic fungus *Trichoderma reesei* (Wood et al., 1986). However, the extracellular exoglucanase activity in the strains of *N. frontalis* studied by Mountfort & Asher (1985) was very low, and little was extractable from the cells. The Avicelase, endoglucanase and β-glucosidase activities of *N. frontalis* had pH and temperature optima of 5–6 and 37–50°C, respectively (Mountfort & Asher, 1985; Pearce & Bauchop, 1985). Endoglucanase production was maximum when the fungus was grown on cellulose

Table 6
Extracellular Hydrolytic Activities of *Neocallimastix frontalis* Strain PN2 after 7 days Growth on Microcrystalline Cellulose (Pearce & Bauchop, 1985)

Enzyme	Maximum activity ($\mu mol/min/ml$)
Exoglucanase	0·012
Endoglucanase	0·51
Amylase	0·066
Xylanase	0·075
Pectinase	0
β-Galactosidase	0·010
β-Glucosidase	0·041
β-Xylosidase	0·006

The substrates used included Avicel, carboxymethyl-cellulose, soluble starch, larch-wood xylan, polygalacturonic acid and *p*-nitrophenyl derivatives of β-D-galactose, β-D-glucose and β-D-xylose.

while synthesis was totally repressed by the addition of glucose, indicating that the enzyme was subject to regulation (Mountfort & Asher, 1985).

The glycanase and glycosidase activities of three separate anaerobic fungi (*Neocallimastix patriciarum*, *Piromonas communis* and an unidentified isolate) have been surveyed by Williams & Orpin (1987*a,b*). Glycanase enzymes able to hydrolyse α- and β-glucans, β-galactans, galactomannan, and arabinoxylan were present in all three, and the activities were dependent upon the carbohydrate source used for growth. Similarly, the broad range of glycosidases shown to be produced by these three fungi was reported to be as extensive as the enzyme profile described for the rumen bacterial and protozoal populations (Williams & Orpin, 1987*b*).

Pectin digestion

The pectolytic enzymes are divided into two main groups: the pectin esterases (pectin pectylhydrolase, EC 3.1.1.11) which catalyse removal of methanol, and the depolymerising enzymes which are either hydrolases or lyases (Prins, 1977; Chesson, 1980). The pectolytic activities of the predominant pectin-degrading bacteria and protozoa have been identified, but little has been published on their properties (Williams, 1986; Wojciechowicz *et al.*, 1982). Pectin-esterase and -lyase activities were demonstrated in *Lachnospira multiparus*, but there was no evidence of polygalacturonase (poly-1,4-α-D-galacturonide glycanohydrolase, EC 3.2.1.15) activity (Silley, 1985). In contrast to the rumen bacteria and protozoa, the anaerobic fungi tested have exhibited rather little hydrolytic activity towards pectin (Pearce & Bauchop, 1985; Williams & Orpin, 1987*b*).

LIMITATIONS TO POLYSACCHARIDE DEGRADATION

Storage polysaccharides

Fructosans are rapidly and fully degraded in the rumen and there appears to be no limitation to their utilisation. Similarly, cereal starches are virtually completely degraded by the rumen microflora (Waldo, 1973). In fact, modern feed-processing methods result in such a finely ground, rapidly fermentable product that there may be a need to slow the rate of utilisation to avoid problems of acidosis (Alderman, 1986; see Chapter 14). Other forms of starch, such as those from maize or sorghum, are generally more slowly degraded than the corresponding cereal products, and significant proportions may escape rumen fermentation to be degraded in the post-ruminal digestive tract. Starches rich in amylose, giving a so-called B-type X-ray diffraction pattern, are poorly degraded in the rumen compared with starches containing normal or low levels of amylose (A-type X-ray diffraction pattern) (Szylit et al., 1978). Methods of feed preparation which insufficiently disrupt the grain, so limiting the accessibility of starch to attacking microorganisms, or the presence of inhibitory compounds which suppress amylolytic activity, can also affect hydrolysis rates and reduce the extent of utilisation in the rumen. Condensed tannins, for example, reduce the value of sorghum starch (particularly that from the bird-resistant varieties) as a ruminant feed (Glennie et al., 1982; Kock et al., 1986). Retrogradation of starch, which undoubtedly occurs during some forms of feed preparation and pelleting, which can restrict utilisation in the foregut of non-ruminants, does not seem to present the same problem to ruminants.

Cell wall polysaccharides

In contrast to the storage carbohydrates, ingested cell wall polysaccharides are rarely completely digested by the microflora, and in most instances a significant proportion escapes fermentation in the rumen and large intestine to be voided with the faeces. However, as indicated previously, plant tissues consist of a heterogeneous population of cells and cell walls which are attacked to differing extents by the rumen microorganisms. Estimates made of total polysaccharide loss within the rumen conceal the fact that polysaccharides from some cell walls are fully degraded within a few hours while similar polysaccharides from other cell wall types remain intact even after exposure for several days to the rumen microflora. The extent to which individual cell walls are attacked depends upon factors both internal (mural) and external (extramural) to the plant cell wall.

Extramural factors
Much of the surface of plant particles entering the rumen may be protected by epicuticular waxes and the cuticle, both of which appear inert to the rumen flora. Invading microorganisms are dependent on broken edges of feed particles or naturally occurring openings such as stomata or lenticels to provide access to

suitable substrates. Inevitably walls of the deeper-lying cells remain protected from attack for longer periods than those of cells forming the more superficial layers. Cuticular material can be very effective in protecting cells from attack. Isolated mesophyll and epidermis cell walls prepared from ryegrass leaf are degraded at exactly the same rate when incubated in the sheep rumen, both being completely digested within 8 hours (Chesson et al., 1986). However, similar observations made with intact ryegrass and other leaves show that, while the leaf mesophyll is lost within the time predicted by incubation of the isolated cell wall, the epidermis is little affected (Akin, 1979, 1986, 1988). The cuticle layer covering the outer surface of the epidermis in the intact leaf appears sufficient to delay degradation substantially, by restricting attack to the inner surfaces of broken cells (Akin, 1986).

Even where bacteria obtain entry into the cell lumen, degradation of the wall may be limited by the presence of a 'warty layer'—a layer lining the inner surface of lignified cell walls which appears resistant to attack and which may have to be mechanically disrupted to allow digestion by adherent bacteria to occur (Engels & Brice, 1985). The nature of the layer remains unclear but its position and staining reactions suggest that it may arise from the plasmalemma of the primary cell which has become progressively suberised during secondary thickening and has adhered to the inner cell wall surface on the death of the cell.

Mural factors

The fine-structure of a polysaccharide can influence both the rate and extent of its microbial degradation. However, much of the available data result either from experiments made with isolated polysaccharides or from those using cell-free enzyme systems, and there is considerable doubt whether such results can be extrapolated to describe the degradation of cell walls by the mixed microbial population of the rumen. The crystallinity of cellulose has been suggested as one possible factor in the control of cell wall digestion. Again, many of the data on the effects of crystallinity on rates of cellulolysis relate to experiments made with highly crystalline samples of isolated cotton- or wood-cellulose and not with the far more amorphous celluloses found in forages and cereal straws. The limited experiments made in the rumen suggest that amorphous and crystalline regions of cellulose are degraded at a common rate (Beveridge & Richards, 1975). It is true that, as Brice & Morrison (1982) found, the degree of side-chain substitution in isolated arabinoxylans is directly related to their degradability. However, as these authors point out, arabinoxylans isolated from mature grass tissue carry a lower level of substitution than is found in younger tissues and so should be more degradable. In fact the reverse is true, with more mature grass (and its arabinoxylan) proving to have a lower degradability than less mature samples. It is evident from such experiments that, in the intact cell wall, factors other than the fine-structure of the polysaccharide have an overriding influence on degradability and that, in lignified tissues, these relate to the phenolic content of the cell wall. Staining reactions confirm that cell walls of tissues that are poorly degraded in the rumen are extensively lignified (Akin, 1979) while delignification substantially enhances cell wall digestion of lignified tissues (Chesson, 1981).

Phenolic compounds of plant cell walls

The phenolic content of primary cell walls is low and, in the monocot, consists largely of monomeric phenolic acids (Fig. 5), aldehydes and low molecular weight phenolic polymers (Harley & Keene, 1984). Phenolic acids can also be extracted from the secondary thickened cell walls of Gramineae where they may account for approximately 1% of dry matter (Salomonsson et al., 1978). Interest in the phenolic acids exists because they are structurally very similar to the precursors of lignin and could represent initiation sites for lignification, and because both ferulic and p-coumaric acids have been shown to be bound to cell wall polysaccharide by, in part at least, alkali-labile linkages (Smith & Hartley, 1983; Kato & Nevins, 1985; Mueller-Harvey et al., 1986). Suggestions that the phenolic acids per se have a role in the control of cell wall degradation (Jung & Fahey, 1983; Jung, 1985; Akin, 1986) seem unlikely. Phenolic acids are not present in sufficient amounts to present major structural restrictions to enzyme action and, although they are released from cell walls during cell wall breakdown (Jung et al., 1983) and have been shown to be mildly toxic to rumen cellulolytic bacteria (Chesson et al., 1982) and fungi (Akin & Rigsby, 1985) in vitro, they are readily metabolised by the rumen microflora (Martin, 1978; Chesson et al., 1982).

Lignin, the major phenolic material present in secondary thickened walls, is formed when its major precursors, p-coumaryl, coniferyl and sinapyl alcohols (Fig. 5), are released into the cell wall and are polymerised in situ to form a three-dimensional macromolecule. The structural elements arising from the three phenylpropane precursors are p-hydroxyphenyl, guaiacyl and syringyl units respectively (Fig. 5). Unlike the polysaccharides in which only the glycosidic linkage is involved in unit bonding, some twenty different types of linkages have been identified within the lignin macromolecule. Most involve either aryl ether linkages or carbon-to-carbon bonds (Fig. 5), the strength of these bonds accounting for the resistance of lignin to both chemical and microbial degradation. It remains uncertain whether lignins are totally random structures, or whether there is a degree of order and reproducibility of structure in lignins within the same plant. However, gross differences relating to the proportions of the three monomer types present can be recognised in lignins of different phylogenic orgins (Alder, 1977; Monties, 1985). Lignins of angiosperms are generally of the guaiacyl–syringyl type and are thus copolymers of coniferyl and sinapyl alcohols. Gramineae lignins differ from other angiosperm lignins in containing a significant proportion of p-hydroxyphenyl units (Erickson et al., 1973). Lignins are extensively bound to the matrix polysaccharides of the cell wall, the lignin–carbohydrate complex so formed varying in structure depending on the plant and plant part examined (Morrison, 1974; Chesson et al., 1983; Ford, 1986). Despite the common use of the term 'lignocellulose', lignins do not appear to be directly bound to cellulose.

Lignin–carbohydrate complexes released from the cell wall during microbial degradation are soluble at rumen pH (Nielson & Richards, 1982), although they may be precipitated by the lower pH of the hindgut. Measurement of the lignin content alone in insoluble material before and after microbial attack can show an

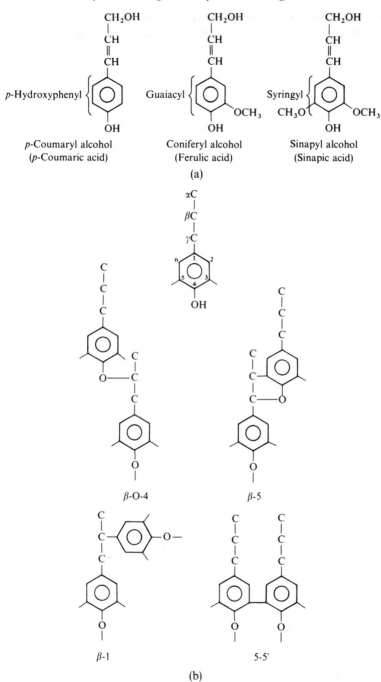

Fig. 5. (a) Structure of the major alcohol precursors of lignin with the equivalent phenolic acid shown in parenthesis, and (b) the four major aryl ether and carbon-to-carbon linkages found after polymerisation of lignin precursors; dimers are shown in skeletal form only.

apparent loss of lignin, implying the ability of rumen microorganisms to metabolise polyphenolic material. Although anaerobic pathways are said to exist for the metabolism of polyphenols, labelled CO_2 release from ^{14}C-labelled lignins held under anaerobic conditions is extremely slow and is measured over a period of months (Benner et al., 1984). Lignin degradation is essentially an aerobic process involving oxidative enzymes or oxidising reagents and is unlikely to occur to any significant extent under the reducing conditions found in the rumen. Minor modifications to the structure of lignin–carbohydrate complexes, such as the removal of sugar units or methoxyl groups may, however, occur. Akin (1980) has described one rumen bacterium apparently able to utilise methoxyl side-chains of phenolic acids as a sole carbon source.

Lignin and the control of cell wall degradation
It is probable that the porosity of lignified cell walls is insufficient to allow the free diffusion of cellulolytic and other enzymes. In consequence, attack is restricted to the inner or outer cell surface, and the chemistry of the surface layer is thus all important in determining the availability of polysaccharides to rumen microorganisms and their enzymes (Chesson, 1986). Observations made by transmission electron microscopy of lignified cell walls undergoing degradation invariably show degradation to be a localised process occurring only in the outermost layer of the wall in closest proximity to the attacking organism. Such micrographs do not show evidence of the more general dissolution which would be expected if enzymes were free to diffuse throughout the hydrated wall.

Residues of lignified tissues or isolated lignified cell walls recovered from the rumen have virtually the same chemical composition as the original material, even after extended incubations (Gordon et al., 1983). All components of the lignified wall, with the exception of lignin itself, are evidently lost at a common rate, and there would appear to be no selective degradation of any one component. Lignin is found in slightly increased proportions in digested residues. Physico-chemical methods applied to the analysis of the surface layers of plant cell walls suggest that lignin may accumulate at the surface because of a selective removal of polysaccharide from the outermost layer. In time, this gradual preferential retention of phenolic material leads to the formation of a protective layer which rumen organisms are unable to degrade and which is capable of protecting the underlying cell wall from further attack. Clearly, the amount of lignin initially present in the wall will dictate the rate at which this layer is formed and hence the extent to which the wall is degraded. Lignin which is exposed at the surface remains bound to the other components of the cell wall and, in theory, should not be free to diffuse away into the rumen liquor. Some undermining of the wall probably occurs, however, and this is likely to be the source of the previously mentioned lignin–carbohydrate complexes found in solution in the rumen.

Forage residues show substantial changes in composition after rumen degradation, with the level of xylose residues greatly increased and cellulose decreased (Gordon et al., 1983). These changes have been widely (and incorrectly) interpreted to suggest that hemicellulose is selectively protected by lignin. Such investigations ignore the previously mentioned heterogeneity of forage samples and the fact that the various types of cells present have walls with different compositions (Gordon *et al.*, 1985) which are degraded at different rates and to different extents (Chesson *et al.*, 1986). Primary cell walls generally are rich in cellulose but have a low hemicellulose content, while the more extensively lignified cells forming the vascular tissue may have walls whose hemicellulose content is two to three times higher than that of primary walls. Most primary tissue is fully degraded within 12 hours in the rumen, so any residues recovered after this time consist solely of the more resistant secondary-thickened tissue with its high hemicellulose content. Cell wall components are lost during digestion at a rate which is common to each cell wall but which varies between cell types.

ACKNOWLEDGEMENT

The authors acknowledge and thank Mr P. J. S. Dewey (Rowett Research Institute) for the analytical data presented in Table 1.

REFERENCES

Adler, E. (1977). Lignin chemistry — past present and future. *Wood Sci. Technol.*, **11**, 169–218.
Akin, D. E. (1979). Microscopic evaluation of forage digestion by rumen microorganisms—a review. *J. Anim. Sci.*, **48**, 701–10.
Akin, D. E. (1980). Attack on lignified grass cell walls by a facultatively anaerobic bacterium. *Appl. Environ. Microbiol.*, **40**, 809–20.
Akin, D. E. (1986). Chemical and biological structure in plants as related to microbial degradation of forage cell walls. In *Control of Digestion and Metabolism in Ruminants*, ed. L. P. Milligan, W. L. Grovum & A. Dobson. Prentice-Hall, Englewood Cliffs, New Jersey, pp. 139–57.
Akin, D. E. (1988). Biological structure of lignocellulose and its degradation in the rumen. *Anim. Sci. Feed Technol.* (in press).
Akin, D. E. & Rigsby, L. L. (1985). Influence of phenolic acids on rumen fungi. *Agron. J.*, **77**, 180–2.
Albersheim, P., Bauer, W. D., Keestra, K. & Talmadge, K. W. (1973). The structure of the wall of suspension-cultured sycamore cells. In *Biogenesis of Plant Cell Wall Polysaccharides*, ed. F. Loewus. Academic Press, London, pp. 117–47.
Alderman, G. (1986). Present and future feed resources for ruminants. In *New Developments and Future Perspectives in Research on Rumen Function*, ed. A. Neimann-Sorensen. EEC, Luxembourg, pp. 1–20.
Åman, P. & Hesselman, K. (1984). Analysis of starch and other main constituents of cereal grains. *Swedish J. Agric. Res.*, **14**, 135–9.
Aspinall, G. O. (1981). Constitution of plant cell wall polysaccharides. In *Plant Carbohydrates*. II. Extracellular Carbohydrates, ed. W. Tanner & F. A. Loewus. Springer-Verlag, Heidelberg, pp. 3–8.
Atkins, E. D. T., Isaac, D. H. & Elloway, H. F. (1979). Conformations of microbial extracellular polysaccharides by X-ray diffraction progress on the *Klebsiella* serotypes. In *Microbial Polysaccharides and Polysaccharidases*, ed. R. C. W. Berkeley, G. W. Gooday & D. C. Ellwood. Academic Press, London, pp. 161–89.

Bailey, R. W. & Gaillard, B. D. E. (1965). Carbohydrases of the rumen ciliate *Epidinium ecaudatum* (Crawley): hydrolysis of plant hemicellulose fractions and β-linked glucose polymers. *Biochem. J.*, **95**, 758–66.

Bailey, R. W., Clarke, R. T. J. & Wright, D. E. (1962). Carbohydrases of the rumen ciliate *Epidinium ecaudatum* (Crawley): action on plant hemicellulose. *Biochem. J.*, **83**, 517–23.

Bauchop, T. (1980). Scanning electron microscopy in the study of microbial digestion of plant fragments in the gut. In *Contemporary Microbial Ecology*, ed. D. C. Ellwood, J. N. Hedger, J. N. Latham, J. M. Lynch & J. H. Slater. Academic Press, London, pp. 305–26.

Bauchop, T. (1981). The anaerobic fungi in rumen fibre digestion. *Agric. Environ.*, **6**, 339–48.

Benner, R., Maccubbin, A. E. & Hodson, R. E. (1984). Anaerobic biodegradation of lignocellulose and synthetic lignin by sediment microflora. *Appl. Environ. Microbiol.*, **47**, 998–1004.

Berghem, L. E. R. & Pettersson, L. G. (1973). The mechanism of enzymatic cellulose degradation: purification of a cellulolytic enzyme from *Trichoderma viride* active on highly ordered cellulose. *Eur. J. Biochem.*, **37**, 21–30.

Beveridge, R. G. & Richards, G. N. (1975). Digestion of polysaccharide constituents of tropical pasture herbage in the bovine rumen. VI. Investigations of the digestion of cell wall polysaccharides of spear grass and of cotton cellulose by viscometry and by X-ray diffraction. *Carbohydr. Res.*, **43**, 167–72.

Biely, P. (1985). Microbial xylanolytic systems. *Trends Biotechnol.*, **3**, 286–90.

Biely, P., Puls, J. & Schneider, H. (1985). Acetyl xylan esterases in fungal cellulolytic systems. *FEBS Lett.*, **186**, 80–4.

Biely, P., MacKenzie, C. R., Puls, J. & Schneider, H. (1986). Cooperativity of esterases and xylanases in the enzymatic degradation of acetyl xylan. *Biotechnology*, **4**, 731–3.

Brice, R. E. & Morrison, I. M. (1982). The degradation of isolated hemicelluloses and lignin-hemicellulose complexes by cell-free, rumen hemicellulases. *Carbohydr. Res.*, **101**, 93–100.

Burke, D., Kaufman, P., McNeil, M. & Albersheim, P. (1974). The structure of plant cell walls. VI. A survey of the walls of suspension cultured monocots. *Plant Physiol.*, **54**, 109–15.

Cheng, K.-J., Stewart, C. S., Dinsdale, D. & Costerton, J. W. (1983/4). Electron microscopy of bacteria involved in the digestion of plant cell walls. *Anim. Feed Sci. Technol.*, **10**, 93–120.

Chesson, A. (1980). Maceration in relation to the post-harvest handling and processing of plant material. *J. Appl. Bacteriol.*, **48**, 1–45.

Chesson, A. (1981). Effects of sodium hydroxide on cereal straws in relation to the enhanced degradation of structural polysaccharides by rumen microorganisms. *J. Sci. Food Agric.*, **32**, 745–58.

Chesson, A. (1986). The evaluation of dietary fibre. In *Feedingstuffs Evaluation: Modern Aspects, Problems, Future Trends*, ed. R. M. Livingstone. Feeds Publication 1, Aberdeen, pp. 18–25.

Chesson, A., Stewart, C. S. & Wallace, R. J. (1982). Influence of plant phenolic acids on growth and cellulolytic activity of rumen bacteria. *Appl. Environ. Microbiol.*, **44**, 597–603.

Chesson, A., Gordon, A. H. & Lomax, J. A. (1983). Substituent groups linked by alkali-labile bonds to arabinose and xylose residues of legume, grass and cereal straw cell walls and their fate during digestion by rumen microorganisms. *J. Sci. Food Agric.*, **34**, 1330–40.

Chesson, A., Gordon, A. H. & Lomax, J. A. (1985). Methylation analysis of mesophyll, epidermis and fibre cell-walls isolated from the leaves of perennial and Italian ryegrass. *Carbohydr. Res.*, **141**, 137–47.

Chesson, A., Stewart, C. S., Dalgarno, K. & King, T. P. (1986). Degradation of isolated grass mesophyll, epidermis and fibre cell walls in the rumen and by cellulolytic rumen bacteria in axenic culture. *J. Appl. Bacteriol.*, **60**, 327–36.

Clarke, A. E., Anderson, R. L. & Stone, B. A. (1979). Form and function of arabinogalactans and arabinogalactan-proteins. *Phytochemistry*, **18**, 521–40.

Coleman, G. S. (1980). Rumen ciliate protozoa. *Advan. Parasitol.*, **18**, 121–73.

Coleman, G. S. (1985). The cellulase content of 15 species of entodiniomorphid protozoa,

mixed bacteria and plant debris isolated from the ovine rumen. *J. Agric. Sci.* (Camb.), **104**, 349–60.
Coleman, G. S. & Laurie, J. I. (1976). The uptake and metabolism of glucose, maltose and starch by the ciliate *Epidinium ecaudatum caudatum*. *J. Gen. Microbiol.*, **95**, 257–64.
Coleman, G. S. & Laurie, J. I. (1977). The metabolism of starch glucose, amino acids, purines, pyrimidines and bacteria by the rumen ciliate *Polyplastron multivesiculatum*. *J. Gen. Microbiol.*, **98**, 29–37.
Coleman, G. S., Laurie, J. I., Bailey, J. E. & Holdgate, S. A. (1976). The cultivation of cellulolytic protozoa isolated from the rumen. *J. Gen. Microbiol.*, **95**, 144–50.
Dahlqvist, A. & Nilsson, U. (1984). Cereal fructosans. Part 1. Isolation and characterization of fructosans from wheat flour. *Food Chem.*, **14**, 103–12.
Engels, F. M. & Brice, R. E. (1985). A barrier covering lignified cell walls of barley straw that restricts access by rumen microorganisms. *Curr. Microbiol.*, **12**, 217–24.
Erickson, von M., Miksohe, G. E. & Somfai, I. (1973). Charakterisierung der lignine von angiospermen durch oxydativen abbau. II. Monokotylen. *Holzforschung*, **27**, 147–50.
Fitt, L. E. & Snyder, E. M. (1984). Photomicrographs of starches. In *Starch*, 2nd edn, ed. R. L. Whistler, J. N. BeMiller & E. F. Paschall. Academic Press, New York, pp. 675–689.
Ford, C. W. (1986). Comparative structural studies of lignin-carbohydrate complexes from *Digitaria decumbens* (Pangola grass) before and after chlorite delignification. *Carbohydr. Res.*, **147**, 101–17.
Forsberg, C. W. (1986). Mechanism of bacterial attachment to dietary fibre in the rumen. In *Proceedings of 13th International Congress of Nutrition*, ed. T. G. Taylor & N. K. Jenkins. Libbey, London, pp. 193–5.
Forsberg, C. W., Beveridge, T. J. & Hellstrom, A. H. (1981). Cellulase and xylanase release from *Bacteroides succinogenes* and its importance in the rumen environment. *Appl. Environ. Microbiol.*, **42**, 886–96.
French, D. (1973). Chemical and physical properties of starch. *J. Anim. Sci.*, **37**, 1048–61.
French, D. (1984). Organisation of starch granules. In *Starch*, 3nd edn, ed. R. L. Whistler, J. N. BeMiller & E. F. Paschall. Academic Press, New York, pp. 183–247.
Glennie, C. W., Daiber, K. H. & Taylor, J. R. N. (1982). Reducing tannin content in sorghum grain by heating. *S. Afr. Food Rev.*, **3**, 51–5.
Gordon, A. H., Lomax, J. A. & Chesson, A. (1983). Glycosidic linkages of legume, grass and cereal cell walls before and after extensive digestion by rumen micro-organisms. *J. Sci. Food Agric.*, **34**, 1341–50.
Gordon, A. H., Lomax, J. A., Dalgarno, K. & Chesson, A. (1985). Preparation and composition of mesophyll, epidermis and fibre cell walls from leaves of perennial ryegrass (*Lolium perenne*) and Italian ryegrass (*Lolium multiflorum*). *J. Sci. Food Agric.*, **36**, 509–19.
Grenet, E. & Demarquilly, C. (1987). Rappels sur la digestion des four dans le rumen (parois) et ses consequences. In *Les Fourages Secs: Recoltes, Traitement, Utilisation*, ed. Demarquilly, C. INRA, Paris, pp. 141–62.
Greve, L. C., Labavitch, J. M. & Hungate, R. E. (1984a). α-L-Arabinofuranosidase from *Ruminococcus albus* 8: purification and possible role in hydrolysis of alfalfa cell wall. *Appl. Environ. Microbiol.*, **47**, 1135–40.
Greve, L. C., Labavitch, J. M., Stack, R. J. & Hungate, R. E. (1984b). Muralytic activities of *Ruminococcus albus* 8. *Appl. Environ. Microbiol.*, **47**, 1141–5.
Griffin, H., Dintzis, F. R., Krull, L. & Baker, F. L. (1984). A microfibril generating factor from the enzyme complex of *Trichoderma reesei*. *Biotechnol. Bioeng.*, **26**, 296–300.
Grouleau, D. & Forsberg, C. W. (1981). Cellulolytic activity of the rumen bacterium *Bacteroides succinogenes*. *Can. J. Microbiol.*, **27**, 517–30.
Grouleau, D. & Forsberg, C. W. (1983). Partial characterization of the extracellular carboxymethyl cellulase activity produced by the rumen bacterium *Bacteroides succinogenes*. *Can. J. Microbiol.*, **29**, 504–17.
Halliwell, G. & Bryant, M. P. (1963). The cellulolytic activity of pure strains of bacteria from the rumen of cattle. *J. Gen. Microbiol.*, **32**, 441–8.

Halliwell, G. & Riaz, M. (1970). The formation of short fibres from native cellulose by components of *Trichoderma koningii* cellulase. *Biochem. J.*, **116**, 35–42.
Hartley, R. D. & Jones, E. C. (1976). Diferulic acid as a component of cell walls of *Lolium multiflorum*. *Phytochemistry*, **15**, 1157–60.
Hartley, R. D. & Keene, A. S. (1984). Aromatic aldehyde constituents of Graminaceous cell walls. *Phytochemistry*, **23**, 1305–7.
Hizukuri, S., Takeda, Y., Yasuda, M. & Suzuki, A. (1981). Multi-branched nature of amylose and the action of debranching enzymes. *Carbohydr. Res.*, **94**, 205–13.
Hobson, P. N., Mann, S. O. & Stewart, C. S. (1981). Growth and rumen function of gnotobiotic lambs fed on starchy diets. *J. Gen. Microbiol.*, **126**, 219–30.
Howarth, R. E., Goplean, B. P., Fesser, A. C. & Brandt, S. A. (1978). A possible role for leaf cell rupture in legume pasture bloat. *Crop. Sci.*, **18**, 129–33.
Howarth, R. E., Cheng, K.-J., Majak, W. & Costerton, J. W. (1986). Ruminant bloat. In *Control of Digestion and Metabolism in Ruminants*, ed. L. P. Milligan, W. L. Grovum & A. Dobson. Prentice-Hall, Englewood Cliffs, New Jersey, USA, pp. 516–27.
Huang, L. & Forsberg, C. W. (1987a). Purification and characterisation of an extracellular chloride-stimulated cellobiosidase from *Bacteroides succinogenes*. *Proc. Can. Soc. Microbiol.* (in press).
Huang, L. & Forsberg, C. W. (1987b). Isolation of a cellodextrinase from *Bacteroides succinogenes*. *Appl. Environ. Microbiol.*, **53**, 1034–41.
Hungate, R. E. (1966). *The Rumen and its Microbes*. Academic Press, New York.
Hungate, R. E. (1975). The rumen microbial ecosystem. *Ann. Rev. Ecol. Syst.*, **6**, 39–60.
Hwa, V. & Ferenci, T. (1984). Binding of *Escherichia coli* K12 to cellulose. *FEMS Microbiol. Lett.*, **25**, 11–15.
Ishii, S. (1984). Cell wall cementing materials of grass leaves. *Plant Physiol.*, **76**, 959–61.
Jung, H. G. (1985). Inhibition of structural carbohydrate fermentation by forage phenolics. *J. Sci. Food Agric.*, **36**, 74–80.
Jung, H. G. & Fahey, G. C. (1983). Interactions among phenolic monomers and *in vitro* fermentation. *J. Dairy Sci.*, **66**, 1255–63.
Jung, H. G., Fahey, G. C. & Merchen, N. R. (1983). Effects of ruminant digestion and metabolism on phenolic monomers of forages. *Br. J. Nutr.*, **50**, 637–51.
Kato, Y. & Nevins, D. J. (1984). Structure of the arabinogalactan from *Zea* shoots. *Plant Physiol.*, **74**, 562–8.
Kato, Y. & Nevins, D. J. (1985). Isolation and identification of (O-ferulolyl-α-L-arabinofuranosyl)-(1–3)-O-β-D-xylopyranose as a component of *Zea* shoot cell-walls. *Carbohydr. Res.*, **137**, 134–50.
Kock, J. L. F., Groenewalld, E. G., Kruger, G. H. J. & Lategan, P. M. (1986). Effect of duration of liquefaction on viscosity and polyphenol content of the mash and subsequent glucose recovery in bird-proof sorghum. *J. Sci. Food Agric.*, **37**, 147–50.
Kosaric, N., Cosentino, G. P., Wieczorek, A. & Duvnjak, Z. (1984). The Jerusalem artichoke as an agricultural crop. *Biomass*, **5**, 1–36.
Krassig, H. (1985). Structure of cellulose and its relation to properties of cellulose fibres. In *Cellulose and its Derivatives*, ed. J. F. Kennedy, G. O. Phillips, D. J. Wedlock & P. A. Williams. Ellis Horwood, Chichester, pp. 3–25.
Kudo, H., Cheng, K.-J. & Costerton, J. W. (1987). Electron microscopic study of the methylcellulose-mediated detachment of cellulolytic rumen bacteria from cellulose fibers. *Can. J. Microbiol.*, **33**, 244–8.
Latham, M. J., Brooker, B. E., Pettipher, G. L. & Harris, P. J. (1978a). *Ruminococcus flavefaciens* cell coat and adhesion to cotton cellulose and to cell walls in leaves of perennial ryegrass (*Lolium perenne*). *Appl. Environ. Microbiol.*, **35**, 156–65.
Latham, M. J., Brooker, B. E., Pettipher, G. L. & Harris, P. J. (1978b). Adhesion of *Bacteroides succinogenes* in pure culture and in the presence of *Ruminococcus flavefaciens* to cell walls in leaves of perennial ryegrass (*Lolium perenne*). *Appl. Environ. Microbiol.*, **35**, 1166–73.

Latham, M. J., Hobbs, D. G. & Harris, P. J. (1979). Adhesion of rumen bacteria to alkali-treated plant stems. *Ann. Rech. Vet.*, **10**, 244–5.
Lau, J. M., McNeil, M., Darvill, A. G. & Albersheim, P. (1985). Structure of the backbone of rhamnogalacturonan I, a pectin polysaccharide in the primary cell walls of plants. *Carbohydr. Res.*, **137**, 111–25.
Leatherwood, J. M. (1965). Cellulase from *Ruminococcus albus* and mixed rumen microorganisma. *Appl. Microbiol.*, **13**, 771–5.
Leatherwood, J. M. (1969). Cellulase complex of *Ruminococcus* and a new mechanism for cellulose degradation. In *Adv. Chem. Ser.*, ed. R. F. Gould. American Chemical Society, Washington, pp. 53–7.
Mackie, R. I., Gilchrist, F. M. C., Robberts, A. M., Hanah, P. E. & Schwartz, H. M. (1978). Microbiological and chemical changes in the rumen during the stepwise adaptation of sheep to high concentrate diets. *J. Agric. Sci.* (Camb.), **90**, 241–54.
Manners, D. J. & Matheson, N. K. (1981). The fine structure of amylopectin. *Carbohydr. Res.*, **90**, 99–110.
Markwalder, H. U. & Neukom, H. (1976). Diferulic acid as a possible crosslink in hemicelluloses from wheat germ. *Phytochemistry*, **15**, 836–7.
Marounek, M. & Bartos, S. (1986). Stoichiometry of glucose and starch splitting by strains of amylolytic bacteria from the rumen and anaerobic digester. *J. Appl. Bacteriol.*, **61**, 81–6.
Martin, A. K. (1978). The metabolism of aromatic compounds in ruminants. In *The Hannah Research Institute*, ed. J. A. Moore and J. A. F. Rook. Hannah Research Institute, Ayr, UK, pp. 148–63.
McGavin, M. J. & Forsberg, C. W. (1987). Purification and characterisation of an endoglucanase from *Bacteroides succinogenes*. *Proc. Amer. Soc. Microbiol.*, K15.
McWethy, S. J. & Hartman, P. A. (1977). Purification and some properties of an extracellular alpha-amylase from *Bacteroides amylophilus*. *J. Bacteriol.*, **129**, 1537–44.
Minato, H. & Suto, T. (1981). Technique for fractionation of bacteria in rumen microbial ecosystem. IV. Attachment of rumen bacteria to cellulose powder and elution of bacteria attached to it. *J. Gen. Appl. Microbiol.*, **27**, 21–31.
Monro, J. A., Bailey, R. W. & Penny, J. (1976). The organisation and growth of primary cell walls of lupin hypocotyl. *Phytochemistry*, **15**, 1193–8.
Monties, B. (1985). Recent advances on lignin inhomogeneity. In *The Biochemistry of Plant Phenolics*, ed. C. F. Van Sumere and P. J. Lea. Oxford University Press, Oxford, pp. 161–81.
Morrison, I. M. (1974). Structural investigations on the lignin-carbohydrate complexes of *Lolium perenne*. *Biochem. J.*, **139**, 197–204.
Mountfort, D. O. & Asher, R. A. (1985). Production and regulation of cellulase by two strains of the rumen anaerobic fungus *Neocallimastix frontalis*. *Appl. Environ. Microbiol.*, **49**, 1314–22.
Mueller-Harvey, I., Hartley, R. D., Harris, P. J. & Curzon, E. H. (1986). Linkage of *p*-coumaroyl and feruloyl groups to cell-wall polysaccharides of barley straw. *Carbohydr. Res.*, **148**, 71–85.
Neilson, M. J. & Richards, G. N. (1982). Chemical structures in a lignin-carbohydrate complex isolated from the bovine rumen. *Carbohydr. Res.*, **104**, 121–38.
Nisizawa, T., Suzuki, H. & Nisizawa, K. (1966). 'Swelling factor' activity of *Trichoderma*-cellulase for absorbent cotton. *J. Ferment. Technol.*, **44**, 659–68.
Ohmiya, K., Shimizu, M., Taya, M. & Shimizu, S. (1982). Purification and properties of cellobiosidase from *Ruminococcus albus*. *J. Bacteriol.*, **150**, 407–9.
Ohmiya, K., Shirai, M., Kurachi, Y. & Shimizu, S. (1985). Isolation and properties of β-glucosidase from *Ruminococcus albus*. *J. Bacteriol.*, **161**, 432–4.
Orpin, C. G. & Bountiff, L. (1978). Chemotaxis in the rumen phycomycete *Neocallimastix frontalis*. *J. Gen. Microbiol.*, **104**, 113–22.
Orpin, C. G. & Letcher, A. J. (1979). Utilisation of cellulose starch, xylan and other

hemicelluloses for growth by the rumen phycomycete, *Neocallimastix frontalis*. *Curr. Microbiol.*, **3**, 121–4.

Pearce, P. D. & Bauchop, T. (1985). Glycosidases of the rumen anaerobic fungus *Neocallimastix frontalis* grown on cellulosic substrates. *Appl. Environ. Microbiol.*, **49**, 1265–9.

Pettipher, G. L. & Latham, M. J. (1979a). Characteristics of enzymes produced by *Ruminococcus flavefaciens* which degrade plant cell walls. *J. Gen. Microbiol.*, **110**, 21–7.

Pettipher, G. L. & Latham, M. J. (1979b). Production of enzymes degrading plant cell walls and fermentation of cellobiose by *Ruminococcus flavefaciens* in batch and continuous culture. *J. Gen. Microbiol.*, **110**, 29–38.

Prins, R. A. (1977). Biochemical activities of gut micro-organisms. In *Microbial Ecology of the Gut*, ed. R. T. J. Clarke & T. Bauchop. Academic Press, London, pp. 73–183.

Prins, R. A. & Van Hoven, W. (1977). Carbohydrate fermentation by the rumen ciliate *Isotricha prostoma*. *Protistologica*, **13**, 549–56.

Puls, J., Schmidt, O. & Granzow, C. (1987). α-Glucuronidase in two microbial xylanolytic systems. *Enzyme Microb. Technol.*, **9**, 83–8.

Reese, E. T., Siu, R. G. H. & Levinson, H. S. (1951). The biological degradation of soluble cellulose derivatives and its relationship to the mechanism of cellulose hydrolysis. *J. Bacteriol.*, **59**, 485–91.

Russell, J. B. (1985). Fermentation of cellodextrins by cellulolytic and noncellulolytic rumen bacteria. *Appl. Environ. Microbiol.*, **49**, 572–6.

Russell, J. B. & Hespell, R. B. (1981). Microbial rumen fermentation. *J. Dairy Sci.*, **64**, 1153–69.

Salomonsson, A. C., Theander, O. & Åman, P. (1978). Quantitative determination by g.l.c. of phenolic acids as ethyl derivatives in cereal straws. *Agric. Food Chem.*, **26**, 830–5.

Setlow, P. & Sacks, L. E. (1983). Cyclic AMP is not detectable in *Clostridium perfringens*. *Can. J. Microbiol.*, **29**, 1228–30.

Silley, P. (1985). A note on the pectinolytic enzymes of *Lachnospira multiparus*. *J. Appl. Bacteriol.*, **58**, 145–9.

Sipat, A., Taylor, K. A., Lo, R. Y. C., Forsberg, C. W. & Krell, P. J. (1987). Molecular cloning of xylanase gene from *Bacteroides succinogenes* and its expression in *Escherichia coli*. *Appl. Environ. Microbiol.*, **53**, 477–81.

Smith, D. (1973). The nonstructural carbohydrates. In *Chemistry and Biochemistry of Herbage*. Vol. 1, ed. G. W. Butler & R. W. Bailey. Academic Press, New York and London, pp. 105–55.

Smith, M. M. & Hartley, R. D. (1983). Occurrence and nature of ferulic acid substitution of cell-wall polysaccharides in Graminaceous plants. *Carbohydr. Res.*, **118**, 65–80.

Smith, W. R., Yu, I. & Hungate, R. E. (1973). Factors affecting cellulolysis by *Ruminococcus albus*. *J. Bacteriol.*, **414**, 729–37.

Stack, R. J. & Cotta, M. A. (1986). Effect of 3-phenyl propionic acid on growth of and cellulose utilization by ruminal bacteria. *Appl. Environ. Microbiol.*, **52**, 209–10.

Stack, R. J. & Hungate, R. E. (1984). Effect of 3-phenyl propanoic acid on capsule and cellulases of *Ruminococcus albus* 8. *Appl. Environ. Microbiol.*, **48**, 218–23.

Stephen, A. M. (1983). Other plant polysaccharides. In *The Polysaccharides*, Vol. 2, ed. G. O. Aspinall. Academic Press, New York, pp. 97–193.

Szylit, O., Durand, M., Borgida, L. P., Atinkpahoun, P. F. & Delort-Laval, J. (1978). Raw and steam pelleted cassava, sweet potato and yam cayenensis as starch sources for ruminant and chicken diets. *Anim. Feed Sci. Technol.*, **3**, 73–87.

Taylor, K. A., Crosby, B., McGavin, M., Forsberg, C. W. & Thomas, D. Y. (1987). Characteristics of the endoglucanase encoded by a *cel* gene from *Bacteroides succinogenes* expressed in *Escherichia coli*. *Appl. Environ. Microbiol.*, **53**, 41–6.

Thomas, G. J. (1960). Metabolism of the soluble carbohydrates of grasses in the rumen of sheep. *J. Agric. Sci.* (Camb.), **54**, 360–72.

Tomasic, J. H., Jennings, H. J. & Glaudemans, C. P. J. (1978). Evidence for a single type of linkage in a fructofuranan from *Lolium perenne*. *Carbohydr. Res.*, **62**, 127–33.

Van der Linden, Y., Van Gylswyk, N. O. & Schwartz, H. M. (1984). Influence of supplementation of corn stover with corn grain on the fibrolytic bacteria in the rumen of sheep and their relation to the intake and digestion of fiber. *J. Anim. Sci.*, **59**, 772–83.

Van Soest, P. J. (1981). *Nutritional Ecology of the Ruminant*. O & B Books, Corvallis, Oregon, pp. 1–373.

Waldo, D. R. (1973). Extent and partition of cereal grain starch digestion in ruminants. *J. Anim. Sci.*, **37**, 1062–74.

Walker, G. J. (1965). The cell-bound α-amylase of *Streptococcus bovis*. *Biochem. J.*, **94**, 289–98.

Walker, G. J. & Hope, P. M. (1964). Degradation of starch granules by some amylolytic bacteria from the rumen of sheep. *Biochem. J.*, **90**, 398–408.

Whistler, R. L. & Daniel, J. R. (1984). Molecular structure of starch. In *Starch*, 2nd edn, ed. R. L. Whistler, J. N. BeMiller & E. F. Paschall. Academic Press, New York, pp. 153–82.

Wilkie, K. C. B. (1979). The hemicelluloses of grasses and cereals. *Adv. Carbohydr. Chem. Biochem.*, **36**, 215–64.

Wilkie, K. C. B. (1983). Hemicellulose. *Chemtech.*, **13**, 306–19, 497.

Williams, A. G. (1986). Rumen holotrich ciliate protozoa. *Microbial Rev.*, **51**, 25–49.

Williams, A. G. & Coleman, G. S. (1985). Hemicellulose-degrading enzymes in rumen ciliate protozoa. *Curr. Microbiol.*, **12**, 85–90.

Williams, A. G. & Orpin, C. G. (1987a). Polysaccharide degrading enzymes formed by three species of anaerobic rumen fungi grown on a range of carbohydrate substrates. *Can. J. Microbiol.*, **33**, 418–26.

Williams, A. G. & Orpin, C. G. (1987b). Glycosidase hydrolase enzymes present in the zoospore and vegetative growth stages of rumen fungi *Neocallimastix patriciarum*, *Piromonas communis* and an unidentified isolate, grown on a range of carbohydrates. *Can. J. Microbiol.*, **33**, 427–34.

Williams, A. G. & Strachan, N. S. (1984). The distribution of polysaccharide-degrading enzymes in the bovine rumen digesta ecosystem. *Curr. Microbiol.*, **10**, 215–20.

Williams, A. G. & Withers, S. E. (1982a). The production of plant cell wall polysaccharide-degrading enzymes by hemicellulolytic rumen bacterial isolates growth on a range of carbohydrate substrates. *J. Appl. Bacteriol.*, **52**, 377–87.

Williams, A. G. & Withers, S. E. (1982b). The effect of the carbohydrate growth substrate on the glycosidase activity of hemicellulose-degrading rumen bacterial isolates. *J. Appl. Bacteriol.*, **52**, 389–401.

Williams, A. G., Withers, S. E. & Coleman, G. S. (1984). Glycoside hydrolases of rumen bacteria and protozoa. *Curr. Microbiol.*, **10**, 287–94.

Williams, A. G., Ellis, A. B. & Coleman, G. S. (1986). Subcellular distribution of polysaccharide depolymerase and glycoside hydrolase enzymes in rumen ciliate protozoa. *Curr. Microbiol.*, **13**, 139–47.

Wojciechowicz, M., Heinrichova, K. & Zioleck, A. (1982). An exopectate lyase of *Butyrivibrio fibrisolvens* from the bovine rumen. *J. Gen. Microbiol.*, **128**, 2661–5.

Wood, T. M. (1985). Properties of cellulolytic enzyme systems. *Biochem. Soc. Trans.*, **13**, 407–10.

Wood, T. M. & McCrae, S. I. (1979). Synergism between enzymes involved in the solubilization of native cellulose. In *Adv. Chem. Ser.* 181, ed. R. D. Brown Jr & L. Jurasek. American Chemical Society, Washington, pp. 181–210.

Wood, T. M. & Wilson, C. A. (1984). Some properties of the endo-(1–4)-β-D-glucanase synthesized by the anaerobic cellulolytic rumen bacterium *Ruminococcus albus*. *Can. J. Microbiol.*, **30**, 316–21.

Wood, T. M., Wilson, C. A. & Stewart, C. S. (1982). Preparation of the cellulase from the cellulolytic anaerobic rumen bacterium *Ruminococcus albus* and its release from the bacterial cell wall. *Biochem. J.*, **205**, 129–37.

Wood, T. M., Wilson, C. A., McCrae, S. I. & Joblin, K. N. (1986). A highly active extracellular cellulase from the anaerobic rumen fungus *Neocallimastix frontalis*. *FEMS Microbiol. Lett.*, **34**, 37–40.

Yu, I. & Hungate, R. E. (1979). The extracellular cellulases of *Ruminococcus albus*. *Ann. Rech. Vet.*, **10**, 251–4.

9
Lipid Metabolism in the Rumen

C. G. Harfoot

Department of Biological Sciences, University of Waikato, Hamilton, New Zealand

&

G. P. Hazlewood

AFRC Institute of Animal Physiology and Genetics Research, Cambridge, UK

Ruminants supply humans with a readily available source of fat in the form of both tissue and milk lipids. It has been known for over 50 years that the compositions of ruminant tissue and milk lipids differ markedly from those of non-ruminant herbivores (Banks & Hilditch, 1931), and much research has been done on ruminant lipids and on the microbial transformations in the rumen which are responsible for the distinctive lipid composition. Two major reviews on lipid metabolism in the rumen are by Viviani (1970) and by Harfoot (1978), and much of the earlier work has been referred to in detail in these reviews, along with briefer mention in more general accounts such as those of Hungate (1966), Prins (1977) and Hobson & Wallace (1982a,b).

The emphasis in this chapter is on aspects of lipid metabolism in the rumen which are purely microbial, i.e. microbial lipid biosynthesis and the lipolysis and biohydrogenation of dietary lipids. For the sake of brevity, emphasis is placed on the researches of the last 10 years or so; for details of earlier work the reader is referred to the reviews listed above, and for details of digestion, absorption and transport of lipids in the ruminant animal the reader is referred to the review by Noble (1978). The use of lipids in manipulation of the rumen fermentation is described in Chapter 13.

ROLE OF MICROORGANISMS IN RUMINANT LIPID METABOLISM

In simple-stomached animals the digestion and absorption of dietary lipid occurs from the small intestine onwards; only negligible changes occur anterior to the small intestine. In the ruminant animal the situation is very different, owing to the activities of microorganisms in the reticulo-rumen. On entering the rumen, dietary acyl lipids are subject to hydrolysis by microbial lipases. Once liberated as free fatty acids, any unsaturated fatty acids are subject to biohydrogenation by rumen bacteria, the end product of this hydrogenation being stearic acid (18:0). Synthesis

Table 1
Fatty Acid Composition (% by weight of total) in Dietary Lipids and in Rumen Liquor[a]

Fatty acid	Dietary lipid	Total fatty acids	Rumen lipid	
			Free fatty acids	Esterified fatty acids
14:0	4·41	2·64	0·79	4·24
16:0	21·48	21·85	23·02	26·50
16:1	1·74	2·11	2·76	traces
18:0	5·01	22·75	48·08	23·94
18:1	5·98	11·35	13·06	14·85
18:2	14·26	11·38	5·67	6·40
18:3	38·46	4·13	—	0·97
Other	8·66	23·79[b]	6·62	23·10[b]

[a] Condensed from Table X, p. 293, of Viviani (1970); — not detected.
[b] These apparently high values include a wide range of individual fatty acids present as the complex lipids of rumen microorganisms and result from the condensation of information given by Viviani (1970).

de novo of microbial lipids also takes place in the rumen, and free acids, both saturated and unsaturated, may be incorporated into microorganisms in the course of cell synthesis. As a result of these microbial transformations, the fatty acids contained in rumen lipids and post-ruminal digesta differ from those present in the diet, being markedly enriched in stearic acid (18:0) at the expense of dietary linoleic (18:2) and linolenic (18:3) acids (see Table 1).

COMPOSITION OF DIETARY LIPIDS

Lipids in the ruminant diet are derived from forage crops and from supplements added to the diet in the form of crushed cereal grains to which are added either crushed oil seeds or their extracts. These supplements are usually referred to as 'concentrates'.

The lipids of forages constitute about 6–7% of the dry weight of leaf tissue (Shorland, 1963) and consist largely of glycolipids and phospholipids. The percentage compositions of a number of the major lipid classes present in forages are shown in Table 2; together these constitute around 95% of the total leaf lipid, the remainder being phosphatidylinositol and diphosphatidylglycerol (cardiolipin).

The fatty acid compositions of the total lipids of various forage plants are shown in Table 3; it can be seen that the compositions are dominated by the unsaturated linolenic (18:3) and linoleic (18:2) fatty acids. In contrast, the seed oils used in concentrates contain predominantly linoleic acid (18:2) and oleic acid (18:1), which are present in triglycerides rather than in glycolipids and phospholipids. It can be seen from Table 3 how the presence of concentrates changes the fatty acid composition of the ruminant diet.

Table 2
Percentage Composition of Some Lipid Classes Present in Common Forage Plants

Plant	PC	PG	PE	MGDG	DGDG	Reference
Medicago sativa[a]	4·6	4·2	5·4	53·5	32·3	Roughan & Batt (1969)
Trifolium repens[a]	4·5	4·1	5·2	46·7	31·2	Roughan & Batt (1969)
T. repens (juvenile)[b]	16·4	11·5	13·1	45·9	13·1	Trémolières (1970)
T. repens (adult)[b]	3·2	4·5	2·3	67·5	22·5	Trémolières (1970)
Lolium perenne[a]	8·0	11·1	6·2	42·1	32·6	Roughan & Batt (1969)
Paspalum sp.[a]	5·6	3·4	4·3	54·2	32·5	Roughan & Batt (1969)
Zea mais[a]	5·1	7·4	3·8	48·1	35·7	Roughan & Batt (1969)

PC, phosphatidylcholine; PG, phosphatidylglycerol; PE, phosphatidylethanolamine; MGDG, monogalactosyl diglyceride; DGDG, digalactosyl diglyceride.
[a] Percentages calculated on weight basis.
[b] Percentages calculated on molar basis.

HYDROLYSIS OF DIETARY LIPIDS

The initial step in the transformation of dietary acyl lipids entering the rumen is the hydrolysis of ester linkages by microbial lipolytic enzymes; this step is a prerequisite for the biohydrogenation of unsaturated fatty acids.

For grazing ruminants, the major dietary lipids reflect the considerable quantity of chloroplast membrane ingested, and they comprise galactolipids (mono- and di-galactosyldiglyceride), sulpholipid (sulphoquinovosyldiglyceride) and phospholipids (mainly phosphatidylcholine, phosphatidylglycerol and phosphatidylethanolamine); in contrast, animals receiving cereals or concentrates containing seed oils will have triglyceride as the major dietary lipid. The observation that triglycerides were rapidly hydrolysed by microbial enzymes when incubated *in vitro* with rumen contents was first made by Garton *et al.* (1958), and has since formed the basis of research papers too numerous to be considered here. Free fatty acids produced were rapidly hydrogenated, and at no stage in the hydrolysis of triglyceride was any mono- or di-glyceride detected (Garton *et al.*, 1959, 1961). At about the same time, Dawson (1959) observed that phosphatidylcholine incubated

Table 3
Fatty Acid Composition (% of total fatty acid) of Various Forages and Ruminant Rations

	12:0	14:0	16:0	16:1	17:0	18:0	18:1	18:2	18:3	Other
Forages										
Mixed pasture grasses[a]	2·9	3·3	9·4	3·0	—	1·5	13·19	20–25	30–39	1·4
Mixed pasture grasses[b]	—	1·1	15·9	2·5	—	3·4	13·2	61·3	0·5	—
Clover pasture[c]	—	—	8·9	7·9	—	2·8	9·5	8·1	58·9	3·9
Rations[d]										
Lucerne + concentrates	0·6	2·0	24·2	4·4	0·9	2·1	9·6	33·3	18·2	4·7
Concentrates	0·1	1·1	15·0	2·9	0·5	0·4	13·2	57·6	6·2	3·0

[a] Garton (1959). [b] Garton (1960). [c] Shorland *et al.* (1955). [d] See Viviani (1970).

with rumen contents or washed rumen microorganisms *in vitro* was metabolised primarily by a combination of phospholipase A, lysophospholipase and GPC-diesterase activities, and to a lesser extent by phospholipase C activity. In view of the quantitative importance of galactolipids in the ruminant diet, it is surprising that the catabolic pathway for these lipids in the rumen was not established until somewhat later. Dawson *et al.* (1974) found that the galactolipids of ^{14}C-labelled pasture grass (*Lolium perenne*), introduced into the ovine rumen, were rapidly hydrolysed with the release of predominantly unsaturated ^{14}C-labelled fatty acids, which were subsequently hydrogenated. In further work (Dawson & Hemington, 1974) it was established that purified ^{14}C-labelled mono- and di-galactosyldiglyceride incubated *in vitro* with rumen contents were rapidly hydrolysed by microbial enzymes with the production, via successive deacylations, of mono- and di-galactosylglycerol respectively. In contrast to the results of Faruque *et al.* (1974), suggesting that the lipolysis of triglycerides and galactolipids in pasture-fed cows was due mainly to the activity of endogenous plant enzymes, it was concluded that the enzymes of rumen microorganisms are largely responsible for the degradation of plant lipids in the rumen (Dawson *et al.*, 1977).

Role of bacteria

Lipolytic bacteria, with the potential for degrading the different lipid components of the diet, have been isolated from rumen contents using anaerobic techniques and a combination of differential and selective media (Hobson & Mann, 1961, 1971; Henderson, 1973a, 1975; Prins *et al.*, 1975; Hazlewood & Dawson, 1975a, 1979). The first isolates from sheep rumen contents were selected from medium containing emulsified linseed oil and were Gram-negative, anaerobic, curved rods which hydrolysed linseed oil triglycerides and tributyrin and fermented glycerol (Hobson & Mann, 1961). A single strain, designated 5S, was subsequently re-isolated in numbers which suggested that it could be an important lipolytic organism *in vivo* (Hobson, 1965), but because of its restricted utilisation of commonly occurring sugars it could not be placed in any of the known genera; Hungate (1966) named the organism *Anaerovibrio lipolytica*. Growth characteristics and lipase production during batch and continuous culture of *A. lipolytica* have been reported (Hobson & Summers, 1966, 1967; Henderson *et al.*, 1969; Henderson, 1971; Henderson & Hodgkiss, 1973; see also Chapter 2), and some of the properties of the purified lipase have been elucidated (Henderson, 1971). The enzyme was entirely extracellular, being associated with cell-surface or extracellular membranous structures. Activity was greatest at pH 7·4 and at 20–22°C, and was enhanced by $CaCl_2$ or $BaCl_2$; $ZnCl_2$ and $HgCl_2$ were inhibitory. Diglycerides were hydrolysed more rapidly than were triglycerides. Phospholipids and galactolipids were not attacked, so, although the organism was shown to be present in rumen contents at around 10^7 ml^{-1} (Prins *et al*, 1975), it probably does not play an important role in lipid digestion in animals receiving forage diets rich in these lipids. In another study, triglycerides up to triolein were hydrolysed by the lipases of 5 Gram-negative, curved rods which had morphological and biochemical properties characteristic of the genus *Butyrivibrio* (Latham *et al.*, 1972).

From the observation that rapid hydrolysis of phosphatidylcholine occurred in the rumens of sheep devoid of ciliate protozoa (Dawson & Kemp, 1969), and subsequent fractionation studies with rumen contents (Hazlewood, 1974), it was concluded that phospholipase activity is largely associated with rumen bacteria. Of 200 isolates selected mainly from medium containing emulsified linseed oil, 3 similar Gram-negative rods had phospholipase activity (Hazlewood & Dawson, 1975a). The most active, strain LM8/1B, was a non-cellulolytic strain of *Butyrivibrio fibrisolvens*. Constitutive phospholipase activity was initially mainly cell-associated, but a variable amount was released into the culture supernatant during autolysis. Phospholipid substrates, phosphatidylcholine, phosphatidylethanolamine and phosphatidylinositol, were catabolised, the former two in a manner characteristic of phospholipase A, lysophospholipase, phosphodiesterase and phosphomonoesterase activities. Phospholipase activity was optimal at pH 6·5–7·5, was stimulated by cysteine, potassium oleate and SDS, and was inhibited by aerobic conditions; divalent cations were not required for activity (Hazlewood & Dawson, 1975a). Surprisingly, appreciable phospholipase activity occurred at low temperatures and the maximum hydrolysis rate at $-10°C$ was higher than that at $39°C$, unless the system at the latter temperature was stimulated by adding oleic acid or sodium dodecyl sulphate (Hazlewood & Dawson, 1976). The low-temperature activity had an absolute requirement for thiol reagents, was stimulated by Ca^{2+}, Mn^{2+} and Mg^{2+}, and was inhibited by EDTA. It seems likely that the activity at temperatures below zero is the direct result of the formation of a solid phase in the incubation medium and is not associated with a phase change in the hydrated phospholipid substrate. A further novel property of the phospholipase system of *Butyrivibrio* LM8/1B was the ability to produce N-acylphosphatidylethanolamine by catalysing an intermolecular transacylation, whereby one molecule of phosphatidylethanolamine acted as acyl donor and another as acyl acceptor (Hazlewood & Dawson, 1975b).

Phospholipase production and the ability to hydrogenate unsaturated fatty acids are properties found individually in only a small proportion of rumen bacteria; that all bacteria originally isolated for their ability to hydrolyse phosphatidylcholine also had a capacity for hydrogenation suggests that the ability to carry out both processes is advantageous to the biochemical economy of an organism (Hazlewood et al., 1976).

Using a selective medium containing grass galactolipids as sole carbon source, Hazlewood & Dawson (1977, 1979) isolated from ovine rumen contents a fatty acid auxotrophic *Butyrivibrio* (strain S2), which was subsequently shown to possess properties consistent with its having a major role in the metabolism of dietary lipids *in vivo*. The long-chain fatty acids necessary for growth of this organism were obtained by hydrolysis of acyl ester linkages in galactolipids, phospholipids, sulpholipid or the non-ionic Tween detergents (Hazlewood & Dawson, 1977, 1979). ^{14}C-labelled mono- and di-galactosyldiglyceride were rapidly degraded with the production of ^{14}C-labelled fatty acids, in a manner similar to that observed *in vivo* (Dawson & Hemington, 1974; Dawson et al., 1974, 1977). Phosphatidylethanolamine, phosphatidylcholine and phosphatidylinositol were all broken down, with phosphatidylethanolamine the preferred substrate; products liberated from the latter indicated the combined presence of phospholipase A, phospholipase C, lysophospholipase and phosphodiesterase activities, again reminiscent of the various activities described for rumen contents (Dawson, 1959). Triglyceride was not hydrolysed, but diglyceride was rapidly and completely catabolised and served as an excellent source of growth-promoting fatty acids. Polyunsaturated fatty acids contained in the various plant lipids were hydrogenated prior to incorporation into bacterial membrane lipids (Hazlewood & Dawson, 1979). Phospholipase activity in batch cultures was predominantly cell-associated but was detectable in culture supernatant from mid-exponential phase onwards; the latter activity was, however, associated with a particulate fraction of molecular weight in excess of 2×10^6 daltons. Subcellular fractionation of *Butyrivibrio* S2 by means of protoplast formation revealed that phospholipases A and C, and galactolipases, were localised in the cytoplasmic membrane (Hazlewood *et al.*, 1983).

Role of ciliate protozoa and anaerobic fungi

Experimental evidence for the involvement of rumen ciliate protozoa in the hydrolysis of dietary lipid is unconvincing, not least because of a paucity of information regarding the role of intracellular and surface-associated bacteria. Wright (1961) found that protozoal extracts hydrolysed tributyrin, but not triolein, linseed oil or olive oil, and on the basis of the action of penicillin in depressing lipolysis under these conditions suggested that protozoa (mainly *Epidinium* spp.) were responsible for 30–40% of the lipolytic activity in the rumen. In other work (Bailey & Howard, 1963) an extract of *Epidinium ecaudatum* was shown to contain α-galactosidase activity capable of liberating galactose from intact galactolipids, even though lipase activity was not demonstrated. Latham *et al.* (1972) also found lipolytic activity in the protozoal fraction of bovine rumen contents, and estimated that about 30% of the total activity could be associated with protozoa. Another ciliate, *Entodinium caudatum*, engulfs particulate matter, including chloroplasts (Hall *et al.*, 1974) and oil droplets (Coleman & Hall, 1969), which it allegedly hydrolyses (Warner, cited by Prins, 1977). Coleman *et al.* (1971) observed that *E. caudatum* catabolises phosphatidylethanolamine via phospholipase A, lysophospholipase, phosphodiesterase and phosphomonoesterase activities, but this process may be more relevant to the internal economy of the cell than to the digestion of dietary lipid in the rumen.

Rumen phycomycete fungi have not, to date, been shown to have a role in the hydrolysis of dietary lipid in the rumen.

BIOHYDROGENATION IN THE RUMEN

Role of microorganisms

Over 50 years ago, Banks & Hilditch (1931) observed that the tissue lipids of ruminants were more saturated than were those of non-ruminants. On the basis of this observation they concluded that biohydrogenation of forage lipids took place in those tissues. We now know this to be untrue; it is now accepted that hydrogenation takes place in the rumen and, to a lesser extent, other regions of the intestinal tract.

Early workers incubated whole rumen contents with either lipid-rich forages (Reiser, 1951), fatty acids representative of those present in forage (Shorland et al., 1955) or, more recently, ^{14}C-labelled unsaturated fatty acids (Ward et al., 1964). Wright (1959, 1960) appears to have been the first worker to investigate whether bacteria or ciliate protozoa were responsible for biohydrogenation; from the results of experiments involving the incubation of fractionated rumen contents, he concluded that both played a part. However, as pointed out by Dawson & Kemp (1969), the results of such experiments may be confused by the fact that bacteria are extensively ingested by many rumen protozoa (see Coleman, 1964, and Chapter 3) and that those ingested bacteria may well have been responsible for the biohydrogenation activity observed in suspensions of washed rumen protozoa. Dawson & Kemp (1969) adopted an alternative approach in which they eliminated the protozoa from the rumens of sheep by using detergent and compared the rates of biohydrogenation of linolenic and oleic acids in these rumens with those in the normal rumen. They concluded that the presence of protozoa was not necessary for biohydrogenation to occur.

Girard & Hawke (1978) using suspensions of washed holotrich protozoa showed that, although [1-^{14}C]linoleic acid was rapidly incorporated into the phospholipids of these organisms, very little biohydrogenation occurred. Bacterial suspensions from the same animals rapidly hydrogenated linoleic acid to *trans*-11-octadecenoic and stearic acids, however.

Experiments done by Singh & Hawke (1979) using fractionated rumen contents incubated with ^{14}C-labelled monogalactosyldiglyceride produced similar results, leading the authors to suggest that the small contribution made by the protozoa to biohydrogenation was due to the activity of ingested bacteria.

At present, the consensus is that it is the bacteria that are largely responsible for biohydrogenation in the rumen; the protozoa are of only very minor importance.

Role of food particles

The contents of the rumen are far from homogeneous; in addition to a bacterial population of some 10^{10}–10^{11} cells ml^{-1} and a protozoal population of some 10^5–10^6 cells ml^{-1} there is at any one time much dietary material present. It is also clear that the bacteria are far from uniformly distributed in the rumen; present opinion suggests that only about half the total bacterial population occurs in free suspension in the liquid, the rest are found adhering to the food particles (Hungate, 1966; see Chapters 2, 8 and 12).

The role of food particles in biohydrogenation is unclear. Experiments using fractionated rumen contents (Viviani & Borgatti, 1967; Hawke & Silcock, 1970) showed that biohydrogenation proceeded more rapidly and to greater completion in the presence of food particles than in their absence. Other experiments involving the incubation of strained rumen contents with linoleic acid *in vitro* and the

Table 4
Rumen Bacterial Isolates and Their Ability to Hydrogenate Linolenic, Linoleic and Oleic Acids

Group[a,b]		Bacterium	End-products of hydrogenation of			Reference
a	b		α-Linolenic acid (cis-9,cis-12,cis-15-Octadecatrienoic)	Linoleic acid (cis-9,cis-12-Octadecadienoic)	Oleic acid (cis-9-Octadecenoic)	
1	A	*Butyrivibrio fibrisolvens*	—[c]	18:1 (46%)	—	Polan *et al.* (1964)
1	A	*B. fibrisolvens* A38		18:2 *cis*-9,*trans*-11 (68%) 18:1 *trans*-9 and *trans*-11 (23%)	Not hydrogenated	Kepler *et al.* (1966)
1	A	*B. fibrisolvens* A38	18:3 *cis*-9,*trans*-11,*cis*-15 18:2 Δ¹¹, Δ¹⁵	18:2 *cis*-9,*trans*-11 18:1 *trans*-11		Kepler & Tove (1967)
1	A	*B. fibrisolvens* S2	18:1 *trans*-11 (into cellular lipids)	18:1 *trans*-11 (into cellular lipids)		Hazlewood & Dawson (1979)
1	A	*Treponema* (*Borrelia*)	Isomerised, then hydrogenated	18:2 *cis*-9,*trans*-11 (5%) 18:2 *trans*-11 (85%)		Sachan & Davis (1969) Yokoyama & Davis (1971)
1	A	*Micrococcus* sp.	Isomerised, then hydrogenated	18:1 *trans*-11 (major intermediate)		Miles *et al.* (1970)
1	A	*Ruminococcus albus* F2/6	18:1 *trans* (95%) 18:1 *cis* (5%) 18:3 *cis*-9,*trans*-11,*cis*-15 (transient) 18:2 *trans*-11,*cis*-15 (transient)	18:2 *cis*-9,*trans*-11 (transient)[d] 18:1 *trans* (95%) 18:1 *cis* (5%)	Not hydrogenated	Kemp *et al.* (1975)
1	A	*Eubacterium* F2/2	18:3 *cis*-9,*trans*-11,*cis*-15 18:2 *trans*-11,*cis*-15 18:1 *trans*-11	18:2 *cis*-9,*trans*-11 18:1 *trans*-11 (95%)	Not hydrogenated	Kemp *et al.* (1975)
1	A	*Eubacterium* W461	18:3 *cis*-9,*trans*-11,*cis*-15 (transient) 18:2 *trans*-11,*cis*-15 (transient) 18:1 *trans* (50%) 18:*trans*-11 (65% of 18:1 *trans*) 18:1 *cis* (50%) 18:1 *cis*-11 (95% of 18:1 *cis*)	18:2 *cis*-9,*trans*-11 (transient) 18:1 *trans* (50%) 18:1 *trans*-11 (95% of 18:1 *trans*)	Not hydrogenated	Kemp *et al.* (1975)
1	A	R8/3 Gram-negative rod	18:2 *trans*-11,*cis*-15 (100%)	18:1 *trans*-11 (96%)	Not hydrogenated	Hazlewood *et al.* (1976)

1	A	LM8/1A; LM8/1B Gram-negative rods	18:2 trans-11,cis-15 100%	18:1 trans-11 (100%)	Not hydrogenated	Hazlewood et al. (1976)
1	A	2/7/2	18:2 trans-11,cis-15 (100%)	18:1 trans-11 (75%) 18:1 cis-9,cis-12 (25% unchanged)	—	Hazlewood et al. (1976)
2	A	EC7/2 Gram-negative rod	18:2 trans-11,cis-15 (70%) 18:1 trans-11 (30%)	18:1 trans-11 (100%)	Not hydrogenated	Hazlewood et al. (1976)
2	A	R7/5 Gram-negative rod	18:1 trans:cis ratio 1:1 (95%) trans-11 (32%) trans-12 (15%) trans-10 (6%) cis-11 (44%) cis-12 (5%)	18:1 trans:cis ratio 1:2 (95%) trans-11 (4%) trans-10 (11%) trans-12 (8%) cis-11 (62%) cis-12 (5%)	Not hydrogenated	Hazlewood et al. (1976)
2	A	2/9/1 Gram-negative vibrio	18:1 trans:cis ratio 2:1 (95%) trans-11 (41%) trans-10 (14%) trans-12 (11%) cis-11 (29%) cis-12 (4%)	18:1 trans:cis ratio 2·5:1 (100%) trans-11 (46%) trans-10 (15%) trans-12 (12%) cis-11 (25%) cis-12 (5%)	Not hydrogenated	Hazlewood et al. (1976)
3	B	Fusocillus babrahamensis P2/2	18:3 cis-9,trans-11,cis-15 (transient) 18:2 trans-11,cis-15 (15%) 18:1 cis-15 (85%)	18:2 cis-9,trans-11 (transient) 18:1 trans-11 (70%) 18:0 (30%)	18:0 80% 18:0 hydroxy (20%)	White et al. (1970) Hazlewood et al. (1976) Kemp et al. (1975)
3	B	Fusocillus T344	18:3 cis-9,trans-11,cis-15 (transient) 18:2 cis-9,trans-11 (transient) 18:1 cis-15 (85%)	18:2 cis-9,trans-11 (transient) 18:1 trans-11 (65%) 18:0 (35%)	18:1 trans-11 (5%) 18:1 cis-9 (5%) 18:0 (90%)	Hazlewood et al. (1976) Kemp et al. (1975)
3	B	R8/5 Gram-negative rod	18:2 trans-11,cis-15 (25%) 18:1 cis-15 (50%) 18:1 trans-15 (25%)	18:1 trans-11 (50%) 18:0 (40%)	18:1 cis-9 (60%) 18:0 hydroxy (40%)	Hazlewood et al. (1976)

[a] Groups in column a after Hazlewood et al. (1976).
[b] Groups in column b after Kemp & Lander (1984).
[c] No data available.
[d] 'Transient' indicates transient appearance only during incubation period.

subsequent removal and fractionation of samples of the incubation mixture into fine food-particles, bacteria, protozoa and cell-free supernatant were done by Harfoot et al. (1973a,b, 1975). They showed that some 80% of the biohydrogenation of the linoleic acid occurred in association with the fine food-particles and that negligible changes occurred in the cell-free supernatant. Of the small changes that occurred in association with the protozoa and bacteria, only an increase in stearic acid associated with the bacterial fraction was of any significance. These findings were attributed to the adsorption of linoleic acid on to the food particles where hydrogenation occurred through the activity of extracellular enzymes produced by the bacteria either attached to the food particles or free in the supernatant. The increase in stearic acid in the bacterial fraction was attributed to either uptake or adsorption by the bacteria. When trilinolein was used as the substrate, lipolysis appeared to take place in the suspending medium; the liberated linoleic acid then rapidly became adsorbed on to plant particles where it was hydrogenated (Harfoot et al., 1975). Conclusive evidence that biohydrogenation is due, at least in part, to extracellular enzymes is lacking; there are no reports of soluble fatty acid-hydrogenating enzymes produced by rumen bacteria grown in either pure or mixed culture (Hazlewood et al., 1976).

Bacterial species responsible for biohydrogenation

For many years, the only bacterial species known to be capable of biohydrogenation was *Butyrivibrio fibrisolvens* (Polan et al., 1964; Kepler et al., 1966; Kepler & Tove, 1967).

More recently, a range of diverse bacteria has been isolated, showing different capacities to biohydrogenate unsaturated fatty acids (Table 4). Their isolation has stimulated research into their contributions to biohydrogenation, both in pure culture and when grown together in mixed culture. In addition, a number of species of bacteria capable of biohydrogenation have been isolated from the caeca and faeces of rats and mice (Eyssen & Verhulst, 1984; Verhulst et al., 1985). Hazlewood et al. (1976) divided the hydrogenating bacteria into three groups, basing the division on the pattern of end-products of hydrogenation and on the isomerisations carried out. A simpler division into two groups, A and B, has been made for some isolates by Kemp & Lander (1984). Both sets of groupings are shown in Table 4. More recent isolates from rats, mice and humans (Eyssen & Verhulst, 1984; Verhulst et al., 1985) fit into one of these categories (group 1 of Hazlewood et al., 1976; group A of Kemp & Lander, 1984). The roles and activities of these different groups of biohydrogenating bacteria are discussed below (see Biochemistry of biohydrogenation; metabolic pathways and the bacteria involved).

It might be considered that the total number of biohydrogenating species described to date is very small. However, it must be remembered that the isolation of such organisms is time-consuming, as biohydrogenators cannot be isolated by using specific selection pressure; rather, a wide range of bacteria must first be isolated and then pure cultures screened for the ability to biohydrogenate unsaturated fatty acids. Kemp et al. (1975) screened some 200 isolates in order to obtain their 5 strains, and Verhulst et al. (1985) screened 100 isolates in order to obtain their 7 strains.

It is not easy to determine whether the strains isolated are representative of the bacteria carrying out biohydrogenation in the rumen, as there is no method for selectively enumerating biohydrogenating bacteria. Kemp et al. (1975) estimated that their 5 biohydrogenating strains were each present in the rumen at between 10^7 and 10^8 ml^{-1}. Similar population densities are reported for *Butyrivibrio*, but this genus includes many strains, of diverse metabolic ability (Hungate, 1966), only a few of which are capable of biohydrogenation. The *Treponema* strain of Yokoyama & Davis (1971) is also reported to occur in about the same numbers. The difficulties associated with estimating the numbers of biohydrogenating bacteria in the rumen have been discussed by Harfoot (1978).

Biochemistry of biohydrogenation

Substrates for biohydrogenation

The major unsaturated fatty acid entering the rumen of an animal grazing on pasture is linolenic acid (*cis*-9,*cis*-12,*cis*-15-octadecatrienoic acid), present largely in glycolipids and to a lesser extent in phospholipids. In a ruminant receiving dietary supplements, appreciable quantities of linoleic acid (*cis*-9,*cis*-12-octadecadienoic acid) enter the rumen, largely as triglyceride. Before these unsaturated fatty acids can be hydrogenated, it is essential that lipolysis takes place and that they be in the form of unesterified fatty acids; the presence of a free carboxyl group is absolutely required for hydrogenation to take place (Kepler et al., 1970; Hazlewood et al., 1976).

Metabolic pathways and the bacteria involved

Incubation of linolenic or linoleic acids with rumen contents *in vivo* and *in vitro*, or with pure cultures of biohydrogenating bacteria, yields a variety of fatty acids in different proportions (see Table 4). A major problem has been to determine which of these fatty acids are true intermediates in the biohydrogenation pathway and which appear as minor products of metabolism (or chemical isomerisation) and are not further metabolised.

The present consensus is that the biohydrogenation of α-linolenic acid proceeds according to the scheme shown in Fig. 1; the corresponding scheme for the biohydrogenation of linoleic acid is shown in Fig. 2. Both pathways involve an initial isomerisation step resulting in the formation of a conjugated *cis*-9,*trans*-11 acid which then undergoes hydrogenation of its *cis* double bond(s) leaving *trans*-11-octadecenoic acid (*trans*-vaccenic acid, 11-elaidic acid) as the penultimate product. Finally this is hydrogenated to stearic acid. Evidence for these pathways comes from three major sources: (a) incubation of unsaturated fatty acids with rumen contents *in vivo* (Wood et al., 1963; Noble et al., 1969); (b) incubations using rumen contents *in vitro* (Ward et al., 1964; Wilde & Dawson, 1966; Dawson & Kemp, 1970); (c) pure-culture studies using biohydrogenating bacteria isolated from the rumen (Table 4) and from non-ruminant sources (Eyssen & Verhulst, 1984; Verhulst et al., 1985).

All bacteria so far isolated appear to belong to the two distinct populations,

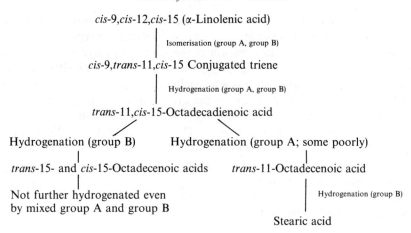

Fig. 1. Scheme for the biohydrogenation of α-linolenic acid; group A and group B refer to the two classes of biohydrogenating bacteria (see text).

group A and group B, of Kemp & Lander (1984) mentioned previously. Members of group A mostly hydrogenate linoleic and α-linolenic acids to *trans*-11-octadecenoic acid, although some isolates produce *trans*-11,*cis*-15-octadecadienoic acid from α-linolenic acid (Hazlewood et al., 1976). They appear to be incapable of hydrogenating octadecenoic acids. In addition to the recent isolates of Kemp et al. (1975), *Butyrivibrio fibrisolvens* belongs to this group, as do a rumen spirochaete (Sachan & Davis, 1969; Yokoyama & Davis, 1971) and some intestinal bacteria from non-ruminants (Eyssen & Verhulst, 1984; Verhulst et al., 1985). Members of group B are capable of hydrogenating a wide range of octadecenoic acids, including *cis*-9 (oleic) and *trans*-11 (*trans*-vaccenic) acids, as well as linoleic acid, to stearic acid. To date, they are the only known hydrogenators of oleic acid to stearic acid. Only 3 isolates of group B are known; they are two species of *Fusocillus* and an unnamed Gram-negative rod, R8/5. Hazlewood et al. (1976) regarded the group A bacteria as comprising two distinct populations: group 1 bacteria capable of hydrogenating only the *cis*-9 double bond of the *cis*-9, *trans*-11 conjugated diene derived from

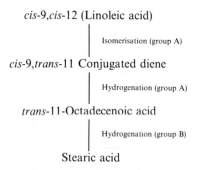

Fig. 2. Scheme for the biohydrogenation of linoleic acid; group A and group B refer to the two classes of biohydrogenating bacteria (see text).

linoleic acid which left the *trans*-11 monoenoic acid as the end-product of hydrogenation, and group 2 bacteria which, in addition, hydrogenate the *cis*-15 double bond of the *cis*-9,*trans*-11,*cis*-15 conjugated triene derived from linolenic acid. Group 3 of Hazlewood *et al.* (1976) corresponds to group B of Kemp & Lander (1984).

From the above, it is clear that the complete hydrogenation of α-linolenic and linoleic acids to stearic can take place only in the presence of members of both group A and group B. Note that, once *trans*-15 and *cis*-15 monoenoic acids have been formed, they are not further hydrogenated by members of either group. Incubations *in vitro* often produce *cis*- or *trans*-15 octadecenoic acids, sometimes as the major end-product (White *et al.*, 1970; Body, 1976); the reasons for this incomplete biohydrogenation route being taken are not understood.

To confirm that members of both group A and group B had to be present before complete hydrogenation of linolenic acid took place, Kemp & Lander (1984) incubated α-linolenic acid with mixed cultures containing pure strains of bacteria belonging to each of these groups. Mixing cultures of *Butyrivibrio fibrisolvens* (group A) with a *Fusocillus* sp. or the Gram-negative rod R8/5 (group B), both previously grown to late exponential phase, resulted in the conversion of approximately two-thirds of the α-linolenic acid substrate into stearic acid; mixing two representatives of group B gave no stearic acid. However, on using a small inoculum of each bacterium and incubating for a longer period of time, the conversion of α-linoleic acid into stearic acid was significantly poorer, presumably because in all instances the group A *Butyrivibrio* outgrew the group B organism. As Kemp & Lander (1984) point out, 'For high yields of stearic acid it must be necessary to strike a balance between the number of each bacterium in the culture and the affinity of each for α-linolenic acid such that the substrate is hydrogenated by the group A bacterium to *trans*-octadec-11-enoic acid which can be hydrogenated to stearic acid by group B bacteria'. The problem of how the *trans*-11-octadecenoic acid is transferred from the group A bacteria to the group B bacteria remains unsolved. It is possible that adsorption on to particles plays a role in this, but at present there is no evidence that the enzymes of rumen biohydrogenation are extracellular (see above).

Biohydrogenation of γ-linolenic acid

The isomer of linolenic acid considered in the above account is the α-form (*cis*-9,*cis*-12,*cis*-15-octadecatrienoic acid), the major fatty acid of grass lipids. The other isomer of linolenic acid, γ-linolenic (*cis*-6,*cis*-9,*cis*-12-octadecatrienoic acid), is present only in small amounts in the lipids of animals and higher plants, but is of frequent occurrence in the lipids of fungi, algae and certain seed oils (Kemp & Lander, 1983).

Kemp & Lander (1983) investigated the biohydrogenation *in vitro* of γ-linolenic acid by *Butyrivibrio* S2 (a group A biohydrogenator) and *Fusocillus babrahamensis* (a group B biohydrogenator). The latter organism was able to hydrogenate γ-linolenic acid completely to stearic acid, and the *Butyrivibrio* only as far as *cis*-6, *trans*-11-octadecadienoic acid. The pathway postulated for the hydrogenation of γ-linolenic acid (Fig. 3) is directly analogous to that for the hydrogenation of

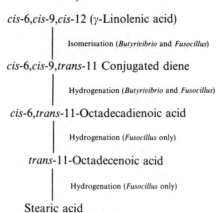

Fig. 3. Scheme for the biohydrogenation of γ-linolenic acid; the *Butyrivibrio* used in the study was a group A biohydrogenator, the *Fusocillus* a group B biohydrogenator. After Kemp & Lander (1983). (Compare Fig. 1 for biohydrogenation of α-linolenic acid.)

α-linolenic acid (Fig. 1), hydrogenation proceeding via a conjugated intermediate. The major difference between the hydrogenation of α-linolenic and γ-linolenic acids is that the former requires two bacteria (one group A, one group B) whereas γ-linolenic acid can be completely hydrogenated by the group B *Fusocillus* acting alone.

Formation of the cis,trans *conjugated diene*
In their studies on the biohydrogenation of linoleic acid by *Butyrivibrio fibrisolvens*, Kepler *et al.* (1966) observed that the first detectable intermediate was a *cis*-9, *trans*-11 conjugated diene. Subsequently, Kepler *et al.* (1970) isolated a Δ^{12}-*cis*,Δ^{11}-*trans* isomerase from this bacterium and tested the substrate specificity of the enzyme against different positional and configurational isomers of octadecadienoic acid. They demonstrated an absolute requirement for the *cis*-9,*cis*-12 diene configuration, along with a free –COOH group at carbon position 1. This enzyme was also shown to have an absolute specificity for *cis*-9,*cis*-12-octadecenoic acids with the final double-bond 6 carbons away from the terminal methyl group (Kepler *et al.*, 1971). Garcia *et al.* (1976) showed that the enzyme from *B. fibrisolvens* was also capable of isomerising *cis*-2,*cis*-5 dienoic acids to the corresponding *trans*-3,*cis*-5 isomers. Since the production of the conjugated diene intermediate is common to all biohydrogenating bacteria of group A so far isolated, it would seem likely that this, or a similar, enzyme is present in all these bacteria.

How is the conjugation carried out? Kepler *et al.* (1971) demonstrated that the isomerase from *B. fibrisolvens* carried out specific protonation of carbon 13 in the D configuration. They incubated the enzyme with substrates of different molecular shape, and concluded from the reaction kinetics that, in the case of linoleic acid, the substrate followed the contours of an infolded, hydrophobic binding site. They postulated that this bonding involved interaction between the active site of the enzyme and the π electrons of the double bond at carbon atom 9, with additional bonding between an electronegative region of the enzyme and the –COOH group of the substrate. Under such conditions, the conformation assumed by the substrate would permit transfer of H^+, thus resulting in the formation of cis-9,trans-11-octadecadienoic acid (Fig. 4).

Range of fatty acids hydrogenated by rumen isolates
The range of isomers of unsaturated fatty acids which can be hydrogenated by isolates such as those shown in Table 4 has been recently investigated; one study looked at the range of positional isomers of acids resembling linoleic acid, while the other study looked at the hydrogenation of ocadecenoic acids.

Hydrogenation of methylene-interrupted cis,cis octadecadienoic acids. These resemble linoleic acid in that the two double bonds are separated by a –CH$_2$– group. Kemp et al. (1984b) investigated the hydrogenation of positional isomers of these fatty acids by bacteria which hydrogenated linoleic acid to stearic acid and bacteria which hydrogenated linoleic acid only as far as the *trans*-11 intermediate. The only isomer which was not hydrogenated by at least one organism was the *cis*-14,*cis*-17 isomer. Of all the isomers tested, only the *cis*-2,*cis*-5 isomer and the *cis*-9,*cis*-12 isomer (i.e. linoleic acid) had to be converted into the corresponding *cis,trans* conjugated diene before further hydrogenation could take place. Those bacteria which had previously been shown to hydrogenate linoleic acid only as far as the *trans*-11 acid gave only octadecenoic acids as the products of hydrogenation; those previously shown to hydrogenate linoleic acid completely to stearic acid gave variable yields of octadecenoic acids and stearic acid when incubated with all

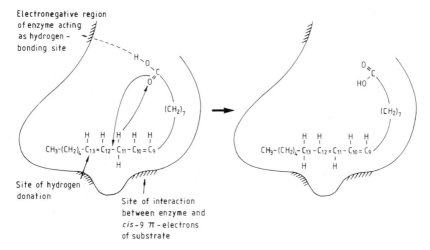

Fig. 4. Proposed mechanism for the interaction between linoleic acid substrate and the hydrophobic binding site of the Δ^{12}-*cis*,Δ^{11}-*trans* isomerase in the conversion of *cis*-9,*cis*-12-octadecadienoic acid into *cis*-9, *trans*-11-octadecadienoic acid. After Kepler et al. (1971).

isomers, except for the *cis*-12,*cis*-15 and the *cis*-18,*cis*-16 isomers from which only octadecenoic acids were produced. Some of the isomers of 18:2 were observed to be inhibitory to growth at the concentrations used (20 μg ml^{-1}); others were stimulatory, but there was neither a clear pattern nor any correlation with the extent to which the isomers were hydrogenated.

Hydrogenation of octadecenoic acid isomers. Kemp et al. (1984a) tested the ability of a rumen *Fusocillus* sp. (a group B biohydrogenator) to hydrogenate octadecenoic acids to stearic acid. Kemp et al. (1984a) used all positional *cis* isomers from *cis*-2 to *cis*-13 except for the *cis*-3 isomer, and all *trans* isomers from *trans*-2 to *trans*-13 except for the *trans*-3 and *trans*-4 isomers. The *cis* and *trans* isomers from positions 5 to 13 inclusive were all hydrogenated to some extent when incubated for 3 hours with late-exponential-phase cultures of the *Fusocillus*. Of the *cis* isomers, *cis*-3 to *cis*-11 isomers were the preferred substrates, with conversions into stearic acid ranging from 73% to 79%. However, only 30% of the *cis*-12 isomer and 5% of the *cis*-13 isomer were hydrogenated. Of the *trans* isomers, about 45% of each of the *trans*-8, *trans*-9 and *trans*-10 isomers were hydrogenated; the rest were poorly hydrogenated.

More extensive hydrogenation occurred when small inocula were used and incubation was extended to 24 hours. In general, these findings indicate that little specificity is shown with regard to the hydrogenation of the methylene-interrupted, unsaturated fatty acids by either the group A or the group B biohydrogenating bacteria. This applies to both the all-*cis* and the all-*trans* isomers. The *cis*-14,*cis*-17 fatty acid in which the double bonds are closest to the methyl end of the molecule appeared to be resistant to hydrogenation. It is of interest that, of those methylene-interrupted, unsaturated fatty acids which were hydrogenated, only those with the *cis*-2,*cis*-5 or *cis*-9,*cis*-12 configuration required the production of a *cis*,*trans* conjugated diene in order that further hydrogenation could take place. Linoleic (*cis*-9,*cis*-12), α-linolenic (*cis*-9,*cis*-12,*cis*-15) and γ-linolenic (*cis*-6,*cis*-9,*cis*-12) acids all possess a *cis*-9,*cis*-12 system, and the production of conjugated intermediates is known to be a prerequisite for the hydrogenation of these fatty acids in the rumen (see above).

Likewise there was little specificity with regard to the hydrogenation of octadecenoic acids by the *Fusocillus* sp. However, *cis* and *trans* acids, in which the double bond occurred towards the middle of the molecule, were hydrogenated preferentially; this was especially the case with the *trans* isomers. Whether these hydrogenations are carried out by one enzyme or a series of enzymes, one for each isomer, is unknown; to date the only studies on hydrogenating enzymes have been done on the *cis*-9,*trans*-12 octadecadienoate reductase of *B. fibrisolvens* (see below), and these studies have not included a survey of substrate specificity.

Mechanism of hydrogenation
Subsequent to the formation of the conjugated diene, each metabolic step involves the transfer of 2 H to each double bond in turn, the *trans*-11 being the last to be hydrogenated. The reduction of the conjugated diene has been much studied by

Tove and his colleagues over some 20 years, both in intact bacteria (Rosenfeld & Tove, 1971) and in cell-free extracts (Hunter et al., 1976) of B. fibrisolvens. However, it is only recently that the probable electron donors have been identified and the reductase enzyme purified.

The electron donors. Yamazaki & Tove (1979) isolated two oxygen-consuming fractions from cell-free extracts of *B. fibrisolvens*. The major fraction (97%) consisted of a mixture of glucose, maltose and dithionite. The minor fraction contained what appeared to be an electron donor for the biohydrogenation of cis-9,trans-11-octadecadienoic acid. Although dithionite could not serve as an electron donor for this reduction, it did supply electrons to the system via an endogenous donor. This electron donor was subsequently identified as α-tocopherolquinol (Fig. 5) (Hughes & Tove, 1980a). Hughes & Tove (1980b) also showed that α-tocopherolquinone (Fig. 5) was present and that, in the presence of NADH in a 1:1 ratio, the α-tocopherolquinone could be reduced to α-tocopherolquinol. The ratio of α-tocopherolquinone produced to fatty acid reduced was 2:1 if aerobic extraction had been used, but 1:1 if anaerobic extraction had been used. They suggested that 2 molecules of α-tocopherolquinol (TQH_2) are oxidised each to the semiquinone (TQH) in order to provide the electrons for the reduction of the *cis* bond of the conjugated diene. Also present in the extracts was a flavin-like compound which appeared to act as an intermediary in the transfer of electrons to the tocopherolquinone. On the basis of these observations the following scheme was proposed for the hydrogenation of cis-9,trans-11-octadecadienoic acid (Hughes & Tove, 1980a):

$$NADH + H^+ \searrow \quad Flavin(ox) \searrow \quad 2TQH_2 \searrow \quad cis\text{-}9,trans\text{-}11\text{-octadecadienoic acid}$$
$$NAD^+ \quad \nearrow \quad Flavin(red) \quad \nearrow \quad 2TQH \quad \nearrow \quad trans\text{-}11\text{-octadecenoic acid}$$

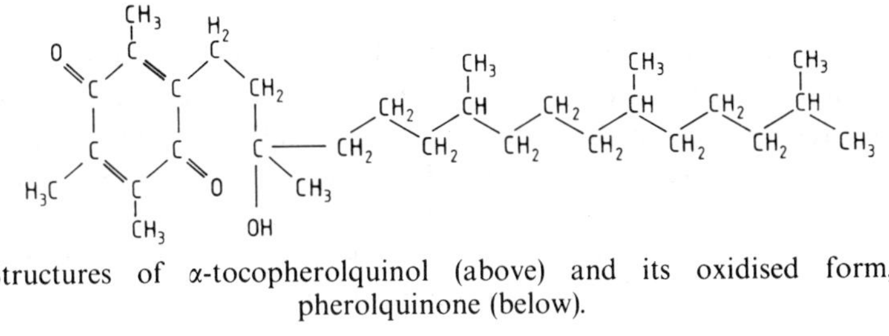

Fig. 5. Structures of α-tocopherolquinol (above) and its oxidised form, α-tocopherolquinone (below).

It is not known whether α-tocopherolquinol is the electron donor for the hydrogenation of the *trans*-11-octadecenoic acid by group B hydrogenators. It is known that both the α-tocopherolquinol and the corresponding quinone are of widespread occurrence among both prokaryotes and eukaryotes (Hughes & Tove, 1982); at present the compounds appear to have no function other than in biohydrogenation.

The cis-9,trans-11 *octadecadienoate reductase*. This has been purified to about 97% homogeneity by Hughes *et al.* (1982). The enzyme has a molecular weight of 60 000 and appears to be a glycoprotein with 10 mol of fucose and 12 mol of galactose per mol of enzyme. Hughes *et al.* suggested that the polysaccharide moiety is linked to the protein via a β-galactosyl linkage. Incubation of *B. fibrisolvens* with ^{3}H-labelled fucose showed incorporation of the label into the reductase. Removal of the carbohydrate did not affect the activity of the enzyme and the authors suggested that the carbohydrate may serve to locate the enzyme at a particular site in the cell membrane. In addition to carbohydrate, the enzyme possesses 2 mol of Fe per mol of enzyme. This iron, present as Fe^{3+}, is absolutely required for enzymic activity and cannot be replaced by Fe^{2+} or by Zn^{2+}, Mg^{2+}, Mn^{2+}, Co^{2+} or Cu^{2+} ions. As Hughes *et al.* (1982) point out, metal-containing enzymes are usually not specific as to the particular ion or to its reduction state; this may indicate that Fe^{3+} is directly involved in the oxidation–reduction reaction, particularly as Hughes & Tove (1980a) propose a one-electron step for biohydrogenation (see above).

Some outstanding problems with regard to biohydrogenation

Unsolved at present are a number of problems related to the mechanism of biohydrogenation, and to the role of biohydrogenation, both in the physiologies of the bacteria which carry out this process, and in the overall economy of the rumen.

Mechanism

When Tove and his colleagues started their studies about 20 years ago, *B. fibrisolvens* was the bacterium of choice as its physiology was reasonably well documented and it could be grown with relative ease (Hungate, 1966). However, it is now known that *B. fibrisolvens* belongs to that group of bacteria which hydrogenate linolenic and linoleic acids only as far as the *trans*-11 intermediate. Questions remaining unanswered at the present time are: does the reductase from *B. fibrisolvens* reduce other *cis,trans* conjugated dienes; how closely does this reductase resemble the corresponding reductases from other group A biohydrogenating bacteria; what enzymes are responsible for the hydrogenation of *cis*-9 and *trans*-11 double bonds by the *Fusocillus* type of biohydrogenating bacteria; what enzyme(s) are responsible for the hydrogenation of the *cis,cis* methylene-interrupted double bonds and the *cis* and *trans* monoenoic acids investigated by Kemp *et al.* (1984a,b)? As regards the control processes involved in biohydrogenation, nothing is known.

Role of biohydrogenation

It is not clear what purpose is served by the biohydrogenation of unsaturated fatty acids. The group A biohydrogenator *Butyrivibrio* S2 readily incorporated *trans*-11-octadecenoic acid, produced from linolenic or linoleic acids, into a range of membrane lipids (Hazlewood & Dawson, 1979), suggesting that hydrogenation has an essential role in the utilisation of dietary fatty acids by fatty acid-auxotrophic bacteria. There is, however, no information regarding the numerical significance of such organisms in the rumen, and it appears that other butyrivibrios, which have group A biohydrogenating activity and are also presumably able to synthesise fatty acids, do not incorporate either the *trans*-11 monoenoic end-product or the *cis*-9, *trans*-11 conjugated intermediate into membrane lipids (Kepler *et al.*, 1970).

Lennarz (1966) suggested that the main function of biohydrogenation was the disposal of reducing power, essential to bacteria living in a reduced environment. From the point of view of the overall economy of the rumen this seems most unlikely; a much more important mechanism for this exists via methanogenesis (see Chapters 6 and 11). Although unsaturated fatty acids are known to inhibit methanogenesis in the rumen (Prins *et al.*, 1972), they do so at millimolar concentration, which is far too low for them to serve as a major competitor for hydrogen. Similarly, the hydrogenator *B. fibrisolvens* possesses hydrogenase (Joyner *et al.*, 1977) and is potentially able to dispose of excess reducing power more efficiently as hydrogen gas than by means of biohydrogenation.

An alternative suggestion for the role of biohydrogenation is that it serves to detoxify fatty acids (Kemp & Lander, 1984; Kemp *et al.*, 1984*b*). It has been known for some time that unsaturated fatty acids are toxic to many microorganisms (Nieman, 1954), including those of the rumen (Prins *et al.*, 1972). Henderson (1973*b*) investigated the effects of fatty acids on pure cultures of rumen bacteria and observed that in some instances long-chain fatty acids were inhibitory at low (0.01–0.1 g l^{-1}; 0.03–0.3 mmol l^{-1}) concentrations. Unfortunately no studies have been done on the effects of fatty acid concentration on hydrogenation by the more recent isolates, but Kemp *et al.* (1984*a*) observed that the presence of starch (which was not utilised) decreased the inhibitory effects observed with the octadecenoic acids used as substrates in their hydrogenation studies. They suggested that this was due to the fatty acids being adsorbed on to the starch and the toxic effects reduced in consequence. Kemp *et al.* (1984*a*) pointed out that, in the rumen, the large amounts of particulate matter permit fatty acid concentrations of 50–100 times higher (i.e. 1–2 mg ml^{-1}) than were used in their experiments (20 μg ml^{-1}) *in vitro*.

Although the role of biohydrogenation is as yet unclear, biohydrogenation is nevertheless of considerable magnitude and has profound effects on the lipid metabolism of the ruminant animal. As Hughes *et al.* (1982) point out, in *B. fibrisolvens* the overall hydrogenation system accounts for nearly 0.5% of the total soluble protein in the cell extract and '... it becomes even more enigmatic to assign such a high proportion of the cellular protein to an enzyme whose benefit to the cell is not apparent. One is free to conclude that biohydrogenation cannot be the only function of this enzyme system in *B. fibrisolvens*, yet we have no inkling, at present, of what other function there may be'.

Factors affecting biohydrogenation

Concentration of fatty acid substrates

It has been known for some time that the concentration of the substrate fatty acid affects the extent to which hydrogenation occurs. Two factors affect this. One is whether the substrate is added as free acid or as triglyceride; the other is whether the experiments are done *in vivo* or *in vitro*. Noble *et al.* (1969) observed that, when emulsions of trilinolein were infused into the rumen, the linoleic acid released by lipolysis was hydrogenated completely to stearic acid, whereas when comparable amounts of free acid were infused, hydrogenation proceeded only as far as the *trans*-11 monoenoic intermediate. Similar experiments done *in vitro* using strained rumen liquor bore out these observations (Harfoot *et al.*, 1973a, 1975). It was also observed that the inhibitory effects of the free acid which resulted in hydrogenation of linoleic acid proceeding only as far as the *trans*-11 intermediate were more pronounced *in vitro* than *in vivo*. The results of these experiments were similar to those done at about the same time, but with rather different aims, by Hawke & Silcock (1970), Viviani & Borgatti (1967) and Wilde & Dawson (1966) in which it was shown that, in the absence of food particles, biohydrogenation of linoleic or linolenic acids proceeded only as far as the *trans*-11 intermediate, if at all. The simplest explanation for these observations is that the food particles compete with the bacteria as sites for adsorption of fatty acids and thus prevent the bacteria from either being coated with a hydrophobic film of lipid or exposed to inhibitory concentrations of unsaturated fatty acid (Henderson, 1973b). Henderson noted that the major inhibitory effects of fatty acids occurred when the amount of fatty acid present was greater than could be adsorbed on to the bacterial cells. It is known that bacteria in the rumen rapidly adsorb fatty acids and that different types of particles compete as sites for this adsorption (Harfoot *et al.*, 1974; Kemp *et al.*, 1984a).

Effects of dietary composition

Recent investigations have shown that components of the diet other than lipid may affect the lipid composition of ruminant tissues and milk. Diets low in roughage (Latham *et al.*, 1972; Leat, 1977) appear to result in decreased rumen lipolysis and biohydrogenation which in turn results in increased proportions of unsaturated fatty acids in cows' milk (Latham *et al.*, 1972) and adipose tissue (Leat, 1977). More recently, Gerson *et al.* (1985) investigated the effects of dietary starch and fibre on the rates of lipolysis and biohydrogenation *in vitro*. They found that addition of readily fermentable carbohydrate did not inhibit either lipolysis or biohydrogenation, but that replacement of fibre (i.e. cellulosic material) with starch resulted in reduction of the rates of lipolysis and biohydrogenation. They suggested that the main biohydrogenators are cellulolytic, although it should be pointed out that most *Butyrivibrio* strains are able to use starch (Hungate, 1966) as can *Ruminococcus albus* F2/6 and *Fusocillus babrahamensis* P2/2 (Table 4; Kemp *et al.*, 1975).

It was observed by Gerson *et al.* (1982) that, when different concentrations of dietary nitrogen were added to diets of constant content of digestible organic material, unsaturated fatty acids comprised between 70% and 80% of the esterified fatty acids of the rumen with 1·1% and 1·4% dietary N present, whereas they comprised between 20% and 30% when low (0·5%) and very high (3.4%) N was present. It was subsequently observed (Gerson et al., 1983) that increased proportions of nitrogen in the diet were accompanied by increased rates of lipolysis and biohydrogenation of triolein and oleic acid by rumen contents *in vitro*. It was concluded that the composition of the microbial population was important in controlling the extent of lipolysis and biohydrogenation, a not unreasonable assumption given the complexity of the rumen microbial population and the presence of different populations of biohydrogenating bacteria.

COMPOSITION OF MICROBIAL LIPIDS

The microbial population of the rumen includes bacteria, ciliate protozoa, flagellate protozoa, phycomycete fungi, amoebae and bacteriophages. The nature of the lipids of these different microorganisms is of great intrinsic interest in relation to microbial function and is of considerable importance in determining the composition of the lipid component available to the host animal in post-ruminal digesta. Available evidence (Keeney, 1970) indicates that, in the ovine rumen, bacterial and protozoal lipids together account for 10–20% of the total lipid present and occur in the ratio of 1:3.

Lipid composition of rumen bacteria

Some of the earliest analytical studies were conducted using mixed bacteria harvested from the fluid phase of rumen contents and freed of protozoa by differential centrifugation. By this approach, Garton & Oxford (1955) determined that lipid accounted for 9% of the dry weight of rumen bacteria and consisted of phospholipids (39%), neutral lipids (38%), steam-volatile fatty acids (12%) and non-saponifiable material (10%). In a subsequent study, Viviani et al. (1968) reported that lipids of mixed rumen bacteria comprised 30% phospholipid and 70% non-phospholipid, of which greater than 40% was unesterified fatty acids. The phospholipids consisted of phosphatidylcholine (1·2%), phosphatidylethanolamine (66·6%) and phosphatidylserine (21·6%) together with unidentified polar lipid (11–12%). Table 5, compiled from the work of several authors, shows the fatty acid composition of total lipids and of phosphatidylethanolamine, phosphatidylserine and the unesterified fatty acid fraction of mixed rumen bacteria. For further data relating to the fatty acid composition of mixed bacteria and of selected species grown in pure culture, readers are referred to Katz & Keeney (1966), Ifkovits & Ragheb (1968), Viviani (1970), Kunsman (1973) and Miyagawa (1982). In addition to the presence of a variety of unusual fatty acids (Ifkovits & Ragheb, 1968), a noticeable feature of the lipids of rumen bacteria is their high proportion of straight-chain and branched-chain saturated fatty acids compared with either total rumen lipids or forage lipids. Although there is evidence to suggest that some rumen bacteria take up and incorporate preformed linoleic acid (Hawke, 1971), it is rare to

Table 5
Fatty Acid Composition (%) of the Lipids of Mixed Rumen Bacteria

Fatty acid	Total lipid[a]	Total lipid[b]	Total lipid[c]	Unesterified fatty acid[c]	Phosphatidyl-ethanolamine[c]	Phosphatidyl-serine[c]
11:0	0·1	n.d.	n.d.	n.d.	n.d.	n.d.
12:0	0·4	4·6	1·2	2·3	0·5	tr
12:0 br	0·7	n.d.	n.d.	n.d.	n.d.	n.d.
13:0	0·3	n.d.	0·8	0·8	0·7	tr
13:0 br	0·7	n.d.	0·6	0·5	0·4	tr
14:0	2·3	3·7	3·9	4·0	4·1	2·4
14:0 br	2·4	n.d.	1·2	1·1	1·6	0·6
15:0	4·4	n.d.	8·0	7·1	10·6	6·0
15:0 br	10·1	n.d.	12·7	9·8	17·8	13·7
16:0	35·2	25·4	31·0	29·7	30·5	28·9
16:0 br	1·0	n.d.	1·2	0·2	1·9	0·9
16:1	—	tr	4·0	2·2	5·2	3·1
17:0	1·8	n.d.	1·6	1·2	1·5	1·8
17:0 br	1·7	n.d.	n.d.	n.d.	n.d.	n.d.
18:0	32·0	20·8	15·0	22·8	7·4	12·5
18:0 br	n.d.	n.d.	0·1	0·1	0·2	0·6
18:1	3·9	19·7	6·0	6·4	7·0	10·5
18:2	3·5[d]	5·6	2·7	2·0	3·4	4·3
18:3	—	tr	1·0	1·1	1·2	1·2
20:0	n.d.	n.d.	0·1	—	—	0·4
Other	—	20·2	8·8	8·8	6·1	13·2

tr, trace; —, not detected; n.d., data not given.
[a] Tweedie et al. (1966). [b] Williams & Dinusson (1973).
[c] Viviani et al. (1968); in the phospholipids, 13:0 br, 14:0 br, 15:0 br and 16:0 br includes 12:1, 13:1, 14:1 and 15:1 straight-chain acids respectively. [d] Includes 18:3.

find polyunsaturated fatty acids in bacteria (Goldfine, 1982), and therefore probable that the low levels of 18:2 and 18:3 found in the total lipids of mixed rumen bacteria originate through non-specific adsorption or by contamination of the bacterial fraction with feed particles; the lack of convincing evidence for the synthesis of 18:2 or 18:3 by pure cultures of rumen bacteria supports this conclusion.

A further striking feature of the lipids of rumen bacteria is their high content of fatty aldehydes, which may be present in both polar and non-polar glycerol ether lipids. Described by Katz & Keeney (1964), this finding was confirmed by Kamio et al. (1969) and was shown to apply to a variety of individual species of anaerobic rumen bacteria (Allison et al., 1962; Wegner & Foster, 1963; Kanegasaki & Takahashi, 1968; Kamio et al., 1970b; Miyagawa, 1982). For *Selenomonas ruminantium* the nature of the fatty aldehydes occurring in plasmalogen lipids was dependent on growth conditions and, in particular, the carbon number of the volatile fatty acids added to the growth medium (Kamio et al., 1970b).

Detailed analysis of the membrane lipids of selected species has shown that anaerobic rumen bacteria can be a rich source of rare and unusual lipids. This is particularly true for members of the genus *Butyrivibrio*, and for *Bacteroides ruminicola* strains, some of which contained over 50% sphingophospholipid

(Kunsman, 1973); an example of a novel lipid is the sphingophospholipid ceramide phosphoryl-3-aminopropane 1,2-diol isolated first from lipids of mixed rumen bacteria and subsequently from a Gram-negative bacterium assigned to the genus *Bacteroides* (Kemp *et al.*, 1972; Hazlewood, unpublished results). Lipid composition of *Butyrivibrio* spp. has been documented in some detail. Three of the strains studied did not contain the phospholipids commonly found in prokaryotes (phosphatidylglycerol, phosphatidylethanolamine and cardiolipin) but instead contained a minor *N*-acylphosphatidylethanolamine component and, in one case, a new lipid, diglyceride galactosylphosphorylethanolamine, as the only N-containing lipids (Clarke *et al.*, 1976). All three strains contained the *n*-butyroyl ester of phosphatidylglycerol, and either monogalactofuranosyldiglyceride or its *n*-butyroyl ester; valeryl, isovaleryl, propionyl and myristyl esters of phosphatidylglycerol were also detected. Two of the three strains contained glycerylphosphorylgalactosyldiglyceride and one also contained what is probably the diacyl derivative of the former. All of the lipids examined were of the plasmalogen type, containing alk-1-enyl chains. Further work with this genus has been confined to a lipolytic, general fatty acid auxotroph, *Butyrivibrio* S2 (Hazlewood & Dawson, 1979), isolated from rumen contents, and has included not just studies of the lipid composition of a bacterium with a defined fatty acid availability but also some biophysical studies of bacterial membranes in the complete absence of acyl chain unsaturation (Hauser *et al.*, 1979; Hazlewood *et al.*, 1980*b*; Hauser *et al.*, 1985). *Butyrivibrio* S2 was unable to synthesise long-chain fatty acids and grew only when provided with exogenous free or esterified long-chain fatty acid (Hazlewood & Dawson, 1977); *n*-saturated acids from C_{13} to C_{18}, positional isomers of *cis*- or *trans*-octadecenoic acid, and C_{18} polyunsaturated fatty acids all stimulated growth (Hazlewood & Dawson, 1979; Hazlewood *et al.*, 1979). The growth-promoting fatty acid was incorporated into a variety of membrane lipids either unchanged, as an ether-linked alkenyl grouping, or as a novel long-chain terminal dicarboxylic acid with mid-chain vicinal dimethyl groups (diabolic acid) (Hauser *et al.*, 1979; Klein *et al.*, 1979). Complex lipids of the organism cultured in the presence of palmitic acid comprised phosphoglycolipids (59%), phospholipids (25%) and glycolipids (13%). Most of the phospholipids contained esterified diabolic acid and an abundance of C_{16} alkenyl groups, but few esterified long-chain fatty acid residues; esterified butyric acid was found in many of the isolated lipids (Hazlewood *et al.*, 1980*a*). The structures of the major phospholipids have been elucidated, and it is evident that diabolic acid functions by linking molecules of plasmalogenic phospholipid and galactolipid (Clarke *et al.*, 1980). Thus, the major lipid (Fig. 6) is the butyroyl ester of *sn*-1-alkenyl-glycero-3-phospho-1'-*sn*-glycerol joined by a molecule of diabolic acid, through esterification of the two vacant 2-hydroxyl groups of the alkenyl-substituted glycerol molecules, to the butyroyl ester of *sn*-1-alkenyl-3-galactosylglycerol. The second major lipid comprises 2 molecules of *sn*-1-alkenyl-glycero-3-phospho-*sn*-1'-glycerol butyroyl ester linked through diabolic acid in a similar manner. Other lipids, differing only in the absence of a butyroyl group, or in having a palmitoyl group in place of one of the butyroyl groups, have also been identified.

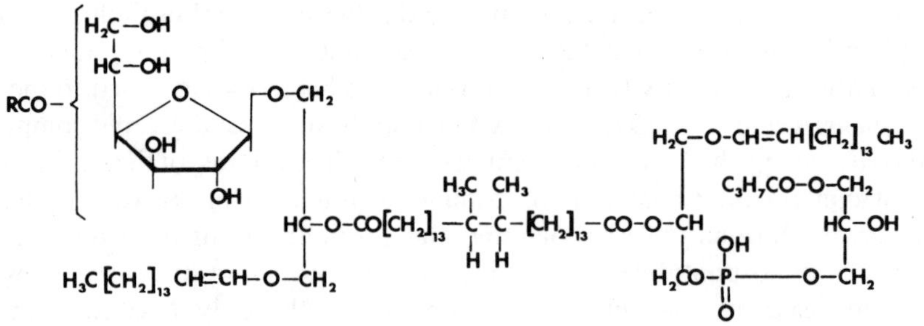

Fig. 6. Structure of the major diabolic acid-containing phospholipid isolated from *Butyrivibrio* S2 grown in the presence of palmitic acid. The R group esterified to the galactose is a butyroyl residue. The butyroyl group on the glycerol residue may be replaced by a palmitol group. Two molecules of *sn*-1-alkenyl-glycero-3-phospho-*sn*-1'-glycerol butyrol ester may also be linked through a diabolic acid.

Two other Gram-negative bacteria studied in some detail are *S. ruminantium* and *Megasphaera elsdenii*. *S. ruminantium* contained ethanolamine phosphatides as the major polar lipid class in its cytoplasmic membrane. These were present in the diacyl, alk-1-enyl acyl, and alkyl acyl forms (Kamio *et al.*, 1970c; Kamio & Takahashi, 1980). For *M. elsdenii*, the plasmalogen analogues of phosphatidylserine and phosphatidylethanolamine were the principal phospholipids (Van Golde *et al.*, 1973).

Lipid composition of rumen protozoa

Early data on the lipid composition of rumen protozoa resulted from the analysis of mixed populations obtained from rumen fluid; see Viviani (1970) and Harfoot (1978) for references. Because of the technical difficulties involved in obtaining protozoal preparations uncontaminated by bacteria or feed particles, a lack of uniformity in analytical procedures and the large variations in the composition and suspension density of the rumen protozoal population which occur both within and between animals (Coleman, 1980), the data should be interpreted with caution. However, results reported by Harfoot (1978), for samples obtained under defined conditions, indicated that mixed protozoa contained a very high proportion of phospholipid (85·5%) and only small quantities of mono-, di- and tri-glycerides. Total fatty acids were less saturated than those of mixed rumen bacteria and contained roughly equal amounts of 18:1 (18%) and 18:2 (16%) and a much smaller amount of 18:3 (< 1%). In general, these results showed reasonable agreement with others obtained for total mixed protozoa, and for mixed holotrich and entodiniomorphid protozoa (see Tables 6 and 7).

Lipid composition of mixed holotrich protozoa was comprehensively investigated by Katz & Keeney (1967), who found 70% phospholipid and 30% non-phospholipid. The phospholipids contained phosphatidylethanolamine (20%), phosphatidylethanolamine plasmalogen (22%), phosphatidylcholine (28%) and unknown phospholipids (29%). Each of the phospholipid fractions contained

Table 6
Fatty Acid Composition of Mixed Rumen Protozoa (% of total fatty acid)

Fatty acid	Percentage of total fatty acids		
	Viviani (1970)	Emmanuel (1974)	Harfoot (1978)
12:0	0·2	<0·1	n.d.
13:0	0·5	<0·1	n.d.
13:0 br	0·2	<0·1	n.d.
14:0	1·5	1·2	1·4
14:0 br	0·5	<0·1	1·5
15:0	4·3	2·5	3·4
15:0 br	4·3	3·0	1·5
16:0	37·8	41·0	43·1
16:0 br	2·3	1·5	—
16:1	6·8	—	1·1
17.0	1·4	1·2	—
17:0 br	—	3·2	3·7
18:0	13·5	10·5	9·3
18:0 br	0·2	<0·1	—
18:1	11·5	15·4	18·4
18:2	6·3	17·0	16·1
18:3	4·7	1·9	0·5
20:0	0·1	<0·1	—
20:0 br	—	—	—
Others	4·2	1·5	—

—, not detected; n.d., data not given.

Table 7
Fatty Acid Composition of the Lipids of Mixed Holotrich and Entodiniomorphid Protozoa (data from Williams & Dinusson, 1973)

Fatty acid	Percentage of total fatty acids	
	Entodiniomorphs	Holotrichs
12:0	0·3	trace
14:0	0·9	2·8
16:0	48·2	37·3
16:1	—[a]	—[a]
18:0	10·4	9·1
18:1	20·6	18·2
18:2	9·7	10·5
18:3	1·1	4·0
Other	8·8	18·1

[a] Not detected.

significant amounts of branched-chain and unsaturated fatty acids preferentially esterified in the 2 position. In ethanolamine plasmalogen, fatty aldehydes rich in branched chains were present in the 1 position. Non-phospholipids comprised mainly unesterified fatty acids (32%), sterol (19%) and diglyceride (10%), with smaller amounts of waxes, hydrocarbons, aliphatic alcohols, monoglycerides and hydroxyacids. In all lipid classes palmitic acid was the major fatty acid.

Total lipids of a pure preparation of the entodiniomorph *Polyplastron multivesiculatum*, obtained from the rumen contents of a sheep fasted for 17 hours to minimise contamination of protozoal lipids by dietary lipid, comprised phospholipids (78·2%), unesterified fatty acids (9·8%), triglycerides (2·0%), diglycerides (1·5%) and sterol esters, waxes etc. (8·6%) (Harfoot, 1978). Phospholipid composition of a second, *in vitro* cultured, entodiniomorph, *Entodinium caudatum*, has been studied in some detail, using chemical hydrolysis to facilitate identification of lipids through the analysis of partial degradation products (Dawson & Kemp, 1967; Broad & Dawson, 1975). The complex pattern of phospholipids revealed was qualitatively similar to that obtained for mixed rumen protozoa (Dawson & Kemp, 1967). Phosphatidylcholine and phosphatidylethanolamine were the major phospholipids (Table 8). Phosphonolipids containing 2-aminoethylphosphonate, or ciliatine, which had been first described in rumen ciliate protozoa by Horiguchi &

Table 8
Phospholipid Composition of Mixed Rumen Protozoa and of *Entodinium caudatum* Cultured *in vitro*.

Phospholipid	Percentage of total lipid P	
	Entodinium caudatum[a]	Mixed protozoa[b]
Phosphoglycerides	83·9	90·0
Phosphatidylcholine	23	36·3
Phosphatidylserine	—	—
Phosphatidylinositol	4	3·1
Phosphatidic acid	—	1·4
Phosphatidylethanolamine + glyceryl ether phospholipids	27	23·2
Diglyceride aminoethylphosphonate	21·5	11·0
Ethanolamine plasmalogen + aminoethylphosphonate plasmalogen	1·7	11·4
Choline plasmalogen	—	0·9
N-(1-Carboxyethyl)phosphatidylethanolamine	6·7	3·6
Sphingophospholipids	16·1	9·1
Ceramide phosphorylethanolamine	9·6	2·2
Ceramide aminoethylphosphonate	6·5	4·1
Sphingomyelin	—	2·0
Unidentified	—	0·8

—, not detected.
[a] Broad & Dawson (1975). [b] Dawson & Kemp (1967).

Kandatsu (1959), occurred primarily in a diglyceride and a sphingophospholipid (ceramide aminoethylphosphonate); mixed rumen protozoa also contained 2-aminoethylphosphonate in a plasmalogen but this could not be positively identified in extracts of *E. caudatum*. Fatty acids esterified in the main phospholipid groups were predominantly octadecenoic acid and palmitic acid; stearic acid was virtually absent. A novel phospholipid, phosphatidyl-*N*-(2-hydroxyethyl)alanine (*N*-(1-carboxyethyl)phosphatidylethanolamine), first detected in the lipids of mixed rumen protozoa (Kemp & Dawson, 1969*a*,*b*), is probably unique to *E. caudatum*, where it accounted for 5–7% of total lipid P (Dawson & Kemp, 1967; Broad & Dawson, 1975).

Lipid composition of rumen fungi

To date, the lipid composition of the anaerobic phycomycete fungi has been elucidated for only two species. Phospholipids of *Piromonas communis* comprised phosphatidylethanolamine (38%), phosphatidylcholine (26%), phosphatidylinositol (12%), phosphatidylserine (6%), phosphatidic acid (5%) and cardiolipin (3%) (Kemp *et al.*, 1984*c*); no sphingolipids, glycolipids, plasmalogens or phosphonyl lipids were detected. In the neutral lipid fraction, free fatty acids (48%), triacylglycerols (10%), 1,2-diacylglycerols (34%) and a variable amount of 1,3-diacylglycerol (up to 8%) were present, as were minor amounts of squalene and a triterpenol, probably tetrahymenol. About half of the total fatty acids were *n*-saturated, even, 12–24 carbon, saturated acids, the remainder being even, 16–24 carbon, monoenoic acids; in all except the 16 carbon acid, double bonds were in the 9 position. Polyenoic acids were not detected. Lipids of *Neocallimastix frontalis* comprised polar lipids (56%) and neutral lipids (44%) (Body & Bauchop, 1985). The phospholipids included phosphatidylethanolamine, phosphatidylcholine and cardiolipin; sphingomyelin was also detected. Triacylglycerols, free fatty acids and sterols were present in the neutral lipid fraction. Total fatty acids included 48% of mainly even, 14–24 carbon, *cis*-monoenoic acids, and were similar in composition to those of *P. communis*.

BIOSYNTHESIS OF MICROBIAL LIPIDS

Cellular lipids of rumen microorganisms in general are generated by synthesis *de novo* and through the direct incorporation of preformed precursor molecules which may be of dietary origin. Aspects of their biosynthesis have been reviewed by Garton (1977) and by Harfoot (1978).

Fatty acids

It is well established that certain branched-chain fatty acids, mainly of the *iso* and *antesio* series, which occur in the tissue- and milk-lipids of ruminants are components of the lipids of rumen bacterial species. These acids are synthesised by chain elongation of branched-chain precursors generated by metabolism of branched-chain amino acids (Garton, 1977). Such a pathway for fatty acid biosynthesis, utilising both branched-chain and straight-chain precursors, has been demonstrated in mixed rumen bacteria (Tweedie *et al*, 1966) and in pure cultures of the selected species *Ruminococcus flavefaciens* (Allison *et al.*, 1961; Keeney *et al.*, 1962), *Ruminococcus albus* (Allison *et al.*, 1962), *Selenomonas ruminantium* (Kanegasaki & Takahashi, 1968; Kanegasaki & Numa, 1970) and *Bacteroides ruminicola* (Keeney, 1970). For some of these organisms, branched-chain and straight-chain fatty acids may be an obligate nutritional requirement (Kanegasaki & Takahashi, 1967). The products of this synthetic pathway are incorporated into cellular lipids as long-chain acids and aldehydes (see, for example, Allison *et al.*, 1962; Wegner & Foster, 1963). In addition to chain elongation, as described above, it is probable that mixed rumen bacteria are also able to synthesise long-chain fatty acids via acetate and glucose, since Patton *et al.* (1970) and Emmanuel *et al.* (1974) noted that significant amounts of radioactivity were incorporated into microbial lipids when ^{14}C-labelled acetate or glucose were incubated with rumen contents *in vitro*. Although there have been no detailed studies of the synthesis *de novo* of long-chain monounsaturated fatty acids by rumen bacteria, an anaerobic pathway for the generation of C_{16} and C_{18} monounsaturated acids exists in some bacteria (Fulco *et al.*, 1964; Gurr, 1974).

A number of studies have been conducted using mixed rumen microorganisms and suspensions of protozoa to evaluate the extent to which preformed fatty acids are utilised as precursors for the synthesis of microbial lipids in the rumen, and it is now apparent that protozoa for certain, and probably also mixed, bacteria can incorporate preformed fatty acids into their cellular lipids (Williams *et al.*, 1963; Patton *et al.*, 1970; Demeyer *et al.*, 1978; Girard & Hawke, 1978; Broad & Dawson, 1975). Among individual bacteria, the fatty acid auxotroph *Butyrivibrio* S2 represents one extreme in this respect, using only exogenous fatty acid (Hauser *et al.*, 1979; Hazlewood & Dawson, 1979), but there is some evidence that other species also incorporate significant amounts of preformed fatty acid into membrane lipids (Hawke, 1971). Although evidence for the existence of polyunsaturated fatty acids in the lipids of rumen bacteria is somewhat limited, it is probable that these acids, if present, also result from the uptake of preformed unsaturated acids (see Harfoot, 1978).

In addition to utilising preformed fatty acids, rumen protozoa synthesise fatty acids *de novo*. Coleman (1969) showed that radioactivity from [U-^{14}C]glucose was incorporated into the lipid fraction of *E. caudatum*. In a subsequent study of mixed protozoa (Emmanuel, 1974), synthesis *de novo* of long-chain fatty acids from propionate, butyrate and isoleucine was demonstrated. Octadecenoic acids, consisting mainly of the Δ^{11} isomer, were synthesised by direct desaturation of saturated acids; available evidence indicated polyunsaturated acids to be of dietary origin.

The results of a recent study (Kemp *et al.*, 1984*c*) indicated that the anaerobic rumen fungus *P. communis* is able to synthesise long-chain fatty acids *de novo*, and to incorporate a range of preformed fatty acids into membrane lipids. Furthermore, the presence of large amounts of monounsaturated fatty acids in the lipids of *N. frontalis* (Body & Bauchop, 1985) and *P. communis* (Kemp *et al.*, 1984c) suggests that a novel anaerobic pathway for the biosynthesis of monoenoic acids, analogous to that currently found only in certain bacteria, may be operative, although for *P. communis* it was suggested that octadecenoic acid was formed by the action of an anaerobic Δ^9 desaturase on stearic acid (Kemp *et al.*, 1984c).

Complex lipids

From the foregoing section it is evident that much information has accrued regarding the origins of the hydrophobic or apolar moieties contained in the cellular lipids of rumen microorganisms. With some exceptions, this is not the case for the polar constituents of complex lipids. Studies conducted with radiolabelled substrates and mixed rumen microorganisms indicated that synthesis *de novo* of phospholipids occurred in the rumen, but gave no indication of the pathways operating in different microbial types (see Harfoot, 1978).

Among studies conducted *in vitro* with pure cultures of rumen bacteria, those reported for *Selenomonas ruminantium* subsp. *lactilytica* are notable. The straight-chain, saturated, volatile fatty acids required for growth by this organism (Kanegasaki & Takahashi, 1967) are elongated and incorporated into the hydrophobic chains of phospholipids consisting mainly of phosphatidylethanolamine, phosphatidylserine and their plasmalogen analogues (Watanabe *et al.*, 1982). By analysing the pattern of labelling of the odd- and even-numbered fatty acids in the phospholipid fraction during growth of *S. ruminantium* with [1-^{14}C]-butyrate, -caproate or -valerate, and the effect of imidazole, an inhibitor of α-oxidation of fatty acids, on growth and fatty acid incorporation, Kamio *et al.* (1970a) concluded that α-oxidation of the odd-numbered short-chain precursor acids is important for the synthesis of phospholipids containing the requisite proportion of even-numbered long-chain fatty acids. In subsequent experiments (Watanabe *et al.*, 1984) a particulate enzyme prepared from *S. ruminantium* subsp. *lactilytica* was shown to catalyse the formation of phosphatidylserine from CDP-diglyceride and serine and phosphatidylethanolamine from phosphatidylserine, indicating that in this organism phosphatidylethanolamine and phosphatidylserine are synthesised via a pathway similar to that which operates in *E. coli* (Raetz, 1978). Pulse-labelling of cells with radioactive precursors of phospholipid revealed that turnover of phosphatidylethanolamine was rapid, while ethanolamine plasmalogen was relatively stable, and it further showed that ethanolamine plasmalogen was formed from serine plasmalogen. Perhaps most significant was the evidence suggesting that the diglyceride moiety of diacyl phospholipids constitutes a large precursor pool from which the 1-*O*-alk-1'-enyl-2-acyl-glycerol moieties of plasmalogens are derived (Watanabe *et al.*, 1984).

A similar approach, using radiolabelled precursors, had been applied to the study of plasmalogen biosynthesis in *Megasphaera elsdenii* which, like *S. ruminantium*, has phosphatidylethanolamine, phosphatidylserine and their plasmalogen analogues as its principal phospholipids (Van Golde *et al.*, 1973). Results obtained suggested that the glycerol moiety of plasmalogens does not have dihydroxyacetone phosphate as a direct precursor, and indicated further that, as with *S. ruminantium*, ethanolamine plasmalogen and phosphatidylethanolamine are derived from serine plasmalogen and phosphatidylserine, respectively, by decarboxylation (Prins *et al.*, 1974). Membrane particles prepared from *M. elsdenii* synthesised phosphatidylserine from CDP-diglyceride and L-serine in a manner analogous to that described for *E. coli* but, despite the substantial amount of phosphatidylethanolamine normally present in the lipids of this organism, phosphatidylserine decarboxylase activity was barely detectable in membranes or whole-cell extracts (Silber *et al.*, 1981).

Lipid biosynthesis has also been studied to a limited extent in the fatty acid auxotroph *Butyrivibrio* S2 (Hazlewood & Dawson, unpublished results). Pulse-labelling of the lipids of this organism resulted in a very rapid incorporation of [1-^{14}C]palmitic acid primarily into phosphatidylglycerol. The appearance of label in the novel plasmalogenic, diabolic acid-containing, phospholipids, which are characteristic of this organism (Clarke *et al.*, 1980), was preceded by a significant lag, suggesting that the condensation reaction which results in diabolic acid formation occurs between fatty acids already esterified in glyceride structures. Furthermore, the suggestion that diacyl phospholipid may serve as the immediate precursor of plasmalogen indicates that there may be similarities between *Butyrivibrio* S2 and *Clostridium butyricum* in this respect (see Koga & Goldfine, 1984).

Current knowledge of the pathways of phospholipid biosynthesis in rumen protozoa derives exclusively from detailed studies conducted with *Entodinium caudatum* cultured *in vitro*. Choline is essential for the growth of this organism (Broad & Dawson, 1976) and is rapidly converted into phosphorylcholine by the action of choline kinase (Broad & Dawson, 1975). Ethanolamine was taken up at a lower rate and converted into phosphorylethanolamine. Phosphatidylcholine and phosphatidylethanolamine were synthesised via a phosphorylated base-CDP-base pathway similar to that found in the tissues of higher animals. Choline phosphate: cytidyl transferase was concluded to be the rate-limiting enzyme in the rapid conversion of choline into phosphatidylcholine (Bygrave & Dawson, 1976), and no evidence was found for base exchange, or for the formation of phosphatidylcholine by methylation of phosphatidylethanolamine (Broad & Dawson, 1975; Bygrave & Dawson, 1976). Pulse-labelling of the phospholipid fraction with $^{32}P_i$ revealed a rapid turnover of phosphatidylinositol and a much slower turnover of phosphatidylethanolamine and phosphatidylcholine (Broad & Dawson, 1975). Phosphatidyl-*N*-(2-hydroxyethyl)alanine, which appears to replace phosphatidylserine in anaerobic protozoa (Kemp & Dawson, 1969*a*,*b*), was synthesised by the substitution of a three-carbon glycolytic intermediate on to the amino group of phosphatidylethanolamine (Coleman *et al.*, 1971); confirmation of the role of phosphatidylethanolamine as the precursor of this novel phospholipid was subsequently obtained by incubating cells with isotopically-labelled phosphatidylethanolamine (Broad & Dawson, 1975). Similarly, phosphatidylethanolamine was shown to be the direct precursor of ceramide phosphorylethanolamine, apparently without the involvement of CDP-ethanolamine as an intermediary (Broad & Dawson, 1973, 1975).

Data derived from a single study of the rumen fungus *Piromonas communis*

(Kemp et al., 1984c) indicated that phosphatidylethanolamine could be converted into phosphatidylcholine via methylation of the base moiety and, further, that phosphatidylserine could be converted into phosphatidylethanolamine by decarboxylation, but the results must be considered inconclusive because of poor incorporation of radiolabelled precursors.

REFERENCES

Allison, M. J., Bryant, M. P., Keeney, M. & Katz, I. (1961). The metabolic fate of isovalerate in *Ruminococcus flavefaciens*. *J. Dairy Sci.*, **44**, 1203.

Allison, M. J., Bryant, M. P., Katz, I. & Keeney, M. (1962). Metabolic function of branched-chain volatile fatty acids; growth factors for ruminococci. II. Biosynthesis of higher branched-chain fatty acids and aldehydes. *J. Bacteriol.*, **83**, 1084–93.

Bailey, R. W. & Howard, B. H. (1963). Carbohydrases of the rumen ciliate *Epidinium ecaudatum* (Crawley). 2. α-Galactosidase and isomaltase. *Biochem. J.*, **87**, 146–51.

Banks, A. & Hilditch, T. P. (1931). The glyceride structure of beef tallows. *Biochem. J.*, **25**, 1168–82.

Body, D. R. (1976). The occurrence of *cis*-octadec-15-enoic acid as a major biohydrogenation product from methyl linolenate in bovine rumen liquor. *Biochem. J.*, **157**, 741–4.

Body, D. R. & Bauchop, T. (1985). Lipid composition of an obligately anaerobic fungus, *Neocallimastix frontalis*, isolated from a bovine rumen. *Can. J. Microbiol.*, **31**, 463–6.

Broad, T. E. & Dawson, R. M. C. (1973). Formation of ceramide phosphorylethanolamine from phosphatidylethanolamine in the rumen protozoon *Entodinium caudatum*. *Biochem. J.*, **134**, 659–62.

Broad, T. E. & Dawson, R. M. C. (1975). Phospholipid biosynthesis in the anaerobic protozoon *Entodinium caudatum*. *Biochem. J.*, **146**, 317–28.

Broad, T. E. & Dawson, R. M. C. (1976). Role of choline in the nutrition of the rumen protozoon *Entodinium caudatum*. *J. Gen. Microbiol.*, **92**, 391–7.

Bygrave, F. L. & Dawson, R. M. C. (1976). Phosphatidylcholine biosynthesis and choline transport in the anaerobic protozoon *Entodinium caudatum*. *Biochem. J.*, **160**, 481–90.

Clarke, N. G., Hazlewood, G. P. & Dawson, R. M. C. (1976). Novel lipids of *Butyrivibrio* spp. *Chem. Phys. Lipids*, **17**, 222–32.

Clarke, N. G., Hazlewood, G. P. & Dawson, R. M. C. (1980). Structure of diabolic acid-containing phospholipids isolated from *Butyrivibrio* sp. *Biochem. J.*, **191**, 561–9.

Coleman, G. S. (1964). The metabolism of *Escherichia coli* and other bacteria by *Entodinium caudatum*. *J. Gen. Microbiol.*, **37**, 209–23.

Coleman, G. S. (1969). The metabolism of starch, maltose, glucose and some other sugars by the rumen ciliate *Entodinium caudatum*. *J. Gen. Microbiol.*, **57**, 303–32.

Coleman, G. S. (1980). Rumen ciliate protozoa. In *Advances in Parasitology*, Vol. 18, ed. W. H. R. Lumsden, R. Muller & J. R. Baker. Academic Press, London and New York, pp. 121–73.

Coleman, G. S. & Hall, F. J. (1969). Electron microscopy of the rumen ciliate *Entodinium caudatum*, with special reference to the engulfment of bacteria and other particulate matter. *Tissue Cell*, **1**, 607–18.

Coleman, G. S., Kemp, P. & Dawson, R. M. C. (1971). The catabolism of phosphatidylethanolamine by the rumen protozoon *Entodinium caudatum* and its conversion into the *N*-(1-carboxyethyl) derivative. *Biochem. J.*, **123**, 97–104.

Dawson, R. M. C. (1959). Hydrolysis of lecithin and lysolecithin by rumen microorganisms of the sheep. *Nature*, **183**, 1822–3.

Dawson, R. M. C. & Hemington, N. (1974). Digestion of grass lipids and pigments in the sheep rumen. *Br. J. Nutr.*, **32**, 327–40.

Dawson, R. M. C. & Kemp, P. (1967). The aminoethylphosphonate-containing lipids of rumen protozoa. *Biochem. J.*, **105**, 837–42.
Dawson, R. M. C. & Kemp, P. (1969). The effect of defaunation on the phospholipids and on the hydrogenation of unsaturated fatty acids in the rumen. *Biochem. J.*, **115**, 351–2.
Dawson, R. M. C. & Kemp, P. (1970). Biohydrogenation of dietary fats in ruminants. In *Physiology of Digestion and Metabolism in the Ruminant*, ed. A. T. Phillipson. Oriel Press, Newcastle-upon-Tyne, pp. 504–18.
Dawson, R. M. C., Hemington, N., Grime, D., Lander, D. & Kemp, P. (1974). Lipolysis and hydrogenation of galactolipids and the accumulation of phytanic acid in the rumen. *Biochem. J.*, **144**, 169–71.
Dawson, R. M. C., Hemington, N. & Hazlewood, G. P. (1977). On the role of higher plant and microbial lipases in the ruminal hydrolysis of grass lipids. *Br. J. Nutr.*, **38**, 225–32.
Demeyer, D. I., Henderson, C. & Prins, R. A. (1978). Relative significance of exogenous and *de novo* synthesised fatty acids in the formation of rumen microbial lipids *in vitro*. *Appl. Environ. Microbiol.*, **35**, 24–31.
Emmanuel, B. (1974). On the origin of rumen protozoan fatty acids. *Biochim. Biophys. Acta*, **337**, 404–13.
Emmanuel, B., Milligan, L. P. & Turner, B. V. (1974). The metabolism of acetate by rumen microorganisms. *Can. J. Microbiol.*, **26**, 183–5.
Eyssen, H. & Verhulst, A. (1984). Biotransformation of linoleic acid and bile acids by *Eubacterium lentum*. *Appl. Environ. Microbiol.*, **47**, 39–43.
Faruque, A. J. M. O., Jarvis, B. D. W. & Hawke, J. C. (1974). Studies on rumen metabolism. IX. Contribution of plant lipases to the release of free fatty acids in the rumen. *J. Sci. Food Agric.*, **25**, 1313–28.
Fulco, A. J., Levy, R. & Bloch, K. (1964). The biosynthesis of Δ^9- and Δ^5-monounsaturated fatty acids by bacteria. *J. Biol. Chem.*, **239**, 998–1003.
Garcia, P. T., Christie, W. W., Jenkins, H. M., Anderson, L. & Holman, R. T. (1976). The isomerization of 2,5- and 9,12-octadecadienoic acids by an extract of *Butyrivibrio fibrisolvens*. *Biochim. Biophys. Acta*, **424**, 296–302.
Garton, G. A. (1959). Lipids in relation to rumen function. *Proc. Nutr. Soc.*, **18**, 112–17.
Garton, G. A. (1960). Fatty acid composition of the lipids of pasture grasses. *Nature* (Lond.), **187**, 511.
Garton, G. A. (1977). Fatty acid metabolism in ruminants. In *Biochemistry of Lipids* II, Vol. 14, ed. T. W. Goodwin. University Park Press, Baltimore, pp. 337–70.
Garton, G. A. & Oxford, A. E. (1955). The nature of bacterial lipids in the rumen of hay-fed sheep. *J. Sci. Food Agric.*, **3**, 142–8.
Garton, G. A., Hobson, P. N. & Lough, A. K. (1958). Lipolysis in the rumen. *Nature*, **182**, 1511–12.
Garton, G. A., Lough, A. K. & Vioque, E. (1959). The effect of sheep rumen contents on triglycerides *in vitro*. *Biochem. J.*, **73**, 46P.
Garton, G. A., Lough, A. K. & Vioque, E. (1961). Glyceride hydrolysis and glycerol fermentation by sheep rumen contents. *J. Gen. Microbiol.*, **25**, 215–25.
Gerson, T., John, A., Shelton, I. D. & Sinclair, B. R. (1982). Effects of dietary N on lipids of rumen digesta, plasma, liver, muscle and perirenal fat in sheep. *J. Agric. Sci.* (Camb.), **99**, 71–8.
Gerson, T., John, A. & Sinclair, B. R. (1983). The effect of dietary N on *in vitro* lipolysis and fatty acid hydrogenation in rumen digesta from sheep fed diets high in starch. *J. Agric. Sci.* (Camb.), **101**, 97–101.
Gerson, T., John, A. & King, A. S. D. (1985). The effects of dietary starch and fibre on the *in vitro* rates of lipolysis and hydrogenation by sheep rumen digesta. *J. Agric. Sci.* (Camb.), **105**, 27–30.
Girard, V. & Hawke, J. C. (1978). The role of holotrichs in the metabolism of dietary linoleic acid in the rumen. *Biochim. Biophys. Acta*, **528**, 17–27.
Goldfine, H. (1982). Lipids of prokaryotes: structure and distribution. In *Current Topics in*

Membranes and Transport, Vol. 17, ed. F. Bronner & A. Kleinzeller. Academic Press, New York and London, pp. 1–43.
Gurr, M. I. (1974). Biosynthesis of fatty acids. In *The Biochemistry of Lipids*, Vol. 4, ed. T. W. Goodwin (MTP International Reviews in Science, Biochemistry Ser. I). Butterworths, London.
Hall, F. J., West, J. & Coleman, G. S. (1974). Fine structural studies on the digestion of chloroplasts in the rumen ciliate *Entodinium caudatum*. *Tissue Cell*, **6**, 243–53.
Harfoot, C. G. (1978). Lipid metabolism in the rumen. *Progr. Lipid Res.*, **17**, 21–54.
Harfoot, C. G., Noble, R. C. & Moore, J. H. (1973a). Food particles as a site for biohydrogenation of unsaturated fatty acids in the rumen. *Biochem. J.*, **132**, 829–32.
Harfoot, C. G., Noble, R. C. & Moore, J. H. (1973b). Factors influencing the extent of biohydrogenation of linoleic acid by rumen microorganisms *in vitro*. *J. Sci. Food Agric.*, **24**, 961–70.
Harfoot, C. G., Crouchman, M. L., Noble, R. C. & Moore, J. H. (1974). Competition between food particles and rumen bacteria in the uptake of long-chain fatty acids and triglycerides. *J. Appl. Bacteriol.*, **37**, 633–41.
Harfoot, C. G., Noble, R. C. & Moore, J. H. (1975). The role of plant particles, bacteria and cell-free supernatant fractions of rumen contents in the hydrolysis of trilinolein and the subsequent hydrogenation of linoleic acid. *Ant. van Leeuwenhoek J. Microbiol. Serol.*, **41**, 533–42.
Hauser, H., Hazlewood, G. P. & Dawson, R. M. C. (1979). Membrane fluidity of a fatty acid auxotroph grown with palmitic acid. *Nature* (Lond.), **279**, 536–8.
Hauser, H., Hazlewood, G. P. & Dawson, R. M. C. (1985). Characterization of membrane lipids of a general fatty acid auxotrophic bacterium by electron spin resonance spectroscopy and differential scanning calorimetry. *Biochemistry*, **24**, 5247–53.
Hawke, J. C. (1971). The incorporation of long-chain fatty acids into lipids by rumen bacteria and the effect on biohydrogenation. *Biochim. Biophys. Acta*, **248**, 167–70.
Hawke, J. C. & Silcock, W. R. (1970). *In vitro* rate of lipolysis and biohydrogenation in rumen contents. *Biochem. Biophys. Acta*, **218**, 201–12.
Hazlewood, G. P. (1974). Metabolism of phospholipids and fatty acids by a rumen bacterium. PhD thesis, University of Cambridge.
Hazlewood, G. P. & Dawson, R. M. C. (1975a). Isolation and properties of a phospholipid-hydrolysing bacterium from ovine rumen fluid. *J. Gen. Microbiol.*, **89**, 163–74.
Hazlewood, G. P. & Dawson, R. M. C. (1975b). Intermolecular transacylation of phosphatidylethanolamine by a *Butyrivibrio* sp. *Biochem. J.*, **150**, 521–5.
Hazlewood, G. P. & Dawson, R. M. C. (1976). A phospholipid-deacylating system of bacteria active in a frozen medium. *Biochem. J.*, **153**, 49–53.
Hazlewood, G. & Dawson, R. M. C. (1977). Acylgalactosylglycerols as a source of long-chain fatty acids for a naturally occurring rumen auxotroph. *Biochem. Soc. Trans.*, **5**, 1721–3.
Hazlewood, G. & Dawson, R. M. C. (1979). Characteristics of a lipolytic and fatty acid-requiring *Butyrivibrio* sp. isolated from the ovine rumen. *J. Gen. Microbiol.*, **112**, 15–27.
Hazlewood, G. P., Kemp, P., Lander, D. & Dawson, R. M. C. (1976). C_{18} unsaturated fatty acid hydrogenation patterns of some rumen bacteria and their ability to hydrolyse exogenous phospholipid. *Br. J. Nutr.*, **35**, 293–7.
Hazlewood, G. P., Reynolds, M. J., Dawson, R. M. C. & Gunstone, F. D. (1979). An automatic colorimeter and its use in evaluating the growth response of an anaerobic general fatty acid auxotroph to *cis*- and *trans*-octadecenoic acids. *J. Appl. Bacteriol.*, **47**, 321–5.
Hazlewood, G. P., Clarke, N. G. & Dawson, R. M. C. (1980a). Complex lipids of lipolytic and general fatty acid-requiring *Butyrivibrio* sp. isolated from the ovine rumen. *Biochem. J.*, **191**, 555–60.
Hazlewood, G. P., Dawson, R. M. C. & Hauser, H. (1980b). The question of membrane fluidity in an anaerobic general fatty acid auxotroph. In *Membrane Fluidity: Biophysical Techniques and Cellular Regulation*, ed. M. Kates & A. Kuksis. Humana Press, Clifton, New Jersey, pp. 191–201.

Hazlewood, G. P., Cho, K. Y., Dawson, R. M. C. & Munn, E. A. (1983). Subcellular fractionation of the Gram negative rumen bacterium *Butyrivibrio* S2 by protoplast formation, and localisation of lipolytic enzymes in the plasma membrane. *J. Appl. Bacteriol.*, **55**, 337–47.

Henderson, C. (1971). A study of the lipase produced by *Anaerovibrio lipolytica*, a rumen bacterium. *J. Gen. Microbiol.*, **65**, 81–9.

Henderson, C. (1973a). An improved method for enumerating and isolating lipolytic rumen bacteria. *J. Appl. Bacteriol.*, **36**, 187–8.

Henderson, C. (1973b). The effects of fatty acids on pure cultures of rumen bacteria. *J. Agric. Sci.* (Camb.), **81**, 107.

Henderson, C. (1975). The isolation and characterisation of strains of lipolytic bacteria from the ovine rumen. *J. Appl. Bacteriol.*, **39**, 101–9.

Henderson, C. & Hodgkiss, W. (1973). An electron microscopic study of *Anaerovibrio lipolytica* (strain 58) and its lipolytic enzyme. *J. Gen. Microbiol.*, **76**, 389–93.

Henderson, C., Hobson, P. N. & Summers, R. (1969). The production of amylase, protease and lipolytic enzymes by two species of anaerobic rumen bacteria. In *Proceedings of the Fourth Symposium on Continuous Cultivation of Micro-organisms*, Prague. Academic Press, London, pp. 189–204.

Hobson, P. N. (1965). Continuous culture of some anaerobic and facultatively anaerobic rumen bacteria. *J. Gen. Microbiol.*, **38**, 167–80.

Hobson, P. N. & Mann, S. O. (1961). The isolation of glycerol-fermenting and lipolytic bacteria from the rumen of the sheep. *J. Gen. Microbiol.*, **25**, 227–40.

Hobson, P. N. & Mann, S. O. (1971). Isolation of cellulolytic and lipolytic organisms from the rumen. In *Isolation of Anaerobes* (SAB Technical Series No. 5), ed. D. A. Shapton & R. G. Board. Academic Press, London, pp. 149–58.

Hobson, P. N. & Summers, R. (1966). Effect of growth rate on the lipase activity of a rumen bacterium. *Nature*, **209**, 736–7.

Hobson, P. N. & Summers, R. (1967). The continuous culture of anaerobic bacteria. *J. Gen. Microbiol.*, **47**, 53–65.

Hobson, P. N. & Wallace, R. J. (1982a). Microbial ecology and activities in the rumen. Part I. *CRC Crit. Rev. Microbiol.*, **9**, 165–225.

Hobson, P. N. & Wallace, R. J. (1982b). Microbial ecology and activities in the rumen. Part II. *CRC Crit. Rev. Microbiol.*, **9**, 253–320.

Horiguchi, M. & Kandatsu, M. (1959). Isolation of 2-aminoethane phosphonic acid from rumen protozoa. *Nature* (Lond.), **184**, 901–2.

Hughes, P. E. & Tove, S. B. (1980a). Identification of an endogenous electron donor for biohydrogenation as α-tocopherolquinol. *J. Biol. Chem.*, **255**, 4447–52.

Hughes, P. E. & Tove, S. B. (1980b). Identification of deoxy-α-tocopherolquinol as another endogenous electron donor for biohydrogenation. *J. Biol. Chem.*, **255**, 11802–6.

Hughes, P. E. & Tove, S. B. (1982). Occurrence of α-tocopherolquinone and α-tocopherolquinol in micro-organisms. *J. Bacteriol.*, **151**, 1397–1402.

Hughes, P. E., Hunter, W. J. & Tove, S. B. (1982). Biohydrogenation of unsaturated fatty acids: purification and properties of cis-9,trans-11 octadecadienoate reductase. *J. Biol. Chem.*, **257**, 3643–9.

Hungate, R. E. (1966). *The Rumen and Its Microbes.* Academic Press, New York and London, pp. 8–90.

Hunter, W. J., Baker, F. C., Rosenfeld, I. S., Keyser, J. B. & Tove, S. B. (1976). Biohydrogenation of unsaturated fatty acids. VII. Hydrogenation by a cell-free preparation of *Butyrivibrio fibrisolvens*. *J. Biol. Chem.*, **251**, 2241–7.

Ifkovits, R. W. & Ragheb, H. S. (1968). Cellular fatty acid composition and identification of rumen bacteria. *Appl. Microbiol.*, **16**, 1406–13.

Joyner, A. E., Winter, W. T. & Godbout, D. M. (1977). Studies on some characteristics of hydrogen production by cell-free extracts of rumen anaerobic bacteria. *Can. J. Microbiol.*, **23**, 346–53.

Kamio, Y. & Takahashi, H. (1980). Isolation and characterisation of outer and inner membranes of *Selenomonas ruminantium*: lipid composition. *J. Bacteriol.*, **141**, 888–98.
Kamio, Y., Kanegasaki, S. & Takahashi, H. (1969). Occurrence of plasmalogens in anaerobic bacteria. *J. Gen. Appl. Microbiol.*, **15**, 439–51.
Kamio, Y., Inagaki, H. & Takahashi, H. (1970a). Possible occurrence of α-oxidation in phospholipid biosynthesis in *Selenomonas ruminantium*. *J. Gen. Appl. Microbiol.*, **16**, 463–78.
Kamio, Y., Kanegasaki, S. & Takahashi, H. (1970b). Fatty acid and aldehyde compositions in phospholipids of *Selenomonas ruminantium* with reference to growth conditions. *J. Gen. Appl. Microbiol.*, **15**, 439–51.
Kamio, Y., Kim, K. C. & Takahashi, H. (1970c). Glyceryl ether phospholipids in *Selenomonas ruminantium*. *J. Gen. Appl. Microbiol*, **16**, 291–300.
Kanegasaki, S. & Numa, S. (1970). Medium-chain fatty acyl-CoA requirement for long-chain fatty acid synthesis in some anaerobic bacteria. *Biochim. Biophys. Acta*, **202**, 436–46.
Kanegasaki, S. K. & Takahashi, H. (1967). Function of growth factors for rumen microorganisms. I. Nutritional characteristics of *Selenomonas ruminantium*. *J. Bacteriol.*, **93**, 456–63.
Kanegasaki, S. & Takahashi, H. (1968). Function of growth factors for rumen microorganisms. II. Metabolic fate of incorporated fatty acids in *Selenomonas ruminantium*. *Biochim. Biophys. Acta*, **152**, 40–9.
Katz, I. & Keeney, M. (1964). The isolation of fatty aldehydes from rumen-microbial lipid. *Biochim. Biophys. Acta*, **84**, 128–32.
Katz, I. & Keeney, M. (1966). Characterization of the octadecenoic acids in rumen digesta and rumen bacteria. *J. Dairy Sci.*, **49**, 962–6.
Katz, I. & Keeney, M. (1967). The lipids of some rumen holotrich protozoa. *Biochim. Biophys. Acta*, **144**, 102–12.
Keeney, M. (1970). Lipid metabolism in the rumen. In *Physiology of Digestion and Metabolism in the Ruminant*, ed. A. T. Phillipson. Oriel Press, Newcastle-upon-Tyne, pp. 489–503.
Keeney, M., Katz, I. & Allison, M. J. (1962). On the probable origin of some milk fat acids in rumen microbial lipids. *J. Am. Oil Chem. Soc.*, **39**, 198–201.
Kemp, P. & Dawson, R. M. C. (1969a). Characterisation of N-(2-hydroxyethyl)alanine as a component of a new phospholipid isolated from rumen protozoa. *Biochim. Biophys. Acta*, **176**, 678–9.
Kemp, P. & Dawson, R. M. C. (1969b). Isolation of a new phospholipid, phosphatidyl-N-(2-hydroxyethyl)alanine from rumen protozoa. *Biochem. J.*, **113**, 555–8.
Kemp, P. & Lander, D. J. (1983). The hydrogenation of γ-linolenic acid by pure cultures of two rumen bacteria. *Biochem. J.*, **216**, 519–22.
Kemp, P. & Lander, D. J. (1984). Hydrogenation *in vitro* of α-linolenic acid to stearic acid by mixed cultures of pure strains of rumen bacteria. *J. Gen. Microbiol.*, **130**, 527–33.
Kemp, P., Dawson, R. M. C. & Klein, R. A. (1972). A new bacterial sphingophospholipid containing 3-aminopropane-1,2-diol. *Biochem. J.*, **130**, 221–7.
Kemp, P., White, R. W. & Lander, D. J. (1975). The hydrogenation of unsaturated fatty acids by five bacterial isolates from the sheep rumen, including a new species. *J. Gen. Microbiol.*, **90**, 100–14.
Kemp, P., Lander, D. J. & Gunstone, F. D. (1984a). Hydrogenation of some *cis* and *trans* octadecenoic acids to stearic acid by a rumen *Fusocillus* sp. *Br. J. Nutr.*, **52**, 165–70.
Kemp, P., Lander, D. J. & Holman, R. T. (1948b). The hydrogenation of the series of methylene-interrupted *cis,cis* octadecadienoic acids by pure cultures of rumen bacteria. *Br. J. Nutr.*, **52**, 171–7.
Kemp, P., Lander, D. J. & Orpin, C. G. (1984c). The lipids of the rumen fungus *Piromonas communis*. *J. Gen. Microbiol.*, **130**, 27–37.

Kepler, C. R. & Tove, S. B. (1967). Biohydrogenation of unsaturated fatty acids. III. Purification and properties of a linoleate Δ^{12}-cis,Δ^{11}-trans isomerase from *Butyrivibrio fibrisolvens*. *J. Biol. Chem.*, **242**, 5686–92.

Kepler, C. R., Hirons, K. P., McNeill, J. J. & Tove, S. B. (1966). Intermediates and products of the biohydrogenation of linoleic acid by *Butyrivibrio fibrisolvens*. *J. Biol. Chem.*, **241**, 1350–4.

Kepler, C. R., Tucker, W. P. & Tove, S. B. (1970). Biohydrogenation of unsaturated fatty acids. IV. Substrate specificity and inhibition of linoleate Δ^{12}-cis,Δ^{11}-trans isomerase from *Butyrivibrio fibrisolvens*. *J. Biol. Chem.*, **245**, 3612–20.

Kepler, C. R., Tucker, W. P. & Tove, S. B. (1971). Biohydrogenation of unsaturated fatty acids. V. Stereospecificity of proton addition and mechanism of action of linoleic acid Δ^{12}-cis,Δ^{11}-trans isomerase from *Butyrivibrio fibrisolvens*. *J. Biol. Chem.*, **246**, 2765–71.

Klein, R. A., Hazlewood, G. P., Kemp, P. & Dawson, R. M. C. (1979). A new series of long-chain dicarboxylic acids with vicinal dimethyl branching found as major components of the lipids of *Butyrivibrio* spp. *Biochem. J.*, **183**, 691–700.

Koga, Y. & Goldfine, H. (1984). Biosynthesis of phospholipids in *Clostridium butyricum*: kinetics of synthesis of plasmalogens and the glycerol acetal of ethanolamine plasmalogen. *J. Bacteriol.*, **159**, 597–604.

Kusman, J. E. (1973). Characterisation of the lipids of six strains of *Bacteroides ruminicola*. *J. Bacteriol.*, **113**, 1121–6.

Latham, M. J., Storry, J. E. & Sharpe, M. E. (1972). Effect of low-roughage diets on the microflora and lipid metabolism in the rumen. *Appl. Microbiol.*, **24**, 871–7.

Leat, W. M. F. (1977). Depot fatty acids of Aberdeen Angus and Friesian cattle reared on hay and barley diets. *J. Agric. Sci.* (Camb.), **89**, 575–82.

Lennarz, W. J. (1966). Lipid metabolism in the bacteria. *Adv. Lipid Res.*, **4**, 175–225.

Miles, S. C., Scott, T. W., Russell, G. R. & Smith, R. M. (1970). Hydrogenation of C_{18} unsaturated fatty acids by pure cultures of a rumen micrococcus. *Austral. J. Biol. Sci.*, **23**, 1109–13.

Miyagawa, E. (1982). Cellular fatty acid and fatty aldehyde composition of rumen bacterium. *J. Gen. Appl. Microbiol.*, **28**, 389–408.

Nieman, C. (1954). Influence of trace amounts of fatty acids on the growth of microorganisms. *Bacteriol. Rev.*, **18**, 147–67.

Noble, R. C. (1978). Digestion, absorption and transport of lipids in ruminant animals. *Progr. Lipid Res.*, **17**, 55–91.

Noble, R. C., Steele, W. & Moore, J. H. (1969). The incorporation of linoleic acid into the plasma lipids of sheep given intraruminal infusion of maize oil or free linoleic acid. *Br. J. Nutr.*, **23**, 709–14.

Patton, R. A., McCarthy, R. D. & Griel, L. C. (1970). Lipid synthesis by rumen microorganisms. II. Further characterization of the effects of methionine. *J. Dairy Sci.*, **53**, 460–5.

Polan, C. E., McNeill, J. J. & Tove, S. B. (1964). Biohydrogenation of unsaturated fatty acids by rumen bacteria. *J. Bacteriol.*, **88**, 1056–64.

Prins, R. A. (1977). Biochemical activities of gut microorganisms. In *Microbial Ecology of the Gut*, ed. R. T. J. Clarke & T. Bauchop. Academic Press, London and New York, pp. 73–183.

Prins, R. A., Van Nevel, C. J. & Demeyer, D. I. (1972). Pure culture studies on inhibitors for methanogenic bacteria. *Ant. van Leeuwenhoek J. Microbiol. Serol.*, **38**, 281–7.

Prins, R. A., Akkermans-Kruyswijk, J., Franklin-Klein, W., Lankhorst, A. & Van Golde, L. M. G. (1974). Metabolism of serine and ethanolamine plasmalogens in *Megasphaera elsdenii*. *Biochim. Biophys. Acta*, **348**, 361–9.

Prins, R. A., Lankhorst, A., Van der Meer, P. & Van Nevel, C. J. (1975). Some characteristics of *Anaerovibrio lipolytica*, a rumen lipolytic organism. *Ant. van Leeuwenhoek J. Microbiol. Serol.*, **41**, 1–11.

Raetz, C. R. H. (1978). Enzymology, genetics and regulation of membrane phospholipid synthesis in *Escherichia coli*. *Microbiol. Rev.*, **42**, 614–59.

Reiser, R. (1951). Hydrogenation of polyunsaturated fatty acids by the ruminant. *Fed. Proc.*, **10**, 236.
Rosenfeld, I. S. & Tove, S. B. (1971). Biohydrogenation of unsaturated fatty acids. IV. Source of hydrogen and stereospecificity of reduction. *J. Biol. Chem.*, **246**, 5025–30.
Roughan, P. G. & Batt, R. D. (1969). Glycerolipid composition of leaves. *Phytochemistry*, **8**, 363–9.
Sachan, D. S. & Davis, C. L. (1969). Hydrogenation of linoleic acid by a rumen spirochaete. *J. Bacteriol.*, **98**, 300–1.
Shorland, F. B. (1963). The distribution of fatty acids in plant lipids. In *Chemical Plant Taxonomy*, ed. T. W. Swain. Academic Press, New York, pp. 253–311.
Shorland, F. B., Weenink, R. O. & Johns, A. T. (1955). Effect of the rumen on dietary fat. *Nature* (Lond.), **175**, 1129.
Silber, P., Borie, R. P., Mikowski, E. J. & Goldfine, H. (1981). Phospholipid biosynthesis in some anaerobic bacteria. *J. Bacteriol.*, **147**, 57–61.
Singh, S. & Hawke, J. C. (1979). The *in vitro* lipolysis and biohydrogenation of monogalactosyldiglyceride by whole rumen contents and its fractions. *J. Sci. Food Agric.*, **30**, 603–12.
Trémolières, A. (1970). Les lipides des tissus photosynthétiques. *Ann. Biol.*, **3–4**, 113–56.
Tweedie, J. W., Rumsby, M. G. & Hawke, J. C. (1966). Studies on rumen metabolism. V. Formation of branched long-chain fatty acids in cultures of rumen bacteria. *J. Sci. Food Agric.*, **17**, 241–4.
Van Golde, L. M. G., Prins, R. A., Franklin-Klein, W. & Akkermans-Kruyswijk, J. (1973). Phosphatidylserine and its plasmalogen analogue as major lipid constituents in *Megasphaera elsdenii*. *Biochim. Biophys. Acta*, **326**, 314–23.
Verhulst, A., Semjen, G., Meerts, U., Janssen, G., Parmentier, G., Asselberghs, S., van Hespen, H. & Eyssen, H. (1985). Biohydrogenation of linoleic acid by *Clostridium sporogenes*, *Clostridium bifermentans*, *Clostridium sordellii* and *Bacteroides* sp. *FEMS Microbiol. Ecol.*, **31**, 255–9.
Viviani, R. (1970). Metabolism of long-chain fatty acids in the rumen. *Adv. Lipid Res.*, **8**, 267–346.
Viviani, R. & Borgatti, A. R. (1967). Micro-organismi del rumine e bioidrogenazione degli acidi grassi poliinsaturi. *Atti Soc. Ital. Sci. Vet.*, **21**, 254–9.
Viviani, R., Borgatti, A. R., Cortesi, P. & Crisetig, G. (1968). Consituenti lipidici dei batteri e dei protozoi del rumine di ovino. *Nuova Vet.*, **44**, 279–83.
Ward, P. F. V., Scott, T. W. & Dawson, R. M. C. (1964). The hydrogenation of unsaturated fatty acids in the ovine digestive tract. *Biochem. J.*, **92**, 60–8.
Watanabe, T., Okuda, S. & Takahashi, H. (1982). Physiological importance of even-numbered fatty acids and aldehydes in plasmalogen phospholipids of *Selenomonas ruminantium*. *J. Gen. Appl. Microbiol.*, **28**, 22–33.
Watanabe, T., Okuda, S. & Takahashi, H. (1984). Turn-over of phospholipids in *Selenomonas ruminantium*. *J. Biochem.*, **95**, 521–7.
Wegner, G. H. & Foster, E. M. (1963). Incorporation of isobutyrate and valerate into cellular plasmalogen by *Bacteroides succinogenes*. *J. Bacteriol.*, **85**, 53–61.
White, R. W., Kemp, P. & Dawson, R. M. (1970). Isolation of a rumen bacterium that hydrogenates oleic acid as well as linoleic acid and linolenic acid. *Biochem. J.*, **116**, 767–8.
Wilde, P. F. & Dawson, R. M. C. (1966). The biohydrogenation of α-linolenic and oleic acid by rumen micro-organisms. *Biochem. J.*, **98**, 469–75.
Williams, P. P. & Dinusson, W. E. (1973). Amino acid and fatty acid composition of bovine ruminal bacteria and protozoa. *J. Anim. Sci.*, **36**, 151–3.
Williams, P. P., Gutierrez, J. & Davis, R. E. (1963). Lipid metabolism of rumen ciliates and bacteria. II. Uptake of fatty acids and lipid analysis of *Isotricha intestinalis* and rumen bacteria with further information on *Entodinium simplex*. *Appl. Microbiol.*, **11**, 260–4.
Wood, R. D., Bell, M. C., Grainger, R. B. & Teekel, R. A. (1963). Metabolism of labelled linoleic-1-^{14}C acid in the sheep rumen. *J. Nutr.*, **79**, 62–8.
Wright, D. E. (1959). Hydrogenation of lipids by rumen protozoa. *Nature* (Lond.), **184**, 875–6.

Wright, D. E. (1960). Hydrogenation of chloroplast lipids by rumen bacteria. *Nature* (Lond.), **185**, 546–7.

Wright, D. E. (1961). Bloat in cattle. XX. Lipase activity of rumen microorganisms. *NZ J. Agric. Res.*, **4**, 216–23.

Yamazaki, S. & Tove, S. B. (1979). Biohydrogenation of unsaturated fatty acids. 8. Presence of dithionite and an endogenous electron donor in *Butyrivibrio fibrisolvens*. *J. Biol. Chem.*, **254**, 3812–7.

Yokoyama, M. T. & Davis, C. L. (1971). Hydrogenation of unsaturated fatty acids by *Treponema* (*Borrelia*) strain B_25, a rumen spirochaete. *J. Bacteriol.*, **107**, 519–27.

10

The Genetics of Rumen Bacteria

G. P. Hazlewood

AFRC Institute of Animal Physiology and Genetics Research, Cambridge, UK

&

R. M. Teather

Animal Research Centre, Research Branch, Agriculture Canada, Ottawa, Ontario, Canada

Investigation of the molecular genetics of rumen microorganisms is currently the most rapidly expanding area of rumen microbiology, and at the same time is the area in which our knowledge is least advanced. As a consequence, findings which would be regarded as modest in the context of more advanced genetic systems represent a significant consolidation of our understanding. There is little doubt that many of the genetic systems which have been shown by classical and *in vitro* genetic techniques to exist in genetically well characterized aerobic microorganisms such as *Escherichia coli*, *Bacillus subtilis* and certain fungi operate also in the anaerobic microorganisms of the rumen. However, their existence is, on the whole, unsupported by experimental evidence and is inferred only by analogy with slightly better studied anaerobic organisms isolated from other ecosystems.

Whatever the general status of molecular genetic studies on anaerobic organisms, there is a further problem to be resolved. Knowledge gained with anaerobes from other environments can probably be applied to their nearest relatives in the rumen, but in many cases the correct classification of common rumen microorganisms is not well established. Recent studies using DNA hybridization, RNA sequence analysis, or analysis of other macromolecular structures have suggested that in a number of cases the classification of rumen bacteria, as well as protozoa, has been incorrect (see Chapters 2 and 3 for other reasons for reclassifications). For example, it has been suggested that *Bacteroides succinogenes* and *Bacteroides* (*Ruminobacter*) *amylophilus*, organisms of major interest to rumen microbiologists, do not belong in the genus *Bacteroides* (Shah & Collins, 1983). Analysis of 16S RNA from 5 strains of *B. succinogenes* confirmed that they were not closely related to other *Bacteroides* spp., and moreover showed such diversity among the 5 strains examined that they would have to be classified into two genera (Stahl, 1987). Attempts to apply techniques developed for gastro-intestinal tract *Bacteroides* spp. to these organisms would probably not be successful. In the case of *Butyrivibrio fibrisolvens*, analysis of cell wall and membrane architecture in a number of strains suggested that isolates conforming to the species description may in fact belong to any of two or three distinct species (Miyagawa, 1982; Dibbayawan et al., 1985). This was confirmed by Mannarelli & Hespell (1987) who examined 31 strains of B. fibrisolvens using DNA–DNA hybridization. The largest grouping that could be identified as a species on the basis of DNA relatedness contained only 5 strains, and it was concluded that the strains comprise a genetically heterogeneous group consisting of a number of species and probably two genera. Perhaps even more surprising, a study of 16S RNA oligonucleotide catalogues in *Selenomonas ruminantium* and *Megasphaera elsdenii* showed that these organisms do not group with any of the 200 Gram-negative bacterial groups catalogued to date, but show a distinct relationship to the Gram-positive *Clostridium* subdivision (Stackebrandt et al., 1985). The successful development and application of techniques for the genetic manipulation of these organisms will require conscious and careful assessment of strains currently in use in different laboratories. Whatever their true phylogenetic position, it can be expected that unique solutions to the technical problems of recombinant-DNA studies will be required. DNA hybridization studies using species-specific probes have found application in the identification of *Bacteroides* spp., both in the characterization of isolates (Coykendall et al., 1980; Johnson, 1980; Salyers et al., 1983) and in estimating the levels of different species within a mixed population (Kuritza et al., 1986). This technique could be usefully applied to the common rumen species.

MOLECULAR GENETICS OF ANAEROBIC BACTERIA

Molecular genetic studies on obligately anaerobic bacteria have lagged somewhat behind those on organisms such as *E. coli* which are more convenient to study. Much of the work that has been carried out has been driven by the need to understand the epidemiology of antibiotic resistance in medically important bacteria such as *Clostridium* and *Bacteroides* spp. Industrial interests have also supported studies on the genera *Clostridium* and *Methanobacterium*. Work in this area has been reviewed by Woods (1982) and Woods & Jones (1984) and will therefore be considered only in general terms here. The reviews showed that there had been reports of intergeneric plasmid transfer by conjugation between *Bacteroides* spp. and *E. coli*, plasmids and bacteriophage had been described in anaerobes, and some systems for protoplast formation and regeneration had been developed. However, no systems for introducing plasmid cloning vectors into cells had been described, and no cloning vectors had been developed. In addition, no natural system for transformation of chromosomal DNA into anaerobic bacteria was known to exist (Saunders et al., 1984; Stewart & Carlson, 1986). Since the preparation of those reviews, however, there have been a number of noteworthy developments in this area.

The presence of a cryptic plasmid was reported in *Methanobacterium thermoautotrophicum* (Meile et al., 1983), and potential shuttle-vectors based on the methanogen plasmid pME2 have been constructed (Meile & Reeve, 1985). Of perhaps more fundamental significance, the expression of genes from archaebacteria (see Chapter 2) in eubacteria has been examined. Wood et al. (1983) showed that *arg*G and *his*A mutations in *E. coli* could be complemented by DNA cloned from *Methanococcus voltae*. Genes from *Methanosarcina barkeri* have been shown to function in both *E. coli* and *Bacillus subtilis* (Morris & Reeve, 1984). The formate dehydrogenase gene from *Methanobacterium formicicum* has also been cloned in *E. coli*. A detailed analysis of the cloned gene's expression and DNA sequence showed that it was probably expressed correctly (or almost correctly) in eubacteria and contained control sequences similar to eukaryotic ones (Shuber et al., 1986). The sequence of the *his*I gene has also been shown to be strongly conserved in the archaebacterium *Methanococcus vannielii*, the eubacterium *E. coli*, and the eukaryote *Saccharomyces cerevisiae* (Beckler & Reeve, 1986). These results suggest that it should be possible to move genetic information from the archaebacteria into genera from other kingdoms. The reverse procedure, however, may not be so straightforward, as there is evidence to suggest that the DNA-dependent RNA polymerase from at least one methanogen, *Methanococcus thermolithotrophicus*, only recognizes promoter sequences on archaebacterial DNA; i.e. it appears that, while archaebacterial control regions contain sequences that function as promoters in eubacteria, eubacterial promoters may not function in archaebacteria (Thomm & Stetter, 1985).

Systems for protoplast formation and regeneration have been described for *Clostridium pasteurianum* (Minton & Morris, 1983), *C. tertium* (Knowlton et al., 1984), and *C. saccharoperbutylacetonicum* (Yoshino et al., 1984). Protoplast fusion has been used to produce recombinants in *C. acetobutylicum* (Jones et al., 1985). Transfection of *C. acetobutylicum* (Reid et al., 1983), polyethylene glycol (PEG)-mediated transformation of *C. perfringens* with a 38·8 kb plasmid conferring tetracycline resistance, pJU124 (Heefner et al., 1984), and transformation of *C. acetobutylicum* with a plasmid cloning vector from *Staphylococcus aureus*, pUB110 (Lin & Blaschek, 1984), have all been reported. Shuttle vectors, combining the *E. coli* plasmid vector pBR322, a small cryptic plasmid from *C. perfringens*, and the tet^r gene from *C. perfringens* plasmids pJU124 or pCW3, have been constructed and been found to function in this transformation system (Squires et al., 1984). Reciprocal transformation experiments between *E. coli* and *C. perfringens* using the shuttle plasmid-vectors have suggested that restriction barriers exist in both directions (10^2- to 10^4-fold reduction in transformation efficiency) but do not provide an insurmountable barrier to intergeneric DNA transfer in this case.

The control of gene expression in the genus *Clostridium* has been studied in a number of laboratories. Regulation of expression of cellulase in *C. thermocellum* (Johnson et al., 1985) and of β-amylase in *C. thermosulphurogenes* (Hyun & Zeikus, 1985) has been examined and appears to be formally analogous to catabolic enzyme regulation in *E. coli*, being both inducible and subject to catabolite repression. Over 20 apparently distinct DNA fragments coding for cellulases have been cloned from *C. thermocellum* into *E. coli* (Millet et al., 1985; Schwarz et al., 1985; Romaniec et al., 1987). The clones containing the genes *cel*A, *cel*B, *cel*C and *cel*D have been extensively characterized and the DNA sequences of the coding and control regions determined (Beguin et al., 1985; Grepinet & Beguin, 1986; Joliff et al., 1986). These studies have shown that *C. thermocellum* genes can be expressed from their own control sequences in *E. coli*. The signal sequence of the *cel*A gene appears to be typical of Gram-positive bacteria. Other features of the control region appear normal, and it is apparent that the DNA transcription and translation systems do not provide a major barrier to intergeneric transfer of genetic information in this case.

The biology and genetics of *Bacteroides* spp. isolated from the gastro-intestinal tract have been reviewed by Salyers (1984), and described in the reviews of Woods (1982) and Woods & Jones (1984). A number of general conclusions from these reviews will be repeated here before discussing more recent developments. The genus *Bacteroides* is phylogenetically distant from most of the eubacteria. Its physiology is relatively poorly understood. This genus plays a major role in polysaccharide degradation in the intestinal tract. To date, all enzyme systems involved in polysaccharide degradation that have been examined have been shown to be regulated, being both inducible by their substrate (or degradation products of the substrate) and repressible by more readily fermentable substrates. This repression is formally similar to the catabolite repression observed in *E. coli*. A number of these enzymes have been isolated and characterized. The intestinal *Bacteroides* represent a major, partially characterized, pool of genetic information that may have applications in the genetic analysis and modification of rumen species.

Cryptic plasmids, and those carrying antibiotic resistance markers, are common among *Bacteroides* spp. Resistance to both clindamycin and erythromycin has been shown to be carried by a 41 kb conjugative plasmid (known as pBF4 or pIP410). Another conjugative plasmid, pBFTM10, is about 14·6 kb and also carries resistance to clindamycin and erythromycin. These plasmids are capable of mobilizing the transfer of other plasmids. There is a large body of evidence which suggests that the clindamycin resistance determinant on these plasmids is part of an element with the properties of a transposon, with the unusual capability of conjugal transfer from a position of chromosomal integration (Mays *et al.*, 1982). The most recent studies of this element support this conclusion, and make use of this property in the construction of *Bacteroides–E. coli* shuttle vectors (Shoemaker *et al.*, 1985, 1986*a,b*). This DNA structure is flanked by a 1·2 kb direct repeat (Rasmussen & Macrina, 1985; Smith, 1985*a,b*; Smith & Gonda, 1985) which has also been shown to be present in the *Bacteroides fragilis* chromosome, suggesting that the element may function as an insertion sequence (Rasmussen & Macrina, 1985). Tetracycline resistance in *Bacteroides* spp. also appears to be carried on a conjugative transposon. This transfer system has the unusual property of being inducible by tetracycline (Privitera *et al.*, 1979). An apparently comparable genetic element has been reported in *Clostridium difficile* (Smith *et al.*, 1981).

There have been a large number of studies reported on the inter- and intra-generic transfer of plasmids among anaerobic bacteria and between anaerobic and aerobic bacteria. A large part of the work reported has focused on the genus *Bacteroides*. Burt & Woods (1976) demonstrated that transfer of an R plasmid from *E. coli* to *Bacteroides fragilis*, *Bacteroides* spp., *Fusobacterium* spp. and other faecal anaerobes was possible, providing the recipient was heat-treated, which was presumed to inhibit restriction systems in the anaerobes. Transfer between *Bacteroides* spp.

(Privitera et al., 1979; Tally et al., 1979; Welch et al., 1979) and from *B. fragilis* to *E. coli* (Mancini & Behme, 1977; Young & Mayer, 1979; Rashtchian & Booth, 1981) has also been reported. Some doubt has been expressed about the interpretation of these experiments (Salyers, 1984). More recent studies have shown that most antibiotic resistance markers are expressed differently in *Bacteroides* and *E. coli*. For example, *E. coli* R plasmid genes for sulphonamide, trimethoprim, ampicillin and tetracycline resistance are not expressed in *Bacteroides* spp., while the clindamycin and erythromycin resistance determinant from *Bacteroides* is not expressed in *E. coli* (Guiney et al., 1984b; Shoemaker et al., 1985). More surprisingly, a cryptic region from a *Bacteroides* plasmid expresses tetracycline resistance in *E. coli* but not in *Bacteroides* (Guiney et al., 1984a,b). These differences in gene expression support the arguments of Salyers (1984) that the occurrence of conjugal R plasmid transfer between *Bacteroides* spp. and *E. coli* has not been established, even while demonstrating that workable shuttle-vectors can be constructed. In the one case where the sequence and expression of a gene from *Bacteroides* has been examined in detail, the pilin gene from *Bacteroides nodosus*, it was found that the gene product retained extensive homology at the N-terminal with similar pilins from *Pseudomonas*, *Mycobacterium* and *Neisseria*. The regulatory sequences upstream of the structural gene were typical of eukaryotes. However, the gene product underwent extensive post-translational processing, which included the removal of the seven N-terminal amino acids, *N*-methylation of the new N-terminal amino acid (phenylalanine), and internal cleavage of the mature 149 amino acid pilin protein between amino acids 72 and 73 (Elleman & Hoyne, 1984; Elleman et al., 1986). While it is not to be expected that such extensive post-translational processing is typical of *Bacteroides* protein synthesis, it does provide an example of some of the ways in which genes transferred between species may fail to produce functional products when gene functions which are not present in the recipient species may be required for successful gene expression.

Bacteriophages have been reported for a number of *Bacteroides* species, but transduction and lysogeny have not been reported. Until recently, all phages capable of lysing intestinal tract *Bacteroides* spp. had been isolated from sewage despite a number of attempts to isolate them from faeces. However, Kai et al. (1985) have reported the isolation of phage capable of infecting *Bacteroides fragilis* and/or *Bacteroides distasonis* from the faeces of only about 5% of the human subjects examined, so there is probably no special significance attached to the previous failures. Bacteriocins have also been reported (Booth et al., 1977), and methods developed to use bacteriocin production and sensitivity in the identification of *Bacteroides* (Riley & Mee, 1981, 1982). However, it has not been reported whether the bacteriocin production genes are plasmid borne. It is possible that plasmids carrying genes for bacteriocins effective against *Bacteroides* spp. could serve as the basis for effective cloning vectors in the rumen environment.

Only one report exists of the successful transformation of *Bacteroides*. The plasmid used was the 14·6 kb plasmid pBFTM10 and the recipient strain was *Bacteroides fragilis* 638 (Smith, 1985c). The method, based on the method of Klebe et al. (1983), does not require the preparation and regeneration of protoplasts.

Transformation and expression of the clindamycin resistance gene required a minimum of 30 minutes, and numbers of transformants continued to increase up to $2\frac{1}{2}$ hours. While transformation efficiency was reasonably high ($4\cdot2 \times 10^3$ transformants/μg plasmid DNA) with this plasmid and bacterial strain, the method did not work with this plasmid and other *Bacteroides fragilis* strains or other *Bacteroides* species. Limited success in transforming other strains of *Bacteroides* was reported with a smaller hybrid-cloning vector.

Genetic systems developed for the aerobic Gram-positive cocci may be applicable to the rumen facultative anaerobe *Streptococcus bovis* or to *Ruminococcus* species. *Streptococcus pneumoniae* and *Streptococcus sanguis* have very well characterized systems for natural transformation by chromosomal or plasmid DNA. This system seems to require either that the incoming DNA is at least partially homologous with existing intracellular DNA or that it consists of multiple copies of a sequence (for recent reviews see Saunders *et al.*, 1984; Stewart & Carlson, 1986). Transformation of protoplasts with plasmid DNA has been reported for *Staphylococcus carnosus* (Gotz *et al.*, 1983), and calcium-shock induced transformation with plasmid DNA has been reported for *Staphylococcus aureus*. A shuttle vector, effective with *Escherichia coli* and *Streptococcus sanguis* or *S. mutans*, has also been developed (Dao & Ferretti, 1985).

DISTRIBUTION AND SIZE CHARACTERISTICS OF PLASMIDS IN RUMEN BACTERIA

Naturally occurring plasmids provide the basis for the development of cloning vectors, as well as providing in some instances a system for DNA transfer (conjugation) and a method for mapping the position of genes on the chromosome (by conjugal transfer of chromosomal DNA initiated by a plasmid integrated in the chromosome, or by co-transfer of chromosomal genes integrated in the plasmid). Plasmids are commonly found in naturally occurring bacterial populations; the functions they encode frequently confer specific competitive advantages upon the host organism. The salient features of the rumen environment, such as rapid turnover, high population density and varied input of organic materials, might be expected to promote the development of plasmid-encoded functions. Although the rumen microbial population has been extensively studied and many different phenotypes have been determined, the occurrence and ecological significance of plasmid-determined traits is not understood. Bacteriocin production, a trait normally carried by plasmids, was first reported in a rumen isolate (*Streptococcus bovis*) in 1976 (Iverson & Millis, 1976c).

Although circumstantial evidence that rumen urease activity may be plasmid-encoded had been presented earlier (Cook, 1976), the first positive report of plasmids in rumen bacteria was made by Teather (1982). Supercoiled plasmid DNA was isolated from a type strain of *Butyrivibrio fibrisolvens* (ATTC 19177), and from 6 newly isolated strains, by lysis of cells at relatively low suspension densities and successive precipitations with SDS–NaCl and PEG 6000. Analysis of the plasmid

DNA by sucrose density gradient centrifugation indicated a molecular weight of 250 megadaltons; this figure was supported by summation of the fragment sizes produced on digestion of the plasmids with HindIII. EcoRI restriction sites occurred at an abnormally low frequency on these large plasmids, indicating possible DNA base modification, but no evidence was found for the presence of either restriction enzymes or periplasmic nucleases with activity against bacteriophage λ or pBR322 DNA. No functions were assigned to these very large plasmids.

Subsequent studies have demonstrated smaller plasmids in a number of strains of Butyrivibrio. Mann et al. (1985, 1986) described the isolation and restriction mapping of a 2·8 kb cryptic plasmid (pOM1) from B. fibrisolvens strain 49 (Bryant & Small, 1956). With the object of producing a shuttle vector capable of replication in B. fibrisolvens and E. coli, pOM1 restricted with PstI or EcoRI was ligated into the EcoRI and PstI sites of pBR325 and into the PstI sites of pAT153 and pHV33. The series of hybrids thus made were transformed into, and stably maintained in, E. coli HB101 and ED8767, but to date no evidence is available to suggest that these plasmids can be transformed into Butyrivibrio spp. More recently (Orpin et al., 1986b and unpublished results), a comprehensive survey was conducted to assess the distribution of plasmids in 150 strains of B. fibrisolvens isolated on a non-selective habitat-simulating medium from domestic cattle, high-arctic Svalbard reindeer (Orpin et al., 1985), captive musk ox, seaweed-grazing Orkney sheep (Greenwood et al., 1983) and domestic Clun Forest sheep. Plasmid DNA was extracted by the method of Keiser (1984). The results are shown in Table 1. Up to 28% of the isolates contained plasmids, of which the majority contained 2–4 plasmid bands which varied in size from 1·8 kb to about 40 kb. At first glance, their distribution seemed dependent both on the species of ruminant from which the bacteria were isolated and, within a given species, on the geographical origin. The function of these plasmids is unknown.

A similar study was conducted with 157 strains of Selenomonas ruminantium isolated from domestic cattle and Svalbard reindeer on a non-selective medium, and from captive musk ox and domestic sheep and goats on selective media with different carbohydrates as sole carbon source (Orpin et al., 1986b and unpublished

Table 1
Distribution and Size Characteristics of Plasmids in Butyrivibrio fibrisolvens

Source	No. of isolates	% with plasmids	No. of plasmid bands (%)	
			1	2–4
Domestic cattle	10	10	—	100
Svalbard reindeer	90	28	16	84
Musk ox	6	0	—	—
Orkney sheep	10	0	—	—
Domestic, Clun Forest sheep	35	14	14	86

results). The results are shown in Table 2. The majority of the strains contained plasmid DNA in the size range 1 kb to greater than 50 kb, with as many as 10 bands of plasmid DNA detectable for each isolate. Striking differences, perhaps characteristic of subpopulations within the species, were apparent in isolates obtained from the selective media containing different carbon sources. All strains utilizing lactate (subsp. *lactilytica*) contained at least 2 bands of plasmid DNA and 60% of the strains contained multiple bands, but it has not proved possible to correlate the presence of a particular band with the ability to use lactate. All strains isolated on xylose harboured plasmid DNA, including 3 bands in excess of 20 kb which were common to this particular phenotype. In the same laboratory, strains isolated on a selective medium containing trehalose as carbon source were shown to carry a common plasmid of about 100 kb. Also, in some lactate-utilizing strains, the loss of the ability to utilize lactate, which occurred after 20 successive transfers in medium containing glucose as carbon source, coincided with the loss of a plasmid of approximately 90 kb, indicating that lactate utilization might be plasmid-encoded; the plasmid in question has not, however, been demonstrated in all strains growing on lactate. There is a tendency for the larger plasmids of selenomonads to fragment, even under mild lysis conditions, so a degree of caution is required in interpreting results obtained with these strains.

The genus *Ruminococcus* has also been examined recently for the presence of plasmids. Kelly & Asmundson (1986, and unpublished results) and Asmundson & Kelly (1987), using mutanolysin to effect cell lysis, found small cryptic plasmids in 6 out of 7 strains of *R. albus* and in all 3 strains of *R. flavefaciens* examined. Again with the object of producing a shuttle vector, a 5·2 kb plasmid (pRf186) from *R. flavefaciens* 186 was ligated into the *Bam*HI site of the *E. coli* plasmid pUC8, and the *Staphylococcus aureus*/*Bacillus* plasmid pUB110. At present it is not known whether the hybrids so formed can be transformed into, and stably maintained in, either *E. coli* or *Bacillus* sp. Using a number of different plasmid isolation techniques, White

Table 2
Distribution and Size Characteristics of Plasmids in *Selenomonas ruminantium* Strains

Source	No. of isolates	% with plasmids	No. of plasmid bands (%)			
			1	2–4	5–7	>8
Domestic cattle[a]	20	100	—	20	50	30
Svalbard reindeer[a]	20	100	—	25	50	25
Musk ox (S)[b]	10	100	—	—	40	60
Sheep (M)	50	65	34	62	4	—
Sheep (X)	8	100	—	63	37	—
Sheep (G)	27	41	—	18	55	27
Goat (L)	22	100	—	40	60	—

[a] Non-selective isolation medium.
[b] The selective isolation medium contained starch (S), mannitol (M), xylose (X), glucose (G) or lactate (L) as sole carbon source.

& Champion (1986) were unable to demonstrate plasmid DNA in 25 *Ruminococcus* strains, including at least 3 strains reported by Kelly & Asmundson (1986) to contain plasmids.

The genus *Bifidobacterium* has also been shown to carry plasmids in a significant proportion (approximately 20%) of isolates (Sgorbati *et al.*, 1982, 1983). Although most of the isolates examined were of human origin, 2 out of 20 isolates of *Bifidobacterium globosum* isolated from the bovine rumen carried moderate sized plasmids.

APPLICATION OF GENE CLONING TECHNIQUES TO RUMEN BACTERIA

The cloning of microbial genes by genetic recombination *in vitro* and their expression in suitable host organisms such as *E. coli* is a powerful technique which, in recent years, has been applied to rumen bacteria as a means of dissecting enzyme complexes and characterizing their functional components, and of investigating the regulation of gene expression in these organisms. Because the cellulolytic capacity of the rumen microflora represents an essential means by which ruminants digest fibrous plant tissue (see Chapter 8), it is perhaps not surprising that gene cloning techniques have been applied, in the first instance, to the cellulase genes of rumen bacteria.

A pseudo-random gene bank was constructed for the cellulolytic bacterium *Bacteroides succinogenes* S85 by ligating 5–15 kb *Sau*3A fragments of chromosomal DNA into the *Bam*HI site of the plasmid vector pUC8 (Crosby *et al.*, 1984*a,b*; Teather *et al.*, 1984). The recombinant plasmids were transformed into *E. coli* RR1, and 15 carboxymethyl cellulase (CMCase)$^+$ clones were detected by means of the Congo Red plate assay (Teather & Wood, 1982). The clones could be subdivided according to the level of production of CMCase activity and were shown by restriction enzyme analysis and DNA hybridization to contain 6 unrelated genes coding for β-glucanase activity. Clone RE3, containing a 2·2 kb insert subcloned from an original high-activity clone BC14, expressed β-glucanase activity to the same level as the original clone and is the subject of further work which will elicit the DNA sequence and provide information on the regulation of transcription and translation and the nature of regulatory sequences in *B. succinogenes*. Clone O-64, which contained a 4·3 kb *Bam*HI fragment subcloned from the original insert of BC14, exhibited enzyme activity of 0·78 units mg^{-1} cell protein (1 unit corresponds to the release of 1 μmol reducing sugar h^{-1}) against CMC, but was inactive against filter paper. CMCase from clone O-64, which in *E. coli* was associated with the cell membrane, did not elute with the major soluble β-glucanase of *B. succinogenes* on DEAE column chromatography, suggesting that during secretion it may undergo post-translational modification in *B. succinogenes* but not in *E. coli*. Expression of the β-glucanase gene was subject to glucose repression in both bacteria. [Apparent catabolite repression has been reported in a number of rumen bacteria, including *Bacteroides ruminicola*, *Streptococcus bovis*, *Selenomonas ruminantium*, *Butyrivibrio* *fibrisolvens*, *Megasphaera elsdenii* and *Ruminococcus albus* (Russell & Baldwin, 1978; Greve *et al.*, 1984).] The characteristics of this endoglucanase as expressed in *E. coli* have been examined more carefully (Taylor *et al.*, 1987). This examination confirmed the apparent catabolite repression of enzyme synthesis in *E. coli*, and showed that enzyme was secreted into the periplasmic space. This implies that the *B. succinogenes* gene product possesses an amino acid sequence which is recognized by the *E. coli* protein export system. The cloning in *E. coli* of DNA fragments expressing a mixed-linkage β-glucanase (Irvin *et al.*, 1986) and a xylanase (Sipat *et al.*, 1987) from *B. succinogenes* has also been described; in these cases as well, there is evidence that the gene products are at least partially exported from *E. coli*. It is once again apparent from these results that the DNA sequences responsible for the regulation of transcription and translation can be recognized even in distantly related organisms. While this is not always the case, there is no systematic barrier to the intergeneric transfer of genetic information between the anaerobic bacteria which have been examined and the more accessible aerobic organisms commonly used in recombinant-DNA work.

Using similar conventional techniques for *in vitro* genetic recombination, genomic libraries have been constructed for the cellulolytic rumen bacteria *Ruminococcus albus* SY3 and *Butyrivibrio* A46 by ligating *Sau*3A and *Eco*RI fragments of chromosomal DNA into plasmid pBR322 and the phage vector λ gt11, respectively (Romaniec *et al.*, 1986 and unpublished results). Both libraries were screened by the Congo Red plate assay and two different CMCase$^+$ clones were recovered from the *R. albus* library; similarly, two CMCase$^+$ clones containing *Butyrivibrio* A46 DNA were found, but on subsequent analysis were shown to contain identical DNA inserts. The two CMCase$^+$ *E. coli* clones containing *R. albus* DNA showed enzyme activity of 5.4 and 0.9 units mg^{-1} cell protein, respectively, against CMC, and caused a significant reduction in viscosity of the substrate. These are characteristics of endo-β-1,4-glucanase. Restriction enzyme mapping showed that the two clones contained different DNA inserts; the more active clone, Ra1, contained a 10.5 kb insert and had some cellobiohydrolase activity, demonstrated by its ability to hydrolyse the aglycone bond of 4-methylumbelliferyl-β-D-cellobioside (MUC). In view of evidence suggesting that the cellulase of *R. albus* may be organized in a multienzyme complex (see Chapter 8), it is likely that more than two *cel* genes are present in the genome, so further work will be necessary to elucidate the genetic basis of cellulase production and regulation in this organism.

CMCase$^+$ clone BA46 recovered from the *Butyrivibrio* A46 gene bank contained a 2.3 kb DNA insert which was subcloned into plasmid pUC12. In common with clone Ra1, the insert from BA46 contained an abnormally high number of *Hin*dIII restriction enzyme sites. A cell-free extract from this clone was active against CMC (1.7 units mg^{-1} cell protein) and MUC, indicating both endoglucanase and cellobiohydrolase activities; CMCase activity amounted to 30% of that found in the culture supernatant of *Butyrivibrio* A46. The CMCase purified from cultures of *Butyrivibrio* A46 was a single protein with a molecular weight of 57 kilodaltons; by comparison, the molecular weight of the recombinant protein isolated from clone BA46 was 36–58 kDa. Sequencing of the 2.3 kb insert from clone BA46 is in progress and *Hin*dIII restriction fragments have been subcloned in a promoter probe vector in order to establish the nature of regulatory sequences in *Butyrivibrio* A46.

GENE TRANSFER IN RUMEN BACTERIA

In common with other natural ecosystems, little is known about the operation of gene transfer mechanisms in the rumen. However, current interest in the genetic manipulation of rumen microorganisms (Smith & Hespell, 1983; Armstrong & Gilbert, 1985; Teather, 1985; Forsberg *et al.*, 1986; Stewart *et al.*, 1986; Orpin *et al.*, 1988), and the consequent need to develop vectors and DNA transfer mechanisms necessary for the application of recombinant DNA techniques to this group of microbes, has greatly stimulated interest in gene transfer in rumen bacteria. In organisms where conjugation or transformation occurs naturally, and where a mechanism for homologous recombination exists, the transfer of genetic material has allowed both recombinant and complementation analysis and has been the basis of many classical experiments in genetics. In the absence of a foundation in classical genetics, the approach to establishing mechanisms of gene transfer in rumen microorganisms has been empirical rather than logical.

The first report of interspecies DNA transfer involving a rumen bacterium resulted from the transfer of the broad host-range conjugative plasmid RP4 from *E. coli* to *B. fibrisolvens* (Teather, 1985). *E. coli* OR28 carrying RP4, which codes for resistance to a number of antibiotics, including ampicillin, was mated with a streptomycin resistant, ampicillin sensitive strain of *B. fibrisolvens* (OR16-1). After 1 hour, clones resistant to both antibiotics appeared and were shown to comprise *B. fibrisolvens* carrying RP4. These clones were not stable. To date this is the only evidence that genes from another bacterial group can be replicated and correctly expressed in a rumen bacterium; it is perhaps all the more surprising since *B. fibrisolvens* is widely regarded as being Gram-positive (see Chapter 2) and, in general, *E. coli* genes do not express well in Gram-positive bacteria.

The same group have achieved limited success in establishing conditions for PEG-induced uptake of plasmid DNA by sphaeroplasts prepared from *B. fibrisolvens* by growth in the presence of the antibiotic vancomycin. Evidence was obtained that pRK248, a 9·6 kb deletion derivative of the broad host-range conjugative plasmid pRK2, was taken up and expressed tetracycline resistance in *B. fibrisolvens* regenerants, but the regeneration frequency was very low and the plasmid appeared to be unstable in its new host. Conditions which allow the preparation of stable sphaeroplasts and L-forms in *Butyrivibrio* have been reported (Hazlewood *et al.*, 1983*a*, 1987) which may, in combination with the plasmid vectors described above, improve results with this species.

Transformation has also been reported for *Selenomonas ruminantium* strain Suc11 (tre$^-$, lact$^-$) isolated on a selective medium containing sucrose as sole carbon source (Orpin *et al.*, 1986*a*). Total DNA from strain Tre6 (tre$^+$, lact$^+$), isolated on a medium containing trehalose, was mixed with growing cells of strain Suc11 on the surface of agar-containing medium. Tre$^+$lact$^+$ Suc11 clones were selected and were subsequently shown to contain a 45 kb plasmid absent from the original Suc11 but present in Tre6. These results suggest that natural transformation may occur in *Selenomonas*, even though the process normally favours uptake of chromosomal rather than plasmid DNA. The prospect for development of a sphaeroplast-based transformation system for this organism also exists, since it is very sensitive to lysozyme and sphaeroplasts produced in this way can be stabilized with 0·4M sucrose (Kamio & Takahashi, 1980). *S. ruminantium* can also be sphaeroplasted by adding to the growth medium difluoromethyl-lysine, an inhibitor of cell wall synthesis in this organism. The effect of difluoromethyl-lysine is reversed by adding lysine to the growth medium (Kamio *et al.*, 1986). The prospect for intergeneric transfer of genetic material between *Selenomonas* and other organisms by transduction or conjugation is relatively poor, however, as the unique peptidoglycan and storage polysaccharide structures found in this organism (Kamio *et al.*, 1980, 1981*a*,*b*) suggest that the macromolecular structures may not provide the necessary common receptor sites.

The prospect exists for developing protoplast transformation systems for other rumen bacterial species as well. A method for the preparation of stable protoplasts of *Bacteroides ruminicola* has been described (Cheng, 1973), and Brigidi *et al.* (1986) have described both protoplast formation and regeneration of viable cells in *Bifidobacterium*.

The quest for a high-efficiency transformation procedure that may be used for selected rumen bacteria has been taken up by numerous groups worldwide, without apparent success; some of the unsuccessful approaches have been discussed previously (Hazlewood *et al.*, 1987). Until very recently this quest was most severely hindered by the lack of native plasmids or lysogenic phage which might serve as the basis for the development of plasmid vectors. Restriction-modification systems of rumen anaerobes are also likely to constitute a potent barrier to transforming DNA but, to date, information regarding the distribution of restriction enzymes or periplasmic nucleases has been sought for only 6 *Butyrivibrio* strains (Teather, 1982) and L-forms generated from *Butyrivibrio* strain PI-7 (Hazlewood, G. P., unpublished results). No activity was identified in these studies. A high level of DNase activity has been demonstrated in culture supernatants of *Bacteroides succinogenes* (Forsberg, C. W., pers. comm). This enzyme (or enzymes) degraded both linear (bacteriophage λ) and supercoiled (pBR322) DNA. There was no apparent sequence specificity.

Various bacteriophages which are now known to occur in the rumen also have the potential for transferring genes between rumen bacteria both *in vivo* and *in vitro*. Phages have been reported in *Bifidobacterium longum* (Sgorbati *et al.*, 1983). Of *S. ruminantium* isolates, 2% were shown to be lysogenic when treated with mitomycin C (Hazlewood *et al.*, 1983*b*), and lysogenic phages have been isolated and characterized. Lysogenic phages have also been described in *Streptococcus bovis* (Iverson & Millis, 1976*a*,*b*; Tarakanov, 1976), and a virulent bacteriophage has been reported for the large bacterium Eadie's Oval (*Magnoovum eadii*) (Orpin & Munn, 1974). To date, phage transduction has not been demonstrated.

MUTAGENESIS IN RUMEN BACTERIA

Mutagenesis, induced by chemical and physical mutagens, has been employed in classical microbial genetics as a means of accelerating the natural process of selection, and of elucidating the mechanisms of regulation of gene expression. It has, however, found little application in the field of rumen microbiology.

Among anaerobic bacteria, it has been reported that *Bacteroides fragilis* is difficult to mutate (Woods & Jones, 1984). Schumann *et al.* (1983) reported evidence to suggest that *Bacteroides fragilis* lacks a *rec*A-type inducible DNA repair and homologous recombination system. The same may be true for *Clostridium acetobutylicum* (Bowring & Morris, 1985). Despite these anomalies, auxotrophic mutants have been isolated in *B. fragilis* and *C. acetobutylicum* by conventional techniques using EMS mutagenesis (Van Tassel & Wilkins, 1978; Bowring & Morris, 1985). Two systems responsible for repair of UV damage (pyrimidine-dimer excision) in *B. fragilis*, one constitutive and one induced by exposure to UV radiation, have been described (Abratt *et al.*, 1986), but a general system for homologous DNA recombination has not yet been demonstrated. If such a system does not in fact exist, then many of the classical approaches to mapping gene position on chromosomal DNA will not be applicable to this group of bacteria.

Biochemical studies of cellulase production by some of the principal species of cellulolytic bacteria in the rumen have indicated that the biosynthesis of these enzymes may be subject to close regulation by substrates or products. For example, cellulase production by *R. albus* is repressed in the presence of cellobiose or a number of alternative energy sources, including polysaccharides (Greve *et al.*, 1984). The genetic basis of such regulation has not been elucidated, but Taya *et al.* (1983) have demonstrated that derepressed *R. albus* mutants showing 2–10 times the cellulolytic activity of the parent strain could be generated by mutation with UV irradiation and N-methyl-N'-nitro-N-nitrosoguanidine.

CONCLUSIONS

It will be apparent from the foregoing that studies of genetic systems and the regulation of gene expression in anaerobic rumen microorganisms are in their infancy. Published work relates solely to rumen bacteria, with apparently no progress having been made with the anaerobic protozoa or fungi. The transfer of fundamental genetic knowledge gained with anaerobic bacteria from other ecosystems to their rumen counterparts is desirable, but it may require further taxonomic studies to establish the true identity of some rumen species. Progress has been made in establishing the distribution and size characteristics of plasmids in rumen bacteria, and a number of hybrid plasmids with potential uses as cloning vectors have been developed. The lack of progress in establishing DNA transfer mechanisms for rumen bacteria reflects the paucity of fundamental genetic information regarding, for example, restriction-modification systems of these organisms and their ability to replicate and express heterologous genes. However, the prospects for developing protoplast transformation in a number of species are good. The genetic basis of cellulase production by three species of rumen bacteria has been successfully investigated via a gene cloning approach, and it is evident that *cel* genes from rumen bacteria may be correctly transcribed from their own regulatory sequences in *E. coli*. In combination with conventional biochemical techniques, gene cloning will play an important role in broadening our understanding of the molecular biology and biochemistry of rumen bacteria.

REFERENCES

Abratt, V. R., Lindsay, G. L. & Woods, D. R. (1986). Pyrimidine dimer excision repair of DNA in *Bacteroides fragilis* wild type and mitomycin C-sensitive UV-sensitive mutants. *J. Gen. Microbiol.*, **132**, 2577–81.

Amstrong, D. G. & Gilbert, H. J. (1985). Biotechnology and the rumen: a mini review. *J. Sci. Food Agric.*, **36**, 1039–46.

Asmundson, R. V. & Kelly, W. J. (1987). Isolation and characterization of plasmid DNA from *Ruminococcus*. *Curr. Microbiol.*, **16**, 97–100.

Beckler, G. D. & Reeve, J. N. (1986) Conservation of primary structure in the *his*I gene of the archaebacterium *Methanococcus vannieli*, the eubacterium *Escherichia coli*, and the eucaryote *Saccharomyces cerevisiae*. *Molec. Gen. Genet.*, **204**, 133–40.

Beguin, P., Cornet, P. & Aubert, J. P. (1985). Sequence of a cellulase gene of the thermophilic bacterium *Clostridium thermocellum*. *J. Bacteriol.*, **162**, 102–5.

Booth, S. J., Johnson, J. L. & Wilkins, T. D. (1977). Bacteriocin production by strains of *Bacteroides* isolated from human feces and the role of these strains in the bacterial ecology of the colon. *Antimicrob. Agents Chemother.*, **11**, 718–24.

Bowring, S. N. & Morris, J. G. (1985). Mutagenesis of *Clostridium acetobutylicum*. *J. Appl. Bacteriol.*, **58**, 577–84.

Brigidi, P., Matteuzzi, D. & Crociani, F. (1986). Protoplast formation and regeneration in *Bifidobacterium*. *Microbiologica*, **9**, 243–8.

Bryant, M. P. & Small, N. (1956). The anaerobic, monotrichous, butyric acid-producing, curved, rod-shaped bacteria of the rumen. *J. Bacteriol.*, **72**, 16–21.

Burt, S. J. & Woods, D. R. (1976). R factor transfer to obligate anaerobes from *Escherichia coli*. *J. Gen. Microbiol.*, **93**, 405–9.

Cheng, K.-J. (1973). Spheroplast formation by an anaerobic Gram-negative bacterium *Bacteroides ruminicola*. *Can. J. Microbiol.*, **19**, 667–9.

Cook, A. R. (1976). The elimination of urease activity in *Streptococcus faecium* as evidence for plasmid-coded urease *J. Gen. Microbiol.*, **92**, 49–58.

Coykendall, A. L., Kazmarek, F. S. & Slots, J. (1980). Genetic heterogeneity in *Bacteroides saccharolyticus* (Holdeman and Moore, 1970) Finegold and Barnes (approved lists 1980) and proposal of *Bacteroides gingivalis* sp. novum and *Bacteroides macacae* (Slots and Genco) comb. nov. *Int. J. Syst. Bacteriol.*, **30**, 559–64.

Crosby, W. L., Collier, B., Thomas, D. Y., Teather, R. M. & Erfle, J. D. (1984a). Cloning and expression in *Escherichia coli* of cellulase genes from *Bacteroides succinogenes*. In *Fifth Canadian Bioenergy R & D Seminar*, ed. S. Hasnain. Elsevier Applied Science, London, pp. 573–6.

Crosby, W. L., Collier, B., Thomas, D. Y., Teather, R. M. & Erfle, J. D. (1984b). Cloning and expression in *Escherichia coli* of cellulase genes from *Bacteroides succinogenes*. *DNA*, **2**, 184.

Dao, M. L. & Ferretti, J. J. (1985). *Streptococcus–Escherichia coli* shuttle vector pSA3 and its use in the cloning of streptococcal genes. *Appl. Environ. Microbiol.*, **49**, 115–19.

Dibbayawan, T., Cox, G., Cho, K. Y. & Dwarte, D. M. (1985). Cell wall and plasma membrane architecture of *Butyrivibrio* spp. *J. Ultrastruct. Res.*, **90**, 286–93.
Elleman, T. C. & Hoyne, P. A. (1984). Nucleotide sequence of the gene encoding pilin of *Bacteroides nodosus*, the causal agent of ovine footrot. *J. Bacteriol.*, **160**, 1184–7.
Elleman, T. C., Hoyne, P. A., McKern, N. M. & Stewart, D. J. (1986). Nucleotide sequence of the gene encoding the 2 subunit pilin of *Bacteroides nodosus* 265. *J. Bacteriol.*, **167**, 243–50.
Forsberg, C. W., Crosby, B. & Thomas, D. Y. (1986). Potential for manipulation of the rumen fermentation through the use of recombinant DNA techniques. *J. Anim. Sci.*, **63**, 310–25.
Gotz, F., Krentz, B. & Schleifer, K. H. (1983). Protoplast transformation of *Staphylococcus carnosus* by plasmid DNA. *Molec. Gen. Genet.*, **189**, 340–2.
Greenwood, Y., Hall, F. J., Orpin, C. G. & Paterson, I. W. (1983). Microbiology of seaweed digestion in Orkney sheep. *J. Physiol.*, **343**, 121P.
Grepinet, O. & Beguin, P. (1986). Sequence of the cellulase gene of *Clostridium thermocellum* coding for endoglucanase B. *Nucleic Acids Res.*, **14**, 1791–9.
Greve, L. C., Labavitch, J. M. & Hungate, R. E. (1984). Muralytic activities of *Ruminococcus albus* 8. *Appl. Environ. Microbiol.*, **47**, 1141–5.
Guiney, D. G., Hasegawa, P. & Davis, C. E. (1984a). Expression in *Escherichia coli* of cryptic tetracycline resistance genes from *Bacteroides* R-plasmids. *Plasmid*, **11**, 248–52.
Guiney, D. G., Hasegawa, P. & Davis, C. E. (1984b). Plasmid transfer from *Escherichia coli* to *Bacteroides fragilis*: differential expression of antibiotic resistance phenotypes. *Proc. Nat. Acad. Sci. USA*, **81**, 7203–6.
Hazlewood, G. P., Cho, K. Y., Dawson, R. M. C. & Munn, E. A. (1983a). Subcellular fractionation of the Gram-negative rumen bacterium, *Butyrivibrio* S2, by protoplast formation, and localization of lipolytic enzymes in the plasma membrane. *J. Appl. Bacteriol.*, **55**, 337–47.
Hazlewood, G. P., Munn, E. A. & Orpin, C. G. (1983b). Temperate bacteriophages of *Selenomonas ruminantium* and a *Fusobacterium* sp. isolated from the ovine rumen. EM1, *Abstracts of Canadian Society of Microbiologists 33rd Annual Meeting*, p. 76.
Hazlewood, G. P., Mann, S. P., Orpin, C. G. & Romaniec, M. P. M. (1987). Prospects for the genetic manipulation of rumen microorganisms. In *Recent Advances in Anaerobic Bacteriology*, ed. S. P. Borriello & J. M. Hardie. Martinus Nijhoff, Dordrecht, pp. 162–76.
Heefner, D. L., Squires, C. H., Evans, R. J., Kopp, B. J. & Yarus, M. J. (1984). Transformation of *Clostridium perfringens*. *J. Bacteriol.*, **159**, 460–4.
Hyun, H. H. & Zeikus, J. G. (1985). Regulation and genetic enhancement of β-amylase production in *Clostridium thermocellum*. *J. Bacteriol.*, **164**, 1162–70.
Irvin, J. E., Wood, P. J., Teather, R. M. & Crosby, W. (1986). Cloning and expression in *Escherichia coli* of a *Bacteroides succinogenes* β-(1→4),(1→3)-D-glucan-specific hydrolase. *Abstracts of 86th Annual Meeting of American Society Microbiology*, p. 273.
Iverson, W. G. & Millis, N. J. (1976a). Characterization of *Streptococcus bovis* bacteriophages. *Can. J. Microbiol.*, **22**, 847–52.
Iverson, W. G. & Millis, N. J. (1976b). Lysogeny in *Streptococcus bovis*. *Can. J. Microbiol.*, **22**, 853–7.
Iverson, W. G. & Millis, N. J. (1976c). Bacteriocins of *Streptococcus bovis*. *Can. J. Microbiol.*, **22**, 1040–7.
Johnson, J. L. (1980). Specific strains of *Bacteroides* species in human fecal flora as measured by deoxyribonucleic acid homology. *Appl. Environ. Microbiol.*, **39**, 407–13.
Johnson, E. A., Bouchot, F. & Demain, A. L. (1985). Regulation of cellulase formation in *Clostridium thermocellum*. *J. Gen. Microbiol.*, **131**, 2303–8.
Joliff, G., Beguin, P. & Aubert, J.-P. (1986). Nucleotide sequence of the cellulase gene *cel* D encoding endoglucanase D of *Clostridium thermocellum*. *Nucleic Acids Res.*, **14**, 8605–13.
Jones, D. T., Jones, W. A. & Woods, D. R. (1985). Production of recombinants after protoplast fusion in *Clostridium acetobutylicum* P262. *J. Gen. Microbiol.*, **131**, 1213–16.
Kai, M., Watanabe, K. & Ozawa, A. (1985). *Bacteroides* bacteriophages isolated from human feces. *Microbiol. Immunol.*, **29**, 895–9.

Kamio, Y. & Takahashi, H. (1980). Outer membrane proteins and cell surface structure of *Selenomonas ruminantium*. *J. Bacteriol.*, **141**, 899–907.

Kamio, Y., Itoh, Y., Terawaki, Y. & Kusano, T. (1980). A new form of structural peptidoglycan in *Selenomonas ruminantium*: existence of polyamine in peptidoglycan. *Agric. Biol. Chem.*, **44**, 2523–6.

Kamio, Y., Itoh, Y. & Terawaki, Y. (1981a). Chemical structure of peptidoglycan in *Selenomonas ruminantium*: cadaverine links covalently to the D-glutamic acid residue of peptidoglycan. *J. Bacteriol.*, **146**, 49–53.

Kamio, Y., Terawaki, Y., Nakajima, T. & Matsuda, K. (1981b). Structure of glycogen produced by *Selenomonas ruminantium*. *Agric. Biol. Chem.*, **45**, 209–16.

Kamio, Y., Poso, H., Terawaki, Y. & Paulin, L. (1986). Cadaverine covalently linked to a peptidoglycan is an essential constituent of the peptidoglycan necessary for the normal growth in *Selenomonas ruminantium*. *J. Biol. Chem.*, **261**, 6585–9.

Keiser, T. (1984). Factors affecting the isolation of ccc DNA from *Streptomyces lividans* and *Escherichia coli*. *Plasmid*, **12**, 19–36.

Kelly, W. J. & Asmundson, R. V. (1986). Genetic studies of cellulolytic anaerobic bacteria from the genus *Ruminococcus*. *XIV International Congress of Microbiology, Abstract P.B17-5*, p. 100.

Klebe, R. J., Harriss, J. V., Sharp, Z. D. & Douglas, M. G. (1983). A general method for polyethylene glycol-induced genetic transformation of bacteria and yeast. *Gene*, **25**, 333–41.

Knowlton, S., Ferchak, J. D. & Alexander, J. K. (1984). Protoplast regeneration in *Clostridium tertium*: isolation of derivatives with high frequency regeneration. *Appl. Environ. Microbiol.*, **48**, 1246–7.

Kuritza, A. P., Shaughnessy, P. & Salyers, A. A. (1986). Enumeration of polysaccharide degrading *Bacteroides* species in human feces by using specific DNA probes. *Appl. Environ. Microbiol.*, **51**, 385–90.

Lin, Y.-L. & Blaschek, H. P. (1984). Transformation of heat-treated *Clostridium acetobutylicum* protoplasts with pUB110 plasmid DNA. *Appl. Environ. Microbiol.*, **48**, 737–42.

Mancini, C. & Behme, R. J. (1977). Transfer of multiple antibiotic resistance from *Bacteroides fragilis* to *Escherichia coli*. *J. Infect. Dis.*, **136**, 597–600.

Mann, S. P., Hazlewood, G. P. & Orpin, C. G. (1985). Possible vector systems for cloning in ruminal bacteria. *J. Appl. Bacteriol.*, **59**, xv–xvi.

Mann, S. P., Hazlewood, G. P. & Orpin, C. G. (1986). Characterization of a cryptic plasmid (pOM1) in *Butyrivibrio fibrisolvens* by restriction endonuclease analysis and its cloning in *Escherichia coli*. *Curr. Microbiol.*, **13**, 17–22.

Mannarelli, M. & Hespell, R. B. (1987). Deoxyribonucleic acid relatedness among strains of the species *Butyrivibrio fibrisolvens*. *Abstracts of 87th Annual Meeting of American Society for Microbiology*, p. 195.

Mays, T. D., Smith, C. J., Welch, R. A., Delfini, C. & Macrina, F. L. (1982). Novel antibiotic resistance transfer in *Bacteroides ruminicola*. *Antimicrob. Agents Chemoth.*, **21**, 110–18.

Meile, L. & Reeve, J. N. (1985). Potential shuttle vectors based on the methanogen plasmid pME2. *Biotechnol.*, **3**, 69–72.

Meile, L., Kiener, A. & Leisinger, T. (1983). A plasmid in the archaebacterium *Methanobacterium thermoautotrophicum*. *Molec. Gen. Genet.*, **191**, 480–4.

Millet, J., Petre, D., Beguin, P., Raynaud, O. & Aubert, J.-P. (1985). Cloning of 10 distinct DNA fragments of *Clostridium thermocellum* coding for cellulases. *FEMS Microbiol. Lett.*, **29**, 145–9.

Minton, N. P. & Morris, J. G. (1983). Regeneration of protoplasts of *Clostridium pasteurianum* ATCC 6013. *J. Bacteriol.*, **155**, 432–4.

Miyagawa, E. (1982). Cellular fatty acid and fatty aldehyde composition of rumen bacteria. *J. Gen. Appl. Microbiol.*, **28**, 389–408.

Morris, C. J. & Reeve, J. N. (1984). Functional expression of an archaebacterial gene from the methanogen *Methanosarcina barkeri* in *Escherichia coli* and *Bacillus subtilis*. In *Microbial Growth on C1 Compounds*, ed. R. L. Crawford & R. S. Hanson. American Society for Microbiology, Washington, DC, pp. 205–9.

Orpin, C. G. & Munn, E. A. (1974). The occurrence of bacteriophages in the rumen and their influence on rumen bacterial populations. *Experientia*, **30**, 1018–20.

Orpin, C. G., Mathiesen, S. D., Greenwood, Y. & Blix, A. S. (1985). Seasonal changes in the ruminal microflora of the high-arctic Svalbard reindeer (*Rangifer tarandus platyrhyncus*). *Appl. Environ. Microbiol.*, **50**, 144–51.

Orpin, C. G., Jordan, D. J., Hazlewood, G. P. & Mann, S. P. (1986a). Genetic transformation of the rumen bacterium *Selenomonas ruminantium*. *J. Appl. Bacteriol.*, **61**, xiv.

Orpin, C. G., Jordan, D. J., Mathiesen, S. D., Veal, N. L., Hazlewood, G. P. & Mann, S. P. (1986b). Plasmid profiles of the rumen bacteria *Selenomonas ruminantium* and *Butyrivibrio fibrisolvens*. *J. Appl. Bacteriol.*, **61**, xiii–xiv.

Orpin, C. G., Hazlewood, G. P. & Mann, S. P. (1988). Possibilities of the use of recombinant DNA techniques with rumen microorganisms. *Anim. Feed. Sci. Technol.* (in press).

Privitera, G., Sebald, M. & Fayolle, F. (1979). Common regulatory mechanism of expression and conjugative ability of a tetracycline resistance plasmid in *Bacteroides fragilis*. *Nature*, **278**, 657–9.

Rashtchian, A. & Booth, S. J. (1981). Stability in *Escherichia coli* of an antibiotic resistance plasmid from *Bacteroides fragilis*. *J. Bacteriol.*, **146**, 212–27.

Rasmussen, J. L. & Macrina, F. L. (1985). Characterization of the directly repeated DNA sequences that flank the clindamycin resistance determinant of pBR4. *Plasmid*, **13**, 225.

Reid, S. J., Allcock, E. R., Jones, D. T. & Woods, D. R. (1983). Transformation of *Clostridium acetobutylicum* protoplasts with bacteriophage DNA. *Appl. Environ. Microbiol.*, **45**, 305–7.

Riley, T. V. & Mee, B. J. (1981). Simple method for detecting *Bacteroides* spp. bacteriocin production. *J. Clin. Microbiol.*, **13**, 594–5.

Riley, T. V. & Mee, B. J. (1982). A bacteriocin typing scheme for *Bacteroides*. *J. Medical Microbiol.*, **15**, 387–91.

Romaniec, M. P. M., Clarke, N. G. E., Orpin, C. G. & Hazlewood, G. P. (1986). Cloning and expression in *Escherichia coli* of cellulase genes from rumen anaerobic bacteria, *Ruminococcus albus* and *Butyrivibrio* sp. *XIV International Congress of Microbiology*, Abstract P.B17-8, p. 101.

Romaniec, M. P. M., Clarke, N. G. & Hazlewood, G. P. (1987). Molecular cloning of *Clostridium thermocellum* DNA and the expression of further novel endo-β-1,4-glucanase genes in *Escherichia coli*. *J. Gen. Microbiol.*, **133**, 1297–1307.

Russell, J. B. & Baldwin, R. L. (1978). Substrate preferences in rumen bacteria: evidence of catabolite regulatory mechanisms. *Appl. Environ. Microbiol.*, **36**, 319–29.

Salyers, A. A. (1984). *Bacteroides* of the human lower intestinal tract. *Ann. Rev. Microbiol.*, **38**, 293–313.

Salyers, A. A., Lynn, S. P. & Gardner, J. F. (1983). Use of randomly cloned DNA fragments for identification of *Bacteroides thetaiotaomicron*. *J. Bacteriol.*, **154**, 287–93.

Saunders, J. R., Docherty, A. & Humphreys, G. O. (1984). Transformation of bacteria by plasmid DNA. *Methods Microbiol.*, **17**, 61–95.

Schumann, J. P., Jones, D. T. & Woods, D. R. (1983). Effect of oxygen and UV irradiation on nucleic acid and protein synthesis in *Bacteroides fragilis*. *J. Bacteriol.*, **156**, 1366–8.

Schwarz, W. H., Bronnenmeier, K. & Staudenbauer, W. L. (1985). Molecular cloning of *Clostridium thermocellum* genes involved in β-glucan degradation in bacteriophage λ. *Biotechnol. Lett.*, **7**, 859–64.

Sgorbati, B., Scardovi, V. & Leblanc, D. J. (1982). Plasmids in the genus *Bifidobacterium*. *J. Gen. Microbiol.*, **128**, 2121–31.

Sgorbati, B., Smiley, M. B. & Sozzi, T. (1983). Plasmids and phages in *Bifidobacterium longum*. *Microbiologica*, **6**, 169–73.

Shah, H. N. & Collins, M. D. (1983). Genus *Bacteroides*: a chemotaxonomical perspective. *J. Appl. Bacteriol.*, **55**, 403–16.

Shoemaker, N. B., Guthrie, E. P., Salyers, A. A. & Gardner, J. F. (1985). Evidence that the clindamycin–erythromycin resistance gene of *Bacteroides* plasmid pBF4 is on a transposable element. *J. Bacteriol.*, **162**, 626–32.

Shoemaker, N. B., Getty, C., Gardner, J. F. & Salyers, A. A. (1986a). Tn4351 transposes in *Bacteroides* species and mediates the integration of plasmid R751 into the *Bacteroides* chromosome. *J. Bacteriol.*, **165**, 929–36.

Shoemaker, N. B., Getty, C., Guthrie, E. P. & Sayers, A. A. (1986b). Regions in *Bacteroides* plasmids pBFTM10 and pB8-51 that allow *Escherichia coli*–*Bacteroides* shuttle vectors to be mobilized by *inc*P plasmids and by a conjugative *Bacteroides* tetracycline resistance element. *J. Bacteriol.*, **166**, 959–65.

Shuber, A. P., Orr, E. C., Reeny, M. A., Schendel, P. F., May, H. D. & Schauer, N. L. (1986). Cloning, expression and nucleotide sequence of the formate dehydrogenase genes from *Methanobacterium formicum*. *J. Biol. Chem.*, **261**, 12942–7.

Sipat, K., Taylor, K. A., Lo, R. Y. C., Forsberg, C. W. & Krell, P. J. (1987). Molecular cloning of a xylanase gene from *Bacteroides succinogenes* and its expression in *Escherichia coli*. *Appl. Environ. Microbiol.*, **53**, 477–81.

Smith, C. J. (1985a). Characterization of the clindamycin resistance region from *Bacteroides* plasmid pBI136. *Plasmid*, **13**, 225.

Smith, C. J. (1985b). Characterization of *Bacteroides ovatus* plasmid pBI136 and structure of its clindamycin resistance region. *J. Bacteriol.*, **161**, 1069–73.

Smith, C. J. (1985c). Polyethylene glycol facilitated transformation of *Bacteroides fragilis* with plasmid DNA. *J. Bacteriol.*, **164**, 466–9.

Smith, C. J. & Gonda, M. A. (1985). Comparison of the transposon-like structure encoding clindamycin resistance in *Bacteroides* R plasmids. *Plasmid*, **13**, 182–92.

Smith, C. J. & Hespell, R. B. (1983). Prospects for the development and use of recombinant deoxyribonucleic acid techniques with ruminal bacteria. *J. Dairy Sci.*, **66**, 1536–46.

Smith, C. J., Markowitz, S. M. & Macrina, F. L. (1981). Transferable tetracycline resistance in *Clostridium difficile*. *Antimicrob. Agents Chemother.*, **19**, 997–1003.

Squires, C. H., Heefner, D. L., Evans, R. J., Kopp, B. J. & Yarus, M. J. (1984). Shuttle plasmids for *Escherichia coli* and *Clostridium perfringens*. *J. Bacteriol.*, **159**, 465–71.

Stackebrandt, E., Pohla, H., Kroppenstedt, R., Hippe, H. & Woese, C. R. (1985). 16S ribosomal RNA analysis of *Sporomusa*, *Selenomonas* and *Megasphaera*: on the phylogenetic origin of Gram positive eubacteria. *Arch. Microbiol.*, **143**, 270–6.

Stahl, D. (1987). Evolutionary tools for microbial ecology and evaluating ecosystem perturbations. *Abstracts of 87th Annual Meeting of American Society for Microbiology*, p. xxxiii.

Stewart, G. J. & Carlson, C. A. (1986). The biology of natural transformation. *Ann. Rev. Microbiol.*, **40**, 211–35.

Stewart, C. S., Gilmour, J. & McConville, M. L. (1986). Microbial interactions, manipulation and genetic engineering. In *New Developments and Future Perspectives in Research on Rumen Function*, ed. A. Niemann-Sorenson. Commission of the European Communities, pp. 143–257.

Tally, F. P., Snydman, D. R., Gorbach, S. L. & Malamy, M. H. (1979). Plasmid mediated transferable resistance to clindamycin and erythromycin in *Bacteroides fragilis*. *J. Infect. Dis.*, **139**, 83–8.

Tarakanov, B. V. (1976). Biological characteristics of *Streptococcus bovis* bacteriophages isolated from lysogenic cultures and sheep rumen. *Mikrobiologica*, **45**, 695–700.

Taya, M., Ohmiya, K., Kobayashi, T. & Shimizu, S. (1983). Enhancement of cellulose digestion by mutants from an anaerobe, *Ruminococcus albus*. *J. Ferment. Technol.*, **61**, 197–9.

Taylor, K. A., Crosby, B., McGavin, M., Forsberg, C. W. & Thomas, D. Y. (1987). Characteristics of the endoglucanase encoded by a *cel* gene from *Bacteroides succinogenes* expressed in *Escherichia coli*. *Appl. Environ. Microbiol.*, **53**, 41–6.

Teather, R. M. (1982). Isolation of plasmid DNA from *Butyrivibrio fibrisolvens*. *Appl. Environ. Microbiol.*, **43**, 298–302.

Teather, R. M. (1985). Application of gene manipulation to rumen microflora. *Can. J. Anim. Sci.*, **65**, 563–74.

Teather, R. M. & Wood, P. J. (1982). Use of Congo Red–polysaccharide interactions in enumeration and characterization of cellulolytic bacteria from the bovine rumen. *Appl. Environ. Microbiol.*, **43**, 777–80.

Teather, R. M., Erfle, J. D., Crosby, B., Collier, B. & Thomas, D. Y. (1984). Cloning and expression in *Escherichia coli* of cellulase genes from the rumen anaerobe *Bacteroides succinogenes*. *Abstracts of 84th Annual Meeting of American Society for Microbiology*, p. 194.

Thomm, M. & Stetter, K. O. (1985). Evidence for specific *in vitro* transcription of the purified DNA-dependent RNA polymerase of *Methanococcus thermolithotrophicus*. *Eur. J. Biochem.*, **149**, 345–51.

Van Tassel, R. L. & Wilkins, T. D. (1978). Isolation of auxotrophs of *Bacteroides fragilis*. *Can. J. Microbiol.*, **24**, 1619–21.

Welch, R., Jones, K. R. & Macrina, F. L. (1979). Transferable lincosamide-macrolide resistance in *Bacteroides*. *Plasmid*, **2**, 261–8.

White, B. A. & Champion, K. M. (1986). Antibiotic resistance and plasmid content of cellulolytic *Ruminococcus* species. *Abstracts of 86th Annual Meeting of American Society of Microbiology*, p. 20.

Wood, A. G., Redborg, A. H., Cue, D. R., Whitman, W. B. & Konisky, J. (1983). Complementation of *arg*G and *his*A mutations of *Escherichia coli* by DNA cloned from the archaebacterium *Methanococcus voltae*. *J. Bacteriol.*, **156**, 19–29.

Woods, D. R. (1982). Molecular genetic studies on obligate anaerobic bacteria. *S. Afr. J. Anim. Sci.*, **78**, 448–50.

Woods, D. R. & Jones, D. T. (1984). Genetics of anaerobic bacteria. In *Herbivore Nutrition in the Subtropics and Tropics*, ed. F. M. C. Gilchrist & R. I. Mackie. Science Press, Pretoria, pp. 297–312.

Yoshino, S., Ogata, S. & Hayasida, S. (1984). Regeneration of protoplasts of *Clostridium saccharoperbutylacetonicum*. *Agric. Biol. Chem.*, **48**, 249–50.

Young, F. E. & Mayer, L. (1979). Genetic determinants of microbial resistance to antibiotics. *Rev. Infect. Dis.*, **1**. 55–62.

11

Microbe–Microbe Interactions

M. J. Wolin & T. L. Miller

Wadsworth Center for Laboratories and Research, New York State Department of Health, Albany, New York, USA

The previous chapters have shown that the rumen harbours a large number of different species of bacteria and protozoa. Metabolic interactions between these different populations are essential for sustaining the microbial community and its collective activities. Products of the metabolism of some species of microorganisms are sources of energy for other species. Similarly, products of vitamin synthesis and nitrogen metabolism of some species become the sources of the vitamins and nitrogen compounds required by other microorganisms. The kinds and extents of these microbial interactions regulate the concentrations and activities of individual species and the qualitative and quantitative nature of the products of the fermentation of dietary substrates. Products that are used by, and are essential for, the ruminant include acetate, propionate, and butyrate and the constituents of microbial cells. The amounts of CH_4 and CO_2 produced by the fermentation are determined by the same microbial interactions. The purpose of this chapter is to describe the major metabolic interactions that define the nature of the microbial community and its contributions to the host. Some of these interactions have, of necessity, been mentioned in connection with the topics of other chapters in the book.

NUTRITIONAL INTERACTIONS

Ruminants do not need B vitamins to be included in their diets. Their requirements are satisfied by microbial synthesis of vitamins in the rumen. Most of the major species of rumen microorganisms also require one or more B vitamins, but their individual requirements differ, and not all of the microorganisms in the rumen synthesize all of the B vitamins. Cross-feeding of vitamins synthesized by some species presumably satisfies the requirements of other species. Table 1 summarizes most of the available information about the B vitamin requirements of rumen bacteria. The number of preeminent species that require biotin and *p*-aminobenzoic

Table 1
Vitamin Requirements of Rumen Bacteria (modified from Hungate, 1966)

Organism	Biotin	Folate	PABA	Pyridoxine	Pantothenate	Thiamin	Riboflavin	Niacin	B_{12}
B. succinogenes	+	−	+	−	−	−	−	−	−
B amylophilus	−	−	−	−	−	−	−	−	−
R. flavefaciens	+	+[a]	+[a]	+	−	+	+	−	−
R. albus	+	−	+	+	−	−	−	−	−
Bu. fibrisolvens	+	+	−	+	−	−	−	−	−
S. bovis	+	−	−	−	+	+	−	+	−
Sel. ruminantium	+	−	+	−	−	−	−	−	−
Su. dextrinosolvens[b]	−	−	+	−	−	−	−	−	−
M. elsdenii[c]	+	−	−	+	+	−	−	−	−

+, required or stimulatory; −, not required.
Abbreviations: PABA, p-aminobenzoate; B_{12}, vitamin B_{12} or cyanocobalamin: B., Bacteroides; R., Ruminococcus; Bu., Butyrivibrio; S., Streptococcus; Sel., Selenomonas; Su., Succinivibrio; M., Megasphaera.
[a] Some strains use either folate or PABA; some strains require tetrahydrofolate (Slyter & Weaver, 1977). [b] Gomez-Alarcon et al. (1982). [c] Forsberg (1978).

acid (PABA) or biotin and folic acid, or biotin or PABA alone, is high. It would be inaccurate to conclude that the species listed as not requiring these vitamins are the only species that synthesize these vitamins. The microbial sources of these vitamins should be investigated, especially since Scheifinger (1974) demonstrated that biotin is required by *Selenomonas ruminantium* for the decarboxylation of succinate to propionate. This decarboxylation, discussed below, is a major reaction in the overall rumen fermentation.

In addition to B vitamins, 1,4-naphthoquinone (Gomez-Alarcon et al., 1982) and haeme (Caldwell et al., 1965; Reddy & Bryant, 1977) are required by important fermentative bacteria and must be produced by microbial species that do not require these compounds. As discussed later, 2-mercaptoethanesulphonic acid, a coenzyme found only in methane-producing bacteria, is required by some, and produced by other, methane-producing bacteria in the rumen.

INTERACTIONS AND N COMPOUNDS

There is extensive bacterial interaction involved in the conversion of dietary protein into microbial nitrogen. Proteolytic species such as *Bacteroides ruminicola, B. amylophilus* and *Butyrivibrio fibrisolvens* degrade dietary protein to peptides, amino acids and NH_4^+ (Hespell & Smith, 1983). *B. ruminicola* uses either peptides or NH_4^+, but not amino acids, as a source of nitrogen (Pittman et al., 1967). A few non-proteolytic bacterial species can use amino acids as a nitrogen source, but a majority of the predominant rumen bacterial species use NH_4^+ as their major source of cellular nitrogen (Bryant, 1974). *Megasphaera elsdenii* and *Bacteroides ruminicola* can degrade branched-chain amino acids to branched-chain fatty acids with one carbon less than the parent amino acid (Hespell & Smith, 1983). Valine, leucine and isoleucine are degraded to form isobutyric, isovaleric and 2-methylbutyric acids, respectively. Many dominant non-proteolytic rumen bacteria require one or more

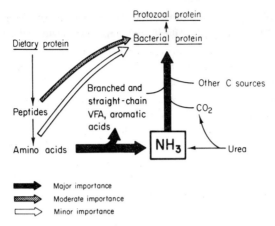

Fig. 1. Protein metabolism in the rumen (VFA, volatile fatty acids). Reproduced from Wolin (1979) with permission from Plenum Publishing Corp.

of these branched-chain fatty acids for synthesis of branched-chain amino acids and long-chain fatty acids (Allison, 1969; Bryant, 1974).

The pathways involved in the metabolism of protein in the rumen are summarized in Fig. 1. The reactions involved in these pathways of protein metabolism have been discussed in detail in Chapter 7; here they will be considered in the context of interactions of particular species of bacteria. Some of the interactions in the pathways have been experimentally demonstrated with *Ruminococcus albus* and *B. ruminicola* grown as a co-culture in a medium with cellulose as the sole energy source and casein as the sole nitrogen source (Bryant & Wolin, 1975). The cellulolytic *R. albus* provided the non-cellulolytic *B. ruminicola* with polysaccharide hydrolysis products for growth (see discussion below of interactions between cellulolytic and non-cellulolytic species). *B. ruminicola* degraded the casein and produced NH_4^+ and isobutyrate or 2-methylbutyrate which are required by *R. albus* for growth. An interesting feature of the study is that it demonstrated interactions involved in energy metabolism, nitrogen metabolism and the production and use of branched-chain acids.

CARBOHYDRATE FERMENTATION

Polysaccharide hydrolysis products

Plant polysaccharides are the primary sources of carbon and energy for ruminants. Rumen microorganisms process the polysaccharides to short-chain volatile fatty acids (VFA) which are then used by the host. The fermentations that produce the VFA provide energy and carbon for the growth and maintenance of the microbial community. Therefore dietary polysaccharides are the primary sources of carbon and energy for sustaining the microorganisms. Ruminant digestion depends entirely on particular microbial populations that specialize in the hydrolysis of plant polysaccharides. The microorganisms and enzymes involved in these polysaccharide degradations have been discussed in Chapter 8.

Polysaccharide hydrolysis and entry of the hydrolysis products into cells is an important area of interaction between species of the microbial ecosystem. Although cellulases are limited to only a few species of bacteria, protozoa and anaerobic phycomycetes, products of cellulose hydrolysis support the growth of many species that are incapable of directly using cellulose. The latter species appear to be sustained mainly by successfully competing with the cellulolytic species for the products of hydrolysis. Similar relationships exist for other polysaccharides. Cellulolytic species do not hydrolyse starch, and vice versa, but some species, e.g. *Bacteroides ruminicola* and *Selenomonas ruminantium*, have the capacity to use different types of polysaccharides although they do not digest cellulose (Wolin, 1979).

There have been few investigations of cross-feeding of products of polymer hydrolysis. Scheifinger & Wolin (1973) first demonstrated that growth of *Selenomonas ruminantium*, a non-cellulolytic species, occurred in co-culture on cellulose with *Bacteroides succinogenes*, a cellulolytic species. They suggested that *S. ruminantium* grew on soluble products of the hydrolysis of cellulose by *B. succinogenes*. Russell (1985) subsequently showed that cellodextrins (primarily cellotetraose and cellopentaose) supported growth of the non-cellulolytic species *Selenomonas ruminantium* and *Bacteroides ruminicola*. The cross-feeding of cellulose hydrolysis products between *Ruminococcus albus* and *Bacteroides ruminicola* was discussed above.

It can be inferred that a species will use products of the hydrolysis of a particular polymer if the species uses small molecular weight forms of the constituent sugars. Examples of small molecular weight forms are maltose (starch), cellobiose (cellulose), and xylobiose (xylan). However, the nature of cross-feeding relationships is not always apparent from available information. For instance, prominent cellulolytic bacteria maintain high numbers in ruminants fed on diets that contain much starch and little cellulose (Latham *et al.*, 1971). However, three significant species of cellulolytic bacteria, *Ruminococcus albus*, *R. flavefaciens* and *Bacteroides succinogenes*, do not ferment maltose, and many strains do not use glucose, so it is not clear how they are supported on high-starch diets. Cross-feeding may not be involved and the diets may contain sufficient cellulose to prevent a severe reduction in the numbers of the cellulolytic species.

Successful competition for soluble products of polymer hydrolysis as substrates for growth by non-hydrolytic species will depend on rates of transport of the soluble compounds into microbial cells and the rates of their conversion into products. Polysaccharide substrates will be in high concentration after feeding and will decrease with time. Since pools of small molecular weight hydrolytic products are never high, the non-hydrolytic microorganisms that use these products as substrates must have relatively low K_m values (high affinities) for the systems used to transport the compounds into the intracellular environments. Differences in K_m between species, and the influence of effectors on transport, could have a significant influence on the composition of the rumen microbial community. Research on competition for products of polymer hydrolysis, including definition of the hydrolytic products formed, determination of the ability of species to use the products as substrates and the kinetics of utilization is necessary for understanding the cross-feeding interactions of the ecosystem.

Food chain and predator–prey relationships

In addition to competition for hydrolysis products, there are two other mechanisms that will allow a non-hydrolytic species to survive in the ecosystem. If a microbial species can obtain carbon and energy from a product of the fermentations of other species that use hydrolysis products, the former can be maintained in the ecosystem. Methanogenic bacteria use H_2 and CO_2 to generate energy, and they use acetate and CO_2 as major sources of cell carbon. Their carbon and energy sources are produced by the organisms that ferment polysaccharide hydrolysis products. This is a food chain interaction: one population produces 'food' used by another population. Secondly, predator–prey interactions can support a species: one population obtains its energy and/or carbon by eating another population. The ingestion of bacteria by ciliate protozoa has been described in Chapter 3. However, the protozoa obtain energy by fermenting dietary polysaccharides and their hydrolysis products. Ingested bacteria are mainly a source of nitrogenous precursors for protein and nucleic acid biosynthesis and may serve as sources of protozoal lipid (Williams, 1986). Although bacteria may not be essential sources of carbon and energy for protozoa, fermentation of reserve polysaccharides of ingested bacteria may contribute to the metabolism of protozoa.

Succinate and propionate relationships

The small molecular weight products of polysaccharide hydrolysis are fermented directly to acetate, butyrate, H_2 and CO_2 by individual species of bacteria and protozoa. The production of propionate and CH_4 requires interactions between species.

The major part of the production of propionate involves the interaction between species that produce succinate and other species that decarboxylate succinate to propionate and CO_2. Pure cultures of several prominent rumen species produce succinate as a major product of carbohydrate fermentation. Species of *Bacteroides* found in high concentration in the human large bowel can produce propionate directly from carbohydrate if they are supplied with cyanocobalamin (Chen & Wolin, 1981). Without cyanocobalamin they produce succinate. In contrast, the rumen *Bacteroides*, *B. ruminicola* and *B. amylophilus*, produce only succinate as a major fermentation product. Neither of these species ferments cellulose, but *B. ruminicola* ferments hemicelluloses and starch and *B. amylophilus* ferments starch, and both species occur in high numbers in the rumen. Other succinate-producing species that are in high concentration in the rumen are the cellulolytic *Ruminococcus*

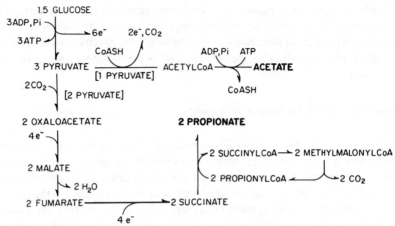

Fig. 2. Pathway of acetate and propionate formation by Selenomonas ruminantium.

flavefaciens and *B. succinogenes* and the non-cellulolytic *Succinivibrio dextrinosolvens*, *Succinomonas amylolytica*, and species of spirochaetes (Hungate, 1966; Wolin, 1979).

Although these major groups of bacteria produce succinate, it does not accumulate in rumen contents (Blackburn & Hungate, 1963). It is rapidly converted into propionate. Decarboxylation of succinate by *Selenomonas ruminantium* appears to be responsible for most of the production of propionate in the bovine rumen *S. ruminantium* produces propionate directly by fermentation of carbohydrates or lactate. Its pathway for forming propionate involves the production of succinate as an intermediate (Fig. 2). Removal of a carboxyl group from succinate (or succinyl-coenzyme A) ultimately leads to the formation of propionate. *S. ruminantium* cells also decarboxylate exogenously added succinate. The organism was co-cultured with *Bacteroides succinogenes*, a cellulolytic species that produces large amounts of succinate from cellulose (Scheifinger & Wolin, 1973). As discussed above, the products of cellulose hydrolysis by *B. succinogenes* supported growth of *S. ruminantium*. Propionate and not succinate was a major product of cellulose fermentation by the co-culture. *S. ruminantium* produced propionate directly from products of cellulose hydrolysis, and it also decarboxylated the succinate produced from cellulose by *B. succinogenes*.

Other than *S. ruminantium*, there appear to be few propionate-forming species in the rumen that might be capable of decarboxylating succinate. *Megasphaera elsdenii* produces propionate as a fermentation product, but the pathway it uses does not include succinate or succinyl-CoA as an intermediate (Paynter & Elsden, 1970). Species of *Veillonella* form propionate and decarboxylate succinate, but they have been isolated only from the sheep rumen (Johns, 1951).

The interactions between succinate-producing and succinate-decarboxylating species in the rumen are extremely important because of the importance of propionate in ruminant physiology; it is the sole gluconeogenic substrate for the ruminant animal. The balance between the two types of microorganisms must usually be adequate, since there are no reports of the accumulation of succinate in the rumen. If *S. ruminantium* is the major decarboxylating species, it must always be present at high enough concentrations to decarboxylate rapidly all of the succinate produced by the other species.

Production of CH_4

None of the carbohydrate-fermenting bacteria and protozoa produce CH_4, but many of them produce formate, H_2 and CO_2 as fermentation products. Methanogenic species of bacteria can then transform the H_2 and CO_2 into CH_4. Formate is converted into H_2 and CO_2 by methanogens that use formate as the primary substrate for the production of CH_4. The methanogenic organisms are not now classified as eubacteria but as part of the kingdom of the 'archaebacteria'. Detailed information about the methanogenic bacteria can be found in comprehensive reviews (Balch *et al.*, 1979; Woese & Wolfe, 1985; Wolin & Miller, 1985); we have briefly summarized the characteristics of methanogens as follows (Wolin & Miller, 1987; see also Chapter 2).

'The composition of cellular constituents of procaryotic archaebacteria and eubacteria are significantly different. The characteristic peptidoglycan polymer of the cell walls of eubacteria is absent from methanogens and other archaebacteria. Archaebacterial lipids are glycerol ethers rather than glycerol esters. Ribosomal RNA nucleotide sequences of archaebacteria and eubacteria indicate early divergence of the two kingdoms during evolution.

Methanogens are a sub-group of the archaebacteria. Although methanogens have many common physiological features, considerable phylogenetic diversity of methanogens is reflected in differences in the macromolecules responsible for the rigidity of the sacculus, differences in the composition of lipids, and differences in rRNA nucleotide sequences. The diversity is also reflected in the representation of cocci, various shaped rods, spirilla, thermophilic and mesophilic species and motile and non-motile species among the different genera and species of methanogens.

Common physiological features include the requirement for anaerobiosis, relatively simple nutritional requirements, and, most importantly, the use of CH_4 formation as the sole energy-generating mechanism of these bacteria. The commonality of CH_4 production is also mirrored by the use of unique common coenzymes for the enzyme reactions involved in the making of CH_4. A very restricted number of substrates are used for methanogenesis. They are H_2–CO_2, formate, acetate, methanol and mono-, di- and tri-methylamine. Almost all methanogens produce CH_4 from H_2 and CO_2. The ability to use the other substrates is restricted to certain species of methanogens. The equations for the production of CH_4 from each of the substrates are shown in Table 2. With H_2 and CO_2 and formate, CH_4 is formed by the reduction of CO_2. In fact, the ability to use formate is based on the ability to convert formate to H_2 and CO_2 and subsequent reduction of CO_2 to CH_4. With all other substrates, CH_4 is made from the methyl groups of the respective compounds.

Table 2
Substrates for Methanogens

Substrates	Equations
H_2 and CO_2	$4H_2 + CO_2 \rightarrow CH_4 + 2H_2O$
Formate	$4HCO_2H \rightarrow CH_4 + 3CO_2 + 2H_2O$
Methanol	$4CH_3OH \rightarrow 3CH_4 + CO_2 + 2H_2O$
Methanol and H_2	$CH_3OH + H_2 \rightarrow CH_4 + H_2O$
Methylamine	$4CH_3NH_2Cl + 2H_2O \rightarrow 3CH_4 + CO_2 + 4NH_4Cl$
Dimethylamine	$2(CH_3)_2NHCl + 2H_2O \rightarrow 3CH_4 + CO_2 + 2NH_4Cl$
Trimethylamine	$4(CH_3)_3NCl + 6H_2O \rightarrow 9CH_4 + 3CO_2 + 4NH_4Cl$
Acetate	$CH_3CO_2H \rightarrow CH_4 + CO_2$

Acetate and H_2–CO_2 are the principal substrates for the CH_4 produced on Earth by methanogenic bacteria. Formate can be an important substrate but it is essentially equivalent to H_2–CO_2 reduction to CH_4. Although methanol and methylamines can be important in special ecosystems, they are quantitatively insignificant in global methanogenesis. Methanol is a product of the fermentation of methylated plant polysaccharides such as pectin, and methylamines are products of the decomposition of choline.'

Hydrogen and CO_2 are the major precursors of CH_4 in the rumen (Hungate, 1967). Conversion of acetate into CH_4 and CO_2, and conversion of propionate and butyrate and other VFA into acetate, H_2 and CO_2, with subsequent conversion into CH_4, does not occur in the rumen. The production of CH_4 from VFA is a very slow process. The turnover of the rumen is too rapid to permit conversion of any of the VFA into CH_4.

Two species of methanogens that have been found to occur in high concentration in the rumen are described in Chapter 2. However, the distribution of these species in the forestomachs of the general population of ruminants is uncertain and there may be as yet unidentified major species of rumen methanogens. Smith & Hungate (1958) isolated from the bovine rumen a methanogen that formed CH_4 from H_2 and CO_2 and was present in high concentrations; it was named *Methanobacterium ruminantium*. A similar methanogen, strain M1, was isolated by Bryant (1965) and became the type strain of the species *Methanobrevibacter ruminantium* (Balch et al., 1979). It has been generally assumed that *Methanobrevibacter ruminantium* is the dominant H_2- and CO_2-using rumen methanogen. It requires 2-mercaptoethane-sulphonate (coenzyme M) and was, until recently, the only methanogen known to have this requirement. However, both coenzyme M-requiring and -nonrequiring *Methanobrevibacter* spp. have been detected at high concentrations in bovine rumen contents (Lovley et al., 1984; Miller et al., 1986). Their taxonomic relationship to *Methanobrevibacter ruminantium* is uncertain (Miller et al., 1986). In one study, *Methanomicrobium mobile* was found at high concentrations in rumen contents from a cow (Paynter & Hungate, 1968), but no additional information regarding its presence in rumen contents has been reported. *Methanomicrobium mobile* and the species of *Methanobrevibacter* grow rapidly with H_2 and CO_2 and do not form CH_4

from methyl groups. *Methanosarcina barkeri*, which has been isolated from cow and sheep rumen contents (Patterson & Hespell, 1979), grows slowly with H_2 and CO_2 and forms CH_4 from methyl groups. It is sustained in the rumen by its ability to produce CH_4 from methanol or methylamines. *Methanosarcina barkeri* does not compete for H_2 with the other methanogens. *Methanobrevibacter* spp. are probably the most significant bovine rumen methanogens, but more investigations are necessary in order to be certain about the importance, and the nature, of the species of *Methanobrevibacter* found in the bovine rumen. There are no reports on the nature of the methanogens in the rumens of other ruminants.

Interspecies H_2 transfer and fermentation

Although there are other H_2-utilizing reactions, essentially all of the H_2 produced by the fermentative microorganisms is used immediately by rumen methanogens to reduce CO_2 to CH_4. The interaction is called 'interspecies H_2 transfer'. While approximately 800 litres of H_2 are produced and used daily for the production of 200 litres of CH_4 in a 500 kg cow, only traces of H_2 accumulate in the gas phase of the rumen (Hungate, 1967). The maintenance of a low partial-pressure of H_2 by methanogenesis has a profound influence on the production of H_2 and other products by the non-methanogenic, fermentative microbial community, as illustrated by the influence of methanogenesis on the fermentation of cellulose by *Ruminococcus albus*. Although *R. albus* produces ethanol, acetate, H_2, CO_2, and sometimes formate, in pure culture, ethanol is not produced in the rumen (Moomaw & Hungate, 1963). If *R. albus* is grown together with an organism that rapidly uses H_2, such as a methanogen, ethanol is not produced.

The scheme of Fig. 3 illustrates how the rapid use of H_2 by methanogens influences the production of ethanol. *R. albus* produces H_2 by cleaving pyruvate to acetyl-CoA, H_2 and CO_2 (Miller & Wolin, 1973). The reaction is not inhibited by H_2.

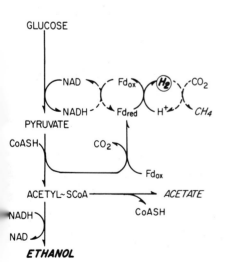

Fig. 3. Carbohydrate fermentation by *Ruminococcus albus* in the absence and presence of a methanogen. In monoculture, *R. albus* ferments glucose to acetate, ethanol, H_2 and CO_2. In co-culture with a methanogen, NADH is used to reduce protons to H_2 and the final products are acetate and CH_4 (Fd, ferredoxin).

The organism also produces H_2 by oxidizing NADH to NAD^+ and H_2 (Glass et al., 1977), and this reaction is inhibited by H_2. When H_2 accumulates, the formation of H_2 from NADH is inhibited and NADH made during glycolysis is re-oxidized by the enzymes used to form ethanol. When H_2 is removed by methanogens or other hydrogen-using species, H_2 is produced from NADH. Acetyl-CoA is then no longer reduced to ethanol; all of the acetyl-CoA produced from pyruvate is transformed into acetate. In effect, the maintenance of a low partial-pressure of H_2 by methanogens increases the production of H_2 and acetate by *R. albus*. Interspecies H_2 transfer couples the oxidative reactions of the fermentation pathway of *R. albus* to the reductive reactions the methanogen uses to reduce CO_2 to CH_4 and thereby eliminates the formation of ethanol.

The ATP yield of the monoculture fermentation of *R. albus* can be estimated by assuming that 2 moles of ATP are produced by the conversion of 1 mole of hexose into 2 moles of pyruvate, and that 1 mole of ATP is produced by the conversion of 1 mole of pyruvate into 1 mole of acetate. The molar yield of acetate per hexose unit fermented increases from 1 for the monoculture fermentation to 2 for the fermentation in the presence of the methanogen. Therefore the molar ATP yields for

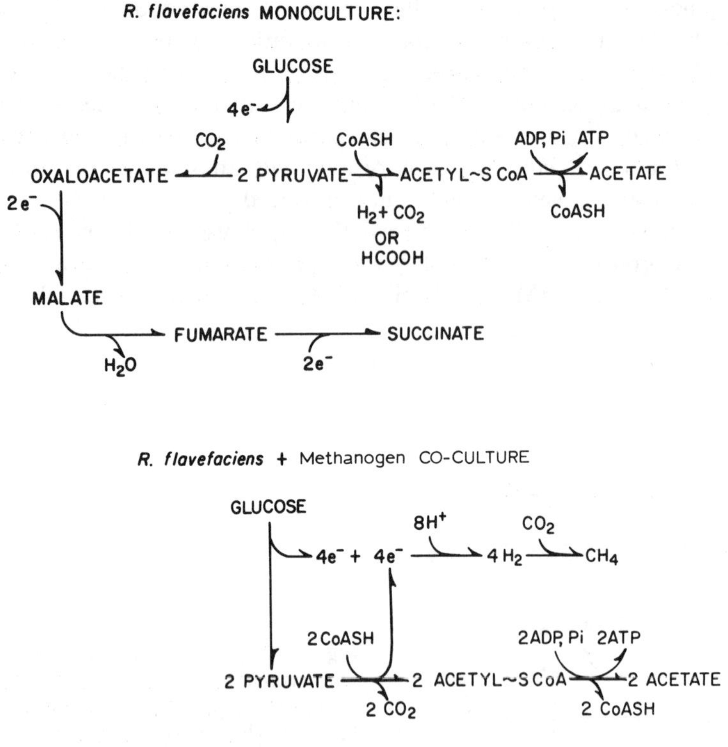

Fig. 4. Carbohydrate fermentation by *Ruminococcus flavefaciens* in the absence and presence of a methanogen. In monoculture, *R. flavefaciens* ferments glucose to acetate, H_2 and CO_2 or formate, and succinate is an electron-sink product. In co-culture with a methanogen, the products are acetate and CH_4.

the monoculture fermentation and with interspecies H_2 transfer are 3 and 4, respectively. In addition to increasing the ATP available to *R. albus*, interspecies H_2 transfer provides additional H_2 for methanogens, and they gain increased energy by using this H_2 to produce extra CH_4.

The fermentations of several other important organisms in the rumen are also influenced by the partial pressure of H_2 and are subject to alteration by interspecies H_2 transfer. *R. flavefaciens*, another cellulolytic species, re-oxidizes NADH by reducing oxaloacetate through a series of reactions to produce succinate. When co-cultured with a methanogen, succinate production by *R. flavefaciens* is dramatically reduced (Latham & Wolin, 1977). The flow of carbon is diverted from succinate to acetate formation. The scheme shown in Fig. 4 illustrates the major features of this fermentation and how interspecies H_2 transfer influences the formation of products.

The fermentation pattern of *Selenomonas ruminantium* is also altered by interspecies H_2 transfer (Scheifinger *et al.*, 1975; Chen & Wolin, 1977). *S. ruminantium* ferments carbohydrates to acetate, propionate and CO_2. It produces only traces of H_2 when it is grown by itself. However, when it is grown with a methanogen propionate formation is significantly decreased and acetate formation increased. The fermentation products of anaerobic, cellulolytic phycomycetes of the rumen ecosystem are also altered by co-culture with methanogens (Bauchop & Mountfort, 1981). Lactate and ethanol, major fermentation products of the phycomycetes, decrease and acetate increases, when the organisms are grown with methanogens (Fig. 5). All the above co-culture fermentations have the potential of providing 4 moles of ATP per mole of hexose fermented, and additional energy for the growth of methanogens.

When grown as monocultures, *R. albus* and the anaerobic phycomycetes form products (ethanol, lactate) that are not major intermediates, or products, of the

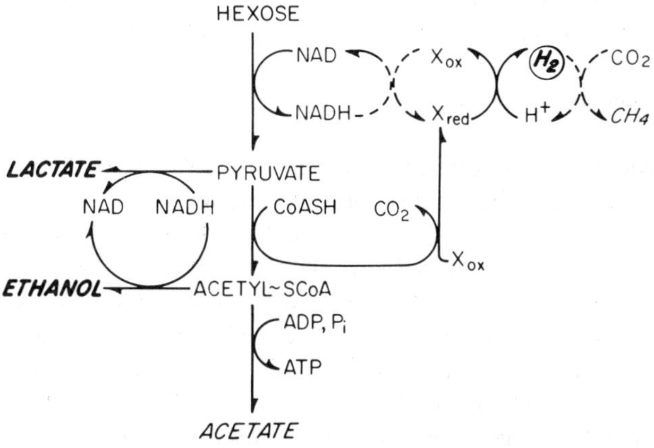

Fig. 5. Carbohydrate fermentation by an anaerobic rumen phycomycete in the absence and presence of a methanogen. In monoculture, the phycomycete produces lactate and ethanol as electron-sink products for NADH oxidation. In co-culture with a methanogen, NADH is oxidized to H_2 and the products of the co-culture fermentation are acetate and CH_4 (X, electron acceptor).

rumen fermentation. Production in the rumen is eliminated by the shift to the production of acetate caused by interspecies H_2 transfer. However, when *R. flavefaciens* and *S. ruminantium* are grown as monocultures, they produce important intermediates and products of the rumen fermentation. With *R. flavefaciens*, interspecies H_2 transfer causes a shift from the production of succinate (an important intermediate in the formation of propionate) to acetate. With *S. ruminantium* there is a shift from the formation of an important product, propionate, to the production of acetate. These shifts probably play a role in regulating the relative amounts of acetate and propionate formed in the rumen.

Although physical associations between fermentative species and methanogens are not essential for interspecies H_2 transfer, close association between the two types of organisms may be an important way to facilitate interspecies H_2 transfer in the rumen. Attachment of methanogens to the surface of rumen protozoa has been reported (Stumm *et al.*, 1982; Krumholz *et al.*, 1983). The relationship of interspecies H_2 transfer to the protozoan fermentations has not yet been elucidated, but the associations suggest that the methanogens might influence the production of fermentation products in the same manner as already described for bacteria and the cellulolytic phycomycetes. However, the fermentations of the protozoa themselves usually result in the production of acetate and butyrate from carbohydrates.

Butyrivibrio fibrisolvens is a major butyrate-producing bacterial species in the rumen. The reduction reactions of the fermentation depend on the production of acetyl-CoA from pyruvate, conversion of acetyl-CoA into acetoacetyl-CoA, reduction of acetoacetyl-CoA to butyryl-CoA and transformation of the latter compound to butyrate and ATP (Miller & Jenesel, 1979). In *Butyrivibrio*, reduced flavins and NADH are electron donors for the reduction reactions. Presumably a similar pathway is operative in the rumen protozoa. A shift from butyrate to acetate formation by interspecies H_2 transfer between a butyrate-producing rumen organism and a methanogen has not yet been demonstrated.

Interactions and the proportions of VFA

There is considerable interest in alterations of the rumen fermentation that would increase the efficiency of production of ruminant animal products. Possible ways of altering the rumen reactions, including the use of feed additives, are described in detail in Chapter 13, but some aspects will be considered here.

Monensin is now a major feed additive. It causes a reduction in CH_4 production and increase in propionate production. These effects on fermentation appear to be translated into an increase in the efficiency of feed utilization by beef cattle. There is evidence that monensin alters the rumen fermentation by inhibiting species that are prominent in producing acetate, butyrate and H_2 (Chen & Wolin, 1979). The uninhibited organisms then make a relatively greater contribution to propionate formation.

In order to clarify the changes caused by monensin and other possible agents that might alter the overall rumen fermentation, it is useful to represent the fermentation as the sum of individual fermentations and the production of CH_4 from H_2 and

CO_2. Analysis of the subsets of activities helps to clarify how a change in one subset will influence other subsets, and how the collective activities of different microbial populations influence the production of the major fermentation products. The subset activities are:

$$\text{Hexose} \rightarrow 1\cdot33 \text{ Propionate} + 0\cdot67\text{Acetate} + 0\cdot67CO_2 \quad (1)$$
$$\text{Hexose} \rightarrow \text{Propionate} + \text{Acetate} + H_2 + CO_2 \quad (1a)$$
$$\text{Hexose} \rightarrow \text{Butyrate} + 2H_2 + 2CO_2 \quad (2)$$
$$\text{Hexose} \rightarrow 2\text{Acetate} + 4H_2 + 2CO_2 \quad (3)$$
$$6H_2 + 1\cdot5CO_2 \rightarrow 1\cdot5CH_4 + 3H_2O \quad (4)$$
$$7H_2 + 1\cdot75CO_2 \rightarrow 1\cdot75CH_4 + 3\cdot5H_2O \quad (4a)$$

Sum $(1)+(2)+(3)+(4)$:

$$3\text{Hexose} \rightarrow 1\cdot33\text{Propionate} + 2\cdot67\text{Acetate} + \text{Butyrate}$$
$$+ 1\cdot5CH_4 + 3\cdot17CO_2 + 3H_2O \quad (5)$$

Sum $(1a)+(2)+(3)+(4a)$:

$$3\text{Hexose} \rightarrow \text{Propionate} + 3\text{Acetate} + \text{Butyrate}$$
$$+ 1\cdot75CH_4 + 3\cdot25CO_2 + 3\cdot5H_2O \quad (5a)$$

The equations for the subset activities are based on knowledge of the fermentations of individual rumen species, information about interactions between species, and general information about bacterial fermentations. For simplicity, all substrate is considered to be composed of hexose units. Equations (1) and (1a) are the sources of all the propionate, and some of the acetate, produced in the rumen. Since two different equations can be used to account for propionate formation, one is numbered (1) and the other is numbered (1a). Equation (2) is the source of all butyrate, and some of the H_2 used to form CH_4. Equation (3) is a major source of the acetate produced in the rumen and the H_2 used to form CH_4. Equation (3) is operative only if equation (4) or (4a) is operative; i.e. it is necessary to use all of the H_2 as it is produced in order to effect the conversion of hexose into acetate and CO_2 through interspecies H_2 transfer.

The relative rates of conversion of hexose into products by populations that use equations (1), (2) and (3), and the activity of rumen methanogens, determine the course of the overall fermentation. Assuming there is no selective absorption of VFA from the rumen, the relative rates determine the proportions of VFA found in the rumen, and they also determine the ratios of CH_4 formed to hexose used.

Calculations based on these equations show that, when the ratio of rates of equations (1), (2) and (3) is 26:26:48, the molar proportions of product (rumen) VFA will be 65 acetate:20 propionate:15 butyrate. To effect a change to 60 acetate:25 propionate:15 butyrate, the ratio of rates of the equations changes to 33:26:41. Therefore a relatively small increase in the activities of the microbial populations that contribute to equation (1) (from 26 to 33) and a small decrease in the activities leading to equation (3) (from 48 to 41), with equation (2) activities remaining constant, will cause the indicated increase in propionate and decrease in acetate. If it is assumed that activities are directly related to concentrations of the responsible microbial populations, then a 27% increase in the concentration of populations responsible for equation (1) and a 15% decrease in populations responsible for equation (3) would cause the alterations in products. Imagine that the rate of the reaction described by equation (1) depends on the concentration of *Bacteroides succinogenes*, a species that would interact with other organisms to ferment cellulose to the products of the reaction. If its contribution to the lower propionate value depended on a concentration *in vivo* of 2×10^9 cells/ml, an increase to $2 \cdot 54 \times 10^9$ cells/ml would be sufficient to effect the change to the higher proportion of propionate. Likewise, if equation (3) depended on *Ruminococcus albus*, and 2×10^9 cells/ml were required for the higher acetate value, a decrease to $1 \cdot 7 \times 10^9$ cells/ml would produce the lower proportion of acetate.

When equation (1a) is substituted for equation (1), the ratio of rates of equations (1a), (2) and (3) is 35:26:39 for molar proportions of 65 acetate:20 propionate:15 butyrate, and 44:26:30 for 60 acetate:25 propionate:15 butyrate. If propionate is produced by a combination of reactions (1) and (1a), intermediate values for rates will be obtained. These rates can be calculated, as well as values for different proportions of VFA, and the associated amounts of CH_4 (equations (4) and (4a) combined with equations (1)–(3)). Table 3 shows the calculated moles of CH_4 produced per mole of hexose fermented for different proportions of VFA in rumen contents. The relationships are the same for any combination of the reactions of equations (1) and (1a).

It is not our purpose to discuss in detail the ramifications of different combinations of rates of the subset fermentations. Our aim is to illustrate how changes in the populations responsible for the subset fermentations can influence the relative amounts of the major VFA formed from dietary polysaccharides. The sensitivity of pure cultures of rumen bacteria to monensin is consistent with inhibition of the bacteria responsible for fermentations (2) and (3) and selection for the bacteria that are responsible for fermentations (1) and (1a) (Chen & Wolin, 1979). This results in a decrease in the production of H_2 and consequently a depression of CH_4 formation. Increases in the concentrations of uninhibited populations that contribute to subset fermentations (1) and (1a) will increase the rate of production of propionate in the rumen.

Table 3
Methane Production and VFA Concentrations in Rumen Contents

Molar ratios acetate:propionate:butyrate	Moles CH_4 produced per mole hexose fermented
65:20:15	0·61
60:25:15	0·54
55:30:15	0·48
70:20:10	0·64
65:25:10	0·57
60:30:10	0·50

Calculated from equations (1), (1a), (2), (3) and (4); see text.

Since production of CH_4 represents a significant loss of energy to the animal, there has been an interest in specific inhibition of the bacteria that produce CH_4. The possible effects on the fermentation of inhibition of methanogenesis will depend on the nature of the species that are the major contributors to equation (3). The equation is valid only in the presence of methanogens. If the major substrate was cellulose and the major fermenting organisms contributing to equation (3) were phycomycetes and *Ruminococcus albus*, then ethanol and lactate would accumulate in the absence of methanogens. The potential for the microbial community to adapt to beneficial use of large amounts of these compounds is unknown. If the major cellulolytic species contributing to equation (3) was *Ruminococcus flavefaciens*, a shift to succinate and thence propionate formation would be expected because interspecies H_2 transfer between *Ruminococcus flavefaciens* and methanogens yields acetate at the expense of succinate. Some H_2 would continue to be produced by equations (1a) and (2) even if the major source of H_2, equation (3), was eliminated by the inhibition of methanogens. Again, the potential for the microbial community to adapt to beneficial use of H_2 is unknown. A possible strategy for shifting the fermentation away from CH_4 that appears to be consistent with known physiological activities of the community would be the use of a specific inhibitor of methanogenesis such as bromoethanesulphonate (Balch & Wolfe, 1979) in the presence of monensin or lasalocid. The former compound would inhibit methanogenesis; the latter compounds inhibit *Butyrivibrio fibrisolvens*, a major butyrate producer, as well as species of *Ruminococcus* (Chen & Wolin, 1979). However, species of *Bacteroides* and *Selenomonas ruminantium* would survive treatment with monensin or lasalocid, and their concentrations should increase because of decreased competition for substrate by the bacteria that are inhibited by these antibiotics. The net effect would be to enhance the fermentations represented by equations (1) and (1a) at the expense of the fermentations of equations (2) and (3). This should result in an increase in the production of propionate at the expense of acetate, butyrate and CH_4.

Another possible strategy would be the inhibition of methanogens with bromoethanesulphonate and the inoculation of the rumen with bacteria that obtain energy by using H_2 to reduce CO_2 to acetate (Ljungdahl, 1986):

$$4H_2 + 2CO_2 \rightarrow CH_3CO_2H + 2H_2O$$

REFERENCES

Allison, M. J. (1969). Biosynthesis of amino acids by ruminal microorganisms. *J. Anim. Sci.*, **29**, 797–807.

Balch, W. E. & Wolfe, R. S. (1979). Specificity and biological distribution of coenzyme M (2-mercaptoethanesulfonic acid). *J. Bacteriol.*, **137**, 256–63.

Balch, W. E., Fox, G. E., Magrum, L. J., Woese, C. R. & Wolfe, R. S. (1979). Methanogens: reevaluation of a unique biological group. *Microbiol. Rev.*, **43**, 260–96.

Bauchop, T. & Mountfort, D. O. (1981). Cellulose fermentation by a rumen anaerobic fungus in both the absence and presence of rumen methanogens. *Appl. Environ. Microbiol.*, **42**, 1103–10.

Blackburn, T. H. & Hungate, R. E. (1963). Succinic acid turnover and propionate production in the bovine rumen. *Appl. Microbiol.*, **11**, 132–5.
Bryant, M. P. (1965). Rumen methanogenic bacteria. In *Physiology of Digestion in the Ruminant*, ed. R. W. Dehority, R. S. Allen, W. Burroughs, N. L. Jacobsen & A. D. McGilliard. Butterworths, Washington, DC, pp. 411–18.
Bryant, M. P. (1974). Nutritional features and ecology of predominant anaerobic bacteria of the intestinal tract. *Am. J. Clin. Nutr.*, **27**, 1313–19.
Bryant, M. P. & Wolin, M. J. (1975). Rumen bacteria and their metabolic interactions. In *Proceedings of First Intersectional Congress of IAMS*, Vol. 2, Developmental Microbiology Ecology, ed. T. Hasegawa. Science Council of Japan, pp. 297–306.
Caldwell, D. R., White, D. C., Bryant, M. P. & Doetsch, R. N. (1965). Specificity of the heme requirement for growth of *Bacteroides ruminicola*. *J. Bacteriol.*, **90**, 1645–54.
Chen, M. & Wolin, M. J. (1977). Influence of CH_4 production by *Methanobacterium ruminantium* on the fermentation of glucose and lactate by *Selenomonas ruminantium*. *Appl. Environ. Microbiol.*, **34**, 756–9.
Chen, M. & Wolin, M. J. (1979). Effect of monensin and lasalocid-sodium on the growth of methanogenic and rumen saccharolytic bacteria. *Appl. Environ. Microbiol.*, **38**, 72–7.
Chen, M. & Wolin, M. J. (1981). Influence of heme and vitamin B_{12} on growth and fermentations of *Bacteroides* species. *J. Bacteriol.*, **145**, 466–71.
Forsberg, C. W. (1978). Nutritional characteristics of *Megasphaera elsdenii*. *Can. J. Microbiol.*, **24**, 981–5.
Glass, T. L., Bryant, M. P. & Wolin, M. J. (1977). Partial purification of ferredoxin from *Ruminococcus albus* and its role in pyruvate metabolism and reduction of nicotinamide adenine dinucleotide by H_2. *J. Bacteriol.*, **131**, 463–72.
Gomez-Alarcon, R. A., O'Dowd, C., Leedle, J. A. & Bryant, M. P. (1982). 1,4-Naphthoquinone and other nutrient requirements of *Succinivibrio dextrinosolvens*. *Appl. Environ. Microbiol.*, **44**, 346–50.
Hespell, R. B. & Smith, C. J. (1983). Utilization of nitrogen sources by gastrointestinal tract bacteria. In *Human Intestinal Microflora in Health and Disease*, ed. D. J. Hentges. Academic Press, New York, pp. 167–87.
Hungate, R. E. (1966). *The Rumen and Its Microbes*, Academic Press, New York, 533 pp.
Hungate, R. E. (1967). Hydrogen as an intermediate in the rumen fermentation. *Arch. Mikrobiol.*, **59**, 158–64.
Johns, A. T. (1951). The mechanism of propionic acid formation by *Veillonella gazogenes*. *J. Gen. Microbiol.*, **5**, 326–36.
Krumholz, L. R., Forsberg, C. W. & Veira, D. M. (1983). Association of methanogenic bacteria with rumen protozoa. *Can. J. Microbiol.*, **29**, 676–80.
Latham, M. J. & Wolin, M. J. (1977). Fermentation of cellulose by *Ruminococcus flavefaciens* in the presence and absence of *Methanobacterium ruminantium*. *Appl. Environ. Microbiol.*, **34**, 297–301.
Latham, M. J., Sharpe, M. E. & Sutton, J. D. (1971). The microbial flora of the rumen of cows fed hay and high cereal rations and its relationship to the rumen fermentation. *J. Appl. Bacteriol.*, **2**, 425–34.
Ljungdahl, L. G. (1986). The autotrophic pathway of acetate synthesis in acetogenic bacteria. *Ann. Rev. Microbiol.*, **40**, 415–50.
Lovley, D. R., Greening, R. C. & Ferry, J. G. (1984). Rapidly growing rumen methanogenic organism that synthesizes coenzyme M and has a high affinity for formate. *Appl. Environ. Microbiol.*, **48**, 81–7.
Miller, T. L. & Jenesel, S. E. (1979). Enzymology of butyrate formation by *Butyrivibrio fibrisolvens*. *J. Bacteriol.*, **138**, 99–104.
Miller, T. L. & Wolin, M. J. (1973). Formation of hydrogen and formate by *Ruminococcus albus*. *J. Bacteriol.*, **116**, 836–46.
Miller, T. L., Wolin, M. J., Hongxue, Z. & Bryant, M. P. (1986). Characteristics of methanogens isolated from bovine rumen. *Appl. Environ. Microbiol.*, **51**, 201–2.

Moomaw, C. R. & Hungate, R. E. (1963). Ethanol conversion in the rumen. *J. Bacteriol.*, **85**, 721–2.
Patterson, J. A. & Hespell, R. B. (1979). Trimethylamine and methylamine as growth substrates for rumen bacteria and *Methanosarcina barkeri*. *Curr. Microbiol.*, **3**, 79–83.
Paynter, M. J. B. & Elsden, S. R. (1970). Mechanism of propionate formation by *Selenomonas ruminantium*, a rumen microorganism. *J. Gen. Microbiol.*, **61**, 1–7.
Paynter, M. J. B. & Hungate, R. E. (1968). Characterization of *Methanobacterium mobilis*, sp. n., isolated from the bovine rumen. *J. Bacteriol.*, **95**, 1943–51.
Pittman, K. A., Lakshmanan, S. & Bryant, M. P. (1967). Oligopeptide uptake by *Bacteroides ruminicola*. *J. Bacteriol.*, **93**, 1499–508.
Reddy, C. A. & Bryant, M. P. (1977). Deoxyribonucleic acid base composition of certain species of the genus *Bacteroides*. *Can. J. Microbiol.*, **23**, 1252–6.
Russell, J. B. (1985). Fermentation of cellodextrins by cellulolytic and noncellulolytic rumen bacteria. *Appl. Environ. Microbiol.*, **49**, 572–6.
Scheifinger, C. C. (1974). Propionate formation from cellulose and soluble sugars through interspecies interaction of *Bacteroides succinogenes* and *Selenomonas ruminantium*. PhD thesis, University of Illinois, Urbana.
Scheifinger, C. C. & Wolin, M. J. (1973). Propionate formation from cellulose and soluble sugars by combined cultures of *Bacteroides succinogenes* and *Selenomonas ruminantium*. *Appl. Microbiol.*, **26**, 789–95.
Scheifinger, C. C., Linehan, B. & Wolin, M. J. (1975). H_2 production by *Selenomonas ruminantium* in the presence and absence of methanogenic bacteria. *Appl. Microbiol.*, **29**, 480–3.
Slyter, L. L. & Weaver, J. M. (1977). Tetrahydrofolate and other growth requirements of certain strains of *Ruminococcus flavefaciens*. *Appl. Environ. Microbiol.*, **33**, 363–9.
Smith, P. H. & Hungate, R. E. (1958). Isolation and characterization of *Methanobacterium ruminantium*. n. sp. *J. Bacteriol.*, **75**, 713–18.
Stumm, C. K., Gijzen, H. J. & Vogels, G. D. (1982). Association of methanogenic bacteria with ovine rumen ciliates. *Br. J. Nutr.*, **47**, 95–9.
Williams, A. G. (1986). Rumen holotrich ciliate protozoa. *Microbiol. Rev.*, **50**, 25–49.
Woese, C. R. & Wolfe, R. S. (1985). *The Bacteria: A Treatise on Structure and Function*. Vol. VIII, Archaebacteria. Academic Press, New York and Washington.
Wolin, M. J. (1979). The rumen fermentation: a model for microbial interactions in anaerobic ecosystems. In *Advances in Microbial Ecology*, Vol. 3, ed. M. Alexander, Plenum, New York and London, pp. 49–77.
Wolin, M. J. & Miller, T. L. (1985). Methanogens. In *Biology of Industrial Microorganisms*, ed. A. L. Demain & N. A. Solomon. Benjamin/Cummings, Menlo Park, California, pp. 189–221.
Wolin, M. J. & Miller, T. L. (1987). Bioconversion of organic carbon to CH_4 and CO_2. *Geomicrobiol. J.*, **5**, 239–59.

12
Compartmentation in the Rumen

J. W. Czerkawski*

Formerly of Hannah Research Institute, Ayr, UK

&

K.-J. Cheng

Research Station, Lethbridge, Alberta, Canada

COMPARTMENTATION IN BIOLOGY

There is little evidence of a truly random or haphazard behaviour within biological systems. The structures of these systems are usually complex and they have evolved to carry out a variety of functions that ensure the existence and propagation of any given species. In general, the function of a given part of a biological system will depend on the requirements of the system as a whole and this, of course, leads to tissue differentiation in complex organisms. If a particular part of a biological system is to accomplish a specific function, it must usually be separated from other parts of the system. In other words, it must have a boundary and thus constitute a compartment. Such a compartment cannot exist in isolation; it interacts with the environment and with other parts of a given biological system. A full understanding of the integrated physiological functions of organs and of their associated bacterial populations can only be achieved in terms of the interrelations between different compartments.

Each cell has a definite boundary and thus forms a discrete compartment, and some of the more complex cells contain structures within their boundaries that are in effect separate compartments (e.g. mitochondria, vacuoles). The cells may exist as free entities (e.g. microorganisms, blood cells) or they may be aggregated to form tissues which in turn may be assembled in more complex functional structures, the organs.

It is unlikely that the more complex biological systems can be readily described in precise molecular terms; instead, compartmental models are set up and tested as part of a normal scientific method. There is often a problem in identifying the nature and number of compartments within a given system. Various organs in the animal body, such as heart, liver or kidney, constitute definite compartments. The gastrointestinal tract can be regarded as a multicompartment system where

---

*Present address: 6 Elm Way, Maybole KA19 8BB, UK.

complex, but nevertheless discrete, compartments can be readily identified. Anatomically the rumen is a complex organ, yet the contents are enclosed in a definite compartment with an inlet and outlet. The rumen and some other parts of the gut (e.g. abomasum, caecum) have distinctly saccular structures and it is not difficult to think of them as compartments. On the other hand, the small intestine consists of a long convoluted tube, but this too can be regarded as a compartment. The shape of a given compartment could be important inasmuch as it could influence the behaviour of the system, but frequently the precise shape need not be known. The vascular system in the animal is very complicated geometrically, with blood vessel diameters ranging from a fraction of a millimetre to several centimetres, and yet blood often behaves as if it was enclosed in a single compartment.

In spite of their complexity in biology, compartments can be represented on paper as simple rectangles; this procedure is helpful in discussions of compartmentation and development of hypothetical models by means of flow diagrams. Some examples of compartmental arrangements are shown diagrammatically in Fig. 1. The one-compartment system could be closed or open. With respect to the environment, there is only one two-compartment closed system, but there are at least three distinct open two-compartment systems, a 'flow-through' type, a flow-through type with reverse flows between compartments, and a system where only one of the compartments communicates with the environment. There are numerous three-compartment systems, and in general the possible number of compartmental systems increases rapidly with the number of compartments. The sizes of compartments and flows are normally described in terms of basic substance, which could be anything from a simple chemical compound to large food particles. If the

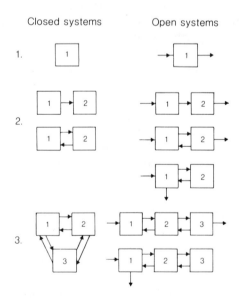

Fig. 1. Diagrams showing some of the types of flow in one-, two- and three-compartment closed and open systems.

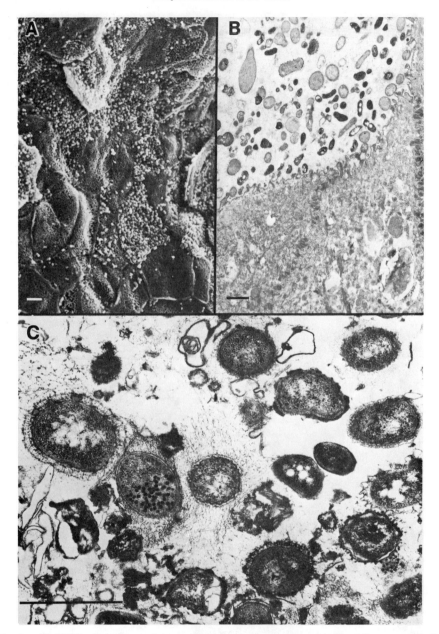

Fig. 2. (a) SEM of the bacterial population of the rumen epithelium from a milk-fed Holstein calf. (b) TEM of a ruthenium red-stained preparation of rumen tissue showing the thick and complex bacterial biofilm that develops on this organ wall. (c) TEM of the rumen fluid bacterial population. The bar in this and subsequent SEM micrographs represent 5 μm and the bar in these and subsequent TEM micrographs represents 1 μm. (SEM, scanning electron micrograph; TEM, transmission electron micrograph.)

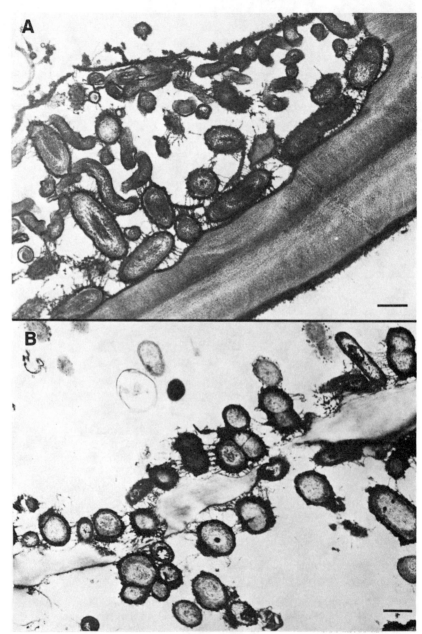

Fig. 3. (a) and (b): TEMs of a ruthenium red-stained preparation of barley straw incubated in a continuous culture of a mixed natural population of bacteria from the bovine rumen.

chemical compound is metabolized, it forms a metabolic pool. Often reference is made to the microbial pool in the rumen, but clearly such a pool consists of many species of microorganisms which are, it will later be shown, metabolically and spatially distributed among many compartments.

The best illustration of a one-compartment closed system is a simple artificial rumen, a batch incubation of a microbial suspension in a closed container, e.g. a syringe (Czerkawski & Breckenridge, 1970). This is initially a one-compartment

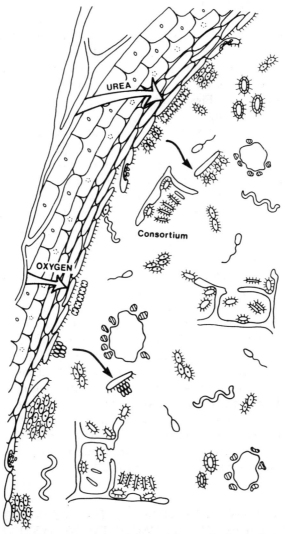

Fig. 4. Diagrammatic representation of bacterial populations of the bovine rumen. Bacteria are seen in the rumen fluid, on food particles, and attached to the rumen epithelium from which distal cells are sloughing into the rumen fluid. The tissue-adherent bacterial population is shown in relation to the urea and oxygen that diffuse through the organ wall, and to the distal epithelial cells, which comprise their nutrient substratum.

system with respect to microorganisms, but it develops a second, non-bacterial compartment with the accumulation of fermentation gases.

The model that has been used most frequently in studies of the rumen is the one-compartment, open system. Hyden (1961) showed that, when an inert, soluble substance was introduced into the rumen, the concentration decreased exponentially with time; this is consistent with a one-compartment, flow-through system. Work in many laboratories has given a great deal of information on the volume of rumen contents and the flows (e.g. Faichney, 1975), but deviations from the simple model can frequently be observed, even if one considers only the inert, soluble marker. The use of additional markers, soluble ones and those that associate with solid digesta, shows quite clearly that the solid and liquid portions of rumen contents are removed at different rates. Since rumen microorganisms are found in suspension, in association with the solid digesta, and adherent to the rumen wall (Cheng & Costerton, 1980) (Figs 2 and 3), there is clearly more than one microbial compartment within this organ.

The rumen has a complex structure and the contents are heterogeneous, consisting primarily of a microbial suspension in free liquid, a solid mass of digesta, and a gas phase. Each of these entities is complex; the properties of microorganisms in suspension deep in the rumen might be different from those of microorganisms close to the rumen wall (Cheng & Costerton, 1980); the solid mass of digesta is certainly heterogeneous, and even the gas phase is complex, consisting of a large gas space (gas-cap), smaller pockets of gas entrapped within the solid mass, and the dissolved gas. Many properties of individual parts of the microbial system of the rumen have been investigated, but often the investigations were governed by the ease of experimentation rather than the importance of the given component. This is why the complex, and relatively inaccessible, semi-solid mass of digesta has received so little attention. In addition, worthwhile investigations were hampered by a lack of a simple conceptual basis for dealing with such a complex system.

The microbial populations of the rumen are effectively subdivided in Cheng's 'three population' theory (Cheng & Costerton, 1980) into those organisms associated with the rumen wall, those free in the rumen fluid, and those associated with feed particles (Figs 2, 3 and 4). This latter population has been subdivided, by Czerkawski (1980), into those organisms that are firmly associated with particulate food (degradative compartment) and those that are more reversibly associated (shuttle compartment). Clearly, in the functional rumen, the compartments are in a dynamic equilibrium because bacteria enter the fluid compartment from the rumen wall and the particle-associated population, and vice versa.

THE RUMEN SIMULATION TECHNIQUE (RUSITEC): A MODEL MULTI-COMPARTMENT SYSTEM

The study of microbial processes in the artificial rumen may shed some light on the integrated physiology of this organ. Techniques have ranged from simple batch incubations to complicated equipment designed to run continuously and even provided with semipermeable membranes for the removal of the end-products of fermentation (see Czerkawski, 1986a). Even with the very complicated techniques, however, little consideration is given to the possibility of microbial compartmentation.

The rumen simulation technique (Rusitec) was developed at the Hannah Research Institute (Czerkawski & Breckenridge, 1977) with compartmentation and the use of fibrous feeds firmly in mind. The rumen and its microbial population evolved to digest natural fibrous feeds, and the concepts discussed here refer mainly to such feeds. However, Rusitec functions well with mixed diets, and it should be possible to develop the concepts of compartmentation to include the digestion of non-structural carbohydrates, e.g. starch.

In Rusitec the solid digesta are contained in coarse nylon gauze bags inside a perforated cage which slides with an amplitude of 5–6 cm at about 8 cycles/min inside a cylindrical reaction vessel. Artificial saliva is infused at the bottom of the vessel and the excess liquid and the fermentation gases are forced through an overflow by a small positive pressure and collected separately. Thus the free microbial suspension is separated physically from the microorganisms associated with the mass of solid digesta, but a continued interaction between these two microbial systems is facilitated by the coarseness of the nylon gauze bags and the perfusion of the solid by the liquid. The experiment is started by placing about 80 g of solid rumen digesta in one nylon bag and a quantity of the animal feed to be used in another bag, inserting these two bags side by side in the perforated cage and filling the reaction vessel with strained, diluted, rumen contents. After 24 hours the solid inoculum is removed and a new bag of food placed in the vessel. Subsequently, each day the older bag is removed, allowed to drain, placed in a larger polythene bag, and washed twice with artificial saliva by pressing gently and squeezing out excess liquid. The combined washings are returned to the reaction vessel and the washed solid is dried, weighed, and analysed. It will be appreciated that the actual residence time of feed in Rusitec is 48 hours, but this can be varied (see Czerkawski, 1986b). The output of end-products of fermentation (methane, volatile fatty acids, etc.) decreases slightly during the first 4–6 days of the experiment, but thereafter, with a balanced diet, a steady fermentation can be maintained indefinitely.

The Rusitec is a rumen without a rumen wall. There is no provision for the removal of substances except by overflow or manual removal of undigested solid. The fermentation patterns obtained and the yield of product per unit mass of food digested, however, closely resemble the fermentation patterns and the composition of rumen contents in sheep kept on the same diet (Czerkawski & Breckenridge, 1977, 1979a,b).

The two 'compartments' separated by the nylon gauze are clearly not a good example of compartmentation as previously defined, because there is a lot of free liquid, as well as the solid feed, inside the bag at any given time. However, at least three, more realistic microbial compartments can be identified in the Rusitec and defined in a simple practical manner (see Fig. 5). Compartment 1 is a free suspension of microorganisms. Its volume and contents can be measured directly, but it is not confined entirely to the space outside the gauze bags. However, the microbial

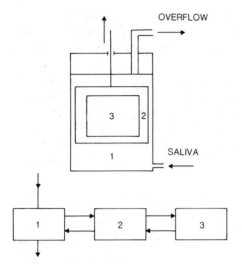

Fig. 5. Relationship between compartments in Rusitec apparatus. Compartment 1 is the free suspension of microorganisms, the microorganisms in compartment 2 are loosely associated with the solid, and those in compartment 3 are firmly bound to the solid (see text).

population associated with the solid mass of digesta is too complex to be regarded as a single compartment.

When the drained undigested material is washed twice as described above, a proportion of microbial matter can be removed, but considerable amounts remain firmly bound to the washed residue. This microbial matter can be measured in terms of microbial markers for bacteria (diaminopimelic acid, DAP) and for protozoa (aminoethylphosphonic acid, AEP), and together these microorganisms account for 5–15% of the dry mass of the washed undigested residue. Because of properties to be discussed later, the fact that these microorganisms are strongly associated with the solid mass of digesta suggests that this population must be considered to be different from that in free suspension (compartment 1). Thus, in Rusitec, compartment 3 is occupied by microorganisms that cannot readily be washed out of the solid digesta. Compartment 3 is easy to define, but it is difficult to study. The volume (ml) of liquid in this compartment is numerically about 2·5 times the weight (g) of dry matter of the feed residue, and the mean microbial concentrations therein are considerably higher than those in compartment 1. As one would expect, a large portion of the cellulolytic activity in the model system is found in this compartment.

Although a large proportion of the microorganisms in the Rusitec are associated with the solid digesta (70–80%), as they are in the actual rumen (Forsberg & Lam, 1977; Craig et al., 1987), part of this population can be removed by squeezing and washing. The properties of the loosely adherent microbial population are different from those of the other two compartments, and the space occupied by these microorganisms is defined as compartment 2. The volume of this compartment can be calculated indirectly (Czerkawski, 1986a) and, since the amount of microbial matter can be determined directly, the microbial concentration can be determined and compared with those of the other two compartments; it is lower than that of compartment 3, but considerably higher than that of compartment 1. Some of the general properties of the three compartments are summarized in Table 1, and the

Table 1
Some of the Compartmental Properties in Rusitec

	Compartment		
	1	2	3
Microbial concentration (mg ml^{-1})	0·7–1·0	20–30	40–60
Protozoal proportion (% total microbial matter)	10–20	30–40	40–50
Microbial distribution (% of total dry mass)	10	56	34
Dilution rate (days^{-1}) (nominal dilution rate 0·9 days^{-1})	1·7	0·8	0·5

Calculations from Czerkawski & Breckenridge (1979b, 1982).

possible interrelation between the compartments can be represented by a three-compartment model (Fig. 3) where compartment 2 is placed between compartments 1 and 3.

Experiments were carried out using the Rusitec apparatus in a simplified manner (Czerkawski & Breckenridge, 1979b) in which a matrix of indigestible hay residues was kept permanently in the vessels and all food was supplied in solution. This resulted in an acceptable fermentation and steady-state conditions that could be described by a relatively simple mathematical model. It was possible to calculate the rates of synthesis of microbial matter in each compartment and the rates of flow of waste and microbial matter between the compartments. The dilution rates were different in each compartment and in general decreased in the passage from 1 to 3 (Table 1). The flow of DAP (bacterial marker) was always greater in the direction compartment 3 to compartment 1 than in the opposite direction. In the rumen, or in Rusitec given solid food, the flow of microbial matter would be in the direction compartment 1 to 3 soon after ingestion of feed, since the flow of soluble food would be in the opposite direction (a possible aid in colonization of solid feed by bacteria). In general, the mean residence time of DAP increased from compartment 1 to 3, an example of the tendency of the bacterial matter to become firmly associated with solid matrices. The proportion of protozoa in the washed undigested residues from the Rusitec was always greater than in the free suspension of microorganisms. When the simplified system was used, with all the food supplied in solution, protozoa disappeared from the effluent liquid. Occasionally a few protozoa could be found in compartment 1, but there were many protozoa in compartment 2 and even greater concentrations (measured in terms of AEP) in compartment 3. This was an example of sequestration *par excellence*, since no protozoa left the Rusitec system and yet a healthy population was being maintained. Since the net synthesis of protozoa under these conditions was zero, the rate of total synthesis must have been equal to the rate of degradation as defined by Van Nevel & Demeyer (1977). Incidentally, under these conditions the protozoa constitute a closed multi-compartment system.

The representation of the microbial population in Rusitec as a three-compartment system is a useful aid in setting up further hypotheses and testing them. Nevertheless, it is an oversimplification.

When solid food is used in Rusitec there is a single compartment 1, but there are two similar compartments 2 and also two compartments 3. This is because at any given time one bag has spent 1 day more in the vessel than the other bag. This might constitute a five-compartment system. It can be shown that, if two dietary constituents are incubated in Rusitec in separate bags (Czerkawski & Breckenridge, 1979a; Krishna et al., 1986), at any given time there are four bags in the vessel and this might be regarded as a nine-compartment system. At all times the Rusitec system lacks the rumen wall-associated microbial population. The digesta in the rumen are not confined to nylon gauze bags and, strictly speaking, the rumen contents might consist of an infinite number of compartments. Nevertheless, the microbial population of the rumen is best described in terms of four basic compartments.

THE RUMEN AS A MULTI-COMPARTMENT SYSTEM

The reticulo-rumen consists of four discernible lobes, two large sacs (dorsal and ventral) in the rumen proper, the reticulum, and a relatively small structure between the reticulum and the main rumen chamber, the cranial sac (see Chapter 1). Although these chambers are interconnected, because of the existence of muscular thickening (the pillars) within this organ, and because of the heterogeneous nature of the contents, the flows of digesta within the rumen are not random and, as often occurs in nature, several conflicting aims are economically achieved.

The effect of rumen movement is that liquid digesta are propelled through the mass of solid digesta. The system can be visualized as consisting of two-particle pools (Czerkawski, 1986b). The fibrous mass of digesta (large-particle pool) is subject to a slow cyclic flow, with small portions being removed at the reticulum end and replenished by spent boli and new food thrown to the other end during reticular contractions. The liquid plus small-particle pool is subject to reciprocating flows. The two pools interact during rumination and during crossflows through the raft of digesta which acts as a renewable double-sided filter.

In Rusitec the liquid digesta are stationary, while the bags of solid digesta (the raft) are propelled through the liquid in such a way that most of the liquid has to pass through the digesta. Thus the mechanism is different from that in the rumen, but the effect is the same: a reciprocating flow of liquid through the solid mass of digesta. Rumination also has a parallel in Rusitec; this is the washing and squeezing of solid digesta. There are several other parallels (see Czerkawski, 1986b), and it is likely that the three microbial compartments identified in Rusitec, and the relations between these compartments, may also apply in the rumen. Because the rumen is bounded by a rumen wall which is well supplied with blood (and therefore oxygen), one would expect the conditions here to be different from those in the interior of the rumen, and hence one would expect to find a different microbial population adherent to the rumen wall (Cheng & Costerton, 1980). Thus in the rumen one finds an additional microbial compartment, named compartment 4 for convenience. The microbial compartments 1, 2 and 3 were defined in a simple manner in Rusitec, and it was possible to describe the properties of these microbial groups. The existence of similar microbial groups has been amply demonstrated in the rumen. For instance,

Merry & McAllan (1983) showed that there were marked differences between the chemical composition of liquid-associated bacteria (compartment 1) and solid-associated bacteria (mainly compartment 2). Craig et al. (1984) compared the N and polysaccharide content of microbial matter in strained rumen contents (compartment 1) and in the microbial matter extracted from the solid digesta with detergent (probably compartment 2 and part of compartment 3). The particle-associated microorganisms had higher polysaccharide and lower N content than those suspended in the liquid phase. Williams & Strachan (1984) determined the glycolytic activity of the microorganisms in the liquid phase of the bovine rumen (compartment 1) and of the microorganisms associated with feed particles incubated in nylon bags in the rumen. The latter were separated into two groups: the microorganisms washed out of the solid (called 'non-adherent population', equivalent to compartment 2) and the microorganisms that could not be removed (called 'attached population', equivalent to compartment 3). The enzymes that degrade non-structural, soluble saccharides were more active in compartments 1 and 2 than in compartment 3 (the mean ratio of activities for compartments 1, 2 and 3 was $1.0:0.9:0.6$). On the other hand, the plant cell-wall polysaccharide-degrading enzymes were considerably more active in the particle-associated populations, particularly in the strongly attached group (compartment 3). The mean ratio of activities for compartments 1, 2 and 3 was $1.0:2.0:4.6$. The above values refer to specific enzyme activities, i.e. per unit weight of protein, and since the microbial concentrations are considerably higher in the inner compartments the actual polysaccharide-degrading activity could be over 100 times greater in compartment 3 than in compartment 1.

The importance of the microbial population of compartment 2 has been stressed by Senshu et al. (1980). They concluded from their work that the strained rumen contents that have been used so frequently in the past (compartment 1) do not truly represent the microbial population of the rumen. Most of the starch fermenters and a proportion of cellulose fermenters could be extracted from the rumen solid by repeated washing and squeezing (compartment 2).

The properties of the four compartments will now be discussed in greater detail.

Compartment 1

The microbial population of compartment 1 is the easiest to sample. It is known as 'strained rumen contents' and it is the most investigated part of the rumen. The volume of this compartment is large, but the microbial concentrations are smaller than in the other compartments.

Because it is a microbial suspension and because the microorganisms rely on soluble substrates, this part of the system is most closely related to a continuous culture. Consequently the population of this compartment will be influenced by the dilution rate and by the supply of soluble food. Since the concentrations of soluble carbohydrates in the rumens of animals on roughage diets are low, except for a short time after feeding, the overall supply of food and the metabolic activity of this population might be low.

The most important role of the liquid occupied by the microbial compartment 1 is as a transport medium for moving solid digesta to different parts of the rumen and for carrying microorganisms and small particles to the omasum. When the food enters the rumen it must be colonized by microorganisms, and therefore the conditions in this compartment must ensure the survival of most types of rumen microorganisms but not necessarily their rapid growth.

Logically the population of compartment 1 should be divided into two main groups. The function of one group is to scavenge the low supply of soluble food. One would expect the members of this group to have a rapid growth rate and very little lysis and resynthesis of microbial matter. These microorganisms would be expected to provide the rumen with protective mechanisms (such as detoxification; Majak & Cheng, 1983; see also Chapter 14). The other part of the population of compartment 1 really belongs to the other compartments (compartments 3 and 4). These microorganisms are in compartment 1 but their contribution to metabolism becomes significant only when they colonize the solid feed. Both groups of microorganisms would be culturable given suitable conditions, but one would expect the latter group to show considerable lag in its growth and metabolism. When hay was incubated with rumen contents *in vitro* (Mehra *et al.*, 1981), the rate of digestion of fibre was negligible for 6 hours and then increased rapidly, reaching a maximum at 24 hours.

Compartment 3

The main function of the rumen is to break down fibrous feeds, and there is no doubt that it is the population of compartment 3 that is primarily responsible. Up to this point our definition of the population of compartment 3 has made no reference to the structure of this compartment, which is probably very complex. The microorganisms may be attached to the solid digesta (Kudo *et al.*, 1986*b*) or simply trapped in the eroded spaces. Functionally, what is important is that these microorganisms are closely associated with the solid and that they cannot be washed off either in the rumen or in the bolus. The bulk of this population will remain in the solid matrix and will leave the rumen only when the undigested particles are small enough.

The mean residence time of solid feed in the rumen is long (2–3 days). The generation times of bacteria depend on the supply of food, and under ideal conditions could be short (hours rather than days); therefore one would expect a great deal of microbial lysis and resynthesis in compartment 3. This could be beneficial to the system as a whole. The release of exogenous enzymes into compartment 1 with the expectation that they would find their way to the fibrous substrate would result in a minimal chance of reaction. It must be concluded that microbial attachment to solid feed or entrapment within the solid is essential for efficient degradation of solid substrate, and indeed there is now a great deal of evidence that such an association does in fact exist (see Akin, 1979; Cheng *et al.*, 1980, 1983). It is possible to see digestive consortia of different microorganisms and almost pure cultures of morphologically similar organisms (Kudo *et al.*, 1986*a*). The shape of some microorganisms is such that they 'conform' to the surface while others secrete polymeric carbohydrates (glycocalyx) which mediates attachment and transfers enzymes to the site of action (Kudo et al., 1986b). The details of attachment of bacteria to fibres and the degradation of the fibre polysaccharides are discussed in Chapter 8.

Plant tissues are not attacked uniformly. The parts that are digested show deep pitting and evidence of attack from within. The heavily lignified structures are not colonized, even though they contain cellulose and hemicellulose (Cheng et al., 1983). This is considered to be a deficiency of the system, and a great deal of effort has gone into the study of the linkage between lignin and carbohydrate in fibrous tissues, and how to break this linkage and thus enable the hard core of lignocellulose to be digested (Chapter 8). It is possible that the uncolonized structures are necessary for the proper functioning of the fibre-digesting system of the rumen. If these were removed the whole structure could collapse (see the 'hotel theory'; Van Soest, 1975).

A single layer of bacteria colonizing the surface of a hay-stem particle (1 cm × 1 mm) in the rumen would account for only a small proportion (1/500) of the actual microbial matter content. The cuticular wax would further reduce this type of colonization, and thus it is not surprising that the entry of microorganisms into the fibrous mass involves the damaged areas of the fibres (Fay et al., 1981). After colonization and the establishment of microbial populations, the actual process of fibre digestion is dependent upon a strongly adherent microbial population attacking the fibre from within.

The population of compartment 3 will contain a large proportion of cellulolytic microorganisms, and it would be expected to have microbial consortia and metabolic couplets, with a great deal of interdependence and specialized metabolic activity in the microorganisms operating in the confined spaces (see Chapter 11). This population would be adapted to dilution rates lower than those in compartment 1, and would be subject to more extensive cell lysis and resynthesis. In a semi-stagnant system, a high microbial recycling (lysis and resynthesis), and proteolysis (see Wallace & Munro, 1986), may be advantageous. It is normally difficult to see *Bacteroides succinogenes* in strained rumen contents (compartment 1), yet these microorganisms are found in profusion inside the fibrous tissues. Since the mean residence of fibrous residues in the rumen is about 2 days, one would expect lysis of trapped cells. Such lysis may result in a complete destruction of cells, but it is possible that certain enzymes survive and continue to digest fibre, for instance in small envelopes. These may contain cellulolytic enzymes and continue the process of cellulolysis (Stewart et al., 1979; Forsberg et al., 1981). It is possible that within compartment 3 the lysis and resynthesis of microbial matter may not be an entirely wasteful process but part of a mechanism for degrading fibrous tissues from within.

Compartment 2

The properties of the population of compartment 2 are in general intermediate between those of compartments 1 and 3, but this compartment has unique features. It appears to be homogeneous with respect to some substances, such as NH_3 and urease (see e.g. Czerkawski & Breckenridge, 1982), and very heterogeneous with respect to individual microenvironments (cellulolytic consortia). The microbial concentrations, and concentrations of common enzymes such as alkaline phosphatase, increase steadily from compartment 1 to compartment 3. The concentrations of glycolytic enzymes increase disproportionately, suggesting that they really belong to compartment 3 and that the boundary region between compartments may not be very distinct.

The existence of a microbial population in compartment 2 with properties different from those in the other compartments has been demonstrated by the work with the Rusitec and by experiments with hay suspended in nylon gauze bags in the rumen (Czerkawski, 1983). It would be difficult to see how two such different microbial populations and their environments as those in compartments 1 and 3 could coexist and interact in the same system, without the intervention of another compartment. The microorganisms in compartment 2 associate with the solid digesta, but they may not be degrading these substrates and they are not as closely bound as those in compartment 3. For instance, cells of the genus *Lachnospira* appear to penetrate the intercellular spaces of forage leaves and yet they use soluble nutrients (Cheng et al., 1980).

The microorganisms in compartment 2 can be removed from the solid by relatively gentle means and almost certainly by rumination. It is also possible that the microbial matter in this compartment can be partly removed by pressures and flows of liquids during rumen movement. After rumination, the squeezed bolus, with a much reduced microbial population of compartment 2, would be sent back into the rumen to be colonized by the microorganisms washed from the bulk of the fibrous mass.

The function of compartment 2 is to transfer microbial matter, nutrients, and end-products of microbial activity between entirely different milieux, and therefore it has been named the 'shuttle' compartment. The strong metabolic activity of this population would create a concentration gradient which would be expected to help with both the supply of nutrients and the removal of end-products. The microorganisms in this compartment would form a link in a food chain.

The heterogeneous nature of compartment 2 and the movement of very active microbial matter, substrates, and end-products within the compartment are consistent with its function. However, if the conditions change by formation of stable froth or a rapid release of soluble materials, its 'shuttle' function could be impaired and serious rumen dysfunction conditions might ensue (e.g. bloat; Howarth et al., 1986; see Chapter 14).

Compartment 4

The microbial population close to the rumen wall forms an interface between well oxygenated tissues (90% of blood flowing to the rumen supplies the epithelium) and an anaerobic system. This population has been studied (Cheng & Costerton, 1980) and, although the microorganisms in this compartment account for no more than about 1% of the total microbial population in the rumen, it may have an important strategic role. The population of bacteria that adheres to the stratified squamous epithelium of the rumen wall occupies an environmental niche that is oxygenated and is very rich in urea which diffuses continuously across the rumen epithelium. This population is ideally suited to this niche because it carries out proteolytic digestion and effective recycling of the protein of sloughed epithelial cells, it is partially facultative in oxygen tolerance, and it produces urease to convert urea into the ammonia that is necessary for the nitrogen metabolism of other compartments. Cheng and his colleagues (Cheng & Wallace, 1979; Wallace *et al.*, 1979) have established that this urease-producing capacity is entirely bacterial and that this important physiological cooperation between the tissue and the colonizing bacteria is established early in the life of the animal (see Chapter 7 for further details of urea metabolism).

SOME CONSEQUENCES OF COMPARTMENTATION IN THE RUMEN

Adequate nitrogen supply

Nitrogen can enter the rumen in animal feed as part of the normal diet or as a supplement. The endogenous nitrogen supply, as urea, can enter the rumen with saliva or through the rumen wall, although there is some disagreement on the proportions supplied by these two paths (Kennedy & Milligan, 1980). The nitrogen in feed, particularly fibrous feed, may not be available to the microorganisms. Although rumen microorganisms can use preformed amino acids, ammonia is considered to be a very important source of nitrogen for microbial synthesis in the rumen (see Chapters 2 and 7).

Experiments with the Rusitec apparatus (Czerkawski & Breckenridge, 1982) showed that the high urease activity of the inoculum declined to very low values and that high urease activity could be induced at will by infusing urea with the artificial saliva. The magnitude of urease activity per unit volume of each compartment was a direct function of urea infusion, but it was considerably higher in compartment 2 than in compartments 1 and 3, and it was not inhibited by high ammonia concentrations (the latter were higher in compartment 2 than in compartment 1). These observations suggest an ingenious mechanism for the efficient use of ammonia nitrogen in the rumen. Since the urease activity in compartment 1 is low, a great deal of the available urea will be hydrolysed in compartment 2, resulting in a large increase in the concentration of ammonia in that compartment. This will facilitate the flow of ammonia to compartment 3 where the ammonia concentrations are low and where it is needed most. This mechanism keeps the concentration of ammonia low where it could result in absorption and toxicity (compartment 1; particularly with salivary input of urea), and it enables the flow of available nitrogen against an apparent concentration gradient (compartment 1 to 2).

Attack from within and the mechanism of fibre digestion

The digestibility of hay chopped to about 1 cm length is similar in the Rusitec and in sheep given the same diet. There is no chewing action in Rusitec, and the washed and

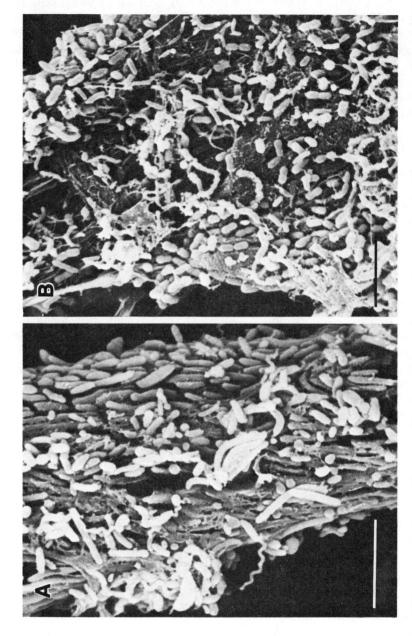

Fig. 6. (a) SEM of a partly digested filter paper fibre from 24-h enrichment cultures inoculated with rumen contents from a steer. (b) SEM of a partly digested filter paper from 24-h enrichment cultures inoculated with rumen contents from a water buffalo.

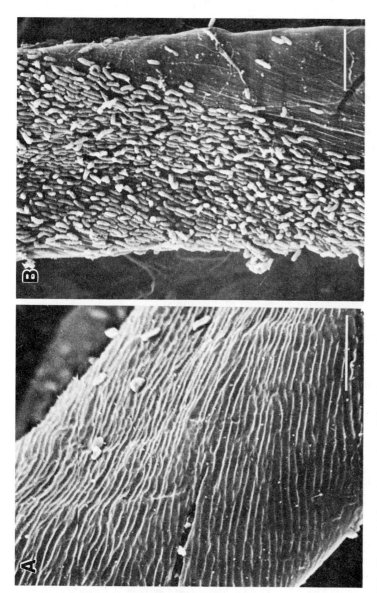

Fig. 7. (a) SEM of bacterial detachment from Whatman No. 1 filter paper. Filter paper was first incubated with *Bacteroides succinogenes* strain E at 39 °C for 20 h, and then treated with 0·1% methylcellulose at 39 °C for 1 h. This SEM shows the very effective detachment of cellulolytic bacteria from their cellulosic substrate. (b) SEM of Whatman No. 1 filter paper incubated with *Bacteroides succinogenes* strain E at 39 °C for 20 h showing extensive bacterial adhesion to the cellulose fibre.

dried fibre matrix becomes discoloured and brittle, but the size and shape of the particles appear to be unchanged. This supports the hypothesis that the fibrous feeds are attacked from within and that it is not necessary to break them down physically. The chewing of food during eating and rumination will break long fibres and expose 'ends' of fibres as well as producing other damage to allow bacteria in, and thus to facilitate initial colonization. The rumination could be important also in the reduction of the size of the already digested particles, so that they can leave the rumen.

The digestion of fibre in the rumen requires a reasonably long time. Fine grinding of feeds could prevent the formation of the desirable 'filter' structure, causing the small particles to flow out too soon. The block structure of contents that cannot flow out with liquid may be necessary for microbial sequestration and for the proper operation of the microbial multi-compartment system.

Microbial attachment

The attachment of rumen microorganisms to fibres and the degradation of fibre polysaccharides have been considered in detail in Chapter 8. Here some aspects of microbial attachment relating to compartment theory are mentioned.

Whatever the mechanism of fibre digestion at the molecular level, the microbial attachment (Fig. 6) to the solid fibrous substance is an important prerequisite of degradation of this substance (Cheng et al., 1977, 1983; Forsberg & Lam, 1977; Kudo et al., 1986b). The attachment of a proportion of the microbial population could be very firm, requiring complete dissolution of the substrate to free the microorganisms—an unlikely happening (see Fig. 7). Any model that aims to describe the mechanism of fibre digestion in the rumen must take both microbial attachment and detachment into account. The characteristic properties of microorganisms in compartment 2 are consistent with these requirements.

When grasses were incubated with a suspension of rumen microorganisms, the encapsulated cocci and the irregularly shaped bacteria accounted for about 80% of all the bacteria attached to the plant cell wall (Amos & Akin, 1978). Apparently the adhesion of the encapsulated cocci was facilitated by extensive extracellular material (Fig. 8), while the irregularly shaped *Bacteroides*-like cells adhered simply because their shape conformed with the shape of the eroded zones and they produced thin, adherent capsules (Costerton et al., 1974). The attachment of bacteria is usually, but not always, associated with digestion of plant tissue and, according to Akin (1979), bacteria must either almost completely adhere, or be in close proximity, to the plant cell wall as a prerequisite to its degradation. This is one of the most important arguments for microbial compartmentation in the rumen.

The attachment of protozoa and subsequent degradation of forage tissue has also been demonstrated (Amos & Akin, 1978; see also Chapter 3). Entodinia were found to degrade mesophyl tissues, leaving the parenchyma bundle sheath intact and creating well-defined cavities, while preserving the three-dimensional structure of the matrix. When lucerne was incubated in nylon bags in the rumen (Bauchop, 1979), large numbers of *Epidinum* became attached to the damaged areas of the stem in a

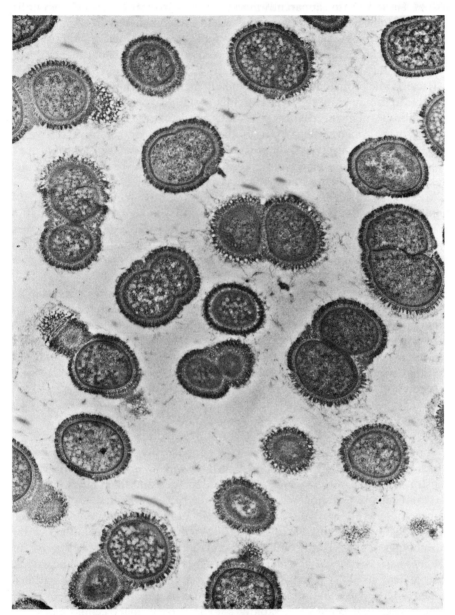

Fig. 8. Cells of *Ruminococcus flavefaciens* isolated from the water buffalo were seen to have their cell walls surrounded by a radially structured zone of electron-dense glycocalyx fibres that was, in turn, surrounded by a less dense, less structured external glycocalyx.

very short time, and the highest concentrations were on the cortex and phloem tissues. These protozoa apparently ingested fibrous tissues even though they utilized only the soluble part of the ingested material. Although these microorganisms were responsible for some physical degradation and removal of underlying tissues, the general structure of the solid matrix, the basis for microbial compartmentation, was preserved.

Bacteria and protozoa are the predominant microbial forms in the rumen, but it is possible to demonstrate appreciable numbers of fungal zoospores which can attach to the fibrous fragments (Bauchop, 1979; Orpin, 1984; see Chapter 4). The life cycle of fungi in the rumen is about 24 hours, and it would appear that the vegetative form of these microorganisms is confined to compartment 3 where they contribute to the degradation of fibrous material; the zoospores would be expected to contribute to the population of compartments 1 and 2 as well as 3.

Microbial attachment, consortia formation (Stanton & Canale-Parola, 1980; Kudo et al., 1986a; Wood et al., 1986) and sequestration are not accidental in a multi-compartment system; they are integral and desirable characteristics of the system. In fact there is a great deal of correspondence between the spatial distribution of microorganisms and their function in the rumen (see Williams & Strachan, 1984).

The very firm attachment of microorganisms to solid surfaces or their entrapment in compartment 3 and their recycling in the stagnant environment seem to implicate the endogenous enzymes released during microbial lysis. It has been suggested that such enzymes would be rapidly destroyed by proteolysis. Since, in compartment 3, the enzymes could be released close to their natural substrates, one would expect that they would become adsorbed onto these substrates, and it is known that the formation of an enzyme–substrate complex can protect the enzyme.

Microbial consortia

Complex substrate transformations commonly involve microbial consortia in which two or more species combine their enzymatic activities to degrade the substrates (see Chapter 11). Pure cultures of bacteria and fungi never achieve the rates of cellulose digestion actually seen in the rumen, and direct examination of rumen contents shows that a spatial hierarchy of microorganisms exists on cellulose surfaces undergoing rapid cellulolysis. When organisms identified in these morphological studies are combined in mixed cultures (e.g. *B. succinogenes* plus *Treponema bryantii* or *B. succinogenes* plus *Butyrivibrio* sp.), the rate and extent of their attack on cellulose approaches that seen in the rumen, and combinations of rumen fungi and methanogenic bacteria (Wood et al., 1986) are similarly effective. It is now clear that cellulose is rapidly colonized by primary cellulose-degrading organisms and subsequently colonized by secondary consortium organisms that do not degrade cellulose but 'drive' the digestive process by end-product removal. These colonization processes may be initiated in compartment 1 and become fully operational in compartments 2 and 3.

Digestive dysfunction due to malfunction in compartmentation

If the overall bacterial reactions in one compartment of the rumen are deranged, by unusual food input or by environmental factors, then the function of the compartment in question may be impaired and overall animal performance compromised (see Chapter 14). When high-energy feeds are presented in fine particle sizes the elevated levels of soluble nutrients available to the bacteria in compartment 1 may cause an overproduction of lactic acid, resulting in acidosis (Brent, 1976; Schwartz & Gilchrist, 1975). Acidosis in compartment 1 may affect compartment 4 to produce lesions through which compartment 4 bacteria enter the bloodstream and accumulate in the liver to produce liver abscesses (Kanoe et al., 1976). Alternatively, the release of large amounts of soluble nutrients from legume forage materials may cause the formation of a frothy foam and result in pasture bloat (Howarth et al., 1986). Thus it is clear that the optimal functioning of the rumen is dependent on the optimal functioning of each compartment and the balanced relationship between these integrated subdivisions of the organ.

Manipulation of compartment processes in the rumen

The four-compartment model of rumen function allows us to target certain processes for optimization of feed digestion. For example, the breeding of slow-nutrient-release legume forages can prevent bloat by slowing the release of soluble nutrients in compartment 1 (Howarth et al., 1986). Similarly, the digestion of certain grasses may be improved by increasing the soluble nutrients available from this slow-release forage. The process of colonization of cellulose by cellulolytic bacteria, which is the main purpose of compartment 3, may be accelerated by the use of a chemical agent (e.g. 3-phenylpropanoic acid) that promotes this specific adhesion process. Similarly, if other chemical agents that promote the formation of effective cellulolytic consortia could be developed, this might accelerate cellulose digestion in compartment 3. The manipulation of the rumen fermentation is discussed in Chapter 13.

CONCLUDING REMARKS

The properties of the four postulated microbial compartments in the rumen are summarized in Table 2 and Figs 2 and 9. Clearly, compartments 1 and 4 are related, inasmuch as they are concerned with inputs and outputs in the whole system. Similarly, compartments 2 and 3 are related because they are both involved in the sequestration of microbial matter, so important in fibre digestion.

The concepts developed here stem directly from the consideration of demands imposed on the rumen as an organ by the need for converting tough fibrous feeds into nutrients that can be used by the host animal. The rumen has an incredibly clever way of solving these problems.

Table 2
Properties of Compartments in the Four-compartment Model of the Rumen

	Compartment			
	4	1	2	3
General purpose	Control of inputs and outputs in the system		Control of sequestration of particular components	
Particular role	Strategic control	Transport medium	Shuttle service	Fibre degradation
Host influence	Strong	Fairly strong	Some influence	Very little influence
Microbial population	Facultatively aerobic	Utilizers of soluble substrates	Complex, links in substrate chains	Fibre digesters[a]
Predominant bacterial species	*Fusobacterium* Gram-positive cocci	*Bacteroides* *Streptococcus* *Megasphaera*	*Lachnospira* *Ruminococcus albus* *Butyrivibrio fibriosolvens*	*Bacteroides succinogenes* *Ruminococcus flavefaciens* *Treponema*

[a] This population will have cellulolytic and hemicellulolytic activities. The attached bacteria may also have pectin- and fructosan-degrading activities and they may be involved in degradation of starch granules.

It is clear that the internal structure of the reticulo-rumen is not designed to stir the solid and liquid contents in the way that one would stir the contents of a round-bottom flask. On the contrary, the raft of digesta in animals on roughage diets is more or less immobilized and the overall structure appears to favour minimal movement of the solid mass of digesta. This must be advantageous, because it is unlikely to disturb the balance of the animal, in which the rumen contents may

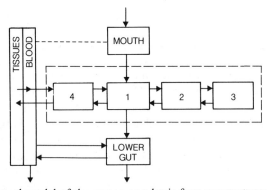

Fig. 9. A conceptual model of the rumen as a basic four-compartment system; particular conditions, not applicable during rumination. Compartments 1, 2, 3 and 4 are defined in the text.

account for up to one-fifth of its weight and whose survival in the wild state depends so much on its agility. The immobilization of the raft of fibrous digesta and its use as a continuously renewable double-sided filter (Czerkawski, 1986b) ensures its controlled and predictable throughput in an incredibly complicated organ. The compartmentation of the microbial population in the rumen is an indispensable and integral part of this system.

In multicellular organisms, the cells differentiate in response to the function of any particular tissue. Some cells remain fixed to form definite functional structures (e.g. muscle cells), while others become mobile (e.g. blood cells). The integrity of this complex system is maintained by connective tissues. Clearly, the microbial system of the rumen is not a simple fermentation vat filled with randomly distributed mixtures of microorganisms. It is a highly structured and functionally compartmented system, *en par* with some of the important organs of the advanced multicellular organism. Little progress will be made in the field of rumen metabolism unless the rumen is viewed in this manner.

REFERENCES

Akin, D. (1979). Microscopic evaluation of forage digestion by rumen microorganisms: a review. *J. Anim. Sci.*, **48**, 701–10.

Amos, H. E. & Akin, D. E. (1978). Rumen protozoal degradation of structurally intact forage tissues. *Appl. Environ. Microbiol.*, **36**, 513–22.

Bauchop, T. (1979). The rumen ciliate *Epidinium* in primary degradation of plant tissues. *Appl. Environ. Microbiol.*, **37**, 1217–23.

Brent, B. E. (1976). Relationship of acidosis to other feedlot ailments. *J. Anim. Sci.*, **43**, 930–5.

Cheng, K.-J. & Costerton, J. W. (1980). Adherent rumen bacteria: their role in the digestion of plant material, urea and epithelial cells. In *Digestive Physiology and Metabolism in Ruminants*, ed. Y. Ruckebush & P. Thivend. MTP Press, Lancaster, pp. 227–50.

Cheng, K.-J. & Wallace, R. J. (1979). The mechanism of passage of endogenous urea through the rumen wall and the role of ureolytic epithelial bacteria in the urea flux. *Br. J. Nutr.*, **42**, 553–7.

Cheng, K.-J., Akin, D. E. & Costerton, J. W. (1977). Rumen bacteria: interaction with particulate dietary components and response to dietary variation. *Fed. Proc.*, **36**, 193–7.

Cheng, K.-J., Fay, J. P., Howarth, R. E. & Costerton, J. W. (1980). Sequence of events in the digestion of fresh legume leaves by rumen bacteria. *Appl. Environ. Microbiol.*, **40**, 613–25.

Cheng, K.-J., Stewart, C. S., Dinsdale, D. & Costerton, J. W. (1983). Electron microscopy of bacteria involved in the digestion of plant cell walls. *Anim. Food Sci. Technol.*, **10**, 93–120.

Costerton, J. W., Damgaard, H. N. & Cheng, K.-J. (1974). Cell envelope morphology of rumen bacteria. *J. Bacteriol.*, **118**, 1132–43.

Craig, W. M., Brown, D. R., Broderick, G. A. & Ricker, D. B. (1984). Polysaccharide levels of fluid and particulate microbes. *Can. J. Anim. Sci.*, **64**(Suppl.), 62–3.

Craig, W. M., Broderick, G. A. & Ricker, D. B. (1987). Quantitation of microorganisms associated with the particulate phase of ruminal ingesta. *J. Nutr.*, **117**, 56–62.

Czerkawski, J. W. (1980). Compartmentation in the rumen. *Hannah Res. Inst. Rep.*, 69–85.

Czerkawski, J. W. (1983). Microbial fermentation in the rumen. *Proc. Nutr. Soc.*, **43**, 101–18.

Czerkawski, J. W. (1986a). The artificial rumen. In *Introduction to Rumen Studies*. Pergamon, Oxford, pp. 51–62.

Czerkawski, J. W. (1986b). A simple model to describe the flow of large and small particles in the rumen. In *Proc. Int. Symp. on Use of Nuclear Techniques in Studies of Animal Production and Health in Different Environments.* International Atomic Energy Agency, Vienna, pp. 485–99.

Czerkawski, J. W. & Breckenridge, G. (1970). Small scale apparatus for studying rumen fermentation *in vitro. Lab. Practice*, **19**, 717–19.

Czerkawski, J. W. & Breckenridge, G. (1977). Design and development of a long-term rumen simulation technique (Rusitec). *Br. J. Nutr.*, **38**, 371–84.

Czerkawski, J. W. & Breckenridge, G. (1979a). Experiments with the long-term rumen simulation technique (Rusitec); response to supplementation of the rations. *Br. J. Nutr.*, **42**, 217–28.

Czerkawski, J. W. & Breckenridge, G. (1979b). Experiments with the long-term rumen simulation technique (Rusitec); use of soluble food and an inert solid matrix. *Br. J. Nutr.*, **42**, 229–45.

Czerkawski, J. W. & Breckenridge, G. (1982). Distribution and changes in urease (EC 3.5.1.5) activity in the rumen simulation technique (Rusitec). *Br. J. Nutr.*, **47**, 331–48.

Faichney, G. J. (1975). The use of markers to partition digestion with gastro-intestinal tract of ruminants. In *Digestion and Metabolism in Ruminants*, ed. I. W. McDonald & A. C. I. Warner. University of New England Publishing Unit, Armidale, Australia, pp. 277–91.

Fay, J. P., Cheng, K.-J., Hanna, M. R., Howarth, R. E. & Costerton, J. W. (1981). A scanning electron microscopy study of the invasion of leaflets of a bloat-safe and a bloat-causing legume by rumen microorganisms. *Can. J. Microbiol.*, **27**, 390–9.

Forsberg, C. W. & Lam, K. (1977). Use of adenosine-5'-triphosphate as an indicator of the microbiota biomass in rumen contents. *Appl. Environ. Microbiol.*, **33**, 528–37.

Forsberg, C. W., Beverage, T. J. & Hellstrom, A. (1981). Cellulose and xylanase release from *Bacteroides succinogenes* and its importance in the rumen environment. *Appl. Environ. Microbiol.*, **42**, 886–96.

Howarth, R. E., Cheng, K.-J., Majak, W. & Costerton, J. W. (1986). Ruminant bloat. In *Control of Digestion and Metabolism in Ruminants*, ed. L. P. Milligan, W. L. Grovum & A. Dobson. Prentice-Hall, Englewood Cliffs, New Jersey, pp. 516–27.

Hyden, S. (1961). Determination of the amount of fluid in the reticulo-rumen of the sheep and its rate of passage to the omasum. *LantbrHogsk. Annlr.*, **27**, 51–79.

Kanoe, M., Imagawa, H., Toda, M., Sato, A., Inoue, M. & Yoshimoto, Y. (1976). Bacteriology of bovine hepatic abscesses. *Jap. J. Vet. Sci.*, **38**, 263–8.

Kennedy, P. M. & Milligan, L. P. (1980). The degradation and utilization of endogenous urea in the gastrointestinal tract of ruminants: a review. *Can. J. Anim. Sci.*, **60**, 205–21.

Krishna, G., Czerkawski, J. W. & Breckenridge, G. (1986). Fermentation of various preparations of spent hops (*Humulus lupulus* L.) using the rumen simulation technique (Rusitec). *Agric. Wastes*, **17**, 99–117.

Kudo, H., Cheng, K. J. & Costerton, J. W. (1986a). Interactions between *Treponema bryantii* and cellulolytic bacteria in the *in vitro* degradation of straw cellulose. *Can. J. Microbiol.*, **33**, 244–8.

Kudo, H., Cheng, K.-J. & Costerton, J. W. (1986b). Electron microscopic study of the methylcellulose-mediated detachment of cellulolytic rumen bacteria from cellulose fibres. *Can. J. Microbiol.*, **33**, 267–72.

Majak, W. & Cheng, K.-J. (1983). Recent studies on ruminal metabolism of 3-nitropropanol in cattle. *Toxicol. Suppl.*, **3**, 265–8.

Mehra, U. R., Singh, U. B., Czerkawski, J. W. & Breckenridge, G. (1981). Convenient and inexpensive method for incubating solid feeds with rumen contents. *Lab. Practice*, **30**, 879–81.

Merry, R. J. & McAllan, A. B. (1983). A comparison of the chemical composition of mixed bacteria harvested from the liquid and solid fractions of rumen digesta. *Br. J. Nutr.*, **50**, 701–9.

Orpin, C. G. (1984). The role of ciliate protozoa and fungi in rumen digestion of plant cell walls. *Anim. Feed Sci. Technol.*, **10**, 121–43.

Schwartz, H. M. & Gilchrist, F. M. C. (1975). Microbial interactions with the diet and the host animal. In *Digestion and Metabolism in the Ruminant*, ed. I. W. McDonald & A. C. I. Warner. University of New England Publishing Unit, Armidale, Australia, pp. 165–79.

Senshu, T., Makamura, K., Sawa, A., Miura, H. & Matsumoto, T. (1980). Inoculum for *in vitro* rumen fermentation and composition of volatile fatty acids. *J. Dairy Sci.*, **63**, 305–12.

Stanton, T. B. & Canale-Parola, E. (1980). *Treponema bryantii* sp. nov., a rumen spirochete that interacts with cellulolytic bacteria. *Arch. Microbiol.*, **127**, 145–56.

Stewart, C. S., Dinsdale, D., Cheng, K.-J. & Paniagua, C. (1979). The digestion of straw in the rumen. In *Straw Decay and its Effect on Disposal and Utilization*, ed. E. Grossbard. Wiley, Chichester, pp. 123–30.

Van Nevel, C. J. & Demeyer, D. I. (1977). Determination of rumen microbial growth *in vitro* from ^{32}P-labelled phosphate incorporation. *Br. J. Nutr.*, **38**, 101–14.

Van Soest, J. (1975). Physico-chemical aspects of fibre digestion. In *Digestion and Metabolism in Ruminants*, ed. I. W. McDonald & A. C. I. Warner. University of New England Publishing Unit, Armidale, Australia, pp. 351–65.

Wallace, R. J. & Munro, C. A. (1986). Influence of the rumen anaerobic fungus *Neocallimastix frontalis* on the proteolytic activity of a defined mixture of rumen bacteria growing on a solid substrate. *Lett. Appl. Microbiol.*, **3**, 23–6.

Wallace, R. J., Cheng, K.-J., Dinsdale, D. & Ørskov, E. R. (1979). An independent microbial flora of the epithelium and its role in the ecomicrobiology of the rumen. *Nature* (London), **279**, 424–6.

Williams, A. G. & Strachan, N. H. (1984). Polysaccharide degrading enzymes in microbial populations from the liquid and solid fractions of bovine rumen digesta. *Can. J. Anim. Sci.*, **64**(Suppl.), 58–9.

Wood, T. M., Wilson, C. A., McCrae, S. I. & Joblin, K. N. (1986). A highly active extracellular cellulase from the anaerobic rumen fungus *Neocallimastix frontalis*. *FEMS Microbiol. Lett.*, **34**, 37–40.

13

Manipulation of Rumen Fermentation

C. J. Van Nevel[a] & D. I. Demeyer[a,b]

[a] *Research Centre for Nutrition, Animal Production and Meat Science, State University of Ghent, Melle, Belgium*

[b] *Institute of Biotechnology, Vrije Universiteit Brussel, Brussels, Belgium*

The previous chapters have considered the various reactions in the rumen, and it has been shown that during fermentation feed components, empirically defined as structural and non-structural carbohydrates, and protein and non-protein nitrogen compounds, are degraded to certain extents by the microbial population. The fermentation end-products formed consist almost exclusively of volatile fatty acids (VFA), methane, carbon dioxide and ammonia. Energy (ATP) is generated during oxidation, mainly by substrate-linked phosphorylation, but anaerobic electron or proton transport systems also operate. Apart from kinetic and osmotic energy needs, generated energy is mainly used in the synthesis of cellular constituents (anabolism), for maintenance purposes of individual cells and populations, and for the increase of biomass, overall referred to as 'microbial growth'. Originally it was thought that the amount of microbial mass formed was primarily dependent on the amount of energy generated during substrate degradation (catabolism), but it has recently been argued that, in the rumen, processes of turnover (predation) and coupling may be more important than energy yield (Thauer & Kröger, 1984; Demeyer & Van Nevel, 1986b; Owens & Goetsch, 1986) (see also Chapter 6).

The rumen fermentation may be characterized by the following parameters: the total amount and relative proportions of VFA formed; the amount of methane formed and organic matter (OM) fermented; the amount of microbial matter synthesized; and the efficiency of the latter process which is most often expressed as g of N incorporated into microbial matter per kg of OM fermented in the rumen ($gN_i/kg\ OM_f$) (Demeyer & Van Nevel, 1986b). Such a description of 'rumen fermentation' is simplifying and ignores important but complex aspects such as recycling of microbial biomass and compartmentation within the rumen (see Chapters 6 and 12). Such aspects have been incorporated in earlier work, however (Demeyer & Van Nevel, 1986a; Demeyer *et al.*, 1986), and this chapter will be based on the simplified overall scheme shown in Fig. 1.

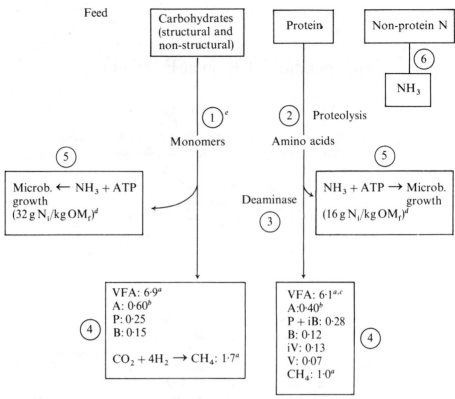

[a] Millimoles per gram monomer (VFA, volatile fatty acids).
[b] Relative proportion of acetic (A), propionic (P), isobutyric (iB), butyric (B), isovaleric (iV) and valeric (V) acid.
[c] 2-Methylbutyric acid excluded.
[d] Grams of nitrogen incorporated per kg of organic matter fermented in the rumen.
[e] ○, sites of modification.

Fig. 1. Important reactions in the rumen and possible sites for modification; data shown are average values derived from Demeyer & Van Nevel (1979).

THE AIM IN MANIPULATING RUMEN FERMENTATION

Considerable research efforts have been devoted to the manipulation of rumen metabolism with the final aim of improving ruminant productivity, and several reviews are available (Demeyer & Van Nevel, 1975; Ørskov, 1975; Chalupa, 1977; Thomas & Rook, 1977; Prins, 1978; Chalupa, 1984; Owens et al., 1984; Demeyer & Van Nevel, 1986a; Demeyer et al., 1986; Thivend & Jouany, 1986). The conviction that modification is sometimes justified is based on the following considerations. Maximal ruminant production using feeds rich in lignocellulosic substances can be achieved only when general fermentation and microbial protein synthesis in the rumen are maximal, provided that most end-products can be used by the animal. Reasons for selective modification of the fermentation pattern can be manifold: whether the efficiency of utilization of the different VFA for growth and fattening may be different or not, a low methane–high propionate pattern in the rumen indicates a fermentation with higher efficiency in terms of gross energy retained in non-gaseous end-products (Ørskov, 1975). However, this pattern is directly related to the low-fat milk syndrome in the dairy cow (Davis & Brown, 1970). Metabolic disorders requiring modification of rumen fermentation are ketosis and acidosis (Kronfeld, 1970; Dirksen, 1970). Manipulation of rumen fermentation can be considered as an optimization process, whereby optimal conditions are sought by maximization and/or minimization of fermentation processes, depending on factors such as kind and level of feeding and animal production. An example of maximization, which seems valid in all circumstances, is rumen microbial protein production and crude-fibre degradation. Microbial protein synthesis can be stimulated by increasing the efficiency of the process or by increasing the amount of OM fermented in the rumen. Processes that should be minimized are methane production, degradation of the true protein fraction of the feed, biohydrogenation of unsaturated fatty acids and, to a certain extent, fermentation of starch.

THE RUMEN AS AN INTEGRATED SYSTEM

It should be mentioned that the considerations used to rationalize manipulation of rumen fermentation are, traditionally, of an analytical character; a reaction within the overall process is considered separately for modification. Over the years, however, it has become apparent that all reactions within the rumen have to be considered as integrated parts of the interactive network of reactions that make up rumen fermentation and, eventually, ruminant metabolism. Because of its integrated character, any intervention within this system will result in a series of interrelated effects. The nature and importance of such interactions have been discussed elsewhere (Demeyer & Van Nevel, 1986a; Demeyer et al., 1986; Jouany et al., 1986) and have been considered in previous chapters. In the present review the subject of manipulation will be mainly discussed by following the analytical approach which has been used during years of experimentation. More detailed descriptions of the reactions involved are given in other chapters; what is considered here is the inducement of changes in rate or extent of these reactions.

MODIFICATION OF THE RUMEN FERMENTATION PATTERN

The most important sites for possible modification of rumen metabolism are indicated in Fig. 1. A first possibility to alter rumen fermentation is an intervention at the dietary level, either for increasing the digestibility of the structural carbohydrates (crude fibre, site 1) or for protecting dietary protein against microbial degradation in the rumen (sites 2 and 3). Chemical (NaOH, KOH, Ca(OH)$_2$, NH$_3$, ozone) or physical (grinding, high-pressure steam) treatments can increase the degradability of crude fibre in the rumen (Capper et al., 1977; Klopfenstein, 1978; Sundstøl et al., 1979; Hart et al., 1981; Shefet & Ben-Ghedalia, 1982; Rangnekar et al., 1982). Sodium hydroxide treatment of grain seemed to be promising (Ørskov, 1979). These treatments can also alter the proportions of VFA formed in the rumen, as they increase fermentation rate (Demeyer, 1981). Feed protein can be protected against microbial attack by treatment with aldehydes or tannins, and by use of protease and/or deaminase inhibitors (Chalupa, 1975; Ferguson, 1975; Brock et al., 1982; Kopecny & Wallace, 1982; Wallace, 1983).

Another intervention is more directly on the microbial level (site 4), where fermentation pattern is altered through action of certain compounds (micro- or macro-quantities in the ration) on the microbes. Net microbial growth (site 5) can also be modified by numerous factors, such as feed additives, residence time of the microbes in the rumen, elimination of protozoa, addition of growth factors and prevention of a lactate-type fermentation (Harrison & McAllan, 1980; Owens & Goetsch, 1986; Demeyer & Van Nevel, 1986b; Vérité et al., 1986). Decreasing the rate of ammonia release from non-protein nitrogen (NPN) sources (site 6) prevents ammonia toxicity and, at the same time, a better coupling between carbohydrate fermentation (ATP production) and ammonia utilization in microbial protein synthesis is obtained (Johnson, 1976; Mizwicki et al., 1980). Such a timed release is observed when NPN substances other than urea are used, and some examples are glycosyl ureides (Merry et al., 1982a,b), isobutylidene diurea (Teller et al., 1978) and formaldehyde-treated urea (Pal & Negi, 1977). For the same purpose, urease has been inhibited by acetohydroxamic acid or phosphoric phenyl ester diamide (Jones, 1968; Makkar et al., 1981; Voigt et al., 1981). A modification of rumen fermentation not shown in Fig. 1 can be done by altering the kinetics of rumen digesta flow, e.g. increasing fractional outflow rate of rumen liquid by infusion of mineral salts (Harrison & McAllan, 1980).

Changes in rumen fermentation caused by altering the composition of the ration and feeding method, or processing ration components, will not be discussed here as they are described in other reviews (Ørskov, 1975; Thomas & Rook, 1977; Sutton, 1980; Thomson & Beever, 1980). Also, possibilities of modification involving genetic manipulation of rumen microorganisms have been discussed in Chapter 10. In this chapter, emphasis will be laid on modification of rumen metabolism by the use of compounds other than major dietary components (additives). Table 1, slightly adapted from Chalupa (1984), shows that such compounds, known as 'rumen manipulators', are numerous. For the reasons discussed above, most of the substances are not very specific and alter simultaneously different sites of rumen fermentation. This renders a strict classification of the compounds based on their site(s) of action impossible. Therefore we decided to use a classification based on the originally chosen action of the compounds.

It should be realized, however, that rumen manipulations should ideally be evaluated in a holistic manner, within the framework of more complex models of rumen metabolism (Demeyer & Van Nevel, 1986b; Demeyer et al., 1986).

Table 1
Compounds Used for Modification of Rumen Fermentation and Their Known Sites of Action (adapted from Chalupa, 1984)

Additive	Site of action[a]
Ionophores	
Monensin	1 2 3 4 5 6 7 8 9 10 11
Lasalocid	1 2 3 4 5 6 7 9 11
Salinomycin	4 5 7 9 11
ICI 139603	2 4
Laidlomycin	4 6
Narasin	11
Polyether A	8
Abiérixin	2 8
Lysocellin	11
Nigericin	11
Lenoremycin	11
X-206	11
Gramicidin	5
Other antibiotics	
Avoparcin	2 3 4 5
Aridicin	2 4 5
Actaplanin	
Tylosin	1 4 6
Chlortetracycline	1 2 7
Penicillin	1 2
Erythromycin	2 6
Chloramphenicol	1 2
Bacitracin	1 6
Capreomycin	4 6
Novobiocin	6
Oxamycin	4 6
Hygromycin B	6
Oxytetracycline	1
Dihydrostreptomycin	1
Kanamycin	1
Thiopeptin	2 4 6
Sulphomycin	4 6
Siomycin	4 6
Taitomycin	4 6
Flavomycin	1 4 5
Virginiamycin	2 3 4 5 6 8 11
Methane inhibitors	
Cl- and Br-methane analogues	3 4 5 8 11
Long-chain fatty acids and derivatives	1 4 5 7 8 9 10
Pyromellitic diimide	4 5
Benzo[1,3]dioxins	4 5
Sulphate-sulphite	4 5
Nitrate-nitrite	5
Bromoethanesulphonic acid	4 5

contd.

Table 1—contd.

Additive	Site of action[a]
Protease/deaminase inhibitors	
Diaryliodonium compounds	3 4 5
p-Chloromercuribenzoate	2
Phenylmethylsulphonyl fluoride	2
Dithiothreitol	2
Pepstatin A	2
N-Tosyl-1-lysine chloromethyl ketone	2
N-Tosylphenylalanine chloromethyl ketone	2
EDTA	2
Iodoacetate	2
Merthiolate	2
Soybean trypsin inh. I and II s	2
Antipain	2
Chymostatin	2
Leupeptin	2
Elastatinal	
Phosphoramidon	2
Cr^{2+}, Sn^{2+}, Ti^{3+}, Zn^{2+}, Ag^{+}, Cd^{2+}, Cu^{2+}, Hg^{2+}	2
Growth factors	
Iso C_4–C_5 vol. fatty acids	8
Niacin	8
Thiamin	4 8
Phenylpropanoic acid	1
Miscellaneous compounds	
Alkylamines	1 4
Captan	1 4
Orotic acid	4
Sarsaponin	1 2 7 9 10
Mineral salts (buffers)	
$NaHCO_3$, Na_2CO_3, MgO	4 8 9
$CaCO_3$, alkaten	
Defaunating agent	
(eliminating rumen protozoa)	1 2 4 5 7 8 9 10
Copper sulphate	
Dioctyl sodium sulphosuccinate (Manoxol OT)	
Alcohol ethoxylate (Teric GN 9)	
Calcium peroxide (Ixper 80 C)	

[a] (1) Crude fibre degradation. (2) Proteolysis. (3) Deaminase. (4) Volatile fatty acid production and relative proportions. (5) Methane production. (6) Lactate production. (7) Protozoal numbers. (8) Microbial growth and its efficiency. (9) Kinetics of digesta in the rumen. (10) Rumen NH_3 level. (11) Conversion of L-tryptophan into 3-methylindole.

MODIFICATION OF RUMEN FERMENTATION BY MICRO-COMPONENTS OF THE DIET

Methane inhibitors

Depending on the type of diet, up to 10% of the gross energy in the diet can be lost through methane formation in the rumen (Czerkawski, 1969). It was soon found that methanogenesis is a process which was easily inhibited by many chemicals, while a simultaneous increase in propionic acid and, in some cases, butyric acid occurred because of the inverse relationship between formation of the gaseous and acid products (Wolin, 1960; Marty & Demeyer, 1973; Van Nevel et al., 1974; Demeyer & Van Nevel, 1975). Examples of compounds toxic for methane bacteria are halogenated (Cl, Br) methane analogues, sulphite, nitrate, unsaturated long-chain fatty acids, trichloroethyl-pivalate and -adipate (Czerkawski & Breckenridge, 1975: Demeyer & Van Nevel, 1975; Van Nevel & Demeyer, 1976). Some of these compounds have been tested *in vivo* and, occasionally, favourable effects on performance of growing ruminants have been found (Van Nevel & Demeyer, 1976; Chalupa, 1984). From all these results, it has become clear that for evaluation of a methane-inhibiting compound, thought to be a very specific agent, a complex number of factors must be taken into account, for example: rate of fermentation in the rumen, especially crude-fibre degradation; extent of hydrogen gas accumulation; effect on protozoa and on microbial protein-synthesis and feed-protein degradation; effect on food intake and a possible adaptation of the microbes to the compound (Trei et al., 1971; Clapperton, 1977). *In vitro*, amichloral (a hemiacetal of chloral and starch), a very potent methane inhibitor, increased fermentation efficiency by 6–8% while amino acid degradation was seriously inhibited (50% of control incubation) (Chalupa et al., 1980), thus demonstrating interaction. Inhibition of methanogenesis with accumulation of hydrogen gas interferes with interspecies hydrogen transfer and alters the NADH/NAD ratio in the microbes. This mechanism can, at least partly, be responsible for the inhibited degradation of reduced amino acids, and more especially branched-chain amino acids, after treatment of the rumen contents with methane inhibitors (Russell & Martin, 1984; Hino & Russell, 1985). In their experiment *in vivo*, Mathers & Miller (1982) confirmed that chloral hydrate (1–4 g/day/sheep) inhibited methane production with no effect on the non-ammonia N (NAN) flow to the duodenum. It is possible that an eventual increase in feed protein escaping rumen fermentation was counterbalanced by a decrease in microbial protein flow, with no overall change as final result. Derivatives of 2,4-bis(trichloromethyl)benzo[1,3]dioxin-6-carboxylic acid were reported to be potent methane inhibitors, remaining persistent throughout a cattle-growing experiment, and favourable effects on animal performance were shown (Davies et al., 1982). We have compared some recent data concerning the effect of methane inhibitors on animal performance, and positive effects were related to the use of high-roughage rations which normally give rise to large quantities of methane and low proportions of propionate in the rumen, while growth rates of the animals are low (Demeyer & Van Nevel, 1986*a*).

A very specific and potent methane inhibitor is 2-bromoethanesulphonic acid (BES), a Br substitute of coenzyme F involved in methyl group transfer during methanogenesis (Wolfe, 1982; Martin & Macy, 1985). Pyromellitic diimide was even more potent, but was rapidly inactivated (degraded) in the rumen when administered *in vivo* or to semi-continuous incubations (Martin & Macy, 1985).

The mechanisms of some methane inhibitors were described earlier (Demeyer & Van Nevel, 1975; Van Nevel & Demeyer, 1976). It is striking that, although in most cases methane inhibition was accompanied by an increased propionic acid proportion, an increase in butyric acid proportion was found in a few cases only (Van Nevel *et al.*, 1969, 1970).

Rumen propionate production is solely of bacterial origin, and the negative relationship between methanogenesis and propionate production has been clearly established from excellent work on bacterial interspecies hydrogen transfer (Wolin, 1975) (see also Chapter 11). Butyrate production is associated with bacterial and protozoal metabolism. However, there is little evidence that hydrogen gas or interspecies hydrogen transfer influences the formation of butyrate in the rumen (Van Nevel *et al.*, 1974; Hino, 1983), although methanogens are physically associated with protozoa, and inhibitors of methanogenesis might be expected to inhibit protozoal metabolism and thus butyrate production (Demeyer & Vervaeke, 1985). From these considerations it is clear, though, that distinction between methane inhibitors and so-called propionate enhancers is artificial; methane inhibitors are normally propionate enhancers and vice versa.

Propionate enhancers

Ionophore antibiotics
  Monensin. Monensin (trade name Rumensin) is a carboxylic polyether antibiotic, produced by *Streptomyces cinnamonensis*, with moderate activity *in vitro* against Gram-positive bacteria and certain mycobacteria (Haney & Hoehn, 1967). It is known as an 'ionophore' antibiotic, and acts as a carrier for monovalent cations for transport through lipid barriers in which it shows a high affinity for sodium (Pressman & de Guzman, 1975; Durand, 1982; Bergen & Bates, 1984). It also has anticoccidial properties. Since 1976, numerous papers discussing the effects of monensin on rumen fermentation and on cattle and sheep performance have been published. Therefore we have divided the literature into type papers, and some of each type will be discussed. Monensin has improved performance of beef cattle and lambs on all kinds of diets, from high concentrates (Raun *et al.*, 1976) to cage-layer manure (Vijichulata *et al.*, 1980) and wilted or unwilted grass silage (Steen, 1984). Better performance was due to a combination of a better weight gain and a decreased feed intake, with an improved feed efficiency as the result (Goodrich *et al.*, 1984). With cattle on pasture, higher daily weight gains were observed (Goodrich *et al.*, 1984). In order to explain the improved feed efficiency, the effect of the drug on rumen metabolism has been investigated very extensively.

(a) Influence of monensin on rumen carbohydrate metabolism. Evidence that monensin increases the molar proportion of propionate in the rumen, both *in vitro* and *in vivo*, is overwhelming (Durand, 1982; Schelling, 1984). In the experiments this was the result of an increased propionate production (Prange *et al.*, 1978; Van Maanen *et al.*, 1978; Rogers & Davis, 1982a; Armentano & Young, 1983; Shell *et al.*, 1983). A lowered acetate and/or butyrate percentage was observed, whereas the concentration of total volatile fatty acids (VFA) remained unchanged or was slightly decreased. In incubations *in vitro* with soluble sugars or ground concentrates, VFA production was stimulated in the presence of low monensin concentrations (Richardson *et al.*, 1976; Van Nevel & Demeyer, 1977a). It was accepted that an increase in propionate production in the rumen resulted in a protein-sparing effect, as in these circumstances the animal used less amino acid for gluconeogenesis (Schelling, 1984). As previously mentioned, higher propionate proportions are normally coupled with a decreased methanogenesis, as formation of these end-products is inversely related (Van Nevel *et al.*, 1974). Papers are quite unanimous: monensin decreased methane production in batch or continuous incubations (Dinius *et al.*, 1976; Van Nevel & Demeyer, 1977a; Bartley *et al.*, 1979; Slyter, 1979; Chalupa *et al.*, 1980; Fuller & Johnson, 1981; Wallace *et al.*, 1981; Martin & Macy, 1985).

Partial adaptation of methanogenesis has been found (Dinius *et al.*, 1976), but a six-month period was necessary before methanogens in anaerobic digesters fed on manure from monensin-treated cattle became adapted (Varel & Hashimoto, 1982). Inhibition of methane production has been confirmed *in vivo*, and was responsible for one-third of the improved energy utilization of monensin-fed animals (Thornton & Owens, 1981; Wedegaertner & Johnson, 1983). Adaptation of methanogenesis *in vivo* was also found (after 12 days), and Na or K concentrations in the ration influenced the final effect of monensin on methane production (Rumpler *et al.*, 1986). A combined use of a typical methane inhibitor, amichloral, with monensin showed, on some occasions, an additive effect both *in vitro* and *in vivo* (Brethour & Chalupa, 1977; Bartley *et al.*, 1979; Chalupa *et al.*, 1980), but *in vitro* results from our laboratory have indicated that combination of both additives can also result in a purely synergistic effect (Table 2). The lack of a response when monensin alone was added to the incubations suggests that formate was not an important intermediate during carbohydrate fermentation (see later).

The effect of monensin on crude-fibre degradation in the rumen or the total digestive tract is not very clear. *In vitro* results are conflicting; e.g. Wallace *et al.* (1981) found a 12% decrease in cellulose and hemicellulose digestion in their Rusitec (see Chapter 12) experiment, which was most emphasized at the initial rate of fibre degradation. A moderate inhibition of cellulolysis *in vitro* was also obtained by Baldwin *et al.* (1982), using rumen fluid from non-adapted animals. In other experiments *in vitro*, no effect on degradation of cotton fibre strips was seen (Dinius *et al.*, 1976; Slyter, 1979). Incubation in a nylon bag suspended in the rumen of cotton fibre or Solka-floc cellulose ('*in sacco*') did not show an effect of the ionophore (Dinius *et al.*, 1976; Lemenager *et al.*, 1978; Ricke *et al.*, 1984; Faulkner *et al.*, 1985). However, incubation of these substances is strictly related to cellulolytic activity in the rumen and gives no information on overall crude-fibre degradability, which is the resultant of degradation and retention time in the rumen of plant cell walls rather than pure cellulose. In Table 3, data on rumen or total digestive-tract

Table 2
Influence of Chloral Hydrate and/or Monensin on Fermentation Pattern and Total Microbial Growth in Incubations with Rumen Fluid[a] (unpublished results)

Additive[b]	Fermentation pattern[c]					2H recov.[d] (%)	Hexose ferm. (μmol)	Microbial growth[e] ($\mu g\, P_{inc}$)
	A	P	B	M	H_2			
None	119±1[f]	43±3	20±1	45±2	0·1±0	85±3	1 141±115	95
CH	88±2	55±4	28±1	11±3	23·1±6·2	68±4	1 121±47	52
M	120±1	42±1	19±1	42±2	0·1±0·03	82±2	1 353±119	36
CHM	88±6	68±1	21±3	6±1	33·3±4·4	72±1	1 117±124	50

[a] 40 ml of rumen fluid were incubated with 250 μmol each of cellobiose and maltose and 10 mg of N as NH_4HCO_3 for 2 hours under CO_2.
[b] CH, chloral hydrate (0·2 mM); M, monensin (25 ppm); CHM, chloral hydrate + monensin.
[c] A, P, B, M, acetic, propionic and butyric acid and methane (μmol per 100 μmol hexose fermented).
[d] 2H recov., metabolic hydrogen recovery.
[e] P incorporation as measure of total microbial growth (one incubation in duplicate).
[f] Mean value ± standard error (3 replicates).

digestibility of crude-fibre and starch are presented. The variability of the final effect is probably explained by the use of adapted or non-adapted animals, as longer adaptation times seemed to decrease the negative effect on crude-fibre degradation in the rumen or total tract, as clearly shown by Poos et al. (1979). Adaptation was also demonstrated by de Jong & Berschauer (1983). Total-tract digestibility of starch was not influenced by monensin.

(b) Influence of monensin on nitrogen metabolism in the rumen. The 3–6% increase in gross energy retained in VFA in the rumen cannot account for the total improvement in feed efficiency observed using monensin (Raun et al., 1976). Therefore the effect of monensin on nitrogen metabolism has been investigated. In incubations in vitro with casein as the sole substrate, and using inocula from unadapted or monensin-adapted animals, protein degradation was significantly lowered after monensin addition (Van Nevel & Demeyer, 1977a; Russell et al., 1981; Whetstone et al., 1981). Deaminase activity was inhibited in incubations with amino acids (Chalupa et al., 1980), and in casein incubations deaminase or transaminase activity was more inhibited than proteolysis. Hino & Russell (1985) showed that monensin, while decreasing methane formation, interfered with interspecies hydrogen transfer and increased the intracellular NADH/NAD ratio, while decreasing the deamination of reduced amino acids. Also, in continuous fermentation systems, more feed protein escaped degradation after monensin treatment (Fuller & Johnson, 1981; Wallace et al., 1981). The effect on microbial growth efficiency in vitro is shown in Table 4, and clear decreases are shown in non-adapted situations. Experiments in vivo investigating the influence on N transactions in the rumen are rather limited and results are presented in Table 5. It is obvious that monensin did not increase NAN (non-ammonia N) flow to the duodenum, and in one experiment a decrease was seen. In certain cases a lowered microbial N flow and sporadic increases in feed protein N were observed, but the

Table 3
Effect of Monensin on Digestibility (dig.)[a] in the Rumen or the Total Digestive Tract of Sheep and Cattle[b]

Reference	Monensin	Adapt. period[c] (days)	Feed component[d]	Rumen dig. (%) Control	Rumen dig. (%) Monensin	Total tract dig. (%) Control	Total tract dig. (%) Monensin
Cottyn et al. (1983) (sheep)	31 ppm	12	OM			79·3	76·2
			NDF			68·9	60·8[e]
			NFE			82·5	80·5
Rogers & Davis (1982[a]) (steers)	33 ppm	16–20	OM			72·6	74·1
			ADF			38·1	39·3
			Starch			93·0	95·5
Horton & Nicholson (1980) (steers)	33 ppm	21	OM			70·3	74·2
			ADF			41·6	48·9[e]
Wedegaertner & Johnson (1983) (steers)	3 mg/kg $BW^{0·75}$	21	DM			73·3	75·9
			NDF			50·5	57·5[e]
			Starch			89·3	90·3
Moore et al. (1980) (sheep)	133 mg/day	20	DM	49·3	54·8		
			ADF	50·0	50·0		
Faulkner et al. (1985) (steers)	36·6 ppm	60	DM			64·5	64·9
			NDF	37·1	33·2	49·8	51·2
Poos et al. (1979) (sheep)	38 ppm	11–17	DM			68·4	60·7[e]
			ADF			61·9	52·7[e]
		40–46	DM			65·5	64·5
			ADF			54·9	52·5
Muntifering et al. (1981) (steers)	33 ppm	31	OM	49·8	39·3[e]		

[a] App. digestibility. [b] Some representative experiments are given. [c] Period allowed for adaptation of flora to monensin feeding. [d] OM, organic matter; NDF, neutral detergent fibre; NFE, N-free extractives; ADF, acid detergent fibre; DM, dry matter. [e] Significantly different from control.

Table 4
Effect of Monensin on Efficiency of Microbial Growth in Rumen Contents *in vitro*

Reference	Inoculum donor	Adaptation period	Monensin dose (ppm)	Microbial growth efficiency	Expressed as
Van Nevel & Demeyer (1977a)	Sheep	Unadapted	0	0·211	mg of N per 100 μmol of hexose fermented
			0·5	0·046[b]	
			0	0·262	
			1·0	0·185	
			0	0·279	
			5·0	0·134[b]	
			25·0	0·052[b]	
Bartley et al. (1979)	Steer	Unadapted	0	16·3[c]	mg of protein per g of substrate fermented
			22	10·5	
			44	11·9	
			88	6·5	
			176	0·7	
Wallace et al. (1981) (Rusitec)	Steer	Partially adapted (9–41 days)	0[a]	0·134	g of DM per g of DM digested
			2	0·137	
			10	0·154	
			50	0·160	
Whetstone et al. (1981) (semicont. culture)	Steer	30 days	0	18·9	mg of DM per 100 mg of substrate degraded
			1	16·6	
			4	19·3	

[a] Milligrams of monensin per day. [b] Statistically significant from control. [c] No statistical analysis.

differences were far from spectacular, while lower concentrations of ammonia in the rumen were frequently noted (Van Nevel & Demeyer, 1977a; Hanson & Klopfenstein, 1979; Poos et al., 1979; Muntifering et al., 1981; Whetstone et al., 1981; Willis & Boling, 1982; Perry et al., 1983; Lebzien et al., 1986). When the animal was not adapted, monensin decreased microbial growth efficiency in the rumen, and it seemed that c. 20 days were needed for such adaptation. The latter was also demonstrated by Mackie et al. (1984) for Gram-positive amd Gram-negative rumen bacteria. A lower microbial N flow to the lower digestive tract can thus be the result of lower microbial growth efficiencies, but decreased organic matter (more specifically crude fibre) degradation in the rumen will lead to the same effect. So the speculation that monensin has a second protein-sparing effect by increasing the amount of protein entering the duodenum has not been confirmed. A positive effect remains possible, if the undegraded feed protein presented to the intestinal tract for further digestion contains more essential amino acids than microbial protein, but experimental data do not support this speculation (Poos et al., 1979; Muntifering et al., 1981; Moore et al., 1980; Faulkner et al., 1985). Nevertheless, the theory of an unknown protein-sparing effect was supported by the fact that a positive influence of monensin was most noticeable when the protein percentage of the diet was low (Hanson & Klopfenstein, 1979; Perry et al., 1983). The existence of such an effect is perhaps related to diet composition and level of intake.

It was suggested that monensin might decrease urease activity in the rumen (Van Nevel & Demeyer, 1977a; Bartley et al., 1979). This has indeed been confirmed, and the effect is due either to the drug inhibiting Ni transport in ureolytic bacteria or to a selection against ureolytic species (Starnes et al., 1984).

From the foregoing, it could be predicted that use of monensin with low-quality rations, high in roughage and urea content, is rather precarious. Experiments have confirmed this assumption (Oltjen et al., 1977; Coombe et al., 1979). Even in these experiments, the ionophore caused the well known effect on VFA pattern in the rumen, but the concentration of total VFA was reduced, indicating that the overall negative effect on animal performance was due to a decreased organic matter digestion rather than to an inhibited ureolysis or microbial growth efficiency.

Monensin was found to decrease 3-methylindole formation from L-tryptophan, thereby preventing acute bovine pulmonary oedema and emphysema, which is sometimes fatal for the animal (Hammond & Carlson, 1980).

(c) Effect of monensin on rumen microbes. The shift in fermentation pattern to propionate is in overall agreement with the effect of monensin on pure cultures of several rumen bacteria: succinate and propionate producers were not affected by $10\,\mu g/ml$ monensin, while acetate-producing organisms were inhibited to a certain extent (Chen & Wolin, 1979; Henderson et al., 1981). However, Dawson & Bolin (1983) suggested that monensin resistance is not reflected *per se* in altered fermentation patterns. Some pure strains of methane bacteria were also inhibited, and this was possibly related to the inhibited Ni transport observed with *Methanobacterium bryantii* (Chen & Wolin, 1979; Henderson et al., 1981; Jarrell & Sprott, 1982). With pure cultures of *Methanobacterium formicicum*, methanogenesis was lowered when formate was the substrate and not when CO_2 plus H_2 was used

Table 5
Effect of Monensin on N Flow in the Abomasum or Duodenum, and Microbial Growth Efficiency in the Rumen

Animal	Monensin	Adaptation (days)	N flow (g/day) NAN[a]	N flow (g/day) Microbial N	Feed N[b]	Microbial growth efficiency	Reference
Steers	0		75·1	44·8	30·3	16·1[e]	Muntifering et al. (1981)
	33 ppm	31	73·3	38·5[d]	34·8	14·7	
Steers	0		140·3	69·5	70·8	19·1[f,h]	Poos et al. (1979)
(BDGsuppl.)[c]	200 mg/day	10	144·4	47·4[d]	97·0[d]	14·4[d]	
Urea	0		123·7	84·2	39·5	23·3[f]	Poos et al. (1979)
suppl.	200 mg/day	10	117·4	56·2[d]	61·2[d]	14·1[d]	
Sheep	0		79·9	55·0	12·4	22·7[f]	Moore et al. (1980)
(CBG)[c]	133 mg/day	20	90·3	48·7	29·1[d]	18·9	
(F-CBG)[c]	0		106·3	52·6	41·2	16·0[f]	Moore et al. (1980)
	133 mg/day	20	104·2	48·0	43·7	20·6	
Cows	0		159·1	135·8	6·0		Lebzien et al. (1986)
(8·2 kg DM/day)[c]	33 ppm	21	135·7[d]	108·7[d]	10·1	1·47 (0)[g]	
(10·0 kg DM/day)	0		206·7	154·7	29·8	1·12[d] (33 ppm)	
	33 ppm	21	152·1[d]	114·6[d]	18·4		
Steers	0		67·4	44·9	22·5	—	Faulkner et al. (1985)
	18·3 ppm	60	71·7	44·9	26·8	—	
	36·6 ppm		67·9	36·3	31·6		
Steers	0		104·0	58·7	44·8	16·8[e]	Darden et al. (1985)
	27·5 ppm	11	102·0	51·8	49·8	13·8	

[a] NAN, non-NH$_3$N. [b] In most cases not corrected for endogenous N. [c] BDG, brewer's dried grains; CBG, coastal bermuda grass; F, formol treated. [d] Significantly different from control. [e] g N$_i$/kg OM digested. [f] g N$_i$/kg OM fermented. [g] g N$_i$/MJME. [h] Calculated from mean values, but probably significant.

(Dellinger & Ferry, 1984), thus confirming earlier work with incubations of mixed rumen bacteria (Van Nevel & Demeyer, 1977a). It seems that methane inhibition can be the result of a decreased hydrogen production from formate through formate lyase. Increased propionate production reflects the shift in flow of electrons from formate and methane to succinate or propionate formation. Interaction of monensin or other ionophores with the protonmotive force driven ATP synthesis of certain methanogens, with lower intracellular ATP content as a result, has been shown (Jarrell & Sprott, 1983).

Monensin also had a lowering effect on growth of lactate-producing bacteria, as well as on lactate production *in vitro*, indicating a possible therapeutic action of the drug for cases of lactic acidosis (Dennis et al., 1981a,b; Nagaraja et al., 1982) (see Chapter 14).

In their excellent review, Bergen & Bates (1984) summarized the effect of monensin on bacteria as follows: the ionophore acts on the flux of ions through membranes, with effects on dissipation of cation and proton gradients and interference with the uptake of solutes and with the primary transport systems in the cells. The organisms try to maintain primary transport by expending metabolic energy, and Gram-negative bacteria able to do electron transport phosphorylation survive better, so there is a shift to these organisms in the rumen. It is this shift which is responsible for the final effect of an ionophore on rumen metabolism.

Lower protozoal counts in the rumen have sometimes been observed after monensin treatment (Poos et al., 1979; Wallace et al., 1981), but unequivocal evidence of a toxic effect on these organisms was given by Hino (1981). The final effect on the rumen fauna is related to the concentration of the drug, which makes it difficult to compare experiments *in vivo* and *in vitro*. Differences between concentration of protozoa and actual total counts, as influenced by the volume of

Table 6
Influence of Monensin on Fractional Outflow Rates of Liquid and Solid Phases in the Rumen

Reference	Monensin	Liquid fraction outflow rate (%/h)		Solids fraction outflow rate (%/h)	
		Control	Monensin	Control	Monensin
Adams et al. (1981)	33 ppm	10.3	9.0	—	—
Rogers & Davis (1982a)	33 ppm	10.6	9.8	—	—
Ricke et al. (1984)	33 ppm	5.1	4.6	4.6	3.9
Bull et al. (1979)	170 mg/day	7.4	5.6	4.9	4.4
Lemenager et al. (1978)	200 mg/day	6.5	4.5^a	2.7	1.5^a
	20 mg/day	5.3	4.8	—	—
	100 mg/day	5.3	4.8	—	—
	200 mg/day	5.3	4.2^a	—	—
Faulkner et al. (1985)	6.1 ppm	4.8	5.2	2.7	3.2^a
	18.3 ppm	4.8	4.6	2.7	2.8^a
	36.6 ppm	4.8	5.1	2.7	2.6^a

^{a} Significantly different from control.

rumen contents, can also be involved. (Such differences have been discussed in Chapter 5.) Effect on protozoa was also the reason for a reduction of the severity of bloat (see Chapter 14) after treating animals with monensin (Katz et al., 1986).

(d) Effect of monensin on some physical parameters in the rumen. Table 6 shows the effect of monensin on fractional outflow rates of liquid and solid phases in the rumen. Slower outflow rates are shown, but differences are rarely statistically significant. The slower rates may be related to a reduced efficiency of rumination through a direct effect on the microbiota, as ruminal contractions were not influenced (Ellis et al., 1985).

Longer retention time of the feed in the rumen may compensate for an eventual lower crude fibre degrading activity, while microbial growth efficiency, fermentation pattern and feed protein degradation can also be affected (Ørskov & McDonald, 1979; Harrison & McAllan, 1980; Demeyer & Van Nevel, 1986b). Rumen osmolality was unaffected, or slightly lowered (from 295 to 280 mOsm/kg), after feeding monensin, due to decreased concentrations of Ca, Mg and K in the rumen fluid (Rogers & Davis, 1982a; Starnes et al., 1984).

Lasalocid. Lasalocid is another carboxylic ionophore antibiotic with anticoccidial properties and produced by *Streptomyces lasaliensis* (Berger et al., 1951). It showed a strong affinity for K, Na, Ca and Mg (Pressman et al., 1980). Like monensin, the drug increased feed efficiency of cattle on different diets and daily gains of steers on pasture were higher (Spears & Harvey, 1984). The influence of lasalocid on rumen fermentation has been investigated *in vitro* and *in vivo* with steers and sheep.

(a) Effect of lasalocid *in vitro*. In batch incubations with strained rumen fluid from an unadapted steer, lasalocid increased propionate production with a slight decrease in acetate and butyrate formation, without effect on total VFA quantities. Methane production and microbial protein synthesis (c. 40%) were inhibited (Bartley et al., 1979). In continuous culture the effect was dependent on the substrate fed: high-grain or high-roughage (Fuller & Johnson, 1981). Only with the high-grain substrate, was propionate production increased, but methane production was lowered (53–66%) with both diets. Bacterial growth did not appear to be influenced, indicating adaptation of the flora.

(b) Effect of lasalocid *in vivo*. A complete model experiment *in vivo* was carried out by Thivend & Jouany (1983), with sheep cannulated in the rumen, duodenum and ileum. With higher doses of lasalocid (43 or 64 ppm), concentration of total VFA in the rumen was lowered, and molar percentages of acetate and butyrate and the percentage of methane in the gas phase were decreased. Propionate proportions were higher, while the concentration of ammonia was unchanged. The amount of NAN entering the duodenum was not influenced by lasalocid, but the proportion of bacterial N decreased from 71% to 56%, due to a lower organic matter digestibility in the rumen with a concomitant increase in the proportion of N of dietary origin (Table 7). Absorption of N in the small intestine was not altered. Efficiencies of rumen bacterial growth showed small and inconsistent changes, according to Darden et al. (1985) and Zorilla-Rios et al. (1985) (Table 7).

Table 7
Effect of Lasalocid on Rumen Digestibility and N Flow at the Duodenum

Reference[a]	Lasalocid	Rumen digestibility (%)[b]						N passage (g/day)[c]						Microbial growth[d]	
		OM[b]		NDF[b]		Starch		NAN		Microbial N		Feed N			
		C[b]	L[b]	C	L	C	L	C	L	C	L	C	L	C	L
(1)	27·5 ppm	60·5	60·2	—	—	55·0	53·8	104	93	58·7	53·1	44·8	40·4	16·8	14·7
(2)	43 ppm	59·5	55·8[i]	—	—	—	—	20·7	21·4	14·2	11·2[i]	5·9	9·2[i]	21·9	19·1
	64 ppm	59·5	50·8[i]	—	—	—	—	20·7	21·8	14·2	11·9[i]	5·9	8·6[i]	21·9	22·5
(3)	21 ppm[e]	69·3	58·8	48·0	51·5	90·4	64·4[i]	—	—	—	—	—	—	—	—
	20 ppm[f]	72·7	71·5	37·3	59·9[i]	86·0	79·2	—	—	—	—	—	—	—	—
(4)	300 mg/day[g]	79·7	78·2	—	—	—	—	89·4	90·6	53·2	54·8	99·6	106·8	8·4	9·2
	300 mg/day[h]	64·6	55·3[i]	—	—	—	—	106·3	101·9	70·8	68·7	91·8	102·1[i]	15·2	20·6[i]

[a] (1) Darden et al., 1985 (steers; 66% corn, 30% alf. hay). (2) Thivend & Jouany, 1983 (sheep; 56% beet pulp, 27·6% barley and corn). (3) Funk et al., 1986 (lambs; 56% corn, 35% alf. hay). (4) Zorrilla-Rios et al., 1985 (steers; wheat pasture, immature and mature).
[b] App. digestibility; C, control; L, lasalocid; OM, organic matter; NDF, neutral detergent fibre. [c] NAN, non-ammonia N. [d] g N/kg OM truly dig.
[e] Control: 1·1% K in diet. L: 2·1% K in diet. [f] Control: 0·9% K. [g] Immature wheat pasture. [h] Mature wheat pasture. [i] Significantly different from control.

The effects on rumen fermentation parameters *in vivo* have been further confirmed by several authors (Thonney *et al.*, 1981; Guttierrez *et al.*, 1982; Paterson *et al.*, 1983; Spears & Harvey, 1984; Funk *et al.*, 1986). It is striking that, in contrast to monensin, lasalocid did not alter rumen ammonia concentration, in spite of the fact that more dietary protein escaped rumen fermentation (Thivend & Jouany, 1986). Furthermore, Ricke *et al.* (1984) found that lasalocid did not affect rumen liquid volume. Whether the drug has an effect on ammonia absorption from the rumen is not known. Adaptation of methanogens to lasalocid occurred (Rumpler *et al.*, 1986). Fractional outflow rate of the liquid phase in the rumen was unchanged or slightly lowered, but the outflow rate of the solid phase was somewhat more decreased (4·6%/h compared to 4·0%/h) (Ricke *et al.*, 1984; Estell *et al.*, 1985). Osmolality and urease activity in the rumen were also reduced (Starnes *et al.*, 1984).

(c) Effect of lasalocid on rumen microbes. The overall effect on pure strains of rumen bacteria was very similar to that brought about by monensin (Chen & Wolin, 1979). Lactate producers were inhibited; resistant organisms were succinate producers or lactate fermenters (Dennis *et al.*, 1981*a*). Lasalocid was more potent in preventing lactic acidosis than monensin or thiopeptin (Nagaraja *et al.*, 1982). Ciliate protozoa were shown to be sensitive to lasalocid, and numbers of Entodinia and *Polyplastron multivesiculatum* were severely decreased by 43 ppm in the feed (Thivend & Jouany, 1983).

Salinomycin. Salinomycin is a third polyether antibiotic. It is produced by *Streptomyces albus* (Kinashi *et al.*, 1973), and is toxic to Gram-positive bacteria.

Feeding the compound to cattle improved weight gain and feeding efficiency (Nakashima *et al.*, 1982; Hicks *et al.*, 1985; Merchen & Berger, 1985; Barclay *et al.*, 1986). Total VFA concentration and pH in the rumen were not changed, but a shift to a higher molar percentage of propionate was observed (Nakashima *et al.*, 1982; Merchen & Berger, 1985; Olumeyan *et al.*, 1986), while methane production was decreased (Webb *et al.*, 1980). Total digestive-tract digestibility of dry matter, organic matter, neutral or acid detergent fibre, or starch, was not affected (5·5–22 ppm in diet) (Merchen & Berger, 1985). It was further shown that salinomycin decreased the fractional outflow rate of the liquid phase in the rumen (by *c.* 20%) (Merchen & Berger, 1985).

After an adaptation period of 8 weeks, it was found that total anaerobic bacterial counts in the rumen were reduced, but no effect on cellulolytic or lactate-utilizing bacteria was apparent, while the proportion of amylolytic organisms was increased. Protozoal counts were initially reduced due to decreased numbers of Entodinia, but adaptation occurred after 4 weeks. Holotrich numbers remained unchanged (Olumeyan *et al.*, 1986).

Other ionophores. Other ionophores interfering with ruminant metabolism are narasin, polyether A, lysocellin, laidlomycin butyrate, ICI 139603 and abiérixin (Bartley *et al.*, 1979; Davies & Broome, 1982; Rowe *et al.*, 1983; Spires & Algeo, 1983; Gill & Owens, 1984; Barao *et al.*, 1985; Preston *et al*, 1985; Thivend & Jouany, 1986). Most of these products are still in a developmental stage and are currently being tested. Where now known, their effect on feedlot performance is comparable with the ionophores of the first generation described above. ICI 139603 and abiérixin decreased feed protein breakdown in the rumen, without significant effect on rumen microbial growth (Rowe *et al.*, 1983; Jouany *et al.*, 1986).

Ionophores: concluding remarks. Never before has an additive for ruminant feeding been more investigated than has monensin. It was soon shown that the effect on rumen VFA pattern and methane production could only partially explain the better feedlot performance of treated animals. It was also surmised that the effect on nitrogen transactions in the rumen did not cause greater amounts of NAN to flow to the duodenum for further digestion, but almost all these experiments were done with animals on low feed intakes.

Although hindgut fermentation has seemed not to be affected (Yokoyama *et al.*, 1985), it is certain that monensin has several effects beyond the rumen (Table 8); an effect on intermediary metabolism and even at the tissue level cannot be excluded. It is becoming clear that monensin is a good example of a compound inducing a series of interrelated changes in the integrated and interactive network of reactions in the rumen and ruminant metabolism. Indeed, the compound may be absorbed and returned to the gut through the entero-hepatic cycle (Owens *et al.*, 1984), and may effect tissue propionate utilization (Wahle & Livesey, 1985). Its interference with ion

Table 8
Effects of Monensin beyond the Rumen

Increased plasma urea levels	Willis & Boling (1982); Pendlum *et al.* (1980)
Increased plasma citrulline levels	Willis & Boling (1982)
Increased levels of plasma Threo, Met, Ser, Glu, glutamine, Ala, Orn, TAA, EAA and NEAA (amino acids)	Pendlum *et al.* (1980)
Increase in blood glucose level	Potter *et al.* (1976)
Reduced plasma insulin concentration	Wahle & Livesey (1985)
Increased plasma propionate concentration	Wahle & Livesey (1985)
Increased serum vitamin B_{12}	Wahle & Livesey (1985)
Increased turnover of glucose in the body	Van Maanen *et al.* (1978)
Reduced total saturated fatty acids in bovine tissue	Marmer *et al.* (1985)
Additive responses with implants	Goodrich *et al.* (1984)
Decreased cadmium content in muscle and liver of steers	Masek *et al.* (1985)
Reduced milk fat content (goats)	Brown & Hogue (1985)
Improved Se status of pregnant ewes and their lambs	Anderson *et al.* (1983)
Decreased faecal and urinary Mg excretion	Greene *et al.* (1985); Kirk *et al.* (1985)
Decreased urinary Ca excretion	Kirk *et al.* (1985)
Increased absorption and retention of P and Zn	Kirk *et al.* (1985)

transport through membranes may affect mineral retention (Starnes et al., 1984; Kirk et al., 1985), as well as quality (Vrchlabsky, 1980) and heavy-metal content (Masek et al., 1985) of meat (see also Table 8). The toxicological aspect of monensin feeding and its fate in the body and environment has been reviewed by Donoho (1984). Of special interest is the fact that no monensin was detected in edible tissues of cattle and chickens. Also, the drug seems to be biodegradable in manure and soil.

Another important point concerning the use of ionophores is their toxicity for horses. The LD_{50} for monensin was 2–3 mg/kg body weight, while for lasalocid 21·5 mg/kg was estimated (Matsuoka, 1976; Hanson et al., 1981). Also, with ruminants higher doses than recommended (33 ppm in feed, 150–200 mg/head) can lead to monensin poisoning (Potter et al., 1984).

Other antibiotics

Avoparcin and other glycopeptide antibiotics. Of the non-ionophore antibiotics, avoparcin, a glycopeptide compound produced by *Streptomyces candidus*, has been most studied, and growth-promoting activities for cattle have been reported (Johnson et al., 1979; Dyer et al., 1980; Cuthbert & Thickett, 1984; Unsworth et al., 1985). Several authors found increased molar percentages of propionate in the rumen fluid (Johnson et al., 1979; Dyer et al., 1980; Cottyn et al., 1983), and Froetschel et al. (1983) found a higher propionate production rate and pool turnover rate, although effects were only significant in sheep fed on a low-fibre diet. They also found slightly decreased rumen ammonia (50 against 45 mg/100 ml) and increased α-amino N levels. This is in line with avoparcin decreasing degradation *in vitro* of amino acids and several feed proteins, the highly soluble ones being most inhibited (Jouany & Thivend, 1986). Casein degradation *in vitro* was also inhibited by 60% (Lindsey et al., 1985). Methane production *in vitro* was 16% lower after avoparcin addition (MacGregor & Armstrong, 1983). No effect on kinetics of the rumen liquid phase was seen (Froetschel et al., 1983). The uptake of amino acids in the small intestine was increased, probably related to enhanced dipeptidase activity in the mucosa (MacGregor & Armstrong, 1984; Parker et al., 1984). Study of avoparcin using pure strains of rumen bacteria revealed toxicity for most Gram-positive organisms (Ruminococci). *Bacteroides succinogenes*, an important producer of succinate (the immediate precursor of propionate), could survive (Stewart et al., 1983; Stewart & Duncan, 1985). However, the drug had little effect on crude-fibre-degrading activities of mixed rumen bacteria *in vitro*. Avoparcin exerts its toxic effect on bacteria through inhibition of incorporation of N-acetylglucosamine into the cell wall peptidoglycan (Speth et al., 1981). Two newer glycopeptide antibiotics are actaplanin and aridicin (Barao et al., 1985; Lindsey et al., 1985). In their effects on fattening parameters and rumen metabolism they are comparable to avoparcin, aridicin inducing higher propionate production *in vitro*, with lower methane formation and casein degradation (Lindsey et al., 1985).

Miscellaneous antibiotics. Tylosin, an antibiotic acting against Gram-positive bacteria, had a positive effect on daily gain and feed efficiency of cattle, while incidence of liver abscesses was reduced (Brown et al., 1973). Purser et al. (1965) found higher percentages of propionate in the rumens of treated sheep, but no change in fermentation pattern occurred in steers (Horton & Nicholson, 1980). In vitro, tylosin was found to act as a strong inhibitor of cellulose digestion and lactate production (Beede & Farlin, 1977a; Baldwin et al., 1982).

Chlortetracycline (aureomycin) gave better growth performance in trials with lambs, with no effect on rumen pH, VFA and ammonia, but cellulolytic activity measured in vitro was greatly reduced (56%) (Purser et al., 1965; Rumsey et al., 1982). Some authors have noted increased protozoal numbers in the rumens of sheep, but no effect on total or viable counts of bacteria (Klopfenstein et al., 1964; Purser et al., 1965). The reason for this is not clear, but a decrease in rumen liquid volume after antibiotic treatment could be involved. Experiments in vitro have also shown that aureomycin, and terramycin, inhibited ammonia production from hydrolysed casein, by 28% and 49% respectively (Broderick & Balthrop, 1979).

In an early experiment by Hogan & Weston (1969), it was shown that rumen ammonia levels were reduced by penicillin, erythromycin and chloramphenicol, while no effect was found with neomycin, oxytetracycline (terramycin) and streptomycin. Large amounts of chloramphenicol (1 g/day) decreased rumen ammonia slightly, with no effect on rumen or total digestive tract digestibility of protein, organic matter and cellulose. With the aim of treating or preventing lactic acidosis (see Chapter 14) in cattle, Beede and Farlin (1977a) tested a series of antibiotics for their ability to inhibit lactate production from wheat in vitro, and bacitracin, capreomycin, novobiocin, oxamycin, erythromycin, hygromycin B and tylosin were effective. VFA production was generally lowered, except with capreomycin, where there was a higher VFA production and a lower acetate/propionate ratio. Oxamycin and capreomycin were further tested in vivo with sheep showing experimentally induced acidosis (Beede & Farlin, 1977b); 77 and 770 mg/kg wheat in the ration were used. With capreomycin, the lower concentration was most effective in increasing rumen pH and decreasing lactate production. Total VFA concentration and pattern were only slightly affected. With oxamycin, the higher concentration increased pH, but lactate inhibition was most effective with the lower concentration. Total VFA were not affected, but acetate/propionate ratio was higher with both concentrations.

The effects of 11 antibiotics and sulphonamides on cellulose digestion in vitro were investigated by Baldwin et al. (1982). Strong inhibitors were chlortetracycline, oxytetracycline, penicillin and tylosin. Two moderate inhibitors were dihydrostreptomycin and monensin, while bacitracin, chloramphenicol, kanamycin, sulphanilamide and sulphathiazole were rather weak agents. The drugs could be divided into four categories from their mode of action on bacteria, but there was no clear relation with their effect on cellulolysis.

Another interesting antibiotic is thiopeptin, a sulphur-containing peptide compound, produced by Streptomyces tateyamensis, with a narrow Gram-positive spectrum (Miyairi et al., 1972). Feedlot performance was improved, with lower acetate/propionate ratios in the rumen (Gill et al., 1979; Nagaraja et al., 1982). Protozoa were not affected by the drug (Dennis et al., 1981c). Lactic acidosis in sheep was completely inhibited (single dose, minimum, 0·18 mg/kg body weight) by an almost completely inhibited lactate production (Muir et al., 1980), due to the effect of the antibiotic on *Streptococcus bovis* (Muir & Barreto, 1979). Similar antibiotics, like sulphomycin, siomycin (sporangiomycin) and taitomycin, had a comparable effect. In recent experiments, the effect of thiopeptin on lactate production from starch *in vitro* depended on the protozoa, as only in their absence was lactate production diminished (Nagaraja et al., 1986). It is known that protozoa can play a buffering role in the rumen by ingestion of starch (Clarke, 1977), but they may degrade lactic acid formed (Newbold et al., 1985). In incubations where proliferation of *Streptococcus bovis* occurred, thiopeptin was more inhibiting to feed protein degradation than was monensin (Russell et al., 1981). So it can be speculated that thiopeptin would be more effective in modifying rumen fermentation in animals receiving highly digestible concentrate rations.

Flavomycin, a phosphorus-containing glycolipid antibiotic produced by several *Streptomyces* strains, is specifically toxic for Gram-positive bacteria, with lower activities on Gram-negatives and no effect on protozoa (see Cafantaris, 1981). Incubations *in vitro* showed that cellulose degradation and propionate production were increased, with inhibition of methanogenesis. Protein degradation was slightly stimulated (Cafantaris, 1981). In the rumens of cattle fed on a concentrate diet supplemented with straw *ad lib* and flavomycin (0·2 and 0·5 mg/kg bodyweight/day), no effect on VFA or ammonia levels was noted, and cellulose degradation (chopped hay) *in sacco* was not changed (Rowe et al., 1982). With sheep, cannulated in the rumen, duodenum and ileum, it was shown that the drug remained active in the whole gastro-intestinal tract, and post-ruminal activities are very probable. An interesting and unusual experiment was recently published by Goetsch & Owens (1986a,b). With cattle on a high-concentrate diet, ileal administration of 2 g of neomycin and 0·25 g of bacitracin not only altered hindgut digestion (a trend to lower organic matter and starch digestion and higher pH) but also increased ruminal escape of feed N, due to an increased fractional outflow rate of particulate matter. On the contrary, on a high-forage diet, disappearance of organic matter in the hindgut was increased, with a trend to higher acid-detergent-fibre digestibility. No significant effects on rumen particulate matter kinetics were found. The mechanism underlying the effect on rumen kinetics with the concentrate diet remains unknown, but it would seem that feedback impulses from the lower digestive tract can influence rumen kinetics and sites of digestion of feed components.

Finally, experiments *in vitro* were done with virginiamycin, which is active against Gram-positive bacteria, and produced by *Streptomyces virginiae* (Van Dijck, 1969). Minor quantities are absorbed, with no residues in tissues. A shift in fermentation pattern, similar to that with most other antibiotics, was found (Van Nevel et al., 1984). Proteolysis and deaminase activity were slightly decreased, but to a lesser extent than with monensin. Results suggested a toxicity towards rumen protozoa. Toxicity towards *Streptococcus bovis* was seen, and lactic acid production from glucose was inhibited (Hedde et al., 1980). *In vivo*, when steers were fed 33 ppm of the drug, no effect on rumen VFA, ammonia or urea was seen, but pH was increased

(Hedde et al., 1980). A positive effect on performance of growing steers has been reported, but differences were only significant during the second 8 weeks of the trial (Parigi-Bini, 1979). Virginiamycin, together with some other antibiotics, decreased formation of 3-methylindole from L-tryptophan *in vitro* (Honeyfield et al., 1985).

Growth factors and miscellaneous compounds

Growth factors

In contrast to antibiotics, growth factors can be considered as mainly organic substances necessary for growth of microbes. Microorganisms cannot synthesize these substances *de novo* in sufficient amounts, the amount being dictated by the growth rate, so they depend on a supply from the feed or from turnover (recycling) of microbial matter within the rumen. These considerations make it clear that a limiting supply of such growth factors can be expected with certain diets and under conditions where turnover, or residence time in the rumen, is low. As an example, stimulatory effects of protein on rumen microbial growth in the animal have been related to rumen retention time and recycling of microbial matter (Demeyer & Van Nevel, 1986b).

Niacin and thiamin. The influence of niacin and thiamin on rumen metabolism and animal performance was recently reviewed (Brent & Bartley, 1984). Niacin supplementation has resulted in increased efficiency of microbial synthesis in incubations *in vitro* with different substrates (Schussler et al., 1978; Bartley et al., 1979; Riddell et al., 1980, 1981; Shields et al., 1983), although on some occasions no effect has been observed (Riddell et al., 1981; Schaetzel & Johnson, 1981; Hannah & Stern, 1985), and when urea was the sole nitrogen source a negative effect was even found (Riddell et al., 1981). It is obvious that, when in control incubations niacin availability was already adequate, supplementation would have no effect. Total VFA production and individual proportions were normally not changed, but the percentage of propionic acid was slightly increased in one experiment with fistulated cows (Riddell et al., 1980). Degradation *in vitro* of high amounts of supplemented niacin was observed (2–8 μg/ml incubation fluid) (Riddell et al., 1985). When cattle received 6 g of niacin per day, high concentrations of the vitamin were found in the rumen and duodenum, while absorption from the small intestine was increased (Riddell et al., 1985). Higher than normal protozoal numbers have been found in the rumens of cattle fed on heated soybean meal plus niacin, indicating that niacin in this protein source is not available to the protozoa (Dennis et al., 1982). Positive effects on the performance of lambs, beef cattle and dairy cows were reported, while niacin could also help to prevent ketosis (Brent & Bartley, 1984).

Thiamin considerably increased microbial growth and its efficiency in incubations *in vitro* with protein-free substrate and rumen fluid from animals fed the same protein-free purified diet (Candau & Kone, 1980). Total VFA production remained unchanged, but propionic acid proportion was almost doubled. Thiamin supplementation of all-concentrate diets for feedlot steers had inconsistent effects on performance, dependent on the animals being all- or not-deficient in thiamin (Grigat & Mathison, 1982, 1983).

Four- and five-carbon VFA and phenylpropanoic acid. Isovaleric or 2-methylbutyric acid and a mixture of isobutyric, valeric, isovaleric and 2-methylbutyric acids increased protein synthesis and its efficiency in incubations of mixed rumen bacteria in a defined medium (Russell & Sniffen, 1984). However, the effect was observed only when the rumen fluid donor received a ration of timothy hay, and not when 60% concentrates–40% hay was fed. Trypticase addition caused a similar effect, but in this case, besides a source of C_4 and C_5 acids, fermentable organic matter was also added. Absence of an effect was a simple indication that in the control incubations the requirement of the bacteria for the higher VFA was already fulfilled. In incubations with different feeds and protein sources, stimulation of microbial nitrogen synthesis and *in vitro* dry matter digestibility was obtained after addition of a mixture of ammonium salts of isobutyric, 2-methylbutyric, valeric and isovaleric acids (Cummins & Papas, 1985). Efficiency of microbial protein synthesis was not calculated. It has been known for a long time that some cellulolytic bacteria require branched-chain C-skeletons for growth (Bryant & Robinson, 1962). Russell & Sniffen (1984) showed that stimulatory effects were also obtained with a mixture of soluble sugars as substrate.

3-Phenylpropanoic acid in the medium stimulated the growth rate and cellulose degradation of *Ruminococcus albus*, but strains of *R. flavefaciens* and *Butyrivibrio fibrisolvens* were not influenced (Stack & Cotta, 1986).

Other substances

The effect, on VFA production by mixed rumen microbes *in vitro*, of orotic acid, an intermediate in pyrimidine biosynthesis, was reported by Bueno & Ralison (1982). Total VFA production was stimulated by orotic acid (4 mg per 20 ml incubation fluid), and this was mainly a result of higher propionic or butyric acid formation, depending on the ration fed to the animal or incubation (supplemented or not with urea). The increased total VFA production could be the result of better fermentation activity, but the shift in proportions was not explained.

Captan, a fungicide, enhanced *in vitro* digestibility of the dry matter, cell walls, acid detergent fibre, cellulose and hemicellulose of alfalfa-brome hay, and a similar effect was found on apparent digestibility in steers fed a 72% chopped alfalfa-brome hay–25% ground-corn ration plus up to 600 ppm captan (Theuninck *et al.*, 1981). Lambs on a corn-based diet with 320 ppm captan had lower acetate and higher propionic acid percentages in the rumen (Smith & Boling, 1983). One interpretation of these results may be that the specific crude-fibre degrading activity of the bacteria and protozoa which replaced the fungi removed by the captan was, at least in these experiments, more intensive than the fungal activity. The increase in the propionate proportion might have been due to replacement by bacteria of fungi in which this acid is not an important end-product of metabolism (Bauchop & Mountfort, 1981). Specific effects of captan on bacteria or protozoa are, however, also possible. A minor role of fungi in rumen metabolism was also observed in the experiments of Windham & Akin (1984) using antibiotics and cycloheximide. Alkylamines have been reported to inhibit cellulolytic activity *in vitro*, while the acetate/propionate ratio was largely increased, as propionate production was more inhibited than acetate (Baldwin et al., 1981). When N,N-dimethyldodecanamine was given to growing lambs (up to 100 ppm), rumen parameters were not altered, although feedlot performance improved (Rumsey et al., 1982). Sarsaponin was found to increase fermentation in anaerobic digesters (L. A. Peekstok, MS thesis. Miami Univ. Oxford, 1979, cited by Valdez et al., 1986), and this naturally occurring steroidal glycoside (yucca plant) has recently been tested as an additive for ruminant feeds. The effects of sarsaponin on rumen digestibility of organic matter and starch were inconsistent and depended on the diet fed (33% concentrates, 50% conc. or 92% conc.) (Geoetsch & Owens, 1985; Goetsch et al., 1985). Rumen ammonia levels were lowered when calves were fed a 92% concentrates diet; this was related by the authors to an inhibition of ureolytic activity with lower transfer of urea through the rumen wall, although a simultaneous inhibition of feed nitrogen degradation was also observed (Ellenberger et al., 1985; Gibson et al., 1985; Goetsch et al., 1985). Only relatively low microbial growth efficiencies (22 g N_i/kg OM_f) tended to increase after sarsaponin supplementation (Goetsch & Owens, 1985; Goetsch et al., 1985). In dairy cows receiving 55% concentrates plus 45% sorghum silage with 77 ppm sarsaponin, no effects on rumen ammonia, total VFA concentration or individual molar proportions were seen (Valdez et al., 1986). Fractional outflow rates of concentrates, silage, fluid or particulate phase from the rumen were increased, although not always significantly (Goetsch & Owens, 1985; Goetsch et al., 1985). In semi-continuous cultures sarsaponin tended to increase bacterial numbers, but protozoa were decreased (Valdez et al., 1986). In the same experiment, acid detergent fibre digestibility was considerably increased, but the control value was extremely low (10·2%). In cows, however, rumen acid detergent fibre digestibility was not altered (Goetsch & Owens, 1985).

Inhibitors of proteolysis and/or deamination

Conversion of feed protein in the rumen into peptides and amino acids (proteolysis) and further to ammonia (deamination) is unfavourable for the host animal, because energy is needed for re-synthesizing microbial protein from ammonia, and not all ammonia is re-synthesized to protein (Hobson & Wallace, 1982). Protection of feed protein against rumen degradation (proteolysis and/or deamination) will result in a higher supply of protein or amino acids in the proximal duodenum (provided there is no decrease in microbial protein production) which can increase animal productivity, providing that the latter is limited by protein supply. Factors controlling the extent of protein degradation in the rumen are the composition and activity of the microbes (proteolytic bacteria and protozoa), the chemical and physical properties of the protein and its residence time in the rumen (Ørskov & McDonald, 1979; Mahadevan et al., 1980; Russell et al., 1981; Wallace & Kopecny, 1983; Wallace & Brammall, 1985; Cotta & Hespell, 1986). Only a limited amount of research has been devoted to protection of dietary protein by chemical manipulation of rumen fermentation, a field which could be very promising (Hobson & Wallace, 1982). Protease inhibitors as well as deaminase inhibitors can be used, as it makes little difference to the animal whether it receives protein or amino acids in the duodenum. Different chemicals and naturally occurring substances are known to inhibit proteolysis, and their effect in vitro is summarized in Table 9. Wallace (1983) showed that several metal ions were also effective, probably due to interaction with SH-enzymes; most potent were Ag^+, Cd^{2+}, Cu^{2+} and Hg^{2+}. The effect(s) of all these compounds on the integrated system of rumen carbohydrate fermentation, microbial growth yields etc., and their influence on digestion in the lower digestive tract, are largely unknown. Diaryliodonium compounds inhibited deaminase activity in the rumen in vitro and in vivo (Chalupa et al., 1983), and greatest protection occurred for valine, methionine, isoleucine, leucine and phenylalanine. The mode of action of these compounds is still unclear, but prevention of amino acids entering the cell is possible, although accumulation of

Table 9

Inhibiting Effect of Chemicals or Naturally Occurring Substances on Proteolytic Activity of Rumen Bacteria or Protozoa in vitro (Casein or Azocasein as Substrate)

Reference	Inhibitor	Concentration	Inhibition (%)	Organism
Kopecny & Wallace (1982)	p-Chloromercuribenzoate	3 mM	82	Mixed rumen bacteria
	PMSF	1 mg/ml	21	
	DTT	4 mM	25	
	Pepstatin A	10 μM	11	
Brock et al. (1982)	PMSF	3 mM	36.8	Mixed rumen bacteria
	TLCK	1 mM	31.1	
	TPCK	1 mM	20.8	
	EDTA	1 mM	19.4	
	EDTA + DDT	2 mM	31.6	
	Cysteine, HCl	2 mM	27.0	
	DTT	2 mM	38.0	
	Iodoacetate	5 mM	31.0	
	Merthiolate	5 mM	71.0	
	Soybean trypsin-inhibitor IS	100 μg/ml	39.4	
	Soybean trypsin-inhibitor IIS	100 μg/ml	40.6	
	Lima bean trypsin-inhibitor	125 μg/ml	31.7	
	Antipain	250 μg/ml	32.6	
	Chymostatin	250 μg/ml	33.6	
	Leupeptin	250 μg/ml	42.8	
Hazlewood & Edwards (1981)	Antipain	8.3 μM	40.0	Bacteroides ruminicola
	Chymostatin	11.7 μM	38.0	
	Elastatinal	8.5 μM	18.0	
	Leupeptin	11.7 μM	42.0	
	Pepstatin A	7.3 μM	32.0	
	Soybean trypsin-inhibitor	11.7 μM	18.0	
	EDTA	20.0 mM	93.0	
Forsberg et al. (1984)	Antipain	50 μg/ml	60.8	Rumen protozoa
	Chymostatin	50 μg/ml	48.0	
	Leupeptin	50 μg/ml	60.7	
	Phosphoramidon	250 μg/ml	11.7	
	Pepstatin A	250 μg/ml	35.4	

PMSF, phenylmethylsulphonyl fluoride
DDT, dithiothreitol
TLCK, N-tosyl-1-lysine chloromethyl ketone ⎫
TPCK, N-tosylphenylalanine chloromethyl ketone ⎭ synthetic protease-inhibitors
EDTA, ethylenediamine tetra-acetate

alanine also suggested inhibition of deaminase or transaminase reactions. In incubations with carbohydrate as substrate, propionate production and overall energetic efficiency of carbohydrate fermentation were increased; *in vivo*, methanogenesis was inhibited (Chalupa *et al.*, 1983). Other strong deaminase inhibitors were hydrazine and analogues, sodium arsenite, and thymol. Diphenyliodonium chloride and hydrazine also inhibited microbial growth (Broderick & Balthrop, 1979). Hydrazine reacts as an antagonist of vitamin B_6, an important factor in amino acid biosynthesis and degradation. A negative aspect of these compounds is their lack of specificity, while some are severely toxic for the host animal. This makes it rather doubtful if they will ever be used on a practical scale. Finally, it would be interesting to search for a substance that lowers the retention time in the rumen of diets with high protein and low crude-fibre contents.

DEFAUNATION

The role of ciliate protozoa in the rumen is under considerable debate and has received much attention recently (Veira, 1986; Jouany *et al.*, 1988); this is also discussed in Chapter 15. The predatory activity of protozoa towards bacteria, and their retention within the rumen, are the main factors contributing to the turnover of microbial N in the rumen, which may amount to half of the microbial N formed. Intensity of turnover determines net growth yields of bacteria, whereas energy yield from fermentation determines total growth yields (Van Nevel & Demeyer, 1977*b*) (see Chapter 6). Defaunation of the rumen through the introduction of chemicals that destroy ciliates into the adult animal's rumen may thus be expected to increase net rumen microbial N yield. Whether such increase in yield (g N_i/kg OM_f) will increase net duodenal microbial N flow (g N_i/day) will depend on the amount of OM fermented in the rumen (g OM_f/day). If defaunation lowers the latter more than it increases the microbial yield, then net bacterial N flow into the duodenum may be decreased by the defaunation. This is one example of the numerous interrelated effects in the rumen and in ruminant metabolism induced by defaunation of the rumen, as discussed in detail by Jouany *et al.* (1988). These authors summarize possible and documented effects in the rumen as follows:

Increase in bacterial N yield
Absence of protozoal N
Inhibition of methanogenesis
Inhibition of rumen fibre digestion
Inhibition of rumen non-soluble protein degradation
Inhibition of higher fatty acid biohydrogenation
Increase in rumen starch degradation
Increases in Cu and trace element supply from the rumen
Increases and changes in rumen bacterial and fungal populations
Changes in kinetics of digesta flow

It should be mentioned that the relative importance of the various effects may change in an interrelated fashion, e.g. as recently suggested for digesta flow kinetics and rumen fibre degradation (Demeyer, 1988). It was suggested that fractional rate of particle outflow (k_p) reflects, rather than determines, rumen fibre degradation. Inhibition of the latter process by defaunation may thus be reflected in increased k_p values. In some experiments such effects are corrected for by an increase in rumen volume. The latter effect may be so important that mean retention time ($1/k_p$) increases, allowing development of a new fibre-hydrolysing (fungal?) flora as alternative to the lost protozoal population. The variability of such interrelated effects is governed by as yet unknown factors, but it helps to explain contradictions in the literature.

Any effect of defaunation on animal production will be a reflection of the balance of all interrelated effects, whose relative importance may change with diet, breed of animal, level of feeding, adaptation to treatment, etc. Again, such complexity renders interpretation of results difficult, and sizable improvements in animal production through defaunation of the rumen have been obtained only with molasses–urea-based diets, although possibilities seem to exist also for fibre-based diets (Preston & Leng, 1986). Further research in this area is certainly indicated, not least in relation to the often observed toxic effects for the animal of chemical defaunation.

MODIFICATION OF RUMEN FERMENTATION BY MACROCOMPONENTS OF THE DIET

Effect of high amounts of lipids in the ration on rumen metabolism

Two decades ago, long-chain fatty acids were found to inhibit methane production in the rumen while, simultaneously, molar proportions of propionate were increased (Czerkawski et al., 1966a,b; Demeyer & Henderickx, 1967; Demeyer et al., 1969). Higher fatty acids can thus be used as manipulators of rumen fermentation. Apart from this aspect, there has been a renewal of interest in the use of high amounts of lipids in the diet of ruminants because of the possibility of increasing the energy density of the diet, thus permitting an improved energy intake of high-productive lactating animals with high energy demands (Palmquist, 1984). In this case, the high grain content of the ration can partly be replaced by lipids, thus abolishing the negative effects of high-starch-containing diets on rumen fermentation (Thomas & Rook, 1977), but only on condition that the fat itself has no undesirable effect on rumen digestion, and this is not always the case. The partial efficiency of milk synthesis can also be increased with high-fat rations, as synthesis of milk fat from higher fatty acids is more efficient than when short-chain fatty acids are used (see Chalupa et al., 1984). It should be mentioned, however, that long-chain fatty acids may inhibit tissue milk-fat synthesis apart from any effect they may have in the

Table 10

Effect of Dietary Lipids on Crude Fibre (CF) Digestibility in the Rumen or Total Digestive Tract of Sheep

Reference	Ration and feeding method	% CF in ration	Lipid added	CF digestibility (%)	
				Rumen	Total tract
Czerkawski (1973)	Hay and cubes (1 and 2×/day)	—	Control		62·3
			40 g LOFA/day[a]		55·2
			80 g LOFA/day		44·9
Sutton et al. (1975)	Hay and conc. (2×/day)	47	Control	44·6	51·6
			CLO (20 g/day)[b]	43·7	51·4
Ørskov et al. (1978)	Dried grass (ad lib.)	22·5	Control		64·2
			Tallow (13%)		44·3[g]
Ikwuegbu & Sutton (1982)	Hay and conc. (2×/day)	16·5	Control	44·0	55·0
			LO (13 ml/day)[c]	28·0[g]	52·0[g]
			LO (40 ml/day)	14·0[g]	41·0[g]
Sutton et al. (1983)	Hay and conc. (2×/day)	39·7	Control	66·0	50·0
			LO (40 g/day)	53·0[g]	19·0[g]
			LO (40 g/day) Prot.[d]	59·0[g]	29·0[g]
			CO (40 g/day)[e]	46·0[g]	12·0[g]
			CO (40 g/day) Prot.	58·0[g]	38·0[g]
Van der Honing et al. (1983)	Hay and conc. (2×/day)	—	Control		75·3
			Tallow (6·8%)		71·5
			Tallow (11·0%)		64·0
Devendra & Lewis (1974a)	Straw, SBM[f] and barley (2×/day)	20–23	Control		46·2
			Maize oil (8%)		37·0[g]
			Sat. lip. (8%)		23·2[g]

[a] LOFA, linseed oil fatty acids. [b] CLO, cod liver oil. [c] LO, linseed oil. [d] Protected form. [e] CO, coconut oil. [f] SBM, soybean meal. [g] Significantly different.

rumen (Banks et al., 1984). The final effect of fat in the ruminant diet will be the result of the effects on rumen fermentation and intermediary metabolism, including mammary gland metabolism. When triglycerides are added to the diet, their effect on rumen fermentation will be dependent on the rate of lipolysis, as the released free fatty acids exert their toxic effects on the rumen microbes (see later). A short survey of the effects of lipids on rumen digestion will be given.

Effect of lipids on carbohydrate metabolism in the rumen and the total digestive tract
Clear differences exist between the digestion of structural and non-structural carbohydrates, and the site of digestion must also be considered.

Degradation of dried grass or cotton thread incubated *in sacco* in the rumen was lowered when the ration of the sheep was supplemented with tallow (Henderson et al., 1977; Kowalczyk et al., 1977; Ørskov et al., 1978). This was the result of a decreased activity of the microbes degrading crude fibre, as the *in sacco* technique does not take into account physical parameters, such as fractional outflow rates or retention time of the feed components in the rumen. The effect of lipids on crude-fibre digestibility in the rumen or total digestive tract of the sheep and cow, which is the resultant of microbial activity, retention time and site of digestion, is summarized in Tables 10 and 11. Only a series of representative experiments are shown. With sheep, rumen digestibility of crude fibre was always negatively influenced by linseed or coconut oil, but not by cod liver oil. With tallow no consistent effect was observed. Organic matter digestibilities are not shown, but followed the crude-fibre pattern. In cow experiments, crude-fibre digestibility in the rumen or total tract was not normally affected by tallow supplementation, the experiment of Rohr et al. (1978) being the exception. All these experiments suggest that the influence of lipids on crude-fibre digestibility is mainly dependent on the nature and amount of fat used (triglycerides, free fatty acids, degree of unsaturation), while the level of intake and/or animal species (sheep or lactating cow) can also be important. Of minor importance are the crude-fibre contents of the basal ration and its digestibility. Unfortunately the rations and feeding methods used in these experiments were very similar, so excluding interpretation of the role these parameters can play. Interpretation of results is also rendered difficult as the effect of the fatty acid fraction of the lipids is dependent on its liberation by lipolysis and hydrogenation in the rumen. The latter processes are certainly affected by diet (Demeyer et al., 1972; Demeyer, 1973). It has been found recently that the short-term replacement of fibre by starch in the diet reduced not only lipolysis but also hydrogenation (Gerson et al., 1985). A shift in crude-fibre digestion from the forestomachs to the hindgut, and an eventual increase in retention time of the feed in the rumen after fat supplementation, are mechanisms that can compensate for the lower activity of crude-fibre-degrading microbes (Devendra & Lewis, 1974b; Ikwuegbu & Sutton, 1982; Sutton et al., 1983; Jenkins & Palmquist, 1984).

Lipids had no effects on rumen or total-tract digestibility of starch or NFE (nitrogen-free extract) fractions of the diet (Devendra & Lewis, 1974a; Van der Honing et al., 1981; Ikwuegbu & Sutton, 1982; McAllan et al., 1983; Tamminga et al., 1983). Several theories exist to explain the negative effect of lipids on crude-fibre

Table 11
Effect of Dietary Lipids on Crude Fibre (CF) Digestibility in the Rumen or Total Digestive Tract of Cows

Reference	Ration and feeding method	% CF in ration	Lipid added[a]	CF digestibility (%) Rumen	CF digestibility (%) Total tract
Van der Honing et al. (1981)	Hay and conc. (2 ×/day)	16·0	Control		55·6
			Soybean oil in conc. (5%)		54·5
			Tallow in conc. (5%)		55·8
Van der Horning et al. (1983)	Hay and conc. (2 ×/day)	14·0	Tallow in conc. (6·8%)		72·3
			Tallow in conc. (11·0%)		70·4
Tamminga et al. (1983)	Hay and conc. (2 ×/day)	—	Control	80·0	72·0
			Tallow in conc. (11%)	92·0	69·0
Palmquist & Conrad (1978)	Alfalfa, corn silage conc. (ab. lib.)	18·0	EE 3·2% (SBM)		35·6
			5·9% (soybeans)		37·4
			5·7% (sat. lip.)		44·3
			10·8% (sat. lip.)		44·0
Palmquist & Conrad (1980)	Alfalfa, corn silage conc. (ad lib.)	15·7–19·5	EE 3·6%		34·6
			8·4% (an. veg. fat)		38·3
			8·7% (tallow)		39·1
			6·0% (an. veg. fat)		37·9
			6·7% (tallow)		48·9
Rohr et al. (1978)	Corn silage dry forage conc.	—	Control (430 g fat/day)		57·7
			Coconut/palmoil (680 g fat/day)		59·1
			Coconut/palmoil (915 g fat/day)		48·7
	Corn silage dry forage conc.		Control (362 g fat/day)		58·1
			Tallow (656 g fat/day)		60·6
			Tallow (950 g fat/day)		50·9
	Corn silage dry forage conc.		Control (478 g fat/day)		50·2
			Soybean oil (953 g fat/day)		55·3

[a] EE, ether extract; SBM, soybean meal.

digestibility (Devendra & Lewis, 1974a). The speculation that inhibition is due to physical coating of the fibre with lipids is no longer valid, as Ørskov et al. (1978) found no effect on degradability in sacco of dried grass coated with tallow. This permitted the authors to conclude that coating with triglycerides cannot be the reason for the negative effect. Recently, Broudiscou et al. (1988) incubated in sacco pure cellulose coated with soybean oil hydrolysate (7% w/w) and also found no effect on degradability. It is now clear that fatty acids are toxic for certain bacteria and protozoa (see later), and cause shifts in the microbial composition which are responsible for the effect on crude-fibre digestion.

Effect of lipids on nitrogen metabolism in the rumen
Incubation of soybean meal in sacco, in the rumens of steers receiving 8% tallow in their diet showed no effect of the fat on the degradation of N compounds (63·3% against 66·5% with tallow), thus indicating no change in the proteolytic activity in the rumen (Olubobokun et al., 1985).

Experiments in vivo with sheep or cows cannulated in the duodenum showed interesting results. The most striking effect was an increase (sometimes more than twofold) in microbial growth efficiency in the rumen when the animals received linseed oil, coconut oil (unprotected form) or tallow, but not when they received cod liver oil (Sutton et al., 1975; Ikwuegbu & Sutton, 1982; Sutton et al., 1983; Tamminga et al., 1983) (Table 12). The increase in NAN (non-ammonia N) entering the duodenum was in most cases entirely the result of the higher microbial N flow (Devendra & Lewis, 1974b; Sutton et al., 1983). From feed N consumed and entering the proximal duodenum, feed N degradation in the rumen was calculated (Table 12). It must be said that such calculation should be interpreted with caution, as endogenous N secretions are not taken into account, and these could also have been altered by lipid supplementation. Furthermore, when total N values were used instead of NAN, no correction was made for eventual change in NH_3 N amounts passing to the lower tract (Ikwuegbu & Sutton, 1982). From Table 12, it seems that linseed oil supplementation had a different effect in the two experiments. Higher efficiencies of microbial growth (g N_i/kg OM_f) were observed, but these may have been affected by the amount of dietary lipid leaving the rumen, a process lowering OM_f by definition. However, substantially higher amounts of microbial N (N_i) entered the duodenum, showing that an increase in microbial growth efficiency reflects an increase in the numerator (N_i) as well as a decrease in the denominator (OM_f).

Taking into account the effect of defaunation of the rumen on bacterial growth efficiency (Ushida et al., 1984; Kayouli et al., 1986), the decreased protozoal numbers after lipid supplementation (see below) are probably the reason for the increased microbial growth efficiency. However, higher amounts of linseed oil can finally become toxic to the bacteria also, thus abolishing the positive effect on growth efficiency (Van Nevel & Demeyer, 1981; Ikwuegbu & Sutton, 1982).

Effect of lipids on the rumen fermentation pattern and kinetic parameters of the digesta
The shift in rumen volatile fatty acid proportions from acetic to propionic after fat

Table 12
Effect of Linseed Oil (LO) or Coconut Oil (CO) on N Metabolism in the Rumen of Sheep

Reference	Ration	N parameter	Control	13 ml LO	26 ml LO	40 ml LO[d]
Ikwuegbu & Sutton (1982)	Hay and concentrates	Feed protein[a] degradation (%)	74	70	42	32
		Bact. N[b] flow in duodenum (g/day)	7.5	8.1	9.7	6.7
		Bact. growth efficiency (g N$_i$/kg OM$_f$)	32	35	59	34

Reference	Ration	N parameter	Control	40 g LO	40 g PLO	40 g CO	40 g PCO[e]
Sutton et al. (1983)	Hay and concentrates	Feed protein degradation (%)	64	70	83	50	78
		Microbial N[c] flow in duodenum (g/day)	9.5	13.9	13.0	12.4	10.3
		Microbial growth efficiency (g N$_i$/kg OM$_f$)	41	94	64	81	44

[a] Calculated from feed N intake and feed N in duodenum (total N or NAN – bact. N or microb. N). [b] DAPA as marker. [c] RNA as marker. [d] ml/day. [e] g/day; P, protected.

supplementation, probably related to the low-fat milk syndrome (Davis & Brown, 1970), is well documented in the literature (Palmquist, 1984). Long-chain fatty acids were in this respect more active than triglycerides (Chalupa et al., 1984). Experiments in vitro showed that higher molar proportions of propionate were due to higher production rates, with concomitant decreases in acetic and butyric acid production (Van Nevel & Demeyer, 1981). Lower butyric acid proportions were generally found in vivo (Demeyer et al., 1969; Czerkawski, 1973; Sutton et al., 1975; Chalupa et al., 1986), but not in the experiment with cows by Palmquist & Conrad, 1978, 1980). Lower butyrate proportions are probably due to the decreased numbers and activities of protozoa, since in defaunated sheep soybean oil hydrolysate had no effect on rumen butyrate proportions (Broudiscou, L. et al., unpublished).

Methane production was inhibited by tallow, unsaturated fatty acids and linseed oil hydrolysate in experiments in vitro and in vivo with cows and sheep (Czerkawski et al., 1966a,b; Demeyer & Henderickx, 1967; Van Nevel & Demeyer, 1981; Van der Honing et al., 1981, 1983). It was clear that the effect of unsaturated fatty acids was not entirely related to the microbial hydrogenation process competing for metabolic hydrogen (Czerkawski et al., 1966a; Demeyer & Henderickx, 1967). It was also shown in incubations with mixed rumen bacteria in vitro that methane inhibition by unsaturated fatty acids was due to a direct toxic effect on methanogens, and highly unsaturated cis isomers were more toxic than trans isomers, possibly related to their ability to form insoluble salts (Jenkins & Palmquist, 1982). Conjugated forms were less active than non-conjugated isomers and a free carboxyl group was necessary (Czerkawski et al., 1966c; Demeyer & Henderickx, 1967; Demeyer et al., 1967; Van Nevel et al., 1971; Prins et al., 1972; Henderson, 1973). Methane inhibition with a simultaneous increase in propionate production is in general agreement with the overall stoichiometry of carbohydrate fermentation by rumen microbes (Hungate, 1966; Demeyer & Van Nevel, 1975), where an increased propionate synthesis acts as an alternative electron sink (Wolin, 1975) (see Chapter 11). Total VFA concentrations remained unchanged when oils were fed (Ikwuegbu & Sutton, 1982; Sutton et al., 1983; Palmquist et al., 1986), but Tamminga et al. (1983) found a 20% decrease when the cows received c. 12% tallow in the concentrates. This was also observed by Chalupa et al. (1986) on feeding oleic acid.

A more consistent effect was the decrease in NH_3 concentrations in the rumen (Henderson et al., 1977; Kowalczyk et al, 1977; Ikwuegbu & Sutton, 1982; Tamminga et al., 1983; Broudiscou, L. et al., unpublished). This effect is probably not related to increased NH_3 absorption, in view of the minor changes in pH after fat addition (Czerkawski, 1973; Kowalczyk et al., 1977; Ikwuegbu & Sutton, 1982; Tamminga et al., 1983; Broudiscou, L. et al., unpublished). On the other hand, the effect on proteolysis is not clear. Besides an increase in bacterial growth efficiency, another explanation for the lower NH_3 values is the increased rumen liquid volume after lipid supplementation which has been found by several workers (Czerkawski et al., 1975; Ikwuegbu & Sutton, 1982; Tamminga et al., 1983; Broudiscou, L. et al., unpublished). This may also explain the eventual lower concentrations of other fermentation end-products. However, correction for rumen liquid volume still gave lower NH_3 pools in the rumens of sheep receiving linseed oil (Ikwuegbu & Sutton,

1982), in contradiction to data calculated from results of Tamminga et al. (1983) with cows. Data on the influence of lipids on fractional outflow rates of liquid and particle phases in the rumen are very scarce. Liquid outflow rate seemed to be only slightly affected or unaffected (Czerkawski et al., 1975; Ikwuegbu & Sutton, 1982; Tamminga et al., 1983; Broudiscou, L. et al., unpublished). In recent work in our laboratory, particle outflow rate (Cr-mordanted hay) was lowered by c. 25% when faunated sheep received soybean oil hydrolysate (80 g/day). A simultaneous increase in particle pool size (c. 60%) was observed (Broudiscou, L. et al., unpublished).

A longer retention time of the feed may also be related to the depression in forestomach motility caused by fatty acids observed by some workers (Titchen et al., 1966; Nicholson & Omer, 1983). Again, longer retention time of the feed may be a reflection of the lower crude-fibre degrading activity of the microbes, whereas an increased rumen fill may be an adaptation to it (see later).

Effect of lipids on rumen microbes
It is obvious that the effects mentioned above are the final results of the action of lipids on rumen bacteria and protozoa. The most consistent effect of lipid supplementation is the considerable decrease in rumen protozoal numbers in sheep and cows (Czerkawski, 1973; Czerkawski et al., 1975; Henderson et al., 1977; Ikwuegbu & Sutton, 1982; Sutton et al., 1983; Tamminga et al., 1983). With 40 ml of linseed oil given per day, the rumen of a sheep was almost completely defaunated (Ikwuegbu & Sutton, 1982). In their experiment with cows, Tamminga et al. (1983) found that Holotrichs and Entodinia were decreased to approximately the same extent by tallow. In certain concentrations, higher fatty acids were directly toxic to rumen protozoa, but the exact mechanism has not yet been elucidated (Girard & Hawke, 1978).

In sheep consuming large amounts of linseed oil (up to 50 g/day) or linseed oil fatty acids (up to 80 g/day), higher than normal total numbers of bacteria (per ml) in the rumen have been noted, and in most cases the number of large bacteria was also increased (Czerkawski, 1973; Czerkawski et al., 1975). This is in agreement with higher counts of bacteria in defaunated animals (Coleman, 1979). Henderson et al. (1977) observed no charges in numbers of cellulolytic, lipolytic or total bacteria in the rumen of sheep fed tallow. The effect of fatty acids on methanogens, in incubations with mixed rumen bacteria, has already been discussed. Pure-strain studies with *Methanobacterium ruminantium* or *Methanobacterium* M.O.H. showed that growth was severely inhibited by long-chain fatty acids (Prins et al., 1972; Henderson, 1973). The most potent inhibitors were the C_{18} unsaturated acids, but toxicity was also found with capric, lauric, myristic and palmitic acids, while stearic acid had a smaller effect. It was again concluded that unsaturation as well as a free carboxyl group were necessary for potent inhibition. It is thought that fatty acids are adsorbed on to the cell wall, thus interfering with nutrient uptake (Galbraith et al., 1971). Results with pure strains of other rumen bacteria are more difficult to summarize. Fatty acids did not affect the growth of most Gram-negative bacteria important in propionate formation (Henderson, 1973; Maczulak et al., 1981).

Another Gram-negative organism, *Butyrivibrio*, was stimulated by low concentrations of oleic, lauric or capric acid (0·01–0·1 g/l), but at higher concentrations inhibition was noted. For this organism, Maczulak *et al.* (1981) found that the effect varied with the strain used. Several strains of Ruminococci have been studied, and all were inhibited by oleic acid, although to different extents (Henderson, 1973; Maczulak *et al.*, 1981). Addition of cellulose reduced the action of fatty acids, due to the competition for adsorption between food particles and bacteria (Harfoot *et al.*, 1974; Maczulak *et al.*, 1981). The defaunating effect of the fatty acids, as well as the results of pure-strain studies, are in rather good agreement with the suggested role of protozoa in plant fibre degradation (Demeyer, 1981) and with the lower crude-fibre degradability and the shift to propionate *in vivo* after fat supplementation.

Use of protected lipids and Ca-soaps
Negative effects of the fatty acids on rumen fermentation can, at least partially, be removed through the use of protected lipids (Scott *et al.*, 1971; Sutton *et al.*, 1983) or insoluble Ca-soaps. Thus the latter compounds did not affect fibre digestion (Jenkins & Palmquist, 1982, 1984; Palmquist, 1984; Olubobokun *et al.*, 1985) and, *in vitro*, Ca-salts of palmitic, stearic or oleic acid had no effect on VFA production or proportions (Chalupa *et al.*, 1984). This was confirmed *in vivo* (Chalupa *et al.*, 1986). The ability of the fatty acids to form insoluble salts in the rumen can thus be an important factor in the final effects on microbes and fermentation.

Some further considerations
We realize that there is certainly a lack of information concerning the effect of lipids on feed protein degradation and fractional outflow rates of the different phases in the rumen, all these being important rumen fermentation parameters. The shift in crude-fibre degradation from the rumen to the hindgut after lipid supplementation can compensate for the loss of VFA production in the rumen. However, bacterial matter synthesized during fermentation in the hindgut is excreted with the faeces, but the higher bacterial N flow in the duodenum can compensate for this loss if both phenomena happen simultaneously.

Finally, it is clear that classical experiments with normally faunated animals do not permit of distinguishing between a primary effect of the lipids on rumen fermentation and a secondary effect caused by the lowered protozoal numbers. This problem can be solved only by comparison of defaunated with refaunated animals. Such an experiment was recently done by Broudiscou *et al.* (1988) and revealed very interesting results. Incubations *in sacco* with wheat straw, pure cellulose, ground corn or soybean meal in the rumens of sheep receiving soybean oil hydrolysate showed that the effects of the fatty acids differed in the defaunated and the refaunated states. In defaunated animals, degradation of the straw was decreased (dry matter), but no effect on the pure cellulose was noted. On the contrary, corn drymatter and soybean N degradations were enhanced by the fatty acids. Refaunation of the sheep changed this picture: wheat straw and cellulose disappearance were decreased, but corn and soybean N degradations remained unaffected. This example is given to illustrate the important role protozoa play in the final overall effect of lipids on rumen fermentation. This final effect is the resultant of the action on bacteria as well as on protozoa.

Use of buffers or artificial saliva

Dietary buffers have been used for controlling pH and for maintaining a normal rumen fermentation in beef steers or lactating cows receiving large amounts of concentrates, thus preventing metabolic disorders like acidosis and the low-fat milk syndrome (Herod et al., 1978; Fulton et al., 1979; Ha et al., 1983; Teh et al., 1985; Coppock et al., 1986; West et al., 1986). On the other hand, artificial saliva (mainly $NaHCO_3$ or mineral salts) has been infused into the rumen or included in the diets of steers, cows and sheep with the aim of increasing the kinetics of rumen digesta, especially the fractional outflow rate of the liquid phase (Harrison et al., 1975; Thomson et al., 1978; Rogers et al., 1979; Hadjipanayiotou et al., 1982; Rogers & Davis, 1982b; Stokes et al., 1985).

Rumen parameters expected to be affected by addition of buffers or artificial saliva are pH, digesta kinetics, fermentation pattern (total VFA and individual proportions), fluid osmolality, methane production, feed protein degradation, microbial protein synthesis and its efficiency. As everywhere in this chapter, interrelation between these various interactive aspects of the integrated system of rumen metabolism renders interpretation of results difficult. Before discussing the overall effect of the addition of buffering substances, the influence of considerable variation of pH in the rumen fluid *in vitro* and *in vivo* caused by the use of mineral acids or alkali will be described.

In incubations *in vitro*, with the substrates wheat or a mixture of corn silage (60%) and concentrates (40%), the pH was varied between 4 and 7 by adding NaOH, HCl, $NaHCO_3$ or Na_2CO_3. With wheat as substrate, lower pH values (4–5) inhibited total VFA production (30–40%), while proportions of the individual acids were also changed; acetate percentages increased, propionate decreased, while butyrate remained constant (Ha et al., 1983). With the mixed ration, lowering the pH also decreased total VFA production, but in this case the percentage of acetate was clearly decreased and propionate increased (Erfle et al., 1982). In a third experiment, no substrate was added and variation of pH of the incubation between 6·2 and 6·8 had no effect on VFA production (Esdale & Satter, 1972). However, decreasing the pH from 6·2 to 5·6 altered the fermentation pattern towards higher propionate and butyrate percentages, in agreement with Erfle et al. (1982). Lower pH values also decreased methane production (pH 5, complete inhibition) while protease and especially deaminase reactions were also inhibited. So it seems that, in rather simple and well controlled *in vitro* systems, manipulation of pH has already shown conflicting results concerning fermentation patterns probably related to the kind of substrate used and the original characteristics (flora composition) of the incubated rumen fluid. Similar experiments *in vivo*, where pH was increased from c. 5·1 to 5·9, showed either small or considerable increases in total VFA concentrations, again depending on the basal diet being fed (Fulton et al., 1979; Ha et al., 1983). The effect on VFA proportions was variable; either no effect was observed (Ha et al., 1983) or a low original value of acetate (44·4%) was increased (to 50·8%) (Fulton et al., 1979). This is a first indication that eventual changes in VFA percentages after buffer treatment will also depend on the original values; rather extreme values (high acetate, low propionate) on a control diet will probably not be changed after feeding buffers. Esdale & Satter (1972) also showed that changes in pH between 6·2 and 6·8 had no effect on the VFA pattern *in vitro* with hay as substrate, and the same was found in the rumens of cows, but very high percentages of acetate (74%) were already found with the basal ration at pH 6·2. Buffering capacity of rumen contents is also highest between these values (Davis, 1979). Adding buffering substances ($NaHCO_3$, MgO, $CaCO_3$, Alkaten, artificial saliva) to the diets of sheep, steers (up to 5%) or cows (up to 288 g/day) showed varying effects on rumen pH, from no effect to clear increases, dependent amongst other things on diet composition (ratio of forage to concentrates; Rogers & David, 1982b; Mees & Merchen, 1985), amount of buffer used (West et al., 1986), sampling time (Mees et al., 1985) and initial pH values. Low original pH values on the control diet showed an increase (Hadjipanayiotou et al., 1982; Rogers & Davis, 1982b; Okeke et al., 1983; Mees et al., 1985), while higher values remained unchanged (West et al., 1986).

Addition of buffers did not usually influence the concentration of total VFA in the rumen, but on some occasions, when the concentrate proportion in the diet was very high (75%), lowered values were observed (Rogers et al., 1979; Rogers & Davis, 1982b). This decrease can be explained by an increased fractional outflow rate of the rumen liquid phase, resulting in lower availability of soluble substrates for microbial activity (Rogers & Davis, 1982b).

When high-concentrate diets were fed, giving low proportions of acetic acid (threshold value c. 60%) and high propionate proportions (threshold value c. 30%), clear increases or decreases were observed after feeding buffers. On the contrary, on high-roughage rations, normally giving acetic acid percentages higher than 60% and propionate lower than 24%, no changes were found (Harrison et al., 1975; Herod et al., 1978; Thomson et al., 1978; Rogers et al., 1979, 1982; Hadjipanayiotou et al., 1982; Rogers & Davis, 1982b; Mees & Merchen, 1985; Mees et al., 1985; Teh et al., 1985). The increase in acetic acid proportion was due to a decreased rate of production of propionic acid, as production rate of acetic acid remained unchanged when buffers were fed (Rogers & Davis, 1982b). Other important parameters are the fractional outflow rates of liquid (D) and particle (k) phase in the rumen. Generally, D was increased after buffer or artificial saliva treatment when, on the basal ration (high concentrates), the original values were low (0.038–$0.60\,h^{-1}$), although some exceptions did exist (Harrison et al., 1975; Thomson et al., 1978; Hadjipanayiotou et al., 1982; Rogers & Davis 1982b; Okeke et al., 1983). In the experiments of Teh et al. (1985), the original D value was $0.066\,h^{-1}$, and was not changed, but the amount of $NaHCO_3$ fed (1% of the diet) was rather low. Two other authors reported D values of 0.059 and $0.055\,h^{-1}$ which were not changed by addition of $NaHCO_3$ (3·5% or 4·5%) (Mees et al., 1985; Kovacik et al., 1986). This was not related to the diet, as in the two cases rations were different: 75% concentrates plus 25% roughage against 50% concentrates plus 50% orchard grass. Higher control D values (0.092–$0.150\,h^{-1}$), normally obtained with higher roughage proportions ($>50\%$) in the diet, were generally not altered after buffer administration (Harrison et al., 1975; Rogers & Davis, 1982b; Mees & Merchen, 1985; Rogers et al., 1985; Stokes et al., 1985), although in one experiment a control D of $0.097\,h^{-1}$ could be increased by infusion of NaCl or $NaHCO_3$ (Rogers et al., 1979). Only a few papers have dealt with the effects of buffers on k, and the results have been rather variable, going from increases (Okeke et al., 1983; Mees et al., 1985) to no change at all (Stokes, 1983; Mees & Merchen, 1985; Kovacik et al., 1986). In one experiment, D and k were even decreased with 4.5% $NaHCO_3$ in the diet (Mees & Merchen, 1985). This variability is difficult to explain, but differences in animals (cow or sheep), diets fed, and marker techniques used could be involved. Buffers slightly increased rumen osmolality and also increased D (Harrison et al., 1975; Rogers et al., 1979; Rogers & Davis, 1982b; Okeke et al., 1983), except in the experiment of Teh et al. (1985) where low amounts of $NaHCO_3$ (1%) and $CaCO_3$ (1.2%) actually decreased osmolality, with no effect on D. Feeding $NaHCO_3$ and NaCl has normally increased water consumption (Rogers & Davis, 1982b; Rogers et al., 1979, 1982, 1985: Stokes, 1983), partially explaining the higher D (Rogers et al., 1979), although infusion of large amounts of water into the rumen of sheep (12 litres/day) had no effect on D or rumen fermentation parameters; only the rumen liquid volume was increased (Harrison et al., 1975).

Data about the influence of the buffers on nitrogen transactions in the rumen are rather scarce; generally, effects have been far from spectacular. Rumen ammonia levels were not influenced in the experiments of Stokes (1983), Mees et al. (1985) and Teh et al. (1985), but increased values were observed by Hadjipanayiotou et al. (1982), although protein degradation in the rumen of the sheep was significantly decreased. Other authors studied the degradation of soybean meal in sacco in the rumen. Buffers increased degradation rate and this positively correlated with rumen pH (Okeke et al., 1983; Kovacik et al., 1986). However, when k was increased simultaneously, the effective degradability remained unchanged (Okeke et al., 1983).

The effects on the flow of N constituents at the proximal duodenum of sheep and on microbial growth efficiency in the rumen are summarized in Table 13. Only in the experiment of Harrison et al. (1975) was total amino acid flow into the duodenum increased after infusion of artificial saliva, but then only in the two animals where D was also highest, and the increase was due to a higher flow of microbial amino acids. Feed protein degradation in the rumen was lower in one experiment (Hadjipanayiotou et al., 1982), but the absolute quantities were much lower than microbial amino acid flow. The efficiency of microbial growth was usually increased, accompanying higher D or k values, although Hadjipanayiotou et al. (1982) found no differences (Table 13). Conflicting results cannot be explained by feeding methods, as continuous feeding as well as two meals per day were used. However, it is striking that increases were obtained only when values on the control ration were low (Harrison et al., 1975; Mees et al., 1985). Another possibly important factor might be the effect on k, as Mees et al. (1985) reported an increase in k, but unfortunately the other authors only measured D. It must also be said that these effects were studied in sheep on low feed intake, and are probably not directly comparable with dairy cows and beef cattle on higher intake levels.

Table 13
Effect of Dietary Buffers on N Fractions in the Proximal Duodenum of Sheep

Reference[a]	Buffer	D (h^{-1})		Total AAN[b]		N fractions (g/day)		Microbial AAN		Microbial growth efficiency[c]	
						Diet. + endog. AAN					
		Control	Buffer	Control	Buffer	Control	Buffer	Control	Buffer	Control	Buffer
(1)[d]	Art. saliva	0·038	0·105	15·42	18·32	4·32	4·83	11·10	13·49	23·9	31·0
	+ 4% PEG	0·087	0·093	17·66	18·02	4·00	4·16	13·66	13·86	30·8	32·3
	(4 litres/day)	0·039	0·091	13·73	16·69	3·33	3·76	10·35	12·93	23·5	31·2
(2)	Art. saliva										
	20 g/kg	0·053	0·059	17·13	20·28	4·33	7·97	12·80	12·31	36·9	36·1
	40 g/kg	0·053	0·077	17·13	17·89	4·33	8·48	12·80	9·41	36·9	31·9
(3)	NaHCO$_3$ (3·5%)	no effect	but k ↑	21·5[e]	22·2[e]	7·7[f]	6·3[f]	13·7[g]	15·9[g]	15·4[h]	17·6[h]

[a] (1) Harrison et al., 1975. (2) Hadjipanayiotou et al., 1982. (3) Mees et al., 1985.
[b] AAN, amino acid nitrogen. [c] g N$_i$/kg OM$_r$. [d] Individual values from 3 sheep. [e] NAN, non-ammonia N. [f] Dietary and endogenous N. [g] Bacterial N. [h] g N$_i$/kg OM truly digested.

Apparent ruminal digestibility of crude fibre was either not affected or increased by addition of $NaHCO_3$ or artificial saliva (Mees et al., 1985; Kovacik et al., 1986; Wedekind et al., 1986), while starch digestion has been decreased or increased, dependent on the crude protein in the diet (Harrison et al., 1975; Mees et al., 1985).

In conclusion, buffers did affect rumen fermentation to a certain extent when high-concentrate diets were fed, but the exact mechanisms underlying the effects were not always clearly determinable. On high-roughage diets, no benefit of buffer addition was noted (Thomas & Hall, 1984). Control of pH and/or changes in D and k certainly are important, but strict separation between these factors is not possible. We showed that alterations in pH have had different effects on fermentation patterns in different experiments. An indication that changes in D are probably most important was given by Mees et al. (1985), where pH values were increased (6·14 to 6·55) by $NaHCO_3$, but neither D nor fermentation pattern was influenced.

The original idea of increasing D or k, with the aim of increasing microbial growth efficiencies, was based on chemostat experiments. In such systems, an increased D was related to higher propionate proportions and higher bacterial growth efficiencies (Isaacson et al., 1975; Van Nevel & Demeyer, 1979). In vivo, however, higher D values have normally resulted in higher acetate proportions. Clear differences exist, however, between an increase in D in a chemostat and in the rumen caused by infusion of hypertonic solutions. In the former, higher D means also that more substrate is available for fermentation per unit of time. In the latter situation, substrate availability remains unchanged. This case was perfectly imitated in a continuous culture of *Selenomonas ruminantium*, where D of the system was increased without change in rate of glucose supply (Silley & Armstrong, 1985). The organism changed its fermentation pattern towards higher proportions of acetate, in agreement with the *in vivo* studies with buffers or artificial saliva. The increase in acetate proportions *in vivo* has been explained by a change in microbial composition (Thomson et al., 1978), resulting from a decreased availability of soluble substrates (higher D) or a proliferation of crude-fibre degrading organisms which form higher relative proportions of acetate. These organisms are mainly associated with the particulate phase, meaning that their residence time in the rumen is less influenced by a changing D than is that of bacteria in the liquid phase. The experiment of Silley & Armstrong (1985) suggests that an increasing D may also change the metabolic activity of a single bacterial species.

A further complication is the fact that when rumen pH was varied between 5·2 and 7·1, in sheep with intragastric casein infusion and VFA infusion into the rumen, lower pH values increased the proportion of acetate, with an opposite effect on the other VFA. This was explained by decreased absorption rates of acetic acid at lower pH values. Osmotic pressure and VFA proportion were not related in this experiment (Macleod et al., 1984).

CONCLUSION

A different susceptibility of the various microbes or biochemical reactions to a particular rumen additive has been the basic principle governing rumen optimization. Compounds resulting in a generally lower microbial fermentation activity in the rumen are not interesting, and a high degree of specificity is desirable. However, the realization that the rumen fermentation is a very complex system of interrelated reactions makes a specific action by a particular additive almost impossible to obtain. A compound may possibly have a very specific direct effect, but the overall effect will be dictated by the changed interrelations within the whole system of rumen and indeed ruminant metabolism. For example, the use of methane inhibitors such as chloral hydrate or BES, specifically toxic only for the unique group of methane bacteria, induces a chain reaction with final effects on the pattern of volatile fatty acid production and even on degradation of certain amino acids.

It is striking that, with practically all additives (rumen manipulators), reducing equivalents (2H) are diverted from methane to propionate, and this is associated with effects on amino acid metabolism. This also explains why, in most manipulated situations, acetic acid increases are rarely seen. The complexity of interrelations within the system of rumen metabolism also includes microbial growth yield and digesta kinetics, which makes the study of rumen optimization a very laborious task. Furthermore, not only the effects on the system of rumen metabolism have to be considered but, because the sites of digestion are often changed, the effects on lower-digestive-tract metabolism should be determined, as well as possible effects at the ruminant tissue level. As adaptation to certain compounds has been observed, so effects *in vivo*, including feedlot performance, should be studied in long-term experiments. When experiments also include toxicological and even ecological aspects, the burden of the experiments for both researcher and experimental animal (multiple cannulae in the digestive tract) becomes considerable. Table 1 shows that there is ample choice of compounds interfering with rumen metabolism; for most compounds, however, experimental data on actions are limited and sometimes contradictory. The complexity of the rumen and the ruminant systems to be manipulated is the main reason for variability and contradiction in experimental results. Much more work to comprehend and rationalize this complexity is needed before safe and efficient application of rumen manipulation can be considered and recommended.

ACKNOWLEDGEMENT

Our research group is financially supported by the IWONL, Brussels, Belgium.

REFERENCES

Adams, D. C., Galyean, M. L., Kiesling, H. E., Wallace, J. D. & Finkner, M. D. (1981). Influence of viable yeast culture, sodium bicarbonate and monensin on liquid dilution rate, rumen fermentation and feedlot performance of growing steers and digestibility in lambs. *J. Anim. Sci.*, **53**, 780–9.

Anderson, P. H., Berrett, S., Catchpole, J., Gregory, M. W. & Brown, D. C. (1983). Effect of monensin on the selenium status of sheep. *Vet. Rec.*, **113**, 498.

Armentano, L. E. & Young, J. W. (1983). Production and metabolism of volatile fatty acids, glucose and CO_2 in steers and the effects of monensin on volatile fatty acid kinetics. *J. Nutr.*, **113**, 1265–77.

Baldwin, K. A., Bitman, J., Thompson, M. J. & Robbins, W. E. (1981). Effects of primary, secondary and tertiary amines on *in vitro* cellulose digestion and volatile fatty acid production by ruminal microorganisms. *J. Anim., Sci.*, **53**, 226–30.

Baldwin, K. A., Bitman, J. & Thompson, M. J. (1982). Comparison of N,N-dimethyldodecanamine with antibiotics on *in vitro* cellulose digestion and volatile fatty acid production by ruminal microorganisms. *J. Anim. Sci.*, **55**, 673–9.

Banks, W., Clapperton, J. L., Girdler, A. K. & Steele, W. (1984). Effect of inclusion of different forms of dietary fatty acid on the yield and composition of cow's milk. *J. Dairy Res.*, **51**, 387–95.

Barao, S. M., Bergen, W. G. & Hawkins, D. R. (1985). Effect of narasin and actaplanin alone or in combination on feedlot performance and carcass characteristics of finishing steers. *J. Anim. Sci.*, **61** (Suppl. 1), 637.

Barclay, R. A., Faulkner, D. B., Fahey, G. C. Jr & Cmarik, G. F. (1986). Effects of salinomycin on performance characteristics and apparent dry matter digestion by grazing beef steers. *Nutr. Rep. Int.*, **33**, 43–53.

Bartley, E. E., Herod, E. L., Bechtle, R. M., Sapienza, D. A. & Brent, B. E. (1979). Effect of monensin|or|lasalocid, with and without niacin or amichloral, on rumen fermentation and feed efficiency. *J. Anim. Sci.*, **49**, 1066–75.

Bauchop, T. & Mountfort, D. O. (1981). Cellulose fermentation by a rumen anaerobic fungus in both the absence and the presence of rumen methanogens. *Appl. Environ. Microbiol.*, **42**, 1103–10.

Beede, D. K. & Farlin, S. D. (1977a). Effects of antibiotics on apparent lactate and volatile fatty acid production: *in vitro* rumen fermentation studies. *J. Anim. Sci.*, **45**, 385–92.

Beede, D. K. & Farlin, S. D. (1977b). Effects of capreomycin disulfate and oxamycin on ruminal pH, lactate and volatile fatty acid concentrations in sheep experiencing induced acidosis. *J. Anim. Sci.*, **45**, 393–401.

Bergen, W. G. & Bates, D. B. (1984). Ionophores: their effect on production efficiency and mode of action. *J. Anim. Sci.*, **58**, 1465–83.

Berger, J., Rachlin, A. T., Scott, W. E., Sternboch, L. H. & Goldberg, M. W. (1951). The isolation of three new crystalline antibiotics from streptomyces. *J. Amer. Chem. Soc.*, **73**, 5295–5303.

Brent, B. E. & Bartley, E. E. (1984). Thiamin and niacin in the rumen. *J. Anim. Sci.*, **59**, 813–22.

Brethour, J. R. & Chalupa, W. V. (1977). Amichloral and monensin in high-roughage cattle rations. *J. Anim. Sci.*, **45** (Suppl.), 222.

Brock, F. M., Forsberg, C. W. & Buchanan-Smith, J. G. (1982). Proteolytic activity of rumen microorganisms and effects of proteinase inhibitors *Appl. Environ. Microbiol.*, **44**, 561–9.

Broderick, G. A. & Balthrop, J. E. Jr (1979). Chemical inhibition of amino acid deamination by ruminal microbes *in vitro*. *J. Anim. Sci.*, **49**, 1101–11.

Broudiscou, L., Van Nevel, C. J., Demeyer, D. I. & Jouany, J. P. (1988). Addition d'hydrolysat d'huile de soja dans la ration de mouton: effet sur la dégradation in sacco de la paille et de la cellulose. *Reprod. Nutr. Dév.*, **28**, 159–60.

Brown, D. L. & Hogue, D. E. (1985). Effects of feeding monensin sodium to lactating goats: milk composition and ruminal volatile fatty acids. *J. Dairy Sci.*, **68**, 1141–7.

Brown, H., Elliston, N. G., McAskill, J. W., Muenster, O. A. & Tonkinson, L. V. (1973). Tylosin phosphate and tylosin urea adduct for the prevention of liver abscesses, improved weight gains and feed efficiency in feedlot cattle. *J. Anim. Sci.*, **37**, 1085–9.

Bryant, M. P. & Robinson, I. M. (1962). Some nutritional characteristics of predominant culturable ruminal bacteria. *J. Bacteriol.*, **84**, 605–14.

Bueno, L. & Ralison, P. (1982). Effects of orotic acid on *in vitro* volatile fatty acid production by sheep rumen fluid. *J. Anim. Sci.*, **55**, 951–6.
Bull, L. S., Rumpler, W. V., Sweeney, T. F. & Zinn, R. A. (1979). Influence of ruminal turnover on site and extent of digestion. *Fed. Proc.*, **38**, 2713–19.
Cafantaris, B. (1981). Uber die Wirkung von Antibiotika-Zusätzen auf die Mikrobielle Gärung im Pansensaft *in vitro*. *Diss. Dr. Agr.*, Univ. Hohenheim.
Candau, M. & Kone, L. (1980). Influence de la thiamine sur la protéosynthèse bactérienne chez le mouton. *Reprod. Nutr. Dév.*, **20**, 1695–9.
Capper, B. S., Morgan, D. J. & Parr, W. H. (1977). Alkali-treated roughages for feeding ruminants: a review. *Tropical Sci.*, **19**, 73–88.
Chalupa, W. (1975). Rumen bypass and protection of proteins and amino acids. *J. Dairy Sci.*, **58**, 1198–1218.
Chalupa, W. (1977). Manipulating rumen fermentation. *J. Anim. Sci.*, **46**, 585–99.
Chalupa, W. (1984). Manipulation of rumen fermentation. In *Recent Advances in Animal Nutrition*, ed. W. Haresign & D. J. A. Cole. Butterworths, London, pp. 143–60.
Chalupa, W., Corbett, W. & Brethour, J. R. (1980). Effects of monensin and amichloral on rumen fermentation. *J. Anim. Sci.*, **51**, 170–9.
Chalupa, W., Patterson, J. A., Parish, R. C. & Chow, A. W. (1983). Effects of diaryliodonium chemicals on rumen fermentation *in vitro* and *in vivo*. *J. Anim. Sci.*, **57**, 186–94.
Chalupa, W., Rickabaugh, B., Kronfeld, D. S. & Sklan, D. (1984). Rumen fermentation *in vitro* as influenced by long-chain fatty acids. *J. Dairy Sci.*, **67**, 1439–44.
Chalupa, W., Vecchiarelli, B., Elser, A. E., Kronfeld, D. S., Sklan, D. & Palmquist, D. L. (1986). Ruminal fermentation *in vivo* as influenced by long-chain fatty acids. *J. Dairy Sci.*, **69**, 1293–1301.
Chen, M. & Wolin, M. J. (1979). Effect of monensin and lasalocid-sodium on the growth of methanogenic and rumen saccharolytic bacteria. *Appl. Environ. Microbiol.*, **38**, 72–7.
Clapperton, J. L. (1977). The effect of a methane-suppressing compound, trichloroethyl adipate, on rumen fermentation and the growth of sheep. *Anim. Prod.*, **24**, 169–81.
Clarke, R. T. J. (1977). Protozoa in the rumen ecosystem. In *Microbial Ecology of the Gut*, ed. R. T. J. Clarke & T. Bauchop. Academic Press, London, New York, San Francisco, pp. 251–75.
Coleman, C. S. (1979). The role of rumen protozoa in the metabolism of ruminants given tropical feeds. *Trop. Anim. Prod.*, **4**, 199–213.
Coombe, J. B., Dinius, D. A., Goering, H. K. & Oltjen, R. R. (1979). Wheat straw–urea diets for beef steers: alkali treatment and supplementation with protein, monensin and a feed intake stimulant. *J. Anim. Sci.*, **48**, 1223–33.
Coppock, C. E., Schelling, G. T., Byers, F. M., West, J. W., Labore, J. M. & Gates, C. E. (1986). A naturally occurring mineral as a buffer in the diet of lactating dairy cows. *J. Dairy Sci.*, **69**, 111–23.
Cotta, M. A. & Hespell, R. B. (1986). Protein and amino acid metabolism of rumen bacteria. In *Control of Digestion and Metabolism in Ruminants*, ed. L. P. Milligan, W. L. Grovum & A. Dobson. Prentice-Hall, Englewood Cliffs, New Jersey, pp. 122–36.
Cottyn, B. G., Fiems, L. O., Boucqué, C. V., Aerts, J. V. & Buysse, F. X. (1983). Effect of monensin-sodium and avoparcin on digestibility and rumen fermentation. *Zeitschr. Tierphysiol. Tierernähr. Futtermittelkde*, **49**, 277–86.
Cummins, K. A. & Papas, A. H. (1985). Effect of isocarbon-4 and isocarbon-5 volatile fatty acids on microbial protein synthesis and dry matter digestibility *in vitro*. *J. Dairy Sci.*, **68**, 2588–95.
Cuthbert, N. H. & Thickett, W. S. (1984). The efficacy of dietary avoparcin for improving the performance of growing-finishing beef cattle. *Anim. Prod.*, **39**, 195–200.
Czerkawski, J. W. (1969). Methane production in ruminants and its significance *World Rev. Nutr. Diet.*, **11**, 240–82.
Czerkawski, J. W. (1973). Effect of linseed oil fatty acids and linseed oil on rumen fermentation in sheep. *J. Agric. Sci.*, **81**, 517–31.

Czerkawski, J. W. & Breckenridge, G. (1975). New inhibitors of methane production by rumen micro-organisms: experiments with animals and other practical possibilities. *Br. J. Nutr.*, **34**, 447–57.

Czerkawski, J. W., Blaxter, K. L. & Wainman, F. W. (1966a). The metabolism of oleic, linoleic and linolenic acids by sheep with reference to their effects on methane production. *Br. J. Nutr.*, **20**, 349–62.

Czerkawski, J. W., Blaxter, K. L. & Wainman, F. W. (1966b). The effect of linseed oil and of linseed oil fatty acids incorporated in the diet on the metabolism of sheep. *Br. J. Nutr.*, **20**, 485–94.

Czerkawski, J. W., Blaxter, K. L. & Wainman, F. W. (1966c). The effect of functional groups other than carboxyl on the metabolism of C_{18} and C_{12} alkyl compounds by sheep. *Br. J. Nutr.*, **20**, 495–508.

Czerkawski, J. W., Cristie, W. W., Breckenridge, G. & Hunter, M. L. (1975). Changes in the rumen metabolism of sheep given increasing amounts of linseed oil in their diet. *Br. J. Nutr.*, **34**, 25–44.

Darden, D. E., Merchen, N. R., Berger, L. L., Fahey, G. C. Jr & Spears, J. W. (1985). Effects of avoparcin, lasalocid and monensin on sites of nutrient digestion in beef steers. *Nutr. Rep. Int.*, **31**, 979–89.

Davies, A. & Broome, A. W. J. (1982). Effects of the ionophoric antibiotic 139603 in lambs. *Proc. Nutr. Soc.*, **42**, 91A.

Davies, A., Nwaonu, H. N., Stanier, G. & Boyle, F. F. (1982). Properties of a novel series of inhibitors of rumen methanogenesis: *in vitro* and *in vivo* experiments including growth trials on 2,4-bis(trichloromethyl)benzo[1,3]dioxin-6-carboxylic acid. *Br. J. Nutr.*, **47**, 565–76.

Davis, C. L. (1979). The use of buffers in the rations of lactating dairy cows. In *Regulation of Acid-Base Balance*, ed. W. H. Hale & P. Meinhardt. Church and Dwight, Piscataway, New Jersey, pp. 51–4.

Davis, C. L. & Brown, R. E. (1970). Low-fat milk syndrome. In *Physiology of Digestion and Metabolism in the Ruminant*, ed. A. T. Phillipson. Oriel Press, Newcastle-upon-Tyne, pp. 545–65.

Dawson, K. A. & Boling, J. A. (1983). Monensin-resistant bacteria in the rumen of calves on monensin-containing and unmedicated diets. *Appl. Environ. Microbiol.*, **46**, 160–4.

De Jong, A. & Berschauer, F. (1983). Adaptation effects of ionophores on rumen fermentation. *S. Afr. J. Anim. Sci.*, **13**, 67–70.

Dellinger, C. A. & Ferry, J. G. (1984). Effect of monensin on growth and methanogenesis of *Methanobacterium formicicum*. *Appl. Environ. Microbiol.*, **48**, 680–2.

Demeyer, D. I. (1973). Lipidstoffwechsel im Pansen. In *Biologie und Biochemie der Mikrobiellen Verdauung*, ed. D. Giesecke & H. Henderickx. BLV Verlaggesellschaft. München, pp. 209–34.

Demeyer, D. I. (1981). Rumen microbes and digestion of plant cell walls. *Agric. Env.*, **6**, 295–337.

Dermeyer, D. (1988). Interdépendance des effets de la défaunation sur l'activité muralytique, le volume et la cinétique du contenu de rumen: résultats préliminaires et hypothèse. *Reprod. Nutr. Dev.*, **28**, 161–2.

Demeyer, D. (1988). Interdépendance des effets de la défaunation sur l'activité muralytique, methane production *in vitro* by mixed rumen bacteria. *Biochim. Biophys. Acta*, **137**, 484–97.

Demeyer, D. I. & Van Nevel, C. J. (1975). Methanogenesis, an integrated part of carbohydrate fermentation, and its control. In *Digestion and Metabolism in the Ruminant*, ed. I. W. McDonald & A. C. I. Warner. University of New England Publ. Unit, Armidale, Australia, pp. 366–82.

Demeyer, D. & Van Nevel, C. (1979). Protein fermentation and growth by rumen microbes. *Ann. Rech. Vet.*, **10**, 275–9.

Demeyer, D. I. & Van Nevel, C. J. (1986a). Chemical manipulation of rumen metabolism. In *Vet. Res. Comm.*, Vol. 1, *The Ruminant Stomach*, ed. L. A. A. Ooms, A.-D. Degryse & R. Marsboom. Janssen Research Foundation, Belgium, pp. 227–50.

Demeyer, D. & Van Nevel, C. (1986b). Influence of substrate and microbial interaction on efficiency of rumen microbial growth. *Reprod. Nutr. Dév.*, **26**, 161–79.

Demeyer, D. I. & Vervaeke, I. J. (1985). Rumen digestion and microbial processes for increasing the feed value of poor quality materials. In *Improved Utilization of Lignocellulosic Materials in Animal Feed*, OECD, Paris, pp. 31–61.

Demeyer, D., Henderickx, H. & Van Nevel, C. (1967). Influence of pH on fatty acid inhibition of methane production by mixed rumen bacteria. *Arch. Int. Physiol. Biochim.*, **75**, 555–6.

Demeyer, D. I., Van Nevel, C. J., Henderickx, H. K. & Martin, J. A. (1969). The effect of unsaturated fatty acids upon methane and propionic acid in the rumen. In *Energy Metabolism of Farm Animals*, ed. K. L. Blaxter. Oriel Press, Newcastle-upon-Tyne, pp. 139–47.

Demeyer, D. I., Van Nevel, C. J. & Henderickx, H. K. (1972). The effect of glucose on lipolysis in the rumen. *Proc. 2nd World Congr. Anim. Feeding*, Madrid, 5, pp. 39–43.

Demeyer, D., Van Nevel, C., Teller, E. & Godeau, J.-M. (1986). Manipulation of rumen digestion in relation to the level of production in ruminants. *Arch. Tierernähr.*, **36**, 132–43.

Dennis, S. M., Nagaraja, T. G. & Bartley, E. E. (1981a). Effects of lasalocid or monensin on lactate-producing or -using rumen bacteria. *J. Anim. Sci.*, **52**, 418–26.

Dennis, S. M., Nagaraja, T. G. & Bartley, E. E. (1981b). Effect of lasalocid or monensin on lactate production from *in vitro* rumen fermentation of various carbohydrates. *J. Dairy Sci.*, **64**, 2350–6.

Dennis, S. M., Nagaraja, T. G. & Bartley, E. E. (1981c). The effects of lasalocid, monensin and thiopeptin on rumen protozoa. *J. Dairy Sci.*, **64**, 132 (abstr.).

Dennis, S. M., Arambel, M. J., Bartley, E. E., Riddell, D. O. & Dayton, A. D. (1982). Effect of heated or unheated soybean meal with or without niacin on rumen protozoa. *J. Dairy Sci.*, **65**, 1643–6.

Devendra, C. & Lewis, D. (1974a). The interaction between dietary lipids and fibre in the sheep. 2. Digestibility studies. *Anim. Prod.*, **19**, 67–76.

Devendra, C. & Lewis, D. (1974b). The interaction between dietary lipids and fibre in sheep. *Malays Agric. Res.*, **3**, 163–76.

Dinius, D. A., Simpson, M. E. & Marsh, P. B. (1976). Effect of monensin fed with forage on digestion and the ruminal ecosystem of steers. *J. Anim. Sci.*, **42**, 229–34.

Dirksen, G. (1970). Acidosis. In *Physiology of Digestion and Metabolism in the Ruminant*, ed. A. T. Phillipson. Oriel Press, Newcastle-upon-Tyne, pp. 612–25.

Donoho, A. L. (1984). Biochemical studies on the fate of monensin in animals and in the environment. *J. Anim. Sci.*, **58**, 1528–39.

Durand, M. (1982). Orientation du métabolisme du rumen au moyen des additifs. *Ann. Zoot.*, **31**, 47–76.

Dyer, I. A., Koes, R. M., Herlugson, M. L., Bola Ojikutu, L., Preston, R. L., Zimmer, P. & Delay, R. (1980). Effect of avoparcin and monensin on performance of finishing heifers. *J. Anim. Sci.*, **51**, 843–6.

Ellenberger, M. A., Rumpler, W. V., Johnson, D. E. & Goodall, S. R. (1985). Evaluation of the extent of ruminal urease inhibition by sarsaponin and sarsaponin fractions. *J. Anim. Sci.*, **61**, 491 (suppl.).

Ellis, W. C., Deswysen, A. G., Pond, K. R. & Jenkins, W. L. (1985). Effects of monensin upon eating and ruminating behavior of cattle fed corn silage. *J. Anim. Sci.*, **61** (Suppl), 492.

Erfle, J. D., Boila, R. J., Teather, R. M., Mahadevan, S. & Sauer, F. D. (1982). Effect of pH on fermentation characteristics and protein degradation by rumen microorganisms *in vitro*. *J. Dairy Sci.*, **65**, 1457–64.

Esdale, W. J. & Satter, L. D. (1972). Manipulation of ruminal fermentation. 4. Effect of altering ruminal pH on volatile fatty acid production. *J. Dairy Sci.*, **55**, 964–70.

Estell, R. E., Hatfield, P. G., Galyean, M. L. & Ross, T. T. (1985). Effects of protein supplementation and lasalocid on intake, digestion and rumen fermentation in ewes fed a mixed hay diet. *J. Anim. Sci.*, **61** (Suppl.), 152.

Faulkner, D. B., Klopfenstein, T. J., Trotter, T. N. & Britton, R. A. (1985). Monensin effects on digestibility, ruminal protein escape and microbial protein synthesis on high-fiber diets. *J. Anim. Sci.*, **61**, 654–60.

Ferguson, K. A. (1975). The protection of dairy proteins and amino acids against microbial fermentation in the rumen. In *Digestion and Metabolism in the Ruminant*, ed. I. W. McDonald & A. C. I. Warner. University of New England Publ. Unit, Armidale, Australia, pp. 448–64.

Forsberg, C. W., Lovelock, L. K. A., Krumholz, L. & Buchanan-Smith, J. G. (1984). Protease activities of rumen protozoa. *Appl. Environ. Microbiol.*, **47**, 101–10.

Froetschel, M. A., Croom, W. J. Jr, Gaskins, H. R., Leonard, E. S. & Whitacre, M. D. (1983). Effects of avoparcin on ruminal propionate production and amino acid degradation in sheep fed high and low fiber diets. *J. Nutr.*, **113**, 1355–62.

Fuller, J. R. & Johnson, D. E. (1981). Monensin and lasalocid effects on fermentation *in vitro*. *J. Anim. Sci.*, **53**, 1574–80.

Fulton, W. R., Klopfenstein, T. J. & Britton, R. A. (1979). Adaptation to high concentrates by beef cattle. 2. Effect of ruminal pH alteration on rumen fermentation and voluntary intake of wheat diets. *J. Anim. Sci.*, **49**, 785–9.

Funk, M. A., Galyean, M. L. & Ross, T. T. (1986). Potassium and lasalocid effects on performance and digestion in lambs. *J. Anim. Sci.*, **63**, 685–91.

Galbraith, H., Miller, T. B., Paton, A. & Thompson, J. K. (1971). Antibacterial activity of long chain fatty acids and the reversal with calcium, magnesium, ergocalciferol and cholesterol. *J. Appl. Bacteriol.*, **34**, 803–13.

Gerson, T., John, A. & King, A. S. D. (1985). The effects of dietary starch and fibre on the *in vitro* rates of lipolysis and hydrogenation by sheep rumen digesta. *J. Agric. Sci.*, **105**, 27–30.

Gibson, M. L., Preston, R. L., Pritchard, R. H. & Goodall, S. R. (1985). Effect of sarsaponin and monensin on ruminal ammonia levels and *in vitro* dry-matter digestibilities. *J. Anim. Sci.*, **61** (Suppl.), 492.

Gill, D. R. & Owens, F. N. (1984). Narasin, actaplanin and tylosin for feedlot steers. *Anim. Sci. Res. Rep., Okl. Agric. Exp. Stn*, 198–203.

Gill, D. R., Owens, F. N., Fent, R. W. & Fulton, R. K. (1979). Thiopeptin and roughage level for feedlot steers. *J. Anim. Sci.*, **49**, 1145–50.

Girard, V. & Hawke, J. C. (1978). The role of holotrichs in the metabolism of dietary linoleic acid in the rumen. *Biochim. Biophys. Acta*, **528**, 17–27.

Goetsch, A. L. & Owens, F. N. (1985). Effects of sarsaponin on digestion and passage rates in cattle fed medium to low concentrate. *J. Dairy Sci.*, **68**, 2377–84.

Goetsch, A. L. & Owens, F. N. (1986a). Effects of dietary nitrogen level and ileal antibiotic administration on digestion and passage rates in beef heifers. 1. High-concentrate diets. *J. Anim. Sci.*, **62**, 830–43.

Goetsch, A. L. & Owens, F. N. (1986b). Effects of dietary nitrogen level and ileal antibiotic adminstration on digestion and passage rates in beef heifers. 2. High-forage diets. *J. Anim. Sci.*, **62**, 844–56.

Goetsch, A. L., Owens, F. N. & Brown, B. E. (1985). Sarsaponin level and digestion with high concentrate diets. *Anim. Sci. Res. Rep., Okl. Agric. Exp. Stn*, 303–5.

Goodrich, R. D., Garrett, J. E., Gast, D. R., Kirick, M. A., Larson, D. A. & Meiske, J. C. (1984). Influence of monensin on the performance of cattle. *J. Anim. Sci.*, **58**, 1484–98.

Greene, L. W., Schelling, G. T. & Byers, F. M. (1985). The effect of monensin and potassium on magnesium balance in lambs. *J. Anim. Sci.*, **61** (Suppl.), 494.

Grigat, G. A. & Mathison, G. W. (1982). Thiamin supplementation of an all-concentrate diet for feedlot steers. *Can. J. Anim. Sci.*, **62**, 807–19.

Grigat, G. A. & Mathison, G. W. (1983). Thiamin and magnesium supplementation of all-concentrate diets for feedlot steers. *Can. J. Anim. Sci.*, **63**, 117–31.

Guttierrez, G. G., Schake, L. M. & Byers, F. M. (1982). Whole plant grain sorghum silage processing and lasalocid effects on stocker calf performance and rumen fermentation. *J. Anim. Sci.*, **54**, 863–8.

Ha, J. K., Emerick, R. J. & Embry, L. B. (1983). *In vitro* effect of pH variations on rumen fermentation, and *in vivo* effects of buffers in lambs before and after adaptation to high concentrate diets. *J. Anim. Sci.*, **56**, 698–706.

Hadjipanayiotou, M., Harrison, D. G. & Armstrong, D. G. (1982). The effects upon digestion in sheep of the dietary inclusion of additional salivary salts. *J. Sci. Food Agric.*, **33**, 1057–62.

Hammond, A. C. & Carlson, J. R. (1980). Inhibition of ruminal degradation of L-tryptophan to 3-methylindole *in vitro. J. Anim. Sci.*, **51**, 207–14.

Haney, M. E. & Hoehn, M. M. (1967). Monensin, a new biologically active compound. 1. Discovery and isolation. *Antimicrob. Agents Chemoth.*, 349–52.

Hannah, S. M. & Stern, M. D. (1985). Effect of supplemental niacin or niacinamide and soybean source on ruminal bacterial fermentation in continuous culture. *J. Anim. Sci.*, **61**, 1253–63.

Hanson, T. L. & Klopfenstein, T. (1979). Monensin, protein source and protein levels for growing steers. *J. Anim. Sci.*, **48**, 474–9.

Hanson, L. J., Eisenbeis, H. G. & Givens, S. V. (1981). Toxic effects of lasalocid in horses. *Am. J. Vet. Res.*, **42**, 456–61.

Harfoot, C. G., Crouchman, M. L., Noble, R. C. & Moore, J. H. (1974). Competition between food particles and rumen bacteria in the uptake of long-chain fatty acids and triglycerides. *J. Appl. Bacteriol.*, **37**, 633–41.

Harrison, D. G. & McAllan, A. B. (1980). Factors affecting microbial growth yields in the reticulo-rumen. In *Digestive Physiology and Metabolism in Ruminants*, ed. Y. Ruckebusch & P. Thivend. MTP Press, Lancaster, pp. 205–26.

Harrison, D. G., Beever, D. E., Thomson, D. J. & Osbourn, D. F. (1975). Manipulation of rumen fermentation in sheep by increasing the rate of flow of water from the rumen. *J. Agric. Sci.*, **85**, 93–101.

Hart, M. R., Walker, H. G., Graham, R. P., Hanni, P. J., Brown, A. H. & Kohler, G. O. (1981). Steam treatment of crop residues for increased ruminant digestibility. 1. Effects of process parameters. *J. Anim. Sci.*, **51**, 402–8.

Hazlewood, G. & Edwards, R. (1981). Proteolytic activities of a rumen bacterium, *Bacteroides ruminicola* R8/4. *J. Gen. Microbiol.*, **125**, 11–15.

Hedde, R. D., Armstrong, D. G., Parish, R. C. & Quach, R. (1980). Virginiamycin effect on rumen fermentation in cattle. *J. Anim. Sci.*, **51** (Suppl. 1), 366–7.

Henderson, C. (1973). The effects of fatty acids on pure cultures of rumen bacteria. *J. Agric. Sci.*, **81**, 107–12.

Henderson, C., Stewart, C. S. & Hine, R. S. (1977). The effect of added tallow on the rumen digestion rate and microbial populations of sheep fed dried grass. *Proc. Nutr. Soc.*, **36**, 148A.

Henderson, C., Stewart, C. S. & Nekrep, F. V. (1981). The effect of monensin on pure and mixed cultures of rumen bacteria. *J. Appl. Bacteriol.*, **51**, 159–69.

Herod, E. L., Bechtle, R. M., Bartley, E. E. & Dayton, A. D. (1978). Buffering ability of several compounds *in vitro* and the effect of a selected buffer combination on ruminal acid production *in vivo. J. Dairy Sci.*, **61**, 1114–22.

Hicks, R. B., Gill, D. R. & Owens, F. N. (1985). Comparison of salinomycin to other ionophores for feedlot steers. *Anim. Sci. Res. Rep., Okl. Agric. Exp. Stn*, 281–5.

Hino, T. (1981). Action of monensin on rumen protozoa. *Jap. J. Zoot. Sci.*, **52**, 171–9.

Hino, T. (1983). Influence of hydrogen on the fermentation in rumen protozoa: *Entodinium* species. *Jap. J. Zoot. Sci.*, **54**, 320–8.

Hino, T. & Russell, J. B. (1985). Effect of reducing-equivalent disposal and NADH/NAD on deamination of amino acids by intact rumen microorganisms and their cell extracts. *Appl. Environ. Microbiol.*, **50**, 1368–74.
Hobson, P. N. & Wallace, R. J. (1982). Microbial ecology and activities in the rumen. *CRC Crit. Rev. Microbiol.*, **9**, 253–320.
Hogan, J. P. & Weston, R. H. (1969). The effect of antibiotics on ammonia accumulation and protein digestion in the rumen. *Austr. J. Agric. Res.*, **20**, 339–46.
Honeyfield, D. C., Nocerini, M. R., Carlson, J. R. & Hammond, A. C. (1985). The effect of ten ionophores on the formation of 3-methylindole from L-tryptophan *in vitro*. *J. Anim. Sci.*, **61** (Suppl. 1), 472.
Horton, G. M. J. & Nicholson, H. H. (1980). Rumen metabolism and feedlot responses by steers fed tylosin and monensin. *Can. J. Anim. Sci.*, **60**, 919–24.
Hungate, R. E. (1966). *The Rumen and Its Microbes*. Academic Press, London and New York, pp. 266–80.
Ikwuegbu, O. A. & Sutton, J. D. (1982). The effect of varying the amount of linseed oil supplementation on rumen metabolism in sheep. *Br. J. Nutr.*, **48**, 365–75.
Isaacson, H. R., Hinds, F. C., Bryant, M. P. & Owens, F. N. (1975). Efficiency of energy utilization by mixed rumen bacteria in continuous culture. *J. Dairy Sci.*, **58**, 1645–59.
Jarrell, K. F. & Sprott, G. D. (1982). Nickel transport in *Methanobacterium bryantii*. *J. Bacteriol.*, **151**, 1195–203.
Jarrell, K. F. & Sprott, G. D. (1983). The effect of ionophores and metabolic inhibitors on methanogenesis and energy-related properties of *Methanobacterium bryantii*. *Arch. Biochem. Biophys.*, **225**, 33–41.
Jenkins, T. C. & Palmquist, D. L. (1982). Effect of added fat and calcium on *in vitro* formation of insoluble fatty acid soaps and cell wall digestibility. *J. Anim. Sci.*, **55**, 957–63.
Jenkins, T. C. & Palmquist, D. L. (1984). Effect of fatty acids or calcium soaps on rumen and total nutrient digestibility of dairy rations. *J. Dairy Sci.*, **67**, 978–84.
Johnson, R. R. (1976). Influence of carbohydrate solubility on non-protein nitrogen utilization in the ruminant. *J. Anim. Sci.*, **43**, 184–91.
Johnson, R. J., Herlugson, M. L., Bola Ojikutu, L., Cordova, G., Dyer, I. A., Zimmer, P. & Delay, R. (1979). Effect of avoparcin and monensin on feedlot performance of beef cattle. *J. Anim. Sci.*, **48**, 1338–42.
Jones, G. A. (1968). Influence of acetohydroxamic acid on some activities *in vitro* of the rumen microbiota. *Can. J. Microbiol.*, **14**, 409–16.
Jouany, J. P. & Thivend, P. (1986). *In vitro* effects of avoparcin on protein degradability and rumen fermentation. *Anim. Feed Sci. Technol.*, **15**, 215–29.
Jouany, J. P., Bertin, G. & Thivend, P. (1986). Influence d'un nouvel antibiotique l'abiérizine, sur la dégradation de l'azote de différentes proteines et sur la formation de produits terminaux dans le rumen, mesurés *in vitro*. *Reprod. Nutr. Dévelop.*, **26**, 295–6.
Jouany, J. P., Demeyer, D. I. & Grain, J. (1988). Effect of defaunating the rumen. *Anim. Feed Sci. Technol.*, in press.
Katz, M. P., Nagaraja, T. G. & Fina, L. R. (1986). Ruminal changes in monensin and lasalocid-fed cattle grazing bloat-provocative alfalfa pasture. *J. Anim. Sci.*, **63**, 1246–57.
Kayouli, C., Van Nevel, C. J., Dendooven, R. & Demeyer, D. I. (1986). Effect of defaunation and refaunation of the rumen on rumen fermentation and N-flow in the duodenum of sheep. *Arch. Tierernähr.*, **36**, 827–37.
Kinashi, H., Otake, N. & Yonehara, H. (1973). The structure of salinomycin, a new member of the polyether antibiotics. *Tetrahedron Lett.*, **49**, 4955–8.
Kirk, D. J., Greene, L. W., Schelling, G. T. & Byers, F. M. (1985). Effects of monensin on Mg, Ca, P and Zn metabolism and tissue concentrations in lambs. *J. Anim. Sci.*, **60**, 1485–90.
Klopfenstein, T. (1978). Chemical treatment of crop residues. *J. Anim. Sci.*, **46**, 841–8.
Klopfenstein, T. J., Purser, D. B. & Tyznik, W. J. (1964). Influence of aureomycin on rumen metabolism. *J. Anim. Sci.*, **23**, 490–5.

Kopecny, J. & Wallace, R. J. (1982). Cellular location and some properties of proteolytic enzymes of rumen bacteria. *Appl. Environ. Microbiol.*, **43**, 1026–33.

Kovacik, A. M., Loerch, S. C. & Dehority, B. A. (1986). Effect of supplemental sodium bicarbonate on nutrient digestibilities and ruminal pH measured continuously. *J. Anim. Sci.*, **62**, 226–34.

Kowalczyk, J., Ørskov, E. R., Robinson, J. J. & Stewart, C. S. (1977). Effect of fat supplementation on voluntary food intake and rumen metabolism in sheep. *Br. J. Nutr.*, **37**, 251–7.

Kronfeld, D. S. (1970). Ketone body metabolism, its control and its implications in pregnancy, toxaemia, acetonemia, and feeding standards. In *Physiology of Digestion and Metabolism in the Ruminant*, ed. A. T. Phillipson. Oriel Press, Newcastle-upon-Tyne, pp. 566–83.

Lebzien, P., Rohr, K., Breves, G. & Höller, H. (1986). Untersuchungen über der Einflusz von Rumensin (Monensin-Natrium) auf die Stickstoffumsetzungen und die Thiaminnettosynthese in den Vormagen von Wiederkäuern. *Zeitschr. Tierphysiol. Tierernähr. Futtermittelkde*, **55**, 177–86.

Lemenager, R. P., Owens, F. N., Shockey, B. J., Lusby, K. S. & Totusek, R. (1978). Monensin effects on rumen turnover rate, twenty-four hour VFA pattern, nitrogen components and cellulose disappearance. *J. Anim. Sci.*, **47**, 255–61.

Lindsey, T. O., Hedde, R. D., Sokolek, J. A., Quach, R. & Morgenthien, E. A. (1985). *In vitro* characterization of aridicin activity in the rumen. *J. Anim. Sci.*, **61** (Suppl. 1), 464.

MacGregor, R. C. & Armstrong, D. G. (1983). The effect of avoparcin on rumen fermentation *in vitro*. *Proc. Nutr. Soc.*, **43**, 21A.

MacGregor, R. A. & Armstrong, D. G. (1984). The feed antibiotic avoparcin and net uptake of amino acids from the small intestine of sheep. *Can. J. Anim., Sci.*, **64** (Suppl.), 134–5.

Mackie, R. I., Bahrs, P. G. & Therion, J. J. (1984). Adaptation of rumen bacteria to sodium and monensin. *Can. J. Anim. Sci.*, **64** (Suppl.), 351–3.

Macleod, N. A., Ørskov, E. R. & Atkinson, T. (1984). The effect of pH on the relative proportions of ruminal volatile fatty acids in sheep sustained by intragastric infusions. *J. Agric. Sci.*, **103**, 459–62.

Maczulak, A. E., Dehority, B. A. & Palmquist, D. L. (1981). Effects of long-chain fatty acids on growth of rumen bacteria. *Appl. Environ. Microbiol.*, **42**, 856–62.

Mahadevan, S., Erfle, J. D. & Sauer, F. D. (1980). Degradation of soluble and insoluble proteins by *Bacteroides amylophilus* protease and by rumen microorganisms. *J. Anim. Sci.*, **50**, 723–8.

Makkar, H. P. S., Sharma, O. P., Dawra, R. K. & Negi, S. S. (1981). Effect of acetohydroxamic acid on rumen urease activity *in vitro*. *J. Dairy Sci.*, **64**, 643–8.

Marmer, W. N., Maxwell, R. J. & Wagner, D. G. (1985). Effects of dietary monensin on bovine fatty acid profiles. *J. Agric. Food Chem.*, **33**, 67–70.

Martin, S. A. & Macy, J. M. (1985). Effects of monensin, pyromellitic diimide and 2-bromoethanesulfonic acid on rumen fermentation *in vitro*. *J. Anim. Sci.*, **60**, 544–50.

Marty, R. J. & Demeyer, D. I. (1973). The effect of inhibitors of methane production on fermentation pattern and stoichiometry *in vitro* using rumen contents of sheep given molasses. *Br. J. Nutr.*, **30**, 369–76.

Masek, J., Gilka, J. & Docekalova, H. (1985). A note on an effect of sodium monensinate supplements on the level of some chemical elements in the organs and muscle of feedlot steers. *Anim. Prod.*, **40**, 511–13.

Mathers, J. C. & Miller, E. L. (1982). Some effects of chloral hydrate on rumen fermentation and digestion in sheep. *J. Agric. Sci.*, **99**, 215–24.

Matsuoka, T. (1976). Evaluation of monensin toxicity in the horse. *J. Amer. Vet. Med. Assoc.*, **169**, 1089–100.

McAllan, A. B., Knight, R. & Sutton, J. D. (1983). The effect of free and protected oils on the digestion of dietary carbohydrates between the mouth and duodenum of sheep. *Br. J. Nutr.*, **49**, 433–40.

Mees, D. C. & Merchen, N. R. (1985). Effects of sodium bicarbonate additions to wheat straw based diets on rumen turnover rates and nutrient digestibility by sheep. *Nutr. Rep. Int.*, **32**, 1067-72.

Mees, D. C., Merchen, N. R. & Mitchel, C. J. (1985). Effects of sodium bicarbonate on nitrogen balance, bacterial protein synthesis and sites of nutrient digestion in sheep. *J. Anim. Sci.*, **61**, 985-94.

Merchen, N. R. & Berger, L. L. (1985). Effect of salinomycin level on nutrient digestibility and ruminal characteristics of sheep and feedlot performance of cattle. *J. Anim. Sci.*, **60**, 1338-46.

Merry, R. J., Smith, R. H. & McAllan, A. B. (1982a). Glycosyl ureides in ruminant nutrition. 2. *In vitro* studies on the metabolism of glycosyl ureides and their free component molecules in rumen contents. *Br. J. Nutr.*, **48**, 287-304.

Merry, R. J., Smith, R. H. & McAllan, A. B. (1982b). Glycosyl ureides in ruminant nutrition. 3. *In vivo* studies on the metabolism of glycosyl ureides and corresponding mixtures of their free component molecules. *Br. J. Nutr.*, **48**, 305-18.

Miyairi, N., Miyoshi, T., Aoki, H., Kohsaka, M., Ikushima, H., Kunugita, K., Sakai, H. & Imanaka, H. (1972). Thiopeptin, a new feed additive antibiotic: microbiological and chemical studies. *Antimicrob. Agric. Chemoth.*, **1**, 192-9.

Mizwicki, K. L., Owens, F. N., Poling, K. & Burnett, G. (1980). Timed ammonia release for steers. *J. Anim. Sci.*, **51**, 698-703.

Moore, C. K., Amos, H. E., Evans, J. J., Lowrey, R. S. & Burdick, D. (1980). Nitrogen balance, abomasal crude protein and amino acids in wethers fed formaldehyde-treated coastalbermuda grass and infused with methionine, glucose or monensin. *J. Anim. Sci.*, **50**, 1145-59.

Muir, L. A. & Barreto, A. Jr (1979). Sensitivity of *Streptococcus bovis* to various antibiotics. *J. Anim. Sci.*, **48**, 468-73.

Muir, L. A., Rickes, E. L., Duquette, P. F. & Smith, G. E. (1980). Control of wheat-induced lactic acidosis in sheep by thiopeptin and related antibiotics. *J. Anim. Sci.*, **50**, 547-53.

Muntifering, R. B., Theurer, B. & Noon, T. H. (1981). Effects of monensin on site and extent of whole corn digestion and bacterial protein synthesis in beef steers. *J. Anim. Sci.*, **53**, 1565-73.

Nagaraja, T. G., Avery, T. B., Bartley, E. E., Roof, S. K. & Dayton, A. D. (1982). Effect of lasalocid, monensin or thiopeptin on lactic acidosis in cattle. *J. Anim. Sci.*, **54**, 649-58.

Nagaraja, T. G., Dennis, S. M., Galitzer, S. J. & Harmon, D. L. (1986). Effect of lasalocid, monensin and thiopeptin on lactate production from *in vitro* rumen fermentation of starch. *Can. J. Anim. Sci.*, **66**, 129-39.

Nakashima, T., Masuno, T., Sakauchi, R. & Hoshino, S. (1982). Effect of salinomycin on feed efficiency, ruminal and blood characteristics of steers. *Jap. J. Zoot.*, **53**, 541-6.

Newbold, C. J., Chamberlain, D. G. & Williams, A. G. (1985). Ruminal metabolism of lactic acid in sheep receiving a diet of sugar beet pulp and hay. *Proc. Nutr. Soc.*, **44**, 85A.

Nicholson, T. & Omer, S. A. (1983). The inhibitory effect of intestinal infusions of unsaturated long-chain fatty acids on forestomach motility of sheep. *Br. J. Nutr.*, **50**, 141-9.

Okeke, G. C., Buchanan-Smith, J. G. & Grovum, W. L. (1983). Effects of buffers on ruminal rate of passage and degradation of soybean meal in steers. *J. Anim. Sci.*, **56**, 1393-9.

Oltjen, R. R., Dinius, D. A. & Goering, H. K. (1977). Performance of steers fed crop residues supplemented with nonprotein nitrogen, minerals, protein and monensin. *J. Anim. Sci.*, **45**, 1442-51.

Olubobokun, J. A., Loerch, S. C. & Palmquist, D. L. (1985). Effect of tallow and tallow calcium soap on feed intake and nutrient digestibility in ruminants. *Nutr. Rep. Int.*, **31**, 1075-84.

Olumeyan, D. B., Nagaraja, T. G., Miller, G. W., Frey, R. A. & Boyer, J. E. (1986). Rumen microbial changes in cattle fed diets with or without salinomycin. *Appl. Environ. Microbiol.*, **51**, 340-5.

Ørskov, E. R. (1975). Manipulation of rumen fermentation for maximum food utilization. *World Rev. Nutr. Diet.*, **22**, 153–82.
Ørskov, E. R. (1979). Recent information on processing grain for ruminants. *Livestock Prod. Sci.*, **6**, 335–47.
Ørskov, E. R. & McDonald, I. (1979). The estimation of protein degradability in the rumen from incubation measurements weighted according to rate of passage. *J. Agric. Sci.*, **92**, 499–503.
Ørskov, E. R., Hine, R. S. & Grubb, D. A. (1978). The effect of urea on digestion and voluntary intake by sheep of diets supplemented with fat. *Anim. Prod.*, **27**, 241–5.
Owens, F. N. & Goetsch, A. L. (1986). Digesta passage and microbial protein synthesis. In *Control of Digestion and Metabolism in Ruminants*, ed. L. P. Milligan, W. L. Grovum & A. Dobson. Prentice-Hall, Englewood Cliffs, New Jersey, pp. 196–223.
Owens, F. N., Weakly, D. C. & Goetsch, A. L. (1984). Modification of rumen fermentation to increase efficiency of fermentation and digestion in the rumen. In *Herbivore Nutrition in the Subtropics and Tropics*, ed. F. M. C. Gilchrist & R. I. Mackie. Science Press, Craighall, South Africa, pp. 435–54.
Pal, R. N. & Negi, S. S. (1977). A note on rumen ammonia and blood urea levels in steers administered urea or sparingly soluble urea derivates as nitrogen supplements. *Anim. Prod.*, **24**, 117–19.
Palmquist, D. L. (1984). Use of fats in diets for lactating dairy cows. In *Fats in Animal Nutrition*, ed J. Wiseman. Butterworths, London, pp. 357–81.
Palmquist, D. L. & Conrad, H. R. (1978). High fat rations for dairy cows: effects on feed intake, milk and fat productions and plasma metabolites. *J. Dairy Sci.*, **61**, 890–901.
Palmquist, D. L. & Conrad, H. R. (1980). High fat rations for dairy cows: tallow and hydrolyzed blended fat at two intakes. *J. Dairy Sci.*, **63**, 391–5.
Palmquist, D. L., Jenkins, T. C. & Joyner, A. E. Jr (1986). Effect of dietary fat and calcium source on insoluble soap formation in the rumen. *J. Dairy Sci.*, **69**, 1020–5.
Parigi-Bini, R. (1979). Researches on virginiamycin supplementation of feeds used in intensive cattle management. In *Performance in Animal Production*, ed. G. Piana & G. Piva. Minerva Medica, Torino, pp. 237–50.
Parker, D. S., MacGregor, R. C., Finlayson, H. J., Stockill, P. & Balios, J. (1984). The effect of including avoparcin in the diet on cell turnover and enzyme activity in the mucosa of the rat small intestine. *Can. J. Anim. Sci.*, **64** (Suppl. 1), 136–7.
Paterson, J. A., Anderson, B. M., Bowman, D. K., Morrison, R. L. & Williams, J. E. (1983). Effect of protein source and lasalocid on nitrogen digestibility and growth by ruminants. *J. Anim. Sci.*, **57**, 1537–44.
Pendlum, L. C., Boling, J. A. & Bradley, N. V. (1980). Nitrogen sources and monensin levels for growing steers fed corn silage. *J. Anim. Sci.*, **50**, 29–34.
Perry, T. W., Shields, D. R., Dunn, W. J. & Mohler, M. T. (1983). Protein levels and monensin for growing and finishing steers. *J. Anim. Sci.*, **57**, 1067–76.
Poos, M. I., Hanson, T. L. & Klopfenstein, T. J. (1979). Monensin effects on diet digestibility, ruminal protein bypass and microbial protein synthesis. *J. Anim. Sci.*, **48**, 1516–24.
Potter, E. L., Raun, A. P., Cooley, C. O., Rathmacher, R. P. & Richardson, L. F. (1976). Effect of monensin on carcass characteristics, carcass composition and efficiency of converting feed to carcass. *J. Anim. Sci.*, **43**, 678–83.
Potter, E. L., Vanduyn, R. L. & Cooley, C. O. (1984). Monensin toxicity in cattle. *J. Anim. Sci.*, **58**, 1499–511.
Prange, R. W., Davis, C. L. & Clark, J. H. (1978). Propionate production in the rumen of Holstein steers fed either a control or monensin supplemented diet. *J. Anim. Sci.*, **46**, 1120–4.
Pressman, B. C. & de Guzman, N. T. (1975). Biological applications of ionophores: theory and practice. *Ann. NY Acad. Sci.*, **264**, 373–86.
Pressman, B. C., Painter, G. & Fahim, M. (1980). Molecular and biological properties of ionophores. In *Inorganic Chemistry in Biology and Medicine*, ed. A. E. Martell. Am. Chem. Soc., Washington, DC, pp. 3–22.

Preston, T. R. & Leng, R. A. (1986). *Matching Livestock Production Systems to Available Resources.* ILCA, Addis Ababa, 331 pp.
Preston, R. L., Pritchard, R. H. & Wolfrom, G. W. (1985). Lysocellin effects on the gain, feed intake and efficiency of growing-finishing cattle. *J. Anim. Sci.*, **61** (Suppl. 1), 493.
Prins, R. A. (1978). Nutritional impact of intestinal drug–microbe interactions. In *Nutrition and Drug Interrelations*, ed. J. N. Hatcock & J. Coon. Academic Press, New York and London, pp. 189–251.
Prins, R. A., Van Nevel, C. J. & Demeyer, D. I. (1972). Pure culture studies of inhibitors for methanogenic bacteria. *Ant. v. Leeuwenhoek J. Microbiol. Serol.*, **38**, 281–7.
Purser, D. B., Klopfenstein, T. J. & Cline, J. H. (1965). Influence of tylosin and aureomycin upon rumen metabolism and the microbial population. *J. Anim. Sci.*, **24**, 1039–44.
Rangnekar, D. V., Badve, V. C., Kharat, S. T., Sobale, B. N. & Joshi, A. L. (1982). Effect of high-pressure steam treatment on chemical composition and digestibility *in vitro* of roughages. *Anim. Feed Sci. Technol.*, **7**, 61–70.
Raun, A. P., Cooley, C. O., Potter, E. L., Rathmacher, R. P. & Richardson, L. F. (1976). Effect of monensin on feed efficiency of feedlot cattle. *J. Anim. Sci.*, **43**, 670–7.
Richardson, L. F., Raun, A. P., Potter, E. L., Cooley, C. O. & Rathmacher, R. P. (1976). Effect of monensin on rumen fermentation *in vitro* and *in vivo*. *J. Anim. Sci.*, **43**, 657–64.
Ricke, S. C., Berger, L. L., Van der Aar, P. J. & Fahey, G. C. Jr (1984). Effects of lasalocid and monensin on nutrient digestion, metabolism and rumen characteristics of sheep. *J. Anim. Sci.*, **58**, 194–202.
Riddell, D. O., Bartley, E. E. & Dayton, A. D. (1980). Effect of nicotinic acid on rumen fermentation *in vitro* and *in vivo*. *J. Dairy Sci.*, **63**, 1429–36.
Riddell, D. O., Bartley, E. E. & Dayton, A. D. (1981). Effect of nicotinic acid on microbial protein synthesis *in vitro* and on dairy cattle growth and milk production. *J. Dairy Sci.*, **64**, 782–91.
Riddell, D. O., Bartley, E. E., Arambel, M. J., Nagaraja, T. G., Dayton, A. D. & Miller, G. W. (1985). Effect of niacin supplementation on ruminal niacin synthesis and degradation in cattle. *Nutr. Rep. Int.*, **31**, 407–13.
Rogers, J. A. & Davis, C. L. (1982*a*). Rumen volatile fatty acid production and nutrient utilization in steers fed a diet supplemented with sodium bicarbonate and monensin. *J. Dairy Sci.*, **65**, 944–52.
Rogers, J. A. & Davis, C. L. (1982*b*). Effects of intraruminal infusions of mineral salts on volatile fatty acid production in steers fed high-grain and high-roughage diets. *J. Dairy Sci.*, **65**, 953–62.
Rogers, J. A., Marks, B. C., Davis, C. L. & Clark, J. H. (1979). Alteration of rumen fermentation in steers by increasing rumen fluid dilution rate with mineral salts. *J. Dairy Sci.*, **62**, 1599–605.
Rogers, J. A., Davis, C. L. & Clark, J. H. (1982). Alteration of rumen fermentation, milk fat synthesis, and nutrient utilization with mineral salts in dairy cows. *J. Dairy Sci.*, **65**, 577–86.
Rogers, J. A., Muller, L. D., Snyder, T. J. & Maddox, T. L. (1985). Milk production, nutrient digestion, and rate of digesta passage in dairy cows fed long or chopped alfalfa hay supplemented with sodium bicarbonate. *J. Dairy Sci.*, **68**, 868–80.
Rohr, K., Daenicke, R. & Oslage, H. J. (1978). Untersuchungen über den Einfluss verschiedner Fettbeimischungen zum Futter auf Stoffwechsel und Leistung von Milchkühen. *Landbauforsch. Völkenrode*, **28**, 139–50.
Rowe, J. B., Morrell, J. S. W. & Broome, A. W. J. (1982). Flavomycin as a ruminant growth promoter: investigation of the mode of action. *Proc. Nutr. Soc.*, **41**, 56A.
Rowe, J. B., Davies, A. & Broome, A. W. (1983). Effect of the novel ionophore ICI 139603 on nitrogen digestion in sheep. In *Protein Metabolism and Nutrition*, ed. R. Pion, M. Arnal & D. Bonin. INRA, Paris, pp. 259–62.
Rumpler, W. V., Johnson, D. E. & Bates, D. B. (1986). The effect of high dietary cation concentration on methanogenesis by steers fed diets with and without ionophores. *J. Anim. Sci.*, **62**, 1737–41.

Rumsey, T. S., Bitman, J., Wrenn, T. R., Baldwin, K.•A., Tao, H. & Thompson, M. J. (1982). Performance, ruminal fermentation and blood constituents of lambs fed N,N-dimethyldodecanamine and chlortetracycline. *J. Anim. Sci.*, **54**, 1040–50.

Russell, J. B. & Martin, S. A. (1984). Effects of various methane inhibitors on the fermentation of amino acids by mixed rumen microorganisms *in vitro. J. Anim. Sci.*, **59**, 1329–38.

Russell, J. B. & Sniffen, C. J. (1984). Effect of carbon-4 and carbon-5 volatile fatty acids on growth of mixed rumen bacteria *in vitro. J. Dairy Sci.*, **67**, 987–94.

Russell, J. B., Bottje, W. G. & Cotta, M. A. (1981). Degradation of protein by mixed cultures of rumen bacteria: identification of *Streptococcus bovis* as an actively proteolytic rumen bacterium. *J. Anim. Sci.*, **53**, 242–52.

Schaetzel, W. P. & Johnson, D. E. (1981). Nicotinic acid and dilution rate effects on *in vitro* fermentation efficiency. *J. Anim. Sci.*, **53**, 1104–8.

Schelling, G. T. (1984). Monensin mode of action in the rumen. *J. Anim. Sci.*, **58**, 1518–27.

Schussler, S. L., Fahey, G. C. Jr, Robinson, J. B., Masters, S. S., Loerch, S. C. & Spears, J. W. (1978). The effect of supplemental niacin on *in vitro* cellulose digestion and protein synthesis. *Int. J. Vit. Nutr. Res.*, **48**, 359–67.

Scott, T. W., Cook, L. J. & Mills, S. C. (1971). Protection of dietary polyunsaturated fatty acids against microbial hydrogenation in ruminants. *J. Am. Oil Chem. Soc.*, **48**, 358–64.

Shefet, G. & Ben-Ghedalia, D. (1982). Effect of ozone and sodium hydroxide treatments on the degradability of cotton straw monosaccharides by rumen microorganisms. *Eur. J. Microbiol. Biotechnol.*, **15**, 47–51.

Shell, L. A., Hale, W. H., Theurer, B. & Swingle, R. S. (1983). Effect of monensin on total volatile fatty acid production by steers fed a high grain diet. *J. Anim. Sci.*, **57**, 178–85.

Shields, D. R., Schaefer, D. M. & Perry, T. W. (1983). Influence of niacin supplementation and nitrogen source on rumen microbial fermentation. *J. Anim. Sci.*, **57**, 1576–83.

Silley, P. & Armstrong, D. G. (1985). Metabolism of the rumen bacterium *Selenomonas ruminantium* grown in continuous culture. *Lett. Appl. Microbiol.*, **1**, 53–5.

Slyter, L. L. (1979). Monensin and dichloroacetamide influences on methane and volatile fatty acid production by rumen bacteria *in vitro. Appl. Environ. Microbiol.*, **37**, 283–8.

Smith, S. I. & Boling, J. A. (1983). Captan administration and nutrient utilization in ruminants. *Nutr. Rep. Int.*, **27**, 969–76.

Spears, J. W. & Harvey, R. W. (1984). Performance, ruminal and serum characteristics of steers fed lasalocid on pasture. *J. Anim. Sci.*, **58**, 460–4.

Speth, J., Greenstein, M. & Maise, W. (1981). The mechanism of action of avoparcin, a glycopeptide antibiotic. *Abstracts 81st ASM Meeting*, A15.

Spires, H. R. & Algeo, J. W. (1983). Laidlomycin butyrate, an ionophore with enhanced intraruminal activity. *J. Anim. Sci.*, **57**, 1553–60.

Stack, R. J. & Cotta, M. A. (1986). Effect of 3-phenylpropanoic acid on growth of, and cellulose utilization by, cellulolytic ruminal bacteria. *Appl. Environ. Microbiol.*, **52**, 209–10.

Starnes, S. R., Spears, J. W., Froetschel, M. A. & Croom, W. J. Jr (1984). Influence of monensin and lasalocid on mineral metabolism and ruminal urease activity in steers. *J. Nutr.*, **114**, 518–28.

Steen, R. W. J. (1984). A comparison of unwilted and wilted grass silages offered to beef cattle without and with monensin sodium. *Grass Forage Sci.*, **39**, 35–41.

Stewart, C. S. & Duncan, S. H. (1985). The effect of avoparcin on cellulolytic bacteria of the ovine rumen. *J. Gen. Microbiol.*, **131**, 427–35.

Stewart, C. S., Crossley, M. V. & Garrow, S. H. (1983). The effect of avoparcin on laboratory cultures of rumen bacteria. *Eur. J. Appl. Microbiol. Biotechnol.*, **17**, 292–7.

Stokes, M. R. (1983). Effect of sodium bicarbonate on rumen turnover in frequently fed sheep. *Can. J. Anim. Sci.*, **63**, 721–5.

Stokes, M. R., Bull, L. S. & Halteman, W. A. (1985). Rumen liquid dilution rate in dairy cows fed once daily: effects of diet and sodium bicarbonate supplementation. *J. Dairy Sci.*, **68**, 1171–80.

Sundstøl, F., Said, A. N. & Arnason, J. (1979). Factors influencing the effect of chemical treatment on the nutritive value of straw. *Acta Agric. Scand.*, **29**, 179–90.
Sutton, J. D. (1980). Digestion and end-product formation in the rumen from production rations. In *Digestive Physiology and Metabolism in Ruminants*, ed. Y. Ruckebusch & P. Thivend. MTP Press, Lancaster, pp. 271–90.
Sutton, J. D., Smith, R. H., McAllan, A. B., Storry, J. E. & Corse, D. A. (1975). Effect of variations in dietary protein and of supplements of cod-liver oil on energy digestion and microbial synthesis in the rumen of sheep fed hay and concentrates. *J. Agric. Sci.*, **84**, 317–26.
Sutton, J. D., Knight, R., McAllan, A. B. & Smith, R. H. (1983). Digestion and synthesis in the rumen of sheep given diets supplemented with free and protected oils. *Br. J. Nutr.*, **49**, 419–32.
Tamminga, S., Van Vuuren, A. M., Van der Koelen, C. J., Khattab, H. N. & Van Gils, L. G. M. (1983). Further studies on the effect of fat supplementation of concentrates fed to lactating dairy cows. 3. Effect on rumen fermentation and site of digestion of dietary components. *Neth. J. Agric. Sci.*, **31**, 249–58.
Teh, T. H., Hemken, R. W. & Harmon, R. J. (1985). Comparison of buffers and their modes in the rumen and abomasum. *Nutr. Rep. Int.*, **32**, 1339–48.
Teller, E., Godeau, J.-M. & De Baere, R. (1978). Het gebruik van ureum en diuredo-isobutaan in de rundveevoeding. 1. Studie van de samenstelling van het pensvocht. *Landbouwtijdschrift*, **31**, 765–74.
Thauer, R. K. & Kröger, A. (1984). Energy metabolism of two rumen bacteria with special reference to growth efficiency. In *Herbivore Nutrition in the Subtropics and Tropics*, ed. F. M. C. Gilchrist & R. I. Mackie. Science Press, Craighall, South Africa, pp. 399–407.
Theuninck, D. H., Goodrich, R. D. & Meiske, J. C. (1981). Influence of captan on *in vitro* and *in vivo* digestibility of forage. *J. Anim. Sci.*, **52**, 377–81.
Thivend, P. & Jouany, J.-P. (1983). Effect of lasalocid sodium on rumen fermentation and digestion in sheep. *Reprod. Nutr. Dév.*, **23**, 817–28.
Thivend, P. & Jouany, J.-P. (1986). New developments and future perspectives in research on rumen function: additives. In *New Developments and Future Perspectives in Research on Rumen Function*, ed. A. Neimann-Sørensen. EEC Publ., Brussels, pp. 199–215.
Thomas, E. E. & Hall, M. W. (1984). Effect of sodium bicarbonate and tetrasodium pyrophosphate upon utilization of concentrate- and roughage-based cattle diets: cattle studies. *J. Anim. Sci.*, **59**, 1309–19.
Thomas, P. C. & Rook, J. A. F. (1977). Manipulation of rumen fermentation. In *Recent Advances in Animal Nutrition*, ed. W. Haresign & D. J. A. Cole. Butterworths, London, pp. 157–83.
Thomson, D. J. & Beever, D. E. (1980). The effect of conservation and processing on the digestion of forages by ruminant. In *Digestive Physiology and Metabolism in Ruminants*, ed. Y. Ruckebusch & P. Thivend. MTP Press, Lancaster, pp. 291–308.
Thomson, D. J., Beever, D. E., Latham, M. J., Sharpe, M.E. & Terry, R. A. (1978). The effect of inclusion of mineral salts in the diet on dilution rate, the pattern of rumen fermentation and the composition of the rumen microflora. *J. Agric. Sci.*, **91**, 1–7.
Thonney, M. L., Heide, E. K., Duhaime, D. J., Hand, R. J. & Perosio, D. J. (1981). Growth, feed efficiency and metabolite concentrations of cattle fed high forage diets with lasalocid or monensin supplements. *J. Anim. Sci.*, **52**, 427–33.
Thornton, J. H. & Owens, F. N. (1981). Monensin supplementation and *in vivo* methane production by steers. *J. Anim. Sci.*, **52**, 628–34.
Titchen, D. A., Reid, C. S. W. & Vlieg, P. (1966). Effects of intra duodenal infusions of fat on the food intake of sheep. *NZ Soc. Anim. Prod. Proc.*, **26**, 36–41.
Trei, J. E., Parish, R. C. & Scott, G. C. (1971). The effect of several methane inhibitors on performance and metabolite concentrations in fattening lambs. *J. Anim. Sci.*, **33**, 1171 (abstr.).

Unsworth, E. F., McCullough, I. I., McCullough, T. A., O'Neill, D. G., Steen, R. W. J. & Titterington, C. I. (1985). The effect of avoparcin and monensin on the performance of growing and finishing cattle offered grass silage-based diets. *Anim. Prod.*, **41**, 75–82.

Ushida, K., Jouany, J. P., Lassalas, B. & Thivend, P. (1984). Protozoal contribution to nitrogen digestion in sheep. *Can. J. Anim. Sci.*, **64** (Suppl.), 20–1.

Valdez, F. R., Bush, L. J., Goetsch, A. L. & Owens, F. N. (1986). Effect of steroidal sapogenins on ruminal fermentation and on production of lactating dairy cows. *J. Dairy Sci.*, **69**, 1568–75.

Van der Honing, Y., Wieman, B. J., Steg, A. & Van Donselaar, B. (1981). The effect of fat supplementation of concentrates on digestion and utilization of energy by productive dairy cows. *Neth. J. Agric. Sci.*, **29**, 79–92.

Van der Honing, Y., Tamminga, S., Wieman, B. J., Steg, A., Van Donselaar, B. & Van Gils, L. G. M. (1983). Further studies on the effect of fat supplementation of concentrates fed to lactating dairy cows. 2. Total digestion and energy utilization. *Neth. J. Agric. Sci.*, **31**, 27–36.

Van Dijck, P. J. (1969). Further bacteriological evaluation of virginiamycin. *Chemotherapy*, **14**, 322–32.

Van Maanen, R. W., Herbein, J. H., McGilliard, A. D. & Young, J. W. (1978). Effects of monensin on *in vivo* rumen propionate production and blood glucose kinetics in cattle. *J. Nutr.*, **108**, 1002–7.

Van Nevel, C. & Demeyer, D. (1976). Remming van de methaanproductie in de pens: een literatuuroverzicht. *Landbouwtijdschrift*, **29**, 431–51.

Van Nevel, C. J. & Demeyer, D. I. (1977a). Effect of monensin on rumen metabolism *in vitro*. *Appl. Environ. Microbiol.*, **34**, 251–7.

Van Nevel, C. J. & Demeyer, D. I. (1977b). Determination of rumen microbial growth *in vitro* from ^{32}P labelled phosphate incorporation. *Br. J. Nutr.*, **38**, 101–14.

Van Nevel, C. J. & Demeyer, D. I. (1979). Stoichiometry of carbohydrate fermentation and microbial growth efficiency in a continuous culture of mixed rumen bacteria. *Eur. J. Appl. Microb. Biotechnol.*, **7**, 111–20.

Van Nevel, C. J. & Demeyer, D. I. (1981). Effect of methane inhibitors on the metabolism of rumen microbes *in vitro*. *Arch. Tierernähr.*, **31**, 141–51.

Van Nevel, C. J., Henderickx, H. K., Demeyer, D. I. & Martin, J. (1969). Effect of chloral hydrate on methane and propionic acid in the rumen. *Appl. Microbiol.*, **17**, 695–700.

Van Nevel, C. J., Demeyer, D. I., Cottyn, B. G. & Henderickx, H. K. (1970). Effect of sodium sulphite on methane and propionate in the rumen. *Zeitschr. Tierphysiol. Tierernähr. Futtermittelkde*, **26**, 91–100.

Van Nevel, C. J., Demeyer, D. I. & Henderickx, H. K. (1971). Effect of fatty acid derivatives on rumen methane and propionate *in vitro*. *Appl. Microbiol.*, **21**, 365–6.

Van Nevel, C. J., Prins, R. A. & Demeyer, D. I. (1974). Observations on the inverse relationship between methane and propionate in the rumen. *Zeitschr. Tierphysiol. Tierernähr. Futtermittelkde*, **33**, 121–5.

Van Nevel, C. J., Demeyer, D. I. & Henderickx, H. K. (1984). Effect of virginiamycin on carbohydrate and protein metabolism in the rumen *in vitro*. *Arch. Tierernährung*, **34**, 149–55.

Varel, V. H. & Hashimoto, A. G. (1982). Methane production by fermentor cultures acclimated to waste from cattle fed monensin, lasalocid, salinomycin or avoparcin. *Appl. Environ. Microbiol.*, **44**, 1415–20.

Veira, D. M. (1986). The role of ciliate protozoa in nutrition of the ruminant. *J. Anim. Sci.*, **63**, 1547–60.

Vérité, R. Durand, M. & Jouany, J.-P. (1986). Influence des facteurs alimentaires sur la protéosynthèse microbienne dans le rumen. *Reprod. Nutr. Dév.*, **26**, 181–201.

Vijichulata, P., Henry, P. R., Ammerman, C. B., Becker, H. N. & Palmer, A. Z. (1980). Performance and tissue mineral composition of ruminants fed cage layer manure in combination with monensin. *J. Anim. Sci.*, **50**, 48–56.

Voigt, J., Piatkowski, B., Krawielitzki, R., Bergner, H. & Görsch, R. (1981). Isobutylidenharnstoff als NPN-Quelle für Wiederkäuer. 1. Pansenfermentation, Verdaulichheit der Nährstoffe und mikrobielle Proteinsynthese bei der Milchkuh. *Arch. Tierernähr.*, **31**, 45–56.

Vrchlabsky, J. (1980). Einfluss des Natriummonensinats auf die Qualität von Rindfleisch. *Fleischwirtschaft*, **60**, 1732–5.

Wahle, K. W. J. & Livesey, C. T. (1985). The effect of monensin supplementation of dried grass or barley diets on aspects of propionate metabolism, insulin secretion and lipogenesis in the sheep. *J. Sci. Food Agric.*, **36**, 1227–36.

Wallace, R. J. (1983). *In vitro* studies of potential inhibitors of protein degradation by rumen bacteria. In *Protein Metabolism and Nutrition*, ed. R. Pion, M. Arnal & D. Bodin. INRA Publ. 16, Vol. 2, pp. 219–22.

Wallace, R. J. & Brammall, M. L. (1985). The role of different species of bacteria in the hydrolysis of protein in the rumen. *J. Gen. Microbiol.*, **131**, 821–32.

Wallace, R. J. & Kopecny, J. (1983). Breakdown of diazotized proteins and synthetic substrates by rumen bacterial proteases. *Appl. Environ. Microbiol.*, **45**, 212–17.

Wallace, R. J., Czerkawski, J. W. & Breckenridge, G. (1981). Effect of monensin on the fermentation of basal rations in the Rumen Simulation Technique (Rusitec). *Br. J. Nutr.*, **46**, 131–48.

Webb, K. E. Jr, Fontenot, J. P. & Lucas, D. M. (1980). Metabolism studies in steers fed different levels of salinomycin. *J. Anim. Sci.*, **51**, 407 (suppl.).

Wedegaertner, T. C. & Johnson, D. E. (1983). Monensin effects on digestibility, methanogenesis and heat increment of a cracked corn-silage diet fed to steers. *J. Anim. Sci.*, **57**, 168–77.

Wedekind, K. J., Muntifering, R. B. & Barker, K. B. (1986). Effects of diet concentrate level and sodium bicarbonate on site and extent of forage fiber digestion in the gastrointestinal tract of wethers. *J. Anim. Sci.*, **62**, 1388–95.

West, J. W., Coppock, C. E., Nave, D. H. & Schelling, G. T. (1986). Effects of potassium buffers on feed intake in lactating dairy cows and on rumen fermentation *in vivo* and *in vitro*. *J. Dairy Sci.*, **69**, 124–34.

Whetstone, H. D., Davis, C. L. & Bryant, M. P. (1981). Effect of monensin on breakdown of protein by ruminal microorganisms *in vitro*. *J. Anim. Sci.*, **53**, 803–9.

Willis, C. M. & Boling, J. A. (1982). Level of protein with and without monensin for finishing steers fed at controlled intake. *Nutr. Rep. Int.*, **25**, 399–407.

Windham, W. R. & Akin, D. E. (1984). Rumen fungi and forage fiber degradation. *Appl. Environ. Microbiol.*, **48**, 473–6.

Wolfe, R. S. (1982). Biochemistry of methanogenesis. *Experientia*, **38**, 198–200.

Wolin, M. J. (1960). A theoretical rumen fermentation balance. *J. Dairy Sci.*, **43**, 1452–9.

Wolin, M. J. (1975). Interactions between the bacterial species of the rumen. In *Digestion and Metabolism in the Ruminant*, ed. I. W. McDonald & A. C. I. Warner. University of New England Publ. Unit, Armidale, Australia, pp. 134–48.

Yokoyama, M. T., Johnson, K. A., Dickerson, P. S. & Bergen, W. G. (1985). Effect of dietary monensin on the cecal fermentation of steers. *J. Anim. Sci.*, **61** (Suppl. 1), 469.

Zorilla-Rios, J., Horn, G. W., Andersen, M. A., Vogel, G. J., Ford, M. J. & Poling, K. B. (1985). Effect of stage of maturity of wheat pasture and lasalocid supplementation on intake, site and extent of nutrient digestion by steers. *Anim. Sci. Res. Rep., Okl. Agric. Exp. Stn*, 175–9.

14
Digestive Disorders and Nutritional Toxicity

K. A. Dawson

Department of Animal Sciences, University of Kentucky, Lexington, Kentucky, USA

&

M. J. Allison

National Animal Disease Center, Ames, Iowa, USA

Most of the compounds entering the rumen are metabolized by ruminal microbes and converted into nutrients which can be used by the host animal, as described in earlier chapters. The ruminal microorganisms may also metabolize many naturally occurring toxins in feeds to substances which do not pose a threat to animal health, and thus the organisms serve as a 'first line of defence' against toxic materials (Reid, 1973). However, some compounds entering the rumen are metabolized to form substances which are toxic to the host animal. Still other compounds in feeds inhibit or stimulate the activities of certain microorganisms; causing dramatic changes in ruminal fermentations. These types of metabolic activities may cause digestive disorders and pathological syndromes that have profound negative effects on the health of the host animal.

DIGESTIVE DISORDERS OF THE RUMEN

The microbial population in the rumen is maintained in balance by a frequent introduction of nutrients and by the physiological regulations provided by the animal (nutrient absorption, flow of saliva and flow out of the rumen). However, sudden changes in diet, or excessive intake of certain types of nutrients, can disrupt the normal microbial balance in the rumen and be detrimental to normal rumen function and to the health of the host animal.

Lactic acidosis

Lactic acidosis is a frequently observed, acute condition that is also known as 'grain engorgement', 'grain overload' and 'acute indigestion'. It occurs when ruminants are overfed on, or abruptly changed to, diets that contain large amounts of starch or other rapidly fermented carbohydrates. A high incidence of acidosis is associated with feedlot cattle which are rapidly changed from forage-based rations to high-concentrate rations, and the incidence can be reduced by management practices which allow time for the rumen microbial population to adapt to increasing amounts of grain.

Initially, when the supply of these carbohydrate substrates is abundant, the rates of acid production by the normal rumen microbes increase (Slyter, 1976). At the same time, some species begin to produce more lactic acid, and some ethanol begins to accumulate (Allison *et al.*, 1964). The production of these metabolites can be related to increases in the growth rates of several groups of rumen bacteria. At low growth rates, acetate and propionate or acetate and ethanol are the main products of *Selenomonas ruminantium* and *Streptococcus bovis*, respectively, but both produce lactic acid at high growth rates. Similarly, several species of ruminal ciliate protozoa produce more lactate and less acetate or butyrate when the supply of carbohydrates is abundant (Williams, 1986). In addition to regulation due to changes in growth rate, the lower pH in the rumen environment also serves to shift the metabolism of *S. bovis* toward lactate by altering cellular regulation of lactate dehydrogenase (Russell & Hino, 1985). In microbial cells, lactate production is advantageous when substrate is in excess, because at such times oxidation of reduced nicotinamide adenine dinucleotide and protection from the accumulation of excess acid (H^+ ions) inside the cell are particularly important.

Lactic acid is a stronger acid (pK_a 3·1) than the volatile acids normally produced in the rumen (average pK_a 4·8). A pH of about 5·5 seems to represent the tolerance limit for many rumen microbial species, but certain species will continue to grow in more acidic environments. Since *S. bovis* is tolerant of the acid conditions produced by the rapid fermentation, and also has the capacity to grow very rapidly, it becomes a major component of the rumen flora and thus establishes a self-perpetuating cycle of lactate production in the rumen. As a result, few bacterial or protozoal species other than the acid-tolerant lactobacilli can complete, and the diverse and predominantly Gram-negative species of the normal rumen are replaced by the Gram-positive *S. bovis* and lactobacilli. Further reduction in pH will inhibit the growth of even *S. bovis*, and a population that is almost a monoculture of lactobacilli can develop. The rate of population change may be influenced by many factors, but it is closely associated with an increase in the supply of readily fermentable substrates. The change may be very rapid and overgrowth by lactobacilli can occur within 24 hours.

Both D(+) and L(−) isomers of lactic acid are produced by the microorganisms in the rumen. At a pH of 6 the D-isomer constitutes around 20% of the total lactate produced, while when the pH is less than 5 and lactate concentrations are above 100 mM the D-isomer may account for 50% of the lactate produced. Enzymes for oxidation of L(−)-lactate (L-lactate dehydrogenase) are widely distributed in the cytosol of all host-animal tissues. The D-lactate must pass the mitochondrial membrane before it can be oxidized by the D-2-hydroxy-acid dehydrogenase (Giesecke & Stangassinger, 1980). Thus the D-lactic acid produced during acidosis is more slowly degraded and is considered more toxic.

Rumen protozoa have the capacity to assimilate starch and soluble sugars and store these energy-rich compounds inside the cell as amylopectin (Chapter 3). This process has survival value for the organisms since it allows metabolism to continue when substrates have been depleted. The process is also important to the host in that the rapid assimilation for storage as amylopectin by the protozoa can help to 'level out' the rate of feed-carbohydrate fermentation and can thus be protective against the burst of acid formation with which acidosis is associated.

Factors other than lactic acid, such as production of histamine and the release of endotoxins from normal rumen bacteria when they are exposed to the low pH, have also been implicated in this condition. However, it seems clear that the marked change in ruminal pH and lactic acid concentration are of major importance.

Bloat

Bloat is a rumen dysfunction which results from the accumulation of gas within the rumen. The trapped gas will produce rumen distension, and the build-up of pressure can be sufficient to cause circulatory impairment and death of the animal. Information concerning the development and physiological characteristics of this syndrome has been summarized in a number of recent reviews (Bartley et al., 1975; Reid et al., 1975; Howarth et al., 1986).

Occasionally bloat may result from the accumulation of free gas in the rumen, but it is most often associated with the formation of a stable froth which is a mixture of gas bubbles and small particles. When in contact with receptors in the cardia region of the reticulum, this froth prevents the opening of the cardia and eructation cannot occur. Although gas formation is definitely the result of microbial activities within the rumen, the development of a stable froth cannot always be related simply to microbial activities.

In 'legume' ('pasture') bloat, soluble plant proteins and saponins are major contributing factors to the formation and stabilization of the froth. Microbes may even provide some protection from this type of bloat by degrading these foam-producing soluble proteins and mucous. Legumes that are rapidly degraded in the rumen, however, have a greater bloat-producing potential than those that are slowly degraded (Howarth et al, 1986). This may be due to a more rapid release of foam-producing soluble proteins by both mechanical (chewing) and microbial actions as well as to more rapid gas production. On the other hand, in 'feedlot' bloat of grain-fed cattle the formation of an extracellular dextran slime by amylolytic bacteria in the rumen is an important factor in the development of a stable foam (Bartley et al., 1975). In this situation, control of the bloat would have to rely on the control of the slime-producing bacteria in the rumen.

There is also evidence that some animals are predisposed to bloat, and in some cases this has been related directly to conditions in the rumen just prior to feeding (Majak et al., 1983). Some of this increased susceptibility to bloat may be associated with the activities and composition of the microbial population in the rumen.

Despite over four decades of intensive research, the complex combinations of animal, dietary and microbial factors which interact in the formation of a stable froth in the rumen are not completely understood. It has not been possible consistently to associate bloat with any particular component in feed materials. In addition, the occurrence of bloat, or increased susceptibility to bloat, has not been definitely related to changes in the numbers or species of ruminal microorganisms. Although oils and detergents can be administered to reduce surface tension and decrease frothing in the rumen, prevention is the most efficient method for treating bloat. Current research is focused on producing bloat-safe pastures and breeding cattle with low susceptibility to bloat.

Thiaminase

The microbial population in the rumen synthesizes B vitamins (Hungate, 1966) to the extent that in most situations mature ruminants do not require exogenous supplies of these nutrients (see Chapter 11). Thiamine biosynthesis in the rumen can account for the resistance of ruminants to thiamine deficiency symptoms when they have been fed on bracken or other feedstuff containing thiaminase. However, there are neurological disorders of ruminants, known as polioencephalomalacia (PEM) or as cerebrocortical necrosis (CCN), that appear to be the result of the destruction of thiamine by thiaminase produced by the rumen bacteria. In a sense, thiaminase, like lactic acid, might be described as a toxic material produced by the rumen microbes, and it is so included in Table 2 (see later). Evidence that the acute symptoms observed are due to decreased levels of thiamine in tissues seems clear, since administration of thiamine is effective in rapidly eliminating the symptoms. Dietary conditions that will consistently lead to PEM and similar neurological disorders have not been established, but an increased incidence of these disorders has been related to the use of high-concentrate rations and dietary patterns that lead to sudden changes in the distribution of microorganisms in the rumen.

Several different types of thiamine-degrading enzymes have been identified in cultures of rumen microorganisms, but most research has been directed toward the destruction of thiamine by thiaminase I (Edwin et al., 1982). This enzyme catalyses a base-exchange reaction leading to the production of competitive thiamine analogues when suitable co-substrates are present. It is clear that affected animals have elevated levels of thiaminase I in their rumens, but thiaminase levels are not a reliable indicator of disease or impending disease symptoms. Production of thiaminase has been associated with a large variety of morphologically separate groups of bacteria in the rumen, but little is known about the major microorganisms responsible for production of this enzyme or about the co-substrates and products of the enzyme (Edwin & Jackman, 1982; Havens et al., 1983; Brent & Bartley, 1984). Work on CCN (PEM) using gnotobiotic lambs is described in Chapter 15.

DETOXIFICATION IN THE RUMEN

Ruminant animals are more resistant to the effects of many toxins than are monogastric animals. This resistance can often be related to the metabolism of these toxic compounds by rumen microbes. Table 1 lists examples of toxic materials and biochemical mechanisms associated with their detoxification in the rumen. Many of

Table 1
Substances Detoxified in the Rumen

Substance	Source	Mechanism of detoxification	Organisms involved	References
Pyrrolizidine alkaloid	Plant materials in feed	Reduction to the methylene derivative	*Peptostreptococcus heliotrinreducans*; unidentified Gram-negative cocci	Russell & Smith (1968); Lanigan (1976)
Nitrite	Reduction of nitrate in feeds	Reduction to ammonia	Unknown	Jones (1972); Allison & Reddy (1983)
Ricinoleic acid	Castor meal	Reduction to fatty acids	Unknown	Robb et al. (1973)
Ochratoxin A (mycotoxin)	Fungal contaminated feed	Hydrolysis to phenylalanine and ochratoxin α	Unknown	Hult et al. (1976)
Botulism toxin	Clostridial contaminated feed	Proteolysis by rumen bacteria	Unknown	Allison et al. (1976)
Oxalate	Plant materials in feeds	Decarboxylation to formate	*Oxalobacter formigenes*	Allison & Reddy (1983); Dawson et al. (1980)
Phyto-oestrogens (biochanin A and genistein)	Subterranean clover and red clover	Degradation to p-ethyl-phenol	Unknown	Nilsson et al. (1967); Braden et al. (1967)
3-Nitropropanol, 3-nitropropanoic acid	Hydrolysis of miserotoxins	Degraded to nitrite and reduction to ammonia	*Coprococcus* sp.; *Megasphaera elsdenii*; *Selenomonas ruminantium*	Majak & Cheng (1981); Majak & Clark (1980)
3-Hydroxy-4(1H)-pyridone	Degradation of mimosine	Unknown	Gram-negative, strict anaerobe (not identified)	Jones & Megarrity (1986); Allison et al. (1987)
Gossypol	Cottonseed meal	Bound to protein in the rumen	Microbial activities may not be required	Reiser & Fu (1962)

the transformations involve reductive reactions or hydrolytic activities of proteases. As might be expected, oxidative degradation and reactions requiring the cleavage of aromatic rings are not generally associated with detoxification in the rumen. In many instances the enzyme systems and the specific microorganisms associated with detoxification have not been identified, but a number of detoxification mechanisms have been studied in detail.

Adaptation or acquired tolerances to toxic compounds

One of the unique characteristics of ruminant animals is their ability to acquire tolerance to increased concentrations of toxic materials in feeds. In some cases this acquired tolerance can be related to changes in populations of rumen microbes that lead to increased rates of toxin degradation. Such population changes should be considered as a means of adaptation to environmental change. In this context, changes in the diet induce changes in the environment for the microbes in the rumen, and the resulting changes in microbial populations are adaptive processes that are important for survival of the animal. In many instances, the mechanisms which bring about these adaptive changes in rumen populations can be traced to selective processes which result in changes in the numbers and activities of specific functional groups of microorganisms. Ecological studies of such population changes have provided a great deal of insight into the dynamics of the microbial populations in the rumen.

Oxalate degradation

Both sodium and potassium oxalate are common in many types of plant material ingested by ruminants, but only a few plants contain amounts sufficient to be considered toxic. Ruminants fed on increasing quantities of oxalate develop the ability to tolerate more oxalate because of the increased rates of oxalate degradation by the rumen microorganisms (Allison & Reddy, 1983). Recently an oxalate-degrading anaerobe was isolated from enrichment cultures of rumen bacteria which had been adapted to increased oxalate levels (Dawson et al., 1980). This organism, *Oxalobacter formigenes*, uses oxalate as a sole energy source and produces carbon dioxide and formate as end-products.

Only small amounts of energy are released during the decarboxylation of oxalate to formate. This accounts for the relatively small cell yield (1·1 g of cells per mole of oxalate degraded) and the low numbers of oxalate-degrading organisms in the rumen (less than 0·5% of the total population) even in animals receiving high levels of oxalate in their feed. The ability to use oxalic acid as an energy source is rare among anaerobic bacteria and thus *O. formigenes* occupies a unique niche in the rumen. The dependence upon oxalate for growth accounts for the substrate-based selection of oxalate-degrading bacteria in the rumen that occurs within four days after increasing the amount of oxalate in the diet. This selection explains the ability of ruminants to adapt to, and tolerate, diets containing amounts of oxalate that would be lethal to non-adapted animals. *O. formigenes* has also been isolated from the intestinal tracts of other animals, and humans. In at least some of these anaerobic habitats, selection of oxalate-degrading organisms occurs along with increased supplies of dietary oxalate (Allison et al., 1985).

Nitrate

In the rumen, nitrate is reduced to nitrite and then to ammonia. This process appears to be the major pathway for dissimilatory nitrate metabolism in the rumen. However, the consumption of plant materials which contain high levels of nitrate will lead to an acute intoxication (Jones, 1972). This syndrome can be related to increased concentrations of the intermediate product, nitrite, in the rumen when the rates of nitrate reduction to nitrite exceed the rates of nitrite reduction to ammonia. Absorption of nitrite from the rumen leads to the production of methaemoglobin and interferes with the ability of blood cells to carry oxygen. Factors which influence the rates of nitrate and nitrite reduction include the availability of substrates which can be used to supply electrons for reduction reactions, the physiological conditions in the rumen environment, and the enzymatic capacity of the microbial population to carry out the reduction of these two compounds. There is evidence that increased dietary nitrate induces changes in the ruminal microbial population that lead to increased rates of nitrate and nitrite reduction and to increased tolerance to nitrate in the diet (Allison & Reddy, 1983).

Increased rates of nitrate reduction have been shown to be accompanied by increases in the relative proportions of nitrate-reducing bacteria in the rumens of nitrate-adapted animals. Nitrate reduction appears to compete with methanogenesis for electrons, and it inhibits methane production in the rumen. Similarly, increased nitrate levels in the diet can be expected to alter other metabolic processes such as end-product formation and microbial protein synthesis. Increased energy availability from dissimilatory nitrate reductase activity may give nitrate-reducing organisms a competitive advantage when nitrate is present in the feed. The selection of such organisms might explain the increasing nitrate reduction in the rumens of animals fed with increasing levels of nitrate.

Increased nitrate and nitrite reduction have been correlated with increased abilities to degrade 3-nitropropanol and 3-nitropropionic acid, the toxic metabolites derived from the miserotoxins that are found in several species of plants eaten by ruminants. Preliminary studies have suggested that nitrate may be used to induce rumen detoxification of 3-nitropropanol (Majak et al., 1982) and that dietary manipulations may be used more effectively to control the degradation of these aliphatic nitrotoxins in the rumen (Majak & Clark, 1980).

Pyrrolizidine alkaloids

Alkaloids are found in a number of plants used in livestock feeds. Some of these compounds are metabolized to reactive pyrrole derivatives by mixed-function oxidases in the liver and thus have pathological effects on the liver, kidney and respiratory system. Two of the potentially toxic alkaloids, heliotrine and lasciocarpine, are reduced to the non-toxic 1-methylene and 7-hydroxy-1-methyl derivatives in the rumen (Russell & Smith, 1968). The degradation of the alkaloids is influenced by alkaloid concentrations in the diet, but the basis for this regulation or other effects of alkaloids on the ruminal ecosystem is not fully understood.

Two anaerobic, heliotrine-degrading bacteria have been isolated from the rumen. One of these organisms, an unidentified Gram-negative coccus, was isolated from an animal which had not been purposely exposed to the alkaloid (Russell & Smith, 1968). This organism uses a number of different carbohydrates as substrates and appears to get very little useful energy from the reductive cleavage of heliotrine. Pure cultures of this organism require a short period to adapt to heliotrine reduction. A second organism, *Peptostreptococcus heliotrinreducans*, was isolated from an animal being fed on a high-alkaloid ration designed to favour the growth of alkaloid-reducing bacteria (Lanigan, 1976). This organism does not ferment carbohydrates, but rather appears to obtain energy from reductive pathways in which formate or hydrogen serve as electron donors and a variety of dissimilar compounds, such as fumarate, nitrate or pyrrolizidines, serve as electron acceptors. The reduction of heliotrine may improve the ability of this organism to compete in the rumens of animals exposed to this alkaloid. However, this organism can also use other ruminal intermediates such as nitrate and fumarate as electron acceptors, and simple selection on the basis of pyrrolizidine reduction may not completely account for increased activities of this organism in the rumen.

TOXIN PRODUCTION IN THE RUMEN

A number of compounds entering the rumen can be transformed into toxic derivatives by the microbial population. Some of the toxic compounds known to be produced by microbial activities in the rumen are listed in Table 2. Many of the toxins produced in the rumen result from the hydrolysis of plant materials. Cyanide, 3-nitropropanol, 3-nitropropanoic acid and goitrin are all produced after hydrolysis of glucoside bonds. Other toxic compounds are the result of a series of reactions which involve deamination, decarboxylation, demethylation and/or reduction. In most instances, the specific organisms which are responsible for the biochemical transformation and release of many of these toxic metabolites have not been identified.

The production of some toxic materials in the rumen can be balanced by degradative pathways which further transform the toxic intermediates into non-toxic products. Examples of such toxic intermediates include 3-nitropropanol and 3-nitropropanoic acid which are derived from miserotoxins (Gustine, 1979), nitrite which is produced from nitrate reduction and is further reduced to ammonia (Jones, 1972), 3-hydroxy-4(1H)-pyridone which is the degradation product of mimosine (Hegarty *et al*, 1964; Tangendjaja *et al.*, 1983), and lactic acid which, when produced at moderate rates in animals that have not been overfed, may be converted into propionate (Slyter, 1976). The concentrations and the ability of these

Table 2
Toxic Substances Produced in the Rumen

Substance	Source	Organisms involved	References
3-Methylindole (skatole)	Tryptophan in feeds	*Lactobacillus* sp.	Yokoyama et al. (1977); Carlson & Breeze (1984)
Nitrite	Reduction of nitrates in feed	*Selenomonas ruminantium*; *Veillonella alcalescens*	Allison & Reddy (1983); DeVries et al. (1974)
Lactic acid	Rapidly degraded carbohydrates (in high-concentrate diets)	*Streptococcus* spp.; *Lactobacillus* spp.	Slyter (1976)
3-Hydroxy-4(1H)-pyridone	Degradation product of mimosine	Unknown	Hegarty et al. (1964); Tangendjaja et al. (1983)
Cyanide	Hydrolysis of cyanogenic glycosides	Gram-negative rods and Gram-positive diplococci	James et al. (1975); Coop & Blakeley (1949)
Dimethyl disulphide	Degradation product of S-methylcysteine sulphoxide (brassica anaemia factor)	*Lactobacillus* spp.; *Veillonella alcalescens*; *Anaerovibrio lipolytica*; *Megasphaera elsdenii*	Barry et al. (1984); Smith (1974)
Equol	Demethylation and reduction of formononetin (a phyto-oestrogen)	Unknown	Nilsson et al. (1967)
Thiaminase	Microbial enzyme	*Clostridium sporogenes*; *Bacillus* spp. (other anaerobes)	Havens et al. (1983); Edwin & Jackman (1982)
3-Nitropropanoic acid and 3-nitropropanol	Hydrolysis of miserotoxins	Unknown	Gustine et al. (1977); Gustine (1979)
Goitrin	Hydrolysis of glucosinolates found in rapeseed meal and other crucifers	Unknown	VanEtten & Tookey (1978)

intermediates to produce toxicity, thus depend not only on the rate of toxin production but also on the rate of detoxification.

Mimosine degradation

Mimosine is a non-protein amino acid found in leaves, pods and seeds of tropical leguminous shrubs that belong to the genus *Leucaena*. Mimosine is hydrolysed in the rumen to produce 3-hydroxy-4(1H)-pyridone (3,4DHP), a potent goitrogen (Hegarty *et al.*, 1976). Diets with high amounts of *Leucaena* are, however, well tolerated by ruminants in some tropical areas. Differences in the susceptibility of ruminants to intoxication has been related to differences in microbial populations in the rumens of animals in different geographic regions. The microbial populations in the rumens of goats in Hawaii have the ability to degrade 3,4DHP, while rumen microbial populations obtained from animals in Australia and central North America do not have this capability (Jones & Megarrity, 1983). Animal-to-animal transfer of 3,4DHP-degrading bacteria occurs readily and it has been possible to transfer a mixed population of 3,4DHP-degrading bacteria from goats in Hawaii to cattle in Australia (Jones & Megarrity, 1986). This transfer conveyed protection from 3,4DHP toxicity to the inoculated animals. However, anaerobic bacteria capable of degrading both 3,4DHP and its isomer 2,3DHP have not yet been characterized.

Examples of geographically limited distribution of functional rumen bacteria such as is found for 3,4DHP-degrading organisms appear to be rare. These results suggest that resident microorganisms may not always be able to occupy the new and unique ecological niches that may accompany the introduction of exotic feedstuffs into the rumen.

Phyto-oestrogens

Infertility in sheep grazing on subterranean clover has been associated with certain isoflavone compounds which exhibit oestrogenic activity. The oestrogenic activity of one of these phyto-oestrogens, formononetin, may be enhanced in the rumen if it is demethylated to from equol which is a more potent oestrogen. Nilsson *et al.* (1967) showed that this conversion of formononetin was greater in the rumens of animals receiving clover diets than in animals receiving a ration containing oats. The microorganisms involved in these conversions have not been identified, but the dissimilar conversion patterns in animals fed on different rations suggest that differences in equol production may be related to differences in the microbial populations that are selected by the different rations.

Other phyto-oestrogens, genistein and biochanin A, are degraded in the rumen to the non-oestrogenic metabolites phenolic acid and *p*-ethylphenol. These processes tend to decrease the oestrogenic activities of these compounds and thus may influence the severity of problems that have been associated with ingestion of many strains of subterranean clover (Braden *et al.*, 1967).

Tryptophan degradation

The microbial population in the rumen plays an important role in the development of pulmonary oedema and emphysema, a disease associated with increased intake of L-tryptophan from lush pastures (Carlson & Breeze, 1984). The toxic metabolite 3-methylindole is the product of a two-step microbial degradation of tryptophan in the rumen and is responsible for the pathological syndrome in cattle (Yokoyama & Carlson, 1974). A number of rumen bacteria degrade tryptophan to indoleacetic acid. Indoleacetic acid is subsequently degraded to 3-methylindole by an anaerobic strain of *Lactobacillus* which does not ferment tryptophan (Yokoyama *et al.*, 1977). Sudden dietary changes induce increased production of 3-methylindole. There appears to be a greater rate of tryptophan conversion into 3-methylindole in animals fed on forage-containing diets than in those fed on concentrate rations (Carlson & Breeze, 1984). This suggests 3-methylindole production may be due both to the increased availability of tryptophan and to diet-induced changes in the rumen microbial population which result in increased rates of tryptophan degradation.

Brassica anaemia factor

Kale and some other *Brassica* species contain substantial levels of *S*-methyl-L-cysteine sulphoxide (SMCO), a free amino acid which is degraded in the rumen to dimethyl disulphide through a series of hydrolytic and reductive reactions. This metabolite is thought to be the primary cause of haemolytic anaemia in ruminants grazing on kale and similar root crops. Four organisms have been shown to have significant *S*-alkyl-cysteine sulphoxide lyase activity and can carry out the hydrolytic cleavage of SMCO. These are strains of *Lactobacillus*, *Veillonella alcalescens*, *Megasphaera elsdenii* and *Anaerovibrio lipolytica* (Smith, 1974). The factors which influence SMCO degradation by mixed populations of rumen microorganisms have not been studied in detail, but the degradation of SMCO appears to be very rapid in the rumens of animals receiving fresh kale, and little of the amino acid can be detected in their rumen contents. The rates of degradation were greater in the rumens of sheep fed on kale than in those fed on fresh lucerne (Barry *et al.*, 1984). This suggests an adaptive change in the rumen population and the selection of specific groups of SMCO-degrading bacteria. Some work with gnotobiotic lambs relating to degradation of SMCO is described in Chapter 15.

NATURALLY OCCURRING COMPOUNDS INHIBITING MICROBIAL ACTIVITIES

A number of naturally occurring compounds inhibit the activities of mixed microbial populations and the growth of pure cultures of rumen bacteria (Table 3). However, details of the mechanisms by which these compounds influence rumen metabolism and digestive processes are not known. For the most part, studies of mixed populations have focused on specific activities, such as cellulose digestion. For example, both oxalate and an alkaloid, perloline, inhibit cellulose degradation by rumen populations *in vitro*, but their abilities to influence other digestive processes in the rumen have not been evaluated (James *et al.*, 1967; Bush *et al.*, 1976). Such inhibitory compounds are found in a variety of plant materials, but they are often at concentrations which would not be expected significantly to influence microbial activities in the rumen.

Similarly there is evidence that some of the phenolic monomers released during microbial degradation of fibrous plant materials may inhibit the growth of certain rumen bacteria and depress cellulose digestion (Chesson *et al.*, 1982; Jung *et al.*, 1983; Jung, 1985). Phenolic compounds which significantly influence cellulose digestion include benzoic acid, cinnamic acid, caffeic acid, ferulic acid and vanillin.

Table 3
Toxins Inhibiting Microbial Activities in the Rumen

Inhibitory substance	Inhibitory activity	References
Alkaloids (perloline)	Inhibits cellulose degradation in the rumen and the growth of several cellulose-degraders in pure culture	Hemken *et al.* (1984); Bush *et al.* (1972)
Oxalate	Inhibits cellulose degradation in the rumen	James *et al.* (1967)
Essential oils	Inhibits growth and activity of mixed populations of rumen bacteria and several bacterial species	James *et al.* (1975); Oh *et al.* (1968)
Phenolic monomers	Inhibits the growth of specific ruminal microorganisms and microbial activities *in vitro*	Jung *et al.* (1983); Chesson *et al.* (1982)
Mycotoxins	Inhibits cellulose digestion and alters volatile fatty acid production	Mertens (1979)

Pure cultures of rumen bacteria differ in their tolerance of phenolic compounds, and specific phenolic compounds have different effects on rates of cellulose digestion by mixed populations of rumen bacteria. In general, studies have tested the inhibitory properties of phenolic compounds at concentrations substantially greater than those found in most forages. However, data from these studies do suggest that phenolic acids influence microbial activities and forage digestion in the rumen (see Chapter 8).

A number of mycotoxins have antimicrobial activities. It is possible that specific mycotoxins may influence the growth of specific microorganisms or influence metabolic processes in the rumen. Several investigators have reported that mycotoxins inhibit fibre digestion and volatile fatty acid formation in the rumen, but it has been difficult to show consistently significant effects of mycotoxins on rumen fermentations (Mertens, 1979).

REFERENCES

Allison, M. J. & Reddy, C. A. (1983). Adaptation of gastrointestinal bacteria in response to changes in dietary oxalate and nitrate. In *Current Perspectives in Microbial Ecology*, ed. M. J. Klug & C. A. Reddy. American Society for Microbiology, Washington, DC, pp. 248–56.

Allison, M. J., Dougherty, R. W., Bucklin, J. A. & Snyder, E. E. (1964). Ethanol accumulation in the rumen after overfeeding with readily fermentable carbohydrate. *Science*, **144**, 54–5.

Allison, M. J., Maloy, S. E. & Matson, R. R. (1976). Inactivation of *Clostridium botulinum* toxin by ruminal microbes from cattle and sheep. *Appl. Environ. Microbiol.*, **32**, 685–8.

Allison, M. J., Dawson, K. A., Mayberry, W. R. & Foss, J. G. (1985). *Oxalobacter formigenes* gen. nov., sp. nov.: oxalate-degrading anaerobes that inhabit the gastrointestinal tract. *Arch. Microbiol.*, **141**, 1–7.

Allison, M. J., Cook, H. M. & Stahl, D. A. (1987). Characterization of rumen bacteria that degrade dihydroxypyridine compounds produced from mimosine. *Proc. 2nd International Symp. Nutr. Herbivores*, Brisbane, Australia.

Barry, T. N., Manley, T. R. & Duncan, S. J. (1984). Quantitative digestion by sheep of carbohydrates, nitrogen and S-methyl-cysteine sulphoxide in diets of fresh kale (*Brassica oleracea*). *J. Agric. Sci.* (Camb.), **102**, 479–86.

Bartley, E. E., Meyer, R. M. & Fina, L. R. (1975). Feedlot or grain bloat. In *Digestion and Metabolism in the Ruminant*, ed. I. W. McDonald & A. C. I. Warner. University of New England Publishers, Armidale, Australia, pp. 551–62.

Braden, A. W. H., Hart, N. K. & Lamgerton, J. A. (1967). The oestrogenic activity and metabolism of certain isoflavones in sheep. *Aust. J. Agric. Res.*, **18**, 335–48.

Brent, B. E. & Bartley, E. E. (1984). Thiamine and niacin in the rumen. *J. Anim. Sci.*, **59**, 813–22.

Bush, L. P., Boling, J. A., Allen, G. & Buckner, R. C. (1972). Inhibitory effects of perloline to rumen fermentation *in vitro*. *Crop Sci.*, **12**, 277–9.

Bush, L. P., Burton, H. & Boling, J. A. (1976). Activity of tall fescue alkaloids and analogues in *in vitro* rumen fermentation. *J. Agric. Food Chem.*, **24**, 869–72.

Carlson, J. R. & Breeze, R. G. (1984). Ruminal metabolism of plant toxins with emphasis on indolic compounds. *J. Anim. Sci.*, **58**, 1040–9.

Chesson, A., Stewart, C. S. & Wallace, R. J. (1982). Influence of plant phenolic acids on growth and cellulolytic activity of rumen bacteria. *Appl. Environ. Microbiol.*, **44**, 597–603.

Coop, I. E. & Blakley, R. L. (1949). The metabolism and toxicity of cyanides and cyanogenic glucosides in sheep. *NZ J. Sci. Technol.*, **39A**, 277–91.

Dawson, K. A., Allison, M. J. & Hartman, P. A. (1980). Isolation and some characteristics of anaerobic oxalate-degrading bacteria from the rumen. *Appl. Environ. Microbiol.*, **30**, 833–9.

DeVries, W., VanWijck-Kapteyn, M. C. & Oosterhuis, S. K. H. (1974). The presence and function of cytochromes in *Selenomonas ruminantium*, *Anaerovibrio lipolytica* and *Veillonella alcalescens*. *J. Gen. Microbiol.*, **81**, 69–78.

Edwin, E. E. & Jackman, R. (1982). Ruminant thiamine requirement in perspective. *Vet. Res. Commun.*, **5**, 237–50.

Edwin, E. E., Jackman, R. & Jones, P. (1982). Some properties of thiaminases associated cerebrocortical necrosis. *J. Agric. Sci.*, **99**, 271–5.

Giesecke, D. & Stangassinger, M. (1980). Lactic acid metabolism. In *Digestive Physiology and Metabolism in Ruminants*, ed. Y. Ruckebusch & P. Thivend. AVI, Westport, Connecticut, pp. 523–39.

Gustine, D. L. (1979). Aliphatic nitro compounds in crownvetch: a review. *Crop Sci.*, **19**, 197–203.

Gustine, D. L., Moyer, B. G., Wangsness, P. J. & Shenk, J. S. (1977). Ruminal metabolism of 3-nitropropanoyl-D-glucopyranoses from crownvetch. *J. Anim. Sci.*, **44**, 1107–11.

Havens, T. R., Caldwell, D. R. & Jensen, R. (1983). Role of predominant rumen bacteria in the cause of polioencephalomalacia (cerebrocortical necrosis) in cattle. *Am. J. Vet. Res.*, **44**, 1451–5.

Hegarty, M. P., Shinckel, P. G. & Court, R. D. (1964). Reaction of sheep to the consumption of *Leucaena glauca* Brenth and to its toxic principle mimosine. *Aust. J. Agric. Res.*, **15**, 153–67.

Hegarty, M. P., Court, R. D., Christie, G. S. & Lee, C. P. (1976). Mimosine in *Leucaena leucocephala* is metabolised to a goitrogen in ruminants. *Aust. Vet. J.*, **52**, 490.

Hemken, R. W., Jackson, J. A. & Boling, J. A. (1984). Toxic factors in tall fescue. *J. Anim. Sci.*, **58**, 1011–16.

Howarth, R. E., Cheng, K. J., Majak, W. & Costerton, J. W. (1986). Ruminant bloat. In *Control of Digestion and Metabolism in Ruminants*, ed. L. P. Mulligan, W. L. Grovum & A. Dobson. Prentice-Hall, Englewood Cliffs, New Jersey, pp. 516–27.

Hult, K., Teiling, A. & Gatenbeck, S. (1976). Degradation of ochratoxin A by a ruminant. *Appl. Environ. Microbiol.*, **32**, 443–4.

Hungate, R. E. (1966). *The Rumen and Its Microbes*. Academic Press, New York.

James, L. F., Street, J. C. & Butcher, J. E. (1967). *In vitro* degradation of oxalate and of cellulose by rumen ingesta from sheep fed *Halogeton glomeratus*. *J. Anim. Sci.*, **26**, 1438–44.

James, L. F., Allison, M. J. & Littledike, E. T. (1975). Production and modification of toxic substances in the rumen. In *Digestion and Metabolism in the Ruminant*, ed. I. W. McDonald & A. C. I. Warner. University of New England Publishers, Armidale, Australia, pp. 576–90.

Jones, G. A. (1972). Dissimilatory metabolism of nitrate by rumen bacteria. *Can. J. Microbiol.*, **18**, 1783–7.

Jones, R. J. & Megarrity, R. G. (1983). Comparative toxicity responses of goats fed on *Leucaena leucocephala* in Australia and Hawaii. *Aust. J. Agric. Res.*, **34**, 781–90.

Jones, R. J. & Megarrity, R. G. (1986). Successful transfer of DHP-degrading bacteria from Hawaiian goats to Australian ruminants to overcome the toxicity of *Leucaena*. *Aust. Vet. J.*, **63**, 259–62.

Jung, H. G. (1985). Inhibition of structural carbohydrate fermentation by forage phenolics. *J. Sci. Agric.*, **36**, 74–80.

Jung, H. G., Fahey, G. C. & Garst, J. E. (1983). Simple phenolic monomers of forages and effects of *in vitro* fermentation on cell wall phenolics. *J. Anim. Sci.*, **57**, 1294–305.

Lanigan, G. W. (1976). *Peptococcus heliotrinreducans* sp. nov., a cytochrome-producing anaerobe which metabolizes pyrrolizidine alkaloids. *J. Gen. Microbiol.*, **94**, 1–10.

Majak, W. & Cheng, K. J. (1981). Identification of rumen bacteria that anaerobically degrade aliphatic nitrotoxins. *Can. J. Microbiol.*, **27**, 646–50.

Majak, W. & Clark, L. J. (1980). Metabolism of aliphatic nitro compounds in bovine rumen fluid. *Can. J. Anim. Sci.*, **60**, 319–25.

Majak, W., Cheng, K. J. & Hall, J. W. (1982). The effect of cattle diet on the metabolism of 3-nitropropanol by ruminal microorganisms. *Can. J. Anim. Sci.*, **62**, 855–60.

Majak, M., Howarth, R. E., Cheng, K. J. & Hall, J. W. (1983). Rumen conditions that predispose cattle to pasture bloat. *J. Dairy Sci.*, **66**, 1683–8.

Mertens, D. R. (1979). Biological effects of mycotoxins upon rumen function and lactating dairy cows. In *Interactions of Mycotoxins in Animal Production*. National Research Council, National Academy of Sciences, Washington, D.C., pp. 118–36.

Nilsson, A., Hill, J. L. & Davies, H. L. (1967). An *in vitro* study of formononetin and biochanin A metabolism in rumen fluid from sheep. *Biochim. Biophys. Acta*, **148**, 92–8.

Oh, H. K., Jones, M. B. & Longhurst, W. M. (1968). Comparison of rumen microbial inhibition resulting from various essential oils isolated from relative unpalatable plant species. *Appl. Microbiol.*, **16**, 39–44.

Reid, C. S. W. (1973). Limitations to productivity of herbage-fed ruminants that arise from the diet. In *Chemistry and Biochemistry of Herbage*, Vol. 3, ed. G. W. Butler & R. W. Bailey. Academic Press, London, pp. 215–62.

Reid, C. S. W., Clarke, R. T. J., Cockrem, F. R. M., Jones, W. T., McIntosh, J. T. & Wright, D. E. (1975). Physiological and genetical aspects of pasture (legume) bloat. In *Digestion and Metabolism in the Ruminant*, ed. I. W. McDonald & A. C. I. Warner. University of New England Publishers, Armidale, Australia, pp. 524–36.

Reiser, R. & Fu, H. C. (1962). The mechanism of gossypol detoxification by ruminant animals. *J. Nutr.*, **76**, 215–18.

Robb, J. G., Laben, R. C., Walker, H. G. & Herring, V. (1973). Castor meal in dairy rations. *J. Dairy Sci.*, **57**, 443–50.

Russell, J. B. & Hino, T. (1985). Regulation of lactate production in *Streptococcus bovis*: a spiraling effect that contributes to rumen acidosis. *J. Dairy Sci.*, **68**, 1712–21.

Russell, G. R. & Smith, R. M. (1968). Reduction of heliotrine by rumen microorganisms. *Aust. J. Biol. Sci.*, **21**, 1277–90.

Slyter, L. L. (1976). Influence of acidosis on rumen function. *J. Anim. Sci.*, **43**, 910–29.

Smith, R. H. (1974). Kale poisoning. *Rep. Rowett Inst.* (Aberdeen), **30**, 112–31.

Tangendjaja, B., Hogan, J. P. & Willis, R. B. H. (1983). Degradation of mimosine by rumen contents: effects of feed composition and *Leucaena* substrates. *Aust. J. Agric. Res.*, **34**, 289–93.

Van Etten, C. H. & Tookey, H. L. (1978). Glocosinolates in cruciferous plants. In *Effects of Poisonous Plants on Livestock*, ed. R. F. Keeler, K. R. VanKampen & L. F. James. Academic Press, New York, pp. 507–20.

Williams, A. G. (1986). Rumen holotrich cilate protozoa. *Microbiol. Rev.*, **50**, 25–49.

Yokoyama, M. T. & Carlson, J. R. (1974). The dissimilation of tryptophan and related indolic compounds by ruminal microorganisms *in vitro*. *Appl. Microbiol.*, **27**, 540–8.

Yokoyama, M. T., Carlson, J. R. & Holdeman, L. V. (1977). Isolation and characteristics of a skatole-producing *Lactobacillus* sp. from the bovine rumen. *Appl. Environ. Microbiol.*, **34**, 837–42.

15
Models, Mathematical and Biological, of the Rumen Function

P. N. Hobson*

*Honorary Research Fellow, Biochemistry Department,
Marischal College, Aberdeen University, Aberdeen, UK*

*Formerly Head, Microbial Chemistry Department,
Rowett Research Institute, Aberdeen, UK*

&

J.-P. Jouany

*Laboratoire de la Digestion des Ruminants, INRA Digestion,
Theix, France*

TYPES OF MODELS

The biological model

In the preceding chapters the functions of the rumen microorganisms as a whole and in the animal, of the microorganisms transferred to a laboratory vessel, and of microorganisms isolated from the rumen mixture acting alone or with one or two others in laboratory apparatus, have been described. It is initially assumed in the experiments *in vitro* that the microbes will react as they do in the rumen. Some comparison of the results with measurements made on the organisms *in vivo* is then made to see how far the initial assumption can be verified. The fact that the reactions *in vitro* do not reproduce those *in vivo* does not necessarily mean that the test organism has no place in the rumen system; it may be that its growth conditions in the rumen have not been properly reproduced in the laboratory, or it may mean that its growth in the rumen is overshadowed by some unisolated organism. On the other hand, the fact that the rumen reactions are apparently reproduced *in vitro* does not always mean that the test organism has a place in the rumen flora and that it is there behaving as in the laboratory experiment. Pure cultures of many organisms may react similarly, as may mixed cultures containing a variety of components.

These cultures may be referred to as 'biological models' of the rumen system. However, they are, probably invariably, reproducing only a part of the rumen reactions. This may be by design; a pure culture or mixed culture of two or three components cannot be expected to reproduce more than the reactions it was

* Present address: 4 North Deeside Road, Aberdeen AB1 7PL, UK.

designed to study. On the other hand, it may not be safe to assume that a complex mixed culture, such as an artificial rumen, is reproducing all features of the natural rumen because it is reproducing the main features.

One of the assumptions inherent in the tests with pure, or simple defined, cultures is that, if the reactions *in vitro* and *in vivo* are very similar, the test organisms must be those which are carrying out the rumen reactions, or else that any other organisms concerned are very similar to the test organisms, and that the rumen reaction requires these organisms alone. As a corollary, it is assumed that if the complex mixed culture, the artificial rumen, functions in its measured parameters as does the rumen although it is known that some organisms present in the natural rumen are not found in the artificial habitat, then these organisms are not necessary to the proper function of the natural system.

Various experiments, of the type just discussed and others, suggested that the functional rumen population was composed of a limited number of organisms and that the others in the complex population that exists in the actual rumen were unnecessary or there only by chance, possibly kept growing by continued inoculation from the animal's surroundings. Undoubtedly there are some organisms, common to soils and waters for instance, that are always present in the rumen in low numbers but to which, because of their properties and the numbers in which they occur, no reasonable function can be assigned (see Chapter 2). Indeed, their properties may make it reasonable to assume that their apparent growth is only the result of continued inoculation. This assumption is probably quite true for organisms present in only tens or hundreds per ml of rumen contents. It is more difficult to prove this hypothesis for organisms which are present in numbers of some millions per ml, even though this number is only a fraction of 1% of the rumen bacterial population of some 10^{10} per ml.

Whether or not the biological model is reproducing *in vitro* the reactions of the rumen flora, it cannot physically reproduce the rumen and its functions in association with the animal. Because of contamination from its surroundings, the normal animal cannot be used to study the hypothesis of the limited functional rumen population. Thus the idea of using ruminants brought up in a sterile environment, with no rumen microbes, to which could be given a defined rumen flora, came about. This is the 'gnotobiotic lamb' concept. The functions of a rumen population *in toto* could be investigated not only by measuring certain parameters but by assessing others in the ability of the flora to keep the animal alive and healthy. The gnotobiotic lamb is still a 'model' in that it seeks to reproduce the rumen function without reproducing the natural rumen flora. It can also be used to model such things as the natural invasion of the rumen by animal-pathogenic organisms, or the involvement of certain bacteria in rumen malfunction or detoxification of certain toxic feeds (Chapter 14). However, unlike the *in vitro* model, the *in vivo* model must reproduce the whole digestive function to keep the animal alive, even if the main objective is to study only part of the system.

To test the functions of the rumen ciliate protozoa, a rather less rigorously controlled biological model than the gnotobiotic lamb can be used, and the production of, and the results obtained with, the two types of model will be described later in this chapter.

The mathematical model

The rumen system is an 'open' microbial culture. A mass of microbial nutrients, in the animal feed, enters the rumen at intervals. In the rumen the nutrients are degraded to various extents to different end-products which may be absorbed from the rumen or pass on with undegraded feed components to the more distal portions of the gut. The microorganisms grow, die, lyse or pass on with the undegraded feed. Liquid, as saliva or drinking water, also enters and leaves the rumen. The system is thus concerned with changes in weights, of microbes and feedstuffs and products, and with rates of degradation and synthesis, and passage into and out of the system. All these changes are related, and it should theoretically be possible to make expressions describing the relationships and use these expressions to determine the effects of changes in one factor on another. This is a 'mathematical' model of quantitative functions of the rumen microbes. The first step in making a mathematical model is to be able to describe the system in words or a diagram. This then should suggest the data which will be needed to evaluate quantitatively the changes of the description. The model may not describe the whole system; this may be impossible or not necessary, but if the whole is not described then it should be remembered that the rest of the system may affect the part described.

Over many years various mathematical expressions have been developed which try to describe the growth and metabolism of bacteria under different conditions. These expressions vary in complexity and the extent to which they describe the bacterial functions (see, for instance, Bazin, 1983; Roels, 1983). This is partly because of theoretical limitations and partly because of the uses to which the expressions have been put. A simple description of oxygen uptake rates which does not concern itself with the mechanisms of oxygen metabolism may be quite sufficient in fermentor control, for instance, and one of the reasons for mathematical modelling is to produce control systems for microbial fermentations. Another of the reasons for mathematical modelling of bacterial growth and metabolism is that the success of the model in describing the actual behaviour of bacteria in culture can help to determine the extent of our knowledge of the processes which go on in the bacterial cell and in populations of bacteria. A further, and more usual, reason is that the model hopes to describe the system with such accuracy that it can replace experiments with the actual system and predict what will happen if some controlling parameters of the system are changed. Thus a rumen model might hope to predict what would happen to production of VFA or other metabolites if feed was changed in quantity or quality. However, the complexity of a rumen model will depend on the use to which the model is to be put as well as the state of knowledge of the microbial system. The mathematical model is like the *in vitro* fermentations with pure or mixed rumen bacteria in that it can describe the whole or only a part of the digestive process, and its success is not dependent on describing accurately the whole.

MATHEMATICAL MODELS

General considerations

Since it is impossible to produce a detailed and accurate model of a pure culture of a bacterium growing on a simple medium, it follows that a complete model of a mixed, natural culture such as the rumen is also impossible. The necessary information to make either model is not available, and the practicalities of formulating the equations and the computing involved if the information were available would be beyond most laboratories. The model must be simplified. The stirred-tank anaerobic digester treating a solids-containing feedstock is similar in many ways to the rumen, although some parameters can be more definitely controlled than for the rumen. Much attention has been paid to modelling digesters, and the rumen modeller can find useful information in papers on digester models. In both systems there is a mixed culture of bacteria performing different reactions and degrading different substrates, but all interacting in some way. One simplification is to assume that some interactions are not taking place, e.g. that the growth of fermentative bacteria is limited by carbohydrate availability and that ammonia as a nitrogen source is in excess. Thus the reactions of the deaminative bacteria can be ignored. Or carbohydrates may be assumed to be fermented directly to VFA, ignoring the possible intermediary formation of lactic or succinic acid. Another simplification is to assume that one, final, reaction in those going on in the mixed culture is the rate-controlling one for all the reactions. Thus methanogenesis, in removing products of fermentative reactions, has been used as such in simplified models of digesters, and digesters have in fact been modelled solely on the basis of a simplified methanogenic reaction. Methanogenesis, removing hydrogen which might inhibit cellulolysis, could be considered as a controlling reaction in the rumen (see Chapter 11). However, before this kind of model can be made the controlling reaction must be definitely determined.

Another simplification, used almost universally, is to consider the reactions of a population of bacteria to be those of a single type of organism. Thus cellulose degradation, in which a number of organisms is involved (see Chapter 8), could be considered to take place by the actions of one organism at rates and to fermentation products which are some mean of those found in the rumen when cellulose or fibre is fed.

A further simplification is to consider the feedstuffs as homogeneous materials. Thus fibre is considered as an entity (e.g. 'cellulose'), degraded at such and such an overall rate, disregarding the chemically heterogeneous makeup of the fibre and the possibility that it will be attacked at different rates as degradation proceeds.

Simplification is necessary for reasons of computation and availability of data, but simplification may make it easier to use the model predictively and generally. Hobson (1985) produced a model of a batch anaerobic digestion of a fibrous feedstock in which a number of different groups of particles were degraded by surface bacteria not only at different overall rates but also at rates which varied as the particles were degraded and different fibre structures were exposed. This model reproduced the different rates of gas production during the course of the digestion, and it could be used as the basis of a continuous-culture model, with a very large amount of computing. However, although the principles could be applied to the degradation of other feedstocks and to other systems such as the rumen, numerical application would need detailed knowledge of the degradation of each particular type of particle in the feedstock. Since surface degradation of particles is also influenced by the shape, and by the size, of the particle (Hobson, 1987), knowledge not only of the rates of degradation of, say, straw particles as attack continued but also of the shapes and sizes of particle used in the feedstock would be needed. So the more detailed the model the more difficult it is to use it predictively for other systems. One way of producing a simplified model is by 'lumping' all data together. For instance, all the available data on rates of degradation of 'pig waste' in anaerobic digesters or of 'grain' or 'wheat' in rumens can be reduced to some mean rate. This rate may well give values within the accuracy of the actual measurements when used to predict from a model the gas production from anaerobic digesters on different farms or the rumen VFA production of animals fed on various amounts of different batches of grains, even though in detail the feeds differ, and this result can be all that is needed.

This is only a brief summary of some problems of mathematical modelling. Overall it is possible to model aspects of a microbial system without real reference to the microbial metabolism, or by considering only a limited part of the microbial activities. It is also possible in this type of model to introduce terms which have no microbiological basis. Such models may, though, calculate some relevant end-product or other feature of the microbial activity sufficiently well for them to be used to predict some property of different systems, such as animals fed on different amounts of hay and grain, within the accuracy of the measurements possible or required. The closer a model comes to biological accuracy the more complex it becomes, and a limitation to such models may be lack of data to give the required rates or other factors needed by the model. Some different rumen models to illustrate these points will be briefly discussed in the next paragraphs.

Steady states

The rumen is a type of continuous-culture, and some early models attempted to simulate the rumen on a similar basis to that used to simulate laboratory fermentations (e.g. Hungate, 1965). Herbert *et al.* (1956) had earlier used Monod kinetics to determine steady-state concentrations of substrate and microbial cells in continuous laboratory fermentations of dissolved substrates. Many others have followed these lines, with additions to the original equations to try to explain experimental variations not described by the simple theories. In these models, rate of change in, say, substrate concentration is described by an input, output, utilisation differential equation which is equated to zero to simulate the steady state when no change with time takes place in the residual substrate concentration or microbial mass or fermentation products in the fermentor. Such differential equations can also be made the basis of more complex models.

The classical continuous-culture models require a theoretically continuous flow of medium into the fermentor, and this can be approached under laboratory conditions. Normally the rumen has an intermittent supply of feed. As Blaxter et al. (1956) appreciated, the rumen approaches closest to the theoretical continuous-culture when the ruminant is given a very small amount of feed at frequent intervals, e.g. on a continuously moving conveyor belt. In these circumstances the rumen contents can attain a near steady-state and continuous-culture theory can reasonably describe the residual concentrations of feed constituents, the amounts of acids produced, and so on, over a period of time. This type of model does not, however, describe the situation leading up to the steady state, for instance when the animal's feed is changed, and as the time between feeds increases the model departs more and more from actuality. So the model cannot describe the usual situation with animals feeding once or twice a day, when a portion of feed can be fermented almost as a batch culture superimposed on a continuous flow system. However, although such a system is not a steady state in the sense used in continuous culture theory, it is a kind of steady state as the 24-hour fermentation pattern is repeated *ad infinitum* if the feeding regime remains unaltered. Such a system might be described as being in a 'dynamic steady state' and it can be modelled.

The earlier continuous-culture theories were developed for dissolved substrates, and the ruminant's feed is usually insoluble particles (although these obviously have to be degraded and dissolved before they can be utilised by microorganisms; see Chapters 7 and 8). This is another factor causing varying degrees of divergence of the rumen from the usual laboratory continuous culture and its theoretical treatment.

The disappearance of feeds in the rumen

The models of Ørskov & McDonald (1979) and of Mertens & Ely (1979, 1982) are examples of the type of model in which no reference is made to bacterial metabolism. The first-named authors examined the breakdown of solid proteins in the rumen. They found that protein enclosed in a nylon bag in the rumen ('*in sacco*'; Chapter 13) of a sheep fed on dried grass disappeared (and was assumed to be degraded) at a rate which decreased with time. The measurements of the amounts degraded (p) after various times (t) could be fitted to an equation of the form

$$p = a + b[1 - \exp(-ct)]$$

This was, as the authors said, purely empirical and the constants were obtained by iterative fitting of the equation to the experimental results. However, a could be reasonably interpreted as a fraction of the protein which is rapidly degraded before the first measurement of the protein remaining in the bag is made, while the rest is degraded exponentially. Of course, a could also be a fraction which is rapidly dissolved out of the bag but not degraded. The degradation of material in a nylon bag does not truly represent the disappearance of a feed from the rumen, as it does not allow for flow of liquid through the rumen and the possibility of particles being washed out of the rumen before they can be fully degraded. The effects of washing out were allowed for by incorporating into the equation measurements of rumen flow rate (k) obtained by determining the rate of disappearance of chromium-labelled protein particles which were indigestible. So an equation

$$P = a + [bc/(c + k)][1 - \exp(-(c + k)t)]$$

was obtained. Thus, if the rate of flow through the rumen is known, the effective degradation (P) of a protein in a feed can be calculated from its rate of degradation in a nylon bag, as in the above equation P tends towards a limit $[a + bc/(c + k)]$ as t increases.

This model tells us nothing about the rumen organisms and the way they degrade protein, or the fate of the degraded protein. It shows only a relationship between the disappearance of protein in a nylon bag, the rumen flow rate and the effective degradation of that protein when it is fed as part of the particular diet used in the nylon bag tests, but that was the objective of the model.

Robinson et al. (1986) examined the ability of simple first-order, second-order, surface and logistic models to describe the disappearance of neutral detergent fibre in four feedstuffs from nylon bags suspended in the rumens of cows eating a feed of 33% long hay and 66% pelleted concentrates at two levels of feed intake. They found that the first-order model (of the form residue at time t = degradable $\times$ $\exp(-kt)$ + undegradable, but with a time lag before degradation started) with, in the case of two of the feeds, a second fraction degradable at a different rate was adequate to describe the disappearance, although obviously not biologically accurate. The other models were no better.

Passage of feed from the rumen is, unlike that from an in vitro fermentor, selective for size, in that particles cannot leave until they have been degraded to a certain minimum size (Chapter 12). In the model above, the protein particles were below this minimum size and so could be degraded or washed out. A chopped-grass feed will, however, contain some fibrous particles of a size which could leave the rumen and some particles which must be degraded before they can leave. The protein was assumed to be homogeneous and always degraded at the same rate. The fit of the data to the model suggested that this was probably true. However, this might just have been an example of another problem in modelling: the accuracy of the experimental measurements may not be sufficient to show up imperfections in the model or to determine which of two models is correct. Although the protein may have been homogeneous, grass fragments are not, and different parts are degraded at different rates and there is always an undegradable, lignified, residue. The model of forage fibre degradation by Mertens & Ely (1979) took these features into account although, like the previous model, it did not involve any explicit concept of microbial metabolism and was not concerned with what were, or what happened to, the products of the feed degradation. It did, however, involve not only the rumen 'compartment' but also a compartment representing the large intestine in which some fibre degradation takes place.

The heterogeneous nature of fibres was simplified by making the feed consist of rapidly, slowly and non-degraded materials. These were separate entities, although in fact one fibre could consist of layers of different degradabilities. Each material existed in large, medium and small particles, although the particles were not explicitly defined. Only small particles could easily flow from the rumen to the intestines, but some passage of medium particles was also allowed. Large particles had to be degraded before they could leave the rumen, while small or medium particles could be degraded or could leave the rumen undegraded. This type of model differs from the ones described previously. The protein model considered the degradation with time of one lot of feed, and this can be described by an equation which gives an asymptotic value for the feed protein digested after some hours. So, given some rates, the digestibility is given by the equation. In the present model, a portion of feed, of defined content in fibre, particle size etc., was given. A computer calculated, from equations giving the relationships between particles of different sizes and rates of degradation and flow from the rumen, the degradation and passage of the components of this feed. After 1 hour, a further portion of feed was given and was degraded along with the residues of the first feed. This process was repeated for 24 daily feeds and for further days. With such a system, a steady state is reached after a number of feeds, in which the feed input balances the feed removal by degradation or passage to the intestine and faeces, and a constant residue remains in the rumen. However, the steady-state values are not given by an equation as in the earlier models of continuous cultures but have to be worked out by repetition of similar calculations. Given suitable values for the rates of degradation of the different fractions of forages, rumen flow rates etc., the model could predict the effects on rumen digestion of altering the particle size of a forage by grinding and pelleting, and so on. By making some assumptions about digestion in the large intestine, some aspects of fibre digestion in the whole animal could be simulated. The model was limited in what it could do both by its structure and by lack of suitable values for the rates and other factors needed. A further discussion and evaluation of the model is given in a later paper (Mertens & Ely, 1982).

These are just examples of models of feed metabolism based on measurements of rate of passage of rumen contents and degradation of the feed which have been proposed by a number of workers. Blaxter et al. (1956) had earlier measured rates of passage and dry-matter digestibilities of dried-grass preparations fed to sheep and proposed an exponential model to describe the passage of the grass from the rumen and so to the intestines and faeces. The time spent in the rumen was related to the digestibility of the grass preparation. The equation gave the rate of passage of a single meal but, by either graphically or mathematically superimposing the curves for passage of a number of meals given at intervals, a picture of the previously mentioned 'dynamic steady state' degradation of feed and production of faeces could be obtained. This work was concerned in consideration of digestibility trials and, again, did not consider the rumen microbes or the products of digestion of the grass. Waldo (1969) modelled degradation of feed on the basis that forages consisted of a degradable fraction which was digested by a process with a first-order rate or could pass from the rumen, and an undegradable fraction that passed through the rumen. The model was tested on a large number of forages (Waldo et al., 1972). Many models have followed these. Of necessity, only a brief description of some models has been given here, but it serves to illustrate the fact that a model of a microbial system need not consider microbial metabolism and that a model may be successful in its aims while simulating only a part of the whole microbial system.

These models are, or can be made to be, dynamic ones in that they can describe, to various extents, changes in the system with time and can predict steady-state conditions of extent of digestion of feeds, rumen fill, and so on, and how these will change when some parameter, such as size of particle, is changed. However, such models are limited in their explanation of mechanisms of microbial activities and in the extent to which they can predict happenings in similar systems. This limitation may be by design, in that the output of the model is sufficient for the purposes of the modeller, or it may be by necessity in that there is not sufficient biological or numerical data on which to base a more detailed model. In the decades from the 1950s to the 1970s a large amount of information on the biochemical pathways and the microorganisms involved in the rumen fermentation was obtained, and this has been described in previous chapters. Data were then available for the construction of more detailed models of the rumen system; also, computers and computer programs able to process the data were becoming more widespread.

Balance models

If the reactions involved are known, differential equations of the type used in continuous-culture theory can be used to describe changes with time in components of the rumen feed or metabolic products. The necessary data can be obtained from 'balance' equations which give absolute values to the reactants and products but have no time component as has a dynamic model.

Substrate and product equations of the balance type are used in analysis of many fermentations. For instance, carbon balance and ATP production are calculated from an equation of the type sPolyglucose + uWater → vGlucose → wAcetic + xPropionic + yCarbon dioxide + zHydrogen + tATP (see Chapters 1 and 11, for instance). Such an equation is in itself a simple balance model and can be used to compute the amounts of VFA and microbial cells (if an equation linking ATP and cell mass is obtained) produced in the rumen from a certain amount of feed carbohydrate.

The growth of knowledge of the rumen reactions referred to earlier allowed many balance equations for rumen reactions to be formulated. Reichl & Baldwin (1975) give an example of a sophisticated balance model of rumen reactions involving many equations. These equations can then be linked to the mathematical equations not only for the rates of breakdown of fibres but also of the constituents of fibres, and to the products of these and other reactions such as protein degradations. Thus the basis of complex dynamic models of the rumen microbial reactions may be built up, but such models, because of mathematical complexity or the amount of data to be fitted, may not be workable by 'manual' mathematical or graphical methods such as were mentioned earlier. However, the equations need not be explicit or solvable mathematically. The power of the modern computer in modelling is its ability to repeat hundreds or thousands of times calculations involving the same equations but different inputs. Thus, if a rate of degradation of a feed component and the products of the degradation are given, the computer can calculate the amount of degradation of an ingested portion of feed and the product formation over a short time, and then repeat the calculation adding more ingested feed to the residual preceding feed, and so on. The whole can then provide a picture of the changes in the rumen contents over a long period of time and show if a steady state or dynamic steady state is reached. Such computer iteration has been extensively used in rumen modelling.

Dynamic models including the microbes

France *et al.* (1982) proposed a model which was more complex than the earlier models of fibre breakdown mentioned here, but which they said was 'more concise and practicable' than the complex models which will be described later. They also said that they had attempted to be more complete in their descriptions of the mathematics and numbers used than was the case with other modelling papers. They considered the rumen inputs to be degradable 'α-hexose' (which also included pectins), degradable and non-degradable 'β-hexose' (including lignin), water-soluble carbohydrates, and degradable and non-degradable nitrogenous compounds. The changes in the rumen pools of substrates were given by differential equations of the type rate of change = inflow + synthesis − outflow − utilisation. The build-up of microbial mass was determined only on the basis of availability of C and N (water-soluble carbohydrate and non-protein nitrogen) and a generalised composition of microbial cells. Microbial catabolism when insufficient substrates for growth were available was also considered, as were salivary and water flows. The various equations were solved numerically for steady states by a computer program, while constants and inputs for grass diets were estimated in various ways or taken from experiments. The modelled rumen outflows of the C and N compounds previously mentioned, and C and N balances for two grasses fed chopped or pelleted, were compared with experimental results. Reasonable agreement with experiment was obtained when the model used continuous feeding, but there were large discrepancies when the normal intermittent feeding was modelled. It was not possible to tell, from the experimental data available, how the model should be modified.

Baldwin and coworkers were one of the first groups to attempt the construction of detailed dynamic models of the rumen based on the increase in knowledge of the system brought about by the previously mentioned work in the 1960s and 1970s (Baldwin *et al.*, 1970).

Baldwin *et al.* (1977a) describe the steps in the development of a complex model of fibre digestion from a simpler model which will be used as a brief illustration of the modelling and some of the general points previously mentioned. They started with the two-compartment model of Waldo *et al.* (1972) in which the cellulosic feed was divided into two, potentially-digestible and indigestible, fractions as mentioned before. Both passage and digestion of the fibres were assumed to be described by first-order equations and no distinction was made between passage of the fractions. Thus the model could be defined by two equations. While the model was successful to a certain extent, it obviously did not take into account many of the factors affecting fibre digestion and it was not generally applicable in that new rate factors had to be experimentally determined for each fibrous feed considered. The model was then expanded on lines similar to those described previously for the model which Mertens and Ely developed about the same time. The digestible and indigestible fractions were divided into large and small particles, forage intake passing from the former pool to the latter, then from the rumen to the intestinal compartment and then to the excreta. This model was defined by ten first-order equations. The best rate constants for these equations were determined by taking an arbitrary set of constants and adjusting these by iteration until they gave the closest fits to the data from twelve experiments done by various workers on digestion of roughage diets by sheep. The model was then tested against seventeen other sets of experimental data on cellulose digestion. Although the model predictions agreed with experimental data in some ways, it was apparent that there was a systematic error in predicted rumen cellulose digestion when this was compared with the experimentally determined apparent cellulose digestibility. It was concluded that cellulose digestion cannot be determined on the basis of particle size and amount of substrate available, and that a model incorporating other factors was necessary.

This model was much more complicated than the previous ones. It brought in the concepts previously mentioned as being missed from earlier models, that the rumen contains a microbial community fermenting and degrading a variety of chemical compounds, and that metabolism or availability of one compound may control the metabolism of another. For instance, nitrogen supply may limit carbohydrate fermentation. The model considered the metabolism of cellulose and hemicellulose and other carbohydrates as well as protein and non-protein nitrogen compounds and other feed constituents. The input to the model was, then, a weight of feed containing various percentages of feed components. The programme could then calculate moles of cellulose input, for instance. The cellulose was, as before, considered to consist of large and small particles which could be degraded or pass from the rumen. However, this cellulose was degraded by microorganisms (bacteria and protozoa were considered as one entity) which 'colonised' the particles during a fermentation lag period. The cellulose was then fermented to VFA and this generated an amount of ATP which was sufficient to grow a certain amount of new bacteria (supposing that nitrogen was available), and so this increased the fermentation rate. Many more factors were taken into consideration, but this description will give an idea of how the model simulated the rumen metabolic activities more closely than the previous models. As in the other models, factors representing degradation rates of feed constituents, ATP yields from fermentations etc. had to be put into the model and these were obtained from various experiments described in the literature.

The concepts of this model were extended and fully described in a paper by the same authors (Baldwin et al., 1977b), and some results of testing the model were given. Testing the model was done by simulating a sheep fed continuously on lucerne chaff. Data on this diet were available in the experiments of different workers. However, there were no data to test the model by simulation of a different, but experimental, feeding system. The model was therefore set to simulate feeding of the lucerne in different amounts and at different frequencies. Thus feeding was done once, twice, four times and twenty-four times a day. Digestion of organic matter, VFA production, microbial growth and other results were obtained. From experimental work on ruminants in general, the overall responses to changes in feeding regime of this type were known, so the success of the model was judged on this basis. In general, the model gave reasonable simulations of rumen digestion, but tests did show up deficiencies in the available data with which the model was supplied. The authors discuss these and, for instance, the conventional methods of feed analysis did not seem accurate enough in determining the amounts of different plant constituents. Equations were included in the model to simulate the break-up of large fibres by rumination and chewing, but data on these processes were inadequate to provide accurate simulations. Since the model was inadequate in some ways it could not only point out deficiencies in knowledge but it could be used to assess some hypotheses about the mechanisms involved in reactions. Thus an apparently imperfect model can provide useful information.

Another model which took into account many of the rumen microbial reactions was described by Black et al. (1980/1) and Beever et al. (1980/1). It gave a simulation of rumen function in which feed intake, saliva flow, feed carbohydrate and protein degradation, production of VFA and microbial cells, and other actions in the rumen digestion were described. The model was tested by simulation of different changes in feeding regimes and feed compositions. Simulations followed experimental observations in many cases. For instance, when the potential degradability of cellulose ('β-hexose') in a simulated subterranean clover diet was increased, the total of β-hexose degraded increased and this led to increases in VFA and ATP production, and so to an increase in the amount of microorganisms in the rumen. This produced an increase in the amount of microbial protein flowing from the rumen to the intestines. However, there were situations that the model could not describe. For instance, the effects of grinding a dried-grass feed were not predicted, apparently because of lack of data for determining the effects of size of particles on rates of outflow from the rumen. The partial failures of the model again pointed to the need for further research on feed analysis, outflow rates from the rumen and other subjects.

Baldwin et al. (1987) attempted to 'develop a model of digestion in the lactating dairy cow which would yield acceptable estimates of rates and patterns of nutrient absorption'. The model was based on the model mentioned in the previous paragraph and on the models of France et al. (1982) and Baldwin et al. (1977b). However, the authors pointed out that the lack of data on details of digestion in the cow governed the model development. The feed intake was divided into soluble and insoluble carbohydrates and proteins and other major components in published analyses of diets, including a reference diet for model construction and diets for comparisons with the model digestion. The model assumed continuous feeding, and Michaelis–Menten or mass-action kinetics were used. The model reacted reasonably to changes in some parameters, although the results again showed the need for more information on some aspects, such as effects of particle size, to make the model reproduce experimental behaviours. The model was deemed satisfactory in simulating the digestion of forage diets and diets with 60% barley or corn, but the simulation of digestion of 90% barley and corn diets was less than satisfactory. The reasons for this were not entirely apparent, but there was need for more data from which the model could be made to react to the rather extreme changes in diet. For instance, no provision was included in the model to simulate the depression in microbial activity which experiment suggests is caused by low rumen pH. Some preliminary results of adjusting the model to reproduce discontinuous feeding were considered realistic.

In the papers describing some of the models, the word 'compartment' is used. The compartment can be a physical one (the rumen or large intestine) or it can be a particular reaction, but it should be noted that the models do not consider the rumen to have compartments in the sense of Chapter 12. It is possible that some of the difficulties encountered in modelling discontinuous feeding regimes are caused by the compartmentalisation of Chapter 12, which is difficult or impossible to reproduce in a model.

Provision of data

Even if all the reactions, and the pathways involved in these, are known, the collection of data for model rates and other values is always a problem in modelling natural systems with natural feeds. Experimental data show animal variations, and the feeds used, while overall of similar composition, can vary in unknown ways because of factors such as method of preparation and differences in physical and chemical structures of materials from plants grown at different times in different areas. Following an earlier paper (Koong et al., 1975), Murphy et al. (1982) described the collection of data from many papers on feeding of roughage and concentrate diets, and the statistical manipulation of these data to determine the best values for VFA productions from the various carbohydrate components, to use in models. The data were used to predict the VFA productions from other diets for which experimental results were available. The predictions were within the normal biological variation in such experiments.

As well as using data from rumen fermentations *in vivo*, the rumen modeller can also use data from experiments *in vitro*. Such data can be from fermentations with mixed organisms or can be from studies with pure cultures of microorganisms. The products of cellulose hydrolysis can, for instance, be assumed to be the acids and gases produced by *Ruminococcus* sp. digesting purified cellulose or cellulose in some fibre, or they can be assumed to be a mixture of the products of the cellulolytic bacteria *Ruminococcus*, *Bacteroides* and *Butyrivibrio*. The bacteria might also be grouped by fermentation products as, for example, acetate- or propionate-producing. The groups can then be considered separately in equations for carbohydrate breakdown. Experiments *in vitro* can also provide data on the requirements of these bacteria or groups of bacteria for nitrogen sources, growth factors etc., and such data can be included in the model to represent interactions, and pathways of product formation. An example of this is given by Reichl &

Baldwin (1976). The obvious problem in using this kind of data is that the fermentation products, growth yields and so on may differ in the rumen from the values obtained in the laboratory. However, if laboratory results do not simulate the rumen, they might be adjusted to provide agreement or they may show a definite gap in knowledge of the behaviour of the organisms *in vivo*. Bacteria have been mentioned here, but the results of protozoal experiments could be used in the same way.

General conclusion

A number of mathematical models of the rumen function have been considered here. These vary in complexity, but all seek to describe quantitatively some aspect of the metabolism of feedstuffs in the rumen. This may vary from a simple statement of the disappearance of a feed which can be related in some way to feed digestibility in animals to a complete description of the conversion of all the components of a feed into fermentation products and microbial cells. The numerical inputs used in construction of a model have to be based on the digestion of one particular feed given in one way, or some mean of the digestions of similar feeds, and ability to predict the digestion processes in other circumstances is best when the model is applied to feeds similar to those on which it was based. This may be because of the simple construction of the model or, in the more complex model, because there is not enough knowledge to programme the model to cope with metabolism under different conditions in the way that a microbial community can. The *raison d'être* and the construction of models vary considerably, and the evaluations of the success of models must be based on different criteria. Whatever their 'success', models help the research worker to comprehend the varied factors that affect the rumen digestion of feeds and the inadequacies in knowledge of the details of the processes which go to make up the life of a complex microbial habitat.

A mathematical model may be deemed 'successful' if it approximates to the natural rumen or predicts one aspect but not another. The models have considered, in more or less detail, the reactions which provide the animal with the energy and nitrogen sources needed for its life, but the animal needs other, known and unknown, microbial products such as vitamins. In the next section a 'biological' model, the gnotobiotic ruminant, will be considered. In this model all rumen functions, major and minor, must be considered and must be correct.

REFERENCES

Baldwin, R. L., Koong, L. J. & Ulyatt, M. J. (1977*a*). The formation and utilization of fermentation end-products: mathematical models. In *Microbial Ecology of the Gut*, ed. R. T. J. Clarke & T. Bauchop. Academic Press, London.

Baldwin, R. L., Koong, L. J. & Ulyatt, M. J. (1977*b*). A dynamic model of ruminant digestion for evaluation of factors affecting nutritive value. *Agric. Systems*, 2, 255–88.

Baldwin, R. L., Lucas, H. L. & Cabrera, R. (1970). Energetic relationships in the formation and utilization of fermentation end-products. In *Physiology and Digestion in the Ruminant*, ed. A. T. Phillipson. Oriel Press, Newcastle-upon-Tyne.

Baldwin, R. L., Thornley, J. H. M. & Beever, D. E. (1987). Metabolism of the lactating cow. II. Digestive elements of a mechanistic model. *J. Dairy Res.*, **54**, 107–31.

Bazin, M. (1983). *Mathematics in Microbiology*. Academic Press, London.

Beever, D. E., Black, J. L. & Faichney, G. J. (1980/1). Simulation of the effects of rumen function on the flow of nutrients from the stomach of sheep. Part 2. Assessment of computer predictions. *Agric. Systems*, **6**, 221–4.

Black, J. L., Beever, D. E., Faichney, G. J., Howarth, B. R. & Graham, N. McC. (1980/1). Simulation of the effects of rumen function on the flow of nutrients from the stomach of the sheep. Part 1. Description of a computer program. *Agric. Systems*, **6**, 195–219.

Blaxter, K. L., Graham, N. McC. & Wainman, F. W. (1956). Some observations on the digestion of food by sheep and on some related problems. *Br. J. Nutr.*, **10**, 69–91.

France, J., Thornley, J. H. M. & Beever, D. E. (1982). A mathematical model of the rumen. *J. Agric. Sci.* (Camb.), **99**, 343–53.

Herbert, D., Elsworth, R. & Telling, R. C. (1956). The continuous culture of bacteria: a theoretical and experimental study. *J. Gen. Microbiol.*, **14**, 601–22.

Hobson, P. N. (1985). A model of anaerobic degradation of solid substrates in a batch digester. *Agric. Wastes*, **14**, 255–74.

Hobson, P. N. (1987). A model of some aspects of microbial degradation of particulate substrates. *J. Ferment. Technol.*, **65**, 431–9.

Hungate, R. E. (1965). *The Rumen and its Microbes*. Academic Press, New York.

Koong, L. J., Baldwin, R. L., Ulyatt, M. J. & Charlesworth, T. J. (1975). Iterative computation of metabolic flux and stoichiometric parameters for alternate pathways in rumen fermentation. *Comput. Programs Biomed.*, **4**, 209–17.

Mertens, D. R. & Ely, L. O. (1979). A dynamic model of fiber digestion and passage in the ruminant for evaluating forage quality. *J. Anim. Sci.*, **49**, 1085–95.

Mertens, D. R. & Ely, L. O. (1982). Relationship of rate and extent of forage utilization: a dynamic model evaluation. *J. Anim. Sci.*, **54**, 895–905.

Murphy, M. R., Baldwin, R. L. & Koong, L. J. (1982). Estimation of stoichiometric parameters for rumen fermentation of roughage and concentrate diets. *J. Anim. Sci.*, **55**, 411–21.

Ørskov, E. R. & McDonald, I. (1979). The estimation of protein degradability in the rumen from incubation measurements weighted according to rate of passage. *J. Agric. Sci.* (Camb.), **92**, 499–503.

Reichl, J. R. & Baldwin, R. L. (1975). Rumen modelling: rumen input–output balance models. *J. Dairy Sci.*, **58**, 879–90.

Reichl, J. R. & Baldwin, R. L. (1976). A rumen linear programming model for evaluation of concepts of rumen microbial function. *J. Dairy Sci.*, **59**, 439–54.

Robinson, P. H., Fadel, J. G. & Tamminga, S. (1986). Evaluation of mathematical models to describe neutral detergent residue in terms of its susceptibility to degradation in the rumen. *Anim. Feed Sci. Technol.*, **15**, 249–71.

Roels, J. A. (1983). *Energetics and Kinetics in Biotechnology*. Elsevier Biomedical Press, Amsterdam.

Waldo, D. R. (1969). Factors influencing the voluntary intake of forages. *Proc. Nat. Conf. Forage Qual. Eval. Util.*, p. E-1.

Waldo, D. R., Smith, L. W. & Cox, E. L. (1972). Model of cellulose disappearance from the rumen. *J. Dairy Sci.*, **55**, 879–90.

BIOLOGICAL MODELS

The '*in vitro*' biological model previously mentioned can never provide a complete answer to the questions of what actually constitutes the rumen flora because it can never reproduce the rumen system as it occurs in the animal. Mechanical means cannot reproduce the rumen mixing, a dialysis membrane the metabolising rumen wall in the absorption of molecules, rumination as an aid to comminution of fibres etc. The only 'container' that reproduces the rumen is the rumen itself. However, in the normal animal, although the rumen is sterile at birth it is rapidly inoculated and ends up with the 'normal' rumen flora and fauna. The inoculum organisms come from the air, feedstuffs and bodily contact with nearby animals (Chapters 2, 3 and 5). While inoculation by ciliate protozoa can be prevented by keeping young animals some feet away from older animals (see Chapter 5), as the relatively large protozoa can travel only a short distance in air currents, rumen bacteria can be carried for long distances in air and can be found in the air of animal houses (Mann, 1963). Spatial isolation can therefore not prevent the development of a rumen bacterial population in young animals. So the only way in which the rumen can be used as a model system is to isolate it, and thus the animal, in a sterile atmosphere from birth. This will keep the rumen uninoculated and so in a position to be inoculated with a population of known bacteria: to become a 'gnotobiotic' ruminant. Experiments described elsewhere (Chapters 5 and 13) have shown that the ciliate protozoa are not necessary in a rumen population that will ensure adequate growth of a young animal. On the other hand, bacteria are essential for the survival of the protozoa (Chapter 3). Since it is difficult or impossible to obtain cultures of protozoa with no, or only defined, bacterial populations, experiments with gnotobiotic ruminants have concentrated on providing a defined rumen bacterial population without protozoa.

However, although the ciliate protozoa are not essential for the digestive functions of the rumen and the life of the ruminant, the results of experiments have been somewhat ambiguous in deciding whether, or to what extent, the protozoa modify the bacterial reactions in the rumen (Chapters 3 and 13). Experiments have thus been done in which defined populations of ciliates have been added to unfaunated animals and the rumen reactions determined. This type of experiment does not necessarily need a gnotobiotic animal kept in a sterile environment, but such work will be described in the last part of this chapter. Prevention of protozoal growth is also a way of manipulating the rumen microbial population, and the subject has thus been mentioned in Chapter 13.

Defined bacterial populations without protozoa

Isolation in a sterile environment is, of course, what has been done in many experiments with 'gnotobiotic' or 'germ-free' animals. There is, however, a big difference between the mice, rats, monkeys, and other animals that are used in the usual gnotobiotic experiments, and ruminants. The former animals digest food by means of enzymes in gut secretions, and microbial digestion is a minor part of the digestion and not strictly necessary for the life of the animal. In the adult ruminant the main digestive process is microbial and the ruminant cannot live without a functioning rumen even if fed on a diet which could support a non-ruminant germ-free animal. Normally, the ruminant at birth has a sterile gut like any other animal, but bacteria begin to colonise the internal and external surfaces of the animal within a few minutes of birth and a gut flora soon develops. This flora is initially similar to that of any other milk-fed animal, and it is not strictly necessary for digestion; so, as with other mammals, a germ-free young ruminant can be reared on milk under sterile conditions. The rumen is here non-functional and rudimentary; function and size develop only as the animal begins to eat solid feed and the adult rumen flora begins to develop (see Chapter 5). The time for which a young ruminant can be kept healthy and growing on a milk diet is limited. So the experimental problem in using germ-free animals as model ruminants to determine the actions and interactions of the rumen flora is that the inoculated bacteria have to be added in a succession which will grow under different conditions as the feed changes from milk to solids, they must produce the fermentation products which will develop the rumen itself, and the final population must ferment solids, to provide the VFA and microbial cells and animal growth factors such as vitamins, at rates sufficient to keep the host animal growing. This is a much bigger problem than establishing a gut flora in an animal which is already growing on products of feed metabolism provided by non-microbial enzymes. The fact that the full animal dietary products must be provided to keep the animal, and so the rumen, functioning means that, contrary to the experiments *in vitro* previously mentioned, it is not possible to do proper tests only on the bacteria concerned in one part of the adult rumen processes by inoculating these bacteria alone. Also, since the normal feed contains carbon, nitrogen and energy sources for the bacteria as relatively undefined polymers and not as the simple, defined or partially defined chemicals of the *in vitro* culture medium, a range of hydrolytic, deaminative and other bacteria must be present to make these polymers available to the bacterial 'culture'. The gnotobiotic ruminant can, by not surviving or growing only poorly, show which bacteria are necessary for proper rumen function; by growing it provides the ultimate test of which bacteria are necessary for the rumen function and which are 'passengers'. If a simplified, functioning rumen flora can be developed, it will obviously simplify biochemical and bacteriological tests on the interactions of bacteria and on biochemical pathways in the rumen. It will also provide a basis for experiments on the factors involved in the stability of mixed cultures and the control of 'invading' pathogenic and non-pathogenic organisms.

In this section 'gnotobiotic' is a general term which denotes both an animal with no known gut or any other body microorganisms and an animal with a known gut flora and no other microorganisms. 'Germ-free' describes the first case, the animal with no known flora. The term 'axenic' has more recently been used to describe the germ-free animal, while the 'axenic' animal inoculated with a well defined flora is said to be 'gnotoxenic'. An animal inoculated with a limited, but imperfectly defined, flora is described as 'meroxenic'. Animals reared normally with others, and having a normal rumen flora, have been described as 'conventional', and gnotobiotic animals transferred from the sterile environment to normal animal housing, and so developing by contact with other animals a normal rumen flora, have been described as being 'conventionalised'.

The experiments to be discussed here are concerned with the actions and interactions of the rumen and intestinal flora in ruminants. It was realised that the introduced rumen bacteria might colonise the lower gut of the animals, and some investigation of the intestinal floras was made. The bacteria, or some of them, might also find habitats in the mouth or external body surfaces of animals, but such interactions were not investigated. Most of the experiments with gnotobiotic lambs described here were planned and carried out 10 or more years ago. The selection of rumen bacteria and other aspects of the work should be viewed with this in mind; for instance, the importance of the anaerobic fungi in rumen metabolism had not then been discovered.

Both gnotoxenic and meroxenic animals have been used in work on the rumen, but the techniques used to obtain and maintain the animals are basically the same in both cases. The procedures were based on those developed for the study of gnotobiotic non-ruminant animals. Some earlier work was done with calves, but the main work has been done with lambs, as these animals can be reared to maturity as ruminants without becoming too big to handle in an isolator. Some lambs have been born naturally and then placed immediately in an isolator, with antibiotic treatment to kill any chance contaminating organisms, but a safer procedure is to deliver the lambs after hysterectomy from a uterus passed through an antiseptic bath into a sterile chamber. The sterile chambers are generally made of plastic film on a metal support, and equipped with glove ports to allow manipulation of the animal from outside the chamber. In the first one, after delivery by hysterectomy the lamb is held until it is breathing and otherwise functioning well. It is then transferred to the rearing chamber equipped with a cage like the usual metabolism cage. The transfer of the lamb and subsequent transfers of feed and debris into and out of the cage are made through an air-lock which can be chemically sterilised. Liquid feeds and equipment such as tubes and hypodermics can be sterilised by autoclaving, and solid feeds by irradiation. The surfaces of the equipment containers, or the actual feed bottles and cans, are sterilised by chemicals while they are in the air-lock before transfer into the isolator. Filter-sterilised air at slight positive pressure is fed to the isolators and they are kept in a clean, temperature-controlled room. Rumen samples can be taken by stomach tube or through a rumen cannula, and rumen inoculations can be done *per os* or via the cannula. The animals can be weighed and blood or other samples taken.

The gnotobiotic status of the animals is monitored throughout the experiments by taking liquid or swab samples from the rumen and body apertures and surfaces, and from the isolator, and examining these by microscope and by culture on a variety of bacteriological media (Elliott *et al.*, 1974). Such tests, of course, only guarantee freedom from contamination with the organisms which will grow in the test media, but a break in the isolator is usually indicated by growth of common bacteria such as staphylococci or bacilli. Freedom from virus contamination is almost impossible to guarantee in these experiments, but is assumed. The apparatus and techniques used in the preparation of feeds, in rearing, and in sampling and testing for gnotobiotic status have been described by Alexander *et al.* (1973*a*,*b*) and by Cushnie *et al.* (1981), as well as in papers describing specific experiments.

The young ruminant

The defined rumen flora. A germ-free ruminant is a contradiction in terms, as the rumen must contain microbes to function. The term is used here to describe the uninoculated lamb. There were some early attempts to rear germ-free ruminants from birth to a few months old on milk or on milk and then solid, semi-purified diets containing a large proportion of milk powder (Kuster, 1915; Smith, 1966; Purser & Bergen, 1969; Soares et al., 1970; Oxender et al., 1971; Trexler, 1971; Ponto & Bergen, 1974). Weaver (1974) fed a milk replacer followed by solid diets containing 24% or 45% wood pulp. The experiments varied in the times that the animals were kept on the feeds and in the data recorded on the animals. However, the general results were that, while the animals grew on milk alone, growth was not as good as that of conventional lambs on the same diets, and the animals either grew poorly or lost weight when fed the solid diets.

The experiments of Lysons and colleagues were the first in which gnotobiotic lambs were reared on milk and then on the type of solid diet fed to conventional early-weaned lambs and in which the inoculation of the rumen with a limited, defined flora designed to carry out the normal rumen solids digestion was attempted (Lysons et al., 1971).

The floras used in these experiments were based on the following species. *Streptococcus bovis* and *Lactobacillus* spp. were chosen as examples of the lactate-producing, sugar-fermenting bacteria found in the rumens of milk-fed animals and in lower numbers in older animals. The *Streptococcus*, as a facultative anaerobe, should also have helped to reduce the rumen Eh to a suitable value for growth of the anaerobes. *Veillonella alcalescens* and *Megasphaera elsdenii* were lactate-fermenting bacteria which would initially remove the lactic acid produced in the rumen of the young animal and later in the adult rumen, and so ensure a pH suitable for the establishment of the bacteria of the adult rumen. *Butyrivibrio fibrisolvens* fermented a range of feed carbohydrates including hemicellulose. *Bacteroides amylophilus* was starch-fermenting and also proteolytic, while *Bacteroides ruminicola* was proteolytic and deaminative. The *Ruminococcus* spp. were highly cellulolytic *in vitro*. *Streptococcus faecium* was ureolytic and *Anaerovibrio lipolytica* was added to give lipolytic activity. A strain of *Methanobacterium ruminantium* provided removal of hydrogen by methane production. A large number of tests involving gnotobiotic lambs inoculated with these bacteria were carried out. Only the main results of the published experiments can be described here.

In the first experiments (Lysons et al., 1976a, 1977) lambs were fed on milk for from 4 to 10 weeks but allowed access to pelleted feed at from about 2 to 7 weeks old, except for three lambs which were given pellets at just over 1 week old. Two pelleted feeds were used. One (no. 1) contained mainly barley straw, barley and wheat feed with groundnut meal, molasses, urea, tallow and minerals. The other (no. 2) was mainly barley with a mixed protein supplement, 10% straw, 5% hay and 5% molasses. These feeds, like the others used later, contained vitamin and trace-metal supplements. The milk fed was a commercial, sterile, concentrated cow's milk suitably diluted.

The results showed that, at least for some weeks, germ-free lambs fed *ad libitum* on milk could grow at a similar rate and with a similar feed conversion efficiency to conventional lambs. The performance of the germ-free and inoculated lambs after weaning from milk is discussed in the next section.

Ureolysis. Experiments on the provision of a complete defined rumen flora which would enable gnotobiotic lambs to function as normal ruminants continued, but some other experiments on young lambs without a full digestive flora were also carried out. In Chapters 7 and 12 the role of the rumen wall flora in ureolysis is described. Barr *et al.* (1980) inoculated two germ-free lambs of 24 days old with ureolytic staphylococci obtained from conventional sheep. One lamb was fed entirely on milk, the other was weaned on to a starch–grass–protein cubed feed. In both cases the ureolytic bacteria had established in the rumen fluid by 9 days after inoculation, and post-mortem examination when the lambs were killed at 51 and 56 days old showed that the staphylococci were also associated with the rumen wall. The establishment of the bacteria reversed the pre-inoculation state of the rumen fluid which had high urea and low ammonia concentrations and absence of urease activity. In the inoculated lambs, urease activity was associated with the fluid and rumen epithelium. The weaned lamb had a lower urease activity than the milk-fed one, although bacterial counts in the rumen fluid and on the rumen wall were similar. Since only the one bacterial species was present in the rumen, the only fermentation product present was lactic acid. Although these experiments seem to support the view of the role of ureolytic staphylococci in the rumen function, they are not conclusive, as it was not shown that these bacteria could successfully compete if they were part of a complete rumen flora.

Escherichia coli *and lactobacilli.* Isolations of pathogenic bacteria can often be made from the vicinity of animals, and strains potentially capable of producing intestinal disturbance may be found in the digestive tracts of healthy animals. One cause of illness and death in young animals is 'scours' caused by pathogenic strains of *Escherichia coli*. However, as with other pathogens, exposure to these strains of coli does not necessarily cause illness, and the severity of the illness can vary from animal to animal. A number of factors have been proposed as controlling the proliferation of pathogens in the digestive tract, and among these is suppression of the pathogen by the commensal flora of the gut. Among the commensal bacteria possibly involved are species of *Lactobacillus*. Gnotobiotic lambs were used by Mann *et al.* (1980) to test the possible role of lactobacilli in controlling *E. coli*.

Four germ-free lambs fed on milk were used. One was inoculated with a pathogenic strain of *E. coli* when 4 days old. At 6 days old it became ill and it died at 8 days, when counts of coli in the faeces had become high. Two lambs were inoculated with *Lactobacillus casei* or *L. acidophilus* at 1 day old and with the *E. coli* at 5 or 7 days old. In these cases, faecal coli did not reach the numbers of the first lamb and the lambs showed no signs of illness. With a fourth lamb, inoculated with coli at 3 days old, the coli proliferated slowly and by 15 days had not reached the numbers found in the faeces of the first lamb when it became ill. Inoculation with lactobacilli at 15 days old reduced the faecal coli numbers and the lamb remained healthy. The three lambs were weaned on to starchy solids over 7 days from 21 or 28 days old and inoculated with a defined rumen flora (see above) for other experiments. The coli remained as part of the rumen flora but caused no illness. These experiments suggested that lactobacilli had a definite role in control of pathogens in the gut of young animals.

Cerebrocortical necrosis. It has been suggested that cerebrocortical necrosis (CCN) in ruminants is associated with higher than normal concentrations of thiaminase in the rumen and low levels of thiamine in the tissues (see Chapter 14). Thiaminase is produced by a number of bacteria, but Shreeve & Edwin (1974) suggested that thiaminase-producing strains of *Clostridium sporogenes* could be the source of the rumen thiaminase and so of CCN. Later work with conventional animals (Edwin *et al.*, 1979) neither confirmed nor disproved this suggestion, while Cushnie *et al.* (1979) approached the problem by using germ-free lambs. Two germ-free lambs were fed on milk and then weaned on to a solid feed. One lamb was kept bacteria-free, the other was given a defined rumen flora (as previously described) at weaning. At about 40 days after inoculation the latter lamb was given orally a large dose of a thiaminase-positive strain of *Cl. sporogenes*. Only at 31 days after this did the clostridia appear in the rumen, when they soon reached about 10^8 per ml. However, when the animal was killed about 7 weeks later there were no clinical or post-mortem signs of CCN. The germ-free lamb was also given a dose of the clostridium at 30 days old. The clostridia established in the rumen after 10 days, and remained at 10^9–10^{10} per ml until the animal was killed at about 8 weeks old. Although the lamb did not grow well because of the lack of rumen flora, it showed no signs of CCN. In both lambs rumen thiaminase was high. Two other lambs were allowed contact for either 2 days or 84 days with conventional animals, to obtain a conventional rumen flora, before being placed in sterile isolators. Repeated oral doses of the clostridium failed to establish it in the rumens of these lambs. The experiments suggested that the clostridium could not compete against the normal rumen flora and was partly suppressed by the defined flora, and cast doubts over the role of thiaminase-producing clostridia in production of CCN.

Antibody production. Blood serum of animals contains circulating antibodies to a wide range of Gram-negative bacteria, particularly enterobacteria. The antibodies are considered to be natural antibodies produced in the absence of overt infection. Sharpe *et al.* (1969) found antibodies against specific strains of bovine rumen bacteria in the blood of cows, sheep, goats and a horse, but not in pigs, rabbits and man. The horse caecum contains bacteria similar to those in the rumen (Davies, 1964). Sharpe *et al.* (1969) concluded that the antibodies against rumen bacteria might afford protection to the animal against salmonella infections, and they investigated this further (Sharpe *et al.*, 1977). The work is fully discussed in the paper just quoted, and here only that related to gnotobiotic lambs is described. It was suggested that the caecum- and, possibly, the colon-walls are the sites of antibody production, and in earlier work (Lysons *et al.*, 1976a) all of the inoculated bacteria, except *Butyrivibrio*, *Ruminococcus* and *Methanobacterium*, were found in the caeca of gnotobiotic lambs, although in different proportions in different lambs. In the work of Sharpe *et al.* (1977) four lambs, reared on milk, a starter ration and then grass cubes, as later described, were used. These were inoculated with strains of *Veillonella*, *Bacteroides ruminicola*, *Ruminococcus*, *Selenomonas*, *Megasphaera*, *Lactobacillus*, *Butyrivibrio* and, in one case, *Escherichia coli*. The *Bacteroides*, *Selenomonas* and *Megasphaera* gave a strong response, antibodies to the former bacteria appearing at 20–40 days after inoculation and to the *Megasphaera* at 28–74 days. Agglutinins to the *Veillonella* and *Ruminococcus* were weak and appeared at 100–136 days after inoculation, and a weak reaction to the *Lactobacillus* appeared at 52–125 days. The *Butyrivibrio* stimulated antibody production at 19–106 days in three lambs, but no antibodies had appeared in the other at 43 days after inoculation. A low titre against *E. coli* appeared in the inoculated lamb after 71 days but it had disappeared 140 days after inoculation. The caecum of one lamb was examined when it was killed at 137 days. *Veillonella* was predominant, with *Megasphaera* and *Bacteroides* about 10% of the *Veillonella* counts. *Lactobacillus*, *Selenomonas* and *Butyrivibrio* numbers were negligible.

Brassica anaemia. When ruminants are fed mainly on kale, cabbage or other brassicas, they can develop a severe haemolytic anaemia (see Chapter 14). The primary toxic factor is *S*-methylcysteine sulphoxide (SMCO) which was found to be converted into dimethyl disulphide in the rumen, and the disulphide is the actual haemolytic factor. One of the proofs of this was provided by gnotobiotic lambs. When SMCO was fed to germ-free lambs, blood levels of SMCO rose, but no symptoms of anaemia were produced. After cultures of rumen bacteria able to convert SMCO into the disulphide in laboratory culture had been introduced into the rumen, the first clinical signs of the anaemia became apparent (Smith, 1980).

The 'adult' ruminant

Feed digestion and growth of lambs. Milk feeding of the young animal naturally continues for months, but on farms it is now the common practice to stop feeding milk when the young ruminant is a few weeks old and to allow it to eat an easily digestible starchy, solid diet before being changed to a fibrous ruminant feed. The reasons for this are economic, in that the young can be brought to a killing weight or to breeding at an earlier age.

Because of difficulties in handling large animals in the isolators, the early-weaning system was used with the gnotobiotic lambs so that a reasonably long period of study of the effects of inoculation with a rumen flora would be available before the animals became too big. Thus the lambs were provided with an adult diet and an adult defined rumen flora even though they were young in age.

The first gnotobiotic animals to be inoculated with a defined rumen flora which should have enabled them to digest a solid diet were those of Lysons *et al.* (1971, 1976*a*, 1977) mentioned in the previous section. The solid feeds and the rumen inocula were described in that section.

On solid diet 1, both germ-free and inoculated lambs grew little, although the inoculated lambs put on rather more weight. Conventional lambs also did not thrive too well on diet 1. It was suggested that this poor growth was due to lack of animal protein in the feed, and so diet 2 was formulated. This was similar to some commercial early-weaning diets. However, the first set of germ-free and inoculated and conventionalised lambs did not do well on this diet, either losing or, in the case of the conventionalised lambs, barely gaining weight. The lack of gain of the animals with rumen floras was suggested as being due to the long period (4 weeks) that elapsed between weaning at 7 weeks and inoculation of the rumen. The animals also had wool balls in the abomasum which could have caused some blockage.

With the conventional animal, some solid feed is taken while milk is the main feed, and the rumen flora develops to at least a certain extent before milk feeding stops. The second set of lambs on diet 2 were given the solid feed for about 4 weeks before milk feeding was stopped at 7 weeks old and the rumen bacteria were given 1 week before the milk was stopped. In this case the germ-free animal continued to grow, although more slowly than it did on milk. This suggested that it was possible for a germ-free lamb to derive some nutrient from a suitable feed (the diet contained fish, meat and vegetable proteins and starch, which could all have been digested by intestinal enzymes). However, the results with this one lamb were not unequivocal as the rumen became increasingly contaminated with a bacillus, a clostridium and a streptococcus as the animal grew older, and it is possible that these bacteria could have given the animal a kind of partially functional rumen flora. The lambs with the defined rumen flora continued to grow at the same rate after weaning as before, but the conventionalised lamb suffered a check at weaning before continuing to grow at the same rate. The presence of a defined flora appeared in this case to give a functional rumen, as the weight gains and feed conversion efficiencies of the inoculated lambs were similar to those of conventional lambs. The inoculation of the defined flora before weaning also seemed to have helped to improve performance. The defined flora also stimulated rumen and intestinal development, but this varied between experiments (Lysons et al., 1976b).

The rumen floras of the inoculated lambs in these experiments were described in a further paper by Lysons et al. (1976a). The species of bacteria given were mentioned earlier. The bacteria were given in the order in which they might be expected to appear in a conventional lamb. *Streptococcus bovis* and the *Lactobacillus* were given first, followed by the lactate-using bacteria to control the pH, and then by the nitrogen-metabolising and fibre-digesting and, in the case of one lamb, methanogenic, ureolytic and lipolytic bacteria. The inoculations were done over periods of about 33 days.

The *Streptococcus* and the *Lactobacillus* established viable populations in 1 or 2 days, and the rumen pH fell as lactate concentration increased. However, the lactate-utilising *Megasphaera* and *Viellonella* also rapidly increased in numbers after inoculation, and lactate then became undetectable and the pH rose. As the other bacteria became established, the rumen pH stabilised in the normal range, about 6·5. All the inoculated bacteria, except *A. lipolytica*, were re-isolated from the lambs. The presence of *A. lipolytica* was inferred from analyses which suggested that lipolysis and hydrogenation of fats had occurred (Leat et al., 1977).

Methane was detected in the rumen gas of the lamb inoculated with *M. ruminantium*, and the bacterium was re-isolated. Hydrogen was found in other lambs. In the lambs which grew best, on diet 2, the viable counts of all bacteria were similar to those in conventional lambs, except for the cellulolytic *Ruminococcus*. This was found in numbers of only 10^3–10^5 per ml, and tests suggested that the lambs had digested only about 15–20% of the dietary cellulose. *Butyrivibrio* predominated in the rumen population and, in conformity, while acetic was the major rumen acid, butyric was in higher concentration than propionic. The total concentration of rumen VFA was lower than in conventional lambs. Although the bacteria had produced rumen populations which could digest the feed to some extent and provide sufficient nutrients for growth of the lambs, it appeared that little nutrient was derived from the straw and hay of the diet and that probably the major part of the VFA came from starch digestion.

Ureolysis. There was no ureolytic activity in the rumen of one lamb with a defined flora lacking the ureolytic *S. faecium*, but at 22 days after inoculation this bacterium was isolated in numbers of 2×10^7 per ml and at 68 days the numbers were 7×10^7. Ureolytic activity was found after inoculation, and by 68 days it was comparable to that in a conventional rumen at 0·33 g urea hydrolysed per 100 ml rumen fluid per hour. Thus the numbers of the bacterium and the urease activity were similar to the values found for conventional sheep, supporting the suggestion that this bacterium was a principal producer of urease in the conventional rumen (Cook, 1976). These experiments were more convincing than the previously mentioned experiments on ureolysis in a gnotobiotic lamb, as in this case the bacterium was competing against other bacteria contained in the conventional rumen flora.

Cellulolysis and amylolysis. The problem of producing a normally active cellulolytic flora in the gnotobiotic lambs was never satisfactorily solved. The experiments of Mann & Stewart (1974) are examples. A gnotobiotic lamb provided with a defined flora similar to those previously mentioned grew well on milk and then on a starter diet of barley, dried grass and fish meal. Thirty days after weaning, this feed was replaced by dried grass alone. Growth continued, but after a further 40 days the lamb lost appetite and began to lose weight, and about 20 days later it was removed from the isolator, when it was about 10 kg lighter than a conventional lamb would be on this diet. The rumen VFA were in fairly normal proportions but in concentration only about half to two-thirds of that for a conventional lamb. It was suggested that lack of cellulolysis resulted in accumulation of undigested fibre in the rumen, and thus poor appetite. A germ-free lamb on the same diets did not grow to any extent after weaning, so the defined flora did convey some benefit. Some of the analyses done implicated *Veillonella*, but not *Selenomonas*, in succinate decarboxylation and propionate formation (see Chapter 11).

In other experiments (Hobson *et al.*, 1981) lambs with a defined flora continued to grow after weaning on to a diet containing 45% barley, 50% grass meal, 5% fish meal and mineral and vitamin supplement. They failed to grow after change to an all-grass feed, but could be made to continue growth by inoculating them with whole rumen contents or (the same thing) exposing them to conventional sheep, or by returning them to the barley–grass diet.

The most nearly normal rumen function was obtained in these experiments when the lambs were weaned straight on to a commercial barley (90%)–white-fish meal (10%)–minerals–vitamins feed. The rumen functions and weight gains of the four gnotobiotic lambs continued to be similar to those of conventional lambs on the same diet for about 120–130 days, when three of the lambs suddenly ceased to grow, and then rapidly lost weight. The fourth continued to gain weight for some days after the others began to lose weight, but the rumen VFA level was found to be low in a sample taken on day 148, so the lamb was removed to a pen with conventional lambs where it continued to thrive. The components of the defined rumen flora continued to be major components of the flora, in the same numbers, after conventionalisation; that is, the same species were found, but whether these were the same strains could not be determined. The main difference was that there was also a varied population of Gram-negative rods and cocci in numbers 100–1000 times less than the principal flora. The flora had become the conventional mixture. Of the first three lambs, one died in the pens and the other two died a few days after being moved in with conventional lambs. The suggestion of these happenings was that the rumen flora suddenly failed, and the rumen flora also appeared to fail after a similar period in the lambs on the barley/grass-meal diet. Conventionalisation either provided some factors from the varied flora that stimulated growth of the defined flora, or provided new strains of the defined species to carry on the rumen function. The lambs that died had evidently become too weak to live long enough to obtain a conventional flora.

The degradation of starch appears to be a relatively simple reaction, and previous work had shown that the *Streptococcus bovis* and the *Bacteroides amylophilus* could attach to starch granules, and they produced α-amylases which could degrade the granules, particularly if these were damaged (Baker & Hobson, 1952; Hobson & MacPherson, 1952; McWethy & Hartman, 1977). The starch granules in the barley would have been damaged to some extent during the pelleting of the feed, so degradation of the barley feed might have been expected. On the other hand, degradation of cellulose, particularly when it is combined with other polysaccharides and lignin in plant fibres, is a complex reaction requiring a number of enzymes for cellulolysis, and possibly the cooperation of bacteria with hemicellulolytic or other activities (see Chapters 8 and 11). The *Ruminococcus* added to the defined flora degraded filter paper in pure culture, and this and the *Butyrivibrio* and *B. ruminicola* would degrade hemicellulose and other plant polysaccharides. It was supposed at the time that the range of enzymes provided would degrade fibres. However, since then the importance of the anaerobic fungi and *Bacteroides succinogenes* in fibre degradation has been shown, and the complete degradation of fibre is possibly more complex than was then thought (see Chapters 2, 4, 8 and 11). The little cellulolysis or fibre digestion that did take place in the gnotobiotic lambs could have been the degradation of the most easily accessible polysaccharides.

Apart from the possibility that the defined flora provided for the lambs did not contain the right organisms for a fibre-digesting flora, it is possible that the limited flora did not provide the correct growth conditions for the ruminococci. Fonty *et al.* (1983*a,b*) had difficulty in establishing *Bacteroides succinogenes* in meroxenic lambs. The *Bacteroides* would establish only in lambs which already had a complex mixture of strict anaerobes and facultative anaerobes derived from inoculations of diluted conventional rumen contents established in the rumen, or which had a flora derived from an inoculum containing 182 types of rumen bacteria.

The rumen flora of the gnotobiotic lambs fed on barley failed, although established in numbers comparable to those in the conventional rumen. However, this flora did not prevent the establishment of a complex conventional flora, and the same species (if not the same strains) became part of a 'permanent' rumen flora when the lamb was conventionalised. These results and those of Fonty *et al.* suggest that, although the rumen reactions may be duplicated by a population of a small number of species of bacteria, for proper feed degradation and stability of the flora a complex mixture of species, and possibly continual renewal with new strains by natural inoculation, is required.

REFERENCES

Alexander, T., Lysons, R. J., Elliott, L. M. & Wellstead, P. D. (1973a). Equipment for rearing gnotobiotic lambs. *Lab. Animals*, 7, 195–217.

Alexander, T., Lysons, R. J., Elliott, L. M. & Wellstead, P. D. (1973b). Techniques for rearing gnotobiotic lambs. *Lab. Animals*, 8, 51–9.

Baker, F. & Hobson, P. N. (1952). The selective staining of intact and damaged starch grains with safranin O and Niagara Blue 4B. *J. Sci. Food Agric.*, 3, 608–12.

Barr, M. E. J., Mann, S. O., Richardson, A. J., Stewart, C. S. & Wallace, R. J. (1980). Establishment of ureolytic staphylococci in the rumen of gnotobiotic lambs. *J. Gen. Microbiol.*, 49, 325–30.

Cook, A. R. (1976). Urease activity in the rumen of sheep and the isolation of ureolytic bacteria. *J. Gen. Microbiol.*, 92, 32–48.

Cushnie, G. H., Richardson, A. J., Lawson, W. J. & Sharman, G. A. M. (1979). Cerebrocortical necrosis in ruminants: effect of thiaminase type 1-producing *Clostridium sporogenes* in lambs. *Vet. Rec.*, 105, 480–2.

Cushnie, G. H., Richardson, A. J. & Sharman, G. A. M. (1981). Procedures and equipment for rearing gnotobiotic lambs. *Lab. Animals*, 8, 51–9.

Davies, M. E. (1964). Cellulolytic bacteria isolated from the large intestine of the horse. *J. Appl. Bacteriol.*, 27, 373–8.

Edwin, E. E., Markson, L. M., Shreeve, J., Jackman, R. & Carroll, P. J. (1979). Diagnostic aspects of cerebrocortical necrosis. *Vet. Rec.*, 104, 4–8.

Elliott, L. M., Alexander, T. J. L., Wellstead, P. D. & Lysons, R. J. (1974). Microbiological monitoring of gnotobiotic lambs. *Lab. Animals*, 8, 51–9.

Fonty, G., Gouet, P., Jouany, J. P. & Senaud, J. (1983a). Ecological factors determining the establishment of cellulolytic bacteria in the rumens of meroxenic lambs. *J. Gen. Microbiol.*, 129, 213–23.

Fonty, G., Jouany, J. P., Thivend, P., Gouet, P. & Senaud, J. (1983b). A descriptive study of rumen digestion in meroxenic lambs according to the nature and complexity of the microflora. *Rep. Nutr. Devel.*, 23, 857–73.

Hobson, P. N. & MacPherson, M. J. (1952). Amylases of *Clostridium butyricum* and a *Streptococcus* isolated from the rumen of sheep. *Biochem. J.*, 52, 671–9.

Hobson, P. N., Mann, S. O. & Stewart, C. S. (1981). Growth and rumen function in gnotobiotic lambs fed on starchy diets. *J. Gen. Microbiol.*, 126, 219–30.

Kuster, E. (1915). Die Gewinnung, Haltung und Aufzucht Keimfreier Tiere und ihre Bedeutung für die Erforschung Natürlicher Lebersvorganger. *Arbeiten K. Gesundheitsamte*, 48, 1–79.

Leat, W. M. F., Kemp, P., Lysons, R. J. & Alexander, T. J. L. (1977). Fatty acid composition of the depot fats from gnotobiotic lambs. *J. Agric. Sci.*, 88, 175–9.

Lysons, R. J., Alexander, T. J. L., Hobson, P. N., Mann, S. O. & Stewart, C. S. (1971). Establishment of a limited rumen microflora in gnotobiotic lambs. *Res. Vet. Sci.*, 2, 486–7.

Lysons, R. J., Alexander, T. J. L., Wellstead, P., Hobson, P. N., Mann, S. O. & Stewart, C. S. (1976a). Defined bacterial populations in the rumens of gnotobiotic lambs. *J. Gen. Microbiol.*, **94**, 257–69.

Lysons, R. J., Alexander, T. J. L., Wellstead, P. D. & Jennings, I. W. (1976b). Observations on the alimentary tract of gnotobiotic lambs. *Res. Vet. Sci.*, **20**, 70–6.

Lysons, R. J., Alexander, T. J. L. & Wellstead, P. (1977). Nutrition and growth of gnotobiotic lambs. *J. Agric. Sci.*, **88**, 597–604.

Mann, S. O. (1963). Some observations on the air-borne dissemination of rumen bacteria. *J. Gen. Micribiol.*, **33**, ix.

Mann, S. O. & Stewart, C. S. (1974). Establishment of a limited rumen flora in gnotobiotic lambs fed on a roughage diet. *J. Gen. Microbiol.*, **84**, 379–82.

Mann, S. O., Grant, C. & Hobson, P. N. (1980). Interactions of *E. coli* and lactobacilli in gnotobiotic lambs. *Microbiol. Lett.*, **15**, 141–4.

McWethy, S. J. & Hartman, P. A. (1977). Purification and some properties of an extra-cellular α-amylase from *Bacteroides amylophilus*. *J. Bacteriol.*, **129**, 1537–44.

Oxender, W. D., Purser, D. B. & Bergen, W. G. (1971). A procedure to obtain germ-free goats by hysterectomy. *Am. J. Vet. Res.*, **32**, 1443–6.

Ponto, K. H. & Bergen, W. G. (1974). Developmental aspects of glucose and VFA metabolism in the germ-free and conventional ruminant. *J. Anim. Sci.*, **38**, 393–9.

Purser, D. B. & Bergen, W. G. (1969). Glucose utilization and hepatic enzyme activities in young gnotobiotic ruminants. *J. Dairy Sci.*, **52**, 790–5.

Sharpe, M. E., Latham, M. J. & Reiter, B. (1969). The occurrence of natural antibodies to rumen bacteria. *J. Gen. Microbiol.*, **56**, 353–64.

Sharpe, M. J., Latham, M. J. & Reiter, B. (1977). The immune response of the host animal to bacteria in the rumen and caecum. In *Digestion and Metabolism in the Ruminant*, ed. I. W. McDonald & A. C. I. Warner. University of New England, Australia, pp. 193–204.

Shreeve, J. E. & Edwin, E. E. (1974). Thiaminase-producing strains of *Cl. sporogenes* associated with outbreaks of cerebrocortical necrosis. *Vet. Rec.*, **94**, 330–5.

Smith, C. K. (1966). The derivation and characteristics of the germ-free ruminant. PhD thesis, University of Notre Dame, Indiana, USA.

Smith, R. H. (1980). Kale poisoning: the brassica anaemia factor. *Vet. Rec.*, **107**, 12–15.

Soares, J. H., Leffel, E. C. & Larsen, R. K. (1970). Neonatal lambs in a gnotobiotic environment. *J. Anim. Sci.*, **31**, 733–40.

Trexler, P. C. (1971). Microbiological isolation of large animals. *Vet. Rec.*, **88**, 15–20.

Weaver, J. M. (1974). The establishment of *Ruminococcus flavefaciens* in gnotobiotic lambs. PhD thesis, University of Maryland, USA.

Defined rumen protozoal populations with undefined bacteria

Preparation of ruminants with controlled fauna

To obtain animals with a controlled fauna, ciliate protozoa have to be totally eliminated (defaunation) before inoculation of a pure washed suspension of protozoa, since there is no known way of selectively eliminating particular ciliate genera.

Methods of defaunating the rumen. Rumen defaunation can be achieved by three different methods. The first consists of separating the young animals from their dam after birth and preventing all physical contact with adult ruminants (Abou Akkada & El Shazly, 1964). It is advisable not to carry out the separation before the age of 48–72 hours (Jouany, 1978) in order to allow the young animal to receive colostrum, which improves its vitality. During this short period of time, the young animal is naturally contaminated with strictly anaerobic bacteria (Fonty et al., 1984) which facilitate weaning (Gouet et al., 1984). This method is reliable, but it is not flexible because animals have to be prepared several months before they can be used.

The second method calls for the use of chemical agents, which are introduced into the rumen by a fistula or an oesophageal tube, to kill protozoa. The treatment is generally more efficient when combined with a fast of 24–72 hours. $CuSO_4$ (50 ml of a 2% solution on two consecutive days) can be used as a chemical additive (Becker et al., 1929/30). This treatment is very traumatising for the animals since about 50% of them die after $CuSO_4$ administration. Death may be partly explained by an increasing animal sensitivity to copper toxicity after elimination of protozoa (Ivan et al., 1986). The use of dioctyl sodium sulphosuccinate (Aerosol OT or Manoxol OT) (Abou Akkada et al., 1968; Orpin, 1977; Coleman & Sandford, 1979) brings about motility inhibition and disintegration of ciliate cells (Wright & Curtis, 1976). The many trials carried out by different workers show that animals have a high sensitivity to this product, since 50–66% of those used in the many trials performed by different workers died after treatment. Autopsies have shown that kidneys, liver and rumen mucosa are greatly altered. According to Lovelock et al. (1982), dioctyl sodium sulphosuccinate brings on pulmonary congestion, electrolyte imbalance and dehydration. Furthermore, total elimination of protozoa is sometimes difficult to achieve. Trials have also been carried out with alcohol ethoxylate (Teric GN9) or with alkanates (3 SL 3) (Bird & Leng, 1978, 1984; Bird et al., 1979). Adding calcium peroxide (Ixper 80 C) (Demeyer, 1982) or oils rich in polyunsaturated fatty acids (Czerkawski et al., 1975; Van Nevel & Demeyer, 1981; Ikwuegbu & Sutton, 1982) can defaunate the rumen, as can ground-barley when animals are fed on this *ad libitum* (Eadie et al., 1970).

The third method consists of emptying the rumen and treating the contents in order to destroy ciliate protozoa before reintroducing the contents into the rumen. This method requires animals that have been fitted with a permanent fistula. As early as 1957, Eadie and Oxford used this principle to eliminate holotrich ciliates by heating the contents to 50°C for 15 minutes. Jouany & Senaud (1979a) suggested that the contents be frozen to kill all the protozoa. Just after emptying the rumen, the mucosa must be washed, then treated with a diluted solution of formaldehyde before being carefully rinsed. The treatment is repeated three times on two consecutive days, during which animals are fasted. At the end of all this operation, the thawed rumen contents are put back into the rumen and animals are fed again. This method is reliable and harmless to the animals if the formaldehyde treatment is carried out rapidly and rinsing is sufficiently thorough. Because the animal fasts for 46 hours after emptying the rumen, a considerable supply of the B vitamins (B_2, B_6, B_{12}) has to be given by intramuscular injections.

It is important to make sure that adding chemicals to the rumen does not bring about the systematic elimination of certain rumen bacteria and fungi. Furthermore, we should ascertain that defaunating chemicals have no direct effect on digestion or digestive-tract physiology. Modifications observed after defaunation must be exclusively due to the elimination of protozoa and not to the antimicrobial effect of chemicals used to defaunate. Eadie & Shand (1981), indeed, have shown that adding nonyl phenol ethoxylate (Synperonic NP9) into the rumen could have a direct effect on the fermentations as well as having an effect on protozoa. By introducing rumen contents that have been previously frozen at $-15°C$ and thawed out, the presence of a 'normal' bacterial population in the rumen can be guaranteed.

Inoculating isolated protozoa into a defaunated rumen

This method consists of introducing a pure washed suspension of protozoa into a previously defaunated rumen. The ciliate inoculum is obtained from rumen contents whose fauna are rich in the genus that is to be isolated. The rumen liquid obtained after filtration through a metallic sieve (1 mm mesh size) is put into a small beaker at 40°C. A drop of this liquid is mixed on a slide with 'Hungate solution' (NaCl, 6 g; $NaHCO_3$, 1 g; H_2KPO_4, 1 g; $MgSO_4$, 0·1 g; $CaCl_2$, 0·1 g; in 1 litre of distilled water) at 40°C under a low-power binocular microscope. The ciliates to be selected are removed manually, one by one, with a curved micropipette fitted with a small teat, and then placed on a new slide with Hungate solution at 40°C. This selection process is repeated three times in order to obtain around 20 cells of one ciliate genus. The pure *Entodinium* spp. inoculum is introduced with a long pipette into the ventral sac of the rumen, through the cannula. *Polyplastron* must be incubated at 40°C for at least 4 hours, in rumen fluid that has previously been defaunated by freezing, before they are again selected and put into the rumen. This supplementary step is done to eliminate the ciliates that are engulfed by *Polyplastron* (see Chapter 3) and which could be released into the medium when they are introduced into the rumen and so give a mixed fauna. The population reaches a minimum concentration for counting (30 cells/ml) about 2 weeks after the genus has been introduced into the rumen.

Changes in the bacterial populations in the rumens of animals with controlled fauna

Defaunated animals. The total elimination of ciliate protozoa brings about an increase in the bacterial population in the rumen (Eadie & Hobson, 1962; Kurihara *et al.*, 1968, 1978; Eadie & Gill, 1971; Teather *et al.*, 1984). This can be explained mainly by the intense predatory activity of protozoa on bacteria (see Chapter 3). According to Coleman (1975), 2·5–45 g of dry bacteria is recycled daily in a sheep rumen containing 50 g of total bacterial dry matter. If prey intake is selective as Coleman (1979) has shown, and bacterial growth rate varies with strain and with the general conditions in the rumen, it can be concluded that the bacterial population will be variously affected by elimination of ciliate protozoa. The modifications will depend on the nature of the animal diet and other factors related to the microbial medium, the rumen fluid. Ingestion of feed particles (starch granules and fibres) and adherent bacteria by protozoa must be taken into consideration to understand the evolution of the new equilibrium of the bacterial population in the defaunated rumen.

Some of the very few studies carried out on the evolution of the composition of the bacterial mixture in the rumen show that defaunation does modify the flora. According to Itabashi & Katada (1976), small Gram-negative rods and selenomona-like bacteria tend to increase when protozoa are eliminated from the rumens of calves fed on a mixed diet (roughage plus concentrate). Results obtained by Kurihara et al. (1968), on a limited number of lambs fed on a similar diet, indicated also that there was an increase in the proportion of Gram-negative rod-type bacteria, while the small Selenomonads decreased. A more detailed analysis of the rumen flora carried out by Kurihara et al. (1978), on sheep fed on a purified diet, showed the positive effect of defaunation on the number of amylolytic bacteria (*Bacteroides amylophilus* and *B. ruminicola*). The cellulolytic flora, on the other hand, is less reduced because the decrease in *Ruminococcus albus* concentrations is compensated for by the increase in *R. flavefaciens*. This suggests that there might be modifications in the balance between species of cellulolytic bacteria. One could predict that defaunated animals would be less able to digest cellulose than faunated animals, because the cellulolytic activity of *R. flavefaciens* is lower than that of *R. albus* (Van Gylswyck & Labuschagne, 1971). The other major cellulolytic bacterium in the rumen, *Bacteroides*, is less affected by defaunation (Whitelaw et al., 1972). Some recent data from Leng (1984) and Soetanto (1985) suggest that defaunation could favour the development of anaerobic fungi and this would compensate for the decreased bacterial cellulolytic activity.

Although it has never been clearly demonstrated, elimination of protozoa does have an effect on some methanogenic bacteria whose close association with the ciliate surface has been described (Imai & Ogimoto, 1978; Vogels et al., 1980; Stumm et al., 1982; Krumholz et al., 1983).

Animals with defined rumen protozoa. Introducing *Entodinium caudatum* or *E. caudatum* plus *Polyplastron multivesiculatum* into a defaunated rumen favours the development of large bacteria at the expense of small bacteria (Fonty, 1984). Itabashi & Kandatsu (1974) observed similar results when comparing defaunated and naturally faunated animals which received a diet based on alfalfa. According to Fonty (1984), the change in the cellulolytic flora is much more difficult to describe. The number of cellulolytic bacteria decreases following inoculation of *E. caudatum* and increases in the presence of *Polyplastron* plus *Entodinium*.

With more details available, Fonty (1984) showed that the establishment of only one ciliate genus (*Entodinium* spp. or *P. multivesiculatum*) in a defaunated lamb rumen containing a simplified bacterial population brings about a decrease in the total number of bacteria, particularly anaerobic bacteria (Table 1). In the presence of *Entodinium* spp., the proportion of amylolytic bacteria falls considerably. This confirms the results of Kurihara et al. (1978). A decreased proportion of bacteria that utilise xylose and pectin is less evident. Conversely, the proportion of cellobiose- and maltose-utilising bacteria increases. By classifying the bacteria according to morphology, respiratory type, Gram-type and end-products of glucose fermentation, Fonty et al. (1983a) showed that the presence of *Entodinium* spp. alone in the rumen provoked a clear decrease in the proportion of bacteria of the group that includes *Butyrivibrio fibrisolvens*, with an increase in cocci (*Peptostreptococcus productis* and *Streptococcus*). *Bacteroides* and *Eubacterium* are not affected by the presence of *Entodinium* spp.

Elimination of protozoa, or the inoculation of some of them into a defaunated

Table 1

Bacterial Counts ($\times 10^{-8}$ g^{-1}) in Rumen Samples from Non-inoculated and *Entodinium* spp.- or *Polyplastron*-inoculated Lambs
(Reproduced with permission from Fonty et al., 1983a)

Lambs	Age (days)	Total aerobic counts	Enterobacteria	Streptococci	Lactobacilli	Total anaerobic counts in medium		Bacterial counts on medium containing individual carbohydrate substrate[a]						
						M2	CCA	Glucose	Starch	Cellobiose	Maltose	Xylose	Pectin	Lactate
1[b]	72	0.1	0.01	0.07	0	280	460	240(52)	400(86)	260(56)	220(47)	350(76)	400(87)	9(2)
	96[d]	0.04	0.02	0.007	0	56	124	116(93)	20(16)	100(80)	120(97)	70(56)	90(72)	5(4)
2[b]	72	0.2	0.02	0.08	0	114	210	170(81)	160(76)	88(42)	100(47)	190(90)	180(88)	6(3)
	96[d]	0.2	0.09	0.017	NT[f]	NT	210	140(66)	20(9)	160(76)	160(76)	100(47)	160(77)	6(3)
3[c]	44	4	0.7	4	0	83	250							
	77[e]	1.7	0.8	1.5	0	46	75							
4[c]	54	4	0.15	4	0	80	230							
	87[e]	1.5	0.5	1.4	0	26	44							

[a] Numbers in parentheses represent percentage of total count in CCA medium.
[b] Lambs 1 and 2 inoculated with *Entodinium* spp. at 80 days.
[c] Lambs 3 and 4 inoculated with *Polyplastron multivesiculatum* at 55 and 69 days.
[d] *Entodinium* sp. present in rumen.
[e] *P. multivesiculatum* present in rumen.
[f] NT, not tested.

rumen, not only acts on the concentrations of bacteria but also modifies the composition of the bacterial mixture in the rumen.

Changes in fermentation parameters in the rumen as a function of the fauna

pH and volatile fatty acids (VFA). Defaunation frequently brings about a decrease in the pH value of the rumen fluid (Conrad *et al.*, 1958; Luther *et al.*, 1966; Whitelaw *et al.*, 1972; Jouany, 1978). The discrepancies between faunated and defaunated animals can be quite considerable 2–3 hours after the beginning of feeding on diets rich in starch (Veira *et al.*, 1983). This situation can be explained by storage of readily fermentable carbohydrates and sugars as amylopectin in the ciliate cells after feed intake (Oxford, 1955; Jouany & Thivend, 1972), and also by the metabolism of rumen lactate by the protozoa (Chamberlain *et al.*, 1983; Newbold *et al.*, 1986). Risks of acidosis in animals that have been fed on cereal-rich diets are thus increased by defaunation (Mackie *et al.*, 1978). Defaunation had no effect on rumen pH when diets rich in plant cell wall carbohydrates (75% wheat straw plus 15% sugar beet pulp) were given (Collombier, 1981).

Defaunation sometimes lowers the VFA concentrations in the rumen although, in general, concentration is not significantly modified. On the other hand, defaunation modifies the molar composition of the VFA mixture. According to many authors, defaunation increases the molar proportion of propionic acid at the expense of butyric acid (Itabashi & Katada, 1976; Jouany *et al.*, 1981; Itabashi *et al.*, 1982; Grummer *et al.*, 1983; Jouany & Senaud, 1983; Ushida *et al.*, 1986*a*), and sometimes at the expense of acetic acid (Eadie & Mann, 1970; Bird & Leng, 1978; Kurihara *et al.*, 1978; Whitelaw *et al.*, 1984). However, increase in propionate does not appear to be a general rule. Some authors have shown diametrically opposite effects which are difficult to relate to the nature of the feed or to the animal (Jouany *et al.*, 1988). Chamberlain *et al.* (1983) showed that, with a grass-silage feed, defaunation of the wether sheep rumen brought about a clear increase of 100–150% in butyrate production when lactic acid was infused into the rumen. At the same time, propionate production decreased, as did acetate production but to a lesser extent. In this case, the elimination of protozoa favoured the bacterial conversion of lactate into butyrate.

The introduction of a controlled fauna into a defaunated rumen only slightly modifies the total concentrations of VFA (Table 2), but it influences the molar proportions of the VFA. An increase in butyrate proportion is found when *Entodinium* spp. or *Polyplastron* are inoculated into the rumens of sheep fed on diets rich in cellulose or starch. A similar change occurs after introducing *P. multivesiculatum* and *Entodinium* spp. into the rumens of meroxenic lambs (Fonty *et al.*, 1983*b*). With diets rich in soluble carbohydrates (sucrose, inulin, lactose), Jouany & Senaud (1983) noted a decrease in the proportion of acetic acid in rumen VFA produced after inoculation with *Polyplastron*. This decrease was compensated for by an increase in propionic acid. With diets based on wheat straw, whether alkali-treated or not, Collombier (1981) observed no modifications in the composition of the VFA mixtures when sheep were inoculated only with *Entodinium* spp. or *Polyplastron*. Contrary to the other previously cited genera, *Isotricha* spp.

Table 2
Volatile Fatty Acids (VFA) and NH_3^- in the Rumens of Animals with Conventional or Controlled Fauna

Authors[a]	Diet (% of concentrate)[b]	VFA (mM/l)[c]				C_2				C_3				C_4				NH_3-N (mg litre^{-1})			
		F	D	E	O	F	D	E	O	F	D	E	O	F	D	E	O	F	D	E	O
(1)	60f	90	89	89	91d	62	64	63	62d	23	23	28	27d	14	11	7	9d	—	—	—	49e
(2)	19g	66	65	62	63e	71	72	72	71e	19	19	19	19e	7	7	6	8e	46	42	39	23e
	19g	93	110	106	92e	69	70	71	71e	19	21	20	19e	10	8	7	8e	26	18	23	12e
	3g	90	78	86	95e	72	73	73	70e	17	19	18	21e	7	5	6	8e	18	14	13	15e
(3)	60f	116	103	88	117e	61	59	61	56e	30	35	25	31e	8	4	13	11e	24	16	25	19e
	70h	111	75	112	134e	53	57	49	58e	27	21	30	22e	19	16	17	17e	26	15	17	19e
	83i	127	99	119	121e	56	61	55	55e	26	19	25	28e	17	17	17	15e	35	21	22	30e
	63j	109	91	71	110e	59	64	54	71e	16	14	21	12e	21	21	23	15e	22	16	14	22e
(4)	49f	91	83	92	102e / 86d	66	66	62	67e / 62d	15	22	28	19e / 25d	16	9	8	11e / 8d	35	15	16	26e / 20d

[a] (1) Williams & Dinusson, 1973. (2) Collombier, 1981. (3) Jouany & Senaud, 1982, 1983. (4) Jouany et al., 1981.
[b] The concentrate is rich in: f, starch; g, protein; h, inulin; i, sucrose; j, lactose.
[c] Rumen contains: F, a conventional fauna; D, no protozoa; E, only *Entodinium* spp.; O, only one other genus which is: d, *Isotricha* spp.; e, *Polyplastron*.
[d] C_2, acetic acid; C_3, propionic acid; C_4, butyric acid.

inoculation into the rumens of animals fed on a mixed diet did not influence the butyric acid content (Jouany et al., 1981).

From the results obtained with animals hosting a simple controlled fauna, it is difficult to relate the presence of protozoa to a generally proven effect on the composition of VFA produced in the rumen (Table 2). Two hypotheses can be entertained: either protozoa play a minor role in the overall production of VFA, or the protozoal metabolism varies according to the nature of the diet.

The net effects of protozoa on the balance between bacterial species of quantitative importance, which can vary depending on the conditions within the rumen, could explain the differences observed in the changes in VFA production when defaunated animals are inoculated with some protozoa. These conditions, which are mainly determined by the nature, form and intake level of the feedstuffs, and by feeding behaviour of the animals, could orient the metabolism of protozoa and their effect on bacteria.

Ammonia. The change in rumen ammonia concentration according to the nature of the rumen fauna is easy to characterise. With the exception of results published by Demeyer et al. (1982), all authors agree that by defaunating the rumen the NH_3-N concentration is greatly decreased (Table 2). Sometimes it reaches 50 mg/litre which, in theory, does not allow optimal microbial synthesis. On the basis of current knowledge, it might be useful to provide more soluble or degradable nitrogen compounds in the diets of defaunated animals.

Inoculating *Entodinium* spp. into a defaunated rumen generally results in a slight increase in NH_3-N concentration. In the presence of *P. multivesiculatum*, the increase in NH_3-N concentration is greater, but the concentration remains lower than that observed in conventional animals. In the presence of *Isotricha prostoma*, the NH_3-N concentration lies between the concentrations noted with the other two genera. According to Itabashi & Kandatsu (1975a,b), the increase in NH_3-N concentration observed in the inoculated rumen could not be due to a lower utilisation of NH_3-N since the half-life of $^{15}NH_3$ was the same in faunated and defaunated animals. The increase may be a result of a higher production of free amino acids from the degradation of feed proteins and peptides in the faunated rumen (Itabashi & Katada, 1976). This is probably a direct effect of the increased proteolytic activity of rumen microbes after protozoal inoculation (Ushida & Jouany, 1985; Ushida et al., 1986a). The amount of recycled microbial nitrogen, measured by the difference between 'total' and 'net' synthesis (see Chapter 6), is higher in the faunated than the defaunated rumen (Demeyer & Van Nevel, 1979), which might also explain some of the changes in the NH_3-N concentration. The increase in recycled N is mainly due to predation of protozoa on bacteria and to a possible increase in mean retention time of rumen contents when protozoa are introduced into a defaunated rumen.

Gases. The few available *in vivo* assessments show that methanogenesis is reduced by 30–45% after defaunation of the rumen (Jouany et al., 1981; Whitelaw et al., 1984). These results, which have been confirmed *in vitro* (Itabashi et al., 1984; Ushida et al., 1986b), would account for a daily saving of 75–100 litres of methane in young bulls weighing about 250 kg (Whitelaw et al., 1984). As far as energy metabolism is concerned, this would be a definite advantage for the defaunated animals. The change in methanogenesis agrees with that of propionic acid if we consider the stoichiometric reactions of rumen fermentation (see Chapters 1 and 11). Animals inoculated with a monospecific fauna or a mixture of two genera of ciliates have a CO_2/CH_4 ratio which is lower when compared with that of defaunated animals and higher than that of conventionally faunated animals (Fig. 1).

The presence of methanogenic bacteria closely associated with the surface of ciliates by an ectosymbiotic attachment (Vogels et al., 1980; Stumm et al., 1982; Krumholz et al., 1983) shows that protozoa play a role in the production of methane precursors. They produce hydrogen (Abou Akkada & Howard, 1960; Prins & Van

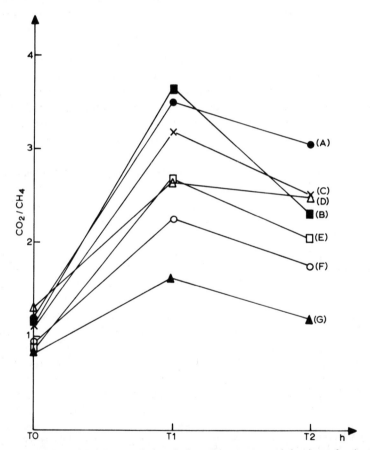

Fig. 1. Change of the CO_2/CH_4 ratio in relation to protozoa and the time after ingestion of a concentrate diet: (A) defaunated animals; (B) animals inoculated with *Entodinium* spp.; (C) animals inoculated with *Isotricha* spp.; (D) animals inoculated with *Polyplastron multivesiculatum*; (E) animals inoculated with *Entodinium* spp. + *Isotricha* spp.; (F) animals inoculated with *P. multivesiculatum* + *Isotricha* spp.; (G) animals inoculated with a conventional fauna.

Hoven, 1977; Van Hoven & Prins, 1977) which is used by methanogenic bacteria (see Chapter 11).

Changes in digestion of plant cell wall carbohydrates in relation to the fauna
  After rumen defaunation. Most studies carried out *in vivo* show that rumen defaunation decreases cell wall carbohydrate digestion by 5–15% when measured over the whole digestive tract. This decrease is much higher when measured only in the rumen (Table 3). Such results indicate that a shift in digestion from the rumen to the hindgut occurs. This shift explains the increased faecal excretion of energy and nitrogen observed by Itabashi *et al.* (1984) when animals were defaunated. The rumen degradability of cell wall carbohydrates measured *in vitro* or *in sacco* decreases by as much as 30% when animals are defaunated (Demeyer, 1981). This result, which corresponds approximately to *in vivo* data, can be partly explained by the presence of cellulolytic enzymes in protozoa (Williams & Coleman, 1985; Williams & Ellis, 1985). Recent results (Broudiscou *et al.*, 1988) indicate that protozoa could be more involved in the digestion of cellulose than in that of hemicelluloses in which bacteria and fungi could have the dominant role. The respective importance of protozoa and bacteria in the cellulolytic activity of the rumen contents can be determined by using ground roughages put into nylon bags and introduced into the rumen ('*in sacco*' method). When the nylon mesh size is

Table 3
Effects of Defaunation on Cell Wall Carbohydrate (ADF) Digestibility in Sheep Digestive Tract or Rumen

Authors[a]	Main source of energy	%	Diet intake (g DM/day)[b]	Digestibility (digested/intake)[c]				
				F[d]	D	P	E	P+E
(1)	Alfalfa hay	100	1 000	(α) 0·65 (a)	0·62	—	—	—
				(β) 0·41 (a)	0·35	—	—	—
	Red clover	100	1 000	(α) 0·76 (a)	0·75	—	—	—
				(β) 0·45 (a)	0·41	—	—	—
(2)	Ray-grass hay	100	1 100	(α) 0·71 (b)	0·68[e]	0·72	0·70[e]	0·72
	Deproteinized whey	42	1 100	(α) 0·59 (b)	0·55[e]	0·58	0·55[e]	0·58
	Barley	50	1 100	(α) 0·60 (b)	0·56[e]	0·61	0·60	0·61
	Jerusalem artichoke	63	1 100	(α) 0·68 (b)	0·66[e]	0·68	0·68	0·68
	Sugar beet	73	1 100	(α) 0·69 (b)	0·62[e]	0·67	0·66	0·69
(3)	Dehydrated alfalfa	49	1 100	(α) 0·48 (b)	0·43[e]	0·47	0·41[e]	—
(4)	Maize silage + shelled corn	95	700	(β) 0·42 (b)	0·34	—	—	—
(5)	Grass hay	50	—	(β) 0·50 (b)	0·36	—	—	—
(6)	Alfalfa hay	65	1 150	(α) 0·43 (b)	0·38[e]	—	—	—
				(α) 0·59 (b)	0·57[e]	—	—	—

[a] (1) Lindsay & Hogan, 1972. (2) Jouany & Senaud, 1982, 1983. (3) Jouany *et al.*, 1981. (4) Veira *et al.*, 1983. (5) Kayouli *et al.*, 1986. (6) Ushida *et al.*, 1986a.
[b] Grams of dry matter per day.
[c] F, faunated animals; D, defaunated animals; P, inoculated with only *Polyplastron*; E, inoculated with only *Entodinium* spp.; P + E, inoculated with *Polyplastron* and *Entodinium* spp.
[d] (α) Digestibility measured in the whole digestive tract. (β) Digestibility measured in the rumen. (a) NDF. (b) ADF.
[e] Significant effect ($P < 0.05$).

≤ 10 μm, only small, non-cellulolytic *Entodinium* spp. can enter the nylon bags, and therefore it is possible to determine the cellulolytic activity of the bacteria only. With mesh size ≥ 100 μm, the cellulolytic activity of large protozoa is involved in plant cell disappearance (Jouany, 1978). Using this methodology, Jouany & Senaud (1979b, 1982, 1983) estimated the cellulolytic activity of bacteria in the faunated and defaunated rumen (Table 4). The authors observed that protozoa had an effect on this parameter and that the effect varied with the nature of the diet. When animals were fed a diet rich in cell wall carbohydrates (97% grass hay), defaunation was found to produce a significant decrease in the cellulolytic activity of the bacteria. This might be due to a lower metabolic activity in the dominant cellulolytic flora as indicated by Yoder *et al.* (1966) and Kurihara *et al.* (1968), or to a smaller number of cellulolytic bacteria in the rumen (Kurihara *et al.*, 1978). When animals were limit-fed on diets rich in easily fermentable carbohydrates (starch, sucrose, lactose, inulin), a high number of ciliates were present (Jouany *et al.*, 1977). The *in sacco* method then showed that rumen defaunation increased the cellulolytic activity of the bacteria. In the latter case, where the amount of cell wall carbohydrates was low, the bacteria probably had more cellulosic material available once the protozoa were eliminated and the nutritional competition between protozoa and bacteria disappeared. The fact that protozoa consume solid particles and adherent bacteria is a limiting factor not to be overlooked when the development of cellulolytic bacteria in faunated animals is studied. In the different experiments it should be noted that no studies have taken fungi into account. In the future they must be taken into consideration even if, as Windham & Akin (1984) pointed out, the amount of cellulose degraded by fungi is quantitatively low compared with that degraded by bacteria. Recent results obtained by Elliott *et al.* (1987) showed that digestibility of straw introduced in nylon bags (64 μm pore size) into the rumen increased by 15–40% when fungi were present. According to Orpin (1983/4), there could be a certain balance between protozoa and fungi, the latter being more numerous in the rumen which hosts a low concentration of protozoa. This suggests that the activity of fungi may substitute for that of protozoa, and could explain some of the positive effects of defaunation on digestion of cell wall carbohydrate from poor roughage (straw).

After inoculation with controlled fauna. Adding *Isotricha* spp. as the only ciliate genus into a defaunated rumen had a negative effect on the digestion of lignocellulose (Jouany *et al.*, 1981). This is explained both by the low hydrolytic activity of *Isotricha* spp. for cell wall carbohydrates, in particular hemicelluloses (Williams & Coleman, 1985), and by a possible negative effect of *Isotricha* spp. on the cellulolytic bacteria and fungi. In all the experiments carried out in France (Jouany & Senaud, 1979b, 1982, 1983; Jouany *et al.*, 1981), inoculating a defaunated rumen with only *Entodinium* spp. did not modify the digestibility of the lignocellulose of a grass-hay diet fed alone or supplemented with 49% of deproteinised whey, or of a diet composed of a mixture of lucerne and barley grain. In the presence of easily fermentable carbohydrates (starch, inulin, sucrose), inoculation of *Entodinium* spp. improved the digestion of lignocellulose to a value close to that measured in conventional animals. Inoculating only *Polyplastron multivesiculatum* had a clear

Table 4

Effects of Rumen Ciliates on Lignocellulose Digestibility (D.ADF) and Cellulolytic Activity of Rumen Bacteria (C.A.) in Sheep Fed on Different Diets

Animals	Cellulose		Starch		Sucrose		Lactose		Inulin	
	D.ADF	C.A.	D.ADF	C.A.	D.ADF	C.A.	D.ADF	C.A.	D.ADF	C.A.
Defaunated	68.8 ± 0.2^a	38.3 ± 1.9^a	55.6 ± 0.7^a	27.3 ± 2.3^a	62.5 ± 1.1^a	28.4 ± 1.5^a	55.4 ± 1.4^a	28.1 ± 1.5^a	66.4 ± 1.9^a	29.9 ± 1.6^a
Contaminated only with										
E	70.1 ± 0.1^b	41.3 ± 3.6^b	60.3 ± 1.8^b	24.5 ± 1.8^b	66.4 ± 3.1^b	26.7 ± 2.5^b	54.7 ± 1.8^a	26.2 ± 1.7^b	68.3 ± 1.2^{ab}	26.3 ± 1.0^b
P	71.6 ± 1.2^c	39.8 ± 1.8^{ab}	61.4 ± 0.6^b	25.4 ± 0.9^b	67.1 ± 4.6^b	26.0 ± 1.6^b	58.6 ± 2.4^b	26.8 ± 0.7^b	68.2 ± 1.3^{ab}	25.6 ± 2.6^b
Faunated	71.1 ± 0.8^c	41.0 ± 0.9^b	60.0 ± 1.2^b	21.1 ± 1.0^c	69.2 ± 0.8^b	26.8 ± 1.4^b	58.9 ± 1.8^b	23.2 ± 0.7^c	68.7 ± 1.2^b	26.2 ± 1.3^b

Means followed by different superscripts in the same column are significantly different ($P \leq 0.05$).
E, *Entodinium* spp.; P, *Polyplastron multivesiculatum*.

positive effect on the lignocellulose digestibility for all the diets studied, digestibility then becoming comparable to that produced by an abundant and mixed fauna. This genus plays an important role in the balance between microorganisms in the rumen, especially the cellulolytic bacteria. *Polyplastron multivesiculatum* also has a direct effect on the digestion of fibres, since *Polyplastron* produces enzymes which are particularly efficient in the hydrolysis of cell wall carbohydrates (Williams & Coleman, 1985; Williams & Ellis, 1985; Bonhomme *et al.*, 1986).

Changes in duodenal nitrogen flow in animals with a controlled fauna
  Flow of undegraded feed nitrogen. The amount of non-degraded feed nitrogen flowing into the duodenum is calculated *in vivo* as the difference between the non-ammonia nitrogen and microbial nitrogen which enters the duodenum of the ruminant. In fact this amount corresponds to the sum of both dietary and endogenous nitrogen. The latter is generally estimated to be 1–2 g N/day for an adult sheep but, according to Harrop (1974) and Harrisson *et al.* (1979), the value varies greatly and it can reach as much as 6 g N/day (Macrae & Reeds, 1980). The endogenous nitrogen flow may be influenced by defaunation, because defaunation alters the mean retention time of feed and decreases fibre digestion in the rumen. Thus most of the comparisons published to date on dietary nitrogen flow in the duodenum obtained on defaunated and faunated animals are rough estimations.

  Furthermore, the accuracy of methods for measuring digestive flow and determining microbial nitrogen is critical when duodenal feed nitrogen has to be quantified. Faichney (1975) showed that the quantification of digestive flow should be done with particular care. The quality of microbial markers, as well as the accuracy of their assay, have also been discussed in many papers (Ling & Buttery, 1978; Smith *et al.*, 1978; Harrisson & McAllan, 1980). Siddons *et al.* (1982) showed that large discrepancies between microbial markers made it impossible to calculate accurately the duodenum feed nitrogen flow. Using defaunated sheep is the only condition in which comparisons between markers for total microbial or for bacterial nitrogen estimations are really valid. Ushida *et al.* (1984, 1986a) and Meyer *et al.* (1986) showed that results obtained with different markers were comparable in defaunated animals, and allowed feed nitrogen degradation to be calculated, although there were some reservations about calculation of endogenous nitrogen, as mentioned before. In spite of these problems, most of the recent experiments show that feed nitrogen degradation is decreased by defaunation (Table 5), hence flow of feed nitrogen to the duodenum is increased. The differences noted between faunated and defaunated animals in the duodenal feed nitrogen flow are sometimes significant (Table 6). Given the limit of accuracy in management and the small number of papers that have dealt with this topic, it is clear that a positive effect of defaunation on the supply of feed nitrogen to the intestine cannot readily be proved in all situations. However, it should be noted that the amount of undegraded feed nitrogen is never lower when animals are defaunated than when they are faunated.

  Flow of microbial nitrogen. Very few authors have, in the same experiment, measured microbial nitrogen by using several markers simultaneously in

Table 5
Influence of Defaunation on Dietary Nitrogen Degradation in the Rumen

Animals	Diet		% dietary N degraded in the rumen		Authors
	Base	Nitrogen	F	D	
Sheep	NaOH-treated straw	Peanut cake + urea (9/1)	63	62	(1)
Sheep	Grass hay		81[a]	56	(2)
			74[a]	70	(2)
Sheep	Corn stover	Corn + fishmeal (30/7)	78	82	(3)
Sheep	Alfalfa hay	Alfalfa	55[b]	36[b]	(4)
Sheep	Alkali-treated straw	Soyabean cake + peanut cake + urea (6/10/1)	69[b]	55[b]	(4)
In vitro		Lupin seed	71	61	(5)
		Peanut cake	71	60	(5)
		Soya cake	48	37	(5)
		Fishmeal	28	17	(5)

(1) Jouany & Thivend, 1983. (2) Kayouli et al., 1986. (3) Meyer et al., 1986. (4) Ushida et al., 1986a. (5) Ushida & Jouany, 1985.
F, faunated animals; D, defaunated animals.
[a] Individual results.
[b] Significant effect of defaunation ($P < 0.05$).

defaunated and faunated animals. By comparing bacterial marker (diaminopimelic acid) flows in both kinds of animals, the effects of protozoa on the amount of bacterial nitrogen flowing into the duodenum can be evaluated. When the same comparison is made with other markers used to estimate the total microbial nitrogen (^{35}S, RNA, purine bases), the amount of protozoal protein entering the duodenum can be calculated by the difference between microbial protein and bacterial protein flows.

Most of the time, defaunation provokes a net increase in overall microbial and bacterial protein flows to the duodenum. The increase can be up to 5% with a roughage diet (Lindsay & Hogan, 1972) and up to 100% with a mixed diet rich in plant cell walls (corn stover) or rich in readily fermentable carbohydrates (corn and molasses) (Meyer et al., 1986). This effect of defaunation on microbial nitrogen flow can be explained mainly by the improvement in bacterial nitrogen flow in defaunated animals, and by the low contribution of protozoal nitrogen to the total microbial nitrogen in the duodenum flow (about 20%) of faunated animals (Ling & Buttery, 1978; Harrisson et al., 1979; Jouany & Thivend, 1983; Cockburn & Williams, 1984; Collombier et al., 1984; Meyer et al., 1986). The differences observed in protozoal/bacterial nitrogen ratio between the rumen and the duodenum illustrate the selective retention of protozoa in the rumen which Hungate et al. (1971), Weller & Pilgrim (1974), Jouany (1978), Coleman et al. (1980), Collombier (1981), Leng et al. (1981), Punia et al. (1984a,b) and Michalowski et al. (1986) have mentioned.

Table 6
Effects of Defaunation on Duodenal Nitrogen Flow (g/day)

Animals	Diets (%)	Feed N[a]		Microbial N		NAN[b]		Authors
		F	D	F	D	F	D	
Sheep	Alfalfa (100)	6	7	12	14	18	21	(1)
	Red clover (100)	11	13	18	19	—	—	(1)
Sheep	Alfalfa + beet pulp (75/15)	9	10	14[d]	18	23[d]	28	(2)
Sheep	NaOH-treated straw + beet pulp (75/15)	9	9	16	16	25	25	(3)
Calves	Corn-silage + corn + urea (48/47/1)	—	—	—	—	16	17	(4)
Sheep	Hay + concentrate (50/50)	7	7	12	15	19[d]	22	(5)
Sheep	Hay + concentrate (50/50)	6[c]	14	16[c]	16	23[c]	30	(6)
		9[c]	9	10[c]	17	49[c]	26	(6)
Sheep	Corn stover + molasses + corn (46/10/30)	7	5	8[d]	16	19	22	(7)
Sheep	Alfalfa hay + barley (65/30)	11[d]	15	12[d]	18	23[d]	33	(8)
Sheep	NaOH-treated straw + beet pulp (83/17)	7[d]	10	15[d]	19	22[d]	30	(8)

(1) Lindsay & Hogan, 1972. (2) Collombier, 1981. (3) Jouany & Thivend, 1983. (4) Veira et al., 1983. (5) Rowe et al., 1985. (6) Kayouli et al., 1986. (7) Meyer et al., 1986. (8) Ushida et al., 1986a.
F, faunated animals; D, defaunated animals.
[a] Feed nitrogen, NAN – microbial N.
[b] NAN, non-ammonia nitrogen.
[c] Individual results.
[d] Significant effect of defaunation ($\bar{P} < 0.05$).

The increased bacterial nitrogen flow in the duodenum of defaunated animals is mainly due to savings in recycled nitrogen in the rumen (see Chapters 3 and 13). Leng (1982) explained that the considerable waste of microbial nitrogen in the faunated rumen is partly due to the death and the lysis of ciliates, although the engulfment of a great number of bacteria, and predation on other ciliates, by protozoa could be the major reason. Half of the microbial nitrogen ingested by protozoa is released into the rumen as amino acids or small peptides (Coleman, 1975). This microbial nitrogen waste, associated with increased ruminal degradation of feed proteins, explains the lower non-ammonia nitrogen flow to the duodenum of faunated animals (Table 6). On the whole, the supply of α-amino nitrogen to the ruminant intestine is positively influenced by defaunation. The introduction of a controlled fauna of *Entodinium* spp. into a defaunated rumen leads to a situation like that of naturally faunated animals when the number of protozoa in the rumen is high. In contrast, no effect on the different forms of nitrogen entering the duodenum was observed when the population of ciliates remained at 10^3/ml, which occurred when animals were inoculated with *Polyplastron* (Table 7).

Effects of controlling the fauna on animal performance
The consequences of eliminating rumen protozoa on animal growth and wool production are not clear (Table 8), as discussed by Veira (1986). The results are influenced both by animal needs and by nutrient supplies (nature of feeds, form of presentation, mode of distribution, level of intake). Unfortunately, in most of the literature data, nutrient supplies have not always been fully described.

The divergent effects observed on animal performances (Table 8) can be explained by the different levels at which the protozoa act in the rumen. Their retention in the rumen and their predation on bacteria are factors which limit the efficiency of microbial synthesis in the rumen by increasing the recycled microbial nitrogen.

Table 7
Effect of Controlled Fauna on Duodenal Nitrogen Flow of Sheep (Jouany & Thivend, unpublished data)

	Defaunated rumen[a]	Rumen inoculated with			S.e.m.
		Polyplastron[b]	Entodinium spp.[b]	Conventional fauna[a]	
Intake (g/day)	35.2	34.5	35.2	34.2	0.6
Duodenal N (g/day)	31.0	30.5	24.6	24.8	2.2
Duodenal NAN (g/day)	28.2	27.2	22.5	22.8	2.0
Bacterial N (g/day)	18.5	17.5	12.1	13.7	1.6
Number of rumen protozoa ($\times 10^{-3}$/ml)	0	9.3	101.1	109.3	3.7

Sheep were fed wheat straw (64%), sugar beet pulp (13%), peanut meal (20%), urea (1%), mineral and vitamins (2%).
[a] 6 sheep were used. [b] 3 sheep were used.

Furthermore, ciliates participate directly in the degradation of feed proteins and could have an indirect effect on protein degradation by the bacteria because ciliates increase feed retention time in the rumen. The increases in both the microbial and the feed proteins that enter the duodenum of a defaunated animal are probably the major cause of improved animal performance when a diet low in non-degradable protein is given to a ruminant with a high requirement for amino acids (young animal with high performance). The supply of a protein supplement that is not degradable in the rumen could limit the differences observed between faunated and defaunated animals, as suggested by Bird *et al.* (1979).

As regards nitrogen metabolism, the increased excretion of faecal nitrogen, which is probably due to a shift in cellulose digestion towards the large intestine, is exactly compensated for by decreased excretion of urinary nitrogen. Therefore defaunation has no effect on nitrogen retention.

The overall effect of protozoa on the production of energy that can be used by the host is much more difficult to describe, because protozoa act on energy metabolism at different levels. They contribute not only by directing the ruminal fermentations, but also by controlling their extent. When animals are given limited amounts of diets rich in cereal or soluble sugars, the entodiniomorphid and Isotrichidae ciliates ingest and then store part of the carbohydrates. This ingestion limits both the production of lactic acid by bacteria and the fall in pH, and consequently improves rumen function (Jouany, 1978). Such a regulating effect of protozoa also appears with diets based on silage which has a high level of lactic acid (Chamberlain *et al.*, 1983). The amount of starch-type carbohydrate stored by ciliates during the 4 hours following feed intake can be between 30% and 40% of their dry matter (Jouany & Thivend, 1972). In these conditions, even a limited flow of protozoa from the rumen could contribute to the quantity of starch entering the intestine, in amounts up to 15–30 g/day in sheep. This is not negligible given the high digestibility of feed starch in the rumen (Thivend & Journet, 1968; MacRae & Armstrong, 1969; Ørskov *et al.*, 1969) and could explain the significant increase in blood glucose levels sometimes found after inoculation of protozoa into a defaunated rumen (Jouany & Senaud, 1982; Itabashi *et al.*, 1984). The lower starch flow into the duodenum of defaunated animals could be partly compensated for by hepatic neoglucogenesis from propionate, and in this case defaunation would no longer have a negative effect on plasma glucose (Whitelaw *et al.*, 1972; Zainab, 1978; Itabashi *et al.*, 1982).

The results of Itabashi *et al.* (1984) showed that the energy fixed by the animals is decreased by defaunation, irrespective of the diet, although the energy excreted in the urine is lowered (Eadie & Gill, 1971). The energy savings, as regards methane production (Whitelaw *et al.*, 1984), are great enough to compensate for the increased energy excreted in the faeces in defaunated animals. The latter is the principal factor responsible for the decrease in metabolisable energy observed in defaunated animals. No data are available on the effects of inoculating a controlled fauna on the energy utilisation of diets by the host.

Lastly it must be noted that protozoa can be beneficial to the host because they limit the toxicity of some plant toxins that ciliates can metabolise (see also Chapter 14). This is the case with ochratoxin-A which is rapidly degraded by protozoa

Table 8
Effects of Rumen Defaunation on Animal Performance

Animals	Age or weight (days or kg)	Diet Based on	Nitrogen source	Growth (g/day) F	Growth (g/day) D	Feed intake (g/day) F	Feed intake (g/day) D	Authors
Lambs	51–100 d	Berseem + concentrate		120	96			(1)
	51–170 d	Berseem + concentrate		96	111			(1)
	51–210 d	Berseem + concentrate		91	106			(1)
	35 kg	Alfalfa hay + corn	Soyabean meal	280	210			(2)
Calves	50 kg	Milk		380	330			(3)
		Milk + cereals then rice shell	Linseed or cottonseed meal	400	310	122^b	122	(3)
				360	240	169^b	175	(3)
		Milk + cereals + peas	Linseed meal	410	320	163^b	150	(3)
Calves	180 kg	Oat straw + molasses	Urea	451	490	159^b	174	(3)
			Soyabean + cottonseed meal + fishmeal, meatmeal			3 760	3 650	(4)
Lambs	—	—	Fishmeal $(0)^a$	530^e	757	4 150	4 230	(4)
			$(4)^a$	0	35			(5)
			$(8)^a$	70	130			(5)
			$(12)^a$	140	170			(5)
				180	150			(5)
	300–375 d	Corn + oats hay + concentrate	—	66^e	46	495	447	(6)

20 kg	Beet pulp+molasses	Urea	181	213	871^e	857	(7)
	NaOH-treated straw+molasses	Urea	102^e	140	878	964	(7)
	Same+tapioca	Urea	239	192	1766	1895	(7)
	Molassine meal	Mozine molasses conc.	135^e	208	1127^e	1530	(7)
16 kg	Oaten chaff+sugar	Fishmeal+urea	122	135	865	890	(8)
			8^d	11^d			(8)
	Oaten chaff+sugar	Fishmeal+urea	122	132	870	930	(8)
			8^d	11^d			(8)
Calves	0–140 d	Hay+concentrate	—	<^{c,e}			(9)
	0–250 d	Hay+concentrate	—	=^{c}			(9)
Lambs	0–35 d	Molasses+conc.	Urea+proteins	88^e	125	1015	1985	(10)
	35–63 d	Molasses+conc.	Urea+proteins	109	99	1166	1189	(10)
	0–63 d	Molasses+conc.	Urea+proteins	100^e	120	1085	1114	(10)

(1) Abou Akkada & El Shazly, 1964. (2) Christiansen et al., 1965. (3) Borhami et al., 1967. (4) Bird & Leng, 1978. (5) Bird et al., 1978. (6) Ramaprasad & Raghavan, 1981. (7) Demeyer et al., 1982. (8) Bird & Leng, 1984. (9) Itabashi & Matsukawa, 1979. (10) Van Nevel et al., 1985.

F, faunated animals; D, defaunated animals.
^{a} Proportion of fishmeal in the diet.
^{b} Total intake of digestible nutrients.
^{c} <, higher growth in faunated animals; =, no difference between faunated and defaunated.
^{d} Wool growth.
^{e} Significant effect of defaunation ($P < 0.05$).

(Kiessling et al., 1984) into phenylalanine and ochratoxin-α which are both non-toxic (Yamazaki et al., 1971). Deacetylation of trichotecenes (diacetoxysirpenol and T-2 toxin) by protozoa partially removes the inhibition of microbial synthesis (Ohta et al., 1978). Ivan et al. (1986) also showed that defaunated rams are more sensitive to copper toxicity than faunated rams. Except for the experiments conducted by Hidiroglou & Jenkins (1974) with selenium, no information is available on the effects that ciliates have on mineral utilisation by the ruminant, although studies in vitro have shown that ciliates need bacteria to use some trace elements (Co) in the rumen (Bonhomme et al., 1982).

REFERENCES

Abou Akkada, A. R. & El Shazly, K. (1964). Effect of absence of ciliate protozoa from the rumen on microbial activity and growth of lambs. Appl. Microbiol., **11**, 384–90.

Abou Akkada, A. R. & Howard, B. H. (1960). The biochemistry of rumen protozoa. III. The carbohydrate metabolism of Entodinium. Biochem. J., **76**, 445–51.

Abou Akkada, A. R., Bartley, E. E., Berube, R., Fina, L. N., Meyer, R. M., Henricks, D. & Julins, F. (1968). Simple method to remove completely ciliate protozoa of adult ruminants. Appl. Microbiol., **16**, 1475–7.

Becker, E. R., Schultz, J. A. & Emmerson, M. A. (1929/30). Experiments on the physiological relationship between the stomach infusoria of ruminants and their hosts, with a bibliography. J. Sci., **4**, 215–51.

Bird, S. & Leng, R. A. (1978). The effects of defaunation of the rumen on the growth of cattle on low-protein, high-energy diets. Br. J. Nutr., **40**, 163–7.

Bird, S. & Leng, R. A. (1984). Further studies on the effects of the presence or absence of protozoa in the rumen on live-weight gain and wool growth of sheep. Br. J. Nutr., **42**, 607–11.

Bird, S., Baigent, D. R., Dixon, R. & Leng, R. A. (1978). Ruminal protozoa and growth of lambs. Proc. Austr. Soc. Anim. Prod., **12**, 137 (Abstr.).

Bird, S., Hill, M. K. & Leng, R. A. (1979). The effects of defaunation of the rumen on the growth of lambs on low-protein, high-energy diets. Br. J. Nutr., **42**, 81–7.

Bonhomme, A., Durand, M., Quintana, C. & Halpern, S. (1982). Effect of cobalt and vitamin B_{12} on the bacterial population. Reprod. Nutr. Devel., **22**, 107–22.

Bonhomme, A., Fonty, G., Foglietti, M. J. & Weber, M. (1986). Endo-1,4-β-glucanase and β-glucosidase of the ciliate Polyplastron multivesiculatum free of cellulolytic bacteria. Can. J. Microbiol., **32**, 219–25.

Borhami, B. E. A., El Shazly, K., Abou Akkada, A. R. & Ahmed, I. A. (1967). Effect of early establishment of ciliate protozoa in the rumen on microbial activity and growth of early weaned buffalo calves. J. Dairy Sci., **50**, 1654–60.

Broudiscou, L., Jouany, J. P., Van Nevel, C. J. & Demeyer, D. I. (1988). Addition d'hydrolysat d'huile de soja dans la ration de mouton; effet sur la dégradation 'in sacco' de la paille et de la cellulose. Reprod. Nutr. Devel., **28**, 159–60.

Chamberlain, D. G., Thomas, P. C. & Anderson, F. J. (1983). Volatile fatty acid proportions and lactic acid metabolism in the rumen in sheep and cattle receiving silage diets. J. Agric. Sci. (Camb.), **101**, 47–58.

Christiansen, W. C., Kawashima, R. & Burroughs, W. (1965). Influence of protozoa upon rumen acid production and liveweight gains in lambs. J. Anim. Sci., **24**, 730–4.

Cockburn, J. E. & Williams, A. P. (1984). The simultaneous estimation of the amounts of protozoal bacterial and dietary nitrogen entering the duodenum of steers. Br. J. Nutr., **51**, 111–32.

Coleman, G. S. (1975). The interrelationships between rumen ciliate protozoa and bacteria. In *Digestion and Metabolism in the Ruminant*, ed. I. W. McDonald & A. C. I. Warner. University of New England Publishing Unit, Armidale, Australia, pp. 149–64.

Coleman, G. S. (1979). Rumen ciliate protozoa. In *Biochemistry and Physiology of Protozoa*, ed. M. Levandowsky & S. M. Kunter, 2nd edn, Vol. 2. Academic Press, New York, pp. 381–408.

Coleman, G. S. & Sandford, D. C. (1979). The uptake and metabolism of bacteria, amino acids, glucose and starch by the spined and spineless forms of the rumen ciliate *Entodinium caudatum*. *J. Gen. Microbiol.*, 117, 411–18.

Coleman, G. S., Dawson, R. M. C. & Grime, D. W. (1980). The rate of passage of ciliate protozoa from the ovine rumen. *Proc. Nutr. Soc.*, 39, 6A (Abstr.).

Collombier, J. (1981). Contribution à l'étude du rôle des protozoaires ciliés du rumen dans l'apport d'azote microbien entrant dans le duodenum du ruminant. Thèse Docteur Ingénieur, Université Clermont II, ordre 105, 85 pp.

Collombier, J., Groliere, C. A., Senaud, J., Grain, J. & Thivend, P. (1984). Etude du rôle des protozoaires ciliés du rumen dans l'apport d'azote microbien entrant dans le duodenum du ruminant, par l'estimation directe de la quantité de ciliés sortant du rumen. *Protistologica*, 20, 431–6.

Conrad, H. R., Hibbs, J. W. & Frank, N. (1958). High roughage system for raising calves based on early development of rumen function. IX. Effect of rumen inoculations and chlortetracycline on rumen function of calves fed high roughage pellets. *J. Dairy Sci.*, 41, 1248–61.

Czerkawski, J. W., Christie, W. W., Breckenridge, G. & Hunter, M. L. (1975). Changes in the rumen metabolism of sheep given increasing amounts of linseed oil in their diet. *Br. J. Nutr.*, 34, 25–44.

Demeyer, D. I. (1981). Rumen microbes and digestion of plant cell walls. *Agric. Environ.*, 6, 295–337.

Demeyer, D. I. (1982). Influence of calcium peroxide on fermentation pattern and protozoa in the rumen. *Arch. Tierernähr.*, 32, 579–93.

Demeyer, D. I. & Van Nevel, C. J. (1979). Effect of defaunation on the metabolism of rumen microorganisms. *Br. J. Nutr.*, 42, 515–24.

Demeyer, D. I., Van Nevel, C. J. & Van de Voorde, G. (1982). The effect of defaunation on the growth of lambs fed three urea-containing diets. *Arch. Tierernähr.*, 32, 595–604.

Eadie, J. M. & Gill, J. C. (1971). The effect of the absence of rumen ciliate protozoa on growing lambs fed on a roughage-concentrate diet. *Br. J. Nutr.*, 26, 155–67.

Eadie, J. M. & Hobson, P. N. (1962). Effect of presence and absence of ciliates on the total rumen bacterial count in lambs. *Nature*, 193, 503–5.

Eadie, J. M. & Mann, S. O. (1970). Development of the rumen microbial population: high starch diets and instability. In *Physiology of Digestion and Metabolism in the Ruminant*, ed. A. T. Phillipson. Oriel Press, Newcastle-upon-Tyne.

Eadie, J. M. & Oxford, A. E. (1957). A simple and safe procedure for the removal of holotrich ciliates from the rumen of an adult fistulated sheep. *Nature*, 179, 485.

Eadie, J. M. & Shand, W. J. (1981). The effect of synperonic NP9 upon ciliate-free and faunated sheep. *Proc. Nutr. Soc.*, 40, 113A (Abstr.).

Eadie, J. M., Hyldgaard-Jensen, J., Mann, S. O., Reid, R. S. & Whitelaw, F. G. (1970). Observations on the microbiology and biochemistry of the rumen in cattle given different quantities of a pelleted barley ration. *Br. J. Nutr.*, 24, 157–77.

Elliott, R., Ash, A. J., Calderon-Cortes, F. & Norton, B. W. (1987). The influence of anaerobic fungi on rumen volatile fatty acid concentration *in vivo*. *J. Agric. Sci.* (Camb.), 109, 13–17.

Faichney, G. J. (1975). The use of markers to partition digestion within gastro-intestinal tract of ruminants. In *Digestion and Metabolism in the Ruminant*, ed. I. W. McDonald & A. C. I. Warner. University of New England Publishing Unit, Armidale, Australia, pp. 277–91.

Fonty, G. (1984). Implantation de la microflore et des protozoaires ciliés dans le rumen d'agneaux holoxéniques et néo-holoxéniques. Facteurs écologiques déterminant l'implantation de *Bactéroïdes succinogènes* (S85) et des protozoaires ciliés dans le rumen d'agneaux holoxéniques isolés, d'agneaux méroxéniques et d'agneaux gnotoxéniques. Evolution des principaux paramètres digestifs. Thèse Doctorat, Université Clermont II, ordre 337, 235 pp.

Fonty, G., Gouet, Ph., Jouany, J. P. & Senaud, J. (1983*a*). Ecological factors determining establishment of cellulolytic bacteria and protozoa in the rumens of meroxenic lambs. *J. Gen. Microbiol.*, **129**, 213–23.

Fonty, G., Jouany, J. P., Thivend, P., Gouet, Ph. & Senaud, J. (1983*b*). A descriptive study of rumen digestion in meroxenic lambs according to the nature and complexity of the microflora. *Reprod. Nutr. Devel.*, **23**, 857–73.

Fonty, G., Jouany, J. P., Senaud, J., Gouet, Ph. & Grain, J. (1984). The evolution of microflora, microfauna and digestion in the rumen of lambs from birth to 4 months. *Can. J. Anim. Sci.*, **64** (Suppl.), 165–6.

Gouet, Ph., Fonty, G., Jouany, J. P. & Nebout, J. M. (1984). Cellulolytic bacteria establishment and rumen digestion in lambs isolated after birth. *Can. J. Anim. Sci.*, **64** (Suppl.), 163–4.

Grummer, R. R., Stapies, C. R. & Davies, C. L. (1983). Effect of defaunation on ruminal volatile fatty acids and pH of steers fed a diet high in dried whole whey. *J. Dairy Sci.*, **66**, 1738–41.

Harrisson, D. G. & McAllan, A. B. (1980). Factors affecting microbial growth yields in the reticulo-rumen. In *Digestive Physiology and Metabolism in Ruminants*, ed. Y. Ruckebusch & P. Thivend. MTP Press, Lancaster, pp. 205–26.

Harrisson, D. G., Beever, D. E. & Osbourn, D. F. (1979). The contribution of protozoa to the protein entering the duodenum of sheep. *Br. J. Nutr.*, **41**, 521–7.

Harrop, C. J. F. (1974). Nitrogen metabolism in the ovine stomach. IV. Nitrogenous components of the abomasal secretions. *J. Agric. Sci.*, **83**, 245–57.

Hidiroglou, M. & Jenkins, K. J. (1974). Influence de la défaunation sur l'utilisation de la seleno-méthionine chez le mouton. *Ann. Biol. Anim. Biochim. Biophys.*, **14**, 157–65.

Hungate, R. E., Reichl, J. & Prins, R. (1971). Parameters of rumen fermentation in a continuously fed sheep: evidence of a microbial rumination pool. *Appl. Microbiol.*, **22**, 1104–13.

Ikwuegbu, D. A. & Sutton, J. D. (1982). The effect of varying the amount of linseed oil supplementation on metabolism of sheep. *Br. J. Nutr.*, **48**, 365–75.

Imai, S. & Ogimoto, K. (1978). Scanning electron and fluorescent microscopic studies on the attachment of spherical bacteria to ciliate protozoa in the ovine rumen. *Jap. J. Vet. Sci.*, **40**, 9–19.

Itabashi, H. & Kandatsu, M. (1974). Effect of rumen ciliate protozoa on the digestibility of food constituents, nitrogen retention and amino acid compositions of the rumen contents. *Jap. J. Zootechn. Sci.*, **45**, 585–91.

Itabashi, H. & Kandatsu, M. (1975*a*). Influence of rumen ciliate protozoa on the concentration of free amino acids in the rumen fluids. *Jap. J. Zootechn. Sci.*, **46**, 600–6.

Itabashi, H. & Kandatsu, M. (1975*b*). Influence of rumen ciliate protozoa on the concentration of ammonia and volatile fatty acids in connection with the utilization of ammonia in the rumen. *Jap. J. Zootechn. Sci.*, **46**, 409–16.

Itabashi, H. & Katada, A. (1976). Studies on nutritional significance of rumen ciliate protozoa in cattle. II. Influence of protozoa on amino acid concentrations in some rumen fractions and blood plasma. *Bull. Tohoku Nat. Agric. Exp. Stn*, **52**, 169–76.

Itabashi, H. & Matsukawa, T. (1979). Studies on nutritional significance of rumen ciliate protozoa in cattle. III. Influence of protozoa on growth rate, food intake, rumen fermentation and various plasma components in calves under different planes of nutrition. *Bull. Tohoku Nat. Agric. Exp. Stn*, **59**, 111–28.

Itabashi, H., Kobayashi, T., Morii, R. & Okamato, S. (1982). Effects of ciliate protozoa on the concentration of ruminal and duodenal volatile fatty acids and plasma glucose and insulin after feeding. *Bull. Nat. Inst. Anim. Ind.*, **39**, 21–32.

Itabashi, H., Kobayashi, T. & Matsumoto, M. (1984). The effect of rumen ciliate protozoa on

energy metabolism and some constituents in rumen fluid and plasma of goats. *Jap. J. Zootechn. Sci.*, **55**, **55**, 248–55.
Ivan, M., Veira, D. M. & Kelleher, C. A. (1986). The alleviation of chronic copper toxicity in sheep by ciliate protozoa. *Br. J. Nutr.*, **55**, 361–7.
Jouany, J. P. (1978). Contribution à l'étude des protozoaires ciliés du rumen: leur dynamique, leur rôle dans la digestion et leur intérêt pour le ruminant. Thèse Doctorat, Université de Clermont II, ordre 256, 2 vol., 195 pp.
Jouany, J. P. & Senaud, J. (1979a). Défaunation du rumen de mouton. *Ann. Biol. Anim. Biochim. Biophys.*, **19**, 619–24.
Jouany, J. P. & Senaud, J. (1979b). Role of rumen protozoa in the digestion of food cellulosic materials. *Ann. Rech. Vet.*, **10**, 261–3.
Jouany, J. P. & Senaud, J. (1982). Influence des ciliés du rumen sur la digestion de différents glucides chez le mouton. I. Utilisation des glucides pariétaux (cellulose et hémicelluloses) et de l' amidon. *Reprod. Nutr. Devel.*, **22**, 735–52.
Jouany, J. P. & Senaud, J. (1983). Influence des ciliés du rumen sur la digestion de différents glucides chez le mouton. II. Régimes contenant de l'inuline, du saccharose et du lactose. *Reprod. Nutr. Devel.*, **23**, 607–23.
Jouany, J. P. & Thivend, P. (1972). Evolution postprandiale de la composition glucidique des corps microbiens du rumen en fonction de la nature des glucides du régime. I. Les protozoaires. *Ann. Biol. Anim. Biochim. Biophys.*, **12**, 673–7.
Jouany, J. P. & Thivend, P. (1983). Influence of protozoa on nitrogen digestion in the ruminant. *IVth Symp. Protein Metabolism and Nutrition*, Clermont-Ferrand, France. Vol. II, Les colloques de l'INRA, 16, pp. 287–90.
Jouany, J. P., Senaud, J., Grain, J., de Puytorac, P. & Thivend, P. (1977) Rumen ciliates of sheep given cellulose, lactose, sucrose and starch diets. *Proc. Nutr. Soc.*, **36**, 72A (Abstr.).
Jouany, J. P., Zainab, B., Senaud, J., Groliere, C. A., Grain, J. & Thivend, P. (1981). Role of the rumen ciliate protozoa *Polyplastron multivesiculatum*, *Entodinium* spp. and *Isotricha prostoma* in the digestion of a mixed diet in sheep. *Reprod. Nutr. Devel.*, **21**, 871–84.
Jouany, J. P., Demeyer, D. I. & Grain, J. (1988). Effect of defaunating the rumen. *Anim. Feed Sci. Technol.*, (in press).
Kayouli, C., Van Nevel, C. J., Dendooven, R. & Demeyer, D. I. (1986). Effect of defaunation and refaunation of the rumen on rumen fermentation and N-flow in the duodenum of sheep. *Arch. Anim. Nutr.* (Berlin), **36**, 827–37.
Kiessling, K. H., Petterson, H., Sandholm, K. & Olsen, M. (1984). Metabolism of aflatoxin, ochratoxin, zearalenone and three trichotecenes by intact rumen fluid, rumen protozoa, and rumen bacteria. *Appl. Environ. Microbiol.*, **47**, 1070–3.
Krumholz, L. R., Forsberg, C. W. & Veira, D. M. (1983). Association of methanogenic bacteria with rumen protozoa. *Can. J. Microbiol.*, **29**, 676–80.
Kurihara, Y., Eadie, J. M., Hobson, P. N. & Mann, S. O. (1968). Relationship between bacteria and ciliate protozoa in the sheep rumen. *J. Gen. Microbiol.*, **51**, 267–88.
Kurihara, Y., Takechi, T. & Shibata, F. (1978). Relationship between bacteria and ciliate protozoa in the rumen of sheep fed on a purified diet. *J. Agric. Sci.* (Camb.), **90**, 373–81.
Leng, R. A. (1982). Dynamics of protozoa in the rumen of sheep. *Br. J. Nutr.*, **48**, 399–415.
Leng, R. A. (1984). The potential of solidified molasses-based blocks for the correction of multinutritional deficiencies in buffaloes and other ruminants fed low-quality agro-industrial byproducts. In *The Use of Nuclear Techniques to Improve Domestic Buffalo Production in Asia*. IAEA, Vienna, pp. 135–50.
Leng, R. A., Gill, M., Kempton, T. J., Rowe, J. B., Nolan, C. V., Stachin, S. J. & Preston, T. R. (1981). Kinetics of large ciliate protozoa in the rumen of cattle given sugar cane diets. *Br. J. Nutr.*, **40**, 371–84.
Lindsay, J. R. & Hogan, J. P. (1972). Digestion of two legumes and rumen bacterial growth in defaunated sheep. *Austr. J. Agric. Res.*, **23**, 321–30.
Ling, J. R. & Buttery, P. J. (1978). The simultaneous use of ribonucleic acid, ^{35}S, diaminopimelic acid and 2-aminoethylphosphonic acid as markers of microbial nitrogen entering the duodenum of sheep. *Br. J. Nutr.*, **39**, 165–79.

Lovelock, L. K. A., Buchanan-Smith, J. G. & Forsberg, C. W. (1982). Difficulties in defaunation of the ovine rumen. *Can. J. Anim. Sci.*, **62**, 299–303.

Luther, R., Trenkle, A. & Burroughs, W. (1966). Influence of rumen protozoa on volatile acid production and ration digestibility in lambs. *J. Anim. Sci.*, **25**, 1116–22.

Mackie, R. I., Gilchrist, F. M. C., Robberts, A. M., Hannah, P. E. & Schwartz, M. (1978). Microbiological and chemical changes in the rumen during the stepwise adaptation of sheep to high concentrate diets. *J. Agric. Sci.* (Camb.), **90**, 241–54.

MacRae, J. C. & Armstrong, D. G. (1969). Studies on intestinal digestion in sheep. II. Digestion of some carbohydrate constituents in hay, cereal and hay–cereal rations. *Br. J. Nutr.*, **23**, 377–87.

Macrae, J. C. & Reeds, P. J. (1980). Prediction of protein deposition in ruminants. In *Protein Deposition in Animals*, ed. B. J. Buttery & D. B. Lindsay. Butterworths, pp. 225–49.

Meyer, J. H. F., Van der Walt, S. I. & Schwartz, H. M. (1986). The influence of diet and protozoal numbers on the breakdown and synthesis of protein in the rumen of sheep. *J. Anim. Sci.*, **62**, 509–20.

Michalowski, T., Harmeyer, J. & Breves, G. (1986). The passage of protozoa from the reticulo-rumen through the omasum of sheep. *Br. J. Nutr.*, **56**, 625–34.

Newbold, C. J., Chamberlain, D. G. & Williams, A. G. (1986). The effects of defaunation on the metabolism of lactic acid in the rumen. *J. Sci. Food Agric.*, **37**, 1083–90.

Ohta, M., Matsumoto, H., Ishii, K. & Ueno, Y. (1978). Metabolism of trichotecene mycotoxins. *J. Biochem.* (Tokyo), **84**, 697–706.

Orpin, C. G. (1977). Studies on the defaunation of the ovine rumen using dioctyl sodium sulfosuccinate. *J. Appl. Bacteriol.*, **43**, 309–18.

Orpin, C. G. (1983/4). The role of ciliate protozoa and fungi in the rumen digestion of plant cell-walls. *Anim. Feed Sci. Technol.*, **10**, 121–43.

Ørskov, E. R., Fraser, C. & Kay, R. N. B. (1969). Dietary factors influencing the digestion of starch in the rumen and small and large intestine of early weaned lambs. *Br. J. Nutr.*, **23**, 217–26.

Oxford, A. E. (1955). The rumen ciliate protozoa: their chemical composition, metabolism, requirements for maintenance and culture, and physiological significance for the host. *Exp. Parasitol.*, **4**, 569–605.

Prins, R. A. & Van Hoven, W. (1977). Carbohydrate fermentation by the rumen ciliate *Isotricha prostoma*. *Protistologica*, **13**, 549–56.

Punia, B. S., Leibholz, J. & Faichney, G. J. (1984a). The flow of protozoal nitrogen to the omasum of cattle. *Proc. Austr. Soc. Anim. Prod.*, **15**, 732 (Abstr.).

Punia, B. S., Leibholz, J. & Faichney, G. J. (1984b). Protozoal numbers in the rumen and omasum of cattle fed kikuyu grass. *Proc. Austr. Soc. Anim. Prod.*, **15**, 733 (Abstr.).

Ramaprasad, J. & Raghavan, G. V. (1981). Note on the growth rate and body composition of faunated and defaunated lambs. *Ind. J. Anim. Sci.*, **51**, 570–2.

Rowe, J. B., Davies, A. & Broome, A. W. S. (1985). Quantitative effects of defaunation on rumen fermentation and digestion in sheep. *Br. J. Nutr.*, **54**, 105–19.

Siddons, R. C., Beever, D. E. & Nolan, J. V. (1982). A comparison of methods for the estimation of microbial nitrogen in duodenal digesta of sheep. *Br. J. Nutr.*, **48**, 377–89.

Smith, R. H., McAllan, A. B., Hewitt, D. & Lewis, P. E. (1978). Estimation of amounts of microbial and dietary nitrogen compounds entering the duodenum of cattle. *J. Agric. Sci.* (Camb.), **90**, 557–68.

Soetanto, H. (1985). Studies on the role of rumen anaerobic fungi and protozoa in fibre digestion. M. Rural Science thesis, University of New England, Armidale, Australia.

Stumm, C. K., Gitzen, H. J. & Vogels, G. D. (1982). Association of methanogenic bacteria with ovine rumen ciliates. *Br. J. Nutr.*, **47**, 95–9.

Teather, R. M., Mahadevan, S., Erfle, J. D. & Sauer, F. D. (1984). Negative correlation between protozoal and bacterial levels in rumen samples and its relation to the determination of dietary effects on the microbial population. *Appl. Environ. Microbiol.*, **47**, 566–70.

Thivend, P. & Journet, M. (1968). Utilisation digestive de l'amidon d'orge chez le ruminant. *Ann. Biol. Anim. Biochim. Biophys.*, **8**, 449–51.
Ushida, K. & Jouany, J. P. (1985). Effect of protozoa on rumen protein degradation in sheep. *Reprod. Nutr. Devel.*, **25**, 1075–81.
Ushida, K., Jouany, J. P., Lassalas, B. & Thivend, P. (1984). Protozoal contribution to nitrogen digestion in sheep. *Can. J. Anim. Sci.*, **64** (Suppl.), 20–1.
Ushida, K., Jouany, J. P. & Thivend, P. (1986a). Role of rumen protozoa in nitrogen digestion in sheep given two isonitrogenous diets. *Br. J. Nutr.*, **56**, 407–19.
Ushida, K., Miyazaki, A. & Kawashima, R. (1986b). Effect of defaunation on ruminal gas & VFA production *in vitro*. *Jap. J. Zootechn. Sci.*, **57**, 71–7.
Van Gylswyck, N. O. & Labuschagne, J. P. L. (1971). Relative efficiency of pure cultures of different species of cellulolytic rumen bacteria in solubilizing cellulose *in vitro*. *J. Gen. Microbiol.*, **66**, 109–13.
Van Hoven, W. & Prins, R. A. (1977). Carbohydrate fermentation by the rumen ciliate *Dasytricha ruminantium*. *Protistologica*, **13**, 599–606.
Van Nevel, L. J. & Demeyer, D. I. (1981). Effect of methane inhibitors on the metabolism of rumen microbes *in vitro*. *Arch. Tierernähr.*, **31**, 141–51.
Van Nevel, C. J., Demeyer, D. I. & Van de Voorde, G. (1985). Effect of defaunating the rumen on growth and carcass composition of lambs. *Arch. Tierernähr.*, **35**, 331–7.
Veira, D. M. (1986). The role of ciliate protozoa in nutrition of the ruminant. *J. Anim. Sci.*, **63**, 1547–60.
Veira, D. M., Ivan, M. & Jiu, P. Y. (1983). Rumen ciliate protozoa: effects on digestion in the stomach of sheep. *J. Dairy Sci.*, **66**, 1015–22.
Vogels, G. D., Hoppe, W. F. & Stumm, C. K. (1980). Association of methanogenic bacteria with rumen ciliates. *Appl. Environ. Microbiol.*, **40**, 608–12.
Weller, R. A. & Pilgrim, A. F. (1974). Passage of protozoa and volatile fatty acids from the rumen of the sheep and from a continuous *in vitro* fermentation system. *Br. J. Nutr.*, **32**, 341–51.
Whitelaw, F. G., Eadie, J. M., Mann, S. O. & Reid, R. S. (1972). Some effects of rumen ciliate protozoa in cattle given restricted amounts of a barley diet. *Br. J. Nutr.*, **27**, 425–37.
Whitelaw, F. G., Eadie, J. M., Bruce, L. A. & Shand, W. J. (1984). Methane production in faunated and ciliate-free cattle and its relationship with rumen volatile fatty acid proportions. *Br. J. Nutr.*, **52**, 261–75.
Williams, A. G. & Coleman, G. S. (1985). Hemicellulose-degrading enzymes in rumen ciliate protozoa. *Curr. Microbiol.*, **12**, 85–90.
Williams, P. P. & Dinusson, W. E. (1973). Ruminal volatile fatty acid concentrations and weight gain of calves reared with and without ruminal ciliated protozoa. *J. Anim. Sci.*, **36**, 588–91.
Williams, A. G. & Ellis, A. B. (1985). Subcellular distribution of glycoside hydrolase and polysaccharide depolymerase enzymes in the rumen entodiniomorphic ciliate *Polyplastron multivesiculatum*. *Curr. Microbiol.*, **12**, 175–82.
Windham, W. R. & Akin, D. E. (1984). Rumen fungi and forage fiber degradation. *Appl. Environ. Microbiol.*, **48**, 473–6.
Wright, D. E. & Curtis, M. W. (1976). Bloat in cattle. XLII. The action of surface chemicals on ciliated protozoa. *NZ J. Agric. Res.*, **19**, 19–23.
Yamazaki, M., Suzuki, S., Sakakibara, Y. & Miyaki, K. (1971). The toxicity of 5-chloro-8-hydroxy-3,4-dihydro-3-methyl-isocoumarin-7-carboxy acid, a hydrolysate of ochratoxin A. *Jap. J. Med. Sci. Biol.*, **24**, 245–50.
Yoder, R. D., Trenkle, A. & Burroughs, W. (1966). Influence of rumen protozoa and bacteria upon cellulose digestion *in vitro*. *J. Anim. Sci.*, **25**, 609–12.
Zainab, B. (1978). Contribution à l'étude des interrelations entre 3 genres de protozoaires ciliés du rumen de mouton: dynamique des populations, rôle dans l'utilisation d'un aliment aggloméré (luzerne/orge: 40/50). Thèse 3ème cycle, Université Clermont II, ordre 545, 68 pp..

16

The Future

P. N. Hobson

*Honorary Research Fellow, Biochemistry Department,
Marischal College, Aberdeen University, Aberdeen, UK*

*Formerly Head, Microbial Chemistry Department,
Rowett Research Institute, Aberdeen, UK*

As was remarked in the Preface, over the last decade there has been a vast publication of information on the rumen microbes and their roles in the rumen reactions. We now know much more about the details of reactions, but the basic reactions remain the same. The reader may then ask 'is there a future for rumen microbiology, or do we know all we need to know about the rumen?'

By and large, the work of the rumen microbiologists has not led to great advances in animal feeding, except in so far as it has emphasised that in feeding a ruminant one is primarily feeding a microbial community. But, this having been realised, it has always been much easier to feed animals on particular materials and see what happens than to try to work out from microbiological experiments what would be the result of such feeding. Because of the time-scale involved in experiments, microbiology has always tended to provide explanations of why things happen, for instance the results of grain over feeding or of antibiotic additions, after the feeding has been done in practice.

There is also the problem that the ruminant is not an industrial fermentor. Small-scale experiments can show that altering a parameter of the running of a fermentor could lead to enhanced formation of a particular product, and this alteration can be put into practice on a large scale. But the suggestions made from small-scale experiments by the rumen microbiologist may not be workable in practice. Altering one parameter may alter others in the complex rumen mixed-culture. Or it may be possible to do the suggested change with experimental animals, where manpower is available and strict control of the environment is possible, but impossible to do this under the economic and environmental constraints of a farm. There is also the problem that the rumen is an open microbial habitat. A flora selected by deletions or additions is open to reversal by environmental microorganisms.

On the other hand, although changes in ruminant feeding practices may not have been initiated by rumen microbiology, the elucidation of the mechanism of action of

Present address: 4 North Deeside Road, Aberdeen AB1 7PL, Scotland, UK.

a feeding system may suggest ways of improving it. Knowledge of the bacteria involved in a particular reaction may lead to formulation of a feed-additive directed to modification of those particular bacteria, and this can be used instead of a broad-spectrum additive selected by feeding trials. The changes in animal feeding practices have previously been brought about by the call for increased human food and increased animal productivity. There has recently been a call for less food production in some temperate, industrialised countries. However, now and in the future, the world will, overall, need more food and better animal productivity. Resources of animal feed are limited. Research can help to overcome this limitation, by modification of wastes and other low-quality feeds to better suit the needs and fermentative abilities of the rumen organisms or, perhaps, by modification of the organisms to make better use of the feeds.

The rumen organisms and the rumen reactions are not peculiar to the rumen. They, or similar organisms and reactions, influence man's life in many ways. They play a part in the natural cycle of organic matter in the world, and so in the continuance of life. On the other hand, they can cause pollution, and their products can be toxic, or cause fires and explosions dangerous to man. They can, though, reduce pollution from wastes from homes, farms and factories, and from wastes and other feedstocks produce fuels and chemicals which can help reduce demands on fossil fuels at the present time and could be one of the replacements of fossil fuels as the supply of these diminishes in the future. Anaerobic microbes can cause sepsis and illness in man and animals. Rumen microbiology was the foundation of research on these microbial systems, and rumen research can still help in our understanding of these processes and in increasing the efficiency of desirable actions while decreasing the undesirable effects of the microbial communities.

There is a call now for research to be 'applied', to bring immediate, financially visible, 'relevant' results. Although it may be difficult to select discrete examples, rumen microbiology has overall been, and is, applied and relevant. There are practical reasons for its continuance. Of course, one cannot foretell all the results of research; today's academic problem may lead to tomorrow's industrial method.

If rumen microbiology is to continue, where should it go? Throughout the book authors have mentioned areas which still need research or where the next step should go, but a short summary of some possible future lines of research could be useful. These final paragraphs are based on the authors' views of the future expressed in their chapters or at other times.

There can be in the future, as in the past, attempts to find rapid solutions to immediate practical problems by use of general principles of rumen microbiology, or there can be a more long-term study of the details of rumen microbes and their reactions which could lead to improvements in ruminant performance but which can also improve our understanding of anaerobic microbiology and have applications beyond the rumen, in biotechnology, medicine and other areas. A possible example of the former is the improving of the feeding value of low-grade vegetation and wastes. It is known that the structure of plant-fibre and the presence of lignin are major determinants of the extent of microbial attack on vegetation, and that these parameters can be affected by treatment of the plant material with alkali or acid. Variations in chemical treatment coupled with simple artificial-rumen type experiments can show which of the more or less *ad hoc* treatments are best. The results of such experiments are immediately applicable. But such methods can be wasteful of the feed they are improving, as indeed can the biological methods, such as incubation of the feed with lignolytic fungi, being advocated. They are wasteful because they provide a generalised solution to the problem. Chemicals can destroy fibre components other than the ones obstructing the rumen degradation; fungi can use up valuable carbohydrate as well as lignin. Further studies on plant structure and the mode of action of the rumen fibre-degrading organisms may show specific alterations which could be made to the structure of the dead plant (or perhaps to the living one by genetic manipulation) to improve its digestion. Or they may show how particular rumen organisms, or consortia of organisms, might be altered to improve fibre digestion. Whether the organisms could be altered might depend on genetic research or research into the growth of mixed cultures, or other aspects of rumen microbiology. Results of such research could have applications in biotechnology and the future production of chemicals and fuels from biomass.

From the point of view of the bacteria themselves, while there may be other genera and species which may have an important role in the rumen to be discovered, application of genetically based methods of taxonomy have shown that the present classifications of anaerobes may be flawed. Taxonomic research can increase our understanding of the relationships of the bacteria and can have application in fields other than the rumen. In an integrated mixed-culture like the rumen, where consortia are all-important, more research is needed into the relationships and interactions of bacteria which are concerned in polysaccharide degradations and cross-provision of energy sources, the provision of growth factors and nitrogen sources, energetic and degradative aspects of inter-species hydrogen transfer, and other matters that control the growth of mixed cultures. Such studies would help to provide a better basis for attempts at manipulation of the rumen and particular reactions and organisms, and again would have application beyond the rumen field. It should not be forgotten, though, that the rumen is only a part of the animal's digestive system; the effects of rumen manipulations on digestion in the lower digestive tract should be examined. Animal growth-promoting substances may also affect the rumen and hind-gut digestions.

In manipulating the rumen, the protozoa and anaerobic fungi must not be forgotten. The ciliates have still to be grown in axenic culture from which their true enzymatic and metabolic activities could be determined. These organisms do not seem to be a vital part of the rumen population, but more work is needed to determine the importance of the ciliates in the normal rumen metabolism and the nutritional value of the digesta leaving the rumen. Aspects of fungal metabolism and their relationships with the other rumen organisms are still to be determined. As with many other studies, these few words could cover a vast amount of work, as the possible interactions between multi-species ciliate, fungal and bacterial populations are immense. Study of the natural variations in rumen microbial populations could help in imposing variations on rumen populations in domesticated animals, and knowledge of the natural development of the functioning rumen could help in bringing farm ruminants into earlier use of solid feeds and, perhaps, in manipulating the rumen population in the adult animal.

The energy-yield of the organisms affects the efficiency of the microbial conversion of feeds into products of use to the animal. Much work remains to be done on the modes of generation of energy for growth and maintenance of the anaerobic bacteria, protozoa and fungi. As well as substrate degradations, the movements of ions and substrates across the cell membrane have been shown to have an effect on yields of non-rumen bacteria; similar effects may be found in rumen organisms.

As with other rumen microbial substrates, although the main pathways of metabolism of nitrogenous substances are known, there are still details to be filled in, particularly in the metabolism of peptides and the interactions in the digestion of nitrogen compounds. Many organisms can potentially use a variety of nitrogen compounds, as they can carbon compounds; the regulation of these activities in a habitat containing a variety of these classes of compound is a further area of study. There is work to be done on the hydrolysis and hydrogenation of lipids, particularly the latter where the bacterial species involved are still relatively unknown.

Genetic manipulation is seen by many as the way to advances in use of microbes. Whether it is desirable for genetically manipulated organisms to be spread in ruminants is part of the general debate on the use of genetic engineering in many fields. However, such debate is rather academic at the moment as, while techniques have been developed for the genetic manipulation of bacteria such as *E. coli*, such techniques are not available for the rumen bacteria. Much more work on the technicalities of genetic engineering is needed before questions about the use of the organisms need to be raised. However, the production of genetically changed rumen organisms is bound up with all aspects of rumen research. What are the rate-controlling reactions in the biochemical pathways, what organisms are involved, can we pick organisms for manipulation in reactions that are carried on by numerous species or interacting consortia? Also, would laboratory-mutated organisms survive in a system which is selecting from a multitude of organisms in the environment and from genetically changing organisms in the system itself?

The gnotobiotic ruminant is an obvious tool for the genetic engineer. If a defined flora properly carrying out the rumen functions can be obtained, this would be a much simpler system for testing the effects and the survival of genetically changed organisms than the normal rumen. Whether such a defined flora can be obtained depends on research in all aspects of rumen microbiology, but the gnotobiotic animal, particularly if protozoa can be used in the microbial populations, offers a means of checking many hypotheses about the rumen functions and the interactions in microbial communities.

Modelling of rumen functions depends on knowledge of the reactions and the microbes involved in them, and it can help to define areas where information is scanty. However, the modelling of fermentation of solid substrates and the integration of the compartmentation theory into models is an area that has hardly yet been touched.

The preceding paragraphs give only an overall view of some areas where more research on the rumen is desirable; those working on particular aspects will see places where more detail needs to be filled in. Although much has been done, there is obviously much more to be done. It has often been said that a pure culture is a laboratory artifact; microbes grow naturally in mixed cultures. Increase in knowledge of the behaviour of microbes in mixed cultures, of which the rumen is a major example, can only help to increase the benefits derived by mankind from the microbes which form a large part of the life of the world.

Index

Abiérixin, 404
Acetohydroxamic acid, 232
Acinetobacter, 25
Adenosine triphosphate (ATP), 10–12, 15–17, 102, 185–93, 196, 197–9, 203, 204, 208, 352–4, 387, 390, 469, 471, 472
ADP, 186, 197
Aerobic bacteria, 151
Aerobic fungi, cellulases of, 262
Alkaligenes faecalis, 25
S-Alkyl-cysteine sulphoxide, 455
Alkylamines, 410
Alysiella filiformis, 25
Amino acids, 192–4, 221–2, 344, 411–13, 428
　metabolism, 228–31
　sources of, 88
　synthesis, 104, 238–40
　uptake of, 87–9
Aminoethylphosphonic acid (AEP), 368–9
Ammonia, 234–8, 387, 407, 411, 420, 493, 494
Amylolysis, 484
Anabolic nitrogen metabolism, 234–40
Anaerobic bacteria, 151
　molecular genetics of, 324–8
　see also Bacteria
Anaerobic fungi, 129–50, 237, 270–1, 490, 515
　attack on plant tissues, 144–6
　chemical composition, 142–4
　culture media, 140
　estimation of population densities and biomass, 138

Anaerobic fungi—*contd.*
　fine structure, 136–7
　genera and species, 131
　intermediary metabolism, 141
　isolation and culture, 139–41
　life cycles *in vivo*, 137–8
　lipid composition of, 311
　nutrition, 140
　role in hydrolysis of dietary lipid, 290
　taxonomy, 132–5
Anaerobic glove-box techniques, 59–60
Anaerobic metabolism, 6–7
Anaerobic microbes, 514
Anaeroplasma abactoclasticum, 50
Anaeroplasma bactoclasticum, 50
Anaerovibrio, 173
Anaerovibrio lipolytica, 36–7, 192, 288, 455, 479, 483
Antibiotics, 171, 394–406
Antibody production, 481
Artificial saliva, 423–7
Aspergillus fumigatus, 129
Aureomycin, 407
Avoparcin, 29, 406

B vitamins, 343, 416
Bacillus, 26, 330
Bacillus macerans, 234
Bacillus megaterium, 86, 87
Bacillus subtilis, 323, 325
Bacitracin, 407
Bacteria, 21–75, 165–75, 221–5
　amylolytic, 490
　carbohydrate contents of, 208

Bacteria—contd.
    characteristics of, 26–60
    cultivation and maintenance, 57–9
    engulfed, 86
    enumeration procedures, 165–7
    factors affecting population, 168–71
    fatty acid composition, 306
    gene cloning techniques, 331–3
    gene transfer in, 333–4
    genetics of, 323–41
    growth yields of, 196–9
    ingestion of, 103–4
    inoculation and establishment of, 167–8
    interrelationships with protozoal populations, 175–6
    large, 50
    lipid composition of, 305–8
    lipolytic, 288
    methanogenic, 347, 349, 490, 495
    mutagenesis in, 335
    phylogenetic relationships, 25–6
    plasmid distribution and size characteristics in, 328–31
    population changes, 489
    sampling and media preparation, 54–7
    uptake, digestion and metabolism, 85
    variation between ruminant species, 174
    wild ruminants, in, 174–5
    see also Anaerobic bacteria
Bacterial flora, 23–5
Bacterial populations, 476–86
Bacterial protein, 86
Bacteroides, 26–33, 171, 173, 232, 239, 307, 324, 326, 327, 357, 473, 481, 482
Bacteroides amylophilus, 26, 31–2, 172, 190, 200, 221–5, 224, 225, 237, 254, 323, 344, 347, 479, 485, 490
Bacteroides fibrisolvens, 190, 196, 204, 225, 233, 239, 263
Bacteroides fragilis, 326, 327, 328, 335
Bacteroides nodosus, 327
Bacteroides ruminicola, 26, 27, 29–31, 85, 172, 190, 192, 194, 196, 199, 201, 202, 203, 221, 222, 224, 225, 228, 230, 231, 239, 240, 265, 306, 312, 331, 334, 344–7, 479, 481
Bacteroides succinogenes, 26, 32–3, 146, 174, 175, 188, 190, 196, 221, 240, 260–6, 268, 323, 331, 332, 346, 348, 356, 380, 406, 485
Balance models, 469–70
BAPNA, 224, 225
BES, 428

Bifidobacterium, 232, 331, 334
Bifidobacterium globosum, 331
Bifidobacterium longum, 334
Bioenergetics, 186
Biohydrogenation, 290–305
    bacteria involved, 295–7
    bacterial species responsible for, 294–5
    biochemistry of, 295–302
    factors affecting, 304–5
    mechanism of, 300–1, 302
    metabolic pathways, 295–7
    problems with regard to, 302–3
    role of, 303
    role of food particles in, 291–4
    role of microorganisms, 290–1
    substrates for, 295
Biological models, 461–3, 475–86
Biological systems, 3–5
Biotin, 343, 344
Blepharoconus krugerensis, 97, 161
Blepharocorys bovis, 97
Blepharocorys ventriculi, 97
Blepharocorythidae, 97
Blepharoprosthium parva, 97, 98
Bloat, 447–8
Bos indicus, 159
Bos taurus, 159
Brassica anaemia, 482
Brassica anaemia factor, 455
Buetschlia, 92
Buetschlia parva, 96
Buetschliidae, 96
Buffers, 423–7
Butyrivibrio, 22, 23, 172, 173, 175, 223, 225, 228, 230–2, 288, 289, 290, 295, 304, 306, 314, 329, 332–4, 380, 473, 481–3, 485
Butyrivibrio fibrisolvens, 35–6, 85, 87, 189, 221–3, 225, 254, 294, 296–8, 301, 303, 323, 324, 328, 329, 331–3, 344, 410, 479, 490
Butyrivibrio ruminicola, 485, 490

Callimastix frontalis, 130
Caloscolex, 81, 161
Capreomycin, 407
Capric acid, 422
Captan, 410
Carbohydrate contents of bacteria, 208
Carbohydrate degradation, 100
Carbohydrate-fermenting groups, 172–4

Carbohydrate fermentation, 141–2, 345–57
  products of, 101–2
Carbohydrate metabolism, 99, 102–3,
  394–402, 416–17
Carbohydrate substrates, 446
Carbohydrate uptake, 99
Carbohydrates, 6–7, 515
Ca-soaps, 422
Catabolite regulatory mechanisms, 205–8
Cellobiose phosphorylase, 196
Cellulases
  aerobic fungi, of, 262
  rumen bacteria, of, 263
Cellulolysis, 484
Cellulolytic activity of rumen bacteria, 498
Cellulose, 2, 89, 258
  mechanism of digestion, 262–5
Cerebrocortical necrosis (CCN), 448, 481
Charon ventriculi, 97
Charonina, 92, 97
Charonina equi, 97, 98
Charonina nuda, 97, 98
Charonina Strand, 97
Charonina ventriculi, 97, 98
Chemical equilibria, 186
Chemical reactions, 3–5
Chemical treatment, 515
Chemical work, 3–5
Chilomastix caprae, 164
Chloral hydrate, 396, 428
Chloramphenicol, 407
Chlorella, 227
Chlortetracycline, 407
Choline, 233, 234
Chytridiomycetes, 135
Ciliate protozoa, 152–63, 230, 413, 487,
  489
  antagonism between species, 162
  counting procedures, 152
  differences within and between domestic
    ruminant species, 159–62
  factors influencing population size and
    composition, 154–9
  faunation, 152–4
  identification, 152
  level of intake, 157
  role in hydrolysis of dietary lipid, 290
  specificity of, 161
  wild ruminants, in, 162–3
Ciliates, 515
Cillobacterium cellulosolvens, 49
Clostridium, 221, 254, 324
Clostridium acetobutylicum, 325, 335
Clostridium amino-valericium, 26

Clostridium butyricum, 314
Clostridium difficile, 326
Clostridium lochheadii, 46
Clostridium longisporum, 46
Clostridium pasteurianum, 236, 325
Clostridium perfringens, 325
Clostridium polysaccharolyticum, 46
Clostridium saccharoperbutylacetonicum,
  325
Clostridium sphenoides, 26
Clostridium sporogenes, 481
Clostridium tertium, 325
Clostridium thermocellum, 196, 325
Clostridium thermosulphurogenes, 325
CO_2, 347, 349–51, 387, 399, 495
Coconut oil, effect on nitrogen
  metabolism, 419
Compartmentation, 361–85, 516
  biology, in, 361–6
  consequences of, 375–81
  digestive dysfunction due to malfunction
    in, 381
  four-compartment model, 382–3
  manipulation of compartment processes,
    381
  multi-compartment model, 366–70
  one-compartment system, 362, 365
  rumen as multi-compartment system,
    370–5
  three-compartment system, 362
  two-compartment system, 362
Corynebacterium, 232
Crude protein content, 217–18
Cyclic $3',5'$-AMP(cAMP), 206
Cycloheximide, 410

Dasytricha, 92, 98, 101, 152
Dasytricha hukuokaensis, 94
Dasytricha ruminantium, 90, 94, 99–103,
  105, 156
Deaminase inhibitors, 411
Defaunation, 413–14, 418, 489, 492, 496,
  501
  effect on animal performance, 502–6
  methods of, 487–8
Detoxification, 448–52
Diaminopimelic acid (DAP), 368–9
Diene, *cis,trans* conjugated 298–9
Dietary effects, 156–7, 169–70, 172–4, 218,
  219
Digestive dysfunction, 445–8
  due to malfunction in
    compartmentation, 381

Dihydrostreptomycin, 407
N,N-Dimethyldodecanamine, 411
Diplodinium, 78, 82, 152, 153, 158, 159, 162
Diplodinium affine, 162, 228
Diplodinium anisacantham, 228
Diplodinium dentatum, 78
Diplodinium minor, 129
Diploplastron, 80
Diploplastron affine, 80, 88, 89, 228
Diurnal variations, 154–6, 168–9, 172
Dynamic models, including microbes, 470–3

EDTA, 223, 226, 227
Electron donors, 301
Electron transport-linked phosphorylation (ETP), 196, 197, 198
Elytroplastron, 81
Elytroplastron bubali, 81
Embden–Meyerhof–Parnas (EMP) pathway, 102, 186–7, 203
Energy-consuming reactions, 194–203
Energy conversion, 3–5
Energy degradation, 3–5
Energy distribution, 201
Energy metabolism, 185
Energy regulation, 203–9
Energy sources, 347, 515
Energy-spilling reactions, 201–3
Energy-yield, 185–94, 516
Enoploplastron, 81
Enoploplastron confluens, 81
Enoploplastron triloricatum, 81, 228
Entodiniomorphid protozoa, 77–90
  characteristics of, 79
  classification, 77–82
  cultivation, 83–4
  division and conjugation, 85
  evolution, 82
  identification, 77–82
  structure, 82
  uptake, digestion and metabolism of dietary components, 85–90
  uptake of particulate matter and soluble compounds, 82
Entodinium, 78, 82, 88, 152–4, 156, 157, 159, 489, 492, 494, 497, 502
Entodinium alces, 163
Entodinium anteronucleatum, 158
Entodinium bursa, 78, 87, 88, 162
Entodinium caudatum, 78, 82–9, 162, 220, 233, 234, 310–12, 490
Entodinium ecaudatum caudatum, 233

Entodinium longinucleatum, 88, 162
Entodinium maggii, 233
Entodinium simplex, 85, 87, 162, 220
Enzymatic digestion of fibre, 16
Enzyme activity, 203–4
Eodinium, 78
Eodinium lobatum, 78
Epidinium, 81, 88, 153, 157, 158, 160, 162
Epidinium ecaudatum, 81, 89, 161, 266, 268, 270
Epidinium ecaudatum caudatum, 228, 230
Epiplastron, 81, 161
Eremoplastron, 78–80
Eremoplastron bovis, 83, 88, 129, 228, 268
Eremoplastron rostratum, 80
Erythromycin, 407
Escherichia coli, 84, 85, 86, 188, 194, 195, 207, 262, 265, 323–33, 336, 480, 481, 516
Ethanolamine, 233
Eubacterium, 49–50, 173, 221, 222, 223, 231
Eubacterium cellulosolvens, 49
Eubacterium oxidoreducens, 49
Eubacterium ruminantium, 49, 254
Eubacterium uniforme, 49
Eubacterium xylanophilum, 49
Eudiplodinium, 80, 153, 162
Eudiplodinium affine, 80
Eudiplodinium maggii, 80, 88, 89, 129, 220, 268
Eudiplodinium medium, 220

Fatty acid substrates, 304
Fatty acids, 28
  hydrogenated by rumen isolates, 299
Fatty aldehydes, 28
Feed-additive formulation, 514
Feed conversion, 516
Feed intake, pasture or range, 171
Feed metabolism, models of, 466–9
Feeding frequency, 157–8, 170
Feeding levels, 170
Feeding materials, 513
Feeding practices, 513, 514
Fermentation
  modelling, 516
  small-scale experiments, 513
Fermentation characterization, 47, 387
Fermentation modification, 387–443
  aim in, 388–9
  compounds used for, 391
  evaluation of, 390
  integrated system, as, 389

Fermentation modification—contd.
   macrocomponents of diet, by, 414–27
   microcomponents of diet, by, 393–413
   sites for, 389–90
Fermentation parameters, 492
Fermentation products
   acids, 12–13
   ATP synthesis, 15–16
   microbes, 13–15
Fibre-degrading organisms, 515
Fibre digestion, 16, 375–8, 515
Flagellate protozoa, 151, 163–5
Flagellates, 131
Flavobacterium spp., 25
Flavomycin, 408
Folic acid, 344
Food chain relationships, 347
Food particles, role in biohydrogenation, 291–4
Free amino acids, uptake of, 87–9
Free-decrements, 10–12
Fructosans, 272
   structure and metabolism, 255
Fructose 1,6-bisphosphate (FBP), 204
Fungi, 151, 226–7
   anaerobic. *See* Anaerobic fungi
   metabolism, 515
Fusobacterium, 221, 326
Fusocillus, 296–8, 300
Fusocillus babrahamensis, 297, 304
Futile cycles, 204–5

Gene cloning techniques, 331–3
Gene transfer in bacteria, 333–4
Genetic manipulation, 516
Geographic differences, 159
Glucose effect, 206
Glucose phosphotransferase system (PTS), 195
Glutamine synthetase-glutamate synthase (GS-GOGAT), 235, 237
Glycanases, 262–71
Glycolysis, 186
Glycopeptide antibiotics, 406
Glycosidases, 268–71
Gnotobiotic animals, 476–7, 482, 484, 516
Gnotoxenic animals, 477–8
Grain engorgement, 445
Grain overload, 445
Growth factors, 409–11
Growth-promoting substances, 515
Gut mutualisms, 5–6

Hemicellulases, 89, 266
Hemicelluloses, 2, 257, 266
Holophryozoon bovis, 97
Holotrich ciliates
   classification of, 93
   occurrence, 91
Holotrich protozoa, 90–105
   classification, 92
   cultivation, 98–9
   morphology, 92–8
   occurrence of, 92–8
   ultrastructure, 92–8
   uptake, digestion and metabolism of dietary components, 99–105
Hydrazine, 413
Hydrogen, 347, 349–51, 399
Hydrogen transfer, 351–4
Hydromycin B, 407
3-Hydroxy-4(1H)-pyridone (3,4DHP), 454

ICI 139603, 404, 405
Indigofera, 25
Ionophore antibiotics, 394–406
Isoleucine, 412
Isotrichia, 98, 99, 101, 152, 154, 156, 492, 497
Isotrichia bubali, 94
Isotrichia intestinalis, 90, 100, 104
Isotrichia intestinalis Stein, 94
Isotrichia prostoma, 90, 94, 100, 102–4, 204, 494
Isotrichidae, 92–4

Kanamycin, 407
Kjeldahl procedure, 217
Klebsiella aerogenes, 85, 86, 89, 201

Lachnospira, 22, 221
Lachnospira multiparus, 48, 175, 271
Lactate dehydrogenase (LDH), 203, 204
Lactic acid, 102, 109
Lactic acidosis, 445–7
Lactobacillus, 46–8, 173, 231, 232, 455, 479, 481–3
Lactobacillus acidophilus, 480
Lactobacillus casei, 480
Lactobacillus ruminis, 48
Lactobacillus vitulinus, 48
Laidlomycin butyrate, 404
Lampropedia, 50

Lasalocid, 29, 402–4
Lauric acid, 422
Leucaena, 454
Leucine, 412
Leucine *p*-nitroanilide (LNA), 224
Lignin–carbohydrate complexes, 274
Lignins, 2, 3, 274, 276–7, 514, 515
Lignocellulose digestibility (DADF), 498
Linoleic acid, 292, 296, 297
Linolenic acid, 292, 295–9
  biohydrogenation of, 297–8
Linseed oil, effect on nitrogen metabolism, 419
Lipid metabolism, 104, 285–322
  effects of dietary composition, 304
  microorganisms in, 285–6
Lipids
  carbohydrate metabolism, effect on, 416–17
  complex, 313–15
  composition of, 286
  crude fibre digestibility, effect on, 415, 417
  fatty acids composition, 311–13
  hydrolysis of, 287–90
  microbial
    biosynthesis, 311–15
    composition of, 305–11
  nitrogen metabolism, effect on, 418–21
  protected, 422
  rumen metabolism, effect on, 414–23
  rumen microbes, effect on, 421–2
Liquid and small particles (LSP), 11, 14, 15
Lolium perenne, 288
Lysocellin, 404

Magnoovum eadii, 50
Maintenance energy, 199–201
Maintenance functions, 199
Mathematical models, 463–74
  feed metabolism, 466–9
  general considerations, 464–5
  provision of data, 473–4
  steady states, 465–6
Megasphaera, 22, 231, 481–3
Megasphaera elsdenii, 26, 39–42, 146, 187, 188, 191, 194, 196, 228, 230, 239, 240, 308, 313, 314, 324, 332, 344, 348, 455, 479
Membrane-transport systems, 194
Meroxenic animals, 478
Metadinium, 80
Metadinium affine, 80

Metadinium medium, 80, 81
Metadinium ypsilon, 80
Methane, 387, 495
  inhibitors, 393–4
  production, 191–2, 349–51
Methanobacterium, 324, 481
Methanobacterium bryantii, 399
Methanobacterium formicicum, 24, 52, 325, 399
Methanobacterium M.O.H., 421
Methanobacterium ruminantium, 350, 421, 479, 483
Methanobacterium thermoautotrophicum, 324
Methanobrevibacter, 351
Methanobrevibacter ruminantium, 52–3, 192, 239, 350
Methanobrevibacter smithii, 142
Methanococcus thermolithotrophicus, 325
Methanococcus vannielii, 325
Methanococcus voltae, 325
Methanogenesis, 494
Methanogens, 50–4
  cultivation of, 59
  substrates for, 350
Methanomicrobium mobile, 52, 350
Methanosarcina barkeri, 52–4, 234, 325, 351
Methanosarcina mazei, 52
Methionine, 231, 412
3-Methylindole, 399, 409
Microbe–microbe interactions, 343–59
Microbial activities, 455–6
Microbial communities, 2
Microbial consortia, 380
Microbial growth rates, 10–12
Microbial populations, 151–83, 515
Microbiology research, 514
Microcetus lappus, 95, 161
Micrococcus, 232
Micrococcus varians, 25
Microorganisms
  attachment to fibres, 378–80
  lipid metabolism, in, 285–6
  role in biohydrogenation, 290–1
Mimosine degradation, 454
Models and modelling, 461–511, 516
  see also Biological models; Mathematical models
Molecular genetics of anaerobic bacteria, 324–8
Monensin, 29, 171, 354, 394–402, 405, 407, 408

Monocercomonas ruminantium, 164, 165
Monocercomonoides caprae, 164
Mucor rouxii, 129
Mutagenesis in bacteria, 335
Mycobacterium, 327
Mycoplasmas, 50, 151
Mycotoxins, 456

NADP-GDH, 235, 237
1,4-Naphthoquinone, 344
Narasin, 404
Neisseria, 327
Neocallimastix 164, 131, 132, 136, 137, 142, 144, 146
  life cycle, 133, 138
Neocallimastix frontalis, 132, 136, 139, 141, 142, 143, 146, 163, 164, 207, 226, 237, 254, 261, 270, 311
Neocallimastix patriciarum, 130, 136, 140–3, 271
Neomycin, 407
Niacin, 409
Nitrate, 233, 451
Nitrogen
  anabolic, 234–40
  degradation in rumen, 500
  fixation, 234
  flow, 499, 501
  metabolism, 103–4, 217–49, 396, 418–21
  microbial, 499–502
  supply, 375
Nitrogenous compounds, 344–5, 516
  breakdown of, 219–34
  flow through rumen, 217–18
  forms of, 218
  metabolism of, 217–49
3-Nitropropanol, 451, 452
3-Nitropropionic acid, 451, 452
Non-protein nitrogen (NPN), 217, 218, 390, 396, 418, 499
Novobiocin, 407
Nucleic acids, 86, 90, 233
Nutritional interactions, 343–4

Octadecadienoate reductase, 302
Octadecenoic acid, 298–300
Oleic acid, 292, 422
Oligoisotricha, 94–5
Oligoisotricha bubali, 95
Ophryoscolex, 81, 82, 152, 159, 220
Ophryoscolex caudatus, 81, 88
Ophryoscolex purkynjei, 81, 85

Ophryoscolex tricoronatus, 162
Opisthotrichum, 81, 159, 161, 163
Oscillospira, 50
Ostracodinium, 80, 162
Ostracodinium dentatum, 80
Ostracodinium mammosum, 80
Ostracodinium obtusum bilobum, 88, 228, 268
Oxalate degradation, 450–1
Oxalobacter formigenes, 450
Oxamycin, 407
Oxygen, metabolic effects of, 105
Oxytetracycline, 407

p-aminobenzoic acid (PABA), 343–4
Parabundleia ruminantium, 97, 161
Paraisotricha, 98
Paraisotrichidae, 98
Parentodinium, 82
Parentodinium africanum, 82, 161
PCMB, 223, 224
Pectic substances, 257
Pectin digestion, 271
Penicillin, 407
Pentatrichomonas hominis, 164
Pentose metabolism, 187
Peptides, 221–2, 411, 516
  metabolism, 227–8
Peptococcus glycinophilus, 26
Peptostreptococcus, 232
Peptostreptococcus elsdenii, 40
Peptostreptococcus heliotrinreducans, 452
Peptostreptococcus productis, 490
pH value, 492
Phenolic acids, 456
Phenolic compounds, 274–7, 456
Phenylalanine, 412
Phenylpropanoic acid, 410
Phosphoenolpyruvate (PEP), 195
Phosphofructokinase, 186
Phosphoketolase, 187
Phyto-oestrogens, 454
Pingius minutis, 97
Piromonas, 131, 135, 136, 138, 142, 144, 146
Piromonas communis, 130, 131, 137, 142, 143, 164, 271, 311–14
Plant cell walls
  adhesion by rumen microorganisms, 260–2
  carbohydrates digestion, 496
  digestion by rumen microorganisms, 260–71

Plant cell walls—*contd.*
  hydrolysis of, 255
  organisation of, 258–9
  phenolic compounds, 274–7
  structure of, 255–9
Plasmids, distribution and size characteristics, in bacteria, 328–31
PMSF, 223, 224
Polioencephalomalacia (PEM), 448
Pollution effects, 514
Polyether A, 404
Polyethylene glycol (PEG), 14
Polymorphella bovis, 97, 161
Polyplastron, 80, 153, 163, 489, 492, 502
Polyplastron alaskum, 163
Polyplastron arcticum, 163
Polyplastron californiense, 163
Polyplastron multivesiculatum, 80, 83, 85, 87–9, 159, 162, 163, 233, 255, 310, 490, 494, 497, 499
Polysaccharides
  degradation, 251–84, 515
  hydrolysis products, 345–7
  limitations to degradation, 272–7
  plant cell wall, 255–8, 272–3
  storage, 272
  structure and digestion of plant storage, 251–5
Predator–prey relationships, 347
Propionate enhancers, 394–409
Propionate production, 347
Propionibacterium, 173, 232
Propionibacterium shermanii, 186
Protein
  degradation, 219–27
  uptake, 87
Proteolysis inhibitors, 411–13
Proteus mirabilis, 85–7
Protonmotive force (PMF), 185
Protozoa, 77–128, 515
  absence of rumen ciliate, effects of, 106
  elimination of, 490–2
  entodiniomorphid. *See* Entodiniomorphid protozoa
  fatty acid composition, 309
  holotrich. *See* Holotrich protozoa
  host digestion of cell constituents, 108
  importance of, 105–110
  inoculation of, 489
  interrelationships with bacterial populations, 175–6
  lipid composition of, 308–11
  rumen metabolism, and, 108–10
Protozoa-free ruminants, 153–4
Protozoal populations, 487–506

Pseudomonas, 327
Pseudomonas aeruginosa, 25
PTS, 206, 207
Pyrophosphate-6-phosphofructokinase, 186
Pyrrolizidine alkaloids, 451–2
Pyruvate metabolism, 187–91

Quin's Ovals, 50

Rumen, 7–16
  anaerobic fungi. *See* Anaerobic fungi
  bacteria. *See* Bacteria
  fermentation. *See* Fermentation
  manipulations, 515
  metabolism, 108–10
  microbiology challenges, 17
  microcosm, 9–10
  organisms, 514
  reactions, 514
Rumensin, 394–406
Ruminant, 7–16
Ruminococcus, 22, 26, 175, 232, 328, 330, 473, 479, 481, 483, 485
Ruminococcus albus, 15, 26, 38, 44–6, 144, 146, 173, 188, 221, 240, 260, 263, 264, 266, 268, 304, 312, 330, 332, 345, 346, 351, 352, 356, 357, 410, 490
Ruminococcus flavefaciens, 26, 44–5, 146, 196, 208, 221, 237, 239, 240, 260, 262–4, 312, 330, 346–8, 352–4, 357, 379, 410, 490
Rusitec, 374, 375
  rumen simulation technique, 366–70

Saccharomyces cerevisiae, 325
Saccharomyces fragilus, 86
Sagittospora cameroni, 129
Salinomycin, 171, 404
Sarsaponin, 411
Seasonal differences, 158
Selenomonas, 22, 221, 228, 231, 232, 334, 481, 482, 484
Selenomonas ruminantium, 22, 26, 33–4, 85, 87, 171, 175, 187, 190–2, 196, 199, 200, 204, 208, 222, 225, 230, 232–4, 237, 239, 254, 265, 306, 308, 312, 324, 329–31, 333, 334, 344, 346, 348, 353, 354, 357, 446
Selenomonas ruminantium subs. *lactilytica*, 313
Semiquinone (TQH), 301

Shock-sensitive systems, 194
Siomycin, 408
S-methyl-cysteine sulphoxide (SMCO), 455, 482
Sphaerita hoari, 129
Sphaeromonas, 131, 135, 136, 137
Sphaeromonas communis, 130, 131, 164
Spirochaetes, 37
Sporangiomycin, 408
Sporormia minima, 129
Staphylococcus, 232
Staphylococcus aureus, 86, 325, 328, 330
Staphylococcus carnosus, 328
Starch, 88–9
  metabolism, 254
  structure, 251–3
Starvation survival, 208
Steroidal glycoside, 411
Streptococcus 46, 173, 231
Streptococcus bovis, 22, 43–4, 84, 167, 175, 196, 204, 221–3, 225, 237, 239, 254, 265, 328, 331, 408, 446, 479, 483, 485
Streptococcus faecalis, 86
Streptococcus faecium, 479
Streptococcus mutans, 328
Streptococcus pneumoniae, 328
Streptococcus sanguis, 328
Streptomyces, 408
Streptomyces albus, 404
Streptomyces candidus, 406
Streptomyces cinnamonensis, 394
Streptomyces tateyamensis, 407
Streptomyces virginiae, 408
Streptomycin, 407
Substrate-level phosphorylation (SLP), 196
Succinate production, 347
Succinimonas, 22
Succinimonas amylolytica, 39, 254
Succinivibrio, 22, 221, 232
Succinivibrio dextrinosolvens, 39, 237, 239, 348
Succinomonas amylolytica, 348
Sulphanilamide, 407
Sulphathiazole, 407
Sulphomycin, 408
Sulphonamides, 407
Syntrophococcus sucromutans, 44

Taitomycin, 408
Taxonomic research, 515
Terramycin, 407
Tetrahymena pyriformis, 105
Tetratrichomonas buttreyi, 164, 165

Thiamin, 409
Thiaminase, 448
Thiopeptin, 407, 408
Three population theory, 366
TLCK, 224, 225, 226
α-Tocopherolquinol (TQH$_2$), 301
Toxic compounds, acquired tolerance to, 450
Toxins
  inhibiting microbial activities, 456
  production, 452–5
TPCK, 224, 225
Transport systems, 194–6
Treponema, 232, 295
Treponema bryantii, 26, 37, 380
Treponema saccharophilum, 26, 37–8
Treponema succinifaciens, 26
Tricarboxylic acid (TCA) cycle, 238–9
Trichoderma reesei, 263, 270
Trifolium repens, 50
Tritrichomonas foetus, 105
L-Tryptophan, 399, 409
Tryptophan degradation, 455
Tylosin, 406, 407

Urea breakdown, 231–3
Urechis caupo, 2
Ureolysis, 480, 484

Valine, 412
Veillonella, 22, 348, 481–4
Veillonella alcalescens, 146, 187, 188, 239, 455, 479
Veillonella parvula, 42
Vibrio succinogenes, 190
Virginiamycin, 408, 409
Vitamin B, 343
Vitamin B$_6$, 413
Volatile fatty acids (VFA), 12–13, 109, 188, 204–5, 227, 228, 231, 345, 387, 389, 395, 396, 399, 407, 409, 410, 422–4, 456, 463–5, 469, 471–3, 477, 484, 485, 493, 494
  interactions and proportions of, 354–7

Wolinella succinogenes, 38–9
Work, definition, 4

Xylanases, 266

Yeasts, 129